Understanding Fiber Optics

Fourth Edition

Jeff Hecht

Upper Saddle River, New Jersey
Columbus, Ohio

Library of Congress Cataloging-in-Publication Data
Hecht, Jeff.
Understanding fiber optics/Jeff Hecht.—4th ed.
p. cm.
ISBN 0-13-027828-9
1. Fiber optics. I. Title.
TA1800.H43 2002
621.36'92—dc21 2001021565

Editor in Chief: Stephen Helba
Assistant Vice President and Publisher: Charles E. Stewart, Jr.
Assistant Editor: Delia K. Uberec
Production Editor: Tricia Rawnsley
Design Coordinator: Robin Chukes
Cover Designer: Becky Kulka
Cover: FPG International
Production Manager: Matthew Ottenweller

This book was set in Garamond by The Clarinda Company. It was printed and bound by R.R. Donnelley & Sons Company. The cover was printed by The Lehigh Press, Inc.

Prentice-Hall International (UK) Limited, *London*
Prentice-Hall of Australia Pty. Limited, *Sydney*
Prentice-Hall Canada, Inc., *Toronto*
Prentice-Hall Hispanoamericana, S.A., *Mexico*
Prentice-Hall of India Private Limited, *New Delhi*
Prentice-Hall of Japan, Inc., *Tokyo*
Prentice-Hall Singapore Pte. Ltd.
Editora Prentice-Hall do Brasil, Ltda., *Rio de Janeiro*

10 9 8 7 6 5 4 3 2
ISBN: 0-13-027828-9

Preface

Fiber optics has come a long way since I wrote the first edition of *Understanding Fiber Optics.* Optical-fiber communications was a radical new technology then, used mostly for high-capacity, long-distance transmission of telephone signals. As I finish the fourth edition, I can look out my office window and see a fiber cable that carries telephone, Internet, and cable-television signals down the street.

Over the years, I have been greatly impressed by the tremendous progress in developing practical fiber-optic equipment. The technology is interesting and elegant, as well as important. I find myself caught up in the advancing field, like a sports writer covering a team blazing its way to a championship. The thrill of technical achievement can be just as tangible to those of us involved with engineering or technology as the thrill of victory is to an athlete.

Although I wrote the first edition mainly for self-study, the book is now used in classroom settings. My goal is to explain principles rather than to detail procedures. When you finish this book you should indeed *understand* fiber optics. You should be able to pick up a trade journal such as *Lightwave* or *Fiberoptic Product News* and understand what you read, just as you should be able to understand the duties of a fiber engineer, a network planner, or a cable installer. You will not be able to do their jobs, but you will be literate in the field. Think of this as Fiber Optics 101, a foundation for your understanding of a growing technology.

To explain the fundamentals of fiber optics, I start with some ideas that may seem basic to some readers. When introducing a relatively new field, it is better to explain too much than too little.

To make concepts accessible, I include drawings to show how things work, limit the mathematics to simple algebra, and step through some sample calculations so you can see how they work. I compare fiber optics with other common technologies and highlight similarities and differences, and I have also organized the book to facilitate cross referencing and review of concepts.

The book is structured to introduce you to basic concepts first, then to dig deeper into fiber hardware and its applications. The chapters are organized as follows:

- The first three chapters present an overview, starting with a general introduction in Chapter 1. Chapter 2 introduces optics, light, and the concept of light guiding. Chapter 3 introduces communication systems and fiber-optic transmission. These chapters assume you have little background in the field, but they are worth reading even if you think your background is adequate.
- Chapters 4 through 8 cover optical fibers, their properties, and how they are assembled into cables. The material is divided into five chapters to make it easier to digest. Chapters 4 through 6 are essential to understanding the fiber concepts found in the rest of the book. Chapter 7 covers special-purpose fibers used in fiber amplifiers, fiber gratings, and a few odd applications such as architectural lighting. Chapter 8 is an overview of cabling.
- Chapters 9 through 12 cover laser and LED light sources, optical transmitters, optical detectors, receivers, optical amplifiers, and electro-optic regenerators. Chapter 12 compares and contrasts the operation of optical amplifiers and electro-optic regenerators.
- Chapters 13 through 16 cover a range of other components used in fiber-optic systems. Chapter 13 covers connectors and splices that join fibers. Chapter 14 covers optical couplers and other passive components used in simple fiber systems. Chapter 15 covers the optics used in wavelength-division multiplexing, which combine and separate signals at different wavelengths sent through the same fiber. Chapter 16 covers optical modulation and optical switching, key components for optical networking.
- Chapter 17 covers the fundamentals of optical and fiber-optic measurements, and describes how optical measurements differ from those of other quantities. Chapter 18 follows, covering fiber-optic test equipment and troubleshooting.
- Chapters 19 through 22 cover principles of fiber-optic communication. Chapter 19 describes the basic principles behind fiber-optic systems and optical networking. Chapter 20 covers major communication standards. Chapter 21 outlines the design of point-to-point single-wavelength systems, with sample calculations, so you can understand how these systems are put together. Chapter 22 extends design concepts, covering wavelength-division multiplexing and the emerging optical network.
- Chapters 23 through 27 cover various aspects of telecommunications, explaining how fiber optics fit into networks used for global and regional telephone and Internet transmission, cable television, and data networks. These chapters focus on different levels and aspects of the global network to keep concepts manageable. Chapter 28 covers special systems that don't fit elsewhere, such as fiber-optic cables for remote control of robotic vehicles, and networks in aircraft and automobiles.

- The final two chapters describe noncommunication applications. Chapter 29 explains the principles and operation of fiber-optic sensors. Chapter 30 covers imaging an illumination with fiber optics.

Most chapters include suggestions for further reading, and a list of resources appears at the back of the book. Links to Web sites are currently being added to my Web site, *http://www.fiberhome.com.* I would welcome any suggestions or comments you might have; please e-mail me at jeff@fiberhome.com or fiber@jeffhecht.com.

The glossary at the back of the book gives you quick translations of specialized terms and acronyms.

This edition also includes appendices that tabulate useful information, such as the values of important physical constants, conversion factors, standard data rates and wavelengths, and a few key formulas.

I have tried to make everything current, but the technology is advancing so fast that some details are bound to become obsolete. When you finish *Understanding Fiber Optics,* you should be prepared to follow the new advances, and perhaps contribute to them as well.

Acknowledgments

Over the years, many people in the fiber-optics industry have given generously of their time to patiently answer my questions. I owe special thanks to Kevin Able, Bill Chang, Erich Dzakler, Jim Hayes, Dennis Horwitz, Jim Masi, Nick Massa, Jim Refi, and Wayne Siddall for reviewing parts of this edition, clearly explaining complex concepts, and pointing me to useful resources. Prentice Hall reviewers Richard J. White, ITT Technical Institute; Stanley M. Krause, St. Philip's College; Kenneth E. Windham, Nash Community College; and Dr. Jalil Moghaddasi, City University of New York Bronx Community College also provided helpful feedback for this edition. David Charlton, Marc Duchesne, Robert Gallawa, Mike Pepper and John Schlager helped with earlier editions. I thank my editors at Prentice Hall, *Laser Focus World,* and *Integrated Communications Design* for patience with me above and beyond the call of duty. I also thank the companies, universities, and individuals who posted papers, application notes, tutorials, standards, and data sheets on the World Wide Web where I could find them when questions arose outside normal working hours. And very special thanks to anyone whose names and contributions may have slipped through my haphazard mental filing systems. Any errors that remain are my own.

This book is dedicated to the memory of Heather Williamson Messenger,
gifted editor, friend, and victim of domestic violence.

Jeff Hecht

Contents

CHAPTER

1

Introduction to Fiber Optics

A Personal View

Light is an old friend that has fascinated me ever since I can remember. As a boy, I bought surplus lenses, prisms, and filters, and used them to play with light; I still have a box of optical toys. In college I studied physics and electronic engineering, but light drew me back, first to lasers and then to fiber optics.

The first optical fibers I saw were in decorative lamps. A bundle of fibers was clamped together at one end and splayed out in a fan at the other. A bulb illuminated the clamped end, and the fibers carried light to the loose ends, where it sparkled like tiny stars. The effect was pretty enough that I bought my sister one as a Christmas present, but useless enough that I wandered away to explore other things.

When I next saw fiber optics, they had improved so much that telephone companies were developing them for communications. Those were the days when phone companies were—with good reason—described as "traditionally conservative" in their use of technology. They probed and tested fiber optics almost as cautiously as a bomb squad investigates a suspicious package. Finally, in 1977, GTE, AT&T, and British Telecom each dared to run live telephone traffic through fiber-optic cables they had pulled through standard underground ducts.

Looking back, the technology they used looks primitive. It was daring then, and it worked. Not only that, it worked flawlessly. The small armies of engineers monitoring those test beds came to countless technical meetings afterward repeating the same monotonous but thrilling conclusion: "It works! Nothing has gone wrong!"

I was at the first fiber-optic trade show in the late 1970s, and have watched the excitement spread since then. Each year the meetings have grown larger. In the early years, breakthroughs were almost routine. The first generation of systems was barely in the ground before a second generation was ready. A third generation followed quickly and, starting in the mid-1980s, became the backbone of the long-distance telephone network. That technology is fading away, and today we're seeing a new revolution in optical networking. State-of-the-art fiber systems can carry billions of bits per second at each of dozens of different wavelengths. Optical devices now switch signals as well as carry them from point to point. The rate of change has varied over the years, but the accomplishments are far beyond what anyone dared dream 20 years ago.

Looking back, it's been an incredible ride. It's a thrill to watch a new technology spring from the laboratory into the real world. Once I heard about fiber optics from research scientists; now I hear about fiber optics from telephone technicians working on the poles down the street. But the fun isn't over yet. The fiber-optics revolution will continue until fiber comes to every home. Today fiber-optic cables run mainly to businesses that need high-speed telecommunications. Fiber goes to only a relative handful of homes today, and it will take time to spread. Yet when it does come, fiber will bring a wealth of new information services, and make today's cable modems and digital subscriber line (DSL) look like yesterday's 1200-bit/second modems. The visionaries who foresaw a wired city were wrong—we will have a fibered society instead. We can all watch it happen.

But that's enough of this visionary stuff. Let's get down to the nuts and bolts—and fiber.

About This Chapter

The idea of communication by light was around long before fiber optics, as were fibers of glass. It took many years for the ideas behind fiber optics to evolve from conventional optics. Even then, people were thinking more of making special optical devices than of optical communications. In this chapter you will see how fiber-optic technology evolved and how it can solve a wide variety of problems in communications.

The Roots of Fiber Optics

Light normally travels in straight lines, but sometimes it is useful to make it go around corners.

Left alone, light will travel in straight lines. Even though lenses can bend light and mirrors can deflect it, light still travels in a straight line between optical devices. This is fine for most purposes. Cameras, binoculars, telescopes, and microscopes wouldn't form images properly if light didn't travel in straight lines.

However, there also are times when people want to look around corners or probe inside places that are not in a straight line from their eyes. Or they may just need to pipe light from place to place, for communicating, viewing, illuminating, or other purposes. That's when they need fiber optics.

Piping Light

The problem arose long before the solution was recognized. In 1881, a Concord, Massachusetts, engineer named William Wheeler patented a scheme for piping light through buildings. Evidently not believing that Thomas Edison's then new incandescent bulb would prove practical, Wheeler planned to use light from a blindingly bright electric arc in the basement to illuminate distant rooms. He devised a set of pipes with reflective linings and diffusing optics to carry light through a building, then diffuse it into other rooms, a concept shown in one of his patent drawings in Figure 1.1.

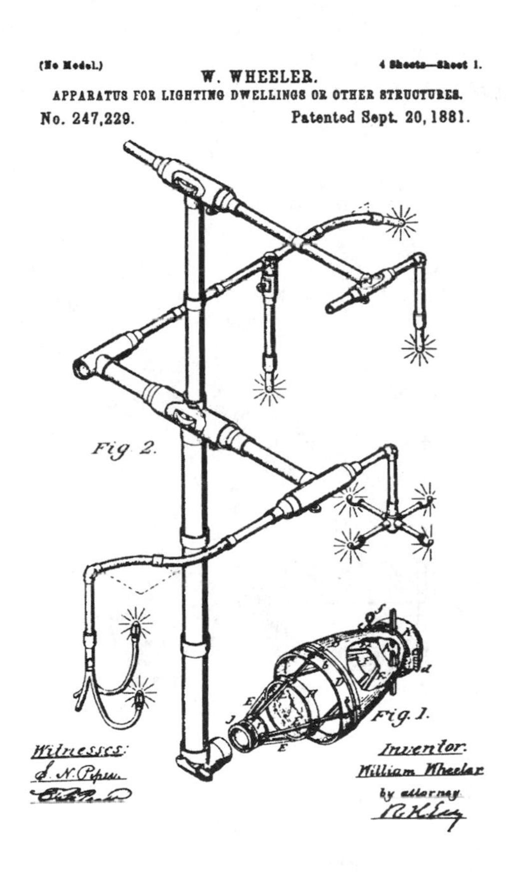

FIGURE 1.1 *Wheeler's vision of piping light (U.S. Patent 247,229).*

Although he was in his twenties when he received his patent, Wheeler had already helped found a Japanese engineering school. He later founded a successful company that made street lamps, and went on to become a widely known hydraulic engineer. Nevertheless, light piping was not one of his successes. Incandescent bulbs proved so practical that they're still in use today. Even if they hadn't, Wheeler's light pipes probably wouldn't have reflected enough light to do the job. However, the idea of light piping reappeared again and again until it finally coalesced into the optical fiber.

Total Internal Reflection

Ironically, the fundamental concept underlying the optical fiber was known well before Wheeler's time. A phenomenon called *total internal reflection,* described in more detail in Chapter 2, can confine light inside glass or other transparent materials denser than air. If the light in the glass strikes the inside surface at a glancing angle, it cannot pass out of the material and is instead reflected back inside it. Glassblowers probably saw this effect long ago in bent glass rods, but it wasn't widely recognized until 1841, when Swiss physicist Daniel Colladon used it in his popular lectures on science.

Colladon's trick, shown in Figure 1.2, worked like this. He shone a bright light down a horizontal pipe leading out of a tank of water. When he turned the water on, the liquid flowed out, with the pull of gravity forming a parabolic arc. The light was trapped within the water by total internal reflection, first bouncing off the top surface of the jet, then off the lower surface, until the turbulence in the water broke up the beam.

The Paris Opera used Colladon's light jet as a special effect in 1853 during performances of a ballet and of Gounod's opera *Faust.* The great Victorian exhibitions of the 1880s borrowed the idea to make illuminated fountains that fascinated fair-goers. But curving water jets had few practical uses. Light inevitably leaked out as the surface of the flowing water grew rough.

FIGURE 1.2
Light guided down a water jet.

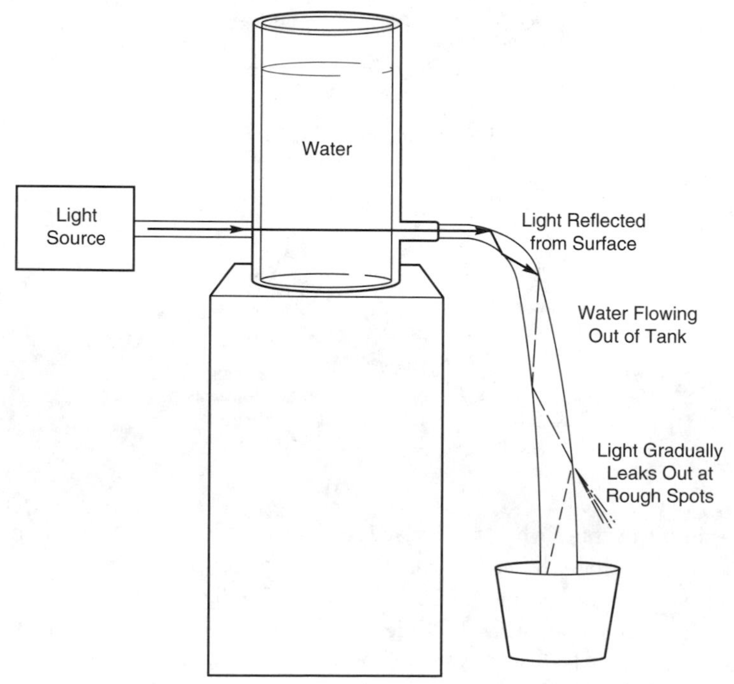

Glass Light Guides and Imaging

Clear glass rods could also guide light. By the early 1900s, they were being used to illuminate microscope slides. Inventors patented schemes for guiding light through bent glass rods to illuminate the inside of the mouth for dentistry. It was better than sticking a gas lamp in the patient's mouth, but it was far from perfect, and was never widely used.

A fine glass fiber is really a very thin rod, so it can guide light in the same way. Glass fibers also are flexible. Assemble a bundle of them, and they can transmit an image from one end to the other, as you will learn in Chapter 30. Clarence W. Hansell, an American electrical engineer and prolific inventor, patented the concept in the late 1920s. Heinrich Lamm, a German medical student, made the first image-transmitting bundle in 1930. However, the images were faint and hazy.

The key development in making optical fibers usable was perfecting a cladding to keep the light from leaking out.

Lamm had to comb the fibers to align them, but he didn't recognize a more important problem. When many bare fibers are bundled together, their surfaces touch, so light can leak from one into the other. The fibers also can scratch each other, and light leaks out at the scratches. Light even leaks out where fingerprint oils cling to the glass. The same problem plagued three men who independently reinvented imaging bundles in the early 1950s: a Danish engineer and inventor, Holger Møller Hansen, and two eminent professors of optics, Abraham van Heel in the Netherlands and Harold H. Hopkins in England.

The solution to that problem seems painfully obvious with 20/20 hindsight. Everyone started by looking at total internal reflection at the boundary between glass and air. However, total internal reflection can occur at any surface where light tries to go from a material with a high refractive index to one with a lower refractive index. Air is convenient, and its refractive index of 1.000293 is much lower than that of ordinary glass, which is 1.5. But total internal reflection occurs as long as the material covering the glass has a refractive index smaller than the glass, as shown in Figure 1.3. Møller Hansen produced total internal reflection by coating glass fibers with margarine, but the results were impractically messy.

Brian O'Brien, a noted American optical physicist, separately suggested the cladding to van Heel in 1951. Van Heel used beeswax and plastic, which were more practical than margarine. In December 1956, Larry Curtiss, an undergraduate student at the University of Michigan, made the first good glass-clad fibers by melting a tube of low-index glass onto a rod of high-index glass. Glass cladding soon became standard, although a few fibers continue to be plastic-clad, and plastic is used to *coat* fibers to protect them mechanically.

Many optical fibers can be bundled together to transmit images.

Glass-clad fibers were the key to making flexible fiber-optic *endoscopes* or *gastroscopes* to look down the throat into the stomach. The fibers are glued in place on each end, but left loose in the middle, forming a flexible bundle that can be bent to follow the natural curves of the throat. Flexible fibers bundles are inserted in the opposite end of the body to examine the colon. The fiber-optic endoscope has become an important medical tool, although some instruments now combine fibers with miniature electronic cameras.

Other imaging applications soon emerged for bundles of optical fibers. Fibers can be melted together and drawn to be extremely fine light guides in rigid bundles. Flexible fiber bundles can "pipe" light for illumination into hard-to-reach spots. A whole industry grew

FIGURE 1.3
Light cannot leak out of clad fibers.

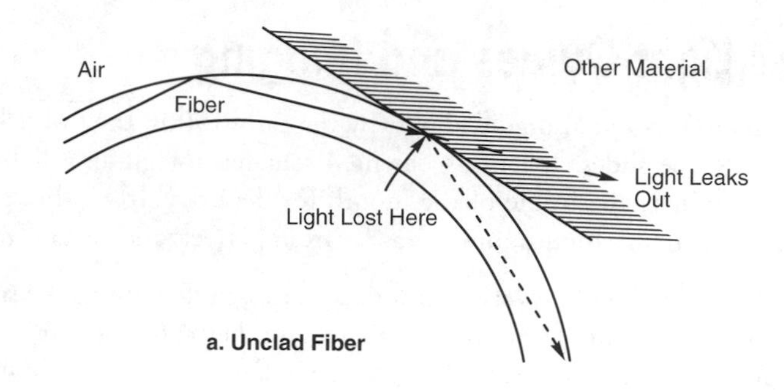

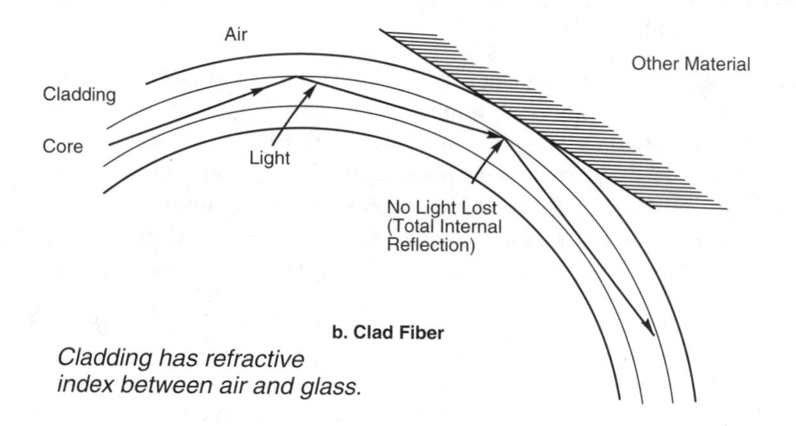

around fiber-optic imaging and light-piping in the 1960s. The technology remains in use, and is covered in Chapter 30, but has largely been eclipsed by fiber-optic communications.

Fibers in Communications

An optical telegraph was invented in France in the 1790s but made obsolete by the electric telegraph.

The idea of communicating by light probably goes back to signal fires on prehistoric hilltops. The ancient Greeks relayed news of the fall of Troy by signal fires; Native Americans used smoke signals. Even the first "telegraph" was an optical one, invented by French engineer Claude Chappe in the 1790s. Operators relayed signals from one hilltop telegraph tower to the next by moving semaphore arms. Samuel Morse's electric telegraph put the optical telegraph out of business, but it left behind countless Telegraph Hills.

In 1880, a young scientist who had already earned an international reputation used beams of light to transmit voices in the first wireless telephone. Alexander Graham Bell had invented the telephone four years earlier, but he considered the photophone his greatest invention. It reproduced voices by detecting variations in the amount of sunlight or artificial light focused through the open air onto a receiver. Bell was elated to hear beams of light laugh and sing, but the Photophone never proved practical in cities, where too many things could get in the way of the light. Wires and radio waves proved more practical for communications.

A few people kept experimenting with optical communications. In the 1930s, an engineer named Norman R. French—who worked for the American Telephone & Telegraph Corp. built around Bell's telephone—patented the idea of communicating by sending light through pipes. But few people took optical communications seriously until Theodore Maiman demonstrated the first laser in 1960.

Invention of the laser stimulated interest in optical communications and led to efforts to reduce light loss in fibers, which was essential for communications.

The laser generates a tightly focused beam of coherent light at a single pure wavelength. It's the optical equivalent of the pure carrier-frequency signal used by a radio or television station. That made it look very promising for communications, and many laboratories started experimenting. They first tried sending laser beams through air, but like Bell soon found that open air was not a very good transmission medium because fog, rain, snow, and haze could block signals. They tried sending light through more modern versions of Wheeler's light pipes and found other troubles.

Optical fibers were available, but they didn't look very promising. The fibers used in endoscopes are much clearer than window glass, but half the light that enters them is lost after 3 meters (10 feet). That's fine for examining the stomach, but not for communications. Go through a mere 20 meters (66 feet) of such fiber, and only 1% of the light remains. Go another 20 meters, and only 1% of that light remains—0.01% of the input. Convinced that transparent solids inevitably absorbed too much light for optical communications, most engineers either gave up or tried to develop new versions of hollow light pipes or better ways to send light through the air.

Two young engineers at Standard Telecommunication Laboratories in England, Charles K. Kao and George Hockham, took a different approach. Instead of asking how clear the best fiber was, Kao asked what the fundamental limit on loss in glass was. He and Hockham concluded that the loss was caused mostly by impurities, not by the glass itself. In 1966, they predicted that highly purified glass should be so clear that 10% of the light would remain after passing through at least 500 meters (1600 feet) of fiber. Their prediction sounded fantastic to many people then, but it proved too conservative.

Publication of Kao and Hockham's paper set off a worldwide race to make better fibers. The first to beat the theoretical prediction were Robert Maurer, Donald Keck, and Peter Schultz at the Corning Glass Works (now Corning Inc.) in 1970. Others soon followed, and losses were pushed down to even lower levels. In today's best optical fibers, 10% of the entering light remains after the light has passed through more than 50 kilometers (30 miles) of fiber. Losses are not quite that low in practical telecommunication systems, but as you will see in Chapter 5, impressive progress has been made. Because of that progress, fiber optics have become the backbones of long-distance telephone networks around the world.

Basic Fiber Concepts

The rest of this book will teach you the details of how fiber optics work and how they are used, particularly in communications. The rest of this chapter is a starting point, to give you a quick snapshot of what's involved in fiber-optic technology. Chapters 2 and 3 go into a bit more depth to lay the foundation for the rest of the book. Their goal is to explain fundamental concepts of light, optics, fibers, communications, and other fiber applications. Later chapters elaborate on those concepts.

Fiber Structures

The basic structure of a fiber is a light-guiding core surrounded by a cladding of lower refractive index.

The basic structure of an optical fiber is quite simple. The light-guiding core of the fiber is made of a material with higher refractive index than the cladding that surrounds it, as shown in Figure 1.4. The difference in refractive index causes the total internal reflection (explained in Chapter 2) that guides light through the core. As we will see in Chapter 4, this is an oversimplified picture, but the core and cladding layers are the fundamental building blocks of optical fibers.

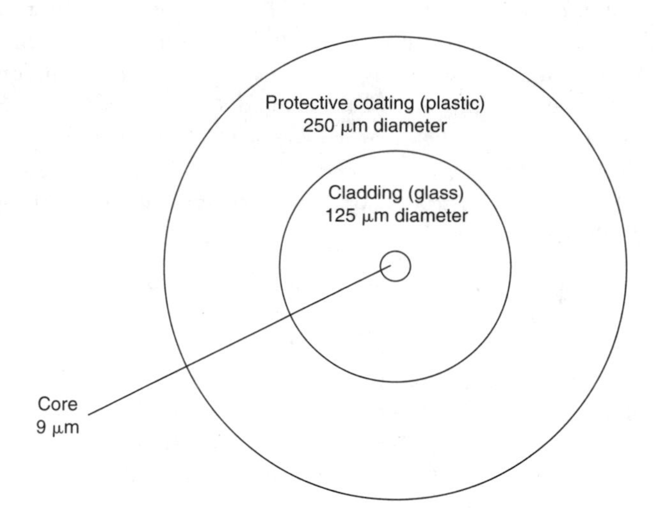

FIGURE 1.4
Cross-section of a typical communications fiber.

Sizes of the core and cladding can vary widely among different types of fibers. Fibers made for high-resolution imaging have thin claddings surrounding small cores. Fibers made for carrying high powers for illumination typically have much larger cores with thin claddings. In contrast, the fibers used for telecommunications have thick cladding layers surrounding small cores. The boundary between core and cladding may be sharp or gradual, with the refractive index changing abruptly or gradually. Some high-performance fibers have multiple layers at the core-cladding boundary.

The standard cladding diameter of telecommunication fibers is 125 micrometers (μm), or 0.005 inches. A plastic coating increases diameter to about 250 μm, easing handling and protecting the glass surface of the fiber from scratches and other mechanical damage. Fibers used for imaging may be as small as several micrometers in diameter; some special-purpose fibers are more than a millimeter (0.04 in.) thick. Fiber dimensions are virtually always given in metric units.

Fiber Materials

Most fibers are made of very pure glass, with small levels of impurities to adjust the refractive index. From a chemical standpoint, the clearest fibers used in telecommunications are essentially pure silicon dioxide, known as *silica* (SiO_2). The fibers used for medical imaging and illumination are made from less pure glass. Some fibers are made from plastic, which is not as clear as glass but is more flexible and easier to handle. A few glass fibers are clad with plastic, but typically plastic is used only as an outer coating for mechanical protection.

Most fibers are nearly pure silica, with dopants added to change the refractive index.

Special-purpose fibers may be made from other materials. For example, fluoride compounds are transparent at longer infrared wavelengths than silica, so they are sometimes used for infrared applications. These fibers may also be called glasses because they are made of materials in a glassy or noncrystalline state, but typically they are identified by material, as, for example, fluoride glass fibers.

Fiber Properties

Mechanically, fibers are stiff but flexible and generally quite strong. Thin fibers are much more flexible than thicker ones. Optical fibers used for communications are often compared to human hairs, but whoever thought of that comparison must have had very stiff hairs or very thin fibers. Communication fibers are stiffer than a man's coarse beard hair of the same length. A better comparison is to monofilament fishing line. Unlike wires, fibers spring back to their original straight form after being bent, and do not stretch permanently when pulled, although glass fibers may break.

Fibers are stiff but flexible and surprisingly strong.

Glass fibers are surprisingly strong, but they can fail if surface cracks propagate through the fiber. Plastic coatings protect fiber surfaces from mechanical damage.

The optical properties of fibers depend on their structure and their composition. The most obvious is loss or signal attenuation, but optical pulses also suffer more subtle effects, which are described in detail in Chapter 5.

Imaging and Bundled Fibers

Optical fibers were first used to transmit images. As described in more detail in Chapter 30, each fiber carries one point of an image from one end of the bundle to the other. If the fibers are arranged in the same way on both ends, this re-creates the original image on the other end of the bundle. Bundles of fibers also can carry light for illumination in which case fiber arrangement is not as critical.

Flexible and rigid bundles are used for imaging and illumination.

Bundles may be flexible or rigid. A flexible bundle consists of many separate fibers, with the two ends fixed together and the fibers loose in the middle (although typically encased in a protective housing). Rigid bundles are made by melting many fibers together into a single rod, which typically is bent to the desired shape during manufacture. Such rigid or fused bundles cost less than flexible bundles, and individual fibers can be thinner than loose fibers, but their inflexibility makes them unsuitable for many applications, such as looking down a patient's throat.

Fiber Optics for Communications

At first glance, fiber-optic imaging might seem more complex than fiber-optic communications. The basic idea of fiber communications is simple, as shown in Figure 1.5. An input signal modulates a light source, producing an optical signal that travels through the fiber to a receiver, which decodes it. You can see the idea if you take a single short large-core fiber, unconnected to any equipment, and look into one end while you point the other end towards and away from a light bulb. (Be sure to keep your eye a safe distance from the fiber end.) Moving the far end of the fiber modulates the light reaching your eye, producing a simple signal.

FIGURE 1.5
Fiber-optic system components.

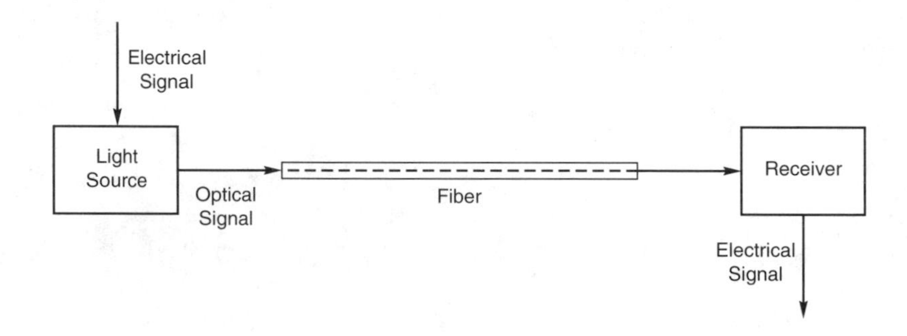

Even the simplest fiber communication systems include a transmitter and receiver as well as the fiber.

Real communication systems are considerably more complex. The example shown in Figure 1.5 is the simplest form of communications, *point-to-point transmission* from a single transmitter to a single receiver. A real fiber-optic system for this simple job is considerably more complex than this simplified drawing. Electronic circuits in a transmitter convert the input signal into a form suitable to modulate output of the light source; they also may monitor the optical output and control the transmitter temperature. A pair of connectors precisely aligns the transmitter with a fiber contained in a cable so its light can enter the fiber. Other connectors may join separate sections of cable. Lengths of fiber may be spliced together to make a longer cable. Amplifiers may boost the strength of the optical signal. Connectors align the output end of the fiber with the input end of the receiver. An optical detector in the receiver generates an electrical signal, which electronic circuits amplify and process to generate the required electrical output signal.

That figure itself is a very simplified version of a communication system. Figure 1.6 shows a sampling of other types of communication systems. Some broadcast signals from a central transmitter to many separate receivers. Others connect mobile phones to the fixed telephone network. Some provide continuous connections among many fixed devices, like an office local area network. The telephone network includes switches that makes temporary connections between pairs of phones.

Traditionally, electronic devices have performed these communication tasks. Fiber-optic components serve the same functions, but work differently. For example, electronic connectors need only to keep a pair of wires touching each other, but fiber-optic connectors

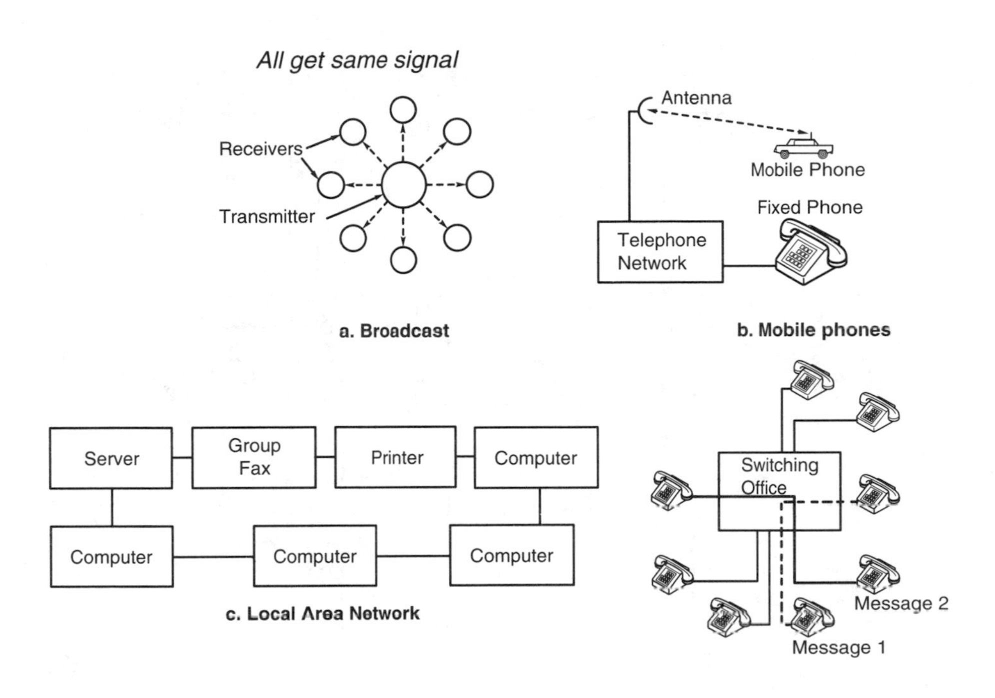

FIGURE 1.6 *Simple types of communication transmission.*

must align the fibers precisely so light leaving one fiber goes straight into the other. Chapters 4 through 16 describe fibers and optical devices and how they work. Chapters 17 and 18 cover optical measurements, troubleshooting, and test equipment. The rest of the book, except for the last two chapters, covers fiber-optic communication systems and the emerging optical network.

The Emerging Optical Network

A communication system performs two basic functions: it transmits and distributes information. You can think of those components as pipes and switches, as shown in Figure 1.7. The job of the pipes is to carry information from its source to its destination. The job of the switches is to direct the information on the proper route.

A communications network consists of signal "pipes" and "switches."

Users sit at the ends of pipes connected to switches. You can connect any two points in Figure 1.7 by drawing lines through the right pipes and switches. Although this is a greatly oversimplified view, you can think of the telephone network as such an array of pipes and switches. The pipes are copper wires, optical fibers, and radio relay links. The switches are electronic boxes—actually special-purpose switching computers—in telephone company facilities. The Internet uses different kinds of switches, but the basic idea is the same. Details vary in other types of communications network, but the basic idea remains the same. You could even extend the idea to the transportation network if you wanted.

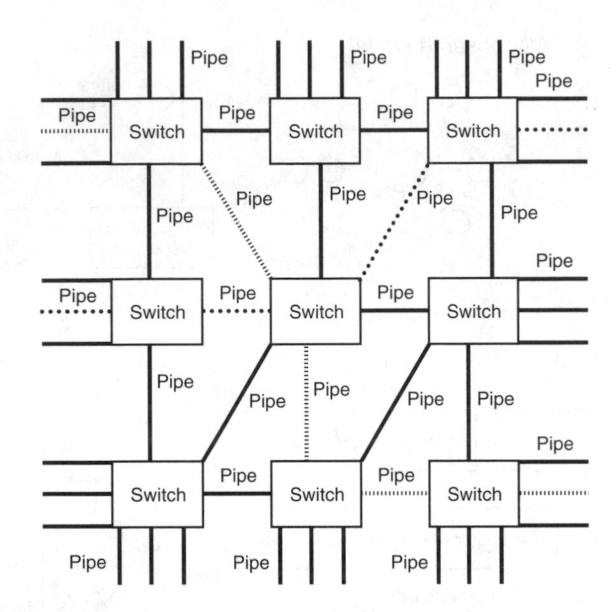

FIGURE 1.7
A communication system consists of "pipes" that transmit signals and "switches" that direct them to their proper destinations.

The first uses of optical fibers were as pipes to carry signals between two points. Until the past few years, fiber-optic systems were simply bigger and better versions of point-to-point transmission links. High-speed long-distance systems carried tremendous volumes of data across continents and under oceans, but they remained only big information pipelines. Electronics converted signals into light on one end, the light passed through the fiber, was amplified where necessary, and emerged at the other end of the fiber pipe to be converted into electrical form and processed electronically.

That changed as fiber capacity increased and it became possible to transmit signals at many separate wavelengths through the same fiber. Each *optical channel* or *lambda* (from the Greek letter λ used to symbolize wavelength) can carry a large volume of independent information. It's much simpler to switch independent optical channels optically than to convert them into electronic form for switching, then back to optical form for the next transmission. An *optical network* is the equipment that handles these optical signals in which the switches as well as the pipes are optical. That technology is still emerging.

Optical networks include optical switches as well as optical transmission.

Optical networking has started at the heart of the telecommunications network, where signals are transmitted at the highest speeds, but is spreading outward. You can think of the optical network as the part of the overall network where the switches as well as the pipes are optical. Later chapters will explain this in more detail.

Transmission Media

The *transmission medium* is the pipe that transmits a signal from point to point. In fiber optics, the signals are transmitted in the form of light through the fiber. In electronic communications, the signals are transmitted as electrons, radio waves, or microwaves. Radio waves and microwaves can pass through open air, through waveguides, or along electrical conductors (such as wires or coaxial cables). The details of those types of transmission are beyond the scope of this book.

Light also can pass through air, and a few communication systems send optical signals short distances through open air or free space. To prevent confusion, I avoid the term "optical communications" in this book. When talking about fiber optics, I specify fiber; when talking about sending light through the air, I talk about *atmospheric* or *free-space laser communications.*

Different media are preferable for different communications jobs. The choice depends on the job and the nature of the transmission medium. Signal distribution is an important factor. If many people in an area receive the same signal—as in broadcast television or radio—the best choice may be nondirectional radio broadcasts through the air. Satellite transmission can distribute television signals to many people over much broader areas. Mobile communications, such as cellular phones, require radio transmission from fixed local antennas or from satellites. Satellites may be the best way to reach remote locations like small mid-ocean islands or polar research stations.

Different media are preferable for different types of communications.

On the other hand, fiber-optic or metal cables typically are used to connect fixed points. Examples include conventional telephones and cable-television networks. Radio transmission can offer similar services, but cables can offer better signal quality, better security, and in many cases more transmission capacity. Cables also do not consume scarce space in the radio-frequency spectrum.

Bandwidth and Distance

Two key factors that determine the choice of transmission medium are how much information the signal carries and how far it must go.

Signal-transmission capacity is called *bandwidth* and, as you will learn in Chapter 3, can be described in various ways depending on the type of signal. It depends on the transmitter and receiver as well as the transmission medium. It's analogous to the amount of water you can get through a pipe.

The maximum transmission distance depends on the transmitter, the receiver, and the transmission medium. For ground-based radio and television stations, it depends on the transmitter power, receiver sensitivity, and atmospheric conditions. You can hear a powerful radio station farther away than a weak one, and a good antenna can pick up weaker stations better than a poor one.

For copper wires, the maximum transmission distance decreases with the signal bandwidth. A pair of telephone wires *can* carry a video signal, but only a very short distance. Coaxial cables can carry video signals much farther, but their transmission also is limited. (As you will learn later, the reason is that attenuation in the cable increases with signal frequency.)

Optical fibers can carry high-bandwidth signals much longer distances than copper.

A big advantage of optical fibers is that they can carry high-bandwidth signals over much longer distances than copper wires or coaxial cables. Fibers also can transmit digital signals very well, an important asset because the global telecommunications network has shifted largely to digital transmission. An additional advantage is that most fiber-optic systems can be upgraded to transmit at higher speeds simply by replacing the transmitter and receiver. That is harder to do with copper cables. Many optical fibers can simultaneously transmit signals at many separate wavelengths, multiplying their total transmission capacity.

Optical fibers have several other advantages that may dictate their selection for certain applications. Compact fiber-optic cables may be the only type that can fit into tight places. For portable systems, fiber cable is lighter than other types. Fiber is inherently secure and difficult to tap, important in applications ranging from military systems to financial data networks. Fiber does not conduct electricity and cannot cause sparks, a must in refineries and chemical plants, where the air may contain traces of explosive gases. Fiber is immune to electromagnetic interference (EMI), which can generate background noise that overwhelms signal transmission on wires.

Some fiber is being installed to "future-proof" communication systems in office buildings and even in homes. Tearing up existing walls is far more expensive than installing cable when a new building is under construction or when an old one is undergoing extensive renovation. Some developers argue that adding fiber during construction is a good investment even if it won't be used immediately, because bandwidth requirements are rising steadily. I was pleasantly surprised to see fiber-optic cable being installed in a home renovated on the Public Broadcasting System's *This Old House* program. It didn't add much to the cost, but it will add a lot to the value when fiber eventually comes to homes.

Fiber-Optic Growth

The bandwidth and other attractions of optical fibers have made them the ideal transmission medium where cables are used. They haven't been installed everywhere because wires got there first and are cheaper for most present home and office applications.

Installation of fiber and construction of new communication systems cost money, and someone has to justify that extra cost. Stereo fanatics with "golden ears" might want to run fiber-optic cables to their speakers, but most of us would never hear the difference. It's cheaper and easier to plug copper cables into existing jacks and sockets. The same is true for phones, cable television, and the wiring on your computer. Fibers don't bring any immediately obvious benefits (because most present services don't need their huge capacity), they aren't readily available, and like any custom equipment, they take more time and money.

Fibers have spread through the global telecommunications network.

However, fibers continue to spread through the global telecommunications network. Today, most telephone calls that go beyond your local community travel on optical fibers. In many communities, fibers run part-way from the telephone company's local switching office to homes and businesses. Fibers carrying cable television signals hang on the poles outside my office window; coaxial cables run to homes from boxes attached to the fiber. Fibers provide the backbone for the Internet, and are just starting to deliver high-speed data services to some homes around the country—at speeds 10 to 100 times higher than cable modems or digital subscriber line (DSL). New high-capacity fiber-optic systems are running from city to city, state to state, and country to country. The fiber-optic industry is growing at an incredible rate. Venture capitalists are virtually throwing money at anything related to optical networking.

It's a remarkable boom time for the fiber-optic industry. Telecommunications users around the world want more bandwidth, and fiber can deliver it. Averaged over the past 20 years, the maximum data rate that could be transmitted through a single fiber has more than doubled every year. That's faster than Moore's law in electronics, where the doubling time is 18 months. There's a lot to learn about fiber optics, as you will see in the chapters that follow, and there's also a lot to do with that knowledge.

Technology Lag Time

The rapid pace of development and the massive scale of the existing telecommunication network create a long lead time in fiber-optic developments. Laboratory developers are quick to announce their latest and greatest work, and trade magazines headline records set by "hero experiment" groups probing the technological frontier. However, the fact that a small army of Ph.D. engineers can transmit 6 terabits (trillion bits) per second through a single fiber in a laboratory equipped with a few million dollars of expensive hardware does not mean this technology is standard. Typically it takes 2 or 3 years before a commercial version is ready for use in the field—although that doesn't stop corporate marketing departments from announcing it. (Such products are called "vaporware" in the computer industry.)

Once a new product is ready for delivery, it still takes time to spread through the communications network. Production capacity is limited, particularly for the highest-performance components. A company building a new telecommunications network may install state-of-the-art equipment, but companies with existing networks don't instantly replace everything when new hardware comes out. They gradually upgrade over a number of years where necessary, but they are likely to leave old cables as long as they're doing the job. Don't be surprised to find a mix of several generations of equipment.

New technology takes time to spread through the network.

Fiber Terms and Terminology

The appendix includes a glossary, tables listing important units, and other important data. Many terms are standardized or widely accepted, but others are not. The communications industry is notorious for its cryptic acronyms and sometimes puzzling buzzwords; I have tried to avoid unclear terminology and all but the most widely accepted acronyms. Because the industry is growing and changing rapidly, the terminology is evolving continuously. Many companies develop their own terminology, and it's not uncommon for different companies to have separate terms for the same technology. I generally avoid proprietary terms, and I particularly despise the meaningless marketing use of "solution" to describe a product or system.

Fiber optics is an international industry, so virtually all units are given in metric units. The only place you're likely to see Imperial units is in giving lengths of cable runs. In the rest of this book, I follow the standard industry practice of using metric units almost exclusively. See Appendix A for a list of metric prefixes.

What Have You Learned?

1. Light rays normally go in straight lines, but optical fibers can guide them around corners.
2. Total internal reflection guides light in optical fibers. The low-refractive-index cladding confines light to the core.
3. Many optical fibers can be bundled together to transmit images.
4. The first optical communications was through the air in the optical telegraph.
5. The invention of the laser stimulated interest in optical communications and led to development of clearer glass fibers.
6. Most fibers are nearly pure silica, with dopants added to create the core-cladding structure.
7. Glass optical fibers are stiff but flexible and surprisingly strong.
8. Telecommunications systems are made of "pipes" and "switches." Initially optical fibers were used only as information pipes.
9. Optical fibers have much higher transmission capacity than copper cables.
10. Optical networking is the use of optical devices to switch and direct optical signals.

What's Next?

In this chapter, we examined the background of fiber-optic technology. In Chapter 2, you'll learn some of the basic physics behind fiber optics, then get a brief introduction to fiber-optic hardware.

Further Reading

On the evolution of fiber optics:

Jeff Hecht, *City of Light: The Story of Fiber Optics* (Oxford University Press, 1999)

On the development of communications in general:

Arthur C. Clarke, *How the World Was One: Beyond the Global Village* (Bantam, 1992)

Irwin Lebow, *Information Highways & Byways: From the Telegraph to the 21st Century* (IEEE Press, 1995)

Laszlo Solymar, *Getting the Message: A History of Communications* (Oxford University Press, 1999)

Questions to Think About for Chapter 1

1. For a bundle of optical fibers to transmit an image, the fibers must be arranged in the same pattern on both ends of the bundle. What limits the size of the smallest details that can be seen?
2. Devise an analogy to show how a bundle of fibers transmits an image using common implements found in a kitchen or cafeteria.
3. Most of the light lost in going through a glass window is reflected at the surface. Ignoring this surface reflection loss, suppose that a one-millimeter window absorbs 1% of the light entering it and transmits 99%. Neglecting reflection, how much light would emerge from a one-meter-thick window?
4. If optical fibers transmit signals so much better than wires, why aren't they used everywhere?
5. Electromagnetic interference occurs in wires because electromagnetic fields cause electrons to move inside the wires, generating currents that interfere with the signal. This is also how antennas work. Why doesn't electromagnetic interference affect signals in an optical fiber?
6. Most switching in the telephone network is done by special-purpose computers. Why do network operators continue to convert most optical signals to electronic form for switching?

Quiz for Chapter 1

1. Light can be guided around corners most effectively in
   a. reflective pipes.
   b. hollow pipes with gas lenses.
   c. clad optical fibers.
   d. bare glass fibers.
2. The first practical use of optical fibers was
   a. for communications via optical telegraph.
   b. in Alexander Graham Bell's Photophone.
   c. to illuminate flowing jets of water.
   d. in bundles to examine the inside of the stomach.
3. What is the principal requirement for a cladding on an optical fiber?
   a. It must have a refractive index lower than the core to produce total internal reflection.
   b. It must be opaque so light doesn't leak out.
   c. It must be made of plastic to keep the fiber flexible.
   d. It must have a lower refractive index than air.

4. Flexible bundles of optical fibers can be used to

a. examine the inside of the stomach without surgery.

b. examine the inside of the colon without surgery.

c. illuminate hard-to-reach machinery.

d. all the above

e. none of the above

5. Today's best optical fibers transmit light so well that 10% of the light entering the fiber remains after

a. 0.5 km.

b. 4 km.

c. 20 km.

d. 50 km.

e. 100 km.

6. All fiber-optic communication systems must include

a. light source, receiver, and fiber.

b. light source and cable.

c. fiber and receiver.

d. fiber only.

e. cable only.

7. A switched telecommunication system makes connections between pairs of terminal devices. Which is the best example of a switched network?

a. Cable television

b. Telephone network

c. A link between a single personal computer and a printer

d. A network that shares one printer among four personal computers

8. The first important role that optical fibers played in modern communications networks was

a. broadcasting signals over wide areas.

b. as pipes for long-distance transmission at high speeds.

c. switching signals among many terminal devices.

d. carrying radio signals in mobile phones.

9. An optical channel is

a. a television channel transmitted optically.

b. all the light transmitted through a single fiber.

c. one of several wavelengths carrying separate signals through an optical fiber.

d. the electrical input to an optical transmitter.

10. Which of the following systems can transmit at the highest speed over the longest distance?

a. An optical fiber carrying a single, pure wavelength

b. A coaxial cable carrying many video signals

c. An optical fiber carrying signals at many separate wavelengths

d. A coaxial cable transmitting light

CHAPTER

2

Fundamentals of Fiber-Optic Components

About This Chapter

Fiber optics is a hybrid field that started as a branch of optics. The basic concept behind a fiber is optical, and some single or bundled optical fibers are used as optical components. However, the most common application of fiber optics is in communications, where many concepts originated in electronic and radio communications. Today, signals shift back and forth between electronic and optical formats as they go through the global telecommunications network. Fiber-optic transmitters and receivers are hybrids that are part optics and part electronics. To understand fiber-optic communications, you need to learn something about three fields: optics, electronics, and communications.

This chapter lays the groundwork for understanding fiber-optic components; Chapter 3 covers communication systems. Later chapters explain particular devices and systems in more detail.

Basics of Optics

The workings of optical fibers depend on basic principles of optics and the interaction of light with matter. The first step in understanding fiber optics is to review the relevant parts of optics. The summary that follows does not cover all of optics, and some parts may seem basic, but you should read it to make sure you understand the fundamentals.

From a physical standpoint, light can be seen either as electromagnetic waves or as photons, quanta of electromagnetic energy. This is the famous wave-particle duality of modern physics. Both viewpoints are valid and valuable. The simplest viewpoint for optics often is to consider light as rays traveling in straight lines between or within optical elements, which can reflect or refract (bend) light rays at their surfaces. The photon viewpoint is useful in electronics and communications, particularly for explaining things like how optical detectors work.

Physically, light is an *electromagnetic wave,* composed of electric and magnetic fields, which vary in amplitude as they move through space. The two fields are perpendicular to each other, and to the direction in which the light travels, as shown in Figure 2.1. The amplitude varies with time *sinusoidally,* like a sine function in trigonometry, rising from zero to a positive peak, going back through zero, hitting a negative peak, then returning to zero. The distance light travels during that complete cycle is called a *wavelength.* The usual symbol for wavelength is the Greek letter λ (lambda), and that's one symbol you want to remember. The number of cycles per second is called the *frequency,* and measured in hertz (after Heinrich Hertz, who discovered electromagnetic waves). Frequency is usually denoted by the Greek letter ν (nu).

FIGURE 2.1
A light wave consists of electric and magnetic fields.

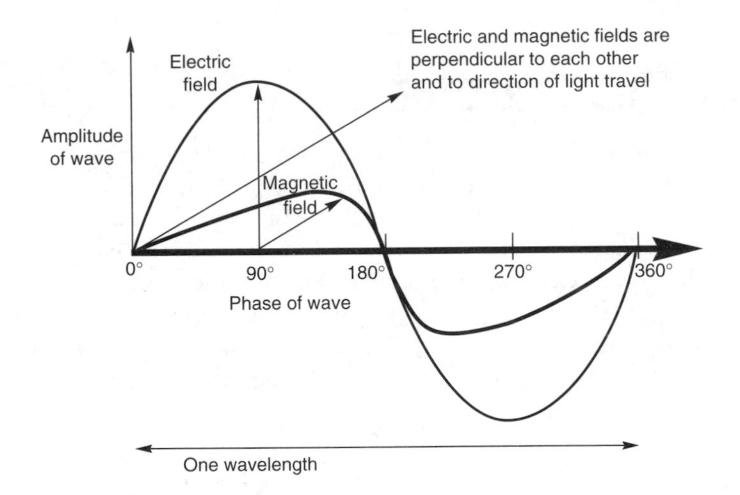

Many light sources such as lasers emit *continuous* light waves, which oscillate steadily at the same frequency. You can think of them as sine waves that go on forever. They represent a continuous stream of photons emitted regularly, each containing a unit of energy. A single photon is a *wave packet,* a series of a few waves that build quickly to a peak amplitude, then fade back to nothing, as shown in Figure 2.2. Like a continuous wave, it has a wavelength and frequency.

The amount of energy in a photon depends on the oscillation frequency or wavelength: The faster the wave oscillates (or equivalently, the shorter the wavelength), the higher the energy.

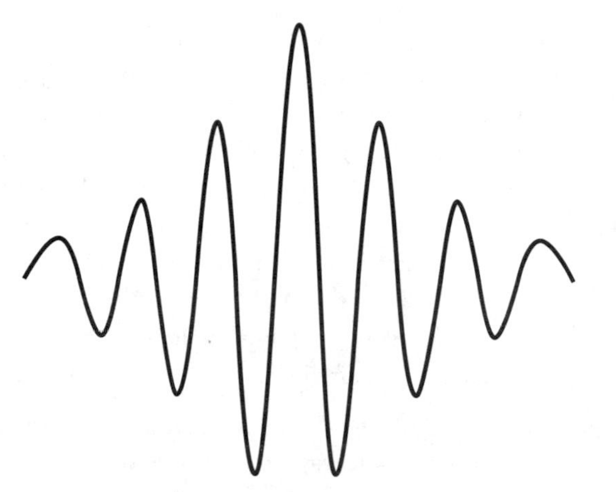

FIGURE 2.2
A single photon is a short packet of waves.

The Electromagnetic Spectrum

What we call "light" is only a small part of the spectrum of electromagnetic radiation. The fundamental nature of all electromagnetic radiation is the same: it can be viewed as photons or waves and travels at the speed of light *(c),* which is approximately 300,000 kilometers per second (km/s), or 180,000 miles per second (mi/s). The difference between radiation in different parts of the electromagnetic spectrum is a quantity that can be measured in several ways: as the length of a wave, as the energy of a photon, or as the oscillation frequency of an electromagnetic field. Figure 2.3 compares these three views.

The light carried in fiber-optic communication systems can be viewed as either a wave or a particle.

Each measurement—wavelength, energy, or frequency—has its own characteristic unit. The preferred unit depends on the part of the spectrum. The optics world usually talks in wavelength, which is measured in metric units—meters, micrometers (μm or 10^{-6} m), and

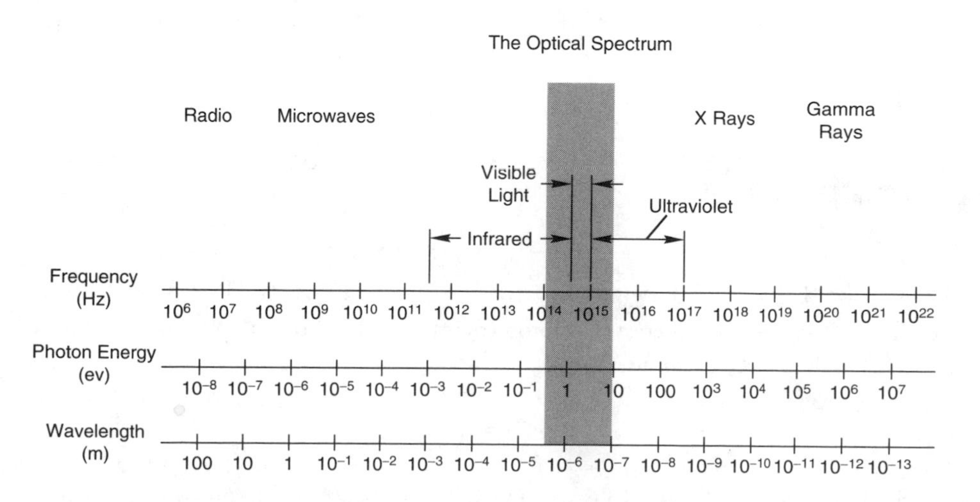

FIGURE 2.3
Electromagnetic spectrum.

nanometers (nm or 10^{-9} m). Wavelength is sometimes measured in angstroms (1Å = 10^{-10} m), but the angstrom is not a standard unit, so it is rarely used. Don't even think of wavelength in inches. (If you absolutely have to know, 1 μm is 0.00003937 in.) Frequency is measured in cycles per second (cps) or hertz (Hz), with megahertz (MHz) meaning a million hertz and gigahertz (GHz) meaning a billion hertz. (The metric system uses the standard prefixes listed in Appendix A to provide different units of length, weight, frequency, and other quantities. The prefix makes a unit a multiple of a standard unit. For example, a millimeter is a thousandth [10^{-3}] of a meter, and a kilometer is a thousand [10^{3}] meters.) Photon energy can be measured in many ways, but the most convenient here is in electron volts (eV)—the energy that an electron gains in moving through a 1-volt (V) electric field.

All the measurement units shown on the spectrum chart are actually different rulers that measure the same thing. There are simple ways to convert between them. Wavelength is inversely proportional to frequency, according to the formula:

$$\text{wavelength} = \frac{c}{\text{frequency}}$$

or

$$\lambda = \frac{c}{\nu}$$

where c is the speed of light, λ is wavelength, and ν is frequency. To get the right answer, all terms must be measured in the same units. Thus c must be in meters per second (m/s), λ must be in meters, and frequency must be in hertz (or cycles per second). Plugging in the approximate value of c, we have a more useful formula for wavelength:

$$\lambda = \frac{3 \times 10^{8}\ \text{m/s}}{\nu}$$

You can also turn this around to get the frequency if you know the wavelength:

$$\nu = \frac{3 \times 10^{8}\ \text{m/s}}{\lambda}$$

Not many people talk about photon energy *(E)* in fiber optics, but a value can be gotten from Planck's law, which states:

$$E = h\nu$$

where h is Planck's constant (6.63×10^{-34} J-s, or 4.14×10^{-15} eV-s) and ν is the frequency. Because most interest in photon energy is in the part of the spectrum measured in wavelength, a more useful formula is

$$E\,(eV) = \frac{1.2399}{\lambda\ (\mu\text{m})}$$

which gives energy in electron volts when wavelength is measured in micrometers (μm).

One important practical consequence of light's wave personality is that light waves have a property called *phase,* which measures the position in the wave's cycle of variation. Look back at Figure 2.1 and you will see one cycle in which the amplitude of the light wave rises, falls, and returns back to the starting point. The light waves emitted by a continuous source repeat this cycle endlessly. Repeating the cycle is like going around a circle, so the phase is measured as an angle between 0° and 360°.

All light waves are not in the same phase. Two waves with the same wavelength may have a phase difference between 0° and 360°. (Larger phase differences don't matter—it's relative position in the cycle that counts, not the number of cycles.) Phase is important because it determines how light waves combine, or more properly *interfere* with each other.

Light waves that are 180° out of phase with each other can cancel each other out.

The light intensity we see is proportional to the square of the wave amplitude, the height of the wave in Figure 2.1. If you turn on two lightbulbs in a dark room, the total intensity is the sum of the two intensities because the light is made up of many different wavelengths. However, if you combine two identical light waves, their amplitudes add or subtract as shown in Figure 2.4. If the peaks of the light waves are neatly lined up in phase with one another (that is, with a 0° phase difference), the amplitudes add and they give a bright spot. This is called *constructive interference.* If the two light waves are aligned so the peaks of one match the troughs of the other, 180° out of phase, their amplitudes cancel each other out in *destructive interference* that gives a dark spot. If the two waves are out of phase by a dif-

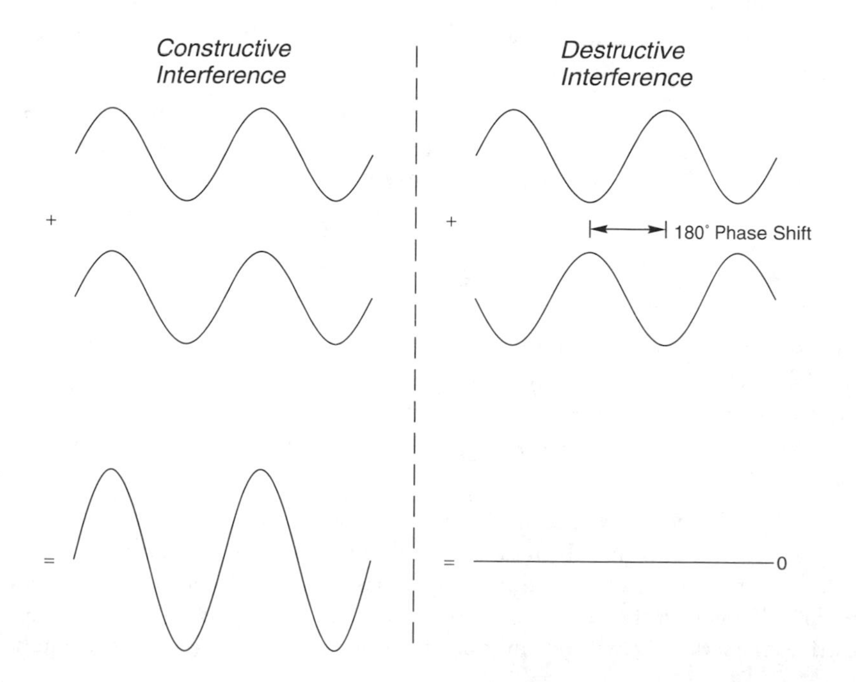

FIGURE 2.4 *Constructive and destructive interference.*

ferent amount, they add together to give an intensity between the maximum and minimum possible.

We are mainly interested in a small part of the spectrum shown in Figure 2.3—the optical region, where optical fibers and other optical devices work. That region includes light visible to the human eye at wavelengths of 400 to 700 nm and nearby parts of the infrared and ultraviolet, which have similar properties. Roughly speaking, this means wavelengths of 200 to 20,000 nm (0.2 to 20 μm).

Fiber-optic communication systems transmit near-infrared light invisible to the human eye.

The wavelengths normally used for communications through silica glass optical fibers are 750 to 1700 nm (0.75 to 1.7 μm) in the near infrared, where silica is the most transparent. Glass and silica fibers can transmit visible light over shorter distances, and special grades of silica (often called fused quartz) can transmit near-ultraviolet light over short distances.

Plastic fibers typically transmit better at visible wavelengths than in the near infrared, so communications through plastic fibers typically is with visible red light. However, plastic fibers are not as transparent as silica glass. Fibers made from certain other materials can transmit at longer infrared wavelengths than silica glass fibers.

Refractive Index

The refractive index of a material is the ratio of the speed of light in a vacuum to the speed of light in the material.

The speed of light *(c)* is often considered the universal speed limit. Nothing is supposed to go faster, although in some special cases light sometimes can go a bit over the speed limit if it carries no information. That universal speed limit is the speed of light *in a vacuum.* Light always travels more slowly when it passes through any transparent material. The degree of slowing down depends on the nature of the material and its density.

This slowing down of light affects how light travels through materials through a parameter called the *refractive index,* denoted by the letter *n* in optics. This is the ratio of the speed of light in vacuum to the speed of light in the material:

$$n = \frac{c_{\text{vac}}}{c_{\text{mat}}}$$

For normal optical materials, the refractive index is always greater than 1.0 in the optical part of the spectrum. In practice, the refractive index is measured by comparing the speed of light in the material to that in air rather than in a vacuum. The refractive index of air at atmospheric pressure and room temperature is 1.000293, so close to 1.0 that the difference is insignificant.

Refraction occurs when light passes through a surface where the refractive index changes.

Although light rays travel in straight lines through optical materials, something different happens at the surface. Light is bent as it passes through a surface where the refractive index changes—for example, as it passes from air into glass, as shown in Figure 2.5. The amount of bending depends on the refractive indexes of the two media and the angle at which the light strikes the surface between them. The angles of incidence and refraction are measured

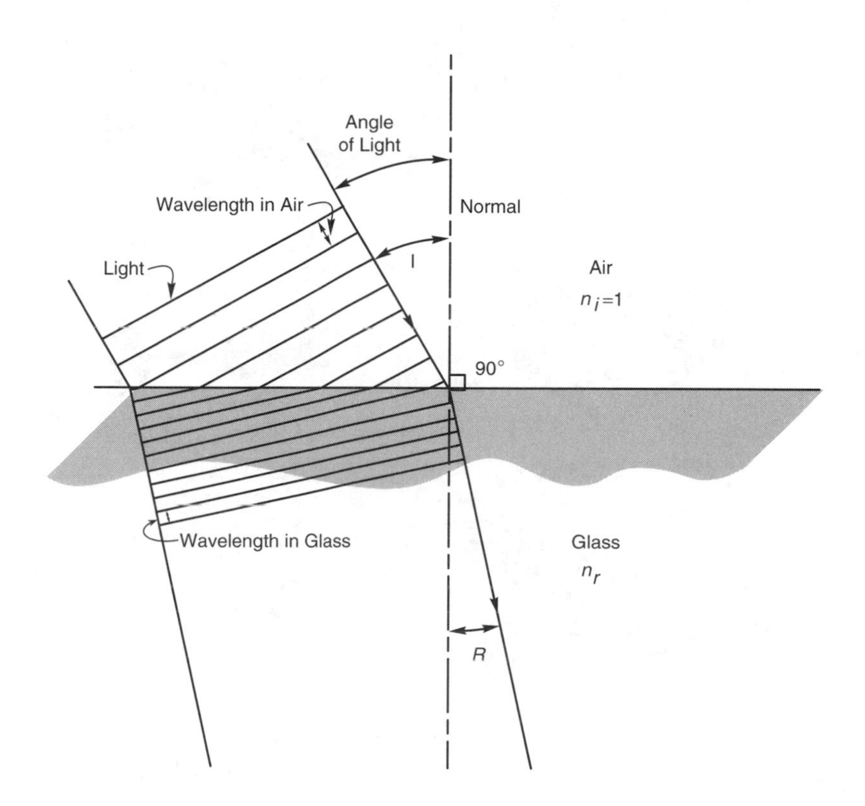

FIGURE 2.5
Light refraction as it enters glass.

not from the plane of the surface but from a line normal (perpendicular) to the surface. The relationship is known as Snell's law, which is written

$$n_i \sin I = n_r \sin R$$

where n_i and n_r are the refractive indexes of the initial medium and the medium into which the light is refracted, and I and R are the angles of incidence and refraction, respectively, as shown in Figure 2.5.

Figure 2.5 shows the standard example of light going from air into glass. The frequency of the wave does not change, but because it slows down in the glass, the wavelength gets shorter, causing the light wave to bend, whether the surface is flat or curved. However, if both front and rear surfaces are flat, light emerges at the same angle that it entered, and the net refraction is zero, as when you look through a flat window. If one or both surfaces are curved, you see a net refraction or bending of the light, as if you were looking through a lens. That is, light rays emerge from the lens at a different angle than they entered. These overall refractive effects are shown in Figure 2.6

What does this have to do with fiber optics? Stop and consider what happens when light in a medium with a high refractive index (such as glass) comes to an interface with a medium

FIGURE 2.6
Refraction through a window and a lens.

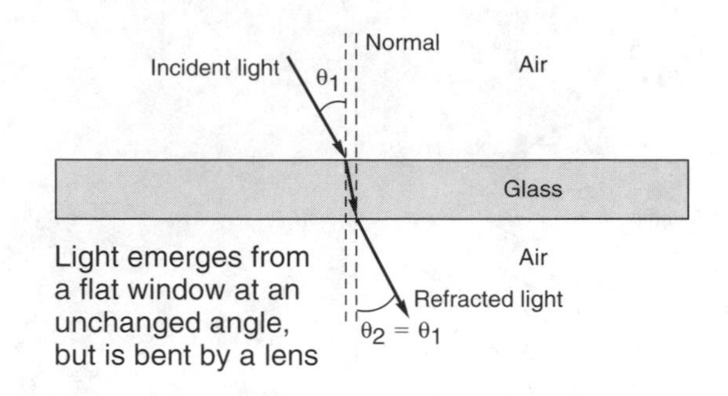

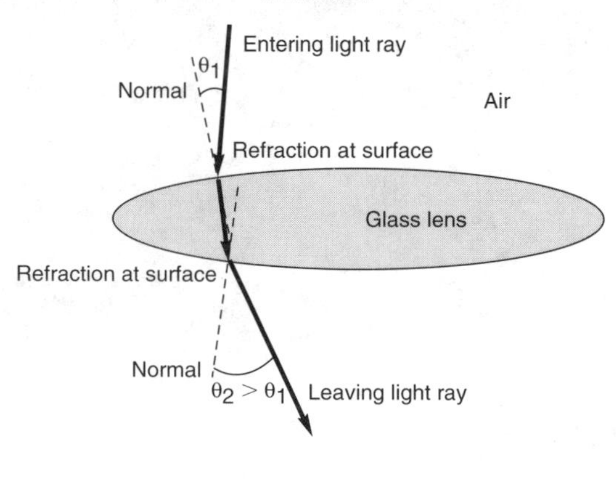

having a lower refractive index (such as air). If the glass has a refractive index of 1.5 and the air an index of 1.0, the equation becomes

$$1.5 \sin I = 1 \sin R$$

If light hits a boundary with a material of lower refractive index at a glancing angle, it is reflected back into the high-index medium. Such total internal reflection is the basic concept behind the optical fiber.

That means that instead of being bent closer to the normal, as in Figure 2.5, the light is bent farther from it, as in Figure 2.7. This isn't a problem if the angle of incidence is small. For $I = 30°$, $\sin I = 0.5$, and $\sin R = 0.75$. But a problem does occur when the angle of incidence becomes too steep. For $I = 60°$, $\sin I = 0.866$, so Snell's law says that $\sin R = 1.299$. Your pocket calculator will tell you this is an error. That angle can't exist because the sine can't be greater than 1.0.

Snell's law indicates that refraction can't take place when the angle of incidence is too large, and that's true. Light cannot get out of the glass if the angle of incidence exceeds a value called the critical angle, where the sine of the angle of refraction would equal 1.0. (Recall from trigonometry that the maximum value of the sine is 1.0 at 90°, where the light would be going along the surface.) Instead, total internal reflection bounces the light back into the glass, obeying the law that the angle of incidence equals the angle of reflection, as shown in

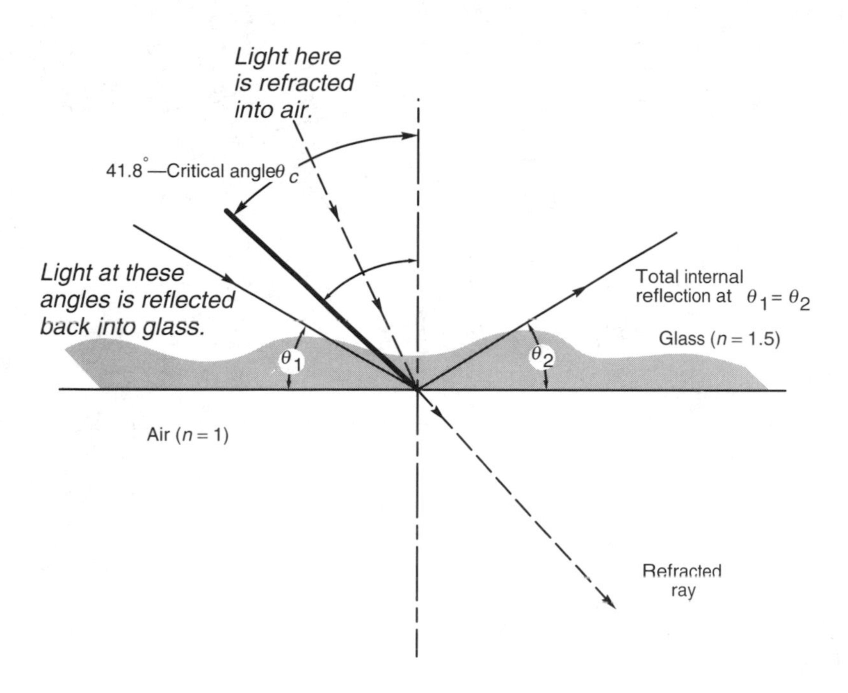

FIGURE 2.7
Refraction and total internal reflection.

Figure 2.7. It is this total internal reflection that keeps light confined in optical fibers, at least to a first approximation. As you will see in Chapter 4, the mechanism of light guiding is more complex in modern communication fibers.

The critical angle above which total internal reflection takes place, θ_c, can be deduced by turning Snell's law around, to give

$$\theta_c = \arcsin (n_r / n_i)$$

For the example given, with light trying to emerge from glass with $n = 1.5$ into air, the critical angle is arcsin (1/1.5), or 41.8°.

Light Guiding

The two key elements of an optical fiber—from an optical standpoint—are its core and cladding. The core is the inner part of the fiber, which guides light. The cladding surrounds it completely. The refractive index of the core is higher than that of the cladding, so light in the core that strikes the boundary with the cladding at a glancing angle is confined in the core by total internal reflection, as shown in Figure 2.8.

Light is guided in the core of an optical fiber by total internal reflection at the boundary of the lower-index cladding.

The difference in refractive index between core and cladding need not be large. In practice, it is only about 1%. This still allows light guiding in fibers. For $n_r / n_i = 0.99$, the critical angle, θ_c, is about 82°. Thus, light is confined in the core if it strikes the interface with the

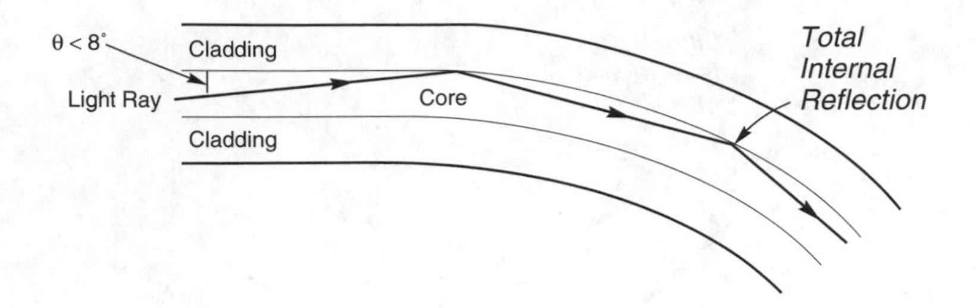

FIGURE 2.8
Light guiding in an optical fiber.

cladding at an angle of 8° or less to the surface. The upper limit can be considered the confinement angle in the fiber.

The angle over which a fiber accepts light depends on the refractive indexes of the core and cladding glass.

Another way to look at light guiding in a fiber is to measure the fiber's acceptance angle—the angle over which light rays entering the fiber will be guided along its core, shown in Figure 2.9. (Because the acceptance angle is measured in air outside the fiber, it differs from the confinement angle in the core.) The acceptance angle normally is measured as the numerical aperture (NA), which for light entering a fiber from air is approximately

$$NA = \sqrt{n_0^2 - n_1^2}$$

where n_0 is the refractive index of the core and n_1 is the index of the cladding. For a fiber with core index of 1.50 and cladding index of 1.485 (a 1% difference), NA = 0.21. An alternative but equivalent definition is the sine of the half-angle over which the fiber can accept light rays: 12° in this example (θ in Figure 2.9). Another alternative definition is NA = $n_0 \sin \theta_c$, where θ_c is the confinement angle in the fiber core (8° in this example). These angles are measured from a line drawn through the center of the core, called the *fiber axis*.

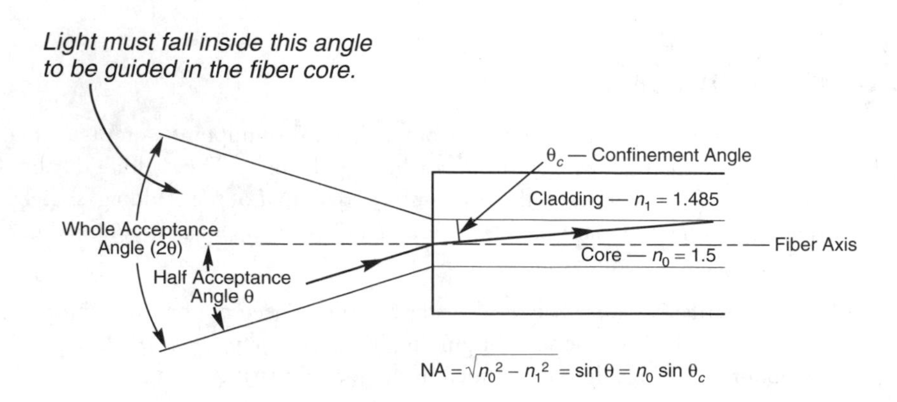

FIGURE 2.9
Measuring the acceptance angle.

Note that the half acceptance angle is larger than the largest glancing angle at which light rays must strike the cladding interface to be reflected, which I said earlier was 8°. What does this mean? Go back and look at Snell's law of refraction again. The difference is the factor n_0, which is the refractive index of the core glass, or 1.5. As you can see in Figure 2.9, refraction bends a light ray entering the fiber so that it is at a smaller angle to the fiber axis than it was in the air. The sine of the angle inside the glass equals that of the angle outside the glass, divided by the refractive index of the core (n_0).

Light Collection Efficiency

An optical fiber will pick up some light from any light source. You can see this if you point a single large-core fiber at a lightbulb. Taking proper precautions for eye safety (see Appendix E), look into the other end of the fiber, bending it so you don't look directly at the light. You should see an illuminated spot in the fiber, showing that the fiber collects some light—but only a very small fraction of the light that the bulb emits.

Developing ways for fibers to collect light efficiently was an important step in developing practical fiber-optic communications. Large-core fibers proved unsuitable for high-speed communications, as discussed in Chapter 4, so light must be concentrated into cores that typically are 8 to 62.5 μm in diameter. Light also must be transferred efficiently from one fiber to another, which requires extremely precise alignment of fiber ends using connectors and splicing techniques covered in Chapter 13.

Light source size and alignment are critical in collecting light in a fiber core.

How efficiently a fiber collects light depends on two critical factors. One is a matter of size: how tightly the light is focused onto the light collecting core. The other is a matter of alignment: how well the light fits into the fiber's collection angle. The two interact to impose demanding requirements.

Simple optics can focus the light from an ordinary bulb so it forms a narrow beam. You can see the results in a flashlight beam or a searchlight. Look carefully, and you can see that the focusing is not perfect, but the beams are strongly directional. However, focusing a large light source into a narrow beam leaves a large spot that spreads far beyond the fiber core. Large light sources can be focused onto small spots with strong magnifying lenses. You've probably used that trick to burn a hole in paper with focused sunlight. However, that normally leaves the light spreading at too large an angle for the fiber to collect it efficiently.

For communication systems, it's generally more efficient to find a light source that is close to the fiber core in size. Generally these are semiconductor lasers, which emit light from a small area on their surface, or optical-fiber amplifiers, which emit light from a doped core, as described later. Light-emitting diodes (LEDs) can be used with some larger-core fibers because they are less expensive and the larger cores can collect more of their light. Larger light sources generally are easier to align with fibers, but their lower intensity delivers less light. Chapter 9 describes light sources in more detail.

Transferring light between fibers requires careful alignment and tight tolerances. Light transfer is most efficient when the ends of two fibers are permanently joined in a splice (de-

Joining the ends of optical fibers requires careful alignment and tight tolerances.

scribed in Chapter 13). Temporary junctions between two fiber ends, made by connectors (also described in Chapter 13) typically have slightly higher losses but allow much greater flexibility in reconfiguring a fiber-optic network. Special devices called couplers (described in Chapter 14) are needed to join three or more fiber ends. One of the most important functional differences between fiber-optic and wire communications is that fiber couplers are much harder to make than their metal-wire counterparts.

Transfer losses must be considered in fiber-optic communication systems.

Losses in transferring signals between wires are so small that they can normally be neglected. This is not so for fiber optics. As you will see in Chapter 21, system designers should account for coupling losses at each connector, coupler, splice, and light source.

Fiber Transmission

Attenuation, dispersion, and nonlinear effects can degrade signals transmitted by optical fibers.

Optical fibers inevitably affect light transmitted through them. The same is true for any material transmitting any kind of signals. You notice these effects most for poor transmitters, like dirty windows or crackling telephone lines. However, they are present even for the tenuous gas dispersed in intergalactic space, which astronomers can spot because it absorbs a tiny fraction of the light passing through it. Generally these effects degrade signals, and if they become large enough, they can make it impossible to receive the signals.

The three principal effects that degrade signals in optical fibers are *attenuation, dispersion,* and *crosstalk.* You can see analogous effects when electronic signals go through copper wires or are broadcast as radio or television signals. These effects are critical to the performance of fiber-optic systems, so I will introduce the concepts here before exploring them in more detail in later chapters.

Fiber Attenuation

Absorption, scattering, and light leakage are the components of fiber attenuation.

Attenuation makes signal strength fade with distance. In some cases, such as broadcast radio, distance alone can cause attenuation because signals spread out through space as they travel. As the signal spreads over a larger volume, the intensity drops.

This is not the case in optical fibers, which are *waveguides* that confine light within the core along their entire length. This prevents signals from spreading over a larger volume, but other effects cause different types of attenuation. The three primary effects are *absorption, scattering,* and *leakage* of light from the fiber core. You will learn more about these later, but the concepts are important to understanding how fiber-optic communications work.

Atoms within the fiber scatter light out of the core.

Although optical fibers are made of extremely pure glass, they absorb a tiny fraction of the light passing through them. The amount depends on the wavelength and the presence of impurities. Certain impurities cause strong absorption, but even pure silica has some absorption. Every transparent object absorbs a little light but transmits most of the light that enters it; opaque materials transmit a little light a little way inside them, but they absorb (or reflect) most of the incident light.

Atoms within the glass also scatter light. The physics are complex, but the atoms act as if they were tiny reflective particles, like droplets in a fog bank. Scattering reflects light off in other directions, so it escapes from the fiber core and is lost from the signal. Like absorption, scattering is inherent in all fiber materials, but generally is small. The amount of scattering increases at shorter wavelengths, so it's higher at visible wavelengths than in the infrared. The physics are the same as for light scattering in the atmosphere, which spreads short-wavelength blue light all over the sky, while allowing longer red wavelengths to reach us as the sun rises and sets.

Light leakage occurs when light escapes from the fiber core into the cladding. It's normally very low unless the fiber is bent sharply, when light can escape by hitting the core-cladding boundary at a steep enough angle to avoid total internal reflection. As you will learn later, fiber installation and the environment can bend fibers in ways that allow light to leak out, but normally this loss is the smallest of the three types. Like leaky plumbing, it's a rare event that indicates something has gone wrong.

Although absorption and scattering are extremely small in optical fibers, total attenuation accumulates when light travels through many kilometers of fiber. Attenuation normally is measured by comparing the strength of the input signal to the output. For example, if 99% of the input light emerges from the other end, a fiber has 1% attenuation.

Attenuation is cumulative, and normally uniform through the entire length of a fiber. Thus every meter of fiber should have the same attenuation as the previous meter. If 99% of the light emerges from the first meter, 99% of that light should emerge from the second meter, and so on. For a 10-meter fiber, the light emerging should be

Attenuation of a fiber is the product of the length times the characteristic loss in decibels per kilometer.

$$\text{Output} = \text{Input} \times 0.99 \times 0.99 \times 0.99 \times 0.99 \times 0.99 \times 0.99 \times 0.99 \times 0.99 \times 0.99 \times 0.99 = 0.904 \times \text{Input}$$

More generally, the output is

$$\text{Output} = \text{Input} \times (\text{transmission/unit length})^{\text{Total length}} = \text{Input} \times (0.99)^{10}$$

These sort of calculations get messy, so generally attenuation is measured in *decibels* (dB), which are very useful units, although peculiar ones. The decibel is a logarithmic unit measuring the ratio of output to input power. (It is actually a tenth of a unit called a *bel* after Alexander Graham Bell, but that base unit is virtually never used.) Loss in decibels is defined as

Decibel losses are easy to underestimate; every 10 dB decreases signal strength by a factor of 10.

$$\text{dB loss} = -10 \times \log_{10}\left(\frac{\text{power out}}{\text{power in}}\right)$$

Thus, if output power is 0.001 of input power, the signal has experienced a 30-dB loss.

The minus sign is added to avoid negative numbers in attenuation measurements. It is not used in systems where the signal level might increase, where the sign of the logarithm indicates if the signal has decreased (minus) or increased (plus).

Each optical fiber has a characteristic attenuation that is measured in decibels per unit length, normally decibels per kilometer. The total attenuation (in decibels) in the fiber equals the characteristic attenuation times the length. To understand why, consider a simple example, with a fiber having the relatively high attenuation of 10 dB/km. That is, only 10% of the light that enters the fiber emerges from a 1-km length. If that output light was sent through another kilometer of the same fiber, only 10% of it would emerge (or 1% of the original signal), for a total loss of 20 dB.

As you can see, the decibel scale simplifies calculations of attenuation. It's widely used in electronics and acoustics as well as optics. You'll learn more about decibels later, but you should realize that they are easy to underestimate. Decibels are really exponents, not ordinary numbers. Every additional 10-dB loss reduces the output a factor of 10. A 20-dB loss is a factor of 100 ($10^{2.0}$), a 30-dB loss is a factor of 1000 ($10^{3.0}$), and a 40-dB loss is a factor of 10,000 ($10^{4.0}$). These numbers can get very big very fast. Appendix B gives some comparisons for decibel units, which you may find surprising.

Bandwidth and Dispersion

Optical fibers are unique in transmitting high-speed signals with low attenuation.

Low attenuation alone is not enough to make fibers invaluable for telecommunications. The thick wires that transmit electrical power also have very low loss, but they cannot transmit information at high speeds. Optical fibers are attractive because they combine loss with high bandwidth to allow high-speed signals to travel over long distances. In a communication system, this becomes high bandwidth, the ability to carry billions of bits per seconds over many kilometers.

Concepts such as *bandwidth* and *information capacity* are crucial in communications, and the next chapter will tell you more about them. They measure the flow of information through a communication system. For example, television signals have more bandwidth than audio signals. In general, the more bandwidth or information, the better.

Attenuation of copper wires increases with signal frequency.

The more information you want to transmit, the faster the signal has to vary, and it's the need for rapidly varying signals that can cause problems in transmitting high-bandwidth signals. Different effects limit different types of communications. The number of dots and dashes an old-fashioned electrical telegraph could transmit was limited by how fast one operator could hit the transmitting key, and how fast another could write down or relay the incoming signals.

Dispersion or pulse spreading limits fiber transmission capacity.

The speed limit on electrical wires comes from the nature of electrical currents. Moving electrons induce currents in the copper around them, so the impedance of a wire increases with the speed at which the signal varies. In practice, that means the higher the frequency, the higher the attenuation. Pairs of copper wires have very low attenuation at the extremely low frequencies used for electrical power transmission, 60 Hz in North America and 50 Hz in Europe, and they can carry audio frequencies over reasonable distances, but not television signals. Coaxial cables can transmit higher frequencies, but their attenuation increases sharply with frequency, as shown in Figure 2.10. In contrast, optical fibers have essentially the same attenuation at all signal frequencies in their normal operating range.

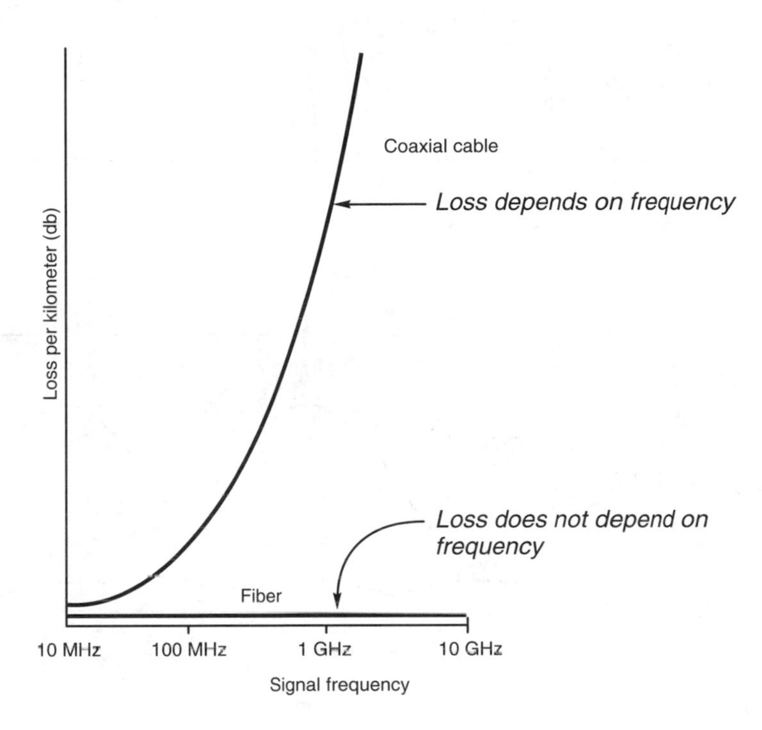

FIGURE 2.10
Loss as a function of frequency.

Other effects do limit fiber-optic transmission capacity. They are easiest to visualize if you consider a series of instantaneous pulses each containing many photons. Like a group of race cars on a track, the photons spread out as they travel, an effect called *dispersion.* Some fibers allow light to travel in many distinct paths called "modes," so they have high dispersion. Slight differences in the photon wavelengths cause chromatic dispersion because the speed of light in glass varies slightly with wavelength.

The effects are small, but like attenuation they build up. The farther the pulses travel, the more they spread out. The photons start in a sharp peak at the same time, but gradually they disperse, with the leading photons from one pulse catching up with the trailing pulses of the previous one. Eventually the pulses have spread out so much they can't be recognized, as shown in Figure 2.11. Chapter 5 tells more about dispersion and how it works.

Most types of dispersion are uniform, causing pulses to stretch out a certain amount for each kilometer they travel. This means that dispersion increases steadily with the distance the signal goes through a fiber. After a certain distance, dispersion will impose a speed limit on transmission. For example, a 2.5-Gbit/s signal might be able to go through 400 kilometers of fiber, but a 10-Gbit/s signal would be limited to only 100 kilometers. This is like the blurring of letters on the page when you look through something hazy; the smaller letters—like the faster pulses—fade away first. System performance can be improved by *dispersion compensation,* which is another concept you'll learn more about later.

FIGURE 2.11
Pulse dispersion.

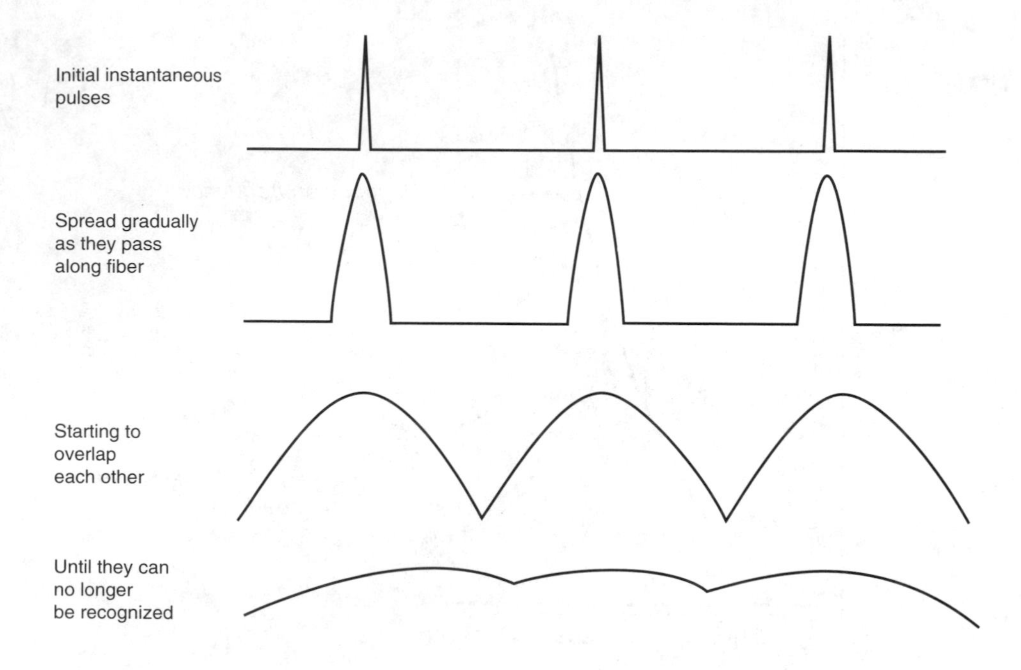

Crosstalk and Nonlinear Effects

Crosstalk is the leakage of signals between nominally independent channels.

Crosstalk occurs when signals cross the barriers that are supposed to separate them from each other. You have crosstalk on the phone if you hear a radio station or another conversation in the background. The different communication channels—phone lines and radio broadcasts—are supposed to be separate from each other. However, a little bit of one can leak into another channel. There are many reasons for electrical crosstalk. Phone wires can act as antennas to pick up strong radio signals. Currents in one pair of wires can induce signals in another pair running beside them. Sometimes other equipment may transmit signals through the air at the same frequency, so your cordless phone might pick up a neighbor's conversation.

Fibers are immune to the usual electronic crosstalk. They don't carry electrical currents, and the light inside them is unaffected by nearby currents. You can run fibers along power lines and never hear a thing, although the 60-cycle hum would overwhelm telephone wires.

Nonlinear interactions between optical channels in the same fiber can cause crosstalk.

However, fibers carrying multiple signals or *optical channels* at different wavelengths—an important technique called *wavelength-division multiplexing*—are vulnerable to crosstalk. Nominally, light signals at different wavelengths passing through the same fiber do not interact because no current flows between them. However, like electrical phone signals passing through parallel wires, there can be secondary interactions, called *nonlinear effects* because they aren't directly proportional to the strength of a single signal. These nonlinear effects are complex, and are the prime cause of crosstalk. You'll learn more about them in Chapter 5.

Other Optical Components

A variety of other optical components are used with fiber-optic systems. Some are standard in all fiber-optic systems, although they may take different forms. Others are used only in certain applications.

Transmitters and Light Sources

Optical *transmitters* generate the signals carried by fiber systems. The standard light sources are *light-emitting diodes (LEDs)* and semiconductor lasers; other lasers may be used in some special cases. Lasers emit much higher powers and can transmit much faster signals, but are more expensive. Several types of light sources are used, as explained in Chapters 9 and 10.

Optical transmitters contain LED or laser light sources, which generate the optical signal.

Visible red LEDs are used with all-plastic fibers, which transmit visible wavelengths better than the near-infrared. Distances are short and speeds are modest; a typical application would be a data link on one floor of an office building.

Near-infrared LEDs and semiconductor lasers made from gallium arsenide (GaAs) and gallium-aluminum arsenide (GaAlAs) emitting at 750 to 900 nm are used with glass optical fibers for somewhat longer distances and speeds. A typical application might be between buildings on a campus. This technology is relatively inexpensive, but limited in performance. LEDs are not common because they offer few advantages over inexpensive GaAs lasers.

Telecommunications systems transmit at 1310 nm or 1530 to 1610 nm.

Semiconductor lasers operating near 1310 nm, or in the range from about 1530 to 1610 nm, are used in telecommunications systems ranging from interbuilding links to transcontinental cables. The lasers are made from another semiconductor compound, indium gallium arsenide phosphide (InGaAsP). LEDs also can be made from this material, but they are rarely used because of their lower power and speed.

The transmitter contains electronics that *modulate* the light source. At low speeds, the transmitter directly modulates light emission by varying the drive current to the light source. High-speed transmitters drive laser light sources at a constant current, producing a steady output beam, which passes through an *external modulator* that controls how much light emerges from the transmitter.

Transmitters may contain multiple laser sources, each operating at different wavelengths. These light sources are modulated separately with different signals. As long as the wavelengths differ sufficiently, a single fiber can carry many of these optical channels.

Receivers and Detectors

A *receiver* converts an optical signal transmitted through an optical fiber into an electrical signal usable by other equipment. The photons enter an opto-electronic *detector* at the end of the fiber, which uses them to generate an electronic current or voltage. Electronics in the receiver amplify and process that signal as needed to replicate the signal that originally drove the transmitter. Chapter 11 covers them in detail.

A receiver converts an optical signal to electronic form.

Optical devices called *demultiplexers* separate optical channels transmitted through a single fiber at different wavelengths before they reach the receiver. This must be done before the detector because detectors respond in the same way to closely spaced wavelengths, so sending all optical channels to the same detector would jumble them.

Amplification and Regeneration

Repeaters and regenerators are three-stage devices that detect a weak optical signal with a receiver, process it to regenerate a clean signal, then drive a separate optical transmitter with the clean signal. They are used where needed to clean up all signal-degrading effects, including dispersion and crosstalk as well as attenuation, before they reach levels that interfere with transmission.

Optical amplifiers merely amplify the strength of weak optical signals, compensating for accumulated attenuation. Because they directly amplify the input signal with no additional processing, they amplify noise and distortion as well as signal. Thus whatever crosstalk or dispersion enters a fiber appears in the output multiplied by the amplification factor. Although this sounds worse than a regenerator, optical amplifiers are much simpler and can simultaneously amplify the optical channels at many separate wavelengths. Electro-optic regenerators or repeaters require separate electronic channels for each optical channel, greatly increasing cost and complexity. Chapter 12 covers both optical amplifiers and electro-optic regenerators.

Some optical devices can clean up optical signals. For example, optics can be added to compensate for chromatic dispersion, by adding an amount of dispersion equal in magnitude but opposite in sign. You will learn more about these and other devices in later chapters.

Noncommunication Fiber Optics

Although communications is the most important application of fiber optics, fibers are used in some other applications. These applications generally require different types of fibers, which Chapters 29 and 30 cover in more detail.

What Have You Learned?

1. Light is part of the electromagnetic spectrum, a wave that has a distinct wavelength, frequency, and photon energy.
2. Wavelength equals the speed of light divided by the frequency.
3. Refractive index *(n)* is a crucial property of optical materials. It equals the speed of light in vacuum divided by the speed of light in the material, so it is always greater than 1.
4. Total internal reflection can trap light inside a material with higher refractive index than its surroundings.

5. The critical angle for total internal reflection at a surface depends on the refractive indexes of the materials on either side. The larger the difference, the larger the angle over which total internal reflection occurs.
6. The core of an optical fiber must have a higher refractive index than the cladding surrounding it.
7. Light is guided through optical fibers by total internal reflection of light entering within an acceptance angle.
8. LEDs and lasers are the light sources for fiber-optic communications.
9. Glass optical fibers transmit signals at 750 to 900, 1310, and 1530 to 1610 nm in the near-infrared. Plastic optical fibers transmit visible light.
10. The small size of optical fibers makes tolerances tight for transferring light into fibers.
11. The low attenuation of optical fibers allows them to carry high-speed signals over long distances.
12. Fiber attenuation is usually measured in decibels per kilometer. Total attenuation of a length of fiber (in decibels) equals that value times the length in kilometers.
13. Transmission capacity of optical fibers depends on the dispersion of light pulses sent through them.
14. Attenuation and dispersion of an optical fiber vary with wavelength.
15. Gallium arsenide LEDs and lasers operate at 750 to 900 nm. InGaAsP lasers emit at 1300 to 1620 nm.
16. Direct modulation of drive current changes how much light a semiconductor laser or LED emits. External modulation changes power in the beam after it leaves the laser.
17. Optical amplifiers, repeaters, and regenerators stretch transmission distance by boosting the strength of weak signals.

What's Next?

In Chapter 3, we will look at how fiber-optic systems are used in communications.

Further Reading

Introductory level:

David Falk, Dieter Brill, and David Stork, *Seeing the Light: Optics in Nature, Photography, Color, Vision and Holography* (Harper & Row, 1986).

B. K. Johnson, *Optics and Optical Instruments* (Dover, 1960)

More advanced:

Eugene Hecht, *Optics* (Addison-Wesley, 1987)

Francis A. Jenkins and Harvey E. White, *Fundamentals of Optics* (McGraw-Hill, 1976)

Questions to Think About for Chapter 2

1. Interference seems to be a strange effect. The total light intensity from two bulbs is the sum of the two intensities. Yet the light intensity is really the square of the amplitudes, and if the two waves are in phase, you double the amplitude, which when squared means the intensity should be four times the intensity of one bulb. Don't these views contradict each other?
2. One photon is a wave packet that doesn't last very long. A continuous light source emits a steady or continuous wave. How is the continuous light source emitting photons?
3. The sun emits an energy of about 3.8×10^{33} ergs per second. A photon with wavelength of 1.3 micrometers has an energy of about 1.6×10^{-12} erg. If you assume the sun emits all its energy at 1.3 μm, how much attenuation in decibels do you need to reduce the sun's entire output to a single 1.3-μm photon per second?
4. If an entire galaxy contains a billion stars, each one as luminous as the sun, how much attenuation does it take to reduce its entire output to a single 1.3-μm photon per second?
5. Suppose a material has attenuation of 10 dB/m at 1.3 micrometers. How thick a block of the material would you need to reduce the sun's entire output to a single photon as in Problem 3?
6. Medical imaging fiber has attenuation of 1 dB/meter at optical wavelengths. If the attenuation is the same at 1.3 μm, and you don't have to worry about the sun's energy melting the fiber, how long a fiber would reduce the sun's output in Problem 3?
7. Atoms and molecules in the atmosphere scatter light in the same way that atoms in glass scatter light in an optical fiber. The shorter the wavelength in the visible spectrum, the stronger the scattering. Where do you think the sky gets its blue color from and why?
8. Diamond has a refractive index of 2.4. What is its critical angle in air and what does that have to do with its sparkle?

Quiz for Chapter 2

1. Which of the following is *not* an electromagnetic wave?
   a. Radio waves
   b. Light
   c. Infrared radiation
   d. X rays
   e. Acoustic waves

2. Optical fibers have minimum loss near 1.5 μm. What is the frequency that corresponds to that wavelength?
   a. 200 MHz
   b. 20 GHz
   c. 200 GHz

d. 20 THz

e. 200 THz

3. An electron-volt is the energy needed to move an electron across a potential of 1 V. Suppose you could convert all the energy from moving an electron across a potential of 1.5 V into a photon. What would its wavelength be?

a. 0.417 μm

b. 0.5 μm

c. 0.827 μm

d. 1.21 μm

e. 1.2399 μm

4. Light that passes from air to glass is

a. reflected.

b. refracted.

c. absorbed.

d. scattered.

5. Light is confined within the core of a simple clad optical fiber by

a. refraction.

b. total internal reflection at the outer edge of the cladding.

c. total internal reflection at the core-cladding boundary.

d. reflection from the fiber's plastic coating.

6. An optical fiber has a core with refractive index of 1.52 and a cladding with index of 1.45. Its numerical aperture is

a. 0.15.

b. 0.20.

c. 0.35.

d. 0.46.

e. 0.70.

7. Zircon has a refractive index of 2.1. What is its critical angle for total internal reflection?

a. 8°

b. 25°

c. 28°

d. 42°

e. 62°

8. The output of a 20-km fiber with attenuation of 0.5 dB/km is 0.005 mW. What is the input power to the fiber?

a. 0.5 mW

b. 0.1 mW

c. 0.05 mW

d. 0.03 mW

e. 0.01 mW

9. What fraction of the input power remains after light travels through 100 km of fiber with 0.3 dB/km attenuation?

a. 0.1%

b. 0.5%

c. 1%

d. 5%

e. 10%

10. If a 1-cm glass plate transmits 90% of the light that enters it, how much light will emerge from a 10-cm slab of the same glass? (Neglect surface reflection.)

a. 0%

b. 9%

c. 12%

d. 35%

e. 80%

11. What happens to light that is scattered in an optical fiber?

a. It escapes from the sides of the fiber.

b. Glass atoms absorb its energy.

c. Glass atoms store the light and release it later.

d. It is reflected back toward the light source.

e. It excites acoustic waves in the glass.

12. What effect does dispersion cause?

a. Scattering of light out the sides of the fiber

b. Stretching of signal pulses that increases with distance

c. Shrinking of signal pulses that become shorter with distance

d. Attenuation of signal pulses

CHAPTER 3

Fundamentals of Communications

About This Chapter

The most widespread applications of fiber optics are in communications. Optical fibers serve as flexible, low-cost "pipes" to carry light signals in environments from climate-controlled offices to the bottom of the ocean. They span distances from across an office to across the Pacific, carrying signals at rates to trillions of bits per second. These fiber-optic systems are part of a global communications network, and to understand how they work, you need to understand the basic concepts behind modern telecommunications.

This chapter introduces basic communications concepts and how fiber optics are used in communication systems. I will go into more detail in the second half of this book, but this chapter will give you the background you need to understand the roles of the components covered in the first half of the book.

Communication Concepts

Communication is the process of transmitting information. It's a big, vague, fuzzy word and broadly applied. The "Communications Department" at a university may include speech teachers, public relations specialists, writers, and broadcasters, who have nothing to do with hardware. By that definition, a writer is in the communications business.

For purposes of this book, we will use a narrower definition that means sending information over a distance by some technical means. A more precise term is *telecommunications,* but we don't need the extra syllables all the time. Traditional dictionaries define telecommunications as transmitting signals a distance using wires or radio, but these days it also means sending signals through fiber optics. In fact, fiber optics have come to

FIGURE 3.1
Home telecommunications.

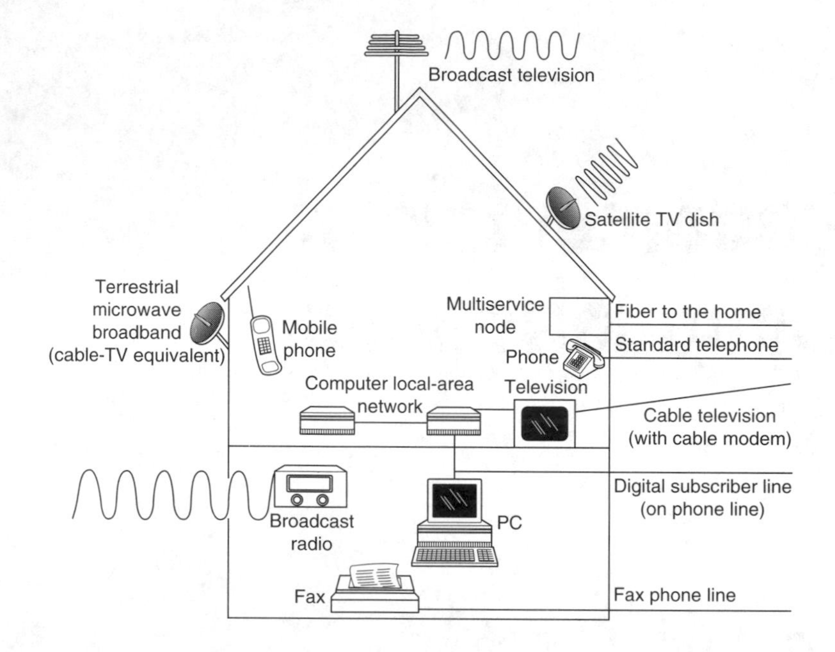

Telecommunications sends signals over a distance by fiber, wire, or radio.

The optical telegraph relayed messages from hilltop to hilltop 200 years ago.

play a crucial role as the high-speed backbone of the global telecommunications network. Today there are many different kinds of telecommunications. Figure 3.1 shows how many you might encounter in the home of a well-off technophile.

Evolution of Communications

The earliest long-distance communications was by signal fires that relayed simple information. One famous example came during the American Revolution, when Paul Revere watched the steeple of Boston's Old North Church for one or two signal lamps. One meant British troops were leaving Boston by land; two meant they were going by sea. Such signals could be seen for miles, but codes giving their meaning had to be prearranged. The presence or absence of a signal light tells the person who sees it only "yes" or "no"—the question being asked had to be known beforehand. It's only a single bit of information, not a detailed message.

A written letter or a human messenger could carry more information. The first systems we might call telecommunications were series of hilltop towers, built by French engineer Claude Chappe in the 1790s. The towers had to be in sight of each other, with an operator in each one. The operator relayed a message by moving arms on top of the tower. The operator of the next tower looked through a telescope to watch the arms move, reading the message, recording it, then moving the arms of his tower to relay it to the next tower. Chappe invented a code that used arm positions to indicate letters or certain common words. His system was called an *optical telegraph.*

Samuel Morse's *electrical telegraph* eventually replaced Chappe's optical one. It also required operators, but they used keys that completed an electrical circuit, sending bursts of current through a wire. The Morse code was a series of dots and dashes (short and long pulses) that represented letters.

The electrical telegraph spread across the continents and in 1866 across the Atlantic. Its wires formed a network running between major cities. Telegraphers received signals, and either sent them for local delivery or relayed them to more distant stations. People did not have telegraphs in their homes, but the stock ticker was invented to serve as a special-purpose telegraph for stockbrokers, relaying a continuous flow of information on stock trades that printed out on a paper tape.

You can think of each dot and dash of the telegraph as a bit of information. Engineers devised ways to make telegraph wires carry two or more signals at once—a process called *multiplexing*—but each signal was still very slow. Alexander Graham Bell's telephone borrowed some principles from the telegraph, but instead of transmitting dots and dashes, it sent a continuously changing electrical current that represented a speaker's voice. Many telegraph companies saw no future in it, but soon telephones started reaching homes and offices.

The telephone network was bigger and more complex than the telegraph network because there were more phones. By the 1890s, thickets of telephone wires stretched between poles in downtown areas. Telephone signals were not as easy to transmit long distances as telegraph signals, because they started out weaker and faded with distance. Voice telephone signals carried much more information than telegraphs. Mechanical devices could regenerate telegraph signals when they became weak, but telephone signals could not be amplified until vacuum-tube circuits were developed.

Radio waves soon carried signals through the air. Radio first carried telegraph signals. Wires worked fine on land, but only radio could send telegrams to ships at sea, and relay urgent messages such as pleas for help. It was radio rescue calls that sent ships to help the survivors of the sinking of the *Titanic*. In the 1920s, radio telephones began to send signals across the Atlantic, something that wires could not do at the time. Radio communicated with ships and airplanes in World War II, and was the only way to get voices across the Atlantic until the first submarine cable was laid in 1956.

Radio communications started at low frequencies, but gradually moved to higher and higher frequencies as electronics improved. The higher the frequency, the more information the signal can carry—a principle we'll explain later. Pictures need much more transmission capacity than sound alone, so television channels are broadcast at higher frequencies than audio radio.

Radio transmission moved to higher frequencies where it could carry more information.

Communications engineers learned how to multiplex many telephone conversations, so they could send many signals over the same path. They developed high-frequency radio relay systems to carry these combined signals long distances. These were chains of towers tens of kilometers apart, repeating signals in an electronic version of Chappe's optical telegraph. Metal coaxial cables like the coax used in cable television systems also could carry these signals.

By the 1970s, satellites were beaming radio signals around the globe. The telephone system had become global, and was generally called the telecommunications network. You could make phone calls to much of the world, but overseas calls cost dollars a minute when dollars were real money, so they were rare except for business. Long-distance calls within the United States were less expensive, but still costly enough to worry about.

Fiber optics spread when the American long-distance telephone market opened in the 1980s.

It was then that fiber optics arrived. When the American long-distance market was opened to competition in the early 1980s, telephone companies built their new high-capacity national backbone systems from fiber. The capacity of fibers has been increasing ever since, and now far exceeds that of any other telecommunications medium.

Signals and Systems: Telecommunications Terminology

That brief account has compressed a large amount of information, and tossed out a few new terms. Telecommunications is full of confusing buzzwords, so let's pause to explain a few important concepts before exploring the field in a bit more detail.

Information is what's transmitted via communications. It may be a very simple message saying "yes" that some anticipated event has happened, like the ancient Greeks announced the fall of Troy with mountaintop signal fires. It can be a huge and complex message, such as the digital files containing an entire book, or a television broadcast of a movie. It doesn't matter if the "information" doesn't contain anything a critic might agree was "information" (such as your least-favorite television program), it counts as information from a telecommunications standpoint.

A signal transmits information.

A *signal* transmits that information. Signals may take many forms, such as optical, acoustic, electronic, or radio-frequency. Signals may be converted from one form into another and still contain the same information. When you make a long-distance telephone call, the sound waves from your mouth are converted to an electrical signal at the telephone, that electrical signal is converted to optical form at the local telephone switching office, back to electrical form at the local switching office on the other end, and back to sound waves at the other person's telephone.

A *system* is the equipment that performs a task, like transmitting signals. We often speak of the telephone system as if it's the whole network of telephone equipment from your phone to the high-capacity fiber-optic cables that carry long-distance calls. However, parts of the telephone system also may be called systems, such as a switching system that directs phone calls. Think of a system as a bunch of stuff designed to work together.

A *solution* is a meaningless marketing buzzword applied to a system or other product that somebody is trying to sell. If the system didn't solve some problem, it wouldn't be useful. The term is annoying because it can conceal the product's function.

An optical channel is a wavelength transmitting a distinct signal.

A *channel* is a distinct signal. One transmission can carry many channels, like a single coaxial cable can deliver many video channels in cable television. An *optical channel* is a signal transmitted at one wavelength; a single fiber may carry one or many optical channels.

Multiplexing is the packaging of multiple channels so they can travel through a single transmission medium. Special equipment on both ends combines and then separates the channels in the multiplexed (combined) signal.

Capacity or *bandwidth* is how much information a system can carry. It can be measured as a data rate (bits per second), a range of frequencies (megahertz), or as a channel count (video or voice channels).

Radio from a telecommunications standpoint describes the frequencies of a large part of the electromagnetic spectrum from about 10 kilohertz to 100 gigahertz. Different parts of that spectrum are used for different purposes. What we usually call "radio" is audio broadcasts at certain radio frequencies. The American AM audio broadcast band is from about 500 to 1700 kilohertz; the American FM audio broadcast band is 87 to 108 megahertz. Broadcast television occupies much (but not all) of the spectrum between about 50 to 1000 megahertz. The radio spectrum is divided into many bands.

Microwaves are radio waves at high frequencies, from about 1 to 100 gigahertz. The higher portion of that band blends into *millimeter waves,* which have wavelengths around a millimeter and are not transmitted well through the atmosphere.

Wireless literally means without wires. In practice, it means signals are sent through the air without a physical connection, such as to a mobile telephone or pager. Typically this is done via radio or microwaves, from the ground or sometimes from satellites. Some household wireless systems, like remote controls, use infrared light.

Satellites orbit above Earth's atmosphere, where they can transmit signals to the ground. *Geosynchronous* satellites circle the earth exactly once every 24 hours in an orbit about 36,000 kilometers (22,000 miles) above the equator, so from the ground they seem to stay in the same place continuously. They include *direct-broadcast satellites,* which transmit many television channels to the ground, as well as other satellites that relay signals between points on the ground for applications such as paging. *Low-orbit satellites* have orbits between a couple hundred and a few thousand kilometers above the surface, so they move quickly across the sky. Arrays of many low-orbit satellites can relay signals as long as the systems keep at least one satellite overhead at all points served, and hand transmission off between satellites.

Coaxial cable is a metal cable with a central wire along its axis that is surrounded by a nonconductive material (usually plastic) and covered by a metallic shield, as shown in Figure 3.2. The central wire carries current while the shield confines the electromagnetic field of the signal. Often called *coax,* it transmits radio and low-frequency microwave signals.

Twisted pair is a pair of thin copper wires covered with insulation and wound around each other in a helical shape, long considered the nominal standard for carrying telephone signals in homes and offices. If you look closely, you'll see that much "twisted pair" may not be twisted and may contain more than two wires, also shown in Figure 3.2. The most common telephone wire in current use is a ribbon of four wires, although only two are needed

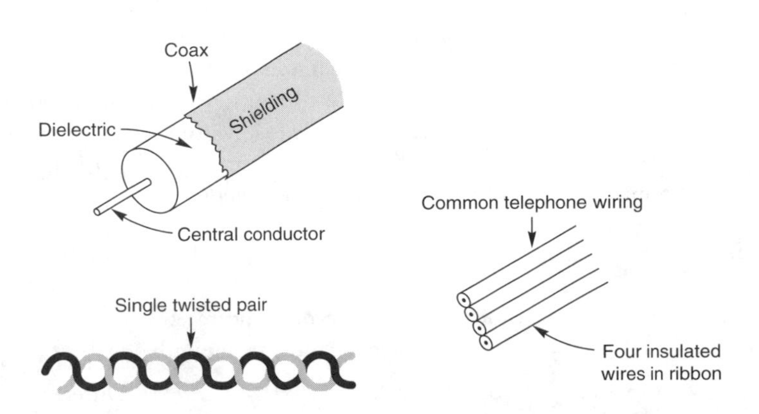

FIGURE 3.2
Types of copper cables.

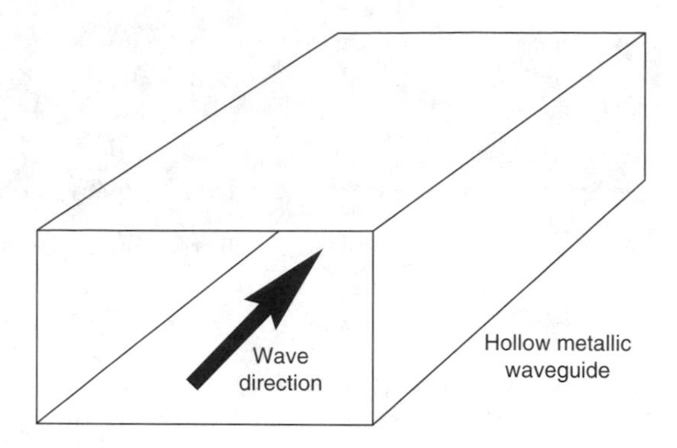

FIGURE 3.3
Hollow metal waveguide.

to carry signals for a single telephone line. Standard twisted pair can carry high-frequency signals only very short distances, but special versions, such as *Category 5* cable, can carry high frequencies farther.

Copper is the generic term for metal telecommunication cables, including both coax and twisted pair.

Waveguides are structures that guide electromagnetic waves. The best-known types are rectangular metal guides that transmit microwaves, as shown in Figure 3.3. Their properties depend on the shape of the guide and the wavelength. Waveguide theory is quite general, and also can give the properties of copper wires and coax. Optical fiber is an example of a *dielectric (nonconductive) waveguide.*

System Functions

Communications may be point-to-point, broadcast, switched, or networked.

Communications systems serve many functions, but it's possible to divide these functions into a few fundamental categories. The simplest are *point-to-point* links, which transmit signals between a pair of nodes, such as from a central computer to a remote instrumentation terminal, or between telephone switching centers in a pair of cities. This was the first use of fiber-optic communications.

Radio and television transmissions are examples of *broadcasting,* which distributes the same signal from a central node to many points. The telephone system is an example of *switched* transmission, which directs unique signals from one sender to one particular recipient. Computer networks are examples of *networks* where many devices are linked together more or less permanently, and can send and receive signals to any point on the network at more or less anytime. These categories are simplified, but they are a good starting point for understanding the complex nature of telecommunication systems.

Broadcasting

Broadcasting sends the same signal to all points.

A broadcast system sends the same signal to everyone who receives it. In its usual simple form, transmission is one way, from the signal source to the individual. Local radio and television transmissions are good examples of pure one-way broadcast systems; signals go from the main antenna to radio and television receivers throughout the community. Satellite television works in the same way; a satellite broadcasts microwaves, which home receivers detect and decode.

Broadcasting doesn't have to be through the air. Cable television networks broadcast signals through optical fibers and coaxial cables. You and your neighbors get most of the same signals, although some of you may not have the electronics you need to decode scrambled premium channels. Cable television systems can carry two-way signals, from you to the point where transmission originates. That's how you order premium pay-per-view programs. If you have a cable modem, the cable carries digital signals both to and from your home.

Merely transmitting signals into the open air doesn't make them a broadcast if everyone can't receive them. Cellular phone systems fill the air with signals radiated from local towers, but they aren't "broadcast" in the sense that anyone can receive them. Only the person with the proper phone is supposed to receive them, although scanners can pick up the signals if they aren't encoded. Pagers work the same way.

Some radio systems do broadcast for everyone to receive, even if the goal is to allow two people to talk to each other. Citizens' Band radio is one example; like talking in an open room, everyone can hear the conversation, and jump in if they feel like it.

Like cable television, some systems may broadcast many signals to the public at large while transmitting other signals only to individuals. Some new radio systems work this way, distributing video signals from local or satellite antennas, while offering individual links at the same time.

Switched Systems

A switched system makes temporary connections between terminals so they can exchange information. The telephone system is a good example. An old-fashioned telephone switchboard made *physical connections* between a pair of phone lines when an operator plugged wires into the corresponding holes. Today, electronic switches do the same thing for local telephone calls, completing a circuit linking your telephone with the phone you're calling.

A switched system makes temporary connections between points.

Long-distance calls are completed by making *logical connections* using *voice circuits* on long-distance lines. These are systems that carry many conversations simultaneously, but reserve slots for each one, to assure everyone on the line has the capacity they need to carry on a conversation. If the capacity isn't available, you get a busy signal.

The key points of a switched system are that the connections are temporary, that they dedicate transmission capacity between a pair of nodes, and that they can be made between any pair of terminals attached to the system, like telephone calls.

Networking

A network connects many terminals that can communicate with each other.

Although the label *network* is used so widely that its meaning often is hard to pin down, *networking* does have a particular meaning for computer communications. A *computer network* consists of many terminals interconnected so they can send signals to each other.

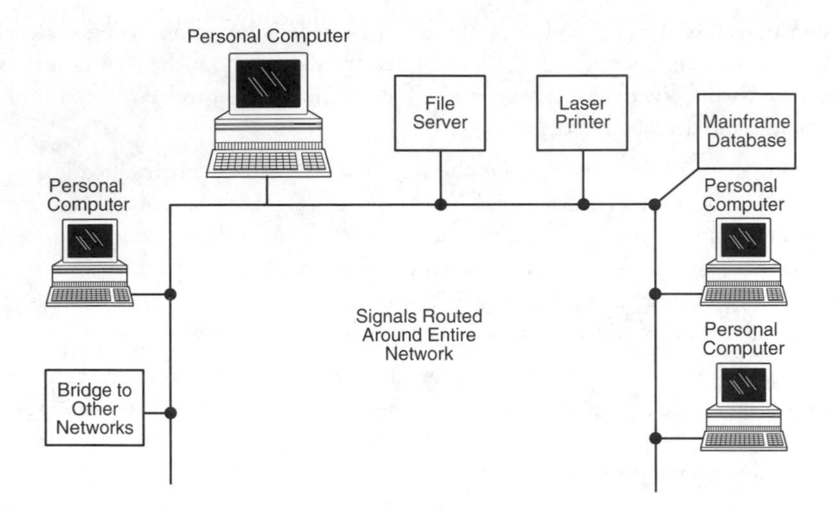

FIGURE 3.4
Local area network (LAN).

Unlike the switched connection of the telephone system, computer network connections are always "on," so each terminal can send data to any other terminal at any time. In addition, each computer terminal is set up to receive only signals directed to it.

An office *local area network (LAN)* is a good example (see Figure 3.4). When one terminal sends a message to another terminal on the network, it may go through many other terminals on the network—but only the intended recipient pays any attention. All the others ignore it. (Otherwise you get into computer networking security issues, which are far beyond the scope of this book.)

Often the results look superficially like the phone system. If you send an e-mail, it should reach only the designated recipient—just like a phone call. However, the internal details differ quite a bit, as you will learn in later chapters. Significantly, there is no circuit capacity reserved specifically for links between any two terminals (unless they're the only terminals attached), so transmission has to be shared with other users.

Pipes and Switches

A telecommunications system consists of "pipes" and "switches."

As I mentioned in Chapter 1, you can think of telecommunications systems as being made of pipes and switches. The pipes transmit information. They may be copper wires, optical fibers, or radio waves going through the atmosphere. They run between points in the system, as you can see in Figure 1.7.

The switches are the devices that make connections and organize the signals being transmitted. When the telephone system was young, its switches were manual—human operators. Mechanical switches were next, then electronic ones, and now optical switches are emerging as we shift to an optical network. Their role is to route and process signals.

The network of pipes and switches can make connections in many ways, as you can see by tracing the possible paths in Figure 1.7. The switching may be only a figurative way of organizing signals so they pass through the pipes efficiently—not just making circuit connections.

Hub-and-spoke airline schedules work somewhat like the pipes and switches of the telecommunications system. Airline flights are the pipes that carry passengers between pairs of points. Airports are the switches, where passengers get off the planes and ground crews direct them to other flights. Fortunately, the telecommunication system rarely leaves you stranded overnight in Chicago while your luggage visits Hong Kong.

Transmission Capacity and Bandwidth

Every communication system has a certain capacity that measures how much information it can transmit or process. Capacity can be measured in various ways, so it deserves a bit of attention.

Bits per second measures digital transmission capacity.

Digital capacity is usually measured as bits per second in communications. (Computer users sometimes think in terms of bytes, with eight bits equaling one byte of data.) For a transmission line, this is the total amount of data that passes a point in a second. For a switching element, it's the number of bits per second that the whole switching node can process.

Transmission line capacity depends on the transmission medium and how it is used. Fiber optics have the highest capacity of any current transmission medium. Typically transmission capacity depends on length of the medium, and declines with increasing distance. The dependence is most obvious in copper wires, which may send tens of megabits per second to an external hard drive on your computer, but only 56,000 bits per second to a telephone switching center 5.5 kilometers (18,000 feet) away. Switching capacity may be higher than transmission line capacity when many transmission lines feed into a single switch.

Bandwidth is a broad term often used to indicate digital transmission capacity. It also can measure the range of frequencies that an analog system can transmit, as described later.

Optical Networking

Combine fiber-optic transmission with optical devices for switching and managing signals, and you have an *optical network*. The term has become a very hot buzzword for the telecommunications industry, and you're sure to encounter it, although its meaning tends to be rather vague.

An optical network uses optics for processing and sending signals.

Ideally, an optical network is a system that transmits signals over fiber optics and switches them with some sort of optical device rather than with the standard electronic switches. That is, optics provide both the pipes and switches. In practice, many nominally optical switches contain some electronics that are vital to their switching function, which makes the meaning of optical networking a little slippery.

It may be more useful to think of an optical network as a system that organizes signals optically. The distinction is subtle but important. New technology makes it possible for fibers to transmit many different optical channels through a single fiber at separate wavelengths. (We'll explain optical channels later.) Systems that process these signals as optical channels rather than break them down into their electronic components can count as optical networks. This means that they use optics to organize the optical channels, treating each signal as a separate channel or *lambda* (after the symbol for wavelength), although they use electronics to process individual channels.

Signal Formats

The signal format is a crucial factor in any communication system. A *transmitter* generates a signal in a form suitable for transmission, while a *receiver* detects the signal and converts it into a usable form. Often the two are packaged together—your telephone includes a transmitter that converts your voice into electronic form, and a receiver that converts incoming electrical signals into sounds you can hear. These signal formats can vary widely.

To understand signal formats, you need to know a bit about general communications theory.

Carriers and Modulation

A signal consists of a modulated carrier wave.

A signal consists of a modulated carrier wave, as shown in Figure 3.5. Both the *carrier* and the *modulation* are important. The concepts are simplest to understand for radio transmission, but are quite general.

A carrier is a nominally pure wave at some frequency chosen by the engineer. For a radio signal, that's the frequency on the dial, such as 570 kilohertz in the AM band or 89.7 megahertz in the FM band. An oscillator at the radio station generates a signal at exactly that frequency.

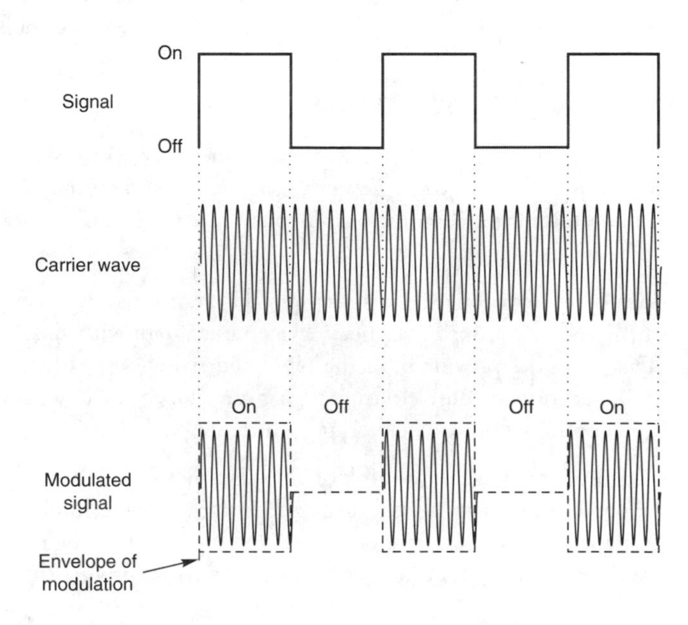

FIGURE 3.5
Signal modulates a carrier wave.

The station then modulates the carrier frequency with the station's audio signal, which is at a much lower frequency. AM stands for *amplitude modulation,* and in an AM radio station the signal modulates the amplitude of the carrier wave. When the signal is strong, so is the carrier frequency. The signal varies much slower than the carrier, forming an *envelope* around the carrier. Figure 3.5 shows digital modulation, where the signal is switched totally off and on, but analog modulation also is possible.

Radio systems also can be modulated in frequency, which is used in the FM (for *frequency modulation*) band. This also is done for conventional analog television broadcast. Its advantage is lower noise than amplitude modulation.

In fiber-optic systems, the carrier wave is an optical wavelength, usually generated by a laser source. In present systems, the transmitted signal modulates the amplitude of the laser light.

Analog and Digital Communications

Communication signals can be transmitted in two fundamentally different forms, analog or digital, as shown in Figure 3.6. The level of an analog signal varies continuously. A digital signal, on the other hand, can be at only certain discrete levels. Most digital signals are coded in binary form, with two levels, either off or on.

Signals can be transmitted in analog or digital formats. Each has its advantages, and both are compatible with fiber optics.

Each format has its advantages. The older analog technology is more compatible with people and much existing equipment. Our ears, for instance, detect continuous variations in the level of sound, not just the presence or absence of sound. Our eyes likewise detect levels of brightness, not simply the presence or absence of light. For that reason, audio and video communications have traditionally been in analog form. Telephone wires deliver a continuously varying signal to a standard telephone handset, which converts those electronic signals into continuously varying sound waves. Standard television sets likewise receive analog video signals, which they decode to display pictures on the screen. In the

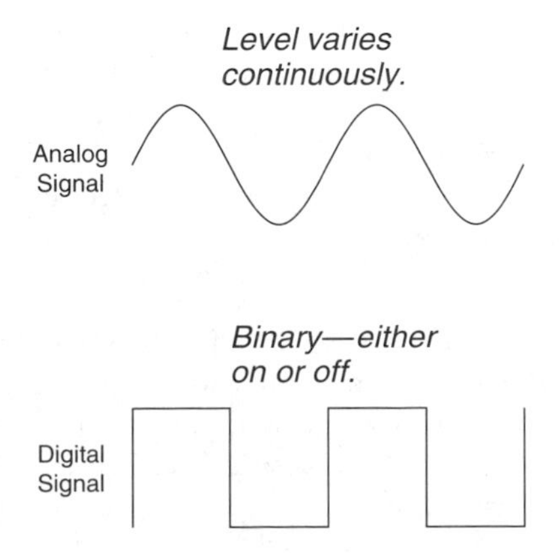

FIGURE 3.6
Analog and digital signals.

future, digital signals will go direct to home electronics, such as digital television (DTV) sets, but the digital signals still must be converted to sounds or pictures for our ears and eyes.

On the other hand, digital signals are easier to process with electronics and optics. It is much simpler and cheaper to design a circuit to detect whether a signal is at a high or a low level (off or on) than to design and build one to accurately replicate a continuously varying signal. Digital signals are also much less prone to distortion, as shown in Figure 3.7. When an analog signal goes through a system that doesn't reproduce it exactly, the result is a garbled signal that can be unintelligible. That happens when you get a distorted voice on the phone. However, when a digital signal is not reproduced exactly, it is still possible to tell the on from the off state, so the signal is clearer. That is one reason digital compact discs reproduce sound much better than analog cassette tapes or phonograph records.

Analog–Digital Conversion

If people need analog signals, but transmission works best with digital, what about converting between the two? That is often done for audio. Compact disc players use a laser to read sound digitized as spots on a rapidly spinning disc, then use internal electronics to convert the digitized sound back to analog form. The telephone network converts the analog signals from a telephone handset into digital code for long-distance transmission, then translates the digital code back to analog form on the other end.

Analog telephone signals are converted to digital format by sampling them 8000 times a second.

The idea of digitization is simple, as shown in Figure 3.8. A circuit called an analog-to-digital converter samples an analog waveform to measure its amplitude. The samples are taken at uniform intervals (8000 times per second in a telephone circuit). The converter assigns the signal amplitude to one of a predetermined number of possible levels. In older telephone circuits, that number was 128 (the number of levels that can be encoded by 7 bits), which converted a 4-kHz analog telephone signal into a stream of 56,000 bits/s (7 bits times 8000 samples). Newer phone circuits use 8-bit coding for 256 levels, generating a data stream at 64,000 bits/s.

Signals require more transmission capacity in digital than in analog form.

The figures in the last paragraph show one disadvantage of digital transmission. Accurate reproduction of an analog signal requires sampling at a rate faster than the highest frequency to be reproduced. In the telephone example, the sampling rate is twice the highest frequency to be reproduced (4000 Hz). Seven or eight bits have to be sent per sampling interval. This requires a large transmission bandwidth. There is no precise equivalence between analog and digital transmission capacity, but the two are comparable—a transmission line capable of handling 10 Mbit/s has an analog capacity of around 10 MHz. That means an analog signal takes only about a tenth of the transmission capacity that it needs in digital form. That is not a problem in telephony, but it has led cable television carriers to stay with analog transmission for most signals.

Fiber-optic systems handle both digital and analog signals.

Fiber optics work well for digital signals and were initially used mainly for digital systems. They have the high transmission capacity needed for digital transmission, and many light sources suffer from nonlinearities that induce distortion in analog signals at high frequencies. However, developers have also succeeded in making highly linear analog fiber systems, which are widely used to distribute signals for cable television.

Output

Input

Unintelligible

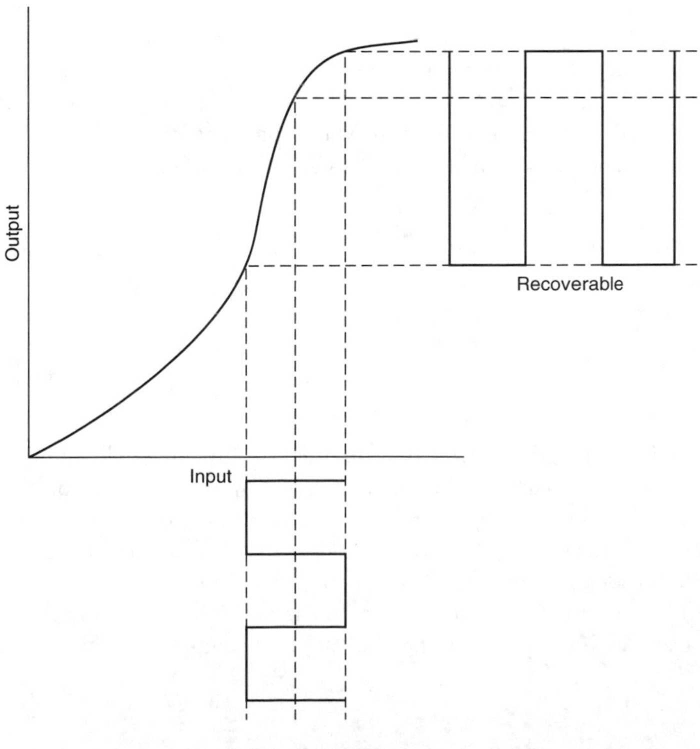

FIGURE 3.7
Distortion of analog and digital signals.

FIGURE 3.8
Digitization of an analog signal.

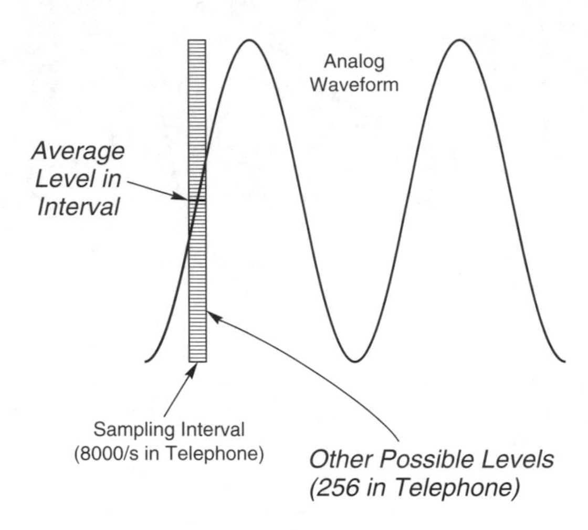

Multiplexing

Multiplexing combines many signals into a single higher-speed signal.

Back in the electrical telegraph era, engineers learned that it costs less to build one high-capacity transmission system than many parallel lower-capacity systems. This led to the invention of *multiplexing,* the combination of many low-speed signals to make a single, unified higher-speed signal that can be transmitted less expensively.

Multiplexing is a general process that can be used in many types of communications. There are several different types of multiplexing, although we don't always recognize them as such.

Frequency-division multiplexing is the transmission of signals on carriers at different frequencies. Radio and television broadcasts, which multiplex transmission through the air, are good examples. Each station broadcasts a signal on its own assigned frequency, so the air transmits every station in the area. You tune your radio or television receiver to pick up the signal you want. Cable television systems likewise assign different television signals to different frequencies.

Wavelength-division multiplexing sends signals through one fiber at separate wavelengths.

Wavelength-division multiplexing is the optical counterpart of frequency-division multiplexing. Separate signals modulate carriers at different optical wavelengths (or, equivalently, frequencies in the optical region). Each wavelength is an *optical channel.* Special optics collect the light and direct it into an optical fiber, which transmits the light. At the other end, *demultiplexing* optics separate the wavelengths, directing each optical channel to its own receiver to regenerate the original signal, as shown in Figure 3.9.

Frequency- and wavelength-division multiplexing are fundamentally similar except that they operate in different parts of the electromagnetic spectrum. Both multiply the transmission capacity by the number of analog or digital channels they carry. The number of possible channels is limited by the transmission medium, channel spacing, and applicable standards. For broadcast radio and television, the main limits come from how the spectrum is assigned for different uses. In optical fibers, the main limits come from the channel spacing and the transmission capacity of the fiber and other components in the system.

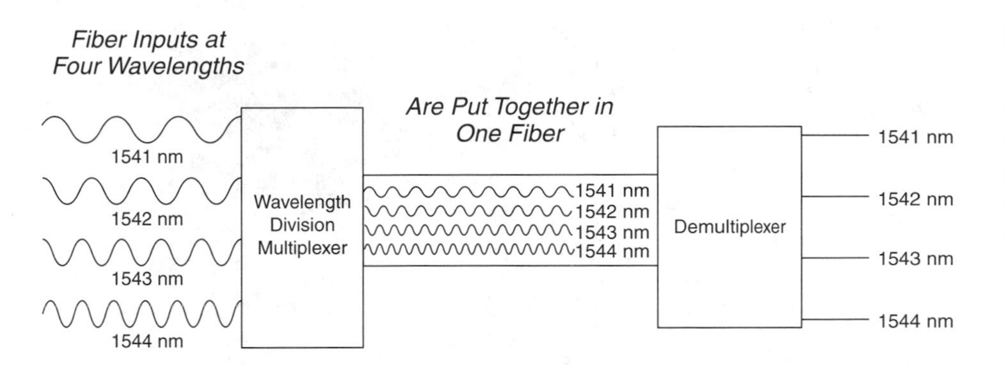

FIGURE 3.9
Wavelength-division multiplexing sends separate signals through one fiber at different wavelengths.

Time-division multiplexing is a different process that is inherently limited to digital signals. It combines bit streams from several different signals to produce a bit streams at a faster data rate, as shown in Figure 3.10. For example, four signals at 10 million bits (megabits) per second can be combined to generate one 40-Mbit/s signal. As you will learn later, a standard set of time division multiplexing rates are used in telecommunications.

Time-division multiplexing combines slow bit streams to make one faster bit stream.

Communications Services

What matters most to us about telecommunication systems is the services that they provide. You're probably familiar in general with telephones, broadcast radio and television, cable television, and the Internet. However, you will benefit from stopping to look under the hood at what goes on behind the scenes in these systems.

The Telephone System

The global telephone network can be loosely divided into a hierarchy of systems, shown in simplified form in Figure 3.11.

Your home or business phone is part of the base of the telephone network, called variously the *subscriber loop,* the *local loop,* or the *access network.* This is the part of the system running from individual telephone subscribers to telephone company switching offices (called *central offices* in the industry) in each community. A typical central office may serve thousands of homes either directly or through feeder cables that carry signals to neighborhood

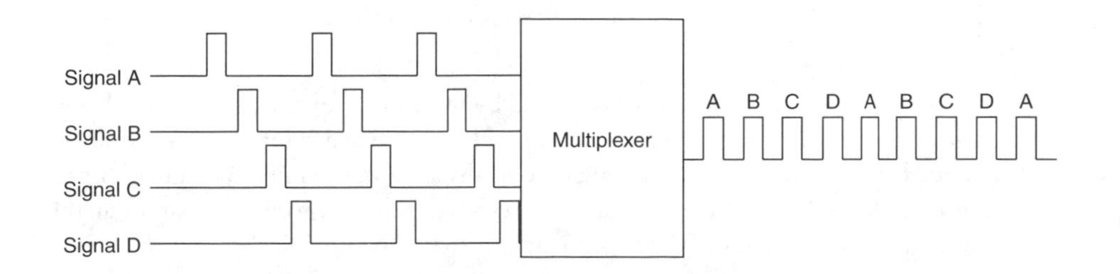

FIGURE 3.10
Time-division multiplexing combines several slow signals into a faster one.

FIGURE 3.11
Parts of the fixed telephone network.

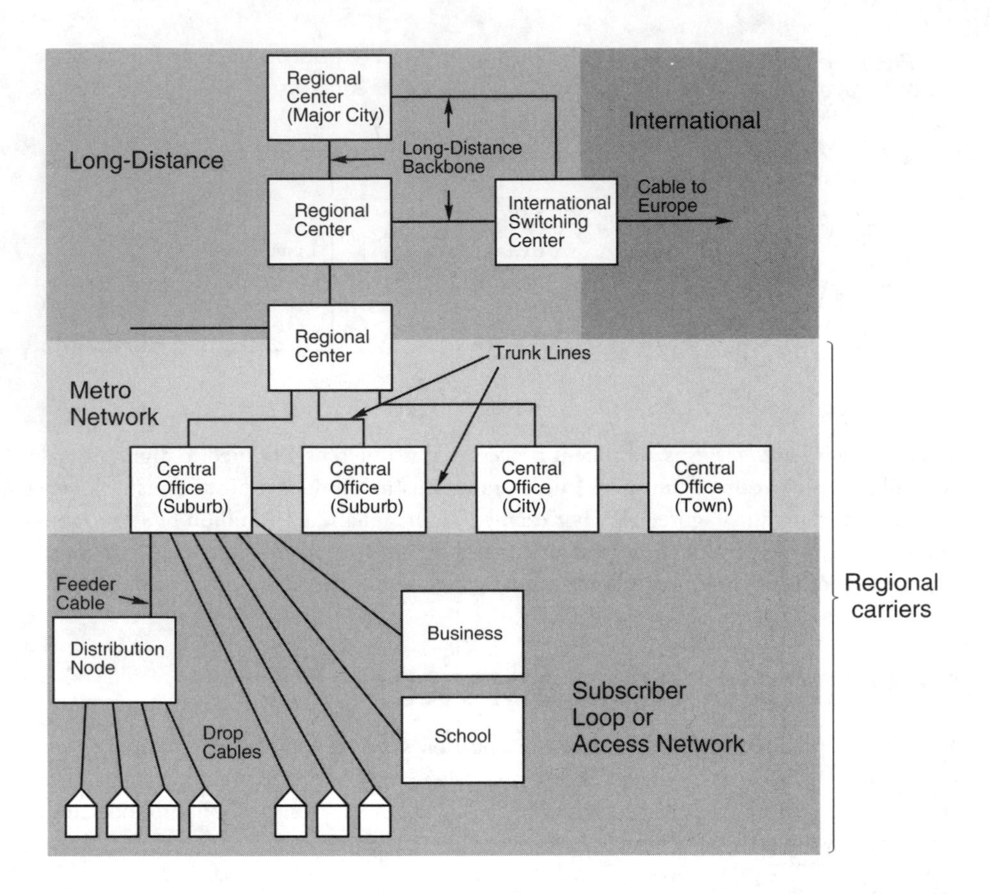

distribution nodes. *Trunk lines* run between central offices, carrying telephone calls between suburbs or towns in an area, or linking towns to nearby urban centers. Together with other links between telephone company facilities and large businesses, they form the *regional* or *metro network* that carries telephone traffic within a region or metropolitan area.

The access and metro networks—or subscriber loop and regional networks—are the domain of *regional carriers* that provide telecommunications service within a region. The best known of these are the old Bell Operating Companies, such as Bell South, SBC, and Verizon (formerly Bell Atlantic). In many areas, competing companies called *competitive local exchange carriers (CLECs)* also offer regional telephone services. Other companies also offer some metro services.

Long-distance service is a separate business offered by separate companies, although changing regulations allow some regional carriers to provide long-distance service in some areas. The difference between what's local and what's long-distance was defined by regulations issued at the time AT&T split from the original regional Bell operating companies at the start of 1984, so it may look rather strange today.

Long-distance carriers such as AT&T, Sprint, and MCI Worldcom pick up signals at regional nodes and transport them around the country via long-distance backbone systems. These backbone systems, in turn, connect with international systems such as submarine cables running to Europe and Asia. The signals pass from carrier to carrier, so things can get quite complex, although mercifully you don't see that in your phone bill.

Cellular phones, pagers, and other wireless mobile communication services link to this global telecommunications system. You can think of these mobile services as the equivalents of regional phone carriers that distribute signals through the air on radio waves rather than through cables. Mobile phones distribute signals from fixed local antennas; pagers and satellite phone systems transmit from satellites.

Optical fibers provide the backbone of the long-distance telephone network, the trunk cables that link central offices with each other, and the cables that link cellular phone towers to the rest of the global phone network. Fibers are spreading through the metro and access networks, although they so far have reached only a few homes.

Optical fibers are the backbone of the long-distance telephone network.

Fiber-optic cables have many advantages besides their high transmission capacity and low signal attenuation. Fiber cables are much smaller than old copper cables, so four fiber cables fit easily into an underground duct built for one copper cable. That can mean big savings because installing new underground ducts in urban areas is far more expensive than replacing old cables with new. Adding more fibers can multiply the potential transmission capacity of new cables at minimal cost, with the extra fibers left as spares (called *dark fibers*) ready for use when their capacity is needed. You can now buy cables containing over 1100 fibers!

Fibers also have other advantages that can be important in particular situations. Fiber cables can be made with no electrical conductors, so they won't carry dangerous current pulses injected by lightning strikes, or conduct ground-loop currents that can disturb sensitive electronics. Signals carried in fibers cannot pick up electromagnetic interference from power lines, generating plants, or other sources—and some run along high-tension power lines.

Cable Television and Video Transmission

Cable-television networks now offer telephone and data transmission services in competition with the telephone network. However, cable-TV (often called CATV, from Community Antenna TeleVision) systems have fundamentally different designs, which stem from their origins. Cable-TV systems were built to offer the same set of video channels to all subscribers, so they lacked the switches that telephone networks use to route conversations to individual phones. Cable also began as purely one-way transmission, unable to carry signals originating in homes. This is changing, but the designs are based on upgrading existing equipment rather than completely rebuilding the entire network.

Design of the cable-television network differs from that of the telephone network.

Video signals require much more bandwidth than sound, so the network that distributes cable-television signals was built with much higher capacity than the subscriber loop that distributes telephone signals. It began with coaxial cables, which can carry dozens of analog video channels, via frequency-division multiplexing, which assigns each channel a different transmission frequency analogous to broadcast frequencies. A central facility, called the *head-end,* sends the same video signals to all homes through cables that spread out in a "tree" architecture. (Premium services are sent to all homes in a coded form that can be viewed only through cable boxes with special decoders rented from the cable company.)

Fiber has replaced coaxial cable in much of the distribution network because it has much lower attenuation, avoiding the need for amplifiers or repeaters in most communities. However, cable systems continue to transmit standard television signals in analog form. Their plans for digital television are well behind those of television broadcasters.

Current cable systems resemble in some ways the telephone subscriber loop. Fibers and coaxial cables carry signals from the head-end to remote nodes from which signals are distributed through coax to individual homes. In modern systems, a small fraction of the bandwidth is set aside for voice and data signals directed to and from individual homes, and some switching capacity has been added. I will describe the details in Chapter 27.

Fiber optics also carry many video signals outside of cable-TV networks. Studios and production facilities use fiber cables because they are smaller, lighter, have higher capacity, and suffer less noise than coaxial cables. Fiber cables are also widely used in temporary installations for electronic newsgathering, such as broadcasting the Olympics. The first fiber-optic Olympic system was installed as a backup for the 1980 Winter Games in Lake Placid, New York, but ended up carrying most video signals because it worked better than the then-standard coaxial cable system.

The Internet and Data Communications

Fibers are used in backbone Internet networks and in high-speed computer networks.

Fiber optics play two distinct roles in computer data communications. Fibers are widely used for Internet backbone transmission and for high-speed connections to and from Internet service providers. Fibers also transmit computer data in networks within companies or linking corporate facilities.

Backbone Internet systems are much like those for other types of telecommunications, and fibers are used for the same reason—their high capacity. The optical hardware is essentially the same; the differences come in data-transmission protocols. I will discuss protocols later, but they are not particularly important here. Think of the Internet backbone as fiber-optic telecommunication hardware owned by companies other than standard long-distance phone companies. The technology is similar, and we won't worry about the differences in ownership and traffic. The same holds for fiber connections linking Internet service providers with telephone company facilities and with the Internet backbone system.

Local data transmission is a different matter because it involves different technology. Typically data travels over a corporate data network that links computers within the same building or in a campus of many buildings. Networks may operate on somewhat different levels. A local-area network (LAN) like the one shown earlier in Figure 3.4 may connect all the devices on one floor of an office building or belonging to one department in a company. Larger *metropolitan-area networks,* or MANs (sometimes called wide-area networks, or WANs) may provide higher-speed links between separate LANs. In a sense, they are networks of networks; a company MAN or WAN may connect separate LANs on different floors of an office building. The Internet is a network of networks on a much larger scale.

Data transmission to and from individual personal computers requires the long-distance, high-speed capacity of fiber. However, other considerations can tilt the scales toward fiber. Fibers do not carry electric currents, so they can be used where voltage isolation is critical, such as in power stations, or to carry signals through explosive atmospheres such as in refineries. Fibers do not pick up electromagnetic interference and power surges, so they can run alongside power cables in the elevator shafts of high-rise buildings—without picking up noise. Fibers also do not radiate electromagnetic signals and are difficult to tap, making them attractive for secure transmission at military facilities and financial institutions.

Fibers do not carry electric currents, so they can be used places where wire transmission would be unsafe or degraded by noise.

The rapid spread of applications requiring high-speed data transmission is pushing fibers into the growing number of high-speed networks, particularly those connecting other networks. For example, copper cables may suffice for a LAN linking all devices in a company's accounting department, but not for the corporate MAN or WAN, which links all the LANs in the company. Schemes for using inexpensive "network computers" or "Internet appliances," which lack hard disks or internal storage, require high-speed access. So do those corporate networks which host software on powerful server computers that must be accessed from individual computers or terminals. The need for fiber increases with the size of software applications and document files, and with the number, size, and detail of graphic images.

The explosive growth of the Internet is driving the expansion of all sorts of data communications in homes, businesses, education, and government. As Web sites install more elaborate graphics and sophisticated software, users demand faster communications so they can access the sites at reasonable speed. In 1985, the best dial-up modems for home use transmitted 1200 bits per second, adequate for text-only electronic bulletin boards. In 2000, the state of the art for homes is a cable modem or digital subscriber line (DSL) delivering hundreds of kilobits per second. High-performance office networks can pump data at even higher speeds.

The trend is not likely to stop. Coaxial cable and twisted-pair copper wires can deliver a megabit over modest distances, but as the demand increases, fiber looks more attractive. Intense data-communications users such as Internet providers, universities, and large businesses already have fiber delivering data that they distribute over copper. Fiber will push farther out into the network as the demand for bandwidth increases.

Fiber will push farther out into the network as the demand for bandwidth increases.

Big pipes won't answer all the problems of data communications. Simple downloads of megabyte files fly on a cable modem, but many Web pages still crawl because they're made of many separate illustrations. Delays accumulate because your computer has to request many separate files, often from separate locations. Some Internet servers respond slowly; others are on the wrong side of slow switches. The networks' switches need to be upgraded to keep pace with the pipes.

Meanwhile, fibers are being installed to "future-proof" new or renovated buildings even when no one plans to use the fibers immediately. It costs much less to install new fiber when the walls are open than when the building is complete.

Special-Purpose Communications

Light weight and EMI immunity encourage the use of fibers on ships and planes.

Optical fibers also have found applications in a variety of special-purpose communication systems because of their high speed, small size, light weight, and immunity to electromagnetic interference. Vehicles ranging in size from automobiles to aircraft carriers require internal communication systems, and optical fibers are attractive for many such applications.

Many newer planes and advanced ships use fiber optics in some of their communications systems. Electromagnetic interference (EMI) is a major concern in control rooms packed with electronic equipment. Fibers do not pick up such noise, and their small size and light weight are important advantages in planes. EMI immunity is particularly important in military ships and planes, which are subject to attack by enemy countermeasures designed to disrupt internal communications or navigation equipment. The automobile industry has long talked about using fiber optics for data networking in cars, but has not been able to cut costs low enough for widespread use of fiber systems.

Portable fiber-optic networks have been developed for deployment on the battlefield, where they would link temporary buildings or tents in a field headquarters. Small size, light weight, and better durability than fat copper cables have been key advantages—but wireless communication systems are taking their place.

Fiber-Optic Communication Equipment

The basic elements of any point-to-point communication system are a transmitter that generates the signal, a transmission medium that carries the signal, and a receiver that detects the signal and converts it into a useful form. We've already looked a bit at communications systems in general. Let's take a brief look at these key elements of fiber-optic communication systems to help you understand the upcoming chapters on optical fibers.

Fiber-Optic Transmitters

Transmitters modulate light intensity to send a signal.

A fiber-optic transmitter modulates the intensity of an optical carrier generated by a light source. This can be done in either of two ways, by directly modulating the input power to the light source or by using a separate optical component that changes the intensity of the light leaving the light source. Each approach has its own advantages, as described in more detail in Chapters 9 and 10.

Direct modulation is simple and inexpensive. It works best for light-emitting diodes (LEDs) and semiconductor lasers, because their light output increases with the drive current passing through the semiconductor device. The input signal modulates the drive current, so the output optical signal is proportional to the input electrical signal. However, LEDs take time to respond to changes in the drive current. Semiconductor lasers are much faster, but their optical properties change slightly with the drive current, causing a slight "chirp" in the wavelength as the signal switches off and on. This becomes a significant problem at speeds above about 622 megabits per second because the wavelength shift can cause significant dispersion in an optical fiber. In addition, pulsing a laser source makes its emission wavelength less stable than if the laser generates a continuous beam, an effect that can reduce the separation between optical channels in wavelength-division multiplexing.

External modulation is more complex, and external modulators cost more than driving the light source directly. However, it generally offers higher performance. The external modulator can change beam intensity without directly affecting the laser's operation, avoiding chirp or any wavelength drift caused by pulsing the laser. External modulators also can work with lasers that can't be modulated directly.

The transmitter serves as an electro-optical interface, converting an input signal into an optical signal that can travel through an optical fiber. In doing so, it also formats the signal in a form required for optical transmission.

Optical Fiber Transmission

The optical fiber is the transmission medium. An ideal transmission medium would have no effect on the signal it carries, but any medium inevitably has some effect.

In a fiber, the two principal limiting effects are attenuation of the signal strength and dispersion of the pulses. Both depend on how far light travels through the fiber. Attenuation weakens the intensity of the optical signal. Dispersion spreads the signal until pulses overlap, essentially blurring it out so it no longer makes sense. Both attenuation and dispersion vary with wavelength, so fibers have certain transmission windows. Long-distance communications requires the low loss and limited dispersion at wavelengths between 1250 and 1650 nanometers. Weaker nonlinear effects can cause crosstalk between optical channels and other types of interference and noise.

Fiber transmission is limited by attenuation and pulse dispersion. Both increase with distance.

Optical amplifiers can boost the strength of optical signals so they can travel farther through optical fibers. They amplify light directly in optical form, without converting the signal to electrical form. Optical amplifiers are not available for all wavelengths, so they, too, can limit transmission windows.

Repeaters and regenerators first convert the optical signal to electronic form, then amplify it and deliver the electronic signal they generate to another transmitter. That transmitter then generates a fresh optical version of the signal. Regenerators can clean up the effects of dispersion and distortion on optical signals. You can think of repeaters and regenerators as a receiver–transmitter pair placed back to back.

Optical Receivers

Receivers are the final elements in any communication systems. They convert a signal transmitted in one form into another form. For example, a radio receiver detects weak radio waves in the air and processes them electronically to generate sound you can hear.

Receivers convert an optical signal to electronic form.

Fiber-optic receivers detect the optical signal emerging from the fiber and convert it to electronic form. A photodetector (usually simply called a *detector*) generates an electric current or voltage from the light it receives. Electronics in the receiver then amplify that signal and process it to decode the signal. Chapter 11 covers the process in more detail.

How well the receiver does its job depends on its sensitivity, speed, and the strength of the signals reaching it. This, in turn, depends on performance of both the detector and the circuits that process the electronic signal it generates. Detectors respond to a relatively broad range of wavelengths, but not to the entire range of visible and infrared wavelengths that fiber systems can transmit. If the power reaching the detector is too low, noise may overwhelm it. Both detectors and receivers also have characteristic rise times, which limits the fastest signals they can detect.

The receiver electronics can enhance the signal generated by the detector. In addition to amplifying it, they clean it up by checking for pulses reaching a threshold value, adjusting signal timing, and sharpening rise times. However, they are inevitably limited, so they can't clean up noise beyond a certain level.

In practice, receivers and transmitters are designed in pairs, with matched properties. They both have to operate at the same wavelength with the same signal format, so the receiver can decode what the transmitter transmits. They also should offer similar levels of performance. It doesn't make sense to spend big bucks on a super-high-speed receiver if the transmitter is much slower—or vice versa.

What Have You Learned?

1. Telecommunications transmits signals over a distance. Transmission media include copper wires, radio waves, and optical fibers.
2. Optical telegraphs were the first form of telecommunications 200 years ago. Electrical telegraphs made them obsolete. Telephones and radio transmission followed.
3. Radio transmission moved to higher frequencies to increase its transmission capacity.
4. Fiber optics became common for long-distance transmission in the 1980s.
5. Telecommunication systems transmit information.
6. Multiplexing combines multiple channels for transmission as a single signal. It can be done optically or electronically, in time, frequency, or wavelength.
7. "Copper" includes twisted-wire pairs and coaxial cables.
8. Point-to-point communication systems link a pair of points, with no other connections.
9. Broadcast communications distribute the same signal to many points.
10. A switched system makes temporary connections between terminals; the telephone network is one example.
11. A network connects many terminals that can communicate with each other continuously.
12. A telecommunication system includes "pipes" and "switches." Fibers were used as pipes with electronic switches; now optical switches are becoming available.
13. An optical network uses optics to process as well as transmit signals.

14. A signal modulates a carrier wave to transmit information.
15. The telephone system includes international, national, regional, and local systems.
16. The cable television network is designed differently from the telephone network, and transmits the same analog video signals to all its customers.
17. Fibers are used for long-distance Internet traffic as well as within corporate networks.
18. Fibers will spread from backbone networks to local networks as bandwidth requirements increase.

What's Next?

Now that you have a general idea how fibers optics and telecommunications work, the rest of the book will help you learn more details about the technology. Chapters 4 through 7 cover optical fibers and their important features.

Further Reading

Roger L. Freeman, *Fundamentals of Telecommunications* (Wiley-Interscience, 1999)

Gil Held, *Voice and Data Internetworking* (McGraw Hill, 2000)

Gary M. Miller, *Modern Electronic Communication* (Prentice Hall, 1999)

Tom Standage, *The Victorian Internet* (Berkeley Books, 1998)

Questions to Think About for Chapter 3

1. How has the tradeoff between speed of communications and the amount of information that can be conveyed changed over the years?
2. Why does multiplex transmission of a combined signal cost less than separate transmission of each signal?
3. How do computer networks and mobile telephones differ from broadcast systems, considering that all three freely distribute signals that many terminals can pick up?
4. How can noise be removed from digital signals? Why doesn't this work for analog signals?
5. What are the main advantages of fiber-optic cables for long-distance transmission?
6. What is the only important telecommunication system that uses fiber to transmit analog signals?
7. Data transmission rates to personal computers have increased from 1200 bits per second with dial-up modems in 1985 to about 400,000 bits per second with a cable modem or DSL in 2000. If bandwidth keeps increasing at the present rate, how fast might transmission be in 2015?

Quiz for Chapter 3

1. Which came first?
   a. The electrical telegraph
   b. The optical telegraph
   c. The telephone
   d. Wireless radio transmission
2. Which of the following are true for analog signals?
   a. They vary continuously in intensity.
   b. They are transmitted in parts of the telephone network.
   c. They are compatible with human senses.
   d. They can be processed electronically.
   e. All of the above
3. Which of the following are true for digital signals?
   a. They can encode analog signals.
   b. They are transmitted in parts of the telephone network.
   c. They can be processed electronically.
   d. They are used in computer systems.
   e. All of the above
4. You digitize a 10-kHz signal by sampling it at twice the highest frequency (i.e., 20,000 times a second) and encoding the intensity in 8 bits. What is the resulting data rate?
   a. 20 kbit/s
   b. 56 kbit/s
   c. 144 kbit/s
   d. 160 kbit/s
   e. 288 kbit/s
5. What part of the telephone network is connected directly to your home telephone if you get your telephone service from a local telephone company?
   a. Subscriber loop
   b. Feeder cable
   c. Trunk line
   d. Backbone system
6. What part of the telephone network carries the highest-speed signals?
   a. Subscriber loop
   b. Feeder cable
   c. Trunk line
   d. Backbone system
7. Time-division multiplexing of 8 signals at 150 Mbit/s each produces
   a. Eight optical channels each carrying 150 Mbit/s.
   b. One channel carrying 120 Mbit/s.
   c. One channel carrying 1.2 Gbit/s.
   d. Eight signals at 150 MHz.
8. Demultiplexing a 10-Gbit/s signal into the signals that were combined to make it *cannot* produce
   a. 10 channels at 1 Gbit/s.
   b. 20 channels at 500 Mbit/s.
   c. 40 channels at 250 Mbit/s.
   d. 100 channels at 50 Mbit/s.
   e. 1 million channels at 10 kbit/s.

9. The carrier signal modulated to produce one optical channel in a fiber-optic system is a(n)

a. Single wavelength of light generated in the transmitter.

b. Radio-frequency signal supplied electronically to the transmitter.

c. Acoustic vibration in the optical fiber.

d. Combination of wavelengths generated by several light sources.

10. The place where a telephone company switches signals in a community is a

a. Local-area network.

b. Central office.

c. Carrier.

d. Backbone network.

e. Head-end.

CHAPTER

4

Types of Optical Fibers

About This Chapter

Optical fibers are not all alike. There are several different types, made for different applications, which guide light in subtly different ways. This chapter describes the basic concepts behind the various types of fibers. It concentrates on fiber design and light guiding. It is closely linked to the two chapters that follow. Chapter 5 describes the important properties of optical fibers. Chapter 6 covers the materials used in fibers, which play a vital role in determining their properties. Together, these three provide an essential groundwork to understanding how most optical fibers work. Chapter 7 covers special fiber types.

Light Guiding

Chapter 2 showed how the total internal reflection of light rays can guide light along optical fibers. This simple concept is a useful approximation of light guiding in many types of fiber, but it is not the whole story. The physics of light guiding is considerably more complex, because a fiber is really a waveguide and light is really an electromagnetic wave with frequency in the optical range.

Like other waveguides, an optical fiber guides waves in distinct patterns called *modes,* which describe the distribution of light energy across the waveguide. The precise patterns depend on the wavelength of light transmitted and on the variation in refractive index that shapes the core, which can be much more complex than the simple, single cores described in Chapter 2. In essence, these variations in refractive index create boundary conditions that shape how electromagnetic waves travel through the waveguide, like the walls of a tunnel affect how sounds echo inside.

It's possible to calculate the nature of these transmission modes, but it takes a solid understanding of advanced calculus and differential equations, which is far beyond the scope of this book. Instead, we'll look at the characteristics of transmission modes,

Total internal reflection is only a rough approximation of light guiding in optical fibers.

which are important in fiber-optic systems. By far the most important is the number of modes the fiber transmits. Fibers with small cores can transmit light in only a single mode. It can be hard to get the light into the fiber, but once it's inside, the light behaves very uniformly. It's easier to get light into fibers with larger cores that can support many modes, but light does not behave the same way in all the modes, which can complicate light transmission, as you will learn later in this chapter.

This chapter covers the many types of optical fibers that have been developed to meet a variety of functional requirements. Their designs differ in important ways. For example, bundles of fibers used for imaging need to collect as much light falling on their ends as possible, so their claddings are made thin compared to their cores. Communications fibers have thicker claddings, both to keep light from leaking out over long distances and to simplify handling of single fibers. Various types of communications have their own requirements. Fibers for short-distance communications typically have large cores to collect as much light as possible. Long-distance fibers have small cores, which can transmit only a single mode, because this well-controlled light can carry signals at the highest speed.

Core-cladding structure and material composition are key factors in determining fiber properties.

The two considerations that affect fiber properties most strongly are the core-cladding structure and the glass composition. The size of the core and cladding and the nature of the interface between them determine the fiber's modal properties and how it transmits light at different wavelengths. The simple types of fiber discussed in Chapter 2 have a *step-index* structure, where the refractive index changes sharply at the abrupt boundary between a high-index core and a low-index cladding. Replacing that abrupt boundary with a gradual transition between core and cladding, or including a series of layers, changes fiber properties. Glass composition, covered in Chapter 6, strongly affects fiber attenuation, as well as influencing pulse spreading.

Combined with other minor factors, these parameters determine important fiber characteristics, including

- Attenuation as a function of wavelength.
- Light-collection capacity.
- Transmission modes.
- Pulse spreading and transmission capacity, as a function of wavelength.
- Tolerances for splicing and connecting fibers.
- Operating wavelengths.
- Tolerance to high temperature and environmental abuse.
- Strength and flexibility.
- Cost.

Figure 4.1 shows selected types of single fibers (as distinct from bundled fibers), along with a plot of refractive index across the core and cladding, called the *index profile.* Only the core and cladding are shown for simplicity; actual fibers have an outer plastic coating 250 μm in diameter to protect them from the environment. I will start with the fiber type that is simplest to explain in terms of total internal reflection, called step-index multimode fiber, because it transmits many modes.

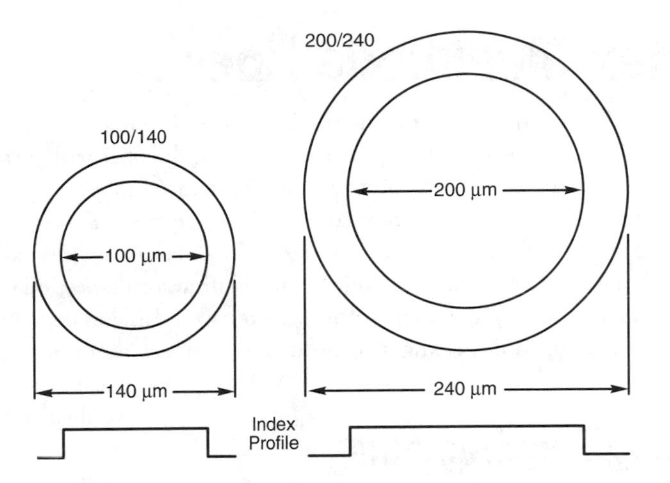

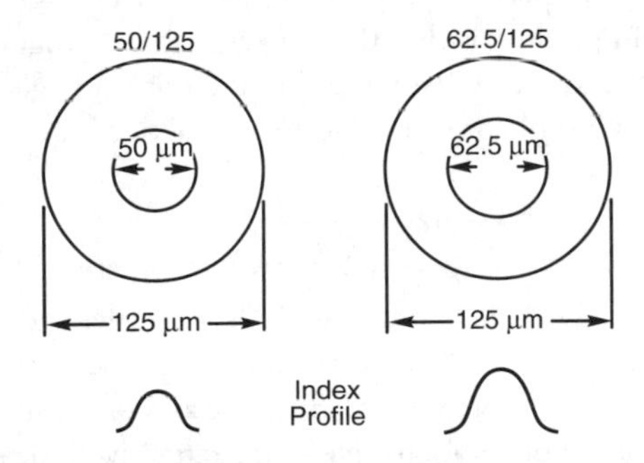

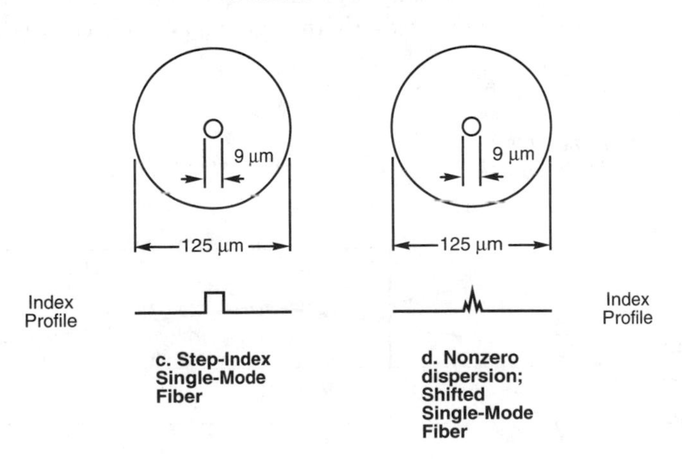

FIGURE 4.1
Common types of optical fiber (to scale).

Step-Index Multimode Fiber

As we saw in Chapter 2, bare, transparent filaments surrounded by air are the simplest type of optical fiber, but they don't work well in practice. Cladding the fiber with a transparent material having lower refractive index protects the light-carrying core from surface scratches, fingerprints, and contact with other cores of the same material, so the light will not escape from the surface. This simple fiber consists of two layers of material, the core and cladding, which have different refractive indexes. If you drew a cross section of the fiber and plotted the refractive index, as in Figure 4.1(a), you would see a step at the core-cladding boundary, where the index changes abruptly.

Light-Guiding Requirements

To guide light, the fiber core must have refractive index higher than the cladding.

As long as the core of a fiber has a diameter many times larger than the wavelength of light it carries, we can calculate fiber properties using the simple model of light as rays. The fundamental requirement for light guiding is that the core must have a higher refractive index than the cladding material. We saw in Chapter 2 that the critical angle for total internal reflection, θ_c, depends on the ratio of core and cladding refractive indexes.

$$\theta_c = \arcsin\left(\frac{n_{\text{clad}}}{n_{\text{core}}}\right)$$

For a typical fiber, the difference is small, about 1%, so the critical angle is arcsin (0.99), or about 82°. Because the critical angle is measured from a line perpendicular to the surface, this means that light rays that are no more than 8° from the axis of the fiber are reflected, as shown in Figure 4.2. This value is not very sensitive to the refractive-index difference. If the difference is doubled to 2%, the critical angle becomes 78.5°, so light rays no more than 11.5° from the axis of the fiber are reflected. Alternatively, you can directly calculate the

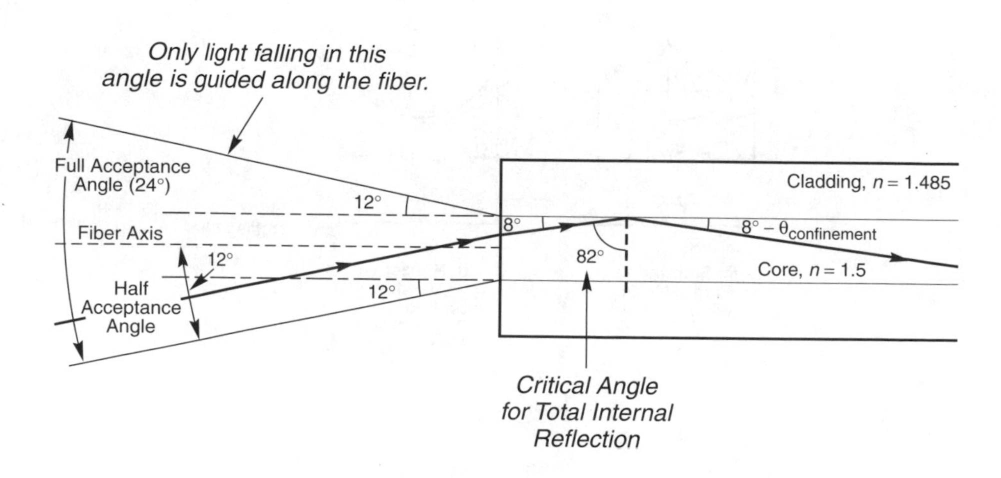

FIGURE 4.2 *Light guiding in a large-core step-index fiber. The confinement angle measures the angle between guided light rays and the fiber axis; the acceptance angle is measured in air.*

confinement angle measured from the core-cladding boundary ($\theta_{confinement}$) by using the arc-cosine:

$$\theta_{confinement} = \arccos\left(\frac{n_{clad}}{n_{core}}\right)$$

The confinement angle gives the maximum angle at which guided light can strike the core-cladding boundary once it's inside the glass. However, refraction occurs when the light enters the glass from air, bending light toward the axis of the fiber. To calculate the acceptance angle, measured in air, you must account for this refraction using the standard law of refraction. As long as the light enters from air, you can simplify this to

The confinement angle is the largest angle at which light rays confined to a fiber core strike the core-cladding boundary.

$$\sin \theta_{half\text{-}acceptance} = n_{core} \times \sin \theta_{confinement}$$

which gives the sine of the largest possible angle from the axis of the fiber, called the half-acceptance angle, $\theta_{half\text{-}acceptance}$. You can calculate the half-acceptance angle directly by juggling the trigonometry a bit more:

$$\theta_{half\text{-}acceptance} = \arcsin\,(n_{core} \times \sin \theta_{confinement})$$

Doubling the half-acceptance angle gives the full-acceptance angle. The confinement angle is small enough that you can roughly approximate the half-acceptance angle by multiplying the confinement angle by the refractive index of the core, n_{core}.

Imaging Fibers

The first clad optical fibers developed for imaging were what we now call step-index multimode fibers. Developers tested a variety of cladding materials with low refractive indexes, including margarine, beeswax, and plastics. However, the key practical development was a way to apply a cladding of glass with lower refractive index than the core.

Step-index multimode fibers were the first fibers developed for imaging.

As we will see in Chapter 6, glass comes in many different formulations with varied refractive indexes. The simplest way to make glass-clad fibers is to slip a rod of high-index glass into a tube with lower refractive index, heat the tube so the softened glass collapses onto the rod, let them fuse together, then heat the whole *preform,* and pull a fiber from the molten end. Figure 4.3 shows the process schematically.

One subtle but crucial requirement is that the core-cladding interface must be smooth and clean on the scale of the wavelength of light. That is possible if the rods are fire-polished but not if they are polished mechanically, a process that leaves grit and fine cracks on the surface. Those flaws remain when rod and tube are melted together and drawn into fiber, and they can scatter or absorb light. (A major reason Larry Curtiss succeeded in making the first good glass-clad fibers was that he used a flame-polished rod.)

The cladding of imaging fibers generally is a thin layer surrounding a thicker core. The reason for this design is that imaging fibers are assembled in bundles, with light focused on one end of the bundle to emerge at the other. Light falling on the fiber cores is transmitted from one end to the other, but light falling on the cladding is lost. The thinner

FIGURE 4.3

A simple way to make step-index multimode fiber is by inserting a glass rod in a tube with lower refractive index, melting the two together to form a preform, and pulling fiber from the hot bottom of the preform.

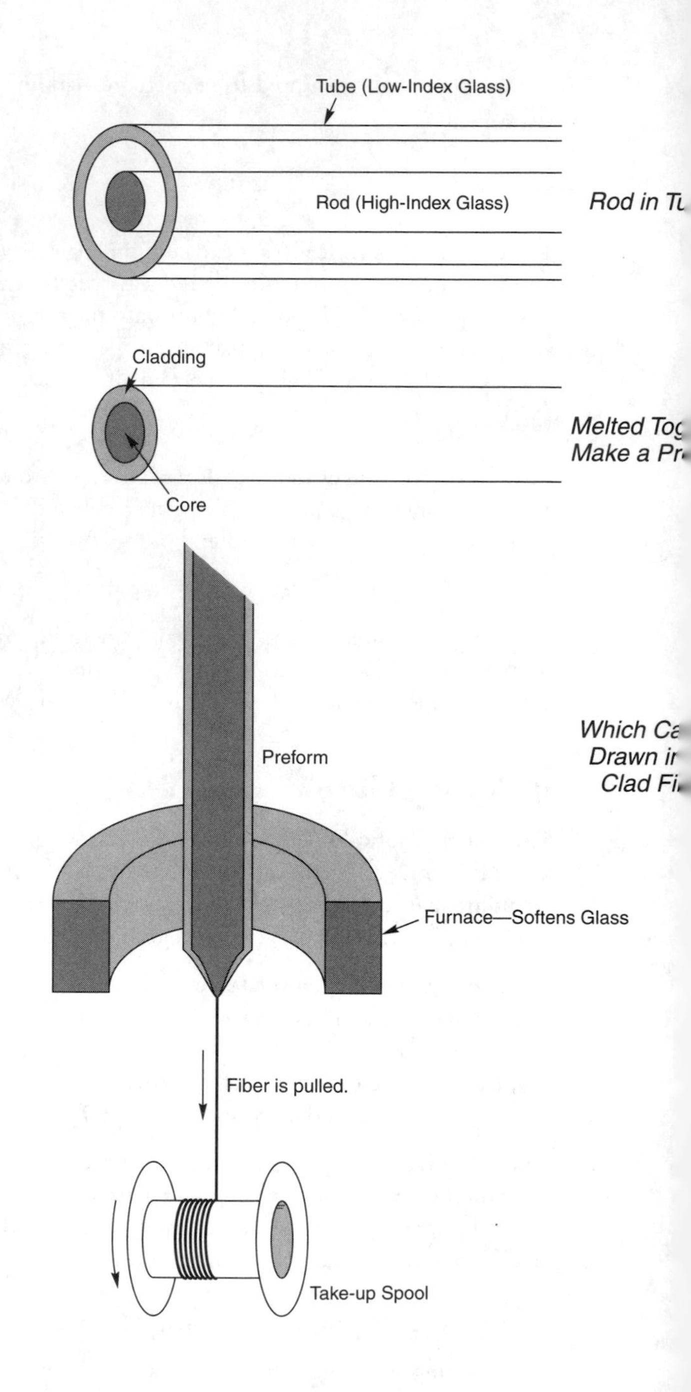

the cladding, the more light falls on the fiber cores and the higher the transmission efficiency.

Reducing the size of individual fibers increases the resolution of images transmitted through a bundle, but very fine fibers are hard to handle and vulnerable to breakage. Typically, the smallest loose fibers used in imaging bundles are about 20 μm (0.02 mm, or 0.0008 in.). Even at this size, they remain large relative to the wavelength of visible light (0.4 to 0.7 μm in air), and you can get away with considering light guiding as determined by total internal reflection of light rays at the core-cladding boundary. (The highest-resolution fiber bundles are made by melting fibers together and stretching the whole solid block.)

Illuminating Fiber

Single step-index fibers with large cores—typically 400 μm to 1 mm—can be used to guide a laser beam from the laser to a target or industrial workpiece. The large diameter serves two purposes. First, it can collect power from the laser more efficiently than a smaller-core fiber. In addition, it spreads the laser power over a larger area at the ends of the fiber and through a larger volume within the fiber. This is important because some laser power inevitably is lost at the surfaces and within the fiber. If the beam must be focused tightly to concentrate it in the fiber, the power density (power per unit area) may reach levels so high it can damage exposed ends of the fiber.

Large-core step-index fibers are used to deliver laser power.

The design of these large-core fibers is similar to those in Figure 4.1(a). The core diameters are proportionally larger, whereas cladding thicknesses do not increase as rapidly. As the fibers become thicker, they also become less flexible.

Communication Fibers

Step-index multimode fibers with cores not quite as large can be used for some types of communications. One smaller type, shown in Figure 4.1(a), has a 100-μm core surrounded by a cladding 20 μm thick, for total diameter of 140 μm. It is typically called 100/140 fiber, with the core diameter written before the overall diameter of the cladding. Typically an outer plastic coating covers the whole fiber, protecting it from mechanical damage and making it easier to handle. The large core is attractive for certain types of communications, because it can collect light efficiently from inexpensive light sources such as LEDs.

Light pulses stretch out in length and time as they travel through large-core step-index fiber.

If you think of light in terms of rays, you can see an important limitation of large-core step-index fibers for communication (see Figure 4.4). Light rays enter the fiber at a range of angles, and rays at different angles travel different paths through the same length of fiber.

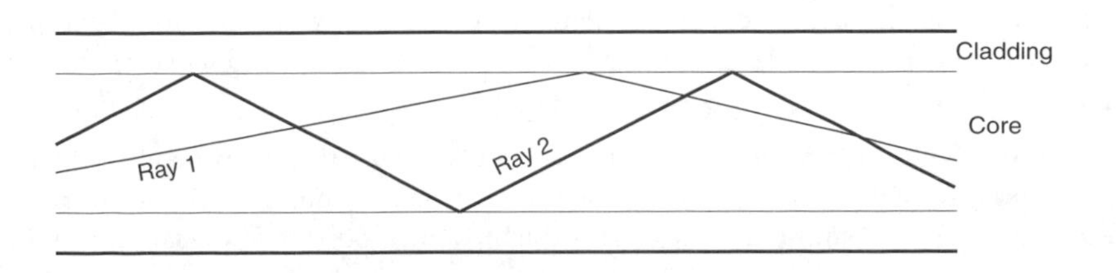

FIGURE 4.4 *Light rays that enter multimode step-index fiber at different angles travel different distances through the fiber, causing pulse dispersion.*

The larger the angle between the light ray and the axis, the longer the path. For example, a light ray that entered at 8° from the axis (the maximum confinement angle in the earlier example) of a perfectly straight 1-m length of fiber would travel a distance of 1.0098 m (1 m/cos 8°) before it emerged from the other end. Thus light just inside the confinement angle would emerge from the fiber shortly after light that traveled down the middle. This pulse-dispersion effect becomes larger with distance and can limit data-transmission speed.

In fact, the ray model gives a greatly simplified view of light transmission down optical fibers. As I mentioned earlier, an optical fiber is a waveguide that transmits lightwaves in one or more transmission modes. Stay tuned for the next section, and I'll explain more about these modes. The larger the fiber core, the more modes it can transmit, so a step-index fiber with a core of 20 μm or more is a multimode fiber. Light rays enter the fiber at different angles, and the various modes travel down the fiber at different speeds. What you have as a result is modal dispersion, which occurs in all step-index fibers that carry multiple modes. It is largely irrelevant for imaging and guiding illuminating beams, but it is a serious drawback for communications. To understand why, we need to take a closer look at modes.

Modes and Their Effects

Waves have distinct propagation modes in a waveguide. A fiber is an optical waveguide.

Modes are stable patterns in which a wave can travel through a waveguide. The wavelength of the wave and the size, shape, and nature of the waveguide determine what modes can propagate. Engineers first developed waveguide theory when they were working with microwaves, and the same theory can be applied to other electromagnetic waves. Thus an optical fiber is merely an optical waveguide.

You don't want to worry about the mathematical details of waveguide theory—and I certainly don't—but it's important to understand their consequences. One is that the number of possible modes increases with the diameter of the waveguide. For a fiber, this means the core diameter. It also depends on the wavelength. In a simple way, the larger the waveguide, measured in wavelengths, the more modes it can carry. In practice, other effects enter the picture.

Types of Waveguides

Optical fibers are dielectric waveguides.

In essence, the walls of a waveguide set boundary conditions for the electric and magnetic fields that make up an electromagnetic wave. Plug the nature of those boundaries into the proper differential equations, and you can calculate the theoretical properties of the waveguide. The most familiar type of microwave waveguide is a rectangular metal tube; its conductive metal walls set up boundaries for the electromagnetic fields of the microwaves passing through it. Another type of microwave waveguide is made of plastic, called a *dielectric* waveguide, because the plastic is an insulator or dielectric. An optical fiber is an optical counterpart to a dielectric waveguide, made of a transparent, nonconductive material (glass or plastic) and with its size closer to the wavelength of light.

The simplest type of microwave dielectric waveguide is a plastic rod suspended in air, similar to an unclad optical fiber. In theory, the waves are guided along the surface through air, not inside the guide. Taking this view can explain some problems of unclad optical fibers.

Surface waveguides work if isolated in air, but the waveguide effect is disrupted if other objects touch their surfaces. Thus an isolated plastic dielectric guide works for microwaves, but unclad glass fibers don't effectively confine light when they touch each other or have fingerprints on their surfaces.

The waveguide surface is the place where its transmission characteristics change, forming a boundary that guides the waves. Cladding a fiber effectively puts that waveguide surface *inside* the optical fiber, where it can't be touched. The communication theorists who first considered optical waveguides considered this a serious drawback, because it meant the light had to travel through the fiber material, which they did not think could be made transparent enough for communications. They later learned otherwise. Adding a cladding also changes the structure of the waveguide and the way it guides light, and this proves to have other important advantages.

Single-Mode Waveguides

One important difference between microwave waveguides and large-core step-index fibers is the number of modes they carry. Microwave guides are less than a wavelength across, and because of those dimensions microwaves can propagate through them in only one mode. In contrast, imaging fibers are many wavelengths across and can carry many modes. That was not attractive to communication researchers, who had learned the hard way that interactions among modes can cause problems in multimode waveguides. They wanted optical waveguides in which only a single mode could propagate, so they wouldn't have to worry about multimode effects.

Small-diameter waveguides carry waves in only a single mode.

The problem with that idea was size. Waveguides restrict propagation to a single mode only if their diameters are below a certain cutoff threshold, which depends on the wavelength. Bare single-mode optical waveguides, designed as scaled-down versions of dielectric microwave waveguides, would have to be less than a wavelength of light thick. That meant their diameters would have to be less than 0.5 μm for visible light, making them practically impossible to handle. The tiny fibers would inevitably have to touch surfaces, so light would leak out—if you could couple any light into something that small.

However, cladding changes the waveguide properties of fibers, because the single-mode cutoff size depends on the difference in refractive index between the core and cladding. The larger the difference, the smaller the fiber must be. The difference is large for a glass fiber (with $n = 1.5$) in air ($n = 1.000293$). It is small—typically less than 1%—for a glass-clad fiber. Although a bare fiber could be no larger than about half the wavelength of light to transmit only a single mode, a clad fiber could have a core diameter several times the wavelength. The larger the core, the more easily it can collect light. The cladding increases fiber diameter, making it easier to handle. (In theory, the cladding could be infinitely thick, but you want fibers thin enough to be flexible.) In addition, the cladding prevents anything from contacting the boundary between core and cladding, which serves a vital function in the waveguide structure.

Clad single-mode fibers have core diameters several times the wavelength.

Clad fibers with larger core diameters can carry multiple modes. The number of possible modes increases rapidly with fiber core diameter.

Modal Properties

Propagation modes are standing waves that travel through the fiber. The details of mode propagation theory are far too complex to discuss here and generally have little relevance to most day-to-day concerns of fiber-optic users. However, there are some exceptions.

Some light penetrates slightly into the fiber cladding.

We saw earlier that waves travel along the surface of an unclad dielectric waveguide. In clad fibers, the core-cladding boundary becomes the "surface" that guides the waves. The cladding changes the structure of the waveguide so much that light travels within the fiber core, but some does penetrate into the cladding, despite the fact that it nominally undergoes total internal reflection. This occurs both in single-mode and multi-mode fibers. It is more significant in single-mode fibers that in practice are characterized by the *mode-field diameter,* which is slightly larger than the core diameter, as shown in Figure 4.5. Technically, the mode-field diameter is the point where light intensity drops to $1/e^2$ (0.135) of the mode's peak intensity. For a single-mode fiber, the peak intensity is at the center of the core.

The leakage of some light into the cladding makes transparency important for cladding material, although not as important as for core material. In single-mode fibers, the guided mode travels mostly in the core but somewhat in the cladding. In multimode fibers, some modes may spend more time in the cladding.

Modes are sometimes characterized by numbers. Single-mode fibers carry only the lowest-order mode, assigned the number 0. Multimode fibers also carry higher-order modes. The number of modes that can propagate in a fiber depends on the fiber's numerical aperture (or acceptance angle) as well as on its core diameter and the wavelength of the light. For a step-index multimode fiber, the number of such modes, N_m, is approximated by

$$\text{Modes} = 0.5\left(\frac{\text{core diameter} \times \text{NA} \times \pi}{\text{wavelength}}\right)^2$$

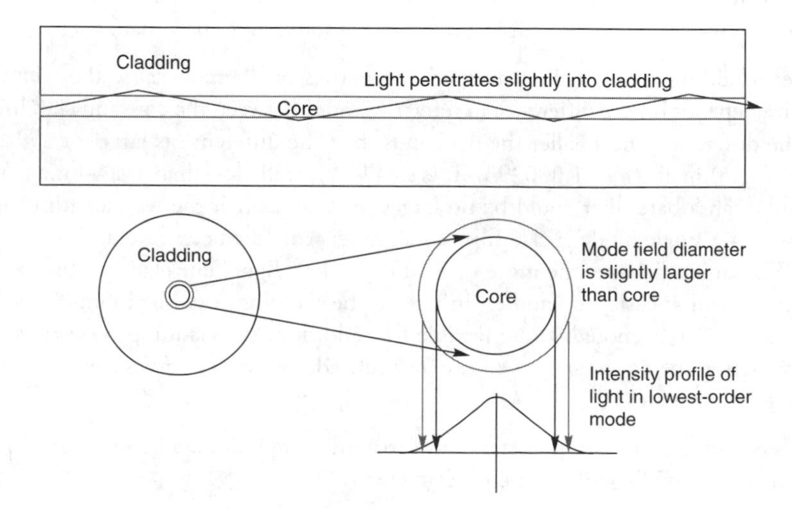

FIGURE 4.5
Light penetrates slightly into the cladding of a single-mode step-index fiber.

or

$$N_m = 0.5\left(\frac{\pi D \times \text{NA}}{\lambda}\right)^2$$

where λ is the wavelength and D is the core diameter. To plug in some representative numbers, a 100-μm core step-index fiber with NA = 0.29 (a typical value) would transmit thousands of modes at 850 nm. This formula is only an approximation and does not work for fibers carrying only a few modes.

Leaky Modes

The difference between the highest-order modes guided in a multimode fiber and the lowest-order modes that are not guided is quite small. Modes that are just beyond the threshold for propagating in a multimode fiber can travel for short distances in the fiber cladding. In this case, the cladding itself acts as an unclad optical fiber to guide those cladding modes.

Some modes can propagate short distances in the cladding of a multimode fiber.

Because the difference between guided and unguided modes is small, slight changes in conditions may allow light in a normally guided mode to leak out of the core. Likewise, some light in a cladding mode may be recaptured. Slight bends of a multimode fiber are enough to allow escape of these leaky modes.

Modal-Dispersion Effects

Each mode has its own characteristic velocity through a step-index optical fiber, as if it were a light ray entering the fiber at a distinct angle. This causes pulses to spread out as they travel along the fiber, in what is called *modal dispersion*. The more modes the fiber transmits, the more pulses spread out.

Modal dispersion in multimode step-index fibers is the largest type of pulse dispersion.

Later we will see that there are other kinds of dispersion, but modal dispersion is the largest in multimode step-index fibers. Precise calculations of how many modes cause how much dispersion are rarely meaningful. However, you can make useful approximations by using the ray model (which works for multimode step-index fibers) to calculate the difference between the travel times of light rays passing straight through a fiber and bouncing along at the confinement angle. For the typical confinement angle of 8° mentioned earlier, the difference in propagation time is about 1%. That means that an instantaneous pulse would stretch out to about 30 ns (30 billionths of a second) after passing through a kilometer of fiber.

That doesn't sound like much, but it becomes a serious restriction on transmission speed, because pulses that overlap can interfere with each other, making it impossible to receive the signal. Thus pulses have to be separated by more than 30 ns. You can estimate the maximum data rate for a given pulse spreading from the equation

$$\text{Data rate} = \frac{0.7}{\text{pulse spreading}}$$

Plug in a pulse spreading of 30 ns, and you find the maximum data rate is about 23 Mbit/s. In practice, the maximum data rate also depends on other factors.

Dispersion also depends on distance. The total modal dispersion is the product of the fiber's characteristic modal dispersion per unit length, D_0, multiplied by the fiber length, L:

$$D_0 \times L$$

Thus a pulse that [illegible] over 1 km will [illegible] m and 300 ns over 10 km. (For very accurate calculations, you should [illegible] where γ is a factor close to 1, which depends on the fiber type. However, γ normally is so close to 1 that it doesn't matter.)

Because total dispersion increases [illegible] distance, the maximum transmission speed decreases. If the maximum data rate for [illegible] of fiber is DR_0, the maximum data rate for L kilometers is roughly:

$$DR = \frac{DR_0}{L}$$

We will learn more about dispersion in Chapter 5. For now, the important thing to remember is that modal dispersion seriously limits transmission speed in step-index multimode fiber.

Graded-Index Multimode Fiber

Replacing the sharp boundary between core and cladding with a refractive-index gradient nearly eliminates modal dispersion.

As communication engineers began seriously investigating fiber optics in the early 1970s, they recognized modal dispersion limited the capacity of large-core step-index fiber. Single-mode fibers promised much more capacity, but many engineers doubted they could get enough light into the tiny cores. As an alternative, they developed multimode fiber in which the refractive index grades slowly from core into cladding. Careful control of the refractive-index gradient nearly eliminates modal dispersion in fibers with cores tens of micrometers in diameter, giving them much greater transmission capacity than step-index multimode fibers.

Optically, graded-index fibers guide light by refraction instead of total internal reflection. The fiber's refractive index decreases gradually away from its center, finally dropping to the same value as the cladding at the edge of the core, as shown in Figure 4.6. The change in refractive index causes refraction, bending light rays back toward the axis as they pass through layers with lower refractive indexes, as shown in Figure 4.7. The refractive index does not change abruptly at the core-cladding boundary, so there is no total internal reflection. (Don't be fooled by the change in slope at the edge of the core in Figure 4.6; it's more like starting up a slow hill than hitting the cliff of a step-index transition.) However, it isn't needed because refraction bends guided light rays back into the center of the core before they reach the cladding boundary. (The refractive-index gradient cannot confine all light entering the fiber, only rays that fall within a limited confinement angle, as in step-index fiber. The refractive-index gradient determines that angle.)

As in a step-index fiber, light rays follow different paths in a graded-index fiber. However, their speeds differ because the speed of light in the fiber core changes with its refractive

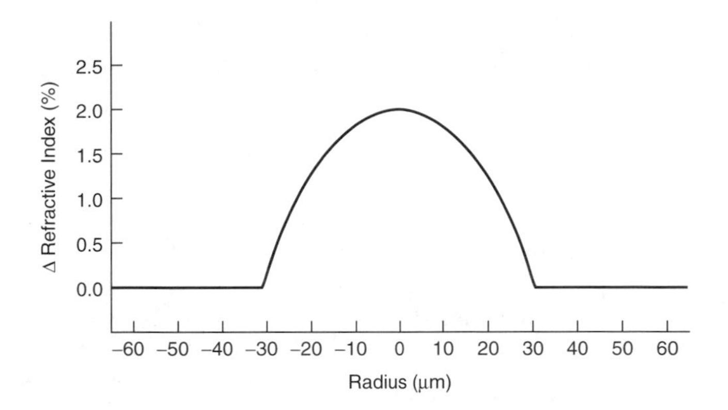

FIGURE 4.6
Refractive-index profile of a graded-index fiber with 62.5-μm core.

index. Recall that the speed of light in a material, c_{mat}, is the velocity of light in a vacuum, c_{vacuum}, divided by refractive index:

$$c_{mat} = \frac{c_{vacuum}}{n_{mat}}$$

Thus the farther the light goes from the axis of the fiber, the faster its velocity. The difference isn't great, but it's enough to compensate for the longer paths followed by the light rays that go farthest from the axis of the fiber. Careful adjustment of the refractive-index profile—the variation in refractive index with distance from the fiber axis—can greatly reduce modal dispersion by equalizing the transit times of different modes.

Practical Graded-Index Fiber

Standard graded-index fibers have 50- or 62.5-μm cores.

Graded-index fibers were developed especially for communications. The long-time standard types have core diameters of 50 or 62.5 μm and cladding diameters of 125 μm; some

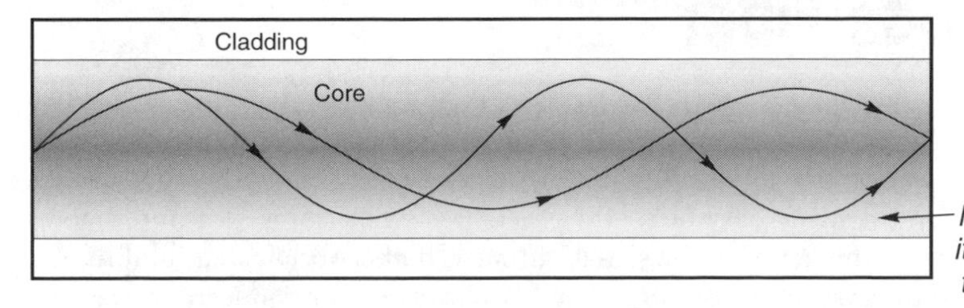

FIGURE 4.7
The refractive-index gradient in a graded-index fiber bends light rays back toward the center of the fiber.

have been made with 85-μm cores and 125-μm cladding diameter. The core diameters are large enough to collect light efficiently from a variety of light sources. The cladding must be at least 20 μm thick to keep light from leaking out.

The graded-index fiber is a compromise, which has much higher transmission capacity than large-core step-index fibers while retaining a core large enough to collect light easily. It was used for some telecommunications until the mid-1980s but gradually faded from use in telephone systems because single-mode fibers worked much better. Graded-index fibers remain in use, mostly for data communications and networks carrying signals moderate distances—typically no more than a couple of kilometers.

Limitations of Graded-Index Fiber

Residual dispersion and modal noise limit performance of graded-index fibers.

Graded-index fibers suffer some serious limitations that ultimately made them impractical for high-performance communications.

Modal dispersion is not the only effect that spreads out pulses going through optical fibers. Other types of dispersion arise from the slight variation of refractive index with the wavelength of light. These remain present in graded-index fibers and became increasingly important as transmission moved to higher speeds. Chapter 5 will describe these dispersion effects.

Multimode transmission itself proved a serious problem. Different modes can interfere with each other, generating what is called *modal noise.* This appears as an uneven distribution of light across the end of the fiber, which continuously changes in response to very minor fluctuations, generating noise. Such modal effects also made it impossible to control precisely how fibers behaved when several were spliced together, because the light in some modes can shift into other modes or leak into the cladding at joints.

In addition, ideal refractive-index profiles are very difficult to realize in practice. The refractive-index gradient must be fabricated by depositing many thin layers of slightly different composition in a precisely controlled sequence. This is expensive, and some fluctuations from the ideal are inevitable.

These limitations do not prevent graded-index fibers from being used in short systems, even at high speeds, as long as dispersion does not accumulate to high enough levels to limit data rates. However, single-mode fibers are standard for long-distance, high-performance systems.

Single-Mode Fiber

The simplest type of single-mode fiber has a step-index profile, with an abrupt boundary between a high-index core and a lower-index cladding.

The basic requirement for single-mode fiber is that the core be small enough to restrict transmission to a single mode. This lowest-order mode can propagate in all fibers with smaller cores (as long as light can physically enter the fiber). Because single-mode transmission avoids modal dispersion, modal noise, and other effects that come with multimode transmission, single-mode fibers can carry signals at much higher speeds than multimode fibers. They are the standard choice for virtually all kinds of telecommunications that involve high data rates or span distances longer than a couple of kilometers, and are often used at slower speeds and shorter distances as well.

The simplest type of single-mode fiber, often called *standard* single mode, has a step-index profile, with an abrupt boundary separating a high-index core and a lower-index cladding. The refractive-index differential is 0.36% for a widely used fiber, and is well under 1% in other standard types. Figure 4.8 shows cross sections of the two principal types of step-index single-mode fiber made from fused silica.

The simplest design is the matched-cladding fiber shown at the top of Figure 4.8. The cladding is pure fused silica; germanium oxide (GeO_2) is added to the core to increase its refractive index.

An alternative design is the depressed cladding fiber shown at the bottom. In this case, the core is fused silica doped with less germanium oxide than is needed for a matched cladding fiber. The inner part of the cladding surrounding the core is doped with fluorine, which *reduces* its refractive index below that of pure fused silica. The outermost part of the core is pure fused silica, without the fluorine dopant.

Both these designs typically are widely used in telecommunications systems operating at 1.31 and 1.5 μm; core diameters are around 9 μm.

Conditions for Single-Mode Transmission

Earlier in this chapter, you saw that the number of modes, N_m, transmitted by a step-index fiber depends on the fiber core diameter, D, the refractive indexes of core (n_0) and cladding (n_1), and the wavelength of light λ. You can write the formula in terms of numerical aperture (NA):

$$N_m = 0.5 \left(\frac{\pi D \times \text{NA}}{\lambda} \right)^2$$

Fiber with a small enough core transmits only a single mode of light.

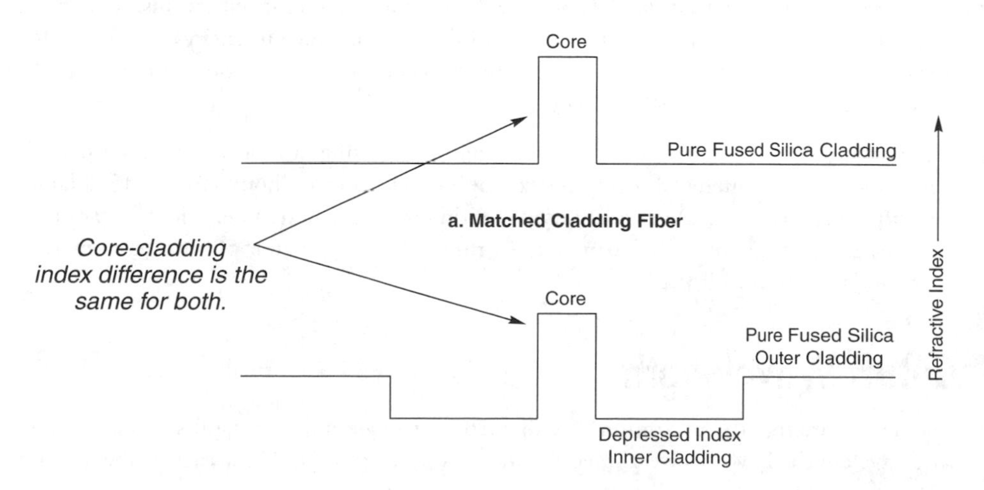

FIGURE 4.8
Two types of step-index single-mode fiber. The difference between core and cladding refractive-index is the same, but in the depressed cladding fiber at the bottom, the inner cladding is doped with fluorine to reduce its refractive index.

You also can replace NA with the core and cladding indexes—useful because NA as acceptance angle isn't very meaningful for single-mode fibers—and reformulate the equation:

$$N_m = 0.5 \left(\frac{\pi D}{\lambda}\right)^2 (n_0^2 - n_1^2)$$

Reducing the core diameter sufficiently can limit transmission to a single mode. By manipulating the mode-number equation and calculating a constant using Bessel functions, you can find the maximum core diameter, D, which limits transmission to a single mode at a particular wavelength, λ:

$$D < \frac{2.4\lambda}{\pi\sqrt{n_0^2 - n_1^2}}$$

If the core is any larger, the fiber can carry two modes.

Note that D is the *maximum* allowable core diameter for single-mode transmission. To allow for the inevitable margins of error, single-mode fibers normally are designed with core diameters somewhat smaller than the maximum value. In practice the refractive-index difference in step-index single-mode fiber is typically less than about 0.5%, and the core diameter is typically several times the wavelength that the fiber is designed to transmit.

Since core area is proportional to the square of core diameter, it varies with the square of wavelength. If all other things are equal, this means that a single-mode fiber designed to transmit a 0.65-micrometer red beam would have a core only one-fourth the area of a fiber made to carry a single mode at 1.3 μm in the near infrared. As a result, coupling light into single-mode fibers gets harder at shorter wavelengths.

Although core diameter is the physical parameter used in the equations for single-mode transmission, the core of a dielectric waveguide does not confine *all* the light. Some light in the guided single mode extends a short distance into the cladding. This spillover is reflected in a number called the *mode field diameter,* which is cited in fiber specifications. It's defined as the distance from the fiber axis (the center of the core) at which intensity drops to a value of $1/e^2$ (0.135) of the peak value in the core. This means that a fraction of the light falls outside the mode-field diameter, as you saw in Figure 4.5.

The mode-field diameter depends on wavelength, increasing at longer wavelengths. Typically mode-field diameter of a step-index single-mode fiber is about 10% to 15% larger than the core diameter. One widely used step-index single-mode fiber with 8.2-μm core has mode field diameter of 9.2 μm at 1310 nm and 10.4 μm at 1550 nm. Its numerical aperture (at 1310 nm) is 0.14.

The cutoff wavelength of a single-mode fiber is the shortest wavelength at which it carries only one mode. At shorter wavelengths it carries two or more modes.

Cutoff Wavelength

We saw before that the maximum core diameter for single-mode transmission depends on the wavelength. If you solve the equation for wavelength, you find that a fiber with a specific core diameter transmits light in a single mode only at wavelengths longer than a value called the *cutoff wavelength,* λ_c, given by

$$\lambda_c = \frac{\pi D \sqrt{n_0^2 - n_1^2}}{2.4}$$

A fiber with diameter D is single-mode at wavelengths longer than λ_c, but as wavelength decreases, it begins to carry two modes at λ_c.

Although core diameter is an important consideration in fiber *design,* cutoff wavelength is important in fiber *use.* If you want a fiber to carry signals in only one mode for a high-performance communication system, you must be sure that all wavelengths transmitted are longer than the cutoff wavelength. To give a safety margin, fibers are designed with their cutoff wavelength somewhat shorter than their shortest operating wavelength. For example, the common step-index single-mode fiber mentioned above, often used at 1310 nm, has a specified cutoff wavelength of 1.26 μm.

What happens at wavelengths shorter than the cutoff? As the wavelength decreases, you first get a second mode, then additional modes. These extra modes can interfere with each other and with the primary mode, causing performance problems. As with any multimode fiber, minor perturbations can affect propagation, and transmission is particularly unpredictable in fibers with only a few modes. Figure 4.9 shows approximate patterns of first- and second-order as modes transmitted in a fiber. As you can see, when the fiber transmits only a single mode, the light intensity is highest at the fiber axis and drops off toward the cladding. The second-order mode is more complex and takes two patterns. One is a pair of intensity peaks above and below the fiber axis; the other is a pair of intensity peaks to the right and left of the fiber axis. Both these patterns appear simultaneously in a dual-mode fiber, superimposed on top of each other and on the first-order mode. The result is a multimodal mess that I can't show clearly in a drawing.

Trade-offs with Single-Mode Fiber

The sheer simplicity of single-mode transmission is one of its primary attractions for fiber-optic communications. By confining light to a single mode, it greatly reduces pulse dispersion. Some dispersion remains, but it depends primarily on the range of wavelengths transmitted in the signal. The smaller the dispersion, the faster pulses can be turned off and on.

Single-mode fiber is a clean and simple transmission system.

Charles Kao recognized the advantages of single-mode fiber in the mid-1960s, but other early developers pointed to a trade-off that seemed inevitable. The smaller the core diameter, the harder it was to couple light into the fiber. Coupling light into single-mode fiber inevitably requires much tighter tolerances than coupling light into the larger cores of multimode fiber. However, those tighter tolerances have proved achievable, and single-mode fibers are widely used. The main applications of multimode fibers today are in systems where connections must be made inexpensively and transmission distances and speeds are modest.

FIGURE 4.9
Modes in single- and dual-mode fibers.

Lowest-order mode is brightest in center, fading outward. This shows only the central zone of the fiber. (The darker the shading, the more light.)

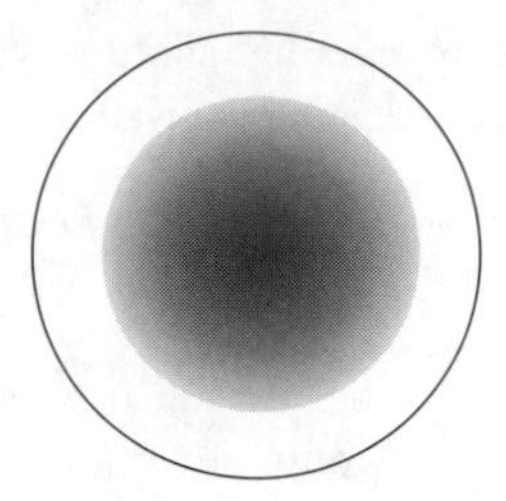

Second mode consists of two patterns (each containing a pair of oval bright regions that fade toward the edges): one with the ovals on the sides of the core and the other with the ovals above and below. The two appear simultaneously, so the actual pattern is an overlay, plus the lowest-order mode.

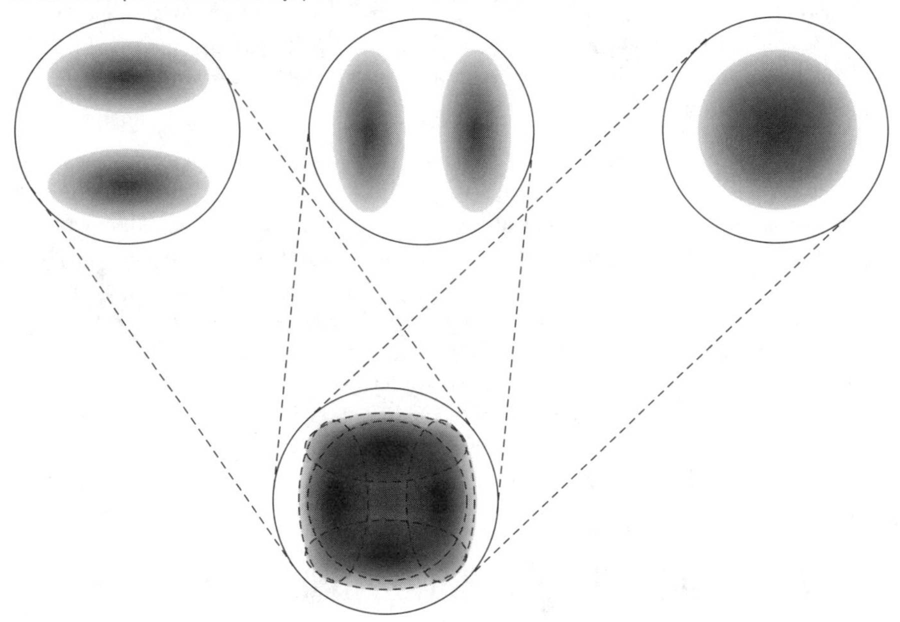

On the other hand, the properties of step-index single-mode fiber are not ideal. Its dispersion is at a minimum at 1.31 μm, but its attenuation has a minimum at 1.55 μm. The best available optical amplifiers, erbium-doped fibers, operate at 1530 to 1610 nm, where dispersion of step-index single-mode fibers is relatively large. These and other limitations have led to development of other single-mode fibers with different structures, which alter their dispersion.

Dispersion-Shifted Single-Mode Fiber

More complex core-cladding designs can shift low dispersion to the 1.5-μm region.

Step-index single-mode fibers have much better properties than developers dreamed were possible 30 years ago. However, they are not ideal. As we will see in Chapter 6, the attenuation of glass fiber has been reduced close to the theoretical minimum, and little improvement is possible without shifting to a new family of materials.

Pulse dispersion is another matter. The major concern in single-mode fiber is spectral or chromatic dispersion, caused by the variation in the speed of light through the fiber with wavelength. Chromatic dispersion is the sum of two quantities, dispersion inherent to the material and dispersion arising from the structure of the waveguide. These two can have opposite signs, depending on whether the speed of light increases or decreases with wavelength. (See Chapter 5 for a more thorough explanation.) Fortuitously, the two cancel each other out near 1.31 μm in standard step-index single-mode fiber, as shown in Figure 4.10.

This is a useful wavelength, but it is not ideal. The loss of glass fibers is lowest at 1.55 μm, and erbium-doped fiber amplifiers operate in that range. Material dispersion is an inherent characteristic of silica fiber that cannot be readily changed without altering glass composition in ways that increase attenuation. However, it is possible to shift the dispersion minimum by changing waveguide dispersion.

Waveguide dispersion arises because light propagation in a waveguide depends on wavelength as well as the waveguide dimensions. The important number is the diameter divided by wavelength. Measured that way, decreasing the wavelength serves to increase the waveguide diameter, whereas increasing wavelength effectively shrinks the waveguide. Thus the distribution of light between core and cladding changes with wavelength.

That change in light distribution affects how fast the light travels through the fiber. The core and cladding have different refractive indexes, which determine the speed of light through them. Because light spends time in both core and cladding, its effective speed through the whole fiber is an average that depends on the distribution of light between core and cladding. A change in wavelength changes that distribution, and thus the average speed, causing waveguide dispersion.

Changing the design of the core-cladding interface can alter waveguide dispersion, shifting the zero point of chromatic dispersion to other wavelengths. There are now several types of dispersion-modified fibers, based on designs that change waveguide dispersion. They are optimized in different ways to meet varying system requirements, particularly the transmission of multiple optical channels for wavelength-division multiplexing.

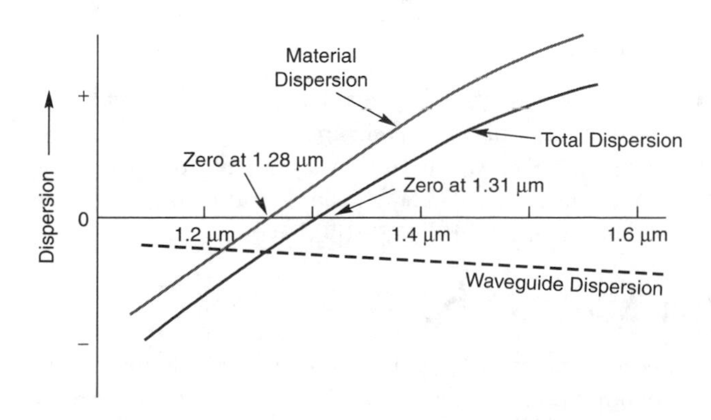

FIGURE 4.10
Waveguide dispersion offsets chromatic dispersion to produce zero dispersion at 1.31 μm in step-index single-mode fiber.

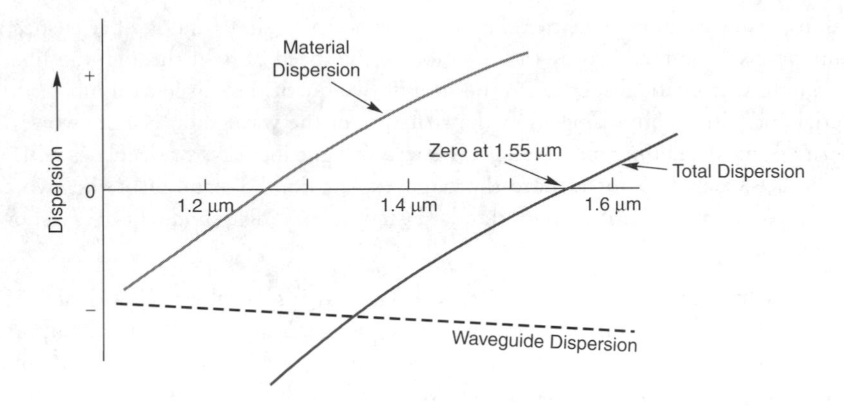

FIGURE 4.11 *A fiber designed with more waveguide dispersion shifts the zero-dispersion wavelength to 1.55 μm.*

Zero Dispersion-Shifted Fiber

Some older fibers had zero dispersion shifted to 1.55 μm.

The first dispersion-shifted fibers had zero dispersion shifted to 1550 nm to match their minimum absorption wavelength. This was done by increasing the magnitude of waveguide dispersion, as shown in Figure 4.11. They were introduced in the mid-1980s and were installed in some systems, but never came into wide use and are no longer manufactured. Originally called simply *dispersion-shifted fibers* they are now known as *zero dispersion-shifted fibers* because their dispersion is zero in the middle of the erbium-doped fiber amplifier band. This type is covered by the International Telecommunications Union G.654 standard, and is sometimes identified by that number.

Designers increased the waveguide dispersion by adapting the layered core design shown in Figure 4.12(a). The *inner core* has a refractive index that decreases with increasing distance from the fiber axis at its center. The next layer, sometimes called the *inner cladding,* has a refractive index that drops as low as that of the outer cladding before starting to rise again. The next layer, called either the *ring* or the *outer core,* has a refractive index that rises to a peak smaller than that of the inner core, then declines to match that of the cladding.

In addition to increasing the waveguide dispersion, this elaborate structure also reduces mode-field diameter to about 8.1 μm at 1550 nm, compared to 10.4 μm for step-index single-mode fiber at the same wavelength.

Although this design worked well for single-channel systems, it proved unsuitable for wavelength-division multiplexing. When multiple optical channels pass through the same fiber at wavelengths where dispersion is very close to zero, they suffer from a type of crosstalk called four-wave mixing, described in Chapter 5. The degradation is so severe that zero dispersion-shifted fiber cannot be used for dense-WDM systems.

Nonzero Dispersion-Shifted Fiber

Zero-dispersion wavelength must lie outside the erbium-fiber band for WDM systems.

The way to avoid four-wave mixing is to move the zero-dispersion wavelength outside the band used for erbium-fiber doped amplifiers. So-called *nonzero dispersion-shifted fibers* do

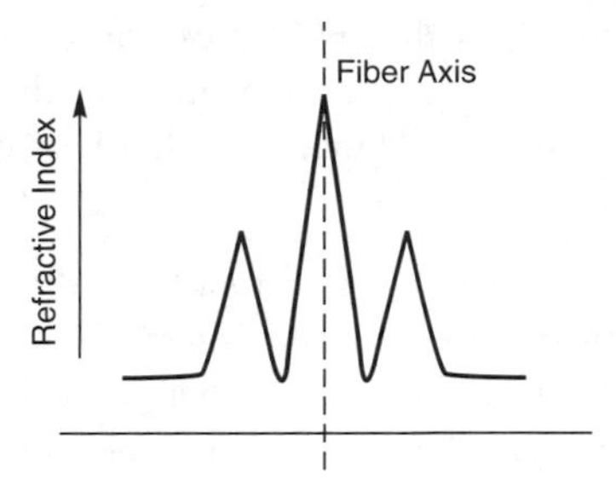

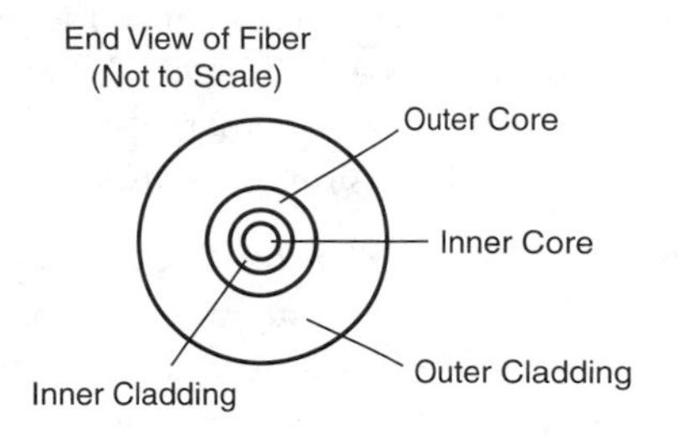

FIGURE 4.12 *Refractive-index profiles of some dispersion-shifted fibers designed for specific applications*

a. **Zero dispersion-shifted fiber.**

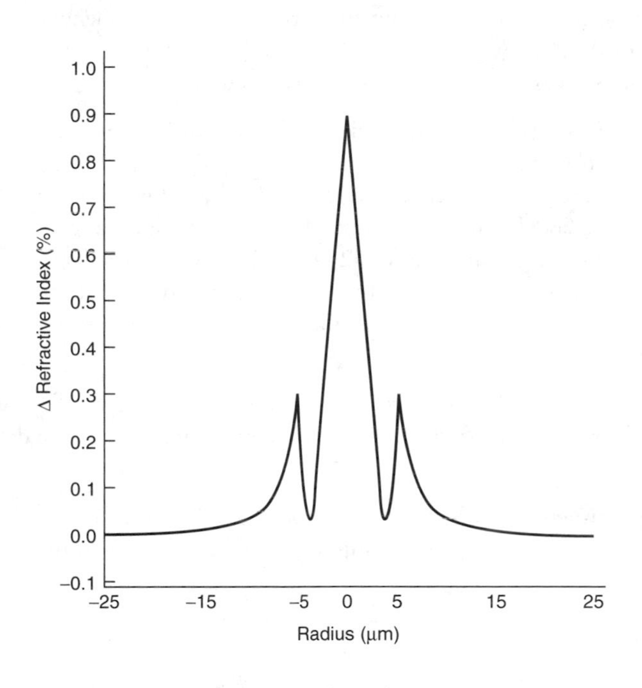

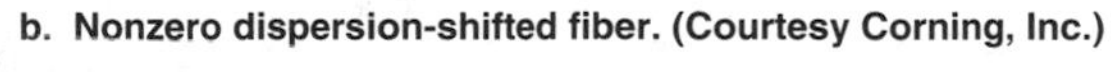

b. Nonzero dispersion-shifted fiber. (Courtesy Corning, Inc.)

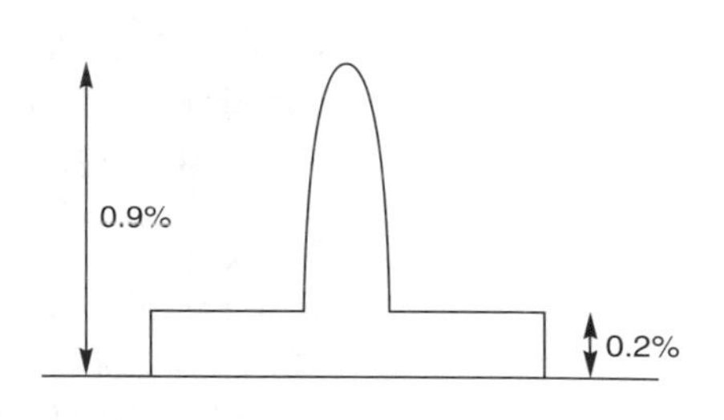

c. Another design for nonzero dispersion-shifted fiber.

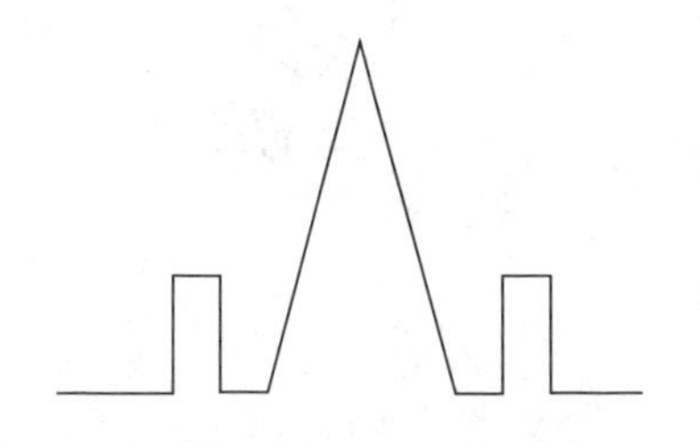

d. Large effective area fiber.

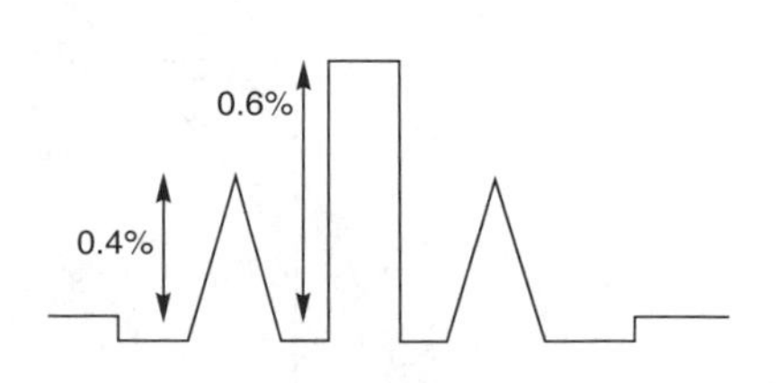

e. Fiber with flattened dispersion slope.

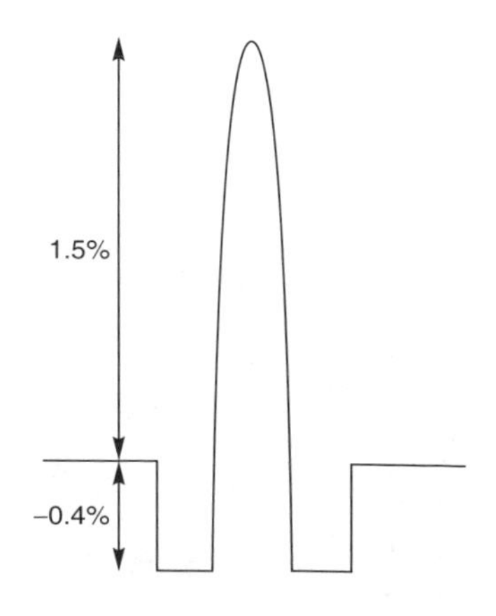

f. A dispersion-compensating fiber.

this by using other layered core structures to adjust the amount of waveguide dispersion differently. Figures 4.12(b) and (c) illustrate two approaches, showing how their refractive-index profiles differ from that of zero dispersion-shifted fiber. As with other designs in this diagram, the dimensions are not exact.

The name comes from the fact that their dispersion is shifted to a value that is low—but not zero—in the 1550-nm band of erbium-fiber amplifiers. The International Telecommunions Union G.655 standard defines nonzero dispersion-shifted fibers as having chromatic dispersion of 0.1 to 6 picoseconds per nanometer-kilometer, but does not specify the sign. (Chapter 5 explains chromatic dispersion.) This small dispersion is enough to keep signals at closely spaced wavelengths from staying in phase over long distances and causing serious crosstalk.

This small dispersion can be provided by moving the zero-dispersion wavelength either above (at shorter wavelengths) or below (at longer wavelengths) the erbium-fiber band, as shown in Figure 4.13. Various types of fibers have been developed.

For dense-WDM applications using erbium-doped fiber amplifiers, the current favorite is a zero-dispersion point at a wavelength of 1500 nm or less. This is shorter than the erbium-amplifier band, and no other optical amplifiers are well developed for this region. In addition, this choice means that the fiber has positive chromatic dispersion (above the X axis on Figure 4.13) in the entire erbium-fiber band from 1525 to 1620 nm. Positive dispersion is an advantage because it is easier to compensate than negative dispersion. In addition, the positive part of the dispersion curve slopes less, so the magnitude of the dispersion is more uniform across the erbium-fiber band.

Some early nonzero dispersion-shifted fibers had zero dispersion at wavelengths of 1580 to 1610 nanometers. Those fibers were dropped when L-band erbium amplifiers were developed for those wavelengths.

Longer-wavelength nonzero dispersion-shifted fibers have been developed with zero dispersion at about 1640 nm, well beyond the erbium-amplifier L-band. This leaves the entire

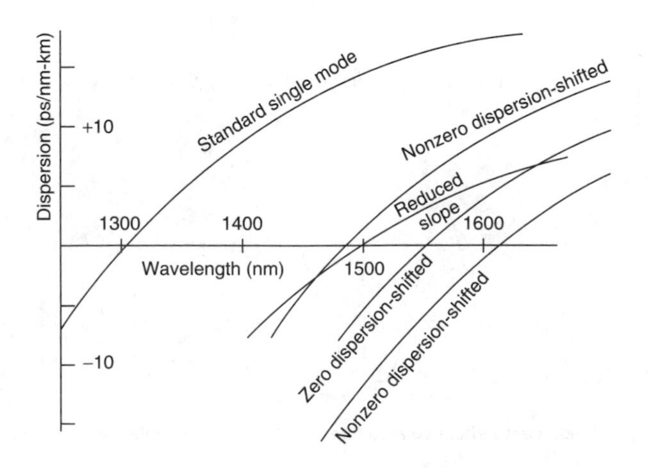

FIGURE 4.13
Dispersion profiles of several single-mode fiber types.

band between 1280 and 1620 nm open, with no zero-crossing point in the middle. Potential uses for such fibers are in the metro network, where transmission distances are too short to require optical amplifiers. An added advantage is that the negative dispersion of these fibers partly offsets the positive wavelength chirp that comes from directly modulating semiconductor laser sources—allowing the use of relatively inexpensive laser transmitters at data rates to 2.5 gigabits per second.

Because nonzero dispersion-shifted fibers are relatively new, their specifications and performance are still evolving.

Reduced Dispersion Slope Fibers

Another way to refine dispersion-modified fibers for dense-WDM systems is to reduce the slope of the dispersion curve. As you can see in Figure 4.13, the dispersion normally changes significantly over the erbium-fiber band. For a typical nonzero dispersion-shifted fiber, the slope is about 0.08 ps/nm^2-km near 1550 nm. That means that dispersion changes about 8 ps/nm-km over a wavelength range of 100 nm.

Reduced slope fibers reduce the change in dispersion with wavelength, but also have smaller effective areas.

This variation with wavelength complicates the task of dispersion compensation for systems with many optical channels. Wavelengths with higher dispersion require more compensation than those with lower dispersion. Proposals to use the entire 1280 to 1650 nm band for metro WDM systems also face problems if the dispersion slope is high. Although they don't require long-distance transmission or optical amplification, the large change in dispersion across the range can cause problems.

Sophisticated multilayer core designs such as the one in Figure 4.12(e) can reduce the dispersion slope below 0.05 ps/nm^2-km, but there are tradeoffs. An important one is that reduced-slope designs tend to have smaller mode-field diameters—about 8.4 μm at 1550 nm, which concentrate optical power in a smaller volume. As you will learn in Chapter 5, raising the power density in fibers increases the strength of nonlinear effects, which can cause crosstalk. Systems with many WDM channels are particularly at risk.

Large Effective Area Fibers

Other nonzero dispersion-shifted fiber designs are intended to maximize the mode-field diameter, which determines the effective area over which optical power is spread in the fiber. This is important because dispersion shifting tends to reduce mode field diameter below that of standard step-index fiber, typically about 9.2 μm at 1310 nm and 10.4 μm at 1550 nm, making dispersion-shifted fiber particularly sensitive to nonlinear effects.

Large effective area fibers reduce nonlinear effects.

Special multilayer core designs like the one in Figure 4.12(d) can spread the mode field over larger areas than in standard dispersion-shifted fiber. In this example, the outer high-index ring draws light outwards, expanding the mode-field diameter. One commercial type has a mode-field diameter of about 9.6 μm at 1550 nm, corresponding to an effective area of 72 square micrometers. Although that doesn't sound much larger than the 8.4 μm mode-field diameter of a reduced-slope fiber, the critical dimension of area is only 55 μm^2 in a reduced-slope fiber. That means the larger fiber has 30% more area, allowing it to carry significantly more power without nonlinear effects.

Fibers can be designed with even larger mode-field diameters, reaching 10.8 μm at 1550 nm. That corresponds to a 100-μm^2 effective area, nearly double that of reduced-slope fiber. However, the tradeoff is dispersion slopes that can reach about 0.11 ps/nm^2-km.

Dispersion-Compensating Fibers

Dispersion-compensating fibers have very high waveguide dispersion.

Some dispersion is inevitable in optical fibers, so engineers have developed *dispersion-compensating fibers,* which have a very high waveguide dispersion. These fibers tend to have a high-index difference between core and cladding, and often have a small effective area; Figure 4.12(f) shows the refractive-index profile of one design.

The overall dispersion of these fibers is opposite in sign and much larger in magnitude than that of standard fibers, so they can be used to cancel out or compensate the dispersion in other single-mode fibers. You'll learn more about dispersion compensation later; for now you only need to remember that special fibers are made for that purpose.

Evolving Fiber Designs

Optical fiber design is continually evolving with changing system requirements. Higher data rates on individual optical channels and increasing numbers of optical channels have pushed the need for better control of dispersion. New single-mode designs already are being fine-tuned and promoted for particular applications. Some fiber types may work better for short-distance, multichannel metro networks, while others fit better into long-distance terrestrial or submarine systems.

Commercial factors also play a role. Companies such as Corning and Lucent Technologies press their own fiber designs, partly to gain advantage in the market. You'll hear different arguments from various companies about what fiber types are best for various systems. There may be no obvious right answer. In fact, different fibers may work best in different parts of a single system, depending on factors such as power levels at various points in the system.

Polarization in Single-Mode Fiber

Light has two orthogonal polarizations.

Light transmission in single-mode fiber also is affected by a property of light that I so far have ignored: polarization. In Chapter 2, we saw that light waves consist of oscillating electric and magnetic fields. The fields are perpendicular to each other and to the direction light travels, as shown in Figure 4.14.

Ordinary unpolarized light is made up of many waves, with their electric and magnetic fields oriented randomly (although always perpendicular to each other for each wave). If all the electric fields (and hence the magnetic fields as well) were aligned parallel to one another, the light would be linearly polarized, which is the simplest type of polarization.

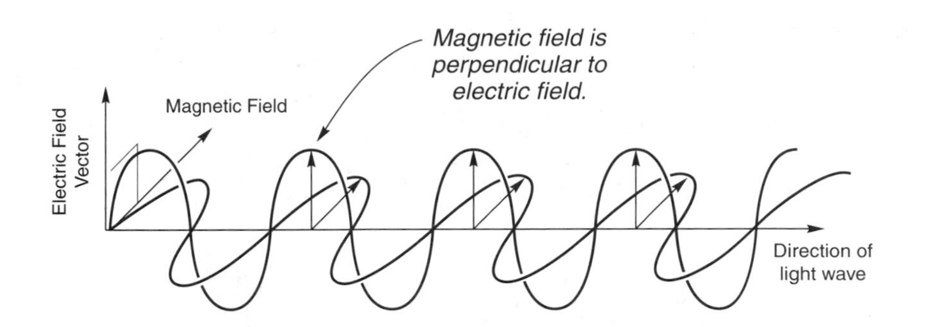

FIGURE 4.14
Electric and magnetic fields in a light wave.

Normal light is considered a combination of two polarizations, vertical and horizontal (determined by the direction of the electric field). A single light wave with its electric field oriented at a different angle is viewed as a combination of waves, one vertically polarized, the other horizontally polarized. Light can also be polarized circularly or elliptically, depending on how electric and magnetic fields oscillate with respect to each other's phase, but that is a matter beyond the scope of this chapter.

Polarization doesn't matter in multimode fibers, but it can be important in single-mode fibers. The reason is that what we call single-mode fibers actually carry two modes with orthogonal polarization. Fibers with circularly symmetric cores can't differentiate between the two linear polarizations. From the standpoint of waveguide theory, the two modes are *degenerate,* meaning they're functionally identical and can't be told apart by the fiber, so light can shift easily between the two polarization modes.

A "single-mode" fiber actually carries two modes with different polarizations.

If the circular symmetry of fibers were perfect, polarization would have little practical impact for communications. However, fiber symmetry is never absolutely perfect. Nor are the forces affecting the fiber applied in perfect symmetry around it. As a result, the two polarization modes may experience slightly different conditions and travel along the fiber at slightly different speeds. This effect is called *polarization mode dispersion,* and it can cause problems in high-performance systems, such as those transmitting time-division multiplexed signals faster than about 2.5 Gbit/s.

Special single-mode fibers can control the polarization of light they transmit. There are two types: true single-polarization fiber and polarization-maintaining fiber. Both intentionally avoid circular symmetry, so they transmit vertically and horizontally polarized light differently. Their cores are asymmetric, and the fiber material may be strained in ways that affect light propagation. The two types have crucial differences in operation.

Single-polarization fiber has different attenuation for light of different polarizations. It transmits light of one polarization well but strongly attenuates light with the orthogonal polarization. Under the proper conditions, a single-polarization fiber attenuates the undesired polarization by a factor of 1000 to 10,000 within a few meters but transmits the desired polarization almost as well as standard single-mode fiber. Thus, only the desired polarization remains at the end.

Polarization-maintaining fiber has internal strain or asymmetry, which effectively splits the input light into two separate polarization modes. This property is called birefringence, which means that the refractive index of the fiber differs for the two polarizations. This prevents the light from shifting between polarizations, as it can while passing through other single-mode fibers. Attenuation of the two polarization modes is similar, but because of the difference in refractive index, they travel at different speeds. Polarization-maintaining fiber will transmit light in a single polarization if the input light is polarized and properly aligned with the polarization direction of the fiber, but otherwise it transmits both polarizations.

Other Fiber Types

Communication fibers generally are classed according to their modal and dispersion properties. However, fibers also can be classified in other ways, such as according to their composition, which are covered in other chapters. The most important examples are:

- Glass fibers with special compositions, such as purified of virtually all hydrogen to eliminate a broad absorption band near 1380 nm, described in Chapter 6.
- Fibers made of nonoxide glasses, which transmit longer infrared wavelengths, also in Chapter 6.
- Plastic fibers covered in Chapter 6.
- Fiber gratings, designed to have specific optical properties, covered in Chapter 7.
- Fibers where light is confined by so-called "photonic bandgap" structures, a new technology in the early research stages, covered in Chapter 7.
- Fibers designed specifically for sensing applications, covered in Chapter 29.
- Fused fiber bundles, covered in Chapter 30.

Planar optical waveguides are not true fibers, but they serve the same function of guiding light waves, and are described briefly in Chapters 7 and 14.

What Have You Learned?

1. There are several different types of optical fibers, with distinct properties.
2. Total internal reflection of light rays only approximates the actual process of light guiding. An optical fiber actually is a dielectric optical waveguide, which propagates light in distinct modes.
3. Fiber properties depend on the core-cladding structure and the materials from which the fiber is made.
4. To guide light, a fiber must have a core with higher refractive index than the cladding.
5. Step-index multimode fibers have a core diameter tens of wavelengths of the light they are guiding. They are used for imaging and illumination, but modal dispersion limits their transmission speed for communications.
6. The number of modes carried by a fiber depends on its core diameter, the refractive indexes of core and cladding, and the wavelength.

7. As dielectric optical waveguides, optical fibers carry light along the core-cladding boundary, with some light in the cladding.
8. Fibers with core diameters only 6 to 10 μm transmit a single mode of light.
9. Grading the refractive-index differential between core and cladding can nearly eliminate modal dispersion in a multimode fiber.
10. Graded-index fibers have standard core diameters of 50 or 62.5 μm. They are used for transmission over distances to a couple of kilometers.
11. Single-mode fibers are used for high-speed communications over distances of more than a kilometer or two. They may be used over shorter distances.
12. Standard single-mode fibers have a step-index profile and zero chromatic dispersion at 1.31 μm.
13. The cutoff wavelength is the shortest wavelength at which a single-mode fiber transmits only one mode.
14. Chromatic dispersion is the sum of material dispersion and waveguide dispersion; all three depend on wavelength. It is the main type of dispersion in single-mode fiber.
15. Dispersion-shifted fiber has waveguide dispersion increased so it cancels material dispersion at a wavelength generally longer than 1.31 μm. Types now in use have dispersion shifted to wavelengths shorter or longer than the erbium-fiber amplifier band, usually to about 1500 nm or about 1640 nm.
16. Reducing dispersion slope makes dispersion change less with wavelength, but tends to decrease the effective area where light is confined in the fiber. Fibers with small effective area are more vulnerable to nonlinear effects.
17. Light can be polarized in vertical or horizontal directions. Normal single-mode fiber carries both polarizations and can suffer polarization-mode dispersion.
18. Single-polarization fibers transmit light in only one polarization. Polarization-maintaining fibers keep light in the same polarization that it had when entering the fiber.

What's Next?

In Chapter 5, you will learn about the most important properties of optical fibers. Chapter 6 will cover fiber materials and fabrication.

Further Reading

Luc B. Jeunhomme, *Single-Mode Fiber Optics: Principles and Applications* (Marcel Dekker, 1990)

Donald B. Keck, ed, *Selected Papers on Optical Fiber Technology* (SPIE Milestone Series Vol. MS38, 1992)

Gerd Keiser, *Optical Fiber Communications,* 3rd ed (McGraw Hill, 2000)

Advanced Treatments:

John A. Buck, *Fundamentals of Optical Fibers* (Wiley-Interscience, 1995)

Ajoy Ghatak and K. Thyagarajan, *Introduction to Fiber Optics* (Cambridge University Press, 1998)

Questions to Think About for Chapter 4

1. A step-index multimode fiber has modal dispersion of about 30 ns/km. Using the formula for maximum data rate for a given dispersion, about how far could it transmit a signal at 1 Gbit/s?
2. Why doesn't dispersion affect imaging or illumination fibers?
3. Graded-index fiber typically is more expensive than step-index single-mode fiber. Yet, it is used to carry Gigabit Ethernet signals several hundred meters. What advantage does it offer?
4. What are the tradeoffs between effective area and dispersion slope?
5. Your system has to transmit wavelength-division multiplexed signals at the 1530 to 1620 nm band of erbium-doped fiber amplifiers. What type of fiber is best?
6. Your cheapskate purchasing department just got a great deal on zero dispersion-shifted fiber. Why can't you use it in the erbium-amplifier system in Question 5?

Quiz for Chapter 4

1. What is the half-acceptance angle for a large-core step-index fiber with core index of 1.5 and cladding index of 1.495?
   a. 4.7°
   b. 7.0°
   c. 9.4°
   d. 11°
   e. 14°
2. Modal dispersion is largest in what type of fiber?
   a. Step-index multimode
   b. Graded-index multimode
   c. Step-index single-mode
   d. Dispersion-shifted single-mode
   e. Polarization-maintaining
3. A fiber has modal dispersion of 20 ns/km. If an instantaneous light pulse traveled through 8 km of such fiber, what would the pulse length be at the end?
   a. 8 ns
   b. 20 ns
   c. 40 ns
   d. 80 ns
   e. 160 ns

4. What is the maximum data rate that the 8-km length of fiber in Problem 3 could carry?

a. 160 Mbit/s

b. 20 Mbit/s

c. 16 Mbit/s

d. 6 Mbit/s

e. 4.4 Mbit/s

5. What guides light in multimode graded-index fibers?

a. Total internal reflection

b. Mode confinement in the cladding

c. Refraction in the region where core grades into the cladding

d. The optics that couple light into the fiber

6. What is the maximum allowable core diameter for a step-index single-mode fiber operating at 1.3 μm, with core index of 1.5 and cladding index of 1.0003 (air)?

a. 0.34 μm

b. 0.89 μm

c. 3.0 μm

d. 4.8 μm

e. 5.5 μm

7. What is the maximum core diameter for a step-index single-mode fiber operating at 1.3 μm, with core index of 1.5 and cladding index of 1.495?

a. 0.89 μm

b. 3.0 μm

c. 4.1 μm

d. 8.1 μm

e. 10.3 μm

8. What is the cutoff wavelength of a single-mode step-index fiber with core diameter of 8 μm, core index of 1.5, and cladding index of 1.495?

a. 0.89 μm

b. 1.15 μm

c. 1.28 μm

d. 1.31 μm

e. 1.495 μm

9. What is the cutoff wavelength of a single-mode step-index fiber with core diameter of 8 μm, core index of 1.5, and cladding index of 1.496?

a. 0.89 μm

b. 1.15 μm

c. 1.28 μm

d. 1.31 μm

e. 1.495 μm

10. What is done to design a dispersion-shifted fiber?

a. Waveguide dispersion is increased to offset material dispersion near 1.55 μm.

b. Material dispersion is reduced at 1.31 μm.

c. Material dispersion is increased to offset waveguide dispersion near 1.55 μm.

d. Core diameter is increased to allow multimode transmission.

e. The fiber core is made asymmetrical to control polarization.

11. For what application is nonzero dispersion-shifted fiber required?

a. Single-wavelength transmission at 1.55 μm
b. Short-distance data communications
c. Single-wavelength transmission at 1.31 μm
d. Dense wavelength-division multiplexing around 1.55 μm
e. Dense wavelength-division multiplexing around 1.31 μm

12. Does single-polarization fiber transmit more or fewer modes than standard step-index single-mode fiber?

a. Both transmit the same number
b. Single-mode fiber transmits fewer because polarization-sensitive fibers distinguish between the two orthogonal polarizations
c. Single-polarization fiber carries fewer because standard step-index fibers do not distinguish between the two orthogonal polarizations
d. Need more information to answer the question

CHAPTER

5

Properties of Optical Fibers

About This Chapter

Now that you have learned about the basic designs of optical fibers, the next step is to understand the properties of fibers important for light transmission. I have already touched upon many properties in Chapter 4; this chapter examines them more thoroughly.

The most important properties for communications are attenuation, light collection and propagation, fiber dispersion, and mechanical strength. Nonlinear effects can be important in some cases, particularly for sensing and high-performance systems. I will start with the property usually at the top of the list—attenuation.

Fiber Attenuation

The attenuation of an optical fiber measures the amount of light lost between input and output. Total attenuation is the sum of all losses. It is dominated by imperfect light coupling into the fiber and absorption and scattering within the fiber. Sometimes other effects can cause important losses, such as light leakage from fibers that suffer severe microbending. Attenuation limits how far a signal can travel through a fiber before it becomes too weak to detect.

Absorption and scattering are both cumulative, with their effects increasing with fiber length. In contrast, coupling losses occur only at the ends of the fiber. The longer the fiber, the more important are absorption and scattering losses, and the less important coupling losses. Conversely, attenuation and scattering may be much smaller than end losses for short fibers.

Loss during fiber transmission is the sum of scattering, absorption, and light-coupling losses.

To briefly review these losses, when you deliver an input power, P_0, to a fiber, a fraction of that light, ΔP, is lost. Thus only the power $P_0 - \Delta P$ gets into the fiber. This light then suffers absorption and scattering loss in the bulk of the fiber. As you learned in Chapter 2, these losses depend on length. If the light lost to absorption per unit length is α and the light lost to scattering per unit length is S, the fraction of light that remains is $(1 - \alpha - S)$. Outside the research laboratory, the quantity that matters is the attenuation per unit length, which is the sum of the absorption and scattering $(\alpha + S)$.

Recall that to calculate the power remaining after a distance D, you raise the fraction of light remaining after attenuation to the power D. This gives a formula for power at a distance D

$$P(D) = (P_0 - \Delta P)\,(1 - [\alpha + S])^D$$

That formula is more useful for looking at the process of light loss than for calculations. It reminds us that absorption and scattering combine to make attenuation, which always takes the same fraction of the light passing through each chunk of fiber. It also reminds us that attenuation acts only on light that gets into the fiber, because some light is lost on entry.

Now let's look at each of these components of loss.

Absorption

Absorption depends on wavelength and is cumulative with distance.

Every material absorbs some light energy. The amount of absorption depends on the wavelength and the material. A thin window of ordinary glass absorbs little visible light, so it looks transparent to the eye. The paper this book is printed on absorbs much more visible light, so it looks opaque. (You can read these words because the blank paper reflects more light than the ink, which absorbs most light striking it and reflects little.) The amount of absorption can vary greatly with wavelength. The clearest glass is quite opaque at an infrared wavelength of 10 μm. Air absorbs so strongly at short ultraviolet wavelengths that scientists call ultraviolet wavelengths shorter than about 0.2 μm the *vacuum ultraviolet* because only a vacuum transmits them.

Absorption depends very strongly on the composition of a substance. Some materials absorb light very strongly at wavelengths where others are quite transparent. For glass, this means that adding small amounts of certain impurities can dramatically increase absorption at wavelengths where glass is otherwise transparent. Removing such impurities was a crucial step to making the extremely transparent fibers used for communications. Typically, absorption is plotted as a function of wavelength. Some absorption peaks can look quite narrow because the material absorbs light in only a narrow range of wavelengths; others spread across a wider range.

Absorption is uniform. The same amount of the same material always absorbs the same fraction of light at the same wavelength. If you have three blocks of the same type of glass, each 1-centimeter thick, all three will absorb the same fraction of the light passing through them.

Absorption also is cumulative, so it depends on the total amount of material the light passes through. That means a material absorbs the same fraction of the light for each unit length.

If the absorption is 1% per centimeter, it absorbs 1% of the light in the first centimeter, and 1% of the *remaining* light the next centimeter, and so on. If the only thing affecting light is absorption, the fraction of light absorbed per unit length is α and the total length is D, the fraction of light remaining after a distance D is

$$(1 - \alpha)^D$$

In our example, this means that after passing through 1 m (100 cm) of glass, the fraction of light remaining would be

$$(1 - 0.01)^{100} = 0.366, \qquad \text{or } 36.6\%$$

Scattering

Atoms scatter a small fraction of passing light.

Atoms and other particles inevitably scatter some of the light that hits them. The light isn't absorbed, just sent in another direction in a process called Rayleigh scattering, after the British physicist Lord Rayleigh, as shown in Figure 5.1. However, the distinction between scattering and absorption doesn't matter much if you are trying to send light through a fiber, because the light is lost from the fiber in either case.

Like absorption, scattering is uniform and cumulative. The farther the light travels through a material, the more likely scattering is to occur. The relationship is the same as for light absorption, but the fraction of scattered light is written S.

$$\text{Remaining light} = (1 - S)^D$$

Scattering depends not on the specific type of material but on the size of the particles relative to the wavelength of light. The closer the wavelength is to the particle size, the more scattering. In fact, the amount of scattering increases quite rapidly as the wavelength λ decreases. For a transparent solid, the scattering loss in decibels per kilometer is given by

$$\text{Scattering} = A\lambda^{-4}$$

where A is a constant depending on the material. This means that dividing the wavelength by 2 multiplies scattering loss (in dB/km) by a factor of 16.

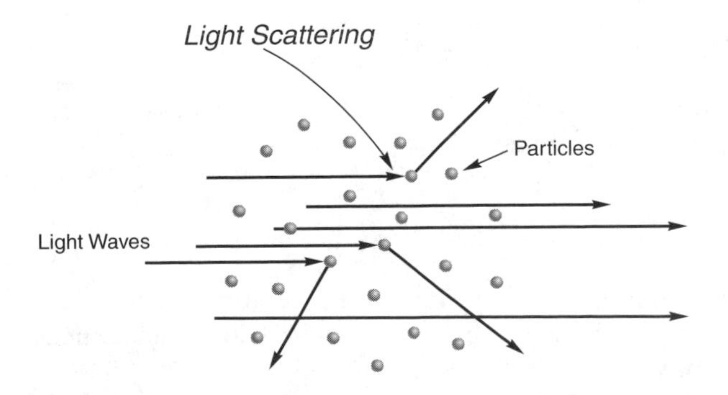

FIGURE 5.1
Rayleigh scattering of light.

Total Loss or Attenuation

Total attenuation, or loss, is the sum of scattering and attenuation; it is measured in decibels per kilometer.

Scattering and absorption combine to give total loss, or attenuation, which is the important number in communication systems. Figure 5.2 plots their contributions across the range of wavelengths used for communications. Attenuation normally is measured in decibels per kilometer for communication fibers. The plot shows small absorption peaks from traces of metal impurities remaining in the glass and other absorption arising from bonds that residual hydrogen atoms form with oxygen in the glass. (I picked this scale to emphasize the peaks, which look lower on other scales, and are smaller in many communication fibers.) The absorption at wavelengths longer than 1.6 μm comes from silicon-oxygen bonds in the glass; as the plot shows, the absorption increases rapidly at longer wavelengths. As a result, silica-based fibers are rarely used for communications at wavelengths longer than 1.65 μm.

Rayleigh scattering accounts for most attenuation at shorter wavelengths. As you can see in Figure 5.2, it increases sharply as wavelength decreases. The space between measured total attenuation and the theoretical scattering curve represents the absorption loss. The closer the two lines, the larger the fraction of total attenuation that arises from scattering. The rapid decrease in scattering at longer wavelengths makes loss lowest in the "valley" around 1.55 μm, where both Rayleigh scattering and infrared absorption are low. Except for the infrared absorption of silica, fiber loss would decrease even more at longer wavelengths.

Total attenuation is what matters for system performance.

The plot in Figure 5.2 compares theoretical scattering and the absorption of pure silica with attenuation measured across the spectrum. It is total attenuation that is important in fiber-optic communications, and that is what is generally measured. Absorption and scattering are hard to separate, and outside the laboratory there is little practical reason to bother. It's most useful to think of the power *(P)* at a distance D along the fiber as defined by

$$P(D) = (P_0 - \Delta P)(1 - A)^D$$

where A is attenuation per unit length, P_0 is initial power, and ΔP is the coupling loss, as before. In practice, it is simpler to make calculations if you first separate fiber attenuation from coupling losses by starting with the power that *enters* the fiber rather than the input power you *attempt* to couple into the fiber.

Calculating Attenuation in Decibels

Attenuation normally is calculated on the logarithmic decibel scale.

As we saw in Chapter 2, attenuation measures the ratio of input to output power: P_{out}/P_{in}. It normally is measured in decibels, as defined by the equation

$$\text{dB (attenuation)} = -10 \log_{10}\left(\frac{P_{out}}{P_{in}}\right)$$

Output power is less than input power, so the result would be a negative number if the equation didn't include a minus sign. You should remember that in some publications decibels are defined so that a negative number indicates loss.

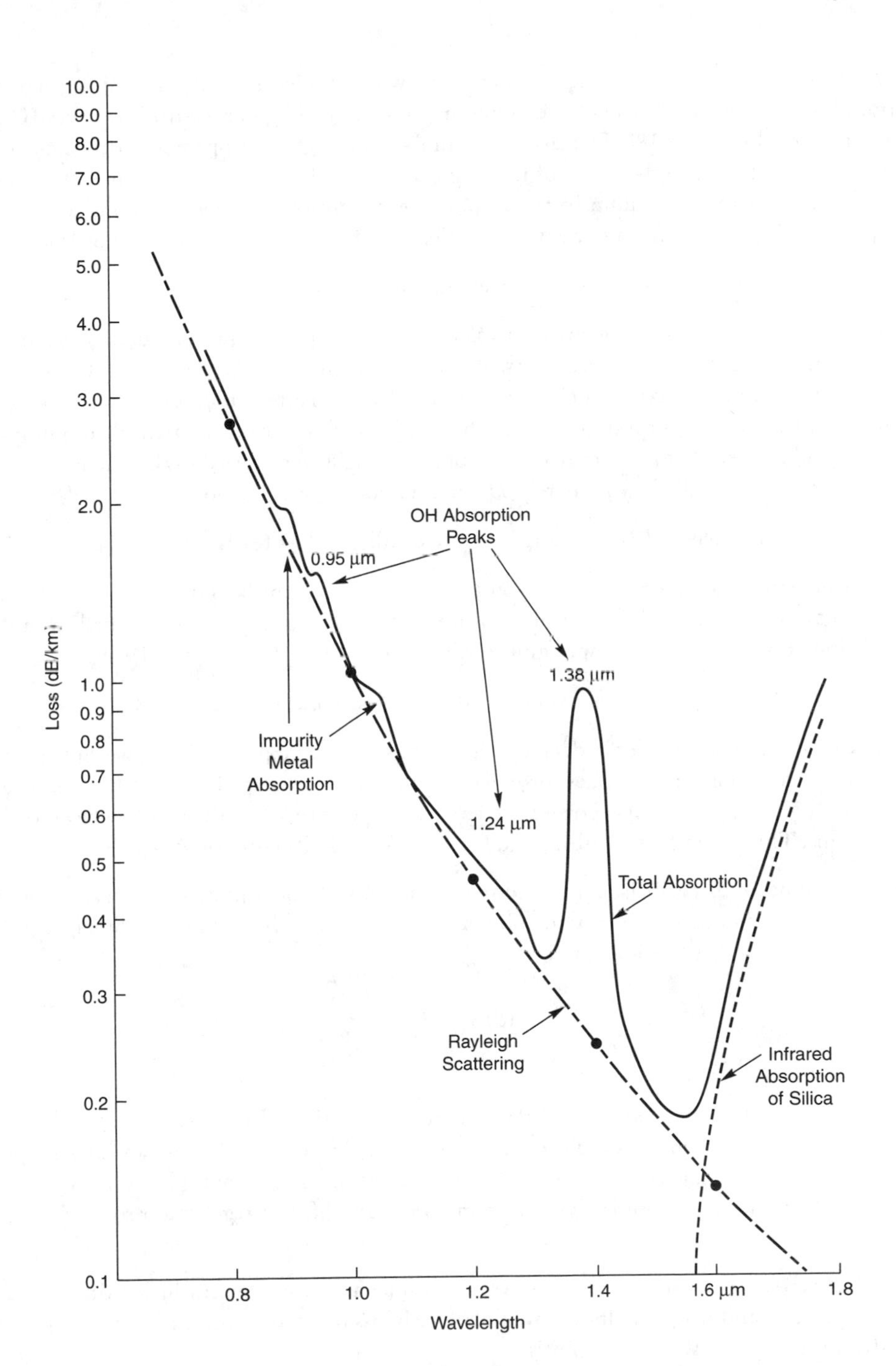

FIGURE 5.2
Total attenuation in a fiber is the sum of absorption and scattering losses.

Decibels may seem to be rather peculiar units, which appear to understate high attenuation. For example, a 3-dB loss leaves about half the original light, a 10-dB loss leaves 10%, and a 20-dB loss leaves 1%. The larger the number, the larger the apparent understatement. A 100-dB loss leaves only 10^{-10} of the original light, and a 1000-dB loss leaves 10^{-100}—a ratio smaller than one atom in the whole known universe. Appendix B translates some representative decibel measurements into ratios. You can also use the simple conversion

$$\text{Fraction of power remaining} = 10^{(-\text{dB}/10)}$$

The decibel scale simplifies calculations of power and attenuation.

Decibels are very convenient units for calculating signal power and attenuation. Suppose you want to calculate the effects of two successive attenuations. One blocks 80% of the input signal, and the second blocks 30%. To calculate total attenuation using fractions, you must convert both absorption figures to the fractions of power transmitted, then multiply them, and convert that number from the fraction of light transmitted to the fraction attenuated. If you use decibels, you merely add attenuations to get total loss.

$$\text{Total loss (dB)} = \text{loss (dB)}_1 + \text{loss (dB)}_2 + \text{loss (dB)}_3 + \cdots$$

The calculations are even simpler if you know the loss per unit length and want to know total loss of a longer (or shorter) piece of fiber. Instead of using the exponential formula mentioned previously, you simply multiply loss per unit length times the distance:

$$\text{Total loss} = \text{dB/km} \times \text{distance}$$

You can also measure power in decibels relative to some particular level. In fiber optics, the two most common decibel scales for power are decibels relative to 1 mW (dBm) and relative to 1 μW (dBμ). Powers above those levels have positive signs; those below have negative signs. Thus 10 mW is 10 dBm, and 0.1 mW is −10 dBm, or 100 dBμ.

If everything is in decibels, simple addition and subtraction suffice to calculate output power from input power and attenuation. You also can write the equation in other ways:

$$P_{out} = P_{in} - \text{loss (dB)}$$

$$\text{Loss (dB)} = P_{in} - P_{out}$$

$$P_{in} = P_{out} + \text{loss (dB)}$$

Note that it is vital to keep track of the plus and minus signs. In this case, we give loss in decibels a positive sign, as we did earlier. If you ever feel confused, you can do a simple truth test, by checking to see if the output power is less than the input. (The only way output can be more than input is if you have an optical amplifier or regenerator somewhere in the system.)

As an example of how the calculations work, consider a fiber system in which 3 dB is lost at the input end and that contains 6 km of fiber with loss of 0.5 dB/km. If the input power is 0 dBm (exactly 1 mW), the output is

$$P_{out} = 0\text{ dBm} - 3\text{ dB (input loss)} - (6\text{ km} \times 0.5\text{ dB/km}) = -6.0\text{ dBm}$$

If you rewrite this as milliwatts, you have 0.25 mW.

Spectral Variation

As we saw before, fiber attenuation is the sum of absorption and scattering, both of which vary with wavelength. The spectral variation depends on the fiber composition. The attenuation curve in Figure 5.2 is fairly typical for single-mode communication fibers, but some types have much lower water peaks at 1380 nm, as described below.

Attenuation varies with wavelength, depending on the material.

Most single-mode communication fibers are used at wavelengths between about 1280 and 1650 nm, where attenuation is generally below 0.5 dB/km except at the water peak where it may reach 1 dB/km. The traditional transmission bands in that region are at 1310 nm, and in the region from about 1530 to 1620 nm where erbium-doped fiber amplifiers are used. Fibers are available that have water content reduced to such low levels that the 1380-nm water peak almost vanishes, allowing them to be used across the entire 1280 to 1650-nm range.

Attenuation generally is higher in commercial graded-index multimode fibers, with typical values about 2.5 dB/km at 850 nm, 0.8 dB/km at 1310 nm, and no more than 3 dB/km at the 1380 nm water peak. As can be seen from Figure 5.2, attenuation is not particularly low at 850 nm; the attraction of that wavelength is its match to the output of gallium-arsenide light sources.

Other materials are used in fibers for other wavelengths. Special grades of quartz are used for ultraviolet-transmitting fibers. Some plastics have relatively even transmission across the visible spectrum. Fluoride compounds are transparent at longer infrared wavelengths than silica glass. Chapter 6 will cover various materials in more detail.

Light Collection and Propagation

Several factors enter into how fibers collect light and propagate it. Most arise from the structure of fibers, described in Chapter 4. This section examines the impact of those considerations.

Core Size and Mode-Field Diameter

Core size is important in coupling light into a fiber. The core must be aligned with the light-emitting region of a laser or LED light source, or the output end of another fiber. To collect light efficiently, the core should be at least as large as the light source. As shown in Figure 5.3, if the light source is larger than the fiber, much of its light goes right into the cladding and quickly escapes from the fiber. The larger the core diameter, the easier it is to align with the light source to collect light. This is purely a matter of geometry, matching the light source to the collecting aperture. Core size does not directly affect the acceptance angle of a fiber, the range of angles over which it collects light.

The larger the core diameter, the easier it is to align with a light source.

Core size is the physical dimension of the core, but light spreads through a slightly larger volume, including the inner edge of the cladding. This *mode-field diameter* or *effective area*

FIGURE 5.3
The match between light-source dimensions and core diameter helps determine light transfer.

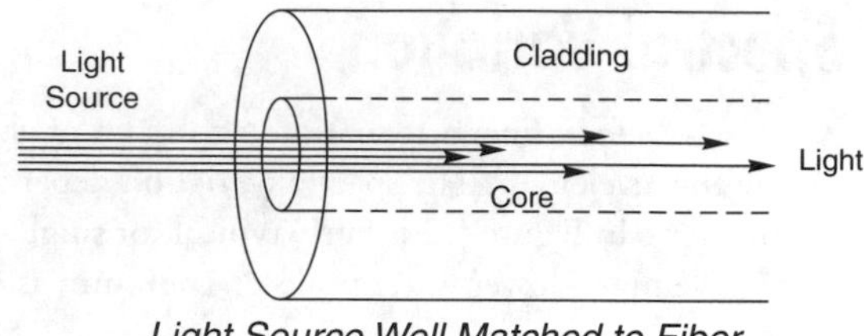

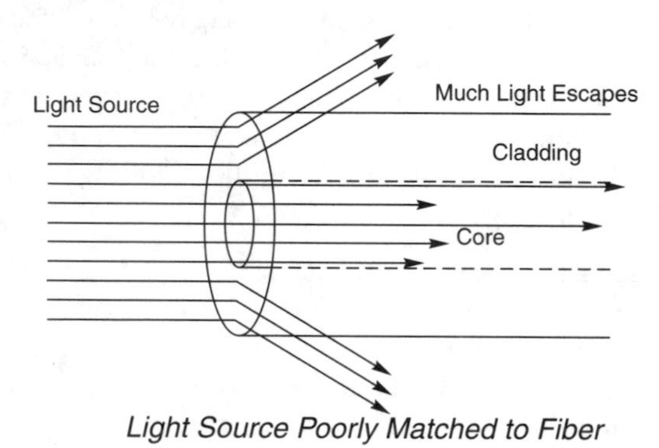

is the critical dimension for light transfer between single-mode fibers. (The difference is not enough to matter in multimode fiber.)

Transferring light between fibers is the business of splices and connectors; details are covered in Chapter 13.

Numerical Aperture

Numerical aperture (NA) measures the fiber's acceptance angle.

A second factor in determining how much light a fiber collects is its acceptance angle, the range of angles over which a light ray can enter the fiber and be trapped in its core. The full acceptance angle is the range of angles at which light is trapped; it extends both above and below the axis of the fiber. The half acceptance angle is the angle measured from the fiber axis to the edge of the cone of light rays trapped in the core; it is shown in Figure 5.4.

The standard measure of acceptance angle is the numerical aperture, NA, which is the sine of the half-acceptance angle, θ, for reasonably small angles. For a step-index fiber, it is defined as

$$\mathrm{NA} = \sqrt{(n_0^2 - n_1^2)} = \sin\theta$$

where n_0 is the core index and n_1 is the cladding index. A typical value for step-index single-mode fiber is around 0.14.

Numerical aperture is not calculated the same way in graded-index fibers; strictly speaking it varies across the core with the refractive index. However, you can measure numerical

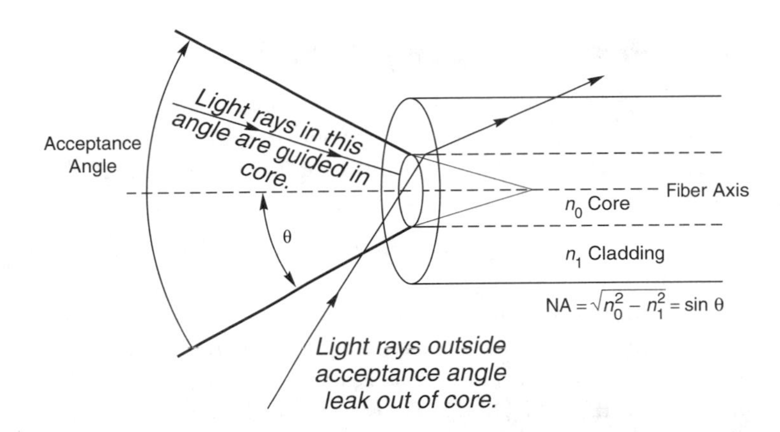

FIGURE 5.4
Light rays have to fall within a fiber's acceptance angle, measured by NA, to be guided in the core.

aperture by monitoring the divergence angle of light leaving a fiber core. As shown in Figure 5.5, the light emerging from a multimode fiber spreads over an angle equal to its acceptance angle. For practical measurements, care must be taken to eliminate modes guided along the cladding, and the *edge* of the beam is defined as the angle where intensity drops to 5% that in the center. NA can be calculated easily from the acceptance angle. Typical NA values are 0.20 for 50/125 graded-index fiber, and about 0.28 for 62.5/125 graded-index fiber.

Core diameter does not enter into the NA equation, but light rays must enter the core as well as fall within the acceptance angle to be guided along a fiber. Large core size and large NA do not have to go together, but in practice larger-core fibers tend to have larger core-cladding index differences and thus larger NAs. For example, step-index multimode fibers typically have NAs of at least 0.3, more than twice the value for single-mode step-index fibers.

The numerical aperture of single-mode fibers is defined by the same equation as for multimode fibers, but light does not spread out from them in the same way. (They carry only a single mode, and their cores are so small that another wave effect called *diffraction* controls how light spreads out from the end.) NA generally is not as important for single-mode fibers as it is for multimode fibers.

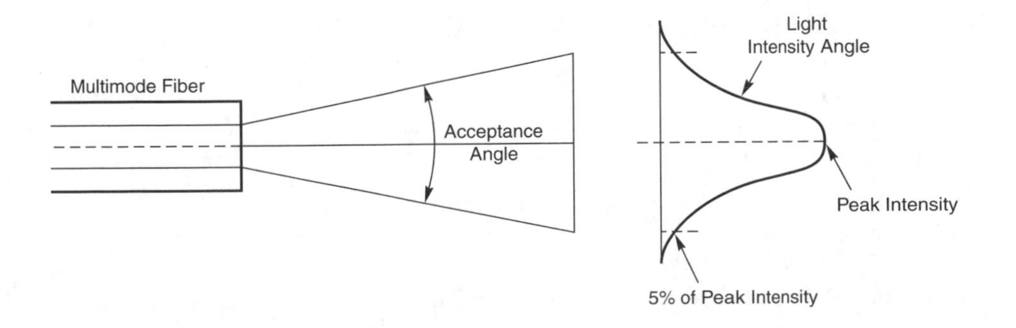

FIGURE 5.5
Intensity of light emerging from a multimode fiber falls to about 5% of peak value at the edge of its acceptance angle.

Cladding Modes and Leaky Modes

Light entering a fiber can be guided along the cladding.

As we have seen earlier, not all light aimed into a fiber is guided along the core. Some enters the cladding at the end of the fiber; other light escapes from the core by hitting the core-cladding boundary at greater than the confinement angle. This light excites *cladding modes,* which can propagate in the cladding.

Total internal reflection at the boundary between cladding and the surrounding material can guide light in the cladding just as it guides light along an unclad fiber. This can happen as long as the surrounding material—air or a plastic coating—has a lower refractive index than the cladding. This might sound like a good way to maximize light transmission, but it's usually undesirable. It can introduce noise in communication fibers and crosstalk between adjacent fibers in an imaging bundle. To prevent this, manufacturers often coat fibers with a plastic having a higher refractive index than the cladding, so light striking the cladding-coating boundary leaks out. The fibers in rigid bundles sometimes are separated by "dark" glass, which absorbs light so it can't pass between claddings.

Leaky modes are not perfectly confined to the core.

In multimode fibers, the boundary between modes guided in the core and modes confined to the cladding is not sharp. Some light falls into intermediate *leaky* modes, which propagate partly in both core and cladding. These modes travel much farther than cladding modes but also are prone to leakage and loss.

Both cladding and leaky modes can lead to spurious results in fiber measurements, so mode strippers have been developed to remove them. These devices work by surrounding part of the fiber with a material having a refractive index equal to or larger than that of the cladding, preventing total internal reflection at the outer boundary of the cladding. Light that leaks into this material is absorbed and lost from the fiber. A long length of fiber also can serve as a mode stripper if the cladding attenuation is much higher than that of the core.

Bending Losses

Bends can cause excess fiber loss.

A variety of outside influences can change the physical characteristics of optical fibers, affecting how they guide light. Typically these effects are modest and must be enhanced or accumulated over long distances to make the kind of sensors described in Chapter 29. However, significant losses can arise if the fiber is bent so sharply that light strikes the core-cladding interface at a large enough angle that the light can leak out.

Bending loss is easiest to explain using the ray model of light in a multimode fiber. When the fiber is straight, light falls within its confinement angle. Bending the fiber changes the angle at which light hits the core-cladding boundary, as shown in Figure 5.6. If the bend is sharp enough, it hits the boundary at an angle outside the confinement angle θ_c, and is refracted into the cladding where it can leak out.

Bend losses fall into two broad categories. *Macrobends* are single bends obvious to the eye, such as a fiber bent sharply where a cable ends at a connector. The case shown in Figure 5.6 is typical. *Microbends* are tiny kinks or ripples that can form along the length of fibers that become squeezed into too small a space. This can happen in a cable when the cabling ma-

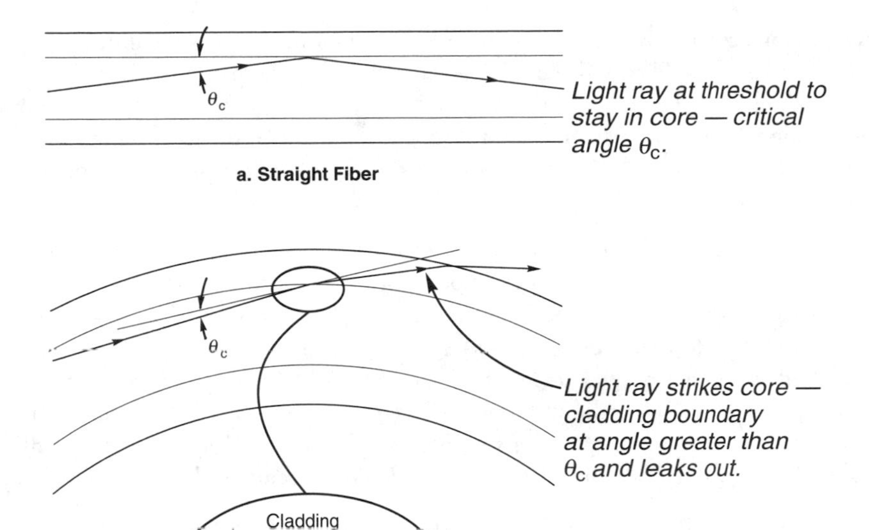

FIGURE 5.6
Light can leak out of a bent fiber.

terial shrinks relative to the fiber, or the fiber stretches relative to the cable. Microbends are smaller, but they cause similar light leakage because they also affect the angle at which light hits the core-cladding boundary.

Dispersion

Dispersion is the spreading out of light pulses as they travel along a fiber. It occurs because the speed of light through a fiber depends on its wavelength and the propagation mode. The differences in speed are slight, but like attenuation, they accumulate with distance. The four main types of dispersion arise from multimode transmission, the dependence of refractive index on wavelength, variations in waveguide properties with wavelength, and transmission of two different polarizations of light through single-mode fiber.

The principal types of dispersion are modal, material, waveguide, and polarization.

Like attenuation, dispersion can limit the distance a signal can travel through an optical fiber, but it does so in a different way. Dispersion does not weaken a signal; it blurs it. If you send one pulse every nanosecond but the pulses spread to 10 ns at the end of the fiber, they blur together. The signal is present, but it's so blurred in time that it is unintelligible.

In its simplest sense, dispersion measures pulse spreading per unit distance in nanoseconds or picoseconds per kilometer. Total pulse spreading, Δt, is

$$\Delta t = \text{dispersion (ns/km)} \times \text{distance (km)}$$

This equation actually gives dispersion in two different forms. One is the unit or characteristic dispersion of the fiber, written as *dispersion* and measured per unit length (in units of time per kilometer). The other is the total pulse spreading in units of time over the entire length. As long as the same fiber is used throughout the cable, the total pulse spreading is simply the characteristic fiber dispersion times the fiber length. If different types of fibers are used, you need to calculate pulse spreading separately for each section, then add them.

The simple equation above holds for modal dispersion, which is the type most important for step-index multimode fibers, where modes travel at different speeds through the fiber. Graded-index fibers nominally equalize the speeds of all transmitted modes, but things don't work that perfectly in the real world. It's functionally impossible to achieve the ideal refractive-index profile needed to make all modes travel at exactly the same speed. That profile depends on wavelength, and fibers carry signals at a range of wavelengths. In practice, you have to rely on manufacturer specifications for the unit dispersion of graded-index fibers, typically specified in units of bandwidth (described below) rather than in time units.

Total pulse spreading is the square root of the sums of the squares of the pulse spreading from modal, chromatic, and polarization-mode dispersion.

Other types of dispersion also add to total pulse spreading. We'll get to them in a minute, but first let's look at how to calculate the total pulse spreading. Material and waveguide dispersion add together to give a wavelength-dependent *chromatic dispersion,* mentioned in Chapter 4. Fibers also experience polarization-mode dispersion. Both quantities are independent of each other and of modal dispersion. That means you have to take the square root of the sum of the squares to get total pulse spreading:

$$\Delta t_{\text{total}} = \sqrt{(\Delta t_{\text{modal}})^2 + (\Delta t_{\text{chromatic}})^2 + (\Delta t_{\text{polarization-mode}})^2}$$

Polarization-mode dispersion doesn't matter in multimode fibers, so for that case the equation becomes

$$\Delta t_{\text{total}} = \sqrt{(\Delta t_{\text{modal}})^2 + (\Delta t_{\text{chromatic}})^2}$$

Likewise, single-mode fibers have no modal dispersion (other than polarization-mode dispersion), so the equation becomes

$$\Delta t_{\text{total}} = \sqrt{(\Delta t_{\text{chromatic}})^2 + (\Delta t_{\text{polarization-mode}})^2}$$

Chromatic Dispersion and Wavelength

Chromatic dispersion depends on the range of wavelengths in the optical signal.

Chromatic dispersion is the pulse spreading that arises because the velocity of light through a fiber depends on its wavelength. It is measured in units of picoseconds (of pulse spreading) per nanometer (of spectral width of the optical signal) per kilometer (of fiber length). The total pulse spreading due to chromatic dispersion, $\Delta t_{\text{chromatic}}$, is calculated by multiplying the fiber's characteristic chromatic dispersion by the range of wavelengths generated by the light source ($\Delta\lambda$) and the fiber length:

$$\Delta t_{\text{chromatic}} = \text{chromatic dispersion (ps/nm-km)} \times \Delta\lambda \text{ (nm)} \times \text{fiber length (km)}$$

The characteristic chromatic dispersion of a fiber is a function of wavelength. It is normally the largest type of dispersion in single-mode fiber systems. As you learned in Chapter 4, chromatic dispersion is the sum of two components, material and waveguide dispersion, which can cancel each other at certain wavelengths. In standard step-index single-mode fiber, material and waveguide dispersion add to zero near 1310 nm. Dispersion shifting moves the zero-dispersion point to other wavelengths, generally longer. To understand chromatic dispersion, we need to look at both material and waveguide dispersion.

Material dispersion arises from the change in a material's refractive index with wavelength. The higher the refractive index, the slower light travels. Thus as a pulse containing a range of wavelengths passes through a material, it stretches out, with the wavelengths with lower refractive index going faster than those with higher indexes. Like absorption, dispersion is a function of the individual material, which changes with wavelength. Communication fibers are nearly pure silica (SiO_2), so their characteristic material dispersion is essentially the same as that of pure fused silica. Figure 5.7 plots both refractive index and material dispersion of fused silica against wavelength.

Material dispersion arises from variations in refractive index with wavelength.

Note that material dispersion has a positive or negative sign, unlike the modal dispersion. You can think of this sign as indicating how the refractive index is changing with wavelength, although that's an oversimplification. The physical meaning of the sign is a bit obscure, but the signs are important in combining material dispersion and waveguide dispersion to calculate total chromatic dispersion. Although chromatic dispersion also has a sign, the calculations for total pulse spreading cancel it out because the formula uses the square of the pulse spreading caused by chromatic dispersion.

As Figure 5.7 shows, the magnitude of material dispersion is large at wavelengths shorter than 1.1 μm. High material dispersion at 850 nm makes chromatic dispersion high at that wavelength, limiting the transmission speed possible even in single-mode fiber. The real benefits of single-mode transmission come from operating at longer wavelengths where the material dispersion is small.

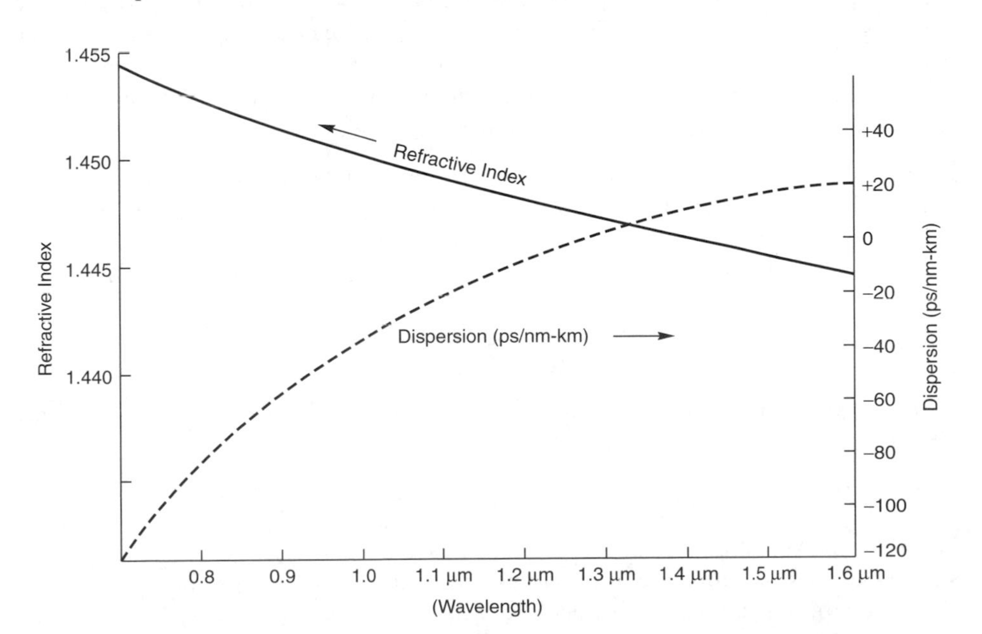

FIGURE 5.7
Material dispersion and refractive index of silica as a function of wavelength.

Waveguide dispersion arises from changes in light distribution between core and cladding.

Waveguide dispersion is a separate effect, arising from the distribution of light between core and cladding. Recall that waveguide properties are a function of the wavelength. This means that changing the wavelength affects how light is guided in a single-mode fiber. For a step-index single-mode fiber, the waveguide dispersion is relatively small, but can be important. More complex refractive index profiles can increase waveguide dispersion, such as the dispersion-compensating fiber in Figure 4.12(f). Like material dispersion, waveguide dispersion has a sign that indicates how changing wavelength affects dispersion.

For most practical purposes, chromatic dispersion is the sum of material and waveguide dispersion.

$$\mathrm{Disp}_{\mathrm{chromatic}} = \mathrm{Disp}_{\mathrm{material}} + \mathrm{Disp}_{\mathrm{waveguide}}$$

Remember that the signs are important. From a physical standpoint, what happens is that the variation with wavelength caused by waveguide dispersion can offset (or add to) that caused by material dispersion. Dispersion shifting is done by designing fibers to have large negative waveguide dispersion, which offsets positive material dispersion at wavelengths longer than 1.28 μm, shifting the region of low chromatic dispersion near the erbium-fiber amplifier band. As mentioned earlier, generally the zero dispersion wavelength is chosen to be a little longer or shorter than the 1530- to 1620-nm erbium-fiber band. Figure 5.8 shows how waveguide and material dispersion combine for step-index single-mode fiber and one type of nonzero dispersion-shifted fiber.

A Closer Look at Chromatic Dispersion

The descriptions of material, waveguide, and chromatic dispersion have been a bit vague because the formal definitions depend on some concepts that require a bit of extra work, a few equations, and a dash of calculus to understand. To delve more deeply, let's consider the case of material dispersion, which is the simplest because it depends only on how the refractive index of the material varies with wavelength. (Chromatic and waveguide dispersion work similarly, but the details are more complex.)

Recall that the velocity of light passing through a material depends on its refractive index. Since the refractive index varies with wavelength, so does the velocity of light in the material. Suppose that the material has a refractive index n_1 at wavelength λ_1 and an index n_2 at wavelength λ_2. The time each wavelength takes to pass through a length of glass L is

$$t = \frac{Ln}{c}$$

If you calculate the difference between the transit times at the two wavelengths, you get what is called the *group delay time,*

$$\text{Group delay} = t_1 - t_2 = \frac{Ln_1}{c} - \frac{Ln_2}{c} = \frac{L}{c}(n_1 - n_2)$$

which measures the difference in travel time for the two wavelengths. This is the same as the pulse spreading through a fiber denoted by Δt.

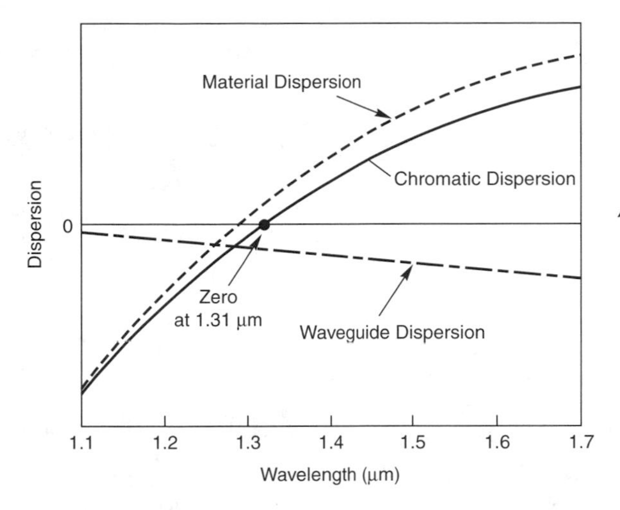

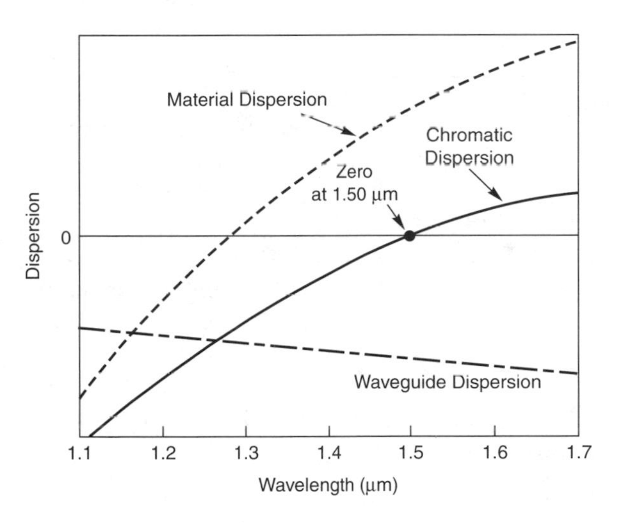

FIGURE 5.8
Different amounts of waveguide dispersion combine with material dispersion to produce different chromatic dispersion.

From a physical standpoint, the group delay is the slope of the curve that plots refractive index as a function of wavelength, shown in Figure 5.9(a) on a different scale that shows its curvature better than Figure 5.7. If you know elementary calculus, that slope is the first derivative of how refractive index n varies with wavelength:

Group delay is the slope of the plot of refractive index versus wavelength.

$$\text{Group delay} = \frac{L}{c}\left(n - \lambda\frac{dn}{d\lambda}\right) = \Delta t$$

This group delay is plotted in Figure 5.9(b). You can think of group delay time as the actual pulse spreading Δt—measured in units of time—caused by the change in refractive index over a range of wavelengths. Remember, however, that this is a time delay, *not* the characteristic material dispersion of the fiber. Characteristic dispersion measures not the *magnitude* of the delay in units of time, but how fast the group delay is *changing* with wavelength

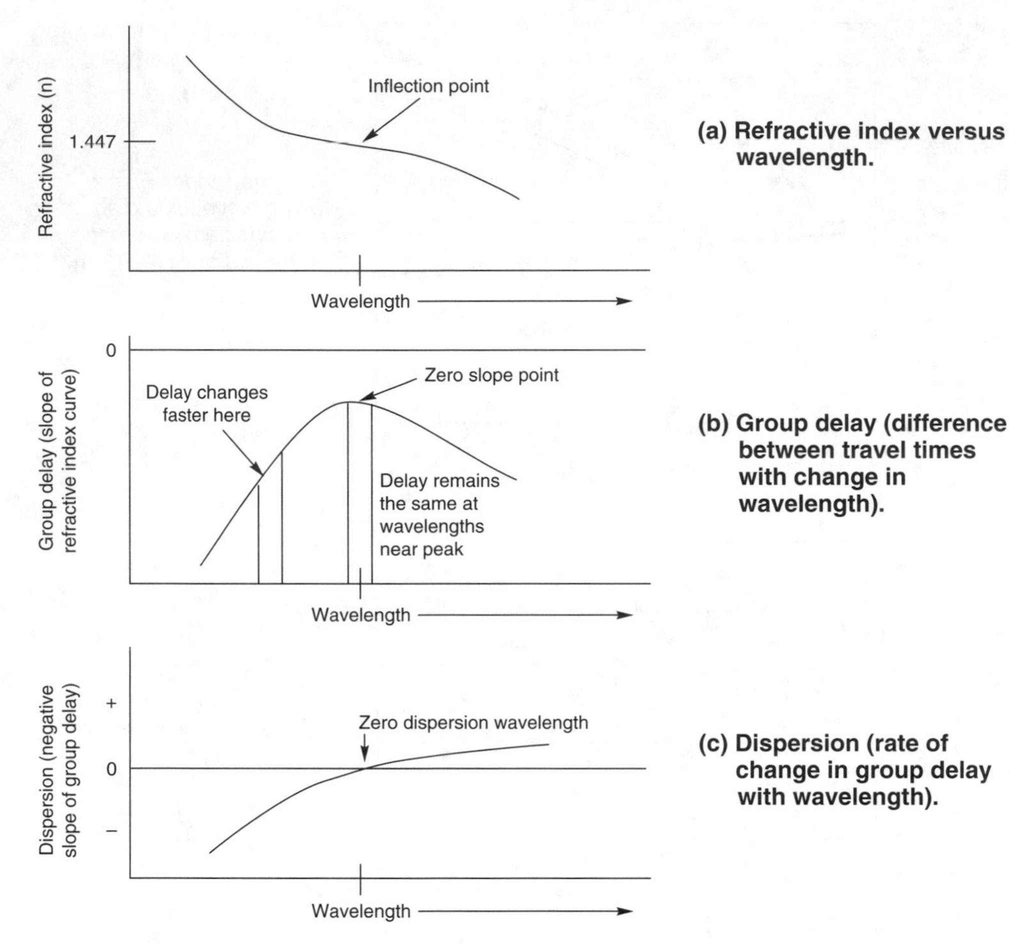

FIGURE 5.9 *Material dispersion is the slope of the slope (or the second derivitive) of a plot of refractive index versus wavelength.*

(generally for a unit length of the fiber rather than for the entire length). This *characteristic material dispersion* is measured in units of picoseconds (of time) per nanometer (of wavelength range) per kilometer (of fiber length). Multiply it by the length of the fiber and the range of wavelengths, and you get the group delay Δt.

You calculate characteristic material dispersion D_{material} as the rate of change of the group delay with wavelength, which is equivalent to measuring the slope of the group delay curve with respect to wavelength. If you divide through by fiber length, and take the differential rate of change in group delay with wavelength, you get

$$D_{\text{material}} = \frac{1}{L} \times \left(\frac{d\,(\text{group delay})}{d\lambda} \right) = \frac{-\lambda}{c} \times \frac{d^2 n}{d\lambda^2}$$

This is the characteristic material dispersion, plotted in Figure 5.9(c), and it represents the slope of the group delay curve. To see what it means graphically, compare it with the plot of group delay in Figure 5.9(b). The group delay is nearly constant at its peak value, so the values are virtually the same at the two wavelengths near the peak (vertical lines). However, at shorter wavelengths the group delay is changing much faster, so the values differ much more at two wavelengths the same distance apart (vertical lines at the left).

You can calculate the total pulse spreading over the length of the fiber, Δt, by multiplying this characteristic dispersion by fiber length L and wavelength range $\Delta\lambda$. This gives:

$$\Delta t = D_{\text{material}} \times L \times \Delta\lambda = \frac{-L\lambda\Delta\lambda}{c} \times \frac{d^2 n}{d\lambda^2}$$

Thus the characteristic material dispersion is proportional to the *second derivative* (or, equivalently, to the slope of the slope) of the plot of refractive index versus wavelength, not directly to the slope of the refractive index curve itself. To reiterate, it's also the slope of the group delay, which measures the travel time through the fiber as a function of wavelength. The *slope* of the group delay curve, in contrast, measures how *fast* the group delay changes with wavelength, which is the characteristic material dispersion. This rate of change of group delay is zero at the peak of the group delay curve, which comes at 1.28 μm in silica fibers. This also is the point where the slope of the refractive-index curve stops decreasing with increasing wavelength and starts increasing again. (Because the refractive index decreases as wavelength increases, the slope is a negative number, plotted below zero on Figure 5.9(b).) Mathematically, the zero material-dispersion wavelength is a maximum of the group velocity curve and a point of inflection in the refractive-index plot.

Dispersion is the slope of group delay, or the second derivative of the plot of refractive index versus wavelength.

Figure 5.9(c) plots characteristic material dispersion. Recall that the formula carries a negative sign, which it gets from the negative value of group delay. The minus sign means that characteristic material dispersion is negative at wavelengths where the group delay curve is rising (i.e., has positive slope), and positive where the middle curve is dropping (i.e., has negative slope).

The components of waveguide dispersion work in a similar way, but the physical relationships are more complex. As you saw earlier, waveguide dispersion has a sign, which matters when adding it to material dispersion to get chromatic dispersion, the number given in product specifications. The sign also matters when compensating for chromatic dispersion to reduce pulse spreading. Chromatic dispersion works like material dispersion; it measures the rate of change of the group delay for all chromatic dispersion, not just for material dispersion.

The signs don't matter when combining the effects of chromatic dispersion with other dispersion, because the pulse spreading enters those equations as squares. As you've probably learned the hard way, it's easy to lose track of signs that don't have an obvious physical meaning. This can happen very easily with material, waveguide, and chromatic dispersion, so don't be surprised if you spot the wrong signs. In normal single-mode fibers, the dispersion should be negative at wavelengths shorter than the zero-dispersion wavelength, and positive at longer wavelengths.

Dispersion Slope and Specifications

In practice, engineers approximate chromatic dispersion by assuming it varies linearly over a defined range of wavelengths. That is, they plot dispersion at a pair of wavelengths, draw a straight line between them, and assume that the dispersion at intermediate wavelengths falls between them, as shown in Figure 5.10. If the two wavelengths are λ_1 and λ_2, and the characteristic chromatic dispersions at those wavelengths are $D(\lambda_1)$ and $D(\lambda_2)$, this means that dispersion $D_{\text{chromatic}}$ at intermediate wavelength λ is

$$D_{\text{chromatic}}(\lambda) = \left(\frac{D(\lambda_2) - D(\lambda_1)}{\lambda_2 - \lambda_1} \times (\lambda - \lambda_2)\right) + D(\lambda_2)$$

Specification sheets often give these equations with the ranges of dispersion and wavelength for which they are valid. Typically there are separate equations for the 1530- to 1565-nm range of C-band erbium-doped fiber amplifiers and the 1565- to 1625-nm L-band.

Dispersion slope gives change in dispersion over a range of wavelengths.

Look closely at the equation, and you can see that it actually multiplies the *dispersion slope* (change in dispersion over a range of wavelength) by the change in wavelength from one endpoint, and adds the dispersion at that endpoint. Thus the equation translates in more descriptive terms to

$$D_{\text{chromatic}}(\lambda) = [(\text{dispersion slope}) \times (\Delta\lambda)] + D(\text{endpoint})$$

Remember this is the slope of *chromatic* dispersion, although the normal term is just "dispersion slope."

Dispersion slope tells how dispersion changes with wavelength. Normally this change is very small over the range of wavelengths generated by a single laser transmitter. However, it is important in wavelength-division multiplexed systems, which carry many optical channels spanning tens of nanometers in wavelength. We will take a closer look later in this chapter.

Specification sheets typically do *not* plot chromatic dispersion directly as a function of wavelength, but give the chromatic dispersion that may be found at a range of wavelengths, such as 2.6 to 6.0 ps/nm-km at 1530 to 1565 nm. These numbers do not mean that the fibers

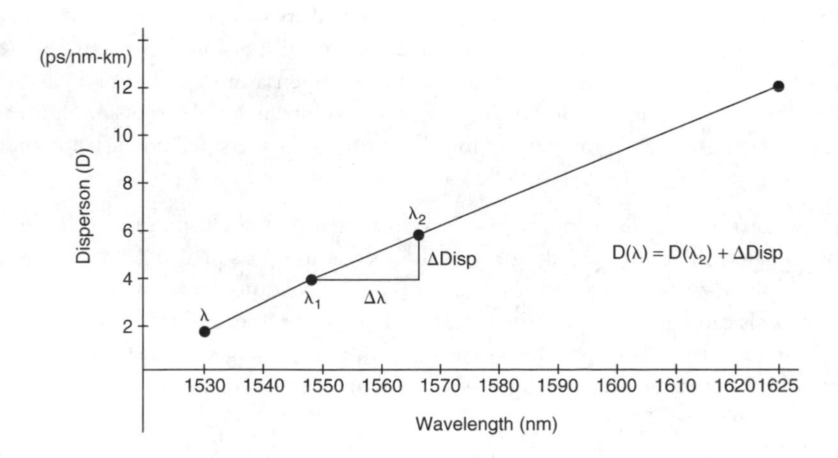

FIGURE 5.10 *Extrapolating fiber dispersion at an intermediate wavelength.*

have 2.6 ps/nm-km dispersion at 1530 nm and 6.0 ps/nm-km at 1565—they mean that the values in this range of wavelengths fall within this "box." As with other specified values, they allow for a range of manufacturing tolerances, so the specified dispersion slope will not always match the slope calculated from the extremes of chromatic dispersion and wavelength.

Source Bandwidth and Chromatic Dispersion

Unlike the pulse spreading caused by other types of fiber dispersion, the spreading caused by chromatic dispersion depends strongly on the light source. If we take $D_{\text{chromatic}}(\lambda)$ as the characteristic dispersion of a unit length (1 km) of fiber, the total pulse spreading from chromatic dispersion $\Delta t_{\text{chromatic}}$ is

$$\Delta t_{\text{chromatic}} = D_{\text{chromatic}}(\lambda) \times \Delta\lambda \times \text{Length}$$

where $\Delta\lambda$ is the range of wavelengths in the optical signal in nanometers and *Length* is the fiber length in kilometers. This means that the spectral bandwidth of the light source is a parameter that system designers can adjust to limit the effects of chromatic dispersion. The higher the data rate, the more important narrow-band sources become, as you will learn in later chapters.

Chromatic Dispersion Compensation and Tailoring

We saw before that pulse dispersion is cumulative, adding up over the entire length of a fiber system. In general that means that adding more fiber only makes things worse. However, it is possible to cancel *chromatic* dispersion if you add fiber with chromatic dispersion of the opposite sign. For example, if you have a fiber with positive chromatic dispersion in the erbium-amplifier band, you can add a fiber with negative chromatic dispersion at the same wavelengths. The idea is similar to having waveguide dispersion offset material dispersion to give low chromatic dispersion at the desired wavelength. In this case, it is two separate fibers, spliced together, that offset each other's dispersion, as shown in Figure 5.11. Dispersion compensation modules can be added to existing fiber systems to upgrade them for higher-speed transmission or operation in the 1550-nm band, or fiber with different dispersion can be integrated into the transmission paths of new systems to "tailor" overall dispersion properties.

Combining fibers with chromatic dispersion of opposite signs can compensate for chromatic dispersion, yielding low overall pulse spreading.

A typical dispersion-compensating fiber has high negative waveguide dispersion that gives it a negative chromatic dispersion that typically is several times the magnitude of the positive chromatic dispersion of the transmission fiber. Thus compensation requires a shorter length of the dispersion-compensating fiber. That is an important consideration because compensating fiber typically has higher attenuation than transmission fibers. Compensating fiber also usually has a small effective area, making it more vulnerable to nonlinear effects, so it is used at the receiving end of the system, where lower power reduces nonlinear effects.

Typically, dispersion is compensated over a range of wavelengths in the erbium-fiber band, but it's easiest to calculate requirements if you look just at one wavelength. Suppose you want to have total chromatic dispersion of +2 ps/nm-km at the 1530-nm short end of the erbium-fiber band over a 1000-km system. You are using nonzero dispersion-shifted transmission fiber with dispersion of +8 ps/nm-km at that wavelength, and you can buy dis-

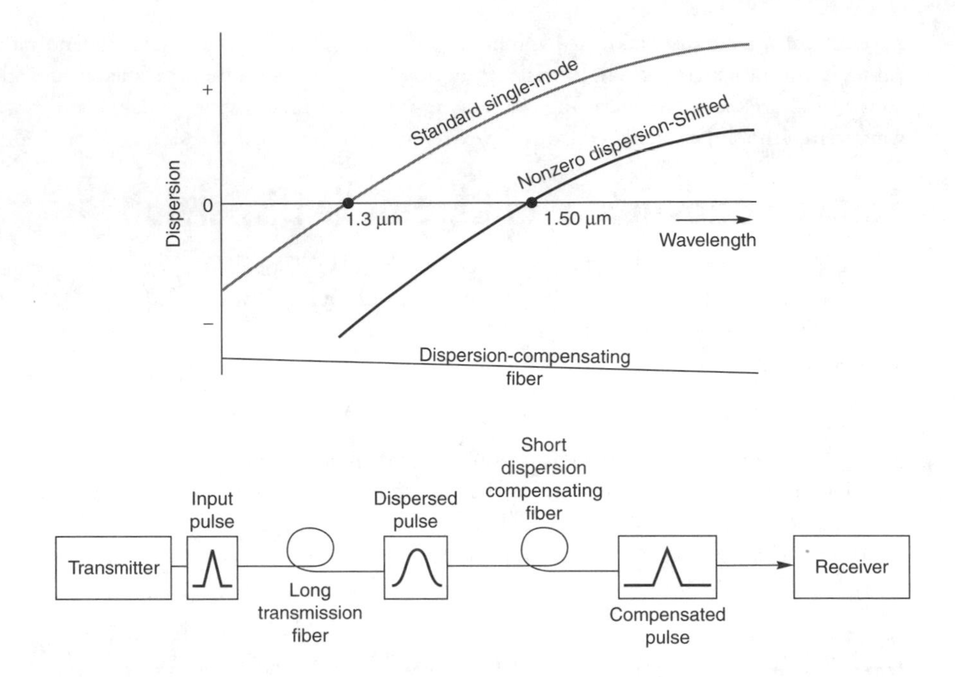

FIGURE 5.11 *Dispersion compensation.*

persion-compensating fiber with dispersion of −100 ps/nm-km at 1530 nm. You can use the general formula

$$D_{net} L_{total} = D_{transmission} L_{transmission} + D_{comp} L_{comp}$$

Where D_{net} is the net dispersion for the entire system, L_{total} is the total length (assuming the compensating fiber is part of the transmission path), $D_{transmission}$ and $L_{transmission}$ are the dispersion and length of the transmission fiber, and D_{comp} and L_{comp} are dispersion and length of the compensating fiber. Plug the numbers in, and you see

$$2000 \text{ ps/ns} = +8\, L_{transmission} - 100\, L_{comp}$$

Since you know that $L_{transmission} + L_{comp} = 1000$ km, you can work out that you need 944 km of nonzero dispersion-shifted transmission fiber and 56 km of compensating fiber. Thus you need about 1 km of compensating fiber for every 17 km of transmission fiber.

You can use the same ideas to calculate the dispersion compensation needed for upgrading existing fiber systems. Other approaches to chromatic dispersion also are possible. One example is an optical delay line that would delay signals a certain amount depending on their wavelength, so the slower signals could catch up. A dispersion-compensating fiber in a box could serve as such a delay line.

WDM transmission requires dispersion compensation over a range of wavelengths.

Multiwavelength Transmission and Dispersion

Dealing with chromatic dispersion is more complex in systems that carry multiple wavelengths. Wavelength-division multiplexing requires management of chromatic dispersion over the entire range of wavelengths that are transmitting optical channels. Typically that can span tens of nanometers in systems with fiber amplifiers, 35 nm in systems with C-band erbium-fiber amplifiers, 55 nm in systems with L-band erbium-fiber amplifiers, or 95 nm in systems with both.

That range of wavelength is large enough for chromatic dispersion to differ significantly among optical channels. Just in the erbium-fiber C-band, the difference can accumulate to 2 ps/nm-km with reduced-dispersion-slope (0.045 ps/nm^2-km) fibers, and to 4 ps/nm-km with other nonzero dispersion-shifted fibers. This becomes important because it means different optical channels may require different amounts of dispersion compensation.

Dispersion management also becomes more complex as the range of wavelengths increases. The dispersion slopes of dispersion-compensating fibers do not match and offset those of transmission fibers, so residual differences remain. These accumulate over distance and can become significant for long-distance, high-speed systems. Additional components or a mix of dispersion-compensating and transmission fibers may be needed.

Polarization-Mode Dispersion

In Chapter 4, you learned that a single mode fiber actually transmits light in two distinct polarization modes. The electric fields of the two modes are perpendicular to each other, or *orthogonal* in the jargon of physics. Normally the two behave just the same in the fiber, so from a physical standpoint they are called *degenerate,* which means they can't be distinguished.

The existence of two polarization modes wouldn't matter if they were perfectly identical, but in the real world even physics is not quite perfect. The polarization modes behave identically only in a fiber that is perfectly symmetrical, and no fiber is that perfect. Stresses within the fiber, and forces applied to it from the outside world, cause the refractive index of glass to differ very slightly for light in the two polarization modes, an effect called *birefringence.*

Fibers have low levels of birefringence that affect light in the two polarization modes differently.

Internal forces make some crystals strongly birefringent; calcite is one example, where the refractive indexes of the two polarizations differ by more than 10%. Look through a calcite prism and you see what looks like a double exposure because the material separates light in the two polarizations. The effect in optical fibers is very tiny—around one part in 10 million (10^{-7}) from manufacturing stresses—but light travels a very long distance in them, so tiny effects can accumulate. If the birefringence was uniform along the whole fiber, light in one polarization mode would get about one wavelength farther ahead of the other every 10 meters, as shown in Figure 5.12. However, the effect called *polarization-mode dispersion* (PMD) isn't that simple.

One reason is that birefringence varies randomly along the length of normal single-mode fibers, because it arises from random fluctuations in manufacturing and environmental stresses distributed along the fiber's length. Another is that the light can shift randomly between polarization modes in normal single-mode fibers (but not in polarization-maintaining or single-polarization fibers, mentioned in Chapter 4). That means that the phase shift

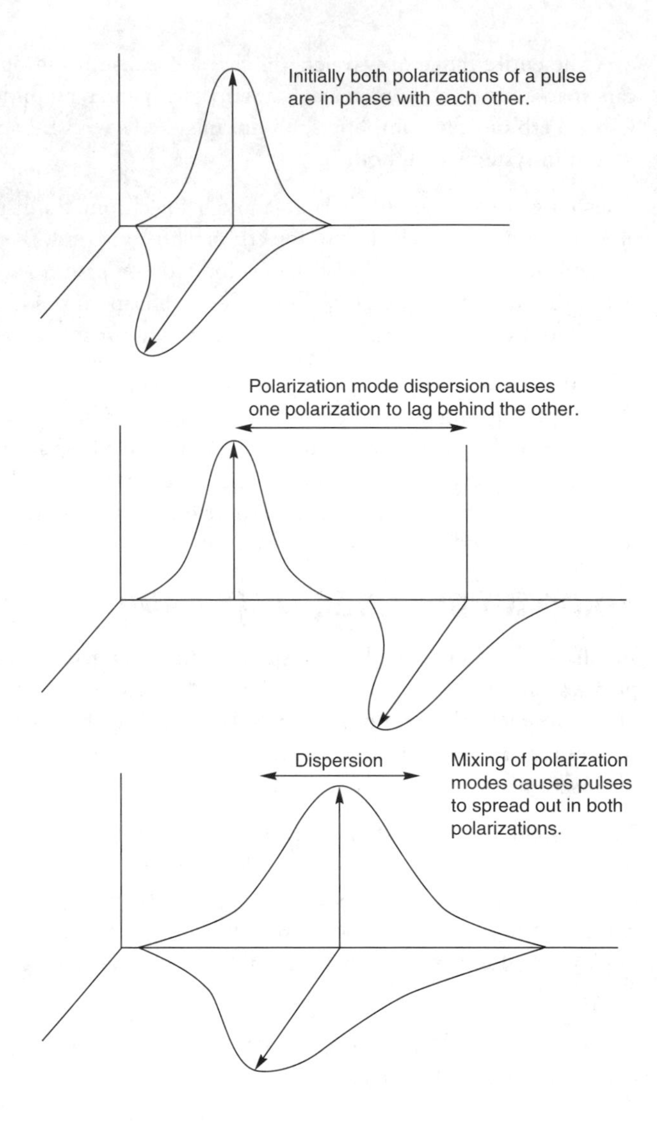

FIGURE 5.12
Polarization-mode dispersion.

between the two modes does not accumulate consistently along the length of the fiber. Instead, the light drifts back and forth between the two modes, and the difference blurs, stretching the pulse duration in time, as you can see at the bottom of Figure 5.12.

Polarization-mode dispersion makes pulse spreading increase with the square root of length.

The random polarization shifts and distribution of birefringence cause the pulse spreading to increase with the square root of fiber length L. The effect is sometimes called *differential group delay,* and is essentially statistical. Each fiber has a characteristic polarization-mode dispersion D_{PMD} when manufactured, but that may change when cabled. Pulse spreading caused by polarization-mode dispersion is Δt_{PMD} is

$$\Delta t_{PMD} = D_{PMD} \times \sqrt{\text{fiber length}}$$

Typical values of existing fiber are 0.05 to 1 picosecond per root kilometer. Fiber now in production has characteristic PMD values of 0.05 to 0.2 ps/km$^{-1/2}$, but cabling can raise the value to 0.5 ps/km$^{-1/2}$. The value may change with time, such as when overhead cables are strained by blowing in the wind, which can increase the instantaneous values of characteristic PMD to over 1 ps/km$^{-1/2}$. The magnitude also can vary with wavelength.

The potential effects of polarization mode dispersion were not considered significant until a few years ago, so manufacturers did not specify PMD values for their earlier fibers. Because cabling effects also are important, the best way to be sure of PMD values in existing cables is to measure them in place. In principle, all optical components can cause polarization mode dispersion, but in practice fiber is the most important source because of the long optical path length.

Polarization-mode dispersion is smaller in magnitude than chromatic dispersion, but technology to compensate for it is still in the early stages of development. In practice, it does not become significant until data rates exceed 2.5 Gbit/s. Careful control allows long-distance transmission at 10 Gbit/s, but PMD poses a challenge to sending higher data rates over long distances.

Dispersion and Transmission Speeds

So far we have considered the effects of dispersion on instantaneous pulses, but real pulses are not instantaneous. In digital systems, the initial pulse starts with a duration Δt_{input}, then experiences spreading due to dispersion of $\Delta t_{\text{dispersion}}$. The output pulse length is not the direct sum of the two pulse durations but the root sum of their squares:

$$\Delta t_{\text{output}} = \sqrt{\Delta t_{\text{input}}^2 + \Delta t_{\text{dispersion}}^2}$$

This gives the pulse length at the end of the system, and it is these pulses that have to be resolved for the system to operate properly.

The degree of overlap at which pulse dispersion causes problems in digital systems depends on the design. One rough guideline for estimating the maximum bit rate is that the interval between pulses should be four times the dispersion, or, equivalently,

Dispersion limits maximum data rate.

$$\text{Maximum bit rate} = \frac{1}{4\,\Delta t_{\text{dispersion}}}$$

Thus, if pulses experience about 1 ns of dispersion, the maximum bit rate is about 250 Mbit/s. It isn't quite this simple in practice because performance depends on other factors as well as dispersion, but it's a useful guideline. For polarization-mode dispersion the usual guideline is more stringent, that the dispersed pulse should be no more than 1/10th as long as the interval between pulses, reflecting the more stringent demands on high-speed systems. Note that these figures consider only dispersion, not the input pulse length, or receiver rise time. Different guidelines relate total system rise time to maximum bit rate, which depend on data transmission format.

Dispersion also affects analog transmission in roughly the same way that it limits bit rates in digital systems. Instead of lengthening digital pulses, dispersion smears out the whole analog waveform, effectively attenuating the highest frequencies in the signal. This limits the analog bandwidth, the frequency at which the detectable signal has dropped 3 dB (50%) compared to lower frequencies.

Transmission capacities of graded-index and step-index multimode fiber often are specified in terms of bandwidth, typically megahertz-kilometers, rather than as dispersion. You can roughly convert that to total system response time Δt_{total} using the formula

$$\text{Bandwidth (MHz)} = \frac{350}{\Delta t_{total}}$$

Typical bandwidths for step-index multimode fiber are around 20 MHz-km. Bandwidths of graded-index multimode fibers depend on wavelength because they suffer both chromatic and modal dispersion. Typical values at 0.85 μm are 200 MHz-km for 62.5/125 fiber and 600 MHz-km for 50/125 fiber. At 1.3 μm, typical bandwidths are 500 MHz/km for 62.5/125 fiber and 1000 MHz/km for 50/125 fiber. Because multimode graded-index fibers are used in local area networks, they also may be specified according to maximum transmission distance using particular Ethernet protocols.

Nonlinear Effects

Nonlinear effects are interactions between light waves, which can cause noise and crosstalk.

Normally light waves or photons transmitted through a fiber have little interaction with each other, and are not changed by their passage through the fiber (except for absorption and scattering). However, there are exceptions arising from the interactions between light waves and the material transmitting them, which can affect optical signals. These processes generally are called *nonlinear effects* because their strength typically depends on the square (or some higher power) of intensity rather than simply on the amount of light present. This means that nonlinear effects are weak at low powers, but can become much stronger when light reaches high intensities. This can occur either when the power is increased, or when it is concentrated in a small area—such as the core of an optical fiber.

Nonlinear optical devices have become common in some optical applications, such as to convert the output of lasers to shorter wavelengths by doubling the frequency (which halves the wavelength). Most nonlinear devices use exotic materials not present in fiber-optic systems in which nonlinear effects are much stronger than in glass. The nonlinearities in optical fibers are small, but they accumulate as light passes through many kilometers of fiber.

Nonlinear effects are weak in optical fibers, but accumulate over long distances.

Nonlinear effects are comparatively small in optical fibers transmitting a single optical channel. They become much larger when dense wavelength-division multiplexing (DWDM) packs many channels into a single fiber. DWDM puts many closely spaced wavelengths into the same fiber where they can interact with one another. It also multiplies the total power in the fiber. A single-channel system may carry powers of 3 milliwatts near the transmitter. DWDM multiplies the total power by the number of channels, so a 40-channel system carries 120 mW. That's a total of 2 mW per square micrometer—or 200,000 watts per square centimeter!

Several nonlinear effects are potentially important in optical fibers, although some have proved more troublesome than others. Some occur in systems carrying only a single optical channel, but others can occur only in multichannel systems. We'll look at each of them in turn, focusing on the more important ones.

Brillouin Scattering

Stimulated Brillouin scattering occurs when signal power reaches a level sufficient to generate tiny acoustic vibrations in the glass. This can occur at powers as low as a few milliwatts in single-mode fiber. Acoustic waves change the density of a material, and thus alter its refractive index. The resulting refractive-index fluctuations can scatter light, called *Brillouin scattering.* Since the light wave being scattered itself generates the acoustic waves, the process is called *stimulated Brillouin scattering.* It can occur when only a single channel is transmitted.

Brillouin scattering scatters light back toward the transmitter, limiting transmitted power.

In fibers, stimulated Brillouin scattering takes the form of a light wave shifted slightly in frequency from the original light wave. (The change is 11 gigahertz, or about 0.09 nanometer for a 1550-nm signal.) The scattered wave goes back toward the transmitter. The effect is strongest when the light pulse is long (allowing a long interaction between light and the acoustic wave), and the laser linewidth is very small, around 100 megahertz. Under such conditions, it can occur at power levels as little as 3 mW in single-mode fibers. However, the power level needed to trigger stimulated Brillouin scattering increases as pulse length decreases, so the effect becomes less severe at higher data rates.

Brillouin scattering directs part of the signal back toward the transmitter, effectively increasing attenuation. The small frequency shift effectively confines the effect to the optical channel generating the effect at present channel spacings, so it does not create crosstalk with other channels. However, it does limit the maximum power a single length of fiber can transmit in one direction. As input power increases, the fraction of power scattered in the opposite direction rises sharply, and the fiber essentially becomes saturated.

Optical signals going in the wrong direction can cause serious problems, so optical isolators must be added to block Brillouin scattering. In general, isolators are placed at transmitters and optical amplifiers, limiting the effects of stimulated Brillouin scattering to a single fiber span between isolators. Special modulation schemes and careful design also can reduce the effects of Brillouin scattering.

Self-phase Modulation

The refractive index of glass varies slightly with the intensity of light passing through it, so changes in signal intensity cause the speed of light passing through the glass. This process causes intensity modulation of an optical channel to modulate the phase of the optical channel that creates it, so the effect is called *self-phase modulation.* As the optical power rises and falls, these phase shifts also effectively shift the frequencies of some of the light; the shifts are in opposite directions at the rising and falling parts of the pulse. The overall result is to spread the bandwidth of the optical channel by an amount that depends on the rate of

An optical channel modulates its own phase by self-phase modulation, which can broaden the range of wavelengths.

change in optical intensity as well as on the nonlinear coefficient of the fiber material. Like stimulated Brillouin scattering, it can occur in a single-channel system.

The spectral broadening caused by self-phase modulation produces dispersion-like effects, which can limit data rates in some long-haul communication systems, depending on the fiber type and its chromatic dispersion. For ultrashort pulses (less than one picosecond) with very high peak powers, self-phase modulation can be very strong, generating a broad continuum of wavelengths. Self-phase modulation also stabilizes pulses called solitons, so they propagate along the fiber with a constant shape, although attenuation reduces their amplitude. This makes soliton transmission an effective way to overcome self-phase modulation.

Cross-phase Modulation

In cross-phase modulation, one channel modulates the phase of other channels.

Systems carrying multiple wavelength channels are vulnerable to *cross-phase modulation* as well as self-phase modulation. In this case, variations in the intensity of one optical channel cause changes in the refractive index affecting other optical channels. These changes modulate the phase of light on other optical channels, in addition to self-phase modulation of the same channel.

The strength of cross-phase modulation increases with the number of channels, and becomes stronger as the channel spacing becomes smaller. There are ways to mitigate this effect, but it can limit transmission speed.

Four-Wave Mixing

Four-wave mixing generates crosstalk among optical channels, noise that can limit WDM systems.

Normally multiple optical channels passing through the same fiber interact with each other only very weakly, making wavelength-division multiplexing. However, these weak interactions in glass can become significant over long fiber-transmission distances. The most important is four-wave mixing (sometimes called four-photon mixing) in which three wavelengths interact to generate a fourth.

Four-wave mixing is one of a broad class of *harmonic mixing* or *harmonic generation* processes. The idea is that two or more waves combine to generate waves at a different frequency that is the sum (or difference) of the signals that are mixed. Second-harmonic generation or frequency doubling is common in optics; it combines two waves at the same frequency to generate a wave at twice the frequency (or, equivalently, half the wavelength). This can happen in optical fibers, but the second harmonic of the 1550 nm band is at 775 nm, far from the communications band, so it doesn't interfere with any signal wavelength.

Four-wave mixing is the strongest nonlinear effect that mixes the frequencies of optical channels in the 1550-nm band to generate noise in that band. As shown in Figure 5.13, three waves combine to generate a fourth frequency. If each frequency is designated by ν, the new frequency, ν_4, is

$$\nu_4 = \nu_1 + \nu_2 - \nu_3$$

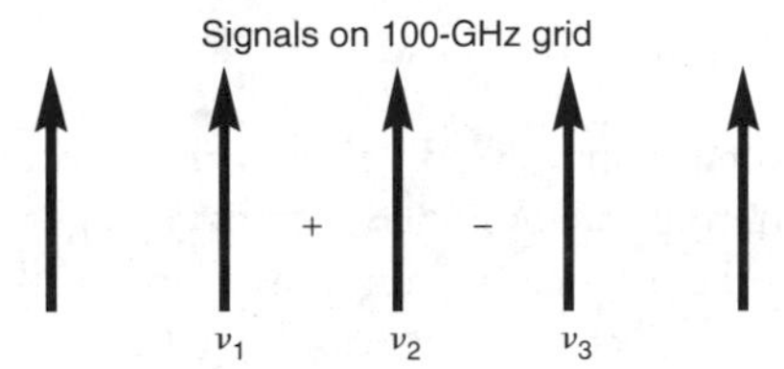

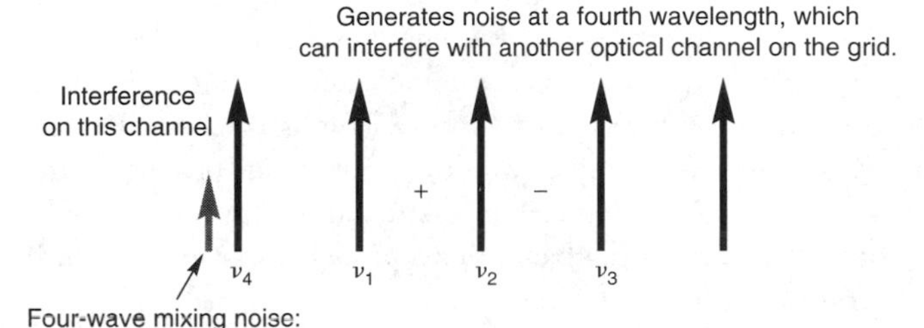

FIGURE 5.13
Four-wave mixing produces noise that interferes with other optical channels in a DWDM grid.

In dense-WDM systems, the optical channels are typically close and spaced on a frequency grid typically separated by 100 or 200 GHz. This means that if ν_1 is the starting point, ν_2 is at a frequency 100 GHz higher, and ν_3 is another 100 GHz higher,

$$\nu_4 = \nu_1 + \nu_1 + 100\text{ GHz} - \nu_1 - 200\text{ GHz} = \nu_1 - 100\text{ GHz}$$

which falls smack on top of another optical channel on the grid. The beating together of two frequencies, ν_1 and ν_2, also can cause four-wave mixing:

$$\nu_4 = \nu_1 + \nu_1 - (\nu_1 + 100\text{ GHz}) = \nu_1 - 100\text{ GHz}$$

Four-wave mixing is a weak effect, but it can accumulate if the signals on the optical channels remain in phase with each other over long distances. This happens when chromatic dispersion is very close to zero. Pulses transmitted over different optical channels, at different wavelengths, stay in the same relative positions along the length of the fiber because the signals experience near-zero dispersion. This amplifies the effect of four-wave mixing, and builds up the noise signal, which interferes with a fourth channel on the grid. (This problem led to abandonment of zero dispersion-shifted fibers.) To overcome this problem, the zero-dispersion point has to be moved out of the erbium-fiber amplifier band at 1530 to 1625 nm. With even modest dispersion, the signals at different wavelengths quickly drift out of phase with each other, reducing four-wave mixing.

Four-wave mixing accumulates if signals remain in phase over long distances, which occurs when dispersion is near zero.

Other techniques also can help control four-wave mixing, such as careful spacing of optical channels. Nonetheless it remains an important fiber property to consider in designing DWDM systems.

Raman Scattering

Stimulated Raman scattering transfers power between signals at different wavelengths.

Stimulated Raman scattering occurs when light waves interact with molecular vibrations in a solid lattice. In simple Raman scattering, the molecule absorbs the light, then quickly re-emits a photon with energy equal to the original photon, plus or minus the energy of a molecular vibration mode. This has the effect of both scattering light and shifting its wavelength.

When a fiber transmits two suitably spaced wavelengths, stimulated Raman scattering can transfer energy from one to the other. In this case, one wavelength excites the molecular vibration, then light of the second wavelength stimulates the molecule to emit energy—at the second wavelength. The Raman shift between the two wavelengths is relatively large, about 13 terahertz (100 nm in the 1550-nanometer window), but it can produce some crosstalk between optical channels. It also can deplete signal strength by transferring light energy to other wavelengths outside the operating band. (This process also can be used to amplify signals, as you will learn in Chapter 12.)

Careful choice of wavelengths can reduce the interference between Raman scattering and other channels. Nonetheless, Raman scattering does impose limits on DWDM systems with many optical amplifiers. Its effects are more serious on the shorter wavelengths in a multiwavelength system.

Mechanical Properties

So far, I have concentrated on the optical properties of fibers. You also need to understand their most important mechanical properties. Although most fibers are assembled into cables with highly automated equipment, installation of cables, connectors, and other components often requires handling individual fibers.

A plastic coating with outer diameter 250 μm covers glass fibers.

Glass fibers are coated with plastic as they are drawn into fiber form. The plastic coating eases handling and protects the fiber's outer surface from physical damage. The standard cladding diameter of communications fibers is 125 μm or 0.125 mm (0.005 in.), thin enough to be difficult to handle or process mechanically. Plastic coating doubles this diameter to 250 μm (0.01 in.), making them easier to pick up and process. Typically this coating consists of two layers, an inner one with outer diameter 245 μm that provides mechanical protection, and a thin outer coating that color-codes the fiber for cabling.

Thin glass fibers are reasonably flexible. Individual telecommunication fibers can be bent into a loop with 5-cm (2-in.) diameter without damage and left that way indefinitely. Equipment used in installation is designed to accommodate that degree of bending. You sometimes can get away with bending fibers more but don't count on it—and you'll never get away with the sort of sharp bend that can be used for copper wires. Thicker glass fibers cannot be bent as tightly. Plastic optical fibers (described in Chapter 6) are more flexible than glass.

In theory, a mechanically perfect glass fiber can withstand a force of 2 million pounds per square inch (14 gigapascals or 14 giganewtons per square meter) pulling along its length.

In practice, inevitable minor surface flaws reduce this to about 500,000 lb/in^2 (3.5 GPa) or less.

The fact that fibers break at randomly distributed surface flaws has some important consequences. The longer the fiber, the more likely it is to contain a flaw that will cause it to break when a certain stress is applied. To weed out the most harmful of these weak points, fiber manufacturers use a simple *proof test* that applies a weight to a length of the fiber as they wind it onto the reel. The weight applies a specified tension along the length of the fiber. If any part of the fiber cannot withstand that tension, it breaks at the weak point.

Fibers are subjected to a proof test to assure they have a minimum strength.

Manufacturers typically proof test fibers under a load of 100,000 lb/in^2 (0.7 GPa), so weaker fibers don't make it out of the plant because they break during the stress test. This doesn't mean you can hold an elephant in the air on a single fiber—fibers are small, and pounds per square inch measures the load applied to a solid square inch of glass, not a thin fiber. (Figure it out yourself and you'll be surprised.) But the numbers do show that glass fibers can withstand reasonable handling, despite our instinctive feeling that glass is fragile.

Pull sharply on a glass fiber and it will snap, apparently without stretching at all. In fact, glass does stretch elastically until it reaches its breaking strength, but it stretches only about 1% beyond its original length before snapping. Copper wires deform plastically, stretching by more than 20% before breaking.

Fiber failure normally starts at a flaw or microcrack in the glass surface. Application of stress spreads the crack, leading quickly to failure if the applied force is beyond the fiber's strength. Flaws are distributed randomly along the surface, and statistics can be used to estimate the chance of failure.

Fibers can suffer from two types of aging. *Dynamic fatigue* arises from the short-lived stresses applied to the fiber either by installation or the temporary environmental effects. Underground cables are affected by the stress of pulling them into a duct, while aerial cables are affected by gusts of wind and snow loading during winter.

Static fatigue is the growth of flaws that occurs while the fiber is maintained under constant conditions. The flaws may grow because of moisture or other environmental factors, or because the cable structure is pulling on the fiber. Moisture is the most common problem because it slowly reacts with silica.

The plastic coating on the fiber is important as a protection against moisture and physical damage to the surface. Experience has shown these coatings to be quite effective.

What Have You Learned?

1. Signal loss is the sum of loss coupling light into a fiber, and of scattering and absorption in the fiber.
2. Material absorption depends on wavelength and is cumulative with distance. Impurities cause absorption peaks.

3. Losses from atomic scattering are higher at shorter wavelengths; scattering losses also are cumulative with distance.
4. Fiber attenuation is the sum of scattering and absorption, measured together in decibels per kilometer.
5. The logarithmic decibel scale is preferred for calculating attenuation and transmission losses. Total attenuation of a fiber is the characteristic loss in decibels per kilometer times the length in kilometers.
6. The larger the fiber core, the more easily it can collect light from a light source.
7. Some light is guided short distances along the cladding in cladding modes. Leaky modes are only partly confined in the fiber core.
8. Dispersion produces pulse spreading that can limit transmission speeds in both analog and digital systems. The main types of dispersion are modal, chromatic, and polarization. Total pulse spreading is the square root of the sums of the squares of the pulse spreading from all three types of dispersion.
9. Material and waveguide dispersion add together to give chromatic dispersion; both can have a positive or negative sign. The pulse spreading caused by chromatic dispersion is proportional to the spectral bandwidth of the transmitter as well as the length of fiber.
10. Material dispersion is a characteristic of the fiber material, which varies with wavelength. It measures the change in group delay with wavelength, which in turn measures the change in refractive index with wavelength.
11. Waveguide dispersion arises from changes in waveguide properties and the distribution of light in the fiber with wavelength.
12. Waveguide and material dispersion can cancel each other to give zero chromatic dispersion if their signs are opposite. Adjusting waveguide dispersion can shift the zero-dispersion wavelength.
13. Fibers with opposite signs of chromatic dispersion can be combined in sequence to compensate for chromatic dispersion in a system.
14. Long wavelength division-multiplexed (WDM) systems require dispersion management over their entire range of wavelengths. Dispersion slope, the change in dispersion with wavelength, is an important consideration.
15. Polarization-mode dispersion arises from low levels of birefringence in fibers, and from the mixing of polarization modes. It is significant at data rates above 2.5 Gbit/s.
16. Nonlinear effects are interactions between light waves, which generate noise and crosstalk. They become important at high power densities that can occur in single-mode fibers.
17. Four-wave mixing among optical channels is the most important potential noise source in WDM systems. It accumulates over long distances when fibers have very low dispersion.
18. Bare glass fibers are coated with plastic to protect their surfaces and ease handling. Glass fibers are quite strong, if they lack surface flaws.

What's Next?

Now that we've talked about fiber structures and characteristics, Chapter 6 will cover fiber materials and manufacture.

Further Reading

Paul Hernday, "Dispersion Measurements," in Dennis Dirickson, ed., *Fiber Optic Test and Measurement* (Prentice Hall, 1998)

Luc B. Jeunhomme, *Single-Mode Fiber Optics: Principles and Applications* (Marcel Dekker, 1990)

Donald B. Keck, ed., *Selected Papers on Optical Fiber Technology* (SPIE Milestone Series Vol. MS38, 1992)

Gerd Keiser, *Optical Fiber Communications,* 3rd ed (McGraw Hill, 2000)

Advanced Treatments:

John A. Buck, *Fundamentals of Optical Fibers* (Wiley-Interscience, 1995)

Ajoy Ghatak and K. Thyagarajan, *Introduction to Fiber Optics* (Cambridge University Press, 1998)

Questions to Think About for Chapter 5

1. The amount of Rayleigh scattering by atoms is proportional to λ^{-4}. How is this related to why the sky looks blue?
2. Can you write a formula that converts loss in decibels into the fraction of power remaining?
3. You have a fiber that transmits a single mode at 850 nm, and a light source with bandwidth of 1 nm. Its chromatic dispersion is about −80 ps/nm-km. What is the maximum data rate that fiber could transmit 100 km, neglecting attenuation, based only on the guideline on page 119 (bit rate = $1/(4 \times \Delta t_{\text{disp}})$)?
4. Estimate the attenuation at 850 nm from Figure 5.2. Assume you need an optical amplifier to boost signal strength after every 30 dB of fiber loss. How far can the fiber transmit signals before it requires an optical amplifier? Which of the limitations in Questions 3 and 4 do you think was the main reason 850-nm systems were never viable for long-distance transmission?
5. Why is it more difficult to compensate for dispersion in a DWDM system than in one transmitting only a single optical channel?
6. Write a formula for how much dispersion-compensating fiber you need to add to an existing system with L_{existing} km of fiber to reduce dispersion to a desired value D_{net}.

7. Four-wave mixing normally occurs only at high powers in glass, yet it can cause significant crosstalk in single-mode fibers. If you have a fiber with effective area of 50 μm^2, what is the power per square centimeter in the fiber if it carries 100 optical channels at 3 mW each?

Quiz for Chapter 5

1. A 1-m length of fiber transmits 99.9% of the light entering it. How much light will remain after 10 km of fiber?
   a. 90%
   b. 10%
   c. 1%
   d. 0.1%
   e. 0.0045%
2. A fiber has attenuation of 0.00435 dB/m. What is the total attenuation of a 10-km length?
   a. 0.0435 dB
   b. 1.01 dB
   c. 4.35 dB
   d. 43.5 dB
   e. Cannot tell without knowing wavelength
3. If 10 mW of light enters the 10-km fiber in Problem 2, how much light remains at the output end?
   a. 0.00045 mW
   b. −33.5 dBm
   c. −3.5 dBμm
   d. All the above are equivalent
   e. None of the above
4. You lose 1.0 dB coupling a 1-mW light source into an optical fiber. You need a signal of 0.1 mW at the other end. How far can you send a signal through fiber with attenuation of 0.5 dB/km?
   a. 1.8 km
   b. 10 km
   c. 18 km
   d. 20 km
   e. 40 km
5. You transmit an instantaneous pulse through a 20-km multimode fiber with total dispersion of 10 ns/km at the signal wavelength. What will the pulse length be at the end?
   a. 200 ns
   b. 100 ns
   c. 50 ns
   d. 20 ns
   e. 10 ns
6. You transmit a 100-ns pulse through the same fiber used in Problem 5. What will the pulse length be at the end?
   a. 300 ns
   b. 224 ns
   c. 200 ns
   d. 150 ns
   e. 100 ns
7. You transmit an instantaneous pulse through a 20-km single-mode fiber with chromatic dispersion of 10 ps/nm-km at the signal wavelength. The spectral width of the input pulse is 2 nm. What is the pulse length at the end of the fiber?

a. 400 ps
b. 250 ps
c. 200 ps
d. 100 ps
e. 32 ps

8. You transmit an instantaneous pulse through a 20-km single-mode fiber with chromatic dispersion of 10 ps/nm-km at the signal wavelength. This time you've spent an extra $2000 for a super-duper laser with spectral width of only 0.002 nm. What is the pulse length at the end of the fiber?

a. 30 ps
b. 20 ps
c. 4 ps
d. 1 ps
e. 0.4 ps

9. A single-mode fiber has material dispersion of 20 ps/nm-km and waveguide dispersion of −15 ps/nm-km at the signal wavelength. What is the total chromatic dispersion?

a. 35 ps/nm-km
b. 25 ps/nm-km
c. 5 ps/nm-km
d. 0 ps/nm-km
e. −35 ps/nm-km

10. You send 200-ps pulses through a 100-km length of the fiber in Problem 9, using a laser with spectral width of 0.002 nm. What is the width of the output pulse?

a. 1 ps
b. 200 ps
c. 250 ps
d. 400 ps
e. 500 ps

11. Your boss says you can't have the extra $2000 for the super-duper narrow-bandwidth laser, so you have to use the cheap model with 2-nm spectral linewidth in the system in Problem 10. What's the width of the output pulse?

a. 200 ps
b. 250 ps
c. 500 ps
d. 1000 ps
e. 1020 ps

12. An optical fiber 125 μm in diameter can withstand a force of 600,000 lb/in^2. What's the heaviest load it could support?

a. A 4-ton elephant
b. A 1/2-ton cow
c. A 95-lb weakling
d. A 10-lb rock
e. A 5-oz. hamster

13. Your job is to send a signal at the highest data rate possible through 2500 km of fiber with polarization-mode dispersion of 1 $ps/km^{-1/2}$. Neglecting all other types of dispersion, what is the best you can do, remembering that polarization-mode dispersion should accumulate to no more than 1/10th the interval between pulses?

a. 10 Gbit/s
b. 5 Gbit/s
c. 2 Gbit/s
d. 1 Gbit/s
e. 100 Mbit/s

14. Suppose you only had to transmit signals 400 km through the same fiber. What is the maximum data rate, again neglecting all other dispersion and remembering that polarization-mode dispersion should accumulate to no more than 1/10th the interval between pulses?

a. 10 Gbit/s
b. 5 Gbit/s
c. 2 Gbit/s
d. 1 Gbit/s
e. 100 Mbit/s

CHAPTER 6

Fiber Materials and Fiber Manufacture

About This Chapter

Materials are the heart of optical fibers. Without ultratransparent materials, fiber-optic communications would be impractical. This chapter describes requirements for fiber-optic materials, the types of materials used, and how they are made into fibers.

Requirements for Making Optical Fibers

The fundamental requirements for making optical fibers sound deceptively simple. You need a material that is transparent and can be drawn into thin fibers with a distinct core-cladding structure that is uniform along the length of the fibers and will survive in the desired working environment. Meeting those requirements turns out to be a challenge, particularly achieving the extreme transparency needed for communications.

Look around and you're sure to see many transparent objects but comparatively few different transparent materials. Ice is transparent, but it doesn't count; it melts at room temperature. Salt and sugar crystals are transparent, but both dissolve too easily in water to be used at normal humidity levels. Virtually all other transparent solids are glass or plastic.

Making thin, uniform fibers is another problem. The usual approach is to heat a material until it softens into a very thick or viscous liquid and then stretch the thick fluid into thin filaments. You can test this for yourself with a glass rod and a flame. Hold both ends, heat the middle until it softens, and then pull the ends apart. The thick liquid

Optical fibers are made by stretching transparent materials into thin filaments.

holds together as you stretch it finer and finer; it cools rapidly to make a thin filament, although simple stretching doesn't make it very uniform. You can do something similar with thick sugar syrup, spinning and pulling it to make cotton candy. Some plastics also work well, but thin liquids don't make fibers, because they tend to fall apart, like water.

Durability is vital. Common sodium chloride is very transparent, but it also soaks up moisture from the atmosphere, so optics made of salt have a distressing tendency to turn into salty puddles unless they are sealed in a dry environment. Some materials are too fragile to survive as long, thin fibers. Plastics and many other materials can't withstand extreme temperatures.

Over the years, silica-based glass and certain plastics have proven the best materials for optical fibers, although you need special glasses and plastics to make low-loss communication fibers. They are most transparent at a limited range of wavelengths in the visible spectrum (0.4 to 0.7 μm) and the near infrared (0.7 to about 2 μm). The clearest window for glass fibers is about 1.2 to 1.7 μm, but they are usable at other wavelengths. Plastic fibers have a window at 0.65 μm and also are reasonably transparent to other visible light.

If you need to transmit wavelengths longer than about 2 μm, you need one of the few exotic compounds that can be made into reasonably transparent fibers, which are described at the end of this chapter.

Glass Fibers

What Is Glass?

Ordinary glass is a noncrystalline compound of silica and other oxides. Many different variations have been developed.

Glass is by far the most common material used in optical fibers, but glass takes many forms, so we should define our terms carefully.

From a scientific standpoint, a glass is a noncrystalline solid—that is, a solid in which the atoms are arranged randomly, not lined up in the neat arrangements of a crystal. You can think of a glass as a sort of liquid with atoms frozen in place by very fast cooling, but it does not flow like a liquid, even over hundreds of years. Typically glasses are compounds such as oxides, but many compounds do not form glasses because they always crystallize. Even compounds such as silica (SiO_2), which readily form good glasses, will crystallize when cooled slowly. Quartz is natural crystalline silica.

The stuff we think of as glass in everyday life is made by melting sand with lime, soda, and some other materials and then cooling the melt quickly. Chemically, the main constituents of ordinary window glass are silica, calcium oxide (CaO), and sodium oxide (Na_2O). Silica accounts for the bulk of the compound. Calcium and sodium compounds improve its properties for glassmaking, notably by reducing its melting temperature. You can make many other types of glass by mixing in other materials. Lead compounds make fine crystal; a dash of cobalt turns the glass a striking deep blue. The glass industry has developed a vast array of glass recipes for different purposes, many going back generations.

Ordinary window glass looks transparent because you don't look through very much glass. Look into the edge of a pane of window glass and you find a strong green color; the wider the pane, the darker the green. The color comes from impurities in the glass. You don't notice their effects when light passes through a few millimeters of window glass, but they add up if you look through the edge of a pane.

Since the 1800s, the optics industry has developed a large family of optical glasses, made of materials that are purer, clearer, and freer of tiny flaws than window glass. Compounds are blended to give glasses with different refractive indexes, important for designers of optical devices. Standard optical glasses have indexes between about 1.44 and 1.8 at visible wavelengths, with pure silica having nearly the lowest refractive index.

Refractive indexes of most optical glasses are between 1.44 and 1.8.

Early fiber-optic developers turned to optical glasses after finding that ordinary glasses absorbed too much light for use in optical fibers. They initially tried coating glass fibers with low-index plastic to serve as the cladding, but when results were poor, they turned to glass cladding.

Rod-in-Tube Glass Fibers

The simplest way to make a glass-clad fiber is by inserting a rod of high-index glass into a tube with lower refractive index. The two are heated so the tube melts onto the rod, forming a thicker solid rod. Then this rod (called a *preform*) is heated at one end and a thin fiber is drawn from the soft tip. The process is shown in Figure 4.3. It is used for image-transmission and illuminating fibers but not for communication fibers.

Simple glass-clad fibers are made by collapsing a low-index tube onto a higher-index rod.

For the fiber to transmit light well, the core-cladding interface must be very clean and smooth. This requires that the rod inserted into the tube must have its surface fire-polished, *not* mechanically polished. Although mechanical polishing gives a surface that looks very smooth to the eye, tiny cracks and debris remain, and if that surface becomes the core-cladding boundary, they can scatter light, degrading transmission.

Another way to draw glass fibers is to pull them from the bottom of a pair of nested crucibles with small holes at their bottoms. Raw glass is fed into the tops of the crucibles, with core glass going into the inner one and cladding going into the outer one. The fiber is pulled continuously from the bottom, with the cladding glass covering the core glass from the inner crucible. The double-crucible process is very rare today, but it has been used in the past and may be used with some special materials.

Limitations of Standard Glasses

Fibers made from conventional optical glasses typically have attenuation of about 1 dB/m, or 1000 dB/km. This is adequate for an image-transmitting bundle to look into a patient's stomach but not for communications from town to town.

Impurities limit transmission of standard glasses.

The main cause of this high loss is absorption by impurities in the glass. Traces of metals such as iron and copper inevitably contaminate the raw materials used in glass manufacture, and those metals absorb visible light. To make extremely clear glass, you need to start with extremely pure silica, which has virtually no absorption at wavelengths from the visible to about 1.6 μm in the near infrared. The concentrations of critical impurities that absorb light at 0.6 to 1.6 μm—including iron, copper, cobalt, nickel, manganese, and chromium—must be reduced to a part per billion (1 atom in 10^9). That level is impractical with standard glass-processing techniques.

Fused-Silica Fibers

Fused silica is the basis for modern communication fibers.

The starting point for modern communication fibers is fused silica, an extremely pure form of SiO_2. It is made synthetically by burning silicon tetrachloride ($SiCl_4$) in an oxyhydrogen flame, yielding chloride vapors and SiO_2, which settles out as a white, fluffy soot. The process generates extremely pure material, because $SiCl_4$ is a liquid at room temperature and boils at 58°C (136°F). Chlorides of troublesome impurities, such as iron and copper, evaporate at much higher temperatures than $SiCl_4$, so they remain behind in the liquid when $SiCl_4$ evaporates and reacts with oxygen. The result is much better purification than you can get with wet chemistry, reducing impurities to the part-per-billion level required for extremely transparent glass fibers.

Dopants, Cores, and Claddings

Silica must be doped to change the refractive index for core and/or cladding.

You cannot make optical fibers from pure silica alone. Optical fibers require a high-index core and a low-index cladding, but all pure silica has a uniform refractive index, which declines from 1.46 at 0.550 μm to 1.444 at 1.81 μm. You need to add dopants to change the refractive index of the silica, but they must be chosen carefully to avoid materials that absorb light or have other harmful effects on the fiber quality and transparency.

Cladding index may be matched to pure silica or depressed by the addition of fluorine.

Most glasses have higher refractive index than fused silica, and most potential dopants tend to increase silica's refractive index. This allows them to be used for the high-index core of the fiber, with a pure silica cladding having a lower refractive index. The most common core dopant is germanium, which is chemically similar to silicon. Germanium has very low absorption, and germania (GeO_2), like silica, forms a glass.

Only a few materials reduce the refractive index of silica. The most widely used is fluorine, which can reduce the refractive index of the cladding, allowing use of pure silica cores. Boron also reduces refractive index, but not as much as fluorine. In practice, step-index silica fibers fall into the three broad categories shown in Figure 6.1. The fiber core may be doped to raise its refractive index above that of pure silica, which is used for the entire cladding. Alternatively, a smaller level of dopant may raise the core index less, but the surrounding inner part of the cladding may be doped—generally with fluorine—to reduce its refractive index. This design is called a *depressed-clad* fiber; normally the fluorine-doped zone is surrounded by a pure silica outer cladding. (Doping at the proper levels complicates processing, so manufacturers prefer to make as much as possible of the fiber from pure silica.) Both designs are used for single-mode fiber. An alternative used for multimode step-index fiber is a pure silica core clad with a lower-index plastic.

As you learned in Chapter 4, the refractive-index profiles of dispersion-shifted fibers are considerably more complex, to provide the extra waveguide dispersion needed to shift the zero-

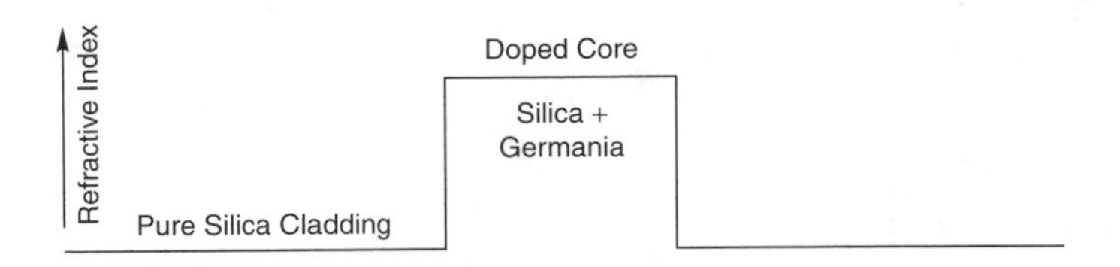

a. Matched-Cladding Fiber

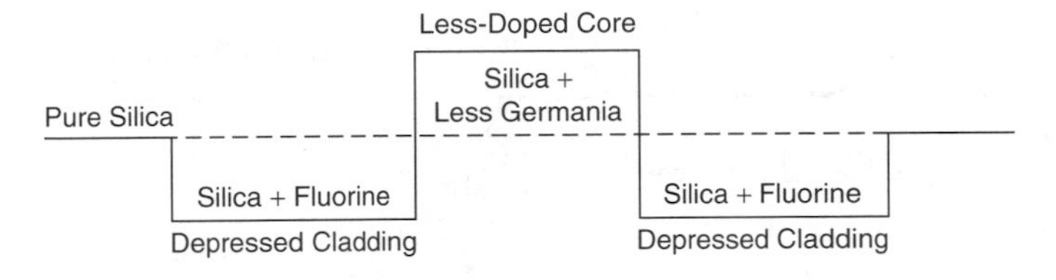

b. Depressed-Clad Fiber

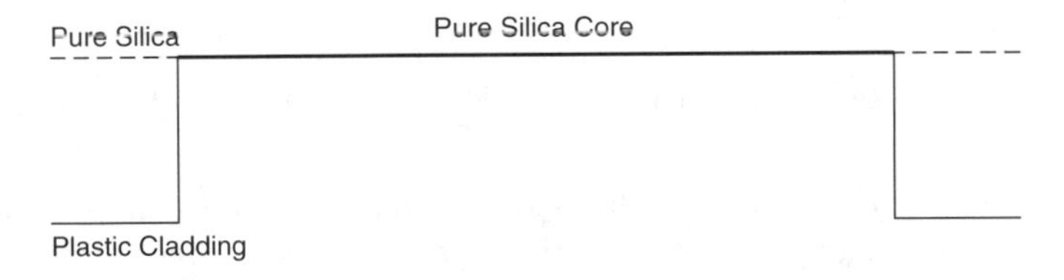

c. Plastic-Clad Silica
(Not Used for Single-Mode Fiber)

FIGURE 6.1
Refractive-index profiles of matched-clad and depressed-clad single-mode fibers and plastic-clad silica multimode fibers.

dispersion point to longer wavelengths. So are the profiles of graded-index multimode fibers. The same dopants are used in these more complex fibers as in simple step-index fibers.

Silica Fiber Manufacture

The trickiest stage in the manufacture of fused-silica optical fibers is making the preform from which the fibers are drawn. Several processes have been developed; they share some common features but have important differences.

The crucial common feature is the formation of fluffy fused-silica soot by reacting $SiCl_4$ (and $GeCl_4$, when it is used as a dopant) with oxygen to generate SiO_2 (and GeO_2 if the silica is doped). The crucial variations are in how the soot is deposited and melted into the final preform.

Fused-silica preforms can be made by depositing glass soot inside a tube of fused silica, which becomes the cladding.

One approach is to deposit the soot on the inside wall of a fused-silica tube, as shown in Figure 6.2. Typically, the tube serves as the outer cladding, onto which an inner cladding layer and the core material are deposited. Variations on the approach are called inside vapor deposition, modified chemical vapor deposition, plasma chemical vapor deposition, and

FIGURE 6.2
Soot deposition inside a fused-silica tube.

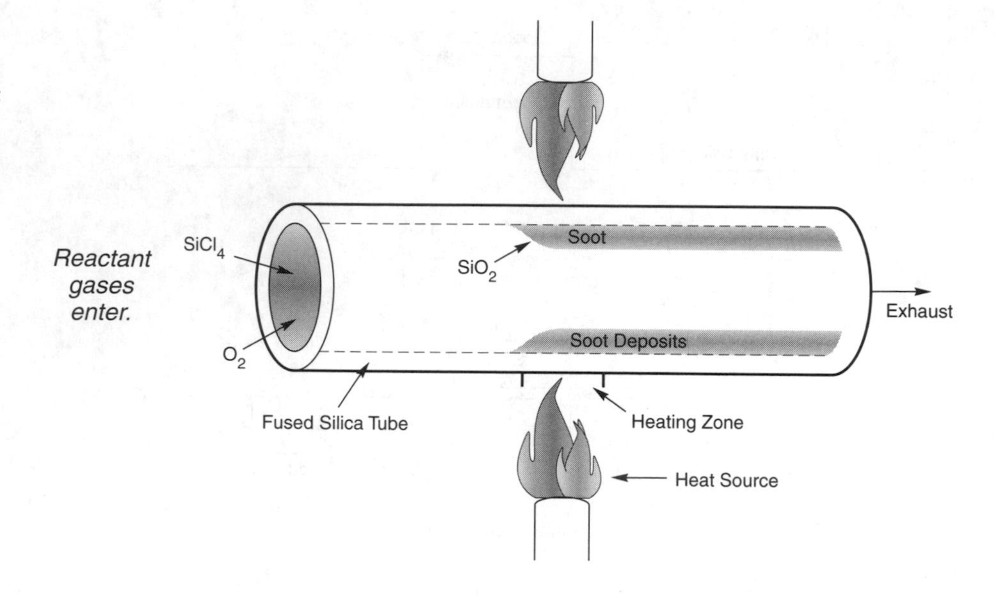

plasma-enhanced chemical vapor deposition. The major differences center on how the reaction zone is heated.

The chemicals react to deposit a fine glass soot, and the waste gas is pumped out to an exhaust. To spread soot along the length of the tube, the reaction zone is moved along the tube. Heating melts the soot, and it condenses into a glass.

The process can be repeated over and over to deposit many fine layers of slightly different composition, which are needed to grade the refractive index from core to cladding in graded-index fibers. The doping of input gases is changed slightly for each deposition step, producing a series of layers with small steps in the refractive index. Step-index profiles are easier to fabricate, because the whole core has the same doping. A final heating step collapses the tube into a preform.

Preforms also can be made by depositing soot on the outside or on the end of a rod.

Another important approach is the outside vapor-deposition process, which deposits soot on the outside of a rotating ceramic rod, as shown in Figure 6.3. The ceramic rod does not become part of the fiber; it is merely a substrate. The glass soot that will become the fiber core is deposited first, then the cladding layers are deposited. The ceramic core has a different thermal expansion coefficient than the glass layers deposited on top of it, so it slips out easily when the finished assembly is cooled before the glass is sintered to form a preform. The central hole is closed either in making the preform or drawing the fiber.

The third main approach is vapor axial deposition, shown in Figure 6.4. In this case, a rod of pure silica serves as a "seed" for deposition of glass soot on its end rather than on its surface. The initial soot deposited becomes the core. Then more soot is deposited radially outward to become the cladding, and new core material is grown on the end of the preform. Vapor axial deposition does not involve a central hole.

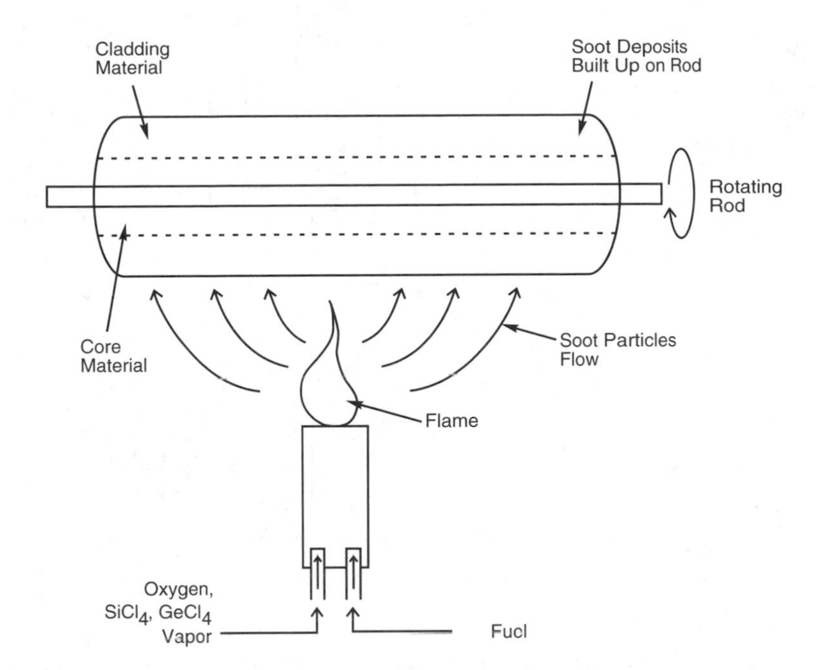

FIGURE 6.3
Outside vapor deposition to make a preform.

All three processes yield long, glass cylinders or rods called *preforms.* They are essentially fat versions of fibers, composed of a high-index fiber covered with a lower-index cladding. They have the same refractive-index profile as the final fiber.

Drawing Fibers

Optical fibers are *drawn* from preforms by heating the glass until it softens, then pulling the hot glass away from the preform. This is done in a machine called a *drawing tower.*

Fibers are drawn from the bottoms of hot preforms.

Drawing towers typically are a couple of stories high and loom above everything else on the floor of a fiber factory. The preform is mounted vertically at the top, with its bottom end in a furnace that heats the glass to its softening point. Initially a blob of hot glass is pulled from the bottom, stretching out to become the start of the fiber. (This starting segment of the fiber normally is discarded.)

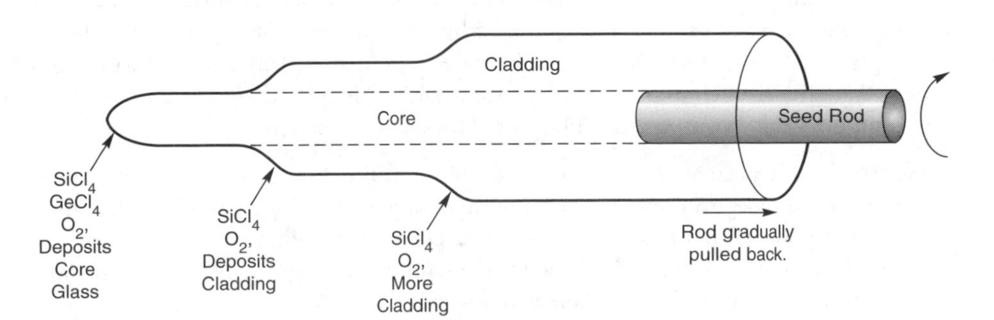

FIGURE 6.4
Vapor axial deposition to make a preform.

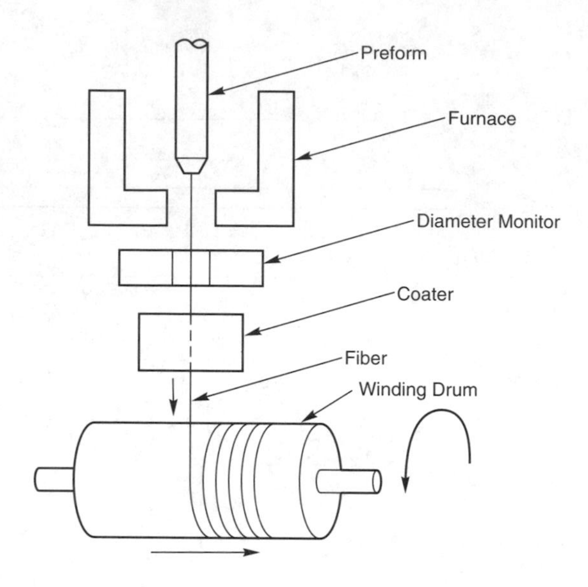

FIGURE 6.5
Drawing glass fibers from preforms. (Courtesy of Corning Inc.)

The hot glass thread emerging from the furnace solidifies almost instantaneously as it cools in open air. As shown in Figure 6.5, the bare glass fiber passes through a device that monitors its diameter, then is covered with a protective plastic coating. The end is attached to a rotating drum or spool, which turns steadily, pulling hot glass fiber from the bottom of the preform and winding plastic-coated fiber onto the drum or spool. The actual length of the draw zone is longer than shown in the figure, to allow the fiber to cool and the plastic coating to cure properly.

Typically the fiber is drawn at speeds well over a meter per second. A single, large preform can yield over 20 kilometers of fiber; smaller preforms yield a few kilometers. After the fiber is drawn, it is proof tested and wound onto final reels for shipping.

Types of Silica Fiber

All-silica fibers are used for communications. Hard-clad silica fibers are used for illumination and beam delivery.

Silica is the standard material for all long-distance communication fibers and for most other fibers used in communications. Except in a few special cases described later, both core and cladding are made of silica, differentiated by different levels of doping. Typically, the cores contain dopants; the claddings may be pure silica (match clad) or doped with index-depressing materials such as fluorine (depressed-clad). This basic design is used for single-mode and graded-index multimode fibers used for communications.

Figure 6.6 shows typical attenuation curves for high-quality dispersion-shifted single-mode fiber and graded-index multimode fiber. The attenuation curve for step-index single-mode fiber is slightly lower than for dispersion-shifted fiber, but the difference is not significant and would not show on this scale. Graded-index fibers have significantly higher loss, but are used only for distances less than a few kilometers.

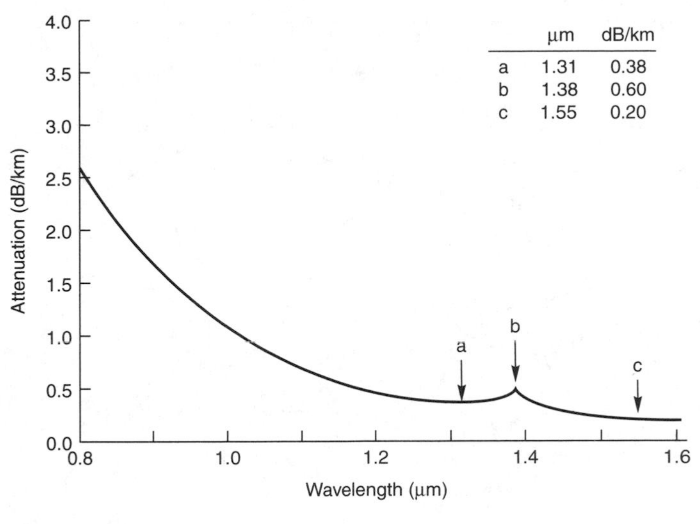

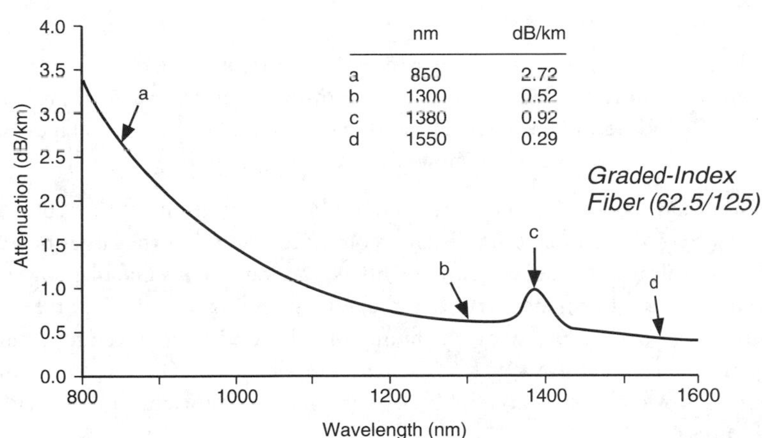

FIGURE 6.6 *Attenuation of non-zero dispersion-shifted fiber (left) and graded-index multimode fiber (below). (Courtesy of Corning Inc.)*

An alternative approach is to clad a pure silica core with silica doped to have a lower refractive index or with a plastic having a lower refractive index. This is mainly done for large-core, step-index multimode fibers. Advantages include simplifying fiber manufacture and avoiding the need for dopants in a large core. (Some large-core step-index fibers are used to deliver high-power laser beams.) Claddings generally are thin compared to the core, with a protective plastic coating 50 to 100 μm thick over the cladding.

Large-core step-index silica fibers are specialty products that come in a variety of configurations. The oldest type is *plastic-clad silica* (PCS), in which the cladding is a silicone plastic. The silicone cladding is fairly easy to strip from the silica core, which is a problem for many applications but an advantage for some uses. *Hard-clad* silica fibers have a tougher plastic cladding, which makes the fibers more durable. Silica-clad fibers can handle higher power levels, important where fibers deliver significant laser powers. The selection of cladding affects attenuation as a function of wavelength, as shown in Figure 6.7.

Large-core silica fibers may be clad with doped silica, hard plastic, or soft plastic; they are used mostly for power transmission.

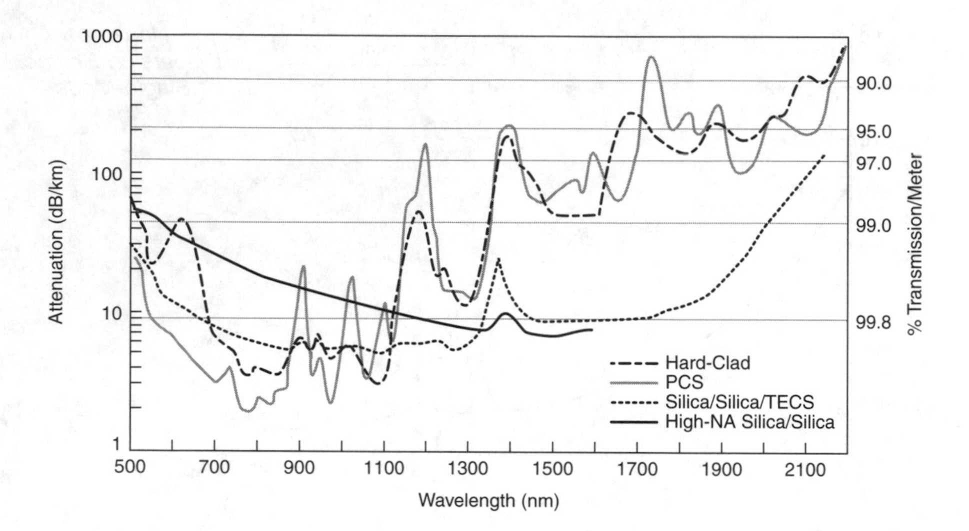

FIGURE 6.7 *Spectral attenuation of various large-core silica fibers. (Courtesy of 3M Specialty Optical Fibers.)*

Composition of the silica core is another important variable. Fibers made in a low-water environment contain little OH and are more transparent in the near-infrared, but those with high OH levels are more transparent in the ultraviolet. (The fibers in Figure 6.7 are all low-OH fibers and the plots do not show ultraviolet wavelengths.)

Typical core diameters of large-core step-index fibers range from 100 to 1000 μm. The smaller fibers may be used for short-distance communication, but the larger fibers are used mostly for illumination. The largest-core fibers can carry considerable power, making them useful for laser-beam delivery, but they are significantly stiffer. For example, the rated continuous power capacity of one family of silica-clad fibers increases from 0.2 kW for 200-μm core fibers to 1.5 kW for 550-μm core fibers, and the rated minimum bend radius increases by a factor of 2.5. Table 6.1 summarizes important optical characteristics of selected fibers.

Although most large-core silica fibers have step-index profiles, some are made with a graded-index core, surrounded by a thin silica cladding and typically a plastic coating and outer buffer layer. Their main application is in delivering high-power laser beams.

Plastic Fibers

Multimode fibers made entirely of plastic have higher loss than silica fibers.

Plastic optical fibers have long been a poor relation of glass. Light, inexpensive, flexible, and easy to handle, plastic has some important attractions. However, these advantages have long been outweighed—especially for communications—by the much higher attenuation of transparent plastics. Years of research have reduced plastic loss considerably, but it still remains much higher than glass. The best laboratory plastic fibers have loss around 50 dB/km. At the 650-nm wavelength preferred for communications using red LEDs, commercial plastic fibers have loss as low as about 150 dB/km. Unlike glass fibers, the loss

Table 6.1 Characteristics of large-core step-index silica fibers. Bandwidths of fibers with cores over 200 μm generally are unrated because they are very rarely used in communications

Fiber Type	*Core/Clad Diameter (μm)*	*Attenuation at 0.82 μm*	*Bandwidth at 0.82 μm*	*NA*
Silica clad	100/120	5 dB/km	20 MHz-km	0.22
Hard clad	125/140	20 dB/km	20 MHz-km	0.48
Plastic-clad, low OH	200/380	6 dB/km	20 MHz-km	0.40
Plastic-clad, high OH	200/380	12 dB/km	20 MHz-km	0.40
Silica clad	400/500	12 dB/km	—	0.16
Hard clad	550/600	12 dB/km	—	0.22
Silica clad	1000/1250	14 dB/km	—	0.16
Plastic-clad, low OH	1000/1400	8 dB/km	—	0.40

of plastic fibers is somewhat lower at shorter wavelengths and is much higher in the near-infrared, as shown in Figure 6.8.

For this reason, plastic optical fibers have found only limited applications. They are used in some flexible bundles for image transmission and illumination, where the light does not have to go far and where lower cost can be important. In communications, plastic fibers are usable only for short links, such as those within an office building or automobile.

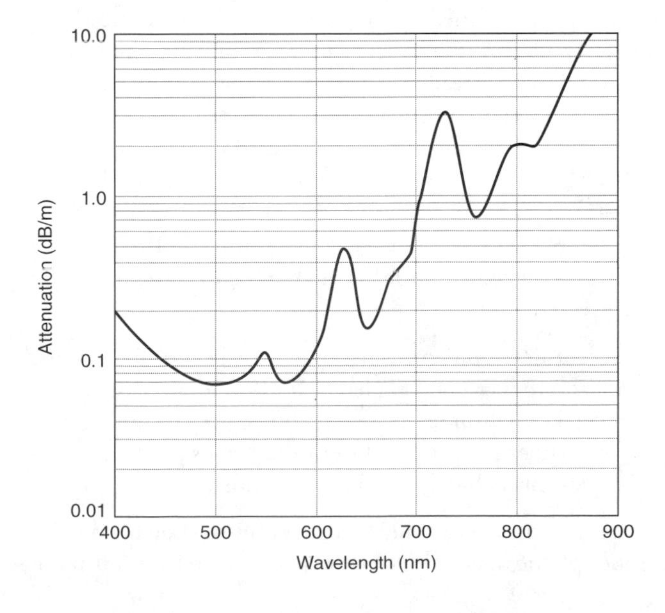

FIGURE 6.8
Attenuation versus wavelength for one commercial PMMA step-index fiber. (Courtesy of Toray Industries Ltd.)

Another important concern with plastic optical fibers is long-term degradation at high operating temperature. Typically plastic fibers cannot be used above 85°C (185°F). This may sound safely above normal room temperature, but it leaves little margin in many environments. The engine compartments of cars, for example, can get considerably hotter. Newer plastics can withstand temperatures to 125°C (257°F), but their optical properties are not as good.

Plastic fibers are designed using the same principles as glass fibers. A low-index core surrounds a higher-index cladding. The refractive-index difference can be large, so many plastic fibers have large numerical apertures. Commercial plastic fibers are multimode types with large cores and step-index profiles, but graded-index fibers have been made in the laboratory. There is little interest in single-mode plastic fibers because the material's high loss makes long-distance transmission impossible.

Step- and Graded-Index Plastic Fibers

Traditional plastic fibers are made of PMMA, with large step-index cores. They are used in bundles and for short data links.

Standard step-index plastic fibers have a core of polymethyl methacrylate (PMMA) and a cladding of a lower-index polymer, which usually contains fluorine. The differences in refractive index typically are larger than in silica or glass fibers, leading to a large numerical aperture. For example, one commercial plastic fiber designed for short-distance communication has a PMMA core with refractive index of 1.492 and a cladding with index of 1.402, giving an NA of 0.47.

Plastic fibers typically have core diameters from about 85 μm to more than 3 mm (3000 μm). You can find larger light-guiding rods of flexible plastic, which sometimes are called fibers, but it's hard to think of something as thick as a pencil as a "fiber." The smallest fibers typically are used only in bundles, but larger fibers are used individually. Typically the claddings are thin, only a small fraction of overall fiber diameter. Large-core plastic fibers cannot carry optical powers as high as those carried by large-core silica fibers, but they are more flexible and less expensive. Plastic fibers with diameters up to around a millimeter are used for some short-distance communication because they are much easier to handle than glass fibers. For example, technicians can splice and connect plastic fibers on site with minimal equipment, instead of the expensive precision equipment required for glass fibers. Figure 6.8 plots attenuation of one PMMA fiber against wavelength, on a scale of decibels per *meter.* The minimum loss, near 500 nm, is equivalent to 70 dB/km, but for communications transmission normally is at the 650-nm wavelength of inexpensive red LEDs. The step-index profile also limits bandwidth, so signals normally are limited to traveling within a building or between adjacent structures.

Graded-index profiles can be made in plastic fibers, increasing bandwidth.

Graded-index plastic fibers are a recent development, which have just begun to reach the market because it had been difficult to produce good graded-index profiles in plastic. Typically a preform is heat-treated to make high-index materials diffuse from a fluorinated plastic core and raise the index of lower-index plastics in the cladding. This plastic preform is then drawn into fiber, much like glass fibers but at much lower temperatures.

As in silica fibers, the advantage of a graded-index profile is broader transmission bandwidth than step-index fibers. Graded-index plastic fibers with core diameters of 50 to 200 μm can transmit 2.5 Gbit/s over distances of 200 to 500 meters, making them attractive for high-speed local area networks. The fluorinated plastic fibers have attenuation around 60 dB/km over a broad range from about 800 to 1340 nm, allowing operation at the 850 and 1300 nm windows. However, attenuation through the entire range is tens of decibels per kilometer, limiting transmission to much shorter distances than with silica fibers.

Issues in Developing Plastic Fibers

High attenuation has been a stubborn problem in plastic optical fibers. Bonds between atoms found in plastics—notably carbon-hydrogen and carbon-oxygen bonds—absorb light at visible and near-infrared wavelengths, even in plastics that look transparent to the eye. Fused silica is much more transparent because these bonds are not present in it.

Attenuation is a key issue in plastic fibers.

Efforts to reduce loss have concentrated on changing the chemical composition of the plastics. One step is to replace normal hydrogen with the heavier (stable) isotope deuterium, which shifts the absorption peaks of carbon-hydrogen bonds to longer wavelengths. Another step is to use fluorinated plastics instead of standard hydrocarbon plastics, because carbon-fluorine bonds have lower attenuation. Figure 6.9 compares attenuation curves for fibers made of standard hydrogen-based PMMA, deuterated PMMA, and one type of fluorinated plastic between 550 and 850 nm. Loss of the fluorinated plastic remains relatively low at wavelengths to 1.3 μm. However, changing composition raises other issues, including the need for more expensive materials.

Another important issue, mentioned earlier, is the durability of plastics, both over time and under extreme conditions. Plastic fibers generally are more flexible than glass, and are easier to cut and install. They generally work fine in a controlled environment such as an office. However, plastics are not as resistant to heat and sunlight as glass. Temperature limitations have proved a particular problem in areas such as the automotive industry, where equipment installed in the engine compartment must withstand frequent temperature cycling and extremes.

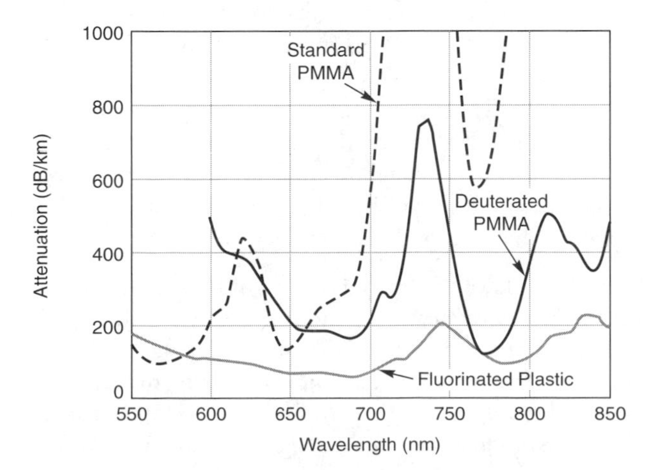

FIGURE 6.9 *Attenuation spectra of graded-index plastic fibers made with regular PMMA, a fluorinated plastic, and deuterated PMMA. (Courtesy of Takaaki Ishigure.)*

Exotic Fibers and Light Guides

From time to time, you may encounter some unusual optical fibers, light guides, or optical waveguides based on novel materials. They presently play little role in communications, but may be used in other applications.

Liquid-Core Fibers (or Light Guides)

In the very early days of fiber-optic communications, developers desperately seeking low-loss materials turned to liquids. They filled thin silica tubes with tetrachloroethylene, a dry-cleaning fluid that is extremely clear and has a refractive index higher than fused silica. The index difference was adequate to guide light, and developers eventually reduced loss to several decibels per kilometer, very good for the time, and better than current plastics.

However, liquid-core fibers were far from a practical communications technology. Filling the tiny capillary tubes took a very long time, but the real problem was thermal expansion. The liquid expanded at a different rate than the tube that held it, so the liquid-core fiber acted like a thermometer, with liquid rising and falling with temperature. If you weren't careful, the liquid could squirt out the ends.

A liquid in a plastic or glass tube can act like a fiber core if its refractive index is higher than the tube.

Now larger-diameter liquid-core light guides are finding a new life transmitting visible light short distances for illumination. Single liquid-core light guides 2 to 10 mm thick are an alternative to standard illuminating bundles. Using suitable fluids, they have lower attenuation than standard bundle fibers, particularly at green and blue wavelengths. The liquid is housed in a plastic tube rather than glass, so the liquid waveguide is more flexible than a large solid fiber. Because lengths are modest—at most 20 m and typically only a few meters—thermal expansion poses little problem.

Midinfrared Fibers

Fibers made of nonsilica glasses transmit infrared wavelengths, which silica absorbs.

The extremely low scattering losses expected at wavelengths longer than 1.55 μm prompted interest in those wavelengths for long-distance communications in the 1980s. The absorption of silica rises rapidly at longer wavelengths, so developers looked to other materials that are transparent in that region. Theorists hoped that extremely low-loss glass fibers could be made from some of those materials. (Recall that glass is a disordered material, not necessarily made from silica.) If other losses could be avoided, the floor set by scattering loss suggested attenuation might be as low as 0.001 dB/km. Such incredibly low loss would allow extremely long transmission distances without amplifiers or repeaters.

Unfortunately, very low-loss infrared fibers have proven exceedingly difficult to make. Purification of the materials is difficult. The raw materials are far more expensive than those for silica fibers. (Despite occasional jokes about desert nations concerning the market on raw materials, silica fibers can't be made from raw sand, as can some glass products.) Infrared materials are harder to pull into fibers because they are much less viscous than silicate glass when molten. The fibers that can be produced are weaker mechanically than silica and suffer other environmental limitations. In short, infrared fibers have been a bust for ultra-long-distance communications.

On the other hand, fibers made from nonsilicate glasses can transmit infrared wavelengths that do not pass through silica fibers. This makes them useful in specialized applications such as infrared instrumentation, although their losses are much larger than the minimum loss of silica fibers at shorter wavelengths.

Fluorozirconate fibers transmit light between 0.4 and 5 μm. Often simply called fluoride fibers, they are made primarily of zirconium fluoride (ZrF_4) and barium fluoride (BaF_2), with some other components added to form a glass compound. The lowest losses for commercial fluorozirconate fibers are about 25 dB/km at 2.6 μm, but loss as low as about 1 dB/km has been reported in the laboratory. A typical transmission curve is shown in Figure 6.10, along with other infrared fibers. Fluoride fibers are vulnerable to excess humidity, so they should be stored and used in low-humidity environments. Fluoride fibers are used in some erbium-doped fiber amplifiers because of desirable optical characteristics. However, they have a refractive index higher than 2, so they have high reflection from their ends.

Fibers made from silver halide compounds (AgBrCl in Figure 6.10) have useful transmission between about 3 and 16 μm in the infrared. They are not a true glass, but a solid made of many small crystals.

Synthetic crystalline sapphire (Al_2O_3) can be drawn into single-crystal fibers that transmit between 0.5 and 3.1 μm. As Figure 6.10 shows, their loss is higher than fluoride fibers, but the material is much more durable.

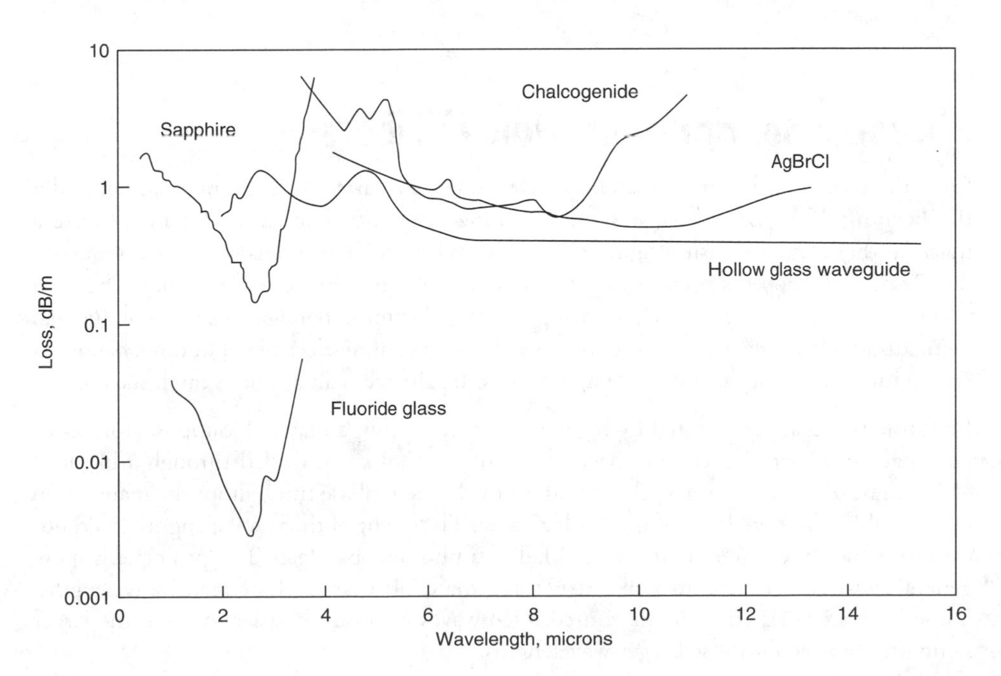

FIGURE 6.10
Attenuation of infrared optical fibers. (Courtesy of James Harrington, Rutgers University)

Hollow Optical Waveguides

Hollow waveguides can transmit longer infrared wavelengths.

Hollow optical waveguides were first developed in the 1960s, after the laser stimulated interest in optical communications. Work on hollow waveguides for the visible and near-infrared stopped shortly after the first low-loss glass fibers were made in 1970s. However, new types of hollow optical waveguides are being developed for infrared wavelengths longer than a few micrometers. I mention them here because they serve the same purpose as infrared fibers, and compete successfully for some infrared applications. There are two basic types of hollow infrared waveguides, metal and hollow glass.

Hollow metal waveguides are coated inside with a nonconductive dielectric material to make them more reflective. The infrared light bounces along the shiny walls, with high reflectivity limiting loss to about 500 dB/km. That isn't bad considering how many reflections the light undergoes. Many hollow glass waveguides work on the same principle; they have the advantage of very smooth surfaces that give loss as low as 0.5 dB/m with suitable coatings, as shown in Figure 6.10.

Other hollow glass waveguides work on a different principle. At certain wavelengths, some materials have an effective refractive index less than 1. Functionally, that means they absorb those wavelengths strongly, but it also means they can serve as a low-index cladding surrounding a hollow core of air, which at that wavelength has a higher refractive index. Silica glass meets those conditions at wavelengths of 7 to 9.4 μm, sapphire does at 10 to 17 μm. These waveguides are called *attenuating total internal reflection* guides because the fraction of the wave in the cladding is absorbed, so loss is over 1 dB/m. However, hollow sapphire guides can be used at the important 10.6-μm wavelength of carbon-dioxide lasers.

Photonic Bandgap or "Holey" Fibers

Novel photonic bandgap materials can confine light, creating an optical waveguide.

Over the past several years, researchers have explored a variety of concepts that fall under the heading of *photonic bandgaps* and *structured materials.* The basic idea is to create a material that cannot transmit light at certain wavelengths. This is analogous to a semiconductor material doped so that it cannot conduct electrons with certain energies. The term *photonic bandgap* comes from an analogy to the electronic bandgap in semiconductors. Semiconductor bandgaps are useful because they can confine electrons. Photonic bandgaps are useful because they can confine light, but so far they remain at the research stage.

Photonic bandgaps are created by regular structures within a material. Suppose, for example, that a uniform and closely spaced array of tiny holes was drilled through a block of glass. Light of certain wavelengths would hit the holes in phase throughout the material, so they would be scattered out of the "holey" zone. Thus light of those wavelengths could not travel through the material, and would fall in a photonic bandgap. The principle is quite general, and can be used to make structured materials that work at many wavelengths. Some laboratory experiments have used microwaves because it's easier to build the larger structures needed for those longer wavelengths.

Changing the spacing of the holes changes the optical characteristics of the material. Extra holes or absences of holes also can change its optical characteristics.

Optical fibers and other optical devices can be made using two or more distinct layers of the structured material instead of layers of glass with different refractive indexes. For example, fibers have been made by stacking together glass rods, with voids intentionally left between them. When the rods are melted together and drawn into fibers, thin air channels remain along the length of the fiber, forming photonic bandgap fibers. You might think of these as hollow optical fibers, with an air core inside a highly reflective flexible tube. We'll talk more about these fibers—called *holey fibers,* or *photonic crystal fibers,* in the next chapter, on special fiber types.

Planar Waveguides

Planar waveguides work on the same principle of total internal reflection as optical fibers, but they come in a different form. A planar waveguide is a thin layer on the surface of a flat material, which has higher refractive index than the bulk material. Typically the high-refractive index is produced by doping the substrate material with something that increases its refractive index. Figure 6.11 shows the basic idea. The boundaries of the doped area form an interface that guides light, like the core-cladding interface in optical fibers. In Figure 6.11, the substrate provides the low-index materials on the sides and bottom, while air is the low-index medium on the top.

Planar waveguides are thin strips on flat substrates that guide light by total internal reflection.

An alternative approach is to deposit a layer of high-index material on a lower-index substrate. In this case, the waveguide is a raised stripe on the substrate, surrounded by air on

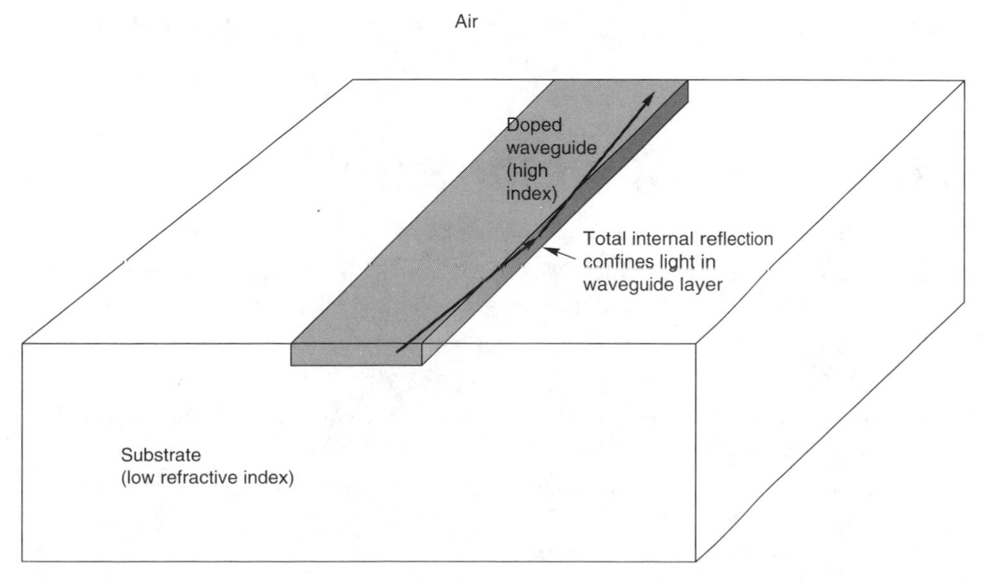

FIGURE 6.11
Planar waveguide.

top and on the sides, and contacting the substrate only on the bottom. As with the doped waveguide, total internal reflection confines light in the waveguide layer.

From a theoretical standpoint, both types of planar waveguides are *dielectric slab waveguides.* That means they are made of nonconducting (dielectric) materials, and are rectangular in cross section, rather than round like a fiber. The theory of such waveguides is quite well developed.

Planar waveguides are used in couplers, lasers, modulators, and switches.

From a practical standpoint, planar waveguides also have some attractions. The technology for making thin stripes of material on flat substrates has been well developed by the semiconductor electronics industry. The technology can be used with a wide variety of materials, including silica glass and other compounds as well as semiconductors. Active optical components such as lasers and photodetectors can be made on the semiconductor materials. So can a wide variety of passive optical components, such as demultiplexers and couplers that divide and combine optical signals. That opens the possibility of integrating optical components on a chip.

On the other hand, planar waveguides also suffer serious practical drawbacks. Their attenuation is much higher than optical fibers, so they can't send signals very far. Their flat, wide geometry differs greatly from the round cores of optical fibers, so light is inevitably lost in transferring from a fiber to a waveguide. Such problems limit the uses of planar waveguides.

Nonetheless, planar waveguide devices can manipulate light in many useful ways. Many semiconductor lasers are planar waveguide devices; you'll learn about them in Chapter 9. Other important planar waveguide devices include couplers, modulators, switches, and wavelength division-multiplexing components; you'll learn about them in Chapters 14, 15, and 16.

For now, what's important is to remember what planar waveguides are, that they can guide light like optical fibers, and that they can serve as the basis for a variety of important components.

What Have You Learned?

1. Fiber-optic materials must be transparent and drawable into thin fibers.
2. Glass is a noncrystalline solid. Most glasses are compounds of silica and other oxides. A wide variety of compositions have been developed for various uses.
3. Silica-based glasses have refractive indexes of 1.44 to 1.8, with pure silica among the lowest.
4. Simple glass-clad fibers are made by collapsing a low-index tube onto a high-index rod, called a preform, heating the tip, and drawing fiber from the soft, hot end.
5. Impurities are the main limit to transmission in standard silica glasses. Synthetic fused silica is the base for communication fibers; it is very clear because impurities are reduced to a part per billion or less.

6. Silica must be doped to form either a high-index core or a low-index cladding for an all-glass fiber. Fluorine can reduce the index of silica; germanium can increase its index.
7. Fused silica preforms are formed by depositing glass soot inside a fused silica tube, on a ceramic rod that is later removed, or on the end of a preform. This soot is melted to make the preform. Fiber is drawn from the bottom of a preform mounted in a drawing tower.
8. Large-core silica fibers are used for illumination and beam delivery. They may be clad with doped silica, hard plastic, or soft plastic.
9. All-plastic fibers have attenuation much higher than silica fibers. They are used for image transmission or short-distance communications.
10. Standard plastic fibers are made from PMMA and have step-index profiles. Graded-index plastic fibers recently became available but are not widely used. Lower-loss plastics are in development, but there are no prospects for reaching the low losses of silica fibers.
11. Silica does not transmit at wavelengths longer than 2 μm, so other materials are used in fibers transmitting at longer wavelengths. None of these are as transparent as silica fibers. Hollow waveguides also are used in the infrared.
12. Research on photonic bandgap materials may lead to new types of fibers.
13. Planar waveguides are thin layers on flat substrates, which guide light by total internal reflection, like optical fibers.

What's Next?

In Chapter 7, you will learn about special types of fibers including fiber Bragg gratings, fibers used in optical amplifiers, and photonic bandgap fibers.

Further Reading

James Harrington, "Infrared Fiber Review," available at *http://irfibers.rutgers.edu/ir_rev_glass.html*

Luc B. Jeunhomme, *Single-Mode Fiber Optics: Principles and Applications* (Marcel Dekker, 1990)

Donald B. Keck, ed, *Selected Papers on Optical Fiber Technology* (SPIE Milestone Series Vol. MS38, 1992)

Yasuhiro Koike and Takaaki Ishigure, "Bandwidth and Transmission Distance Achieved by POF," *IECE Transactions Electronics,* Vol. E82-D, No. 98, August 1999, available on the Web at *http://www.ieice.org/eng/index.html*

University of Bath, "Photonic Crystal Fibre," *http://www.bath.ac.uk/Departments/Physics/groups/opto/pcf.html*

Questions to Think About for Chapter 6

1. Window glass looks very clear when you look straight through a pane, but when you look into the edge it looks green. What causes this color?
2. Why is it easy to make fibers from glass but impossible to make them from ice, even in Antarctica?
3. What are the main trade-offs in picking dopants for the core and cladding of fused-silica fibers?
4. The large-core step-index fibers listed in Table 6.1 have cores that are nominally pure silica and generally carry higher laser powers than telecommunication fibers. Yet their attenuation is higher than for single-mode telecommunication fibers at the same wavelengths. Why should this happen?
5. How would you compare the advantages of plastic and glass fibers? What are the best features of each? Where might plastic have an advantage?
6. Roughly how many times higher is the minimum loss of plastic fibers than glass fibers, measured in dB/km?
7. If you aimed a 1.55-μm laser beam straight up through clear air, about 90% of the light would escape into space, with 10% scattered or absorbed. The atmosphere becomes more tenuous at higher altitudes, so assume that sending a beam into space is equivalent to transmitting it through 10 km of air. What is the equivalent attenuation in dB/km? How does this compare with the best optical fibers at that wavelength? Does that imply a limit on the transparency of hollow photonic bandgap fibers?

Quiz for Chapter 6

1. What is the most essential property of all glass?
   a. It is a noncrystalline solid.
   b. It is a crystalline solid.
   c. It must be transparent.
   d. It must be made of pure silica.
   e. It must have a refractive index of 1.5.
2. What type of fiber is drawn from a preform made by fusing a low-index tube onto a higher-index rod?
   a. Step-index single-mode
   b. Graded-index multimode
   c. Dispersion-shifted
   d. Short-distance imaging and illumination
   e. All of the above
3. What impurity levels are required in fused silica for communications fibers?
   a. Less than 0.1%
   b. Less than 0.001%
   c. One part per million

d. Ten parts per billion

e. One part per billion

4. What is done to make a depressed-clad fiber?

a. The fiber is flattened by rollers to depress it before the cladding is applied.

b. The refractive index in the core is depressed by adding germanium.

c. The refractive index in the inner part of the cladding is depressed by adding fluorine.

d. The fiber is clad with a low-index plastic.

e. The entire fiber is made of pure silica because it has the lowest refractive index of any glass.

5. How are preforms for communications fibers made?

a. By the rod-in-tube method

b. By soot deposition in a fused silica tube

c. By soot deposition on the outside of a ceramic rod

d. By vapor axial deposition on the end of a rod

e. By methods b, c, and d

6. What is *not* used as a cladding for silica fiber?

a. Silica with refractive index depressed by adding fluorine

b. Silica with refractive index increased by adding germanium

c. Hard plastic

d. Soft plastic

e. Pure silica cladding on a core doped to have higher refractive index

7. What type of fiber could transmit the highest laser power?

a. Step-index silica fiber with a 550-μm core

b. Hard-clad silica fiber with a 100-μm core

c. All-plastic fiber with a 1000-μm core

d. Single-mode fiber

e. Plastic-clad silica fiber with a 200-μm core

8. What is the lowest loss of laboratory all-plastic fibers?

a. 1 dB/km

b. 50 dB/km

c. 150 dB/km

d. 500 dB/km

e. 1 dB/m

9. At what wavelength does PMMA plastic fiber have the lowest loss?

a. 500 nm

b. 650 nm

c. 850 nm

d. 1.3 μm

e. 1.55 μm

10. Why would you use fluorozirconate fibers?

a. You couldn't find any other fibers

b. Because their attenuation is 0.001 dB/km

c. To transmit near-infrared wavelengths of 2–5 μm where silica fibers have high loss

d. To transmit infrared wavelengths near 10 μm

e. To compensate for losses in plastic fibers

11. What types of fibers can be used at wavelengths longer than 2 μm?
    a. Fused silica, fluoride, and silver halide
    b. Plastic, fused silica, and fluoride
    c. Fluoride, chalcogenide, and silver halide
    d. Plastic, fused silica, and chalcogenide
    e. Only plastic-clad silica

12. What is the main advantage of graded-index plastic fiber over other plastic fibers?
    a. Higher bandwidth
    b. Uses less plastic
    c. More flexible
    d. As clear as graded-index glass fiber
    e. Larger core diameters

13. What is a holey fiber?
    a. A fiber made from a flawed preform that contains tiny air bubbles, which scatter light from the sides
    b. A fiber that guides light through holes in a plastic with a very low refractive index
    c. A photonic bandgap fiber in which the glass contains tiny holes that block light transmission at certain wavelengths
    d. A theoretical proposal, which has yet to be demonstrated
    e. Another name for a hollow optical waveguide used to transmit the 10.6-μm wavelength of carbon-dioxide lasers

14. In what way is a planar waveguide like an optical fiber?
    a. It guides light through a region of high refractive index by total internal reflection.
    b. It has attenuation below 0.5 dB/km at 1.3 to 1.6 μm.
    c. It is a flexible structure.
    d. It has a plastic coating to protect it from scratches.

CHAPTER

7

Special-Purpose Fibers

About This Chapter

Chapters 4 through 6 covered standard optical fibers whose main function is to guide light, whether for communications or imaging. This chapter covers special-purpose fibers designed for other purposes, particularly fiber gratings and fiber amplifiers.

Fiber gratings are passive optical components, which selectively reflect and transmit light of certain wavelengths. So far they are the most important example of optical fibers designed to serve the function of optical components. Fiber amplifiers are optical fibers which have their cores doped with erbium so they can amplify weak optical signals. This chapter will describe the structure of amplifying fibers; you will learn about their operation later, in Chapter 12. This chapter also covers some other special-purpose fibers that have been developed or are still at the laboratory stage. Chapter 4 covered polarizing fibers, and Chapter 29 will describe fiber sensing.

Fiber Gratings

Normal optical fibers are uniform along their lengths. A slice taken from any one point on the fiber would look very much like a slice taken from any other part of the fiber, neglecting any tiny imperfections. However, it is possible to make the refractive index of the core glass vary periodically along the length of a fiber, rising then falling, then rising again. Such fibers are called *fiber gratings,* because the refractive index variations scatter light passing through the fiber, an effect similar to how grooves etched in bulk optical devices called *diffraction gratings* scatter light hitting their surfaces. Both processes depend on the wavelength of light.

A diffraction grating is a row of fine parallel lines or grooves, usually on a reflective surface. The lines reflect or scatter light waves, and interference among light of different wavelengths spreads out a rainbow of colors, as shown in Figure 7.1. You can see the

FIGURE 7.1
Diffraction grating spreads light out by wavelength; angle depends on the wavelength and line spacing.

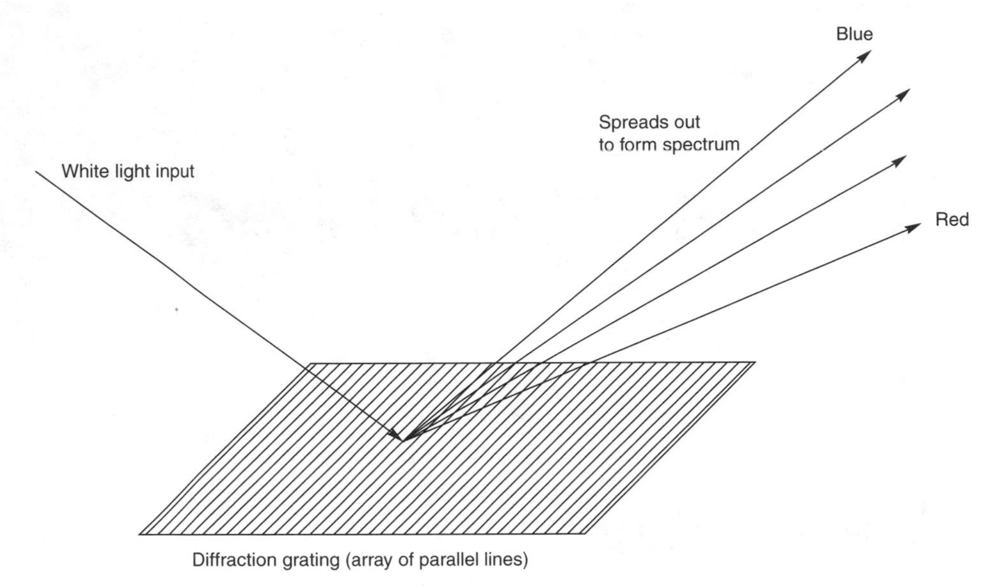

Refractive index varies periodically along the length of a fiber grating, allowing the grating to selectively reflect certain wavelengths.

same effect if you reflect light from a CD, where the tiny pits of recorded data are wound in a tight spiral, and the parallel curved lines they form act like the straight lines in a diffraction grating. Although diffraction gratings depend on different optical effects than prisms, the two may be used interchangeably to disperse the spectrum and separate light at different wavelengths.

The Bragg effect selectively reflects light with wavelengths that match the grating period.

In a fiber grating, the "lines" are not grooves etched on the surface. Instead, they are uniformly spaced regions in the fiber where the refractive index has been raised from that of the rest of the core. These variations scatter light by what is called the *Bragg effect.* Bragg scattering isn't exactly the same as scattering from a diffraction grating, but a Bragg grating does selectively reflect a narrow range of wavelengths. Each time the light hits a region of higher refractive index, a bit is scattered backwards. If the wavelength matches the spacing of the high-index zones in the fiber, the waves scattered from each high-index zone interfere constructively, producing strong reflection, as shown in Figure 7.2.

The high-index regions also scatter light at other wavelengths, but the scattered waves differ in phase so they cancel each other by destructive interference. Thus these nonresonant wavelengths are transmitted through the grating with low loss.

Fabrication of Fiber Gratings

Ultraviolet light forms gratings by affecting bonds in a fiber.

Ultraviolet light creates fiber gratings by breaking atomic bonds in the germania-doped silica glass of the fiber core. (Glass composition is adjusted to maximize the effect.) Typically, an external ultraviolet laser illuminates the fiber through a thin, flat slab of silica with a pattern of fine parallel troughs etched on its bottom, as shown in Figure 7.3. The slab is called a *phase mask.* It diffracts most of the light in two directions, as shown, where they generate

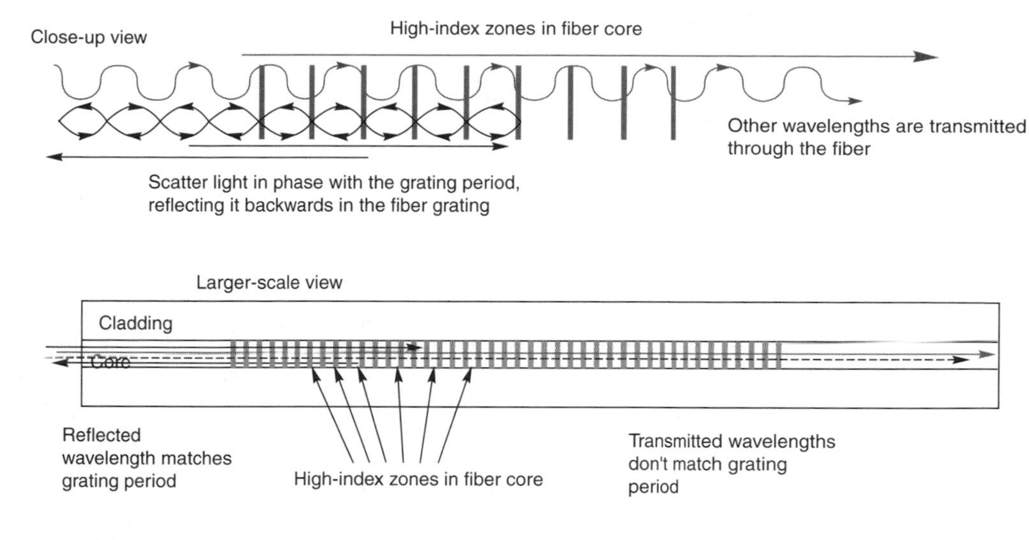

FIGURE 7.2

Fiber Bragg grating reflects light at wavelengths that match the grating period and transmits other wavelengths.

an interference pattern covering the fiber. Regions of high and low intensity alternate. In the regions of high intensity, the ultraviolet light breaks bonds in the glass, changing its refractive index and forming a grating. Because of the geometry, the grating lines in the fiber are half as far apart as the parallel lines in the phase mask. If the phase mask spacing is D, the spacing of the fiber grating is $D/2$. The laser wavelength does not affect line spacing, but it does affect the strength of the grating.

The amount of change in the refractive index depends on the extent of ultraviolet irradiation, the glass composition, and any special processing before treatment. Typically, pulsed ultraviolet lasers illuminate the fibers for a few minutes at high intensities. This can increase the refractive index of germania-doped silica by a factor of 0.00001 to 0.001. Treating the fiber with hydrogen before illuminating it can increase the sensitivity, so the refractive index increases up to 1%. The higher levels of change are comparable to the difference in refractive index between core and cladding, which typically does not reach 1% in single-mode fiber.

Reflection and Transmission in Fiber Gratings

What happens to the light traveling through the fiber depends on its wavelength. Each line reflects a little bit of the light at all wavelengths. If the wavelength in the glass is exactly twice the spacing of the lines written in the fiber, all the scattered light is in phase, so the light waves interfere constructively. That wavelength is reflected. The more lines, the more uniform the spacing, and the more strongly they are written, the stronger the reflection.

Line spacing and refractive index determine the wavelength reflected by a fiber grating.

The wavelength selected is twice the distance between the lines written into the fiber because the light wave has to go through the region between them twice, once into the fiber

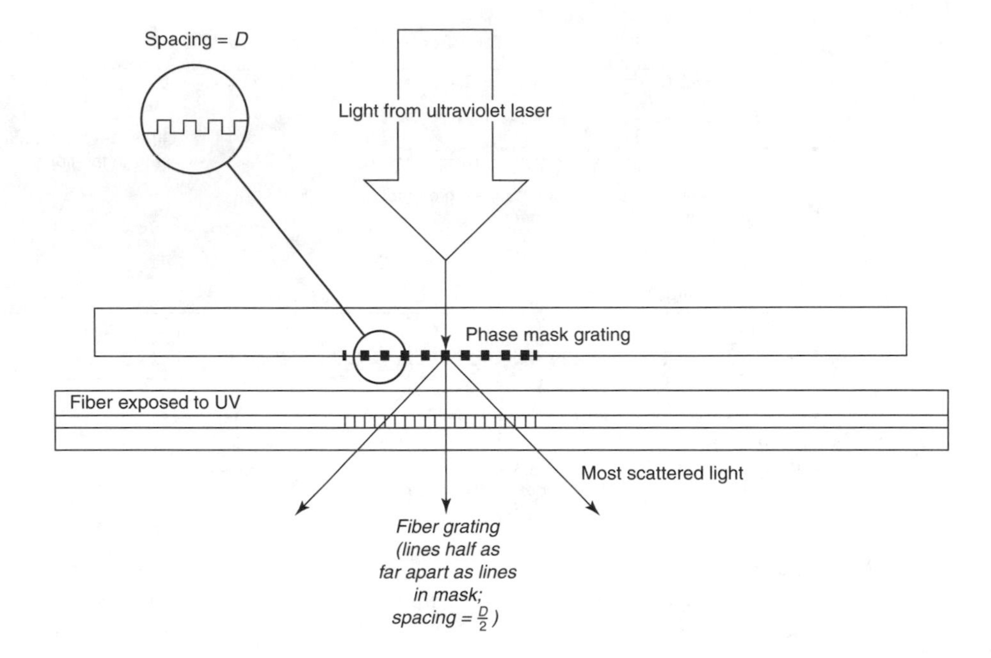

FIGURE 7.3 *Ultraviolet light writes grating in sensitive fiber core.*

grating, and once when reflected back out. Remember also that the wavelength that counts is the wavelength in the glass, which is shorter than the wavelength in air. This means the refractive index enters the equation if we want the results in terms of the wavelength in air (which is the usual way of denoting wavelength). If D is the grating spacing and n the refractive index of the glass, the reflected wavelength (measured in air) is

$$\lambda_{\text{grating}} = 2nD$$

For example, if the grating spacing is 0.500 μm and the refractive index is 1.47, the selected wavelength is 1.47 μm. You also can flip the equation to calculate the grating spacing needed to reflect a particular wavelength. Note that in order to select a precise wavelength, you must know the exact refractive index and grating spacing.

Other wavelengths that do not meet this criterion are not reflected in phase, so the scattered light waves do not add constructively. The reflected light averages out to zero, so they are transmitted essentially unaffected. The result is a simple line-reflection filter, which reflects the selected wavelength and transmits other wavelengths. No optical device behaves perfectly, and in practice reflection increases strongly over a range of wavelengths, with peak reflection at the selected wavelength. Fiber Bragg gratings can be made to have peak reflection across a narrow band, with nearly square sides, as shown in Figure 7.4 for a grating with peak reflection in a band centered at 1538.19 nm. The reflection curve shows the fraction of light (in dB) reflected at that wavelength; the transmission plot shows the transmission loss at that wavelength. This filter—a composite based on typical products—

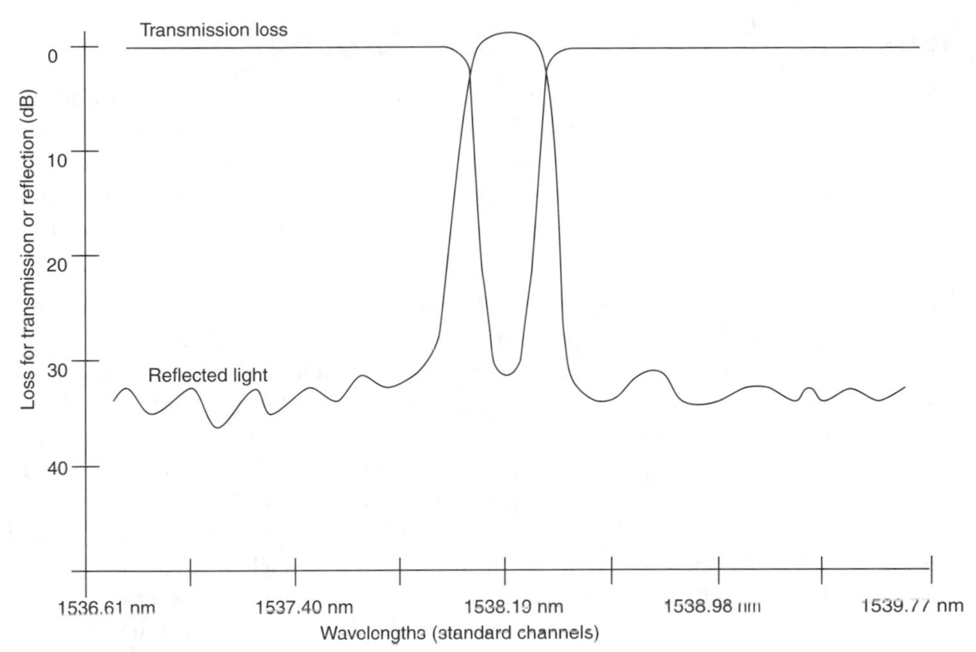

FIGURE 7.4
Reflection and transmission in a fiber grating.

reflects -30 dB (10^{-3}) of the incident light at wavelengths outside the selected band, which is 100 GHz wide (about 0.8 nm). The rest of the light outside the selected band passes through unaffected.

The variation of the reflectivity with the wavelength depends on the nature of the grating. Fine, thin, evenly spaced lines tend to concentrate reflection at a narrow range of wavelengths. Turning up exposures to make a stronger grating will increase reflectivity and broaden the range of reflected wavelengths. Commercial devices using this design select a range of wavelengths as narrow as a few tenths of a nanometer and ranging up to several nanometers wide. The narrow ranges are well matched to the requirements of wavelength-division multiplexing in the 1.55-μm band.

Complex and Graded Gratings

Fiber gratings also can be made in which the regions with high-refractive index are not uniformly spaced or are not perpendicular to the length of the fiber. For example, a grating can be "chirped," so the spacing changes along the length of the fiber.

The details of these designs are worthy of a (rather complicated) book of their own, and you need not worry about them here. That technology is still evolving. What you do need to remember is that fiber gratings can be made into a range of optical devices that have different effects on different wavelengths. We'll talk about the most important examples later, but more devices are likely to emerge from the laboratory in coming years.

Wavelength Selection by Fiber Gratings

Fiber gratings can select one wavelength from many carried by a fiber.

The ability of fiber gratings to select one or more wavelengths is important in wavelength-division multiplexing, or where pump and signal wavelengths must be combined or separated. Other optical devices can do the same thing, but fiber gratings select a narrow range of wavelengths and fit naturally into fiber-optic systems. To see the uses of a fiber grating, consider a system carrying signals at eight wavelengths: 1546 nm, 1548 nm, 1550 nm, 1552 nm, 1554 nm, 1556 nm, 1558 nm, and 1560 nm.

You already saw how a fiber grating selectively reflects one wavelength. Suppose you want to pick out one of the eight wavelengths, 1552 nm, and drop it at a node without affecting the optical channels at the other seven wavelengths. The simplest approach is to couple all eight input optical channels to a fiber grating through a device called an optical circulator, as shown in Figure 7.5. All eight wavelengths start down the fiber grating, and all but 1552 nm pass straight through it. However, the grating reflects 1552 nm back to the optical circulator. The circulator is a directional device described in more detail in Chapter 15. You can think of it as an optical one-way street, that is, it transmits light in one direction but not the other. The circulator transmits light that enters through one input port to the Bragg grating. Light reflected back from the Bragg grating is routed to a different port, called the "drop" port in Figure 7.5. That port acts as a filter, which selects the 1552-nm signal and directs it to the desired location.

The other wavelengths can be transmitted to another destination or directed to another fiber-grating filter. A second fiber grating can pick off another wavelength, such as 1554 nm, and the remaining wavelengths can be transmitted or sent through another filter stage. This is part of a process called wavelength-division demultiplexing, essential to separate signals in multiwavelength systems.

Fiber-grating devices can do more complex functions, as shown in Figure 7.6, where a communication system carries four wavelengths: 1550, 1552, 1554, and 1556 nm. The three shorter wavelengths must go from town A to town C, but the system has to send signals at 1556 nm from A to B and from B to C. This is called an *add-drop multiplexer,* because it drops a signal at an intermediate location and replaces it with another signal at the same wavelength.

In this case, the add-drop multiplexer first extracts the 1556-nm channel going from A to B by reflecting it from a fiber grating that transmits 1550, 1552, and 1554 nm. The optical

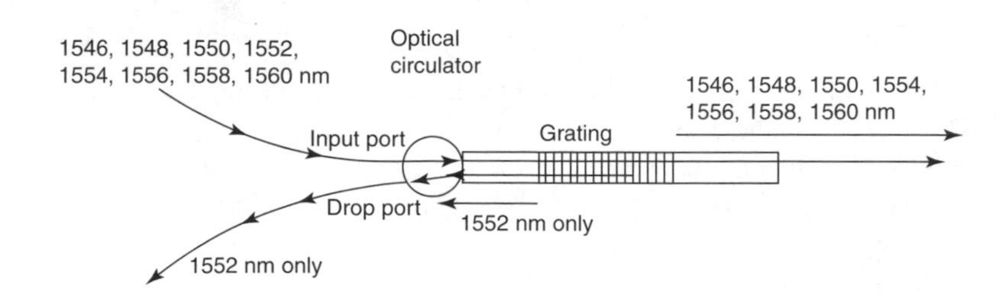

FIGURE 7.5
Fiber Bragg grating selects one wavelength.

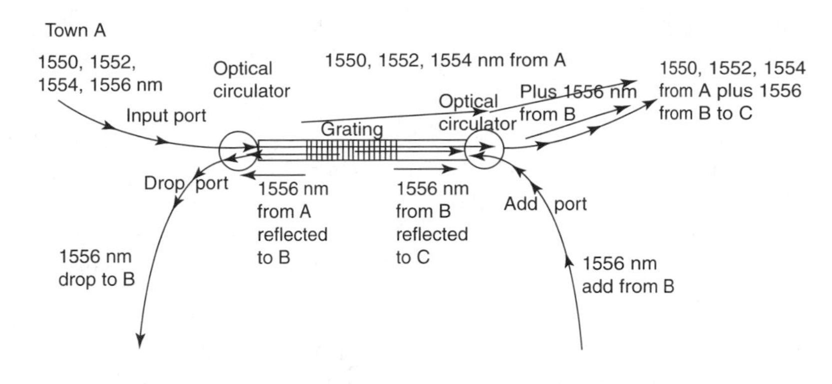

FIGURE 7.6
Add–drop multiplexer uses fiber gratings to drop one wavelength at an intermediate point and to replace it with another signal on the same optical channel.

circulator drops the 1556 nm channel at B. A transmitter at B sends its own 1556-nm signal, which enters the fiber grating from the other end, and is also reflected. The optical circulator sends all signals passing through the fiber grating to C, adding the 1556-nm channel from B to the 1550, 1552, and 1554 nm channels from A.

Dispersion Compensation

Another use of fiber gratings is to compensate for chromatic dispersion in an optical fiber. The grating serves as a selective optical delay line, which adjusts the transit times of different wavelengths in a pulse so they are approximately equal.

A series of gratings with different spacings can serve as a delay line to compensate for dispersion.

Suppose, for example, the longer wavelengths in a pulse arrived first, and the shorter wavelengths arrived last. As shown in Figure 7.7, the grating is made so segments which reflect different wavelengths are in different positions along the length of the grating. In this case, the longest wavelengths (λ_4 in Figure 7.7) arrive first and are transmitted through to the

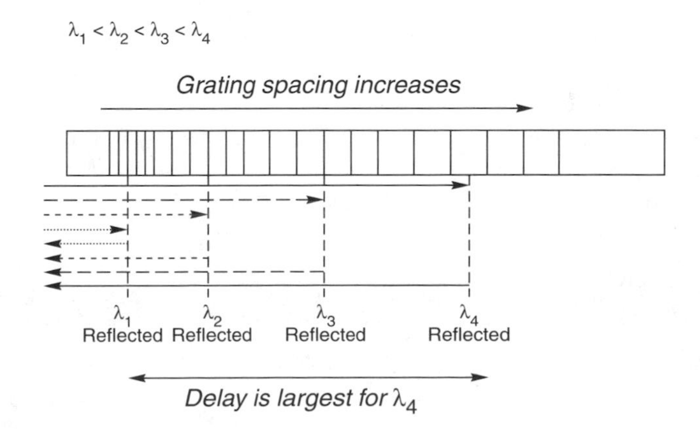

FIGURE 7.7
Fiber grating works as a delay line.

last part of the grating. The first part of the grating reflects the shortest wavelengths, which arrive last. The longer wavelengths have to travel a longer distance, so they are delayed, allowing the shorter wavelengths to catch up.

To offset 100 ps of dispersion, the distance between the two grating segments should be equivalent to 50 ps of travel time along the grating. Intermediate gratings would produce intermediate delays for other wavelengths. If the average refractive index was 1.5, this would require only about 10 mm of fiber grating. The reflected wavelengths would all return to the entrance of the grating at the same time, where an optical circulator would collect the dispersion-compensated signal.

This simple compensation scheme could correct dispersion across a single optical channel. Gratings to compensate for dispersion on other optical channels in a WDM system could be stacked at different points along the fiber grating, or the signal could be demultiplexed and the wavelengths could be routed to separate compensators.

Other Fiber-Grating Filters

Adjusting the spacing and strength of fiber gratings allows them to reflect a broader range of wavelengths, or reflect a controlled fraction of the light to reduce intensity at certain wavelengths. Some fiber gratings are designed specifically for sensing applications, either directly to serve as sensors, or to serve as components within sensing systems.

Important applications include:

- Mirrors that reflect a narrow range of wavelengths to stabilize laser emission at those wavelengths. Fiber gratings are a natural complement to lasers in which the active medium is an optical fiber, described in Chapter 9.
- Filters that reduce the intensity of light by a fraction across a range of wavelengths. This can offset uneven transmission or amplification of light by other components in fiber-optic systems. Examples are *gain-equalization filters,* which offset the uneven amplification of wavelengths by erbium-doped fiber amplifiers, described later in this chapter and in Chapter 12.

Other types of fiber Bragg gratings are in development.

Fiber gratings can be tuned by adjusting effective distance between high-index zones.

The wavelength selected by a fiber Bragg grating can be tuned by changing the refractive index and/or the grating spacing, the two factors that determine reflected wavelength. One possibility is to change the grating temperature, which directly affects the refractive index. Temperature change also causes thermal expansion or contraction of the glass, shrinking or stretching the grating period. Other mechanisms also are possible. Tunable fiber gratings remain in development, and are not in wide use.

Fiber cores are doped to make them act as lasers or optical amplifiers.

Doped Fibers for Amplifiers and Lasers

The cores of optical fibers can be doped with special materials to make the fibers act as optical amplifiers or lasers. You will learn more about optical amplifiers in Chapter 12, and more about lasers in Chapter 9. This chapter will teach you about the special fibers used in fiber amplifiers and lasers.

The Fiber Amplifier Concept

At its simplest, an optical fiber amplifier is merely a very thin, solid-state laser. Like all lasers, it contains a material that emits light. In an optical-fiber amplifier, that material is an element such as erbium that is mixed with the glass of the fiber core. Lasers are made to generate light on their own, as you'll learn in Chapter 9. Fiber amplifiers are made to amplify weak optical signals passing through the fiber.

The fiber amplifier gets its energy from a pump laser, which illuminates the light-emitting atoms in the fiber. The pump laser light transfers energy to the atoms, putting them in an excited state. An optical signal passing through the fiber can stimulate the excited atoms to emit their energy in the form of light that is in phase with the optical signal and at the same wavelength. This process amplifies a weak optical signal. Each photon produced as the signal is amplified can itself stimulate the emission of more photons. Figure 7.8 shows the process schematically.

A pump laser excites atoms in a fiber amplifier so it can amplify light.

The fiber's core-cladding structure concentrates the input signal and the pump light, increasing the intensity in a small area and improving the efficiency of stimulated emission and amplification. To concentrate pump and signal energy, fiber amplifiers designed for communications systems have single-mode cores of roughly the same size as single-mode communication fibers.

The wavelengths absorbed by and emitted by the light-emitting atoms are key parameters for operation of fiber amplifiers. These absorption and emission lines depend primarily on the type of light-emitting atom doped into the core, and secondarily on the composition of the glass in the fiber. That is, the type of atom selects the wavelength region, while the interaction between the atom and the glass molecules surrounding it essentially adjusts where the light falls in that wavelength range.

Another important parameter is gain, which measures the amplification per unit length of fiber. Gain depends on both the materials and the operating conditions, and varies with wavelength for all materials. For low input powers, the output power is proportional to the gain times the length:

$$P_{\text{output}} = P_{\text{input}} \times \text{Gain} \times \text{Length}$$

Gain is amplification per unit length at low powers. At high powers, gain may saturate.

At high input powers, the gain may *saturate,* so each increment of input power produces less and less output power, as shown in Figure 7.9. Essentially the optical amplifier has run out of the power it needs to generate more output.

If the pump laser is off, a fiber amplifier typically becomes an absorber. This is inherent in the physics of optical amplification. Atoms emit light at characteristic wavelengths, which depend on their internal structure, but they also can absorb light at those same wavelengths. They emit when the pump laser gives them extra energy to release, but if the pump laser is off, they tend to absorb the same wavelengths.

FIGURE 7.8
Signal amplification in an erbium-doped fiber amplifier.

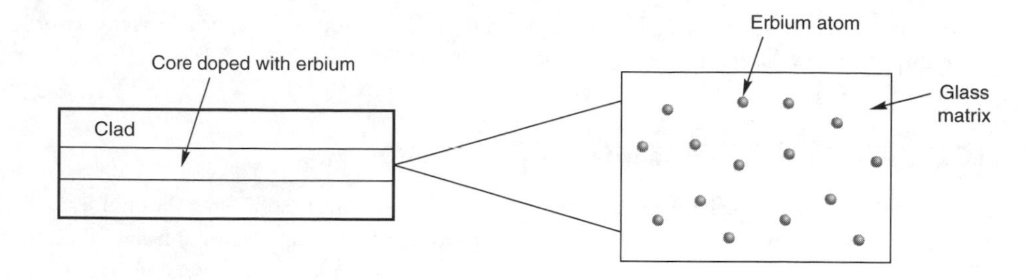

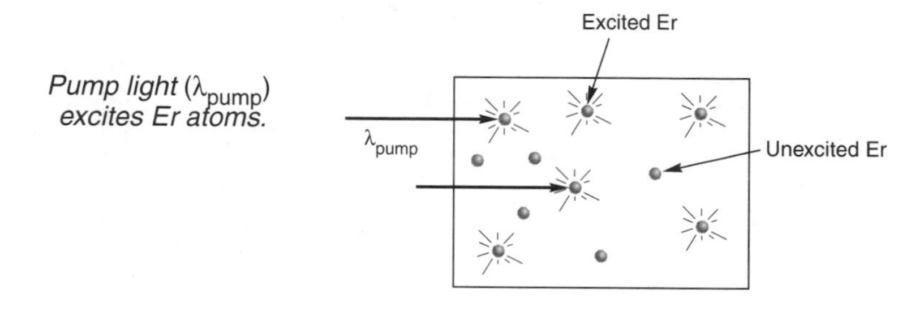

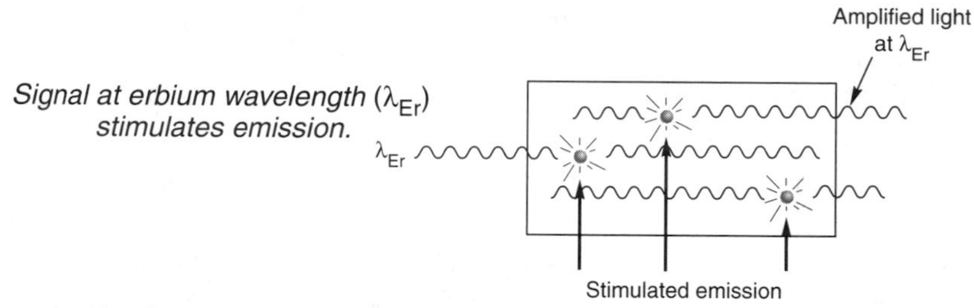

What communication system sees.

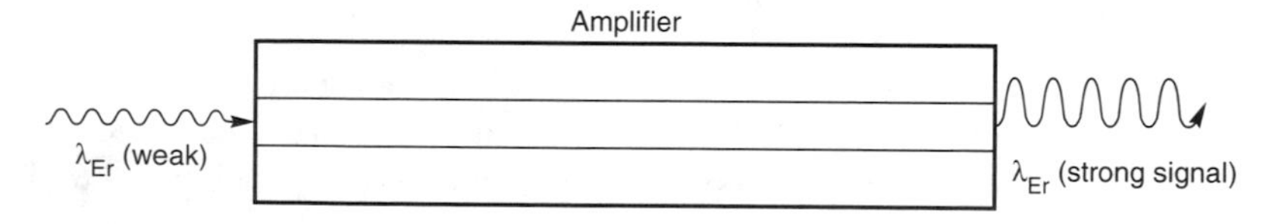

Most fiber amplifiers are doped with erbium, which amplifies wavelengths from 1530 to 1625 nm.

Erbium-Doped Fiber Amplifiers

The standard optical amplifier used in modern telecommunication systems is the *erbium-doped fiber amplifier,* often abbreviated EDFA. You'll learn more about its properties as an amplifier in Chapter 12, but here we will concentrate on its fiber characteristics.

Erbium has gained wide acceptance because it has useful gain at wavelengths between about 1530 and 1625 nm, a range where silica fibers have their lowest attenuation. It can

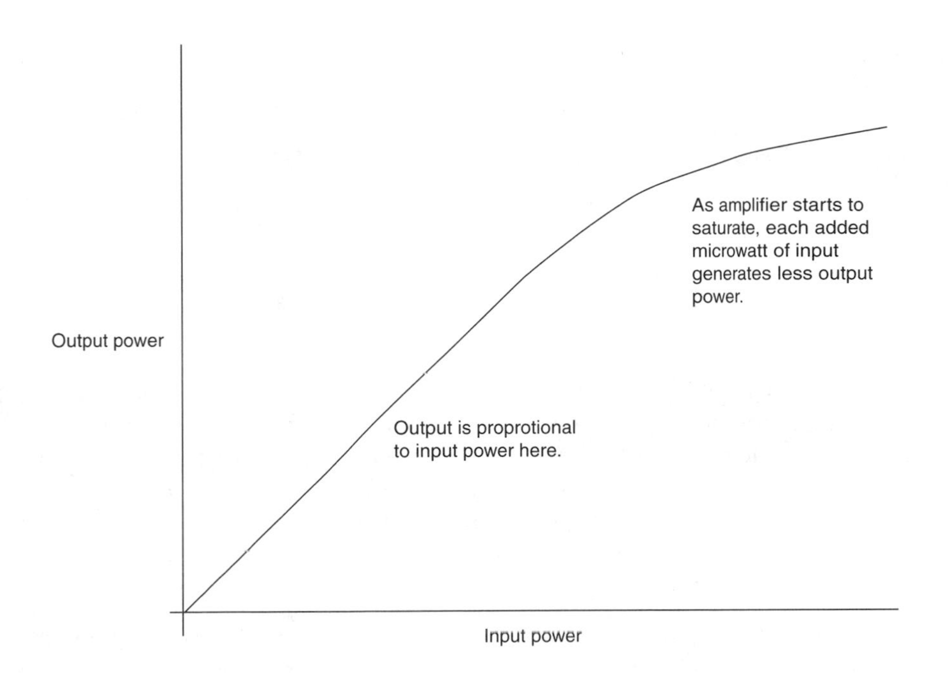

FIGURE 7.9
Saturation in a fiber amplifier.

be excited by wavelengths of 980 or 1480 nm, readily available from semiconductor lasers.

Small amounts of erbium are added to the core of amplifying fibers of various compositions, with concentration depending on the amplifier design. Normally the fiber is a typical fused-silica composition, but for some applications erbium may be doped into fibers made from fluoride compounds or other materials. Tellurite fibers—containing tellurium glasses—have been developed because they have higher gain at longer wavelengths.

Other Amplifying Fibers

Many other fiber amplifiers are designed on the same principles as erbium-doped fiber amplifiers. An element that emits light in the desired range is doped into the core of an optical fiber, and illuminated by light from a pump laser directed along the fiber. Praseodymium-doped fibers have been developed for use at 1310 nm, but have found few applications. Thulium-doped fibers are in development for use at 1470 to 1500 nm. Other types are being investigated.

Raman-fiber amplifiers work on a different principle, described briefly in Chapter 6. Nonlinear interactions between light waves and atoms in the fiber shift some energy from the wavelength of a strong pump beam to the longer wavelength of a weak optical signal.

Raman amplification does not require special dopants, and typically is done in single-mode communication fibers.

Dual-Core Fiber for High-Power Lasers

Dual-core lasers are used in high-power fiber lasers.

A different fiber structure is used to make fiber lasers that generate continuous beams delivering a watt or more of optical power. By laser standards these are high-power devices, particularly since their output is generated in a fiber core. These fibers are optimized to convert pump energy from a pump laser into energy from the fiber laser. (The same design also can be used for high-power amplifiers.)

A fiber laser resembles a fiber amplifier, but with mirrors on both ends of the fiber and no input signal to be amplified. The fiber laser is an *oscillator,* which generates its own signal with no outside input other than energy from the pump laser. Some excited atoms in the fiber release their light energy spontaneously, and it is this energy that is amplified as it passes along the fiber. The mirrors reflect the light back and forth through the fiber, with one of them transmitting a small fraction of the light each time the light bounces off it. (You'll learn more about lasers in Chapter 9.)

To optimize power transfer, many fiber lasers use special fibers with dual inner and outer cores, as shown in Figure 7.10. The inner core has the highest refractive index, the outer core has an intermediate refractive index, and the surrounding cladding has the lowest index. The pump light is directed into the outer core, which has a large and asymmetric

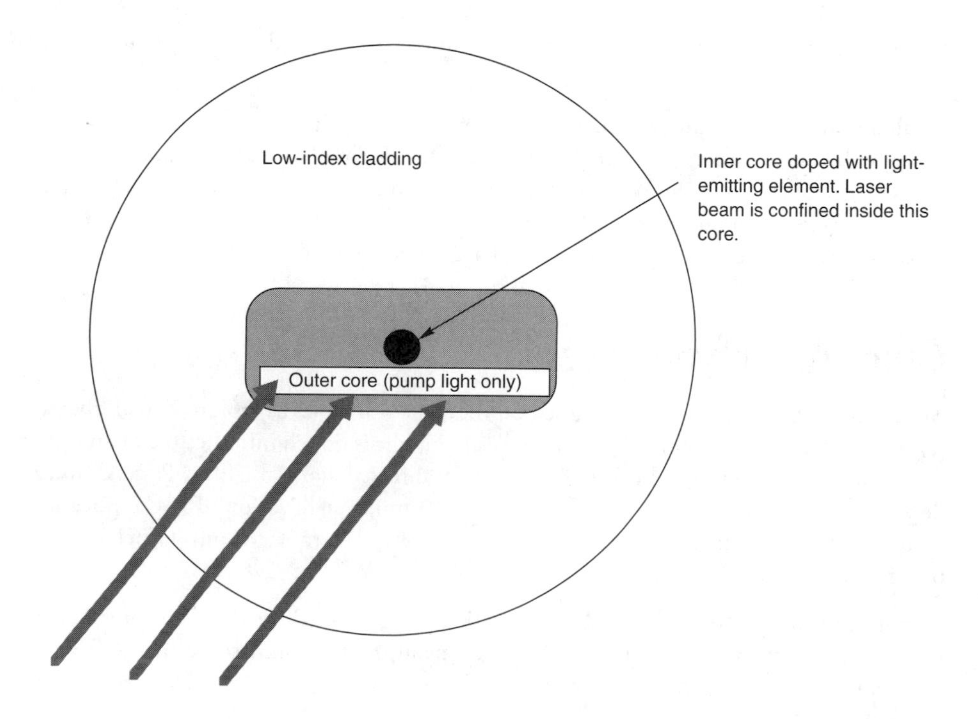

FIGURE 7.10 *Dual-core fiber for high-power amplifiers and fiber lasers. Outer core collects pump light; the inner core is doped with the light-emitting element. The outer core can take various shapes; this one is simple to draw.*

cross-section. The large size allows it to collect pump light efficiently from powerful pump lasers that emit from large apertures. The refractive index differential is large enough to confine pump light in the outer core by total internal reflection, but the asymmetric shape makes it follow an irregular path through the fiber. Ideally, the pump light follows a path that takes it through the inner core repeatedly.

Only the inner core is doped with the light-emitting element excited by the pump laser. The refractive-index difference between the inner and outer cores confines stimulated emission to the inner core, where the high density of excited atoms and stimulated emission produces high gain and high power.

In erbium-doped fiber lasers, the light-emitting element is erbium. However, so far the highest powers—exceeding 100 watts in a continuous beam—have been produced by fibers with a single-mode inner core doped with ytterbium, which emits around 1120 nm.

Side-Glowing Decorative Fibers

Normal optical fibers guide light from end to end. The glass or plastic scatters a tiny amount out the sides of the fiber, but that light is undetectable. Essentially all the light emerging from the fiber exits at their ends. This creates glittering points of light at the ends of fibers in displays, but produces no light at all along their sides.

Fibers containing scattering material glow from their sides like neon tubes.

Special types of fibers can be made to scatter light out their sides for decorative lighting. The trick is to add light-scattering materials to the core of a fiber, as shown in Figure 7.11. Light passing through the core bounces off fine grains of the material, making the side appear to glow. The fiber, the scattering particles, or the illuminating light may be colored; you could even add fluorescent materials. The scattered light makes the fiber look like a glowing neon tube.

Typically these side-glowing fibers are thick plastic or liquid-core fibers, often a few millimeters thick. Some may have diameters of more than a centimeter, which would be better called flexible rods rather than fibers. However, they retain the core-cladding structure of optical fibers, which they use to guide light along the scattering regions. Because so much light leaks out the side, their attenuations are quite high.

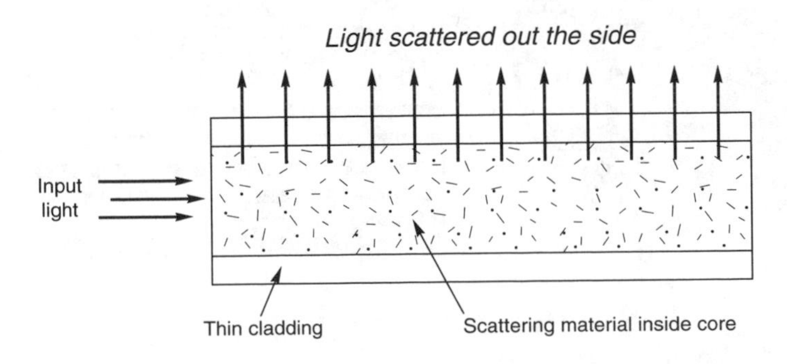

FIGURE 7.11
Side-glowing fiber contains material in the core that scatters light out the side, so the fiber glows along its length like a neon tube.

Photonic Bandgap or "Holey" Fibers

A photonic bandgap material excludes light at certain wavelengths; it can be shaped to create a waveguide.

Chapter 6 mentioned photonic bandgaps or structured materials in which the structure of the material blocks light from passing through the material at certain wavelengths. When these materials are shaped into fibers, they can produce a waveguide effect. Several groups are developing these materials, and predictably the competing teams have different names for concepts that are fundamentally similar. Some examples include *photonic bandgap fibers, photonic crystal fibers, "holey" fibers,* and *omniguides.*

Figure 7.12 shows one example, a fiber drawn from an array of rods with spaces left between them so the cladding contains an array of regularly spaced holes. These regularly spaced holes create a photonic bandgap, a zone in which light cannot travel at certain wavelengths. The details are complex, but essentially the holes combine to reflect light at wavelengths selected by the refractive index and the hole spacing, like a fiber grating. Just as light at the reflected wavelength can't pass through a fiber grating, light at the selected wavelength can't pass through the photonic bandgap.

A layer of this "holey" material can effectively trap light of the wavelength it selects. If the holey material forms the cladding of an optical fiber, it can trap light in the core. In Figure 7.12, a cladding of holey photonic bandgap material traps light in the core, which is solid glass lacking holes. Unlike conventional fibers, this guide transmits only a single mode regardless of the wavelength, and its core diameter can be larger than in a normal single-mode fiber.

Other variations are possible. A hollow core can be surrounded by a photonic bandgap layer that confines the light to travel through air or free space inside the fiber, which is impossible without a photonic bandgap. A fiber-optic version of a coaxial cable has been designed, with a photonic-bandgap wire in the middle surrounded by a transmissive layer and an outer photonic bandgap.

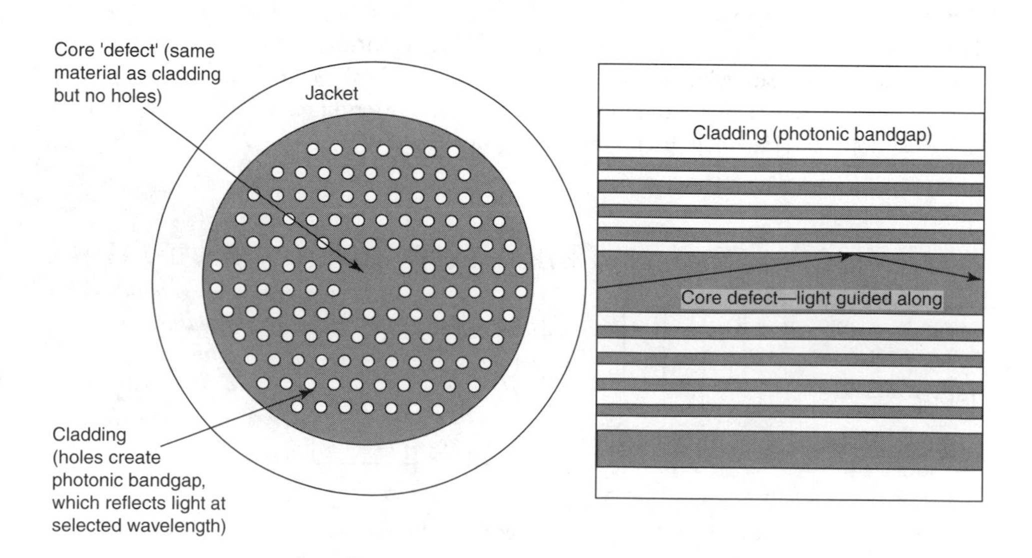

FIGURE 7.12
A photonic bandgap fiber, in cross-section. The central core zone lacks the holes present in the cladding, so it guides light reflected by the photonic bandgap cladding.

At this writing, these devices remain in the laboratory. Their likely applications remain unclear. Initially, they may be used in situations that require special optical properties such as control of polarization modes. The emergence of optical networking may open new opportunities.

Graded-Index Fiber Lenses

In certain cases, short lengths of graded-index fibers can act like tiny lenses to focus light. These fiber-optic microlenses are rarely used in fiber-optic systems, but they have found applications in some systems that manipulate images point by point, such as photocopiers, scanners, and facsimile machines. Note that they are different from the other types of fiber-optic imaging bundles described in Chapter 30.

Graded-index fibers can focus light and act as lenses.

In Chapter 4, you saw that light follows a sinusoidal path through graded-index fiber. When you looked at a cone of light entering a long graded-index fiber, you saw light leaving from the fiber and spreading at the same angle as when it entered. Now look instead at the path of an individual ray through a short segment of graded-index fiber, shown in Figure 7.13, and compare that with the path of a light ray in a step-index multimode fiber.

Total internal reflection from the step-index boundary keeps light rays at the same angle to a step-index fiber axis all along the fiber. The output angle equals the input angle. Graded-index fibers refract light rays, so the angle between the ray and the fiber axis changes constantly as the light ray bends back and forth following a sinusoidal path. If you cut a graded-index fiber after the light ray has gone through 180° or 360° of the sinusoidal curve, the light emerges at the same angle that it entered. If you cut the fiber at some other point, the light can emerge at a different angle, as shown in Figure 7.13. Making light rays emerge at a different angle is equivalent to focusing them with a lens, so a segment of graded-index fiber can serve as a lens and focus light that enters its core.

The key parameter of graded-index fiber lenses (usually sold under the trade name Selfoc) is the fraction of a full sinusoidal cycle that light goes through before leaving the fiber. That fraction is called the *pitch*. A 0.23-pitch lens, for instance, has gone through 0.23 of a cycle, or 82.8°. The value of the pitch depends on factors including the refractive-index gradient

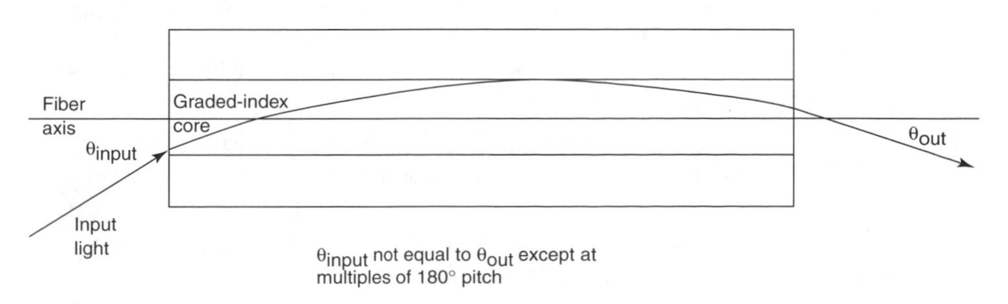

FIGURE 7.13
Graded-index fiber lens.

in the fiber, the fiber refractive index, its core diameter, and the wavelength. Typical graded-index fiber lenses are a few millimeters long—very short by fiber standards.

Sensing Fibers

Normal optical fibers are designed to be insensitive to their environment, but sensing fibers must change their optical properties in response to their surroundings. Chapter 29 covers the special fibers used in sensing applications.

What Have You Learned?

1. Fiber gratings are periodic variations in refractive index along the length of a fiber, produced by illuminating the fiber with ultraviolet light.
2. A uniformly spaced fiber grating selectively reflects light at a narrow range of wavelengths selected by the spacing of the grating. It transmits other wavelengths.
3. One important feature of a fiber grating is that it comes in the form of a fiber, making it easy to transfer light between it and another fiber.
4. A fiber grating can select one wavelength from among many transmitted by an optical fiber, reflecting the selected wavelength and transmitting the rest. This can be used as an add-drop multiplexer for wavelength-division multiplexing, together with a directional component called an optical circulator.
5. The wavelength reflected by fiber gratings can be tuned by adjusting effective distance between high-index zones.
6. Fibers with their cores doped with a light-emitting element can be used in optical fiber amplifiers and fiber lasers. Most optical fiber amplifiers are based on fibers doped with erbium, a rare earth element that amplifies light at 1530 to 1625 nm.
7. Light-emitting atoms in an optical-fiber amplifier are excited by a pump laser. The gain of the fiber is its amplification per unit length at low power; gain may saturate at high power.
8. Fiber amplifiers boost the strength of a weak input signal.
9. Fiber lasers have mirrors on each end and generate their own beam with no external signal input. The highest-power fiber lasers have a dual core in which the pump laser directs light into a large outer core, and the beam is generated in an inner single-mode core doped with the light-emitting element. The highest powers come from ytterbium-doped fibers.
10. Praseodymium- and thulium-doped fibers also are being investigated for fiber amplifiers. Praseodymium can amplify signals near 1300 nm, and thulium can amplify at 1470 to 1500 nm.
11. Fiber cores can be doped with scattering material so they glow from their sides like neon tubes; such fibers are used as decorations.

12. Photonic bandgap or "holey" fibers surround the fiber core with a photonic bandgap material that excludes light at certain wavelengths. The photonic bandgap material guides light at the reflected wavelengths.
13. Short lengths of graded-index fiber can focus light and act as lenses.

What's Next?

In Chapter 8, we move on to learn about the cables that contain optical fibers in communication systems.

Further Reading

P. C. Becker, et al. *Erbium Fiber Amplifiers: Fundamentals and Technology* (Academic Press, 1999)

University of Bath, "Photonic Crystal Fibre," *http://www.bath.ac.uk/Departments/Physics/groups/opto/pcf.html*

K. O. Hill and G. Meltz, "Fiber Bragg grating technology fundamentals and overview," *Journal of Lightwave Technology,* Vol. 15, pp. 1263–1276, August 1997

Raman Kashyap, *Fiber Bragg Gratings* (Academic Press, 1999)

Gerd Keiser, *Optical Fiber Communications* (McGraw Hill, 2000), Chapter 11 on optical amplifiers

Questions to Think About for Chapter 7

1. A fiber grating reflects light waves that are in phase with the grating, with a wavelength $\lambda = 2nD$ where D is the spacing between the gratings. This wavelength is the longest that would make exactly one round-trip between a pair of grating lines. A wavelength exactly half that value also would be resonant with that grating spacing, because exactly two waves would make one round-trip. Why isn't that light also reflected?
2. Why is refractive index always measured in air rather than in glass?
3. Optical circulators are complex and expensive. Why bother using one for optical demultiplexing when a fiber grating can reflect the desired wavelength?
4. What is the advantage of doping erbium only in the fiber core?
5. The gain of an erbium-doped fiber varies as a function of wavelength. It is highest from about 1530 to 1565 nm, and lower at longer wavelengths. Recalling how gain is defined, what is one way to make an erbium-doped fiber amplifier with a high amplification factor at longer wavelengths?
6. Think of a way to demonstrate the effect seen in light scattering from the sides of a fiber designed for decorative lighting. Start with a laser pointer.

7. An erbium-doped fiber amplifier is operating at a power high enough that its output has saturated at 10 mW/channel for 20 optical channels with 100-GHz spacing. You add 20 more channels at intermediate wavelengths. Assuming the amplifier is so saturated it can't generate any more total output, what is the new output per channel?

Quiz for Chapter 7

1. What creates the grating effect in a fiber grating?
   a. Lines etched on the fiber surface by high-power ultraviolet pulses
   b. Changes in the refractive index of the fiber core induced by ultraviolet light
   c. Interference among several modes in a multimode fiber
   d. Variations in glass composition caused by changes in doping during preform fabrication
   e. Optical white magic

2. A grating with period of 0.5 μm is made in a glass fiber with core refractive index of 1.5. What is the wavelength of light reflected most strongly?
   a. 1300 nm
   b. 1500 nm
   c. 1550 nm
   d. 1600 nm
   e. none of the above

3. Which wavelengths are transmitted by the grating in Problem 2?
   a. None; all light is reflected or scattered from the fiber
   b. 1300 and 1600 nm only
   c. 1300, 1500, and 1600 nm
   d. 1300, 1550, and 1600 nm
   e. 1300, 1500, and 1550 nm

4. You want to make an add-drop multiplexer that drops optical channels at 1541 and 1543 nm, while transmitting six other channels at 1540, 1542, 1544, 1545, 1546, and 1547 nm. What type of fiber gratings do you need?
   a. One reflecting 1541 nm and one reflecting 1543 nm
   b. One reflecting 1543 nm and one transmitting 1541 nm
   c. One reflecting 1541 nm and one transmitting 1543 nm
   d. Separate gratings reflecting 1540, 1542, 1544, 1545, 1546, and 1547 nm
   e. One reflecting 1540, 1542, and 1544 nm, and one reflecting 1545, 1546, and 1547 nm

5. You are using a fiber grating in a WDM system with channel spacing of 100 GHz (0.8 nm). Which of the following features are required?
   a. The grating's reflection bandwidth should be narrower than 100 GHz.
   b. The grating's transmission bandwidth should be narrower than 100 GHz.
   c. The grating's reflection bandwidth should be wider than 100 GHz.

d. The grating should transmit a 100-GHz bandwidth centered on the optical channel selected.
e. None of the above

6. What determines the wavelengths that can be amplified in a fiber amplifier?

a. The number of modes guided in the fiber
b. Attenuation of the fiber
c. Properties of the light-emitting ion doped into the core
d. The fiber core diameter
e. Dispersion of the fiber

7. What wavelengths are amplified by an erbium-doped fiber amplifier?

a. 980–1480 nm
b. 1250–1350 nm
c. 1300–1700 nm
d. 1470–1530 nm
e. 1530–1625 nm

8. Which of the following signals can an erbium-doped fiber amplifier amplify?

a. 2.5 Gbit/s signal at 1551 nm
b. 100 Mbit/s signal at 1540 nm
c. 622 Mbit/s signal at 1540 nm
d. 10 Gbit/s signal at 1546 nm
e. All of the above

9. Amplification in an erbium-doped fiber amplifier depends on which of the following?

a. Length of the doped fiber
b. Erbium concentration in the core
c. Pump power
d. All of the above
e. None of the above

10. How does a fiber laser differ from a fiber amplifier?

a. A fiber laser has mirrors on the ends and no input signal.
b. Only a fiber amplifier requires a pump laser.
c. A fiber amplifier generates much more power.
d. A fiber laser amplifies an external signal.
e. They are the same thing sold under different brand names.

CHAPTER

8

Cabling

About This Chapter

Cabling is not glamorous, but it is a necessity for virtually all communication uses of fiber optics. A cable structure protects optical fibers from mechanical damage and environmental degradation, eases handling of the small fibers, and isolates them from mechanical stresses that could occur in installation or operation. The cable makes the critical difference in determining whether optical fibers can transmit signals under the ocean or just within the confines of an environmentally controlled office building.

This chapter discusses the major types of fiber-optic cable you are likely to encounter. You will see what cables do, where and why different types are installed, what cables look like on the inside, how cables are installed, and what happens to fibers in cables.

Cabling Basics

Fiber-optic cables resemble conventional metal cables externally, and they use some materials and jacketing technology borrowed from copper wire cables. Polyvinyl chloride (PVC) sheaths are common on both fiber-optic cables and coaxial cables used inside buildings, but fiber cables are often brightly colored, whereas coax usually has a black jacket. Polyethylene (PE) is used for both metal and fiber outdoor cables to protect against the environmental rigors of underground burial or aerial installation.

Some important differences can be subtle. Because optical fibers are not conductive, they do not require electrical insulation to isolate circuits from each other. Optical cables can be made nonconductive by avoiding use of metals in their construction, which produces all-dielectric cables that are immune to ground-loop problems and resistant to lightning strikes. Fiber-optic cables tend to be smaller because one fiber has the same capacity as many wire pairs and because fibers themselves are small.

Fiber-optic cables resemble metal-wire cables but differ because signals are transmitted as light, not electricity.

Some major differences in cable design are necessary because glass fibers react differently to tension than copper wires. Pull gently on a fiber and it will stretch slightly and then spring back to its original length. Pull a fiber hard enough and it will break (starting at a weak point or surface flaw). Pull a copper wire, applying less stress than you did to break the fiber, and it will stretch by more than 20% and not spring back to its original length. In mechanical engineering terminology, fiber is elastic (because it contracts back to its original length), and copper is inelastic (because it stays stretched out).

Fibers must be isolated from tension, which can cause breakage or long-term reliability problems.

As you learned in Chapter 5, glass fibers are strong. Manufacturers proof-test fibers under stress of 100,000 pounds per square inch, or 0.7 giganewton per square centimeter. (The usual units are thousands of pounds per square inch, kpsi, one of the few cases where Imperial units are widely used in fiber optics.) The test normally is performed as the plastic-coated fiber is wound onto the shipping reel, so weaker fibers should not make it out of the plant. Cable manufacturers may perform their own proof-test before cabling the fiber to make sure it retains its original strength.

Although the strength per unit area is very high for glass fibers, you should remember that a strong tug applies a large force per unit area across the small diameter of a fiber. What counts in assessing strength is the cross-section of the glass, not the thicker diameter of the plastic-coated fiber. A standard 125-μm fiber has a cross-sectional area of only 0.000019 square inch (0.00012 square centimeter), so a 2-lb (1-kg) force applied along the fiber corresponds to the 100 kpsi stress test.

Fibers are plastic coated as well as proof-tested before cabling; the coating protects the fiber surface from handling damage and environmental moisture. The cable structure isolates the fibers from tension, both during installation and in service, such as when they are hanging from outdoor poles. Instead the cable structures apply tension to strength members that run the length of the cable, either through the center or in another layer. As described later, these strength members may be metallic or nonmetallic.

Reasons for Cabling

Cabling is the packaging of optical fibers for easier handling and protection. Uncabled fibers work fine in the laboratory and in certain applications such as sensors and a fiber-optic system for guiding missiles, which I will describe later. However, like wires, fibers must be cabled for most communications uses.

Ease of Handling

Cables make fibers easier to handle.

One reason for cabling fibers is to make them easier to handle. Physically, single glass optical fibers resemble monofilament fishing line, except the fibers are stiffer. Protective plastic coatings raise the outer diameter of communication fibers to 250–900 μm, but they are still so small that they are hard to handle. Fibers without an outer color-coding layer are transparent enough to be hard to see on many surfaces. Try to pick up one loose fiber with your fingers and you'll soon appreciate one virtue of cable.

Cabling also makes multiple fibers easy to handle. Most communication systems require at least two fibers, one carrying signals in each direction. Some require many fibers, and some cables used on high-traffic routes contain hundreds of fibers. Cabling puts the fibers in a single easy-to-see and easy-to-handle structure.

Cables also serve as mounting points for connectors and other equipment used to interconnect fibers. If you take that function too much for granted, try butting two bare fibers together with your hands and finding some way to hold them together permanently.

Protection from Damaging Forces

Another major goal of cabling is to prevent physical damage to the fiber during installation and during use by forces applied intentionally or unintentionally. The two major concerns are stress or tension applied along the length of the cable, and crushing forces applied across the cable's diameter. The ability of cables to withstand these forces varies widely with cable design.

Cables prevent physical damage to fibers during installation and use.

The most severe stresses intentionally applied along the lengths of cables come during installation. Many cables are pulled into place through outdoor underground ducts or through conduits within buildings. Pulling gear is attached directly to strength members or cable jackets on one end, isolating the fibers from the applied forces. Then enough force is applied slowly to pull the entire cable into place.

Cables encounter much less static stress once they are installed, although aerial cables hanging from supports must be able to support their own weight. In cold environments, the cables must also be able to support snow and ice adhering to them.

Static fatigue is a significant issue because glass fibers age very quickly if stretched much. This makes it important to isolate fibers from mechanical loads, so elongation during manufacture and installation is no more than 0.1% to 0.2%.

Cables also can experience short-duration dynamic forces along their lengths. Most of these are unintentional, such as tree limbs falling on aerial cables. Cables can provide reasonable protection against light branches or overweight crows landing on them, but they can be broken by severe shocks. Falling trees and telephone poles snap aerial cables; careless backhoe operators dig up and break underground cables. Even cables inside buildings are vulnerable to damage if someone yanks or trips on them.

Crush resistance is another important cable specification, measuring how well they can withstand force applied from the sides. Requirements differ widely. Ordinary intrabuilding cables are not made to be walked on, but a few have been made for installation under carpets. Deep-sea submarine cables must withstand the pressure of several kilometers of seawater above them. Cable structures that leave no internal cavities, such as embedding the fibers in thick polyethylene, allow the cables to withstand such high static pressures.

Crush resistance is how well cables withstand crushing force applied from the side.

Cables are armored to withstand unusual stresses. For example, the portions of submarine cables near shore are armored to protect them against damage from fishing trawlers and boat anchors. Buried cables must withstand a different type of crushing force applied in a small area: the teeth of gophers, who gnaw anything they can get their teeth around. The front teeth of gophers and other rodents grow continually, so they instinctively gnaw on objects they find underground. This is one case where the small size of fiber cables is undesirable, because it makes them just bite-sized for gophers. To prevent such damage, cables buried in areas where burrowing rodents live typically are sheathed in steel armor and built to larger sizes than gophers like to munch.

Cables are made stiff to keep fibers from being bent too tightly. This practice also helps prevent fibers from developing tiny microcracks, caused by surface nicks, which can lead to fiber breakage.

Figure 8.1 gives a sampling of cable cross sections and shows their applications. The light-duty office cable at the top looks like electric zip cord for a lamp; the indoor-outdoor cable resembles indoor coaxial cable; and the armored cable is lighter than armored electrical cable. The deep-sea cable is about an inch thick and feels as hefty as a policeman's night-stick, but in shallow water it requires extra layers of heavy steel armor to prevent damage from shipping and fishing operations.

Environmental Protection

Cabling helps protect fibers against degradation caused by moisture.

Cabling also protects fibers from more gradual degradation caused by the environment. Long-term exposure to moisture can degrade fiber strength and optical properties. Most cables designed for use in uncontrolled (i.e., outdoor or underground) environments include barriers to keep moisture out. Aerial cables must withstand extremes of temperature—from heating to high temperatures on a hot, sunny day in the summer to freezing in the winter. The combination of cold and moisture presents an added danger—freezing of moisture in the cable. Because water expands when it freezes, it can apply forces on the fiber that produce microbends, increase losses, and even cause microcracks. Cables are designed to prevent the types of degradation important in the environments where they are used; for example, water-blocking materials are used to prevent water from entering loose-tube cables.

A significant long-term concern for fibers carrying signals at 1300 to 1650 nm is the possible influx of hydrogen into the fiber. As you learned in Chapter 5, hydrogen causes a large absorption peak centered on 1380 nm. This poses two potential problems.

Hydrogen absorption builds up with hydrogen level in the cable.

The strength of the absorption depends on hydrogen content. At high hydrogen levels, weak absorption at the fringes of the peak can build up to significant levels across a broad range of wavelengths. This can impair transmission throughout the long-wavelength band. Hydrogen is rare in the open air, but it can accumulate in cable structures where electric currents cause the decomposition of water molecules or some plastics. At high hydrogen pressures, the gas can diffuse into the glass, causing absorption, so modern cables avoid structures that might produce hydrogen.

Even modest levels of hydrogen can impair transmission near 1380 nm. That band has not been used in the past, but the development of extremely low-water fibers opens that window for use over short distances where optical amplifiers are not needed. Early tests have showed no major problems, but the issue remains open.

In all other ways, fiber has shown no sign of serious environmental degradation if properly cabled. Nearly a quarter-century has passed since the first fiber-optic systems were installed.

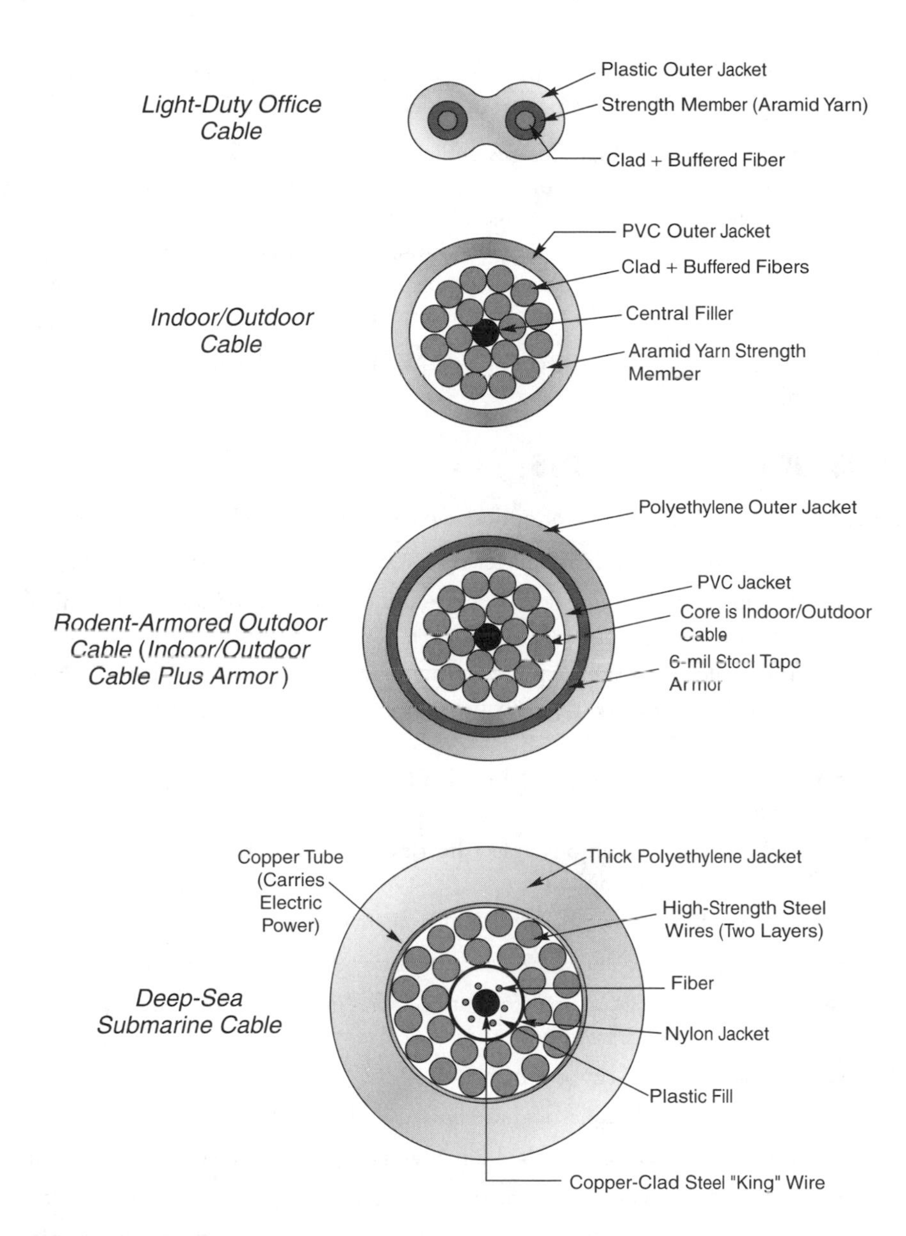

FIGURE 8.1
Cross sections of four grades of cable, from light-duty indoor to deep-sea submarine cable (not to scale).

Fiber and cabling technologies have improved greatly since then, but careful examination of those early systems has not shown serious fiber degradation.

Types of Cable

The same optical fiber may be used in many different environments, but this is not so for cable. Cables are designed to provide a controlled environment for the fibers they contain under particular conditions. They must meet fire and electrical safety codes, which ban

Cables are designed for particular environments.

indoor installation of materials that produce toxins or catch fire easily. Thus, choice of a cable design depends on the environment in which it is to be installed.

Cable manufacturers use modular designs, which they adapt and assemble into cables for particular needs. They have families of cables, which can include all the major types of fiber used for communications. They often assemble cables from subunits, which can contain various numbers of fibers. They may adapt an indoor-outdoor cable for direct burial in the ground by adding metal armor and a waterproof plastic jacket to a product normally used indoors. In short, they build to fit customer requirements.

A variety of factors enter into their choices. We'll start by looking at environmental considerations.

Types of Environments

The major types of environments for optical cable can be loosely classified as follows:

- Inside devices (e.g., inside a telephone switching system or computer).
- Intraoffice or horizontal (e.g., across a room, usually to individual terminals or work groups).
- Intrabuilding or riser (e.g., up wiring risers or along elevator shafts between floors in a structure; typically between distribution nodes on each floor that serve multiple users).
- Plenum installations (i.e., through air spaces in a building; must meet special codes).
- Interbuilding or campus links (short exterior connections; link distribution nodes in separate buildings).
- Chemical-resistant cables for industrial environments.
- Temporary light-duty cables (e.g., remote news gathering at sports events).
- Aerial cables (e.g., strung from utility poles outdoors). May be supported by lashing to support wires or other cables.
- All-dielectric self-supporting cables.
- Cables installed in plastic ducts buried underground.
- Direct-burial cables (i.e., laid directly in a trench or plowed into the ground).
- Submarine cables (i.e., submerged in ocean water or sometimes fresh water).
- Instrumentation cables, which may have to meet special requirements (e.g., withstand high temperatures, corrosive vapors, or nuclear radiation).
- Composite cables, which include fibers and copper wires that carry signals (used in buildings). Note the differences from hybrid cables, as follows.
- Hybrid power-fiber cables, which carry electric power (or serve as the ground wire for an electric power system) as well as optical signals.

These categories are not exhaustive or exclusive, and some are deliberately broad and vague. Instrumentation, for example, covers cables used to log data collected while drilling to explore for oil or other minerals. Special cables are needed to withstand the high temperatures and severe physical stresses experienced within deep wells. There is some overlap among categories; composite cables, for example, may also be classed as intraoffice cables.

Cable Design Considerations

Cables are designed to meet the special requirements of particular environments. A variety of considerations go into cable design, starting with the physical environment and the services being provided. They lead to a wide variety of cable types on the market. The most important considerations are summarized next.

Cables used inside buildings must meet fire and electrical codes.

- Intradevice cables should be small, simple, and low in cost, because the device containing them protects the cables.
- Intraoffice and intrabuilding cables must meet the appropriate fire and electrical codes. The National Electric Code (issued by the National Fire Protection Association) covers fiber cables that contain only fibers and cables that contain both fibers and copper wires. The primary concern is fire safety, because many cable materials are flammable, and some release toxic gases when they burn. Table 8.1 lists cable types and fire-safety tests specified by Underwriters Laboratories. Outdoor cables that do not meet these requirements can run no more than 50 ft (15 m) within a building before terminating in a cable box or being spliced to an approved indoor cable. Indoor/outdoor cables are available that meet indoor requirements and can withstand outdoor conditions, although they are not as rugged as heavy-duty outdoor cables.

Table 8.1 Cable specifications under U.S. National Electric Code

Cable Type	*Description*	*Designation*	*UL Test*
General-purpose (horizontal)—fiber only	Nonconductive optical fiber cable	OFN	Tray/1581
General-purpose (horizontal)—hybrid (fiber/wire)	Conductive optical fiber cable	OFC	Tray/1581
Riser/backbone—fiber only	Nonconductive riser	OFNR	Riser/1666
Riser/backbone—hybrid	Conductive riser	OFCR	Riser/1666
Plenum/overhead—fiber only	Nonconductive plenum	OFNP	Plenum/910
Plenum/overhead—hybrid	Conductive plenum	OFCP	Plenum/910

- Plenum cables are special intrabuilding cables made for use within air-handling spaces, including the spaces above suspended ceilings, as well as heating and ventilation ducts. They are made of materials that retard the spread of flame, produce little smoke, and protect electronic equipment from damage in fires, called "little smoke, no halogen" (LSNH) materials. (Halogens produce toxins.) Cables meeting the UL 910 specification can be run through air spaces without special conduits. The special materials are expensive and are less flexible and less abrasion-resistant than other cable materials, but installation savings and added safety offset the extra cost.
- Fiber count depends on the number of terminals served. Individual terminals may be served by a two-fiber duplex cable that looks like the zip cord used for electric lamps. Other types are round or oval in cross section and the largest contain over a thousand fibers. Multifiber cables often terminate at patch panels or communications "closets" where they connect to cables serving individual terminals.

Breakout cables are intrabuilding cables with fibers packaged into subcables.

- Breakout or fanout cables are intrabuilding cables in which the fibers are packaged as single- or multifiber subcables. This allows users to divide the cable to serve users with individual fibers, without the need for patch panels. Figure 8.2 shows an example.
- Composite or hybrid cables include both fibers and copper wires to deliver different communication services or communication and power to the same point. For example, the fiber may connect a workstation to a local area network while the wires carry voice telephone service to the same user. Figure 8.3 shows an example.
- Temporary light-duty cables are portable and rugged enough to withstand reasonable wear and tear. They may contain only a single fiber (e.g., to carry a

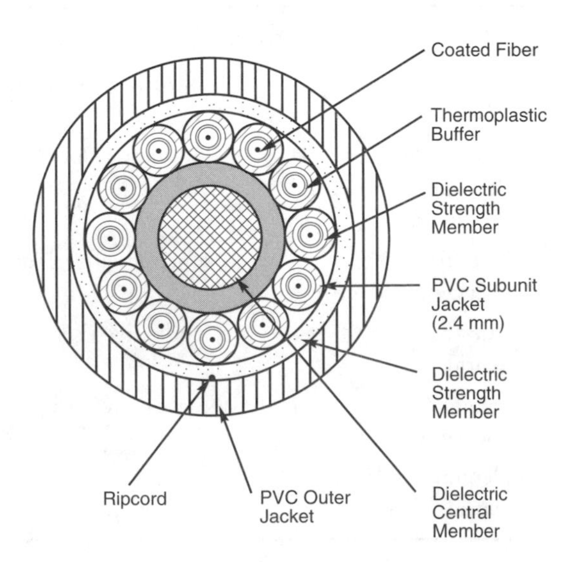

FIGURE 8.2
Breakout cable. (Courtesy of Corning Cable Systems)

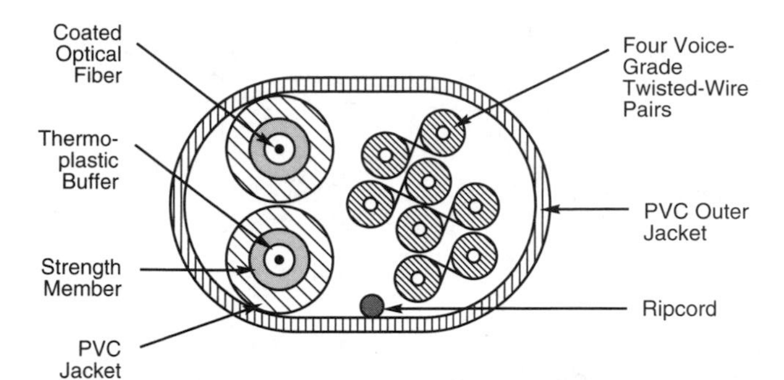

FIGURE 8.3
Composite cable contains both copper wires and fibers. (Courtesy of Corning Cable Systems)

video feed from a camera) and should be durable enough to be laid and reused a few times.

- Outdoor cables are designed to survive harsh outdoor conditions. Most are strung from overhead poles, buried directly in the ground, or pulled through underground tubes called ducts, but a few in protected areas are exposed to surface conditions. Most have polyethylene jackets, which keep out moisture and withstand temperature extremes and intense sunlight. However, polyethylene does not meet indoor fire codes, so such cables can run only short distances indoors, typically to equipment bays from which indoor cables fan out.

Outdoor cables can withstand harsher environments than intrabuilding cables but do not meet the same fire and building codes.

- Aerial cables are made to be strung from poles outdoors and typically also can be installed in underground ducts. They normally contain multiple fibers, with internal stress members of steel or synthetic yarn that protect the fibers from stress. Figure 8.4 shows two types of aerial installations, which normally use different types of cables. One suspends the cable between adjacent poles, supported by internal strength members or by a strength member packaged parallel to the fiber unit, which hangs below in what is called a "figure-8" cable. The other approach is to run a strong "messenger wire" between poles and lash the fiber cable to it by winding a supporting filament around both. Lashing supports the fiber cable at more frequent intervals and reduces the stress applied along its length, which can be large if the only supports are at the poles. Many aerial fiber cables are designed only for lashing, not to withstand the high stress of suspension between poles.
- All-dielectric cables contain no metal elements, either to conduct electricity or to serve as strength members. These cables use nonconductive strength members, such as aramid yarns or Kevlar. The all-dielectric construction prevents lightning surges and ground-loop problems, so they are widely used outdoors, particularly in lightning-prone areas.
- Armored cables are similar to outdoor cables but include an outer armor layer for mechanical protection and to prevent rodent damage. Steel or all-dielectric central members may be used. They can be installed in ducts or aerially, or directly buried underground (which requires extra protection

FIGURE 8.4
Aerial cable installations.

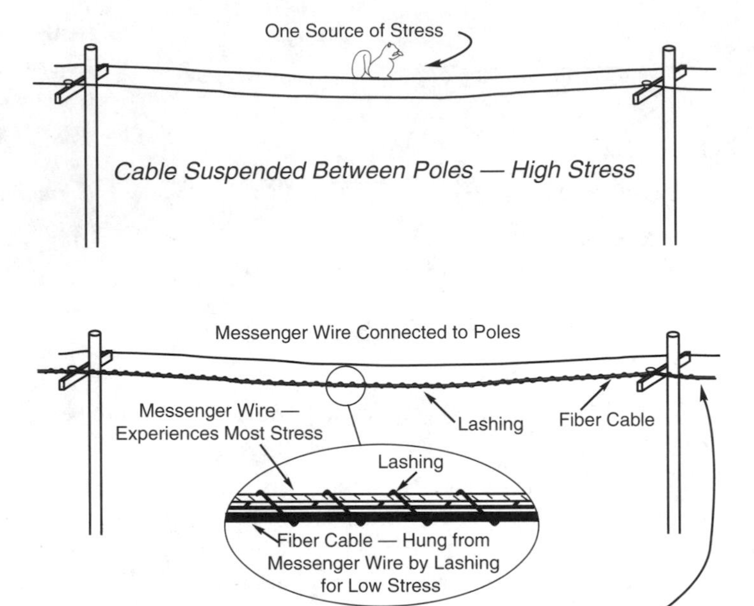

against the demanding environment of dirt). Normally, the armor is surrounded inside and out with polyethylene layers that protect it from corrosion and cushion the inside from bending damage.

- Submarine cables can operate while submerged in fresh or salt water. Those intended to operate over relatively short distances—no more than a few kilometers—are essentially ruggedized and waterproof versions of direct-burial cables. Cables for long-distance submarine use are more elaborate, as shown in Figure 8.5. Some parts of submarine cables are buried under the floor of the river, lake, or ocean, largely to protect them from damage by fishing trawlers and boat anchors. The multilayer design shown in Figure 8.5 can withstand ocean floor pressures; the outer armor is not needed on the deep-sea bed, where no protection is necessary against fishing trawlers and other boat damage.

Elements of Cable Structure

All fiber-optic cables are made up of common elements.

As you learned earlier, fiber-optic cables are diverse in nature, reflecting the diverse environments where they are used. Nonetheless, these diverse cables are built from similar structural elements. Individual fibers are housed in one of three different elements—a tight buffer, a loose tube, or a ribbon. These structural elements can house one or more fibers,

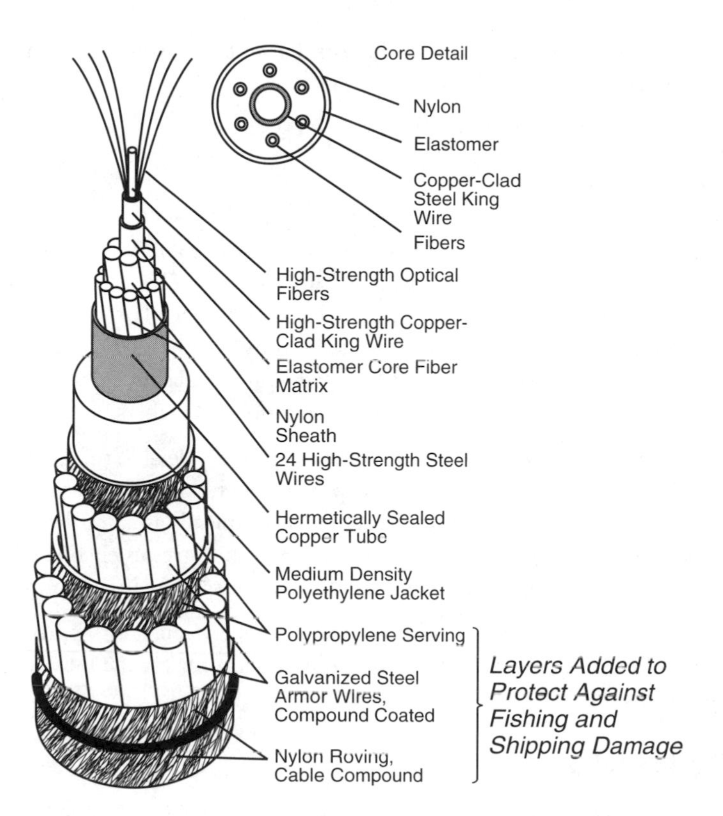

FIGURE 8.5
Fiber-optic submarine cable. (Courtesy of TyCom Ltd.)

and they, in turn, are grouped together within the cable. The more fibers, the more levels of grouping are likely. Some of these internal assemblies may be called *subcables* or *breakouts* because they can be split from the main cable and strung to another point. Individual fibers within each group may be color coded, with the groups or subcables themselves, in turn, color coded so each fiber can be identified.

Multifiber tight-buffer and loose-tube cables have central members around which the cable is built. Often these are strength members, but sometimes they may be fillers. Strength members also may be wrapped around the internal assembly of fibers or subcables. Ribbon fibers are assembled differently, with flat fiber ribbons stacked in the center of the cable or subcable and strength members surrounding them. Single-fiber or dual-fiber *(duplex)* cables also typically have surrounding strength members.

A *jacket* encloses the entire cable structure, sealing it from the environment. *Armor* provides crush resistance, particularly where cables might be exposed to burrowing rodents or mechanical hazards. *Rip cords* embedded in the cable structure can split the jacket to access individual fibers or subcables. Electrical conductors deliver power where needed, or allow one cable to carry both electronic and fiber-optic signals.

Let's look at these structures in more detail.

Fiber Housings

Fibers can be housed in tight buffers or loose tubes.

The two fundamental types of fiber housing are rigid jackets and loose tubes, both shown in Figure 8.6. Each type has its own advantages and its own niches. *Tightly buffered* cables cover a plastic-coated fiber with a thick buffer layer of harder plastic, making it easier to handle. These buffered fibers then can be packed tightly together in a compact cable structure, an approach widely used indoors. *Ribbon cables* are similar in many ways, but encase several parallel fibers in a flat, outer buffer to form a ribbon of fibers.

Loose tube cables leave the plastic-coated fiber loose inside a flexible tube, which usually is filled with a gel that prevents moisture from seeping through the hollow tube. Multiple fibers can fit inside each tube. The loose-tube structure mechanically isolates the fibers from the cable structure, an advantage in handling thermal and other stresses encountered outdoors. Although the gel can be messy it's usually considered an advantage in outdoor applications.

Each cable type has its advocates, and the boundaries between their applications are not rigid. Tightly buffered cables are built for outdoor applications, while loose-tube cables are made for indoor use.

LOOSE-TUBE CABLE

In the simplest loose-tube design, a single plastic-coated fiber is contained in a long tube, with inner diameter much larger than the fiber diameter. The fiber is installed in a loose helix inside the tube, so it can move freely with respect to the tube walls. This design protects the fiber from stresses applied to the cable in installation or service, including effects of changing temperature. Such stresses can cause bending losses as well as damage the fiber.

There are several variations on the loose-tube approach. Multiple fibers can run through the same tube, either separately or assembled into one or more ribbons. Usually the fibers are color-coded. The tube does not have to be a physically distinct cylinder running the length of the cable. It can be formed by running grooves along the length of a solid cylindrical structure encased in a larger tube, or by pressing corrugated structures together and running fibers through the interstices. The end result is the same: the fiber is isolated from stresses applied to the surrounding cable structure.

Loose-tube cables are filled with a gel for use outdoors.

Loose tubes can be used without any filling. However, if they are to be used outdoors, they are normally filled with a jellylike material. The gel acts as a buffer, keeping out moisture and letting the fibers move in the tube.

Loose-tube cables are used outdoors because they effectively isolate the fibers from external stresses such as changes in temperature, preventing damage and resulting in lower fiber loss. They can be installed from poles, in ducts, or by direct burial. A single tube can contain many fibers, making it possible to achieve high fiber densities in a compact cable. The cables can be made of flame-retardant materials to meet codes for indoor use, especially where high fiber counts are needed.

FIGURE 8.6
Tightly buffered and loose tube structures for cables.

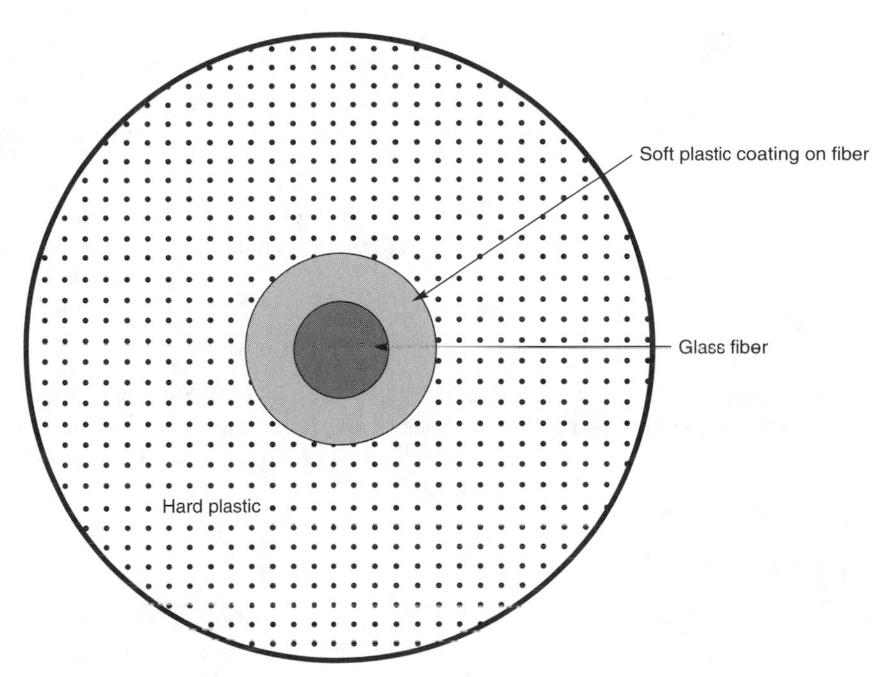

(a) Tightly buffered fiber, encased in hard plastic buffer.

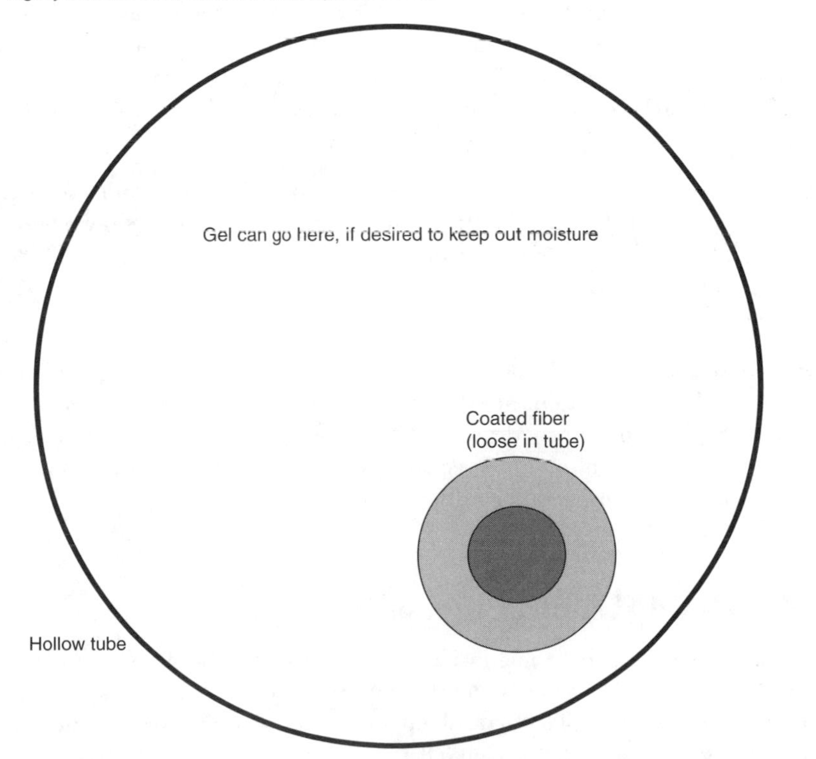

(b) Loose tube, with fiber inside a hollow tube, which may be filled with gel.

TIGHTLY BUFFERED FIBER

Tightly buffered fibers are typically used indoors.

A tightly buffered fiber is encased (after coating) in a plastic layer. The coating is a soft plastic that allows deformation and reduces forces applied to the fiber. The harder outer buffer provides physical protection.

Tight buffering tolerances assure that the fibers are in predictable positions, making it easier to install connectors. The tight-buffer structure creates subunits that can be divided among many terminals, without using patch panels. Tight-buffer cables are smaller for small fiber counts than loose-tube cables, but the ability to pack many fibers into a single loose tube makes that advantage disappear as the fiber count increases.

A major advantage of tight-buffered cable for indoor use is its compatibility with materials that meet fire and electrical codes. Although losses may be somewhat higher than in loose-tube cables, indoor transmission distances are short enough that it's not a problem.

RIBBON CABLE

Many parallel fibers can be encased in plastic to form a ribbon around which cables can be built.

The arrangement of fibers in a ribbon cable shown in Figure 8.7 is in some ways a variation on the tightly buffered cable. Each ribbon is made by aligning several coated fibers parallel and touching each other, then coating them with a plastic buffer or jacket to form a single multifiber ribbon. While tightly buffered fibers are individual structures, the ribbon is a unit containing 4 to 12 fibers, which resembles the flat 4-wire cables used for household telephones. Figure 8.7 shows two 12-fiber ribbons stacked on top of each other; up to a dozen 12-fiber ribbons can be stacked to make an extremely dense block of fibers.

Ribbon cables can be used by themselves as the basis of a fiber cable, with a stack of ribbons at the core surrounded by a cable structure. Alternatively, one or more ribbons can be stacked inside loose tubes and assembled into a loose-tube fiber cable. If the ribbon-containing tubes are used as subcables, this can yield very high fiber counts, such as the cable shown in Figure 8.8.

The simple structure makes a ribbon cable easy to splice in the field because each fiber is in a precisely predictable position. Multifiber connectors also can be installed easily on ribbon cables. The ribbons allow very dense packing of fibers, important for some applications. However, installation can cause uneven strain on different fibers in the ribbon, leading to unequal losses and other potential problems.

Fiber Arrangements in Cable

Fibers can be arranged in a cable in many different ways. The simplest cables are round with a single fiber at their center. Duplex (two-fiber) cables may either be circular or oval in cross section or be made like electrical zip cord, with two single-fiber structures bonded together along their length, as in Figure 8.1.

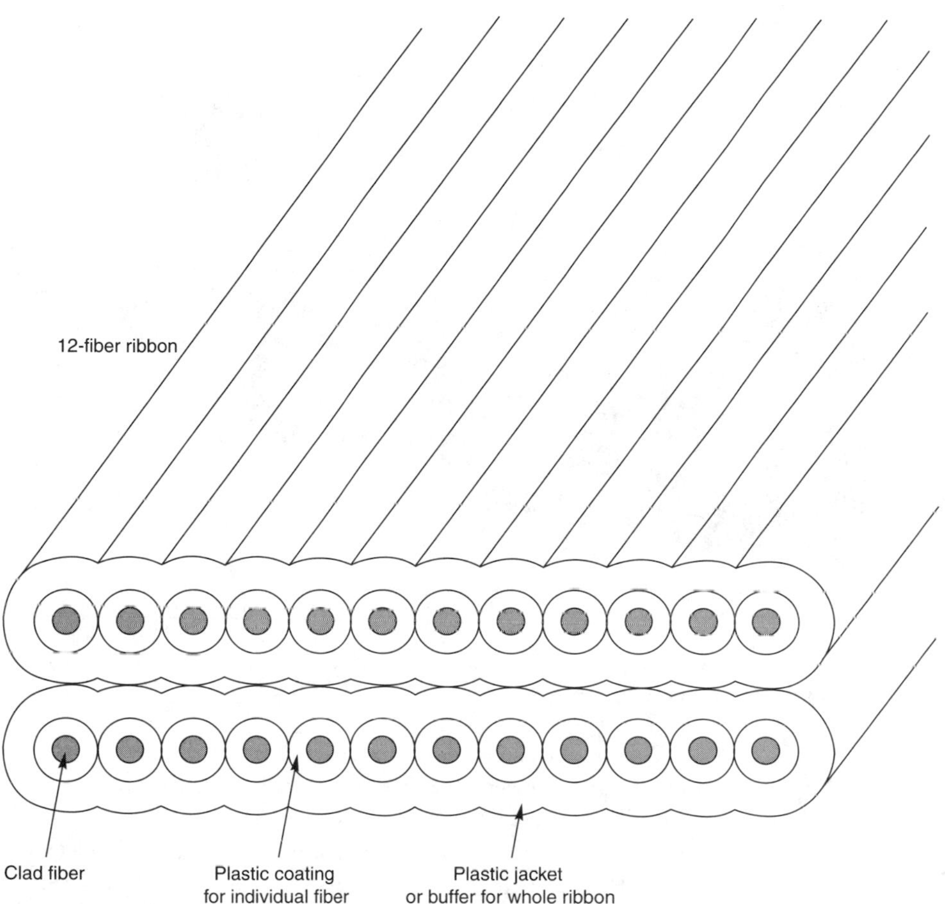

FIGURE 8.7
Fibers in a ribbon cable.

The more fibers in the cable, the more complex the structure. One common cable structure has six tightly buffered fibers wound loosely around a central member. The buffered fibers are wound so that they don't experience torsion in the cable. In loose-tube cables, the fiber count can be raised by putting multiple fibers in each tube. Groups of 8 or 12 fibers may also be wound around a central member.

Cables with more fibers are built up of modular structures. For example, a 36-fiber cable can be made from six loose-tube or tightly buffered modules containing six fibers each, or from three 12-fiber ribbons. A dozen 12-fiber ribbons make a 144-fiber cable. Putting 12 fibers in each loose tube and adding a second ring of 12 loose tubes gives a 216-fiber cable, as shown in Figure 8.9. Design details depend on the manufacturer.

864-fiber cables are in use, and cables with many more fibers are in development.

High fiber counts were rare in early installations, but they have become popular as fibers are used to distribute signals to more customers in large metropolitan areas. Cables with several hundred fibers are in use, and manufacturers now offer cables with up to 1,132 fibers.

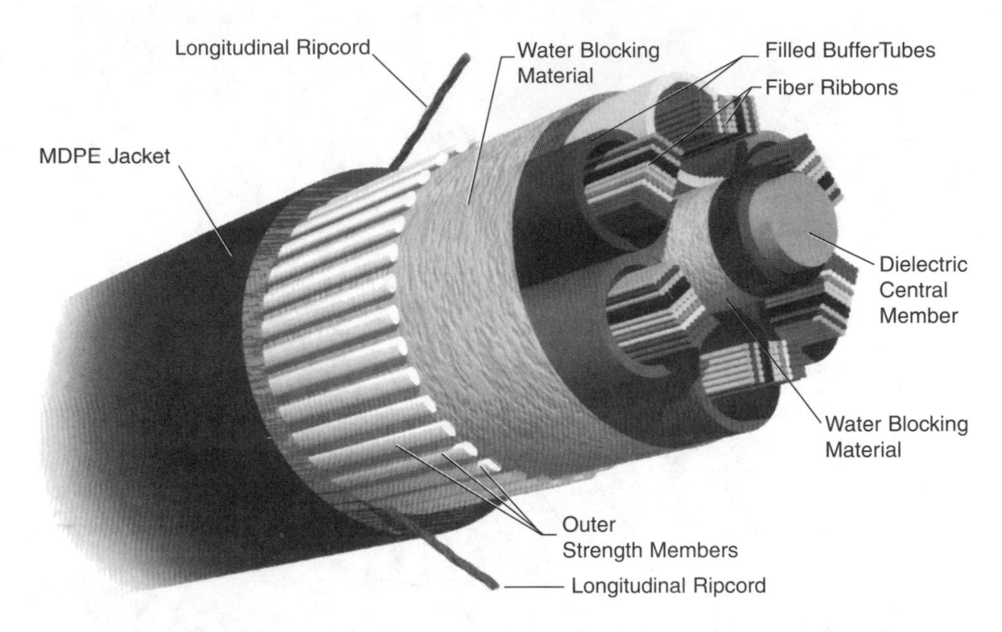

FIGURE 8.8 *Stacking ribbons inside loose tubes can give very high fiber counts. (Courtesy of Pirelli Cable)*

Cabling Materials

Materials play an important role in cable properties.

The choice of materials plays a crucial role in determining characteristics of a cable. Designers usually face trade-offs among several factors. Fire safety is crucial for indoor cables, particularly those that run through air spaces, because some compounds used in outdoor cables produce toxins or catch fire easily. Moisture resistance and temperature tolerance is critical in most outdoor environments. Aerial cables must survive severe temperature extremes, sunlight, and wind loading; cable pulled through ducts must withstand surface abrasion as well as tension along its length. The major materials used in cabling are as follows:

- Polyethylene: Standard for outdoor cables because it resists moisture, it is stable over a wide temperature range, and resists abrasion. It does not meet indoor fire-safety rules, so only short runs are used indoors to reach service panels.
- Polyvinyl chloride (PVC): The most common material for indoor cables, it is available in different grades for various requirements. It is flexible, fire-retardant, and can be extruded easily during cabling, but it is not as durable or moisture-resistant as polyethylene.
- Polyvinyl difluoride (PVDF): A plastic used for plenum cables, it retards fire better than polyethylene and produces less smoke. It is less flexible and harder to extrude.
- Low smoke-no halogen (LSNH) plastics: Produce little smoke and no toxic halogen compounds, they are safest for use in enclosed spaces. They also

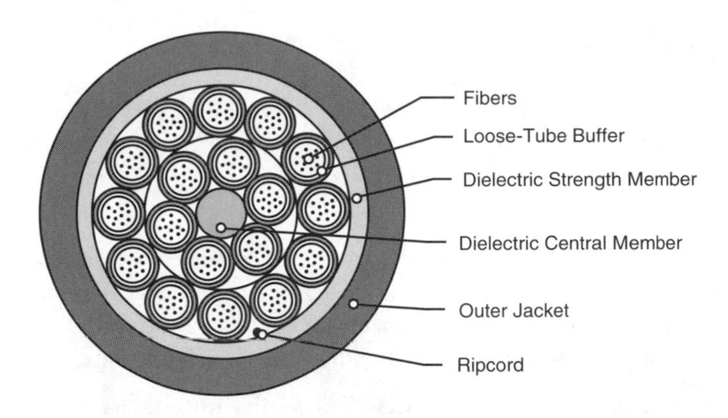

FIGURE 8.9
Modular cable containing 216 fibers, with 12 in each of 18 loose tubes. (Courtesy of Corning Cable Systems)

protect electronic equipment from corrosion damage during fires but are costly and less durable than PVC.

- Aramid yarn: Dielectric strength members, they are known by the trade name Kevlar.

Other Structural Elements

Fibers and their buffers are not the only structural elements of cables. Many—but not all—fiber-optic cables include other components to provide strength and rigidity.

Many cables contain central members to make them rigid and strength members to withstand tensile forces.

Many cables are built around central members made of steel, fiberglass, or other materials. These run along the center of the cable and provide the rigidity needed to keep it from buckling, as well as a core to build the cable structure around. A central member may be overcoated with plastic or other material to match cable size requirements and to prevent friction with other parts of the cable. Small indoor cables containing few fibers generally lack central members, but they are common in outdoor cables, and in indoor cables with high fiber counts.

Strength members in general are distinct from central members. Their role is to provide tensile strength along the length of the cable during and after installation. The usual strength members are strands of dielectric aramid yarn (better known under the trademark Kevlar) wound around the core of the cable. In some cases, tight-buffered fibers may be wound around them to form subunits of the cable. When a cable is pulled into a duct, the tension is applied directly to the strength member.

The structure containing the fibers normally surrounds the central member and, in turn, is surrounded by the strength member and one or more outer jacketing layers of plastic. The composition depends on the application.

Buried and underwater cables require armor.

Underwater and buried cables are among the types that require one or more layers of protecting armor. Typically for buried cables, steel is wound around an inner plastic sheath. An outer plastic sheath is then applied over the armor to prevent corrosion. The metal armor helps protect against crushing damage, rocks, and rodents. Underwater cables in shallow waters may have multiple layers to protect against damage from shipping and fishing operations, as shown in Figure 8.5.

Blown-in Fibers

Fibers can be blown through microducts instead of installing normal cables.

Traditionally, cables are installed as finished units, containing both fibers and protective structures. An alternate approach is to install hollow microducts—typically about 5 mm in diameter—then to blow fibers through them. Forcing air through the microduct carries the fiber along with it.

The process uses conventional single- or multimode fibers coated with a blowable coating, designed to be dragged along by air forced through the microducts. The air can carry the coated fibers around bends, and over distances of more than 1000 feet (300 meters), usually within a building or between adjacent buildings.

The use of blown fibers is a two-stage process. The flexible microducts that carry the fibers must be installed first, then the fiber is blown through them. The major attraction is that the fibers need not be blown in at once, and that fibers can be replaced easily, like cables in underground ducts. Thus you could install the microducts when renovating a building, and blow in the fibers later, when you knew transmission requirements. If a fiber was damaged, you could pull it out and blow in a replacement. With the proper equipment, fibers can be blown into place very quickly.

From a functional standpoint, blow-in fibers behave like loose-tube cables without a filler in the tubes. Materials used in the plastic tubes are chosen to meet the appropriate fire and electrical codes. While blown fibers are not widely used, the technology is available.

Cable Installation

Special techniques are used to install different types of cable.

Cable installation is a specialized task, and the detailed procedures are beyond the scope of this book. Most methods for installing optical cables have been adapted from those used for copper cables. Outdoor cables are laid along rights of way leased or owned by telecommunications carriers, such as along a railroad or highway, which are well marked after the cables are installed.

The basic approaches depend on the type of cable being installed.

- Submarine cables are laid from ships built for that purpose. Typically they are buried in a trench dug on the sea floor at depths of less than about 200 meters (600 feet), and laid directly on the ocean floor in deep ocean basins.
- Direct-buried cables normally are laid in a deep, narrow trench dug with a cable plow, which is then covered with dirt.
- Cable ducts are plastic tubes laid in trenches dug for the purpose, then covered over. Duct sizes and flexibility vary; some are only an inch or two and can be wound on large spools; others are a few inches in diameter and rigid. The ducts typically are directly covered by soil, but sometimes may be encased in concrete to add structural integrity and prevent service disruptions. The ducts are installed without cables inside. Duct routes may be direct

between endpoints, or may be routed through a series of underground access points at manholes.

- Cables are installed in ducts by threading a pull line through the duct, attaching it to the cable, then pulling the cable through the ducts. If manholes or other access points are available on the route, cable runs are pulled between them.
- Self-supporting aerial cables may be suspended directly from overhead poles. Other aerial cables can be suspended from messenger wires, strong steel wires strung between poles. If a messenger wire is used, the cable is lashed to it with a special lashing wire running around both the cable and the messenger wire, or sometimes wound around it. This is a common installation for many overhead fiber cables because it minimizes strength requirements.
- Plenum cables are strung through interior air spaces.
- Interior cables may be installed within walls, through cable risers, or elsewhere in buildings. Installation is easiest in new construction. Only special cables designed for installation under carpets should be laid on the floor where people walk.
- Temporary light-duty cables are laid by people carrying mobile equipment that requires a broadband (typically video) connection to a fixed installation.

Cable Changes and Failure

Cabling can cause minor changes in fiber properties, particularly attenuation. The major reason for these changes is microbending, which depends on the fiber's local environment and on stresses applied to it. These stresses generally are negligible for cables in protected or stable environments, such as inside buildings or in buried duct. Aerial cables are subject to the most extreme variations, because they are exposed to conditions from summer sun to winter ice. Repeated heating and cooling can cause microbending because the glass fiber and the cable materials do not expand and contract with temperature at the same rate. Tightly buffered fibers generally are considered more vulnerable to microbending losses, but in well-made cables the differences are quite small.

Microbending can cause changes in fiber attentuation after cabling

Telephone companies were very cautious before beginning their massive conversion to fiber-optic cables, conducting extensive tests and field trials to evaluate the reliability of fiber. Engineers were instinctively wary of glass, but many tests have shown that fiber cables are very durable under normal conditions. Although poor installation can damage fibers, fibers or cables rarely fail in properly installed systems unless physical damage occurs.

Most cable failures are due to physical damage.

Gophers and other burrowing rodents are natural enemies of all buried cables. Their front teeth grow continually, and they instinctively gnaw on anything they can get their incisors around, including cables. Gophers are a serious problem in some areas in the United States, and cables buried in those areas should be armored and otherwise designed to deter gopher attack. The U.S. Fish and Wildlife Service even developed a standardized "gopher test" in which a cable was run through a gopher cage and left in place for seven days to test its resistance to the gnawing rodent. Cables buried in some areas must meet these rodent-proof standards. (Cables containing obsolete fibers are sometimes called "gopher bait" because they are of little use for telecommunications.)

Human activity is responsible for most other damage to buried cables. The archetypical example is a backhoe digging up and breaking a buried cable, an event industry veterans call "backhoe fade." Construction damage by careless contractors is a serious problem but the subject of some jokes. One telecommunications company runs its long-distance cables alongside natural-gas pipelines, where it posts warnings, which one top manager summarized as "you dig, you die."

Aerial cables are vulnerable to many other types of damage, from errant cranes to falling branches and heavy loading with ice and snow. All kinds of cables can be—and are—cut by mistake by maintenance crews. Light-duty indoor cables can be damaged by closing doors or windows on them; the fibers can be broken without causing obvious damage to the outside of the cable. Tripping over a cable is not likely to break the cable, but it could snap the fibers inside, or jerk the cable out of a connector at one end. (In general, connectors are the weakest points of short cables.)

Fibers do not always break at the exact point where the cable is damaged. When stressed along their lengths, fibers tend to break at weak points, which may be a short distance away from the obvious damage to the cable. The break points can differ among fibers in a multifiber cable, so you might have to go a ways from the break before you find the ends of all fibers. As with copper cables, physical damage to the cable does not invariably sever fibers.

What Have You Learned?

1. Fibers can break at inherent flaws and develop microcracks under tension, so they must be protected from stretching forces.
2. Cabling packages fibers for protection and easier handling.
3. Cables must resist crushing as well as isolate the fiber from tension along its length.
4. Cables protect fibers from heat, cold, and moisture. The designs chosen vary widely, depending on environmental conditions.
5. Indoor cables must meet fire and electrical safety codes. The major impact of these codes is on composition of the plastics used in the cable structure.
6. Cables can carry from one to hundreds of fibers.
7. Important structural elements in a cable are the housing for the fiber, strength members, jacketing, and armor. Armor is used only in certain environments where physical damage is a threat.
8. Outdoor cables may be hung from poles, pulled through ducts, or buried directly in the ground.
9. Fibers can be enclosed in a loose tube, a tight plastic buffer, or a ribbon.

10. The physical arrangement of fibers in the cable depends on the number of fibers and how they are to be distributed in the installation.
11. Much cable damage occurs when a sudden force is applied, such as when a backhoe digs up a buried cable.

What's Next?

In Chapter 9, we will examine the light sources used with fiber-optic cables.

Further Reading

Bob Chomycz, *Fiber Optic Installer's Field Manual* (McGraw Hill, 2000)

Jim Hayes, *Fiber Optics Technician's Manual* (Delmar Publishers, 1996)

Questions to Think About for Chapter 8

1. The design of fiber-optic cables evolved from the design of electrical cables for similar applications. What is one common problem with aging electrical cables that does not affect fiber-optic cables?
2. What is one problem with fiber-optic cables *not* present with electronic cables?
3. Why make one cable with 1000 fibers rather than 10 with 100 fibers?
4. A cable weighs 300 kilograms per kilometer, with poles placed 50 meters apart on the road. A flock of twenty 0.5-kilogram crows perch on the cable to watch the scenery. By what percentage do they increase the weight of the cable span?
5. An ice storm coats the cable with a 1 cm layer of ice. Assume for simplicity that the cable is 2 cm in diameter, and that ice weighs 1 gram/cubic centimeter. How much weight does the ice add to a 50-meter span of the cable?
6. What is the main difference between the structures of indoor and outdoor cables?

Quiz for Chapter 8

1. What part of a cable normally bears stress along its length?
   a. The fiber
   b. The plastic coating of the fiber
   c. A strength member
   d. Metallic armor
   e. All of the above
2. Cables cannot protect fibers effectively against
   a. gnawing rodents.
   b. stresses during cable installation.
   c. careless excavation.
   d. static stresses.
   e. crushing.

3. Light-duty cables are intended for use
   a. within office buildings.
   b. in underground ducts.
   c. deep underground where safe from contractors.
   d. on aerial poles where temperatures are not extreme.
4. The special advantages of plenum cables are what?
   a. They are small enough to fit in air ducts.
   b. They meet stringent fire codes for running through air spaces.
   c. They are crush resistant and can run under carpets.
   d. They have special armor to keep rodents from damaging them.
5. Outdoor cables are *not* used in which of the following situations?
   a. Suspended overhead between telephone poles
   b. Tied to a separate messenger wire suspended between overhead poles
   c. Inside air space in office buildings
   d. Pulled through underground ducts
6. A loose-tube cable is
   a. a cable in which fibers are housed in hollow tubes in the cable structure.
   b. a cable for installation in hollow tubes (ducts) underground.
   c. cable for installation in indoor air ducts.
   d. used underwater.
   e. none of the above
7. Which of the following are usually present in direct-burial cables but *not* in aerial cables?
   a. Strength members
   b. Outer jacket
   c. Armor
   d. Fiber housing
8. Which type of cable installation requires pulling the cable into place?
   a. Direct burial
   b. Underground duct
   c. Military field systems
   d. Submarine cable
9. The main cause of differences in properties of a fiber before and after cabling is
   a. microbending.
   b. temperature within the cable.
   c. application of forces to the fiber.
   d. damage during cabling.
10. The most likely cause of failure of cabled fiber is
    a. hydrogen-induced increases in attenuation.
    b. corrosion of the fiber by moisture trapped within the cable.
    c. severe microbending losses.
    d. physical damage to the cable.

CHAPTER 9

Light Sources

About This Chapter

Many types of light sources are used in fiber-optic systems, from cheap red LEDs to sophisticated and expensive lasers made to emit infrared light at an extremely precise wavelength. Lasers and LEDs generate the light transmitted through the fiber; they may modulate the signal directly, or come with an external modulator that controls the beam intensity. Some operate at telephone-like speeds over short distances, but others may carry trillions of bits per second through thousands of kilometers of fiber. Optical amplifiers also are light sources, but instead of generating their own signals, they amplify weak input light signals.

Light sources are actually parts of transmitters, but to approach the subject systematically I have separated the two. This chapter covers light sources alone, while Chapter 10 covers transmitters, modulation, and related topics. In Chapter 11, you will look at receivers, which convert the optical signals back into electronic form at the other end of the fiber. In Chapter 12 you will learn about the workings of optical amplifiers.

Light Source Considerations

Several factors determine the choice of light sources for fiber-optic systems. The wavelength of a signal source must fall within a transmission window of the optical fiber being used. The power must be high enough to span the distance to the first optical amplifier or the receiver—but not so high it causes nonlinear effects in the fiber or overloads the receiver. The range of wavelengths emitted by the source should not be so broad that dispersion limits transmission speed. The source also must transfer light efficiently into the transmitting fiber.

The same conditions apply for optical amplifiers, but in addition they must be matched to the wavelengths that carry the original signal. In wavelength-division multiplexed systems, they must amplify *all* the wavelengths to the same extent, not just some of them. It wouldn't do if the shortest or longest wavelengths got lost because they fell outside the amplification band. Because optical amplifiers are analog devices, care must be taken that they don't distort input signals when amplifying them.

Wavelength, spectral width, and power are important in fiber-optic light sources.

The primary light sources for fiber-optic communications are semiconductor lasers (often called diode lasers) and light-emitting diodes (LEDs). As semiconductor devices, they fit well with other electronics in transmitters. Other types of lasers are used in some cases, particularly fiber lasers. The most common optical amplifiers are erbium-doped fiber amplifiers, described briefly in Chapter 7. Other fiber amplifiers and semiconductor optical amplifiers have found limited use or are still in development.

Operating Wavelength

Source wavelength affects attenuation and dispersion.

Both signal attenuation and pulse dispersion are functions of operating wavelength. The standard transmission windows in glass fibers are 780 to 850 nm for short systems using gallium-arsenide light sources, and 1280 to 1620 nm for telecommunications. Most present telecommunication systems operate either at 1310 nm or in the 1530 to 1565 nm C-band of erbium-doped fiber amplifiers; many links shorter than a kilometer or two transmit at 780 to 850 nm. The erbium-amplifier band has become standard for long-distance transmission; 1310 nm is used for unamplified systems and in some older long-distance systems. Other wavelengths in the low-loss band from 1280 to 1620 nm are available but are rarely used because standard sources are less expensive.

Plastic fibers require different transmission wavelengths. As you learned in Chapter 6, most in common use have lowest loss around 650 nm in the red.

Wavelength-division multiplexing complicates the picture by sending many optical channels through the same fiber at different wavelengths. Each optical channel has a distinct central wavelength. In addition, each optical channel contains a range of wavelengths, called the *spectral width,* which may be measured in units of wavelength or frequency. Lasers and LED sources normally generate individual optical channels. Several optical channels can be combined to make a composite WDM signal. The distinction between optical channels and composite WDM signals is important; remember that individual light sources produce single optical channels, while optical amplifiers typically amplify composite WDM signals.

Single-frequency lasers are needed for telecommunications at rates above 1 Gbit/s.

Chromatic dispersion increases with the range of wavelengths in the signal on one optical channel. This spectral width is 30 to 50 nm for standard LEDs, compared to 0.5 to 3 nm for modest-priced diode lasers, as shown in Figure 9.1. Operation at speeds above about 1 Gbit/s over long distances requires narrow-line or *single-frequency* lasers with typical spectral widths a small fraction of 1 nm. Typically their spectral widths are measured in frequency units rather than wavelength. If a single-frequency laser has a spectral width of 10 GHz, that corresponds to a wavelength range of about 0.056 nm at 1300 nm or 0.08 nm at 1550 nm.

Composition of the semiconductor material determines the range of wavelengths over which light sources can emit. The structure of the device determines what wavelengths it emits, including the spectral width. A laser and an LED made of the same material may have the same central wavelength, but an LED has a much broader spectral width. Different laser structures give different spectral widths. Conversely, lasers with the same structures made of different materials have comparable spectral widths but different center wavelengths.

The output wavelengths of an optical amplifier depend on the input signal as well as on the amplifier itself. Optical amplifiers work only over a limited range of wavelengths, so signals

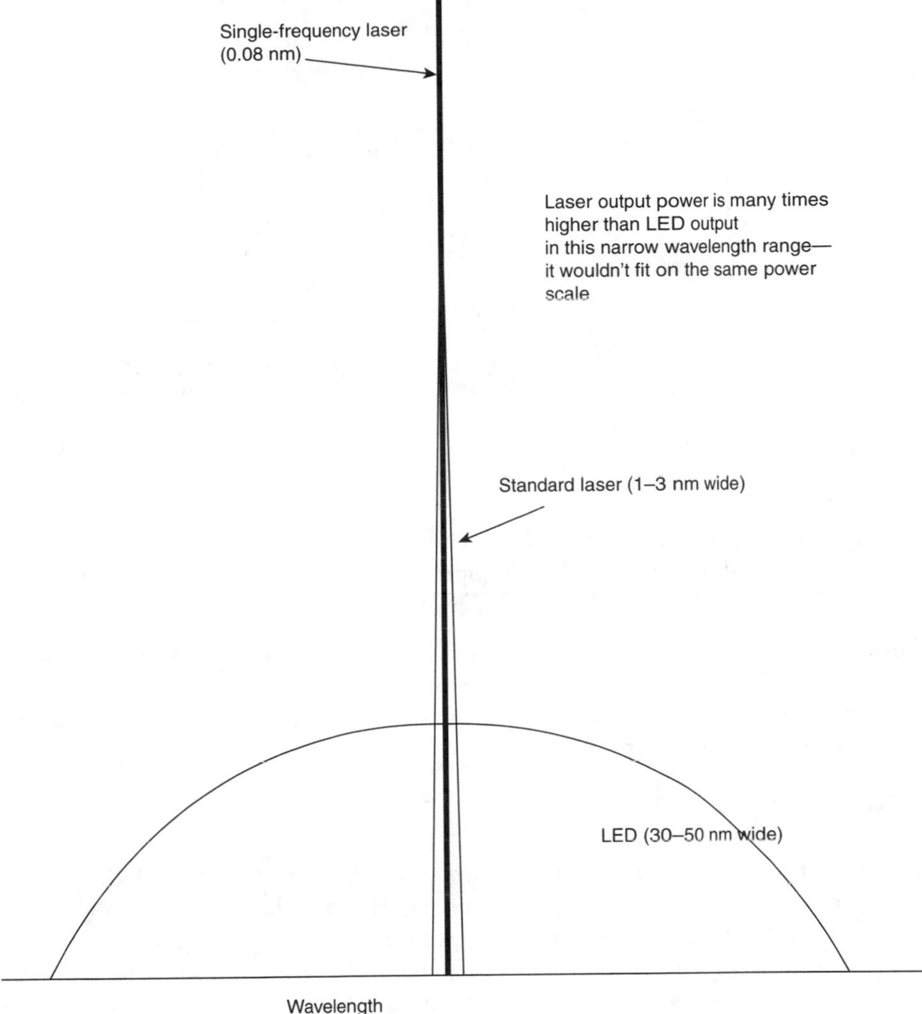

FIGURE 9.1
Comparison of LED and laser spectral widths.

outside that range are not amplified. As long as the input signals fall into that range, the wavelengths and spectral widths of each optical channel in the output are the same as those in the input.

The rest of this chapter will explain these concepts in more detail.

Output Power and Light Coupling

Power from communication light sources can range from more than 100 mW for certain lasers to tens of microwatts for LEDs. Not all that power is useful. For fiber-optic systems, the relevant value is the power delivered into an optical fiber. That power depends on the angle over which light is emitted, the size of the light-emitting area, the alignment of the source and fiber, and the light-collection characteristics of the fiber, as shown in Figure 9.2.

FIGURE 9.2
Light transfer from an emitter into a fiber.

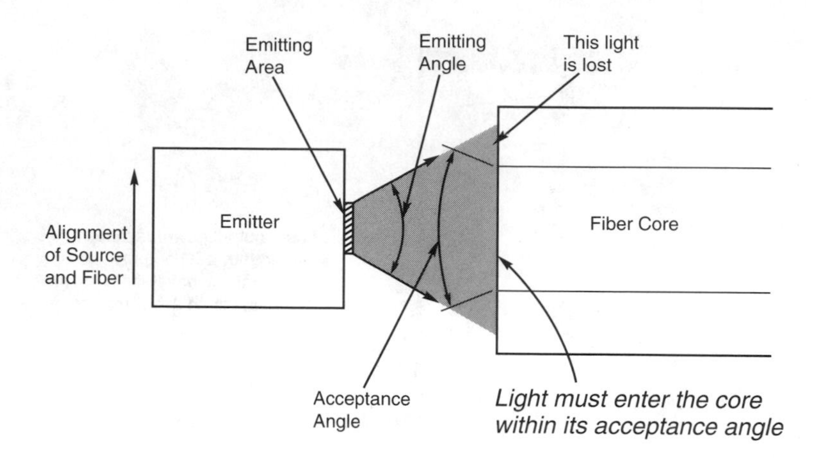

The light intensity is not uniform over the entire angle at which light is emitted but rather falls off with distance from the center. Typical semiconductor lasers emit light that spreads at an angle of 5° to 20°; the light from LEDs spreads out at larger angles.

Losses of many decibels can easily occur in coupling light from an emitter into a fiber, especially for LEDs with broad emitting areas and wide emitting angles. This makes it important to be sure you know if the power level specified is the output from the device or the light delivered into the fiber.

Modulation

Diode lasers and LEDs are modulated directly.

The output power from semiconductor lasers and LEDs varies directly with the drive current. This allows *direct modulation* in which the transmitted signal modulates the intensity of the drive current, and the output light signal varies along with the drive current. Other lasers generate steady beams or fire a series of pulses with even spacing between them. These beams require *external modulation* in which they pass through a modulator, which changes its transparency in response to a separate electronic input signal.

You can think of direct modulation as flicking a light switch off and on, and external modulation as opening and closing a shutter in front of the light. External modulation is more expensive but offers better performance. A laser source operated with a steady drive current has a narrower spectral width than a directly modulated (pulsed) laser.

Several factors are important in modulation. One is speed. Lasers can change their output intensity faster than LEDs, and external modulators are even faster. Another factor is *wavelength chirp,* which arises when changes in the current passing through a directly modulated laser change the semiconductor's refractive index and thus its output wavelength. The linearity of modulation also is important, particularly for analog communications. Depth of modulation and modulation format affect performance. Many of these details are covered in the description of transmitters in Chapter 10.

Cost/Performance Trade-offs

As any student of engineering reality would expect, light sources with the most desirable characteristics cost the most. The cheapest light sources are LEDs with slow rise times, large emitting areas, and relatively low output power. Diode lasers with narrow bandwidths in the 1530- to 1620-nm band, where optical fibers have their lowest losses, are the most expensive. The higher-power and narrower-line emission of lasers comes at a marked price premium, with the narrowest-line lasers costing the most. The only real performance advantage of LEDs is generally longer lifetime than some lasers.

LED Sources

An LED emits light when a current flows through it.

LEDs that emit red or near-infrared light are common light sources for short fiber systems. The basic concept of a light-emitting diode is shown in Figure 9.3. The diode is made up of two regions, each doped with impurities to give it the desired electrical characteristics. The *p* region is doped with impurities having fewer electrons than atoms they replace in the semiconductor compound, which create "holes" where there is room for electrons in the crystalline lattice. The *n* region is doped with impurities that donate electrons, so extra electrons are left floating in the crystalline matrix. Applying a positive voltage to the *p* region and a negative voltage to the *n* region causes the electrons and holes to flow toward the junction of the two regions. If the voltage is above a certain (low) level, which depends on the material in the thin junction layer, the electrons drop into the holes in a process called recombination. As long as that voltage is applied in the same direction, electrons keep flowing through the diode and recombination continues at the junction.

In many semiconductors, notably silicon and germanium, the released energy is dissipated as heat—vibrations of the crystalline lattice. (Light emission from porous silicon is a special case that depends on the microstructure of the silicon crystal.) In other materials, usable in LEDs, the recombination energy is released as a photon of light, which can emerge from the semiconductor material. The most important of these semiconductors, gallium arsenide and related materials, are called III-V compounds because they are made of elements from the IIIa and Va columns of the periodic table:

IIIa	*Va*
Aluminum (Al)	Nitrogen (N)
Gallium (Ga)	Phosphorus (P)
Indium (In)	Arsenic (As)
	Antimony (Sb)

A new technology is developing around semiconducting polymers. These are plastics that have the electrical characteristics of semiconductors. (Conventional plastics are insulators, because they bond electrons tightly.) As with their crystalline counterparts, semiconducting

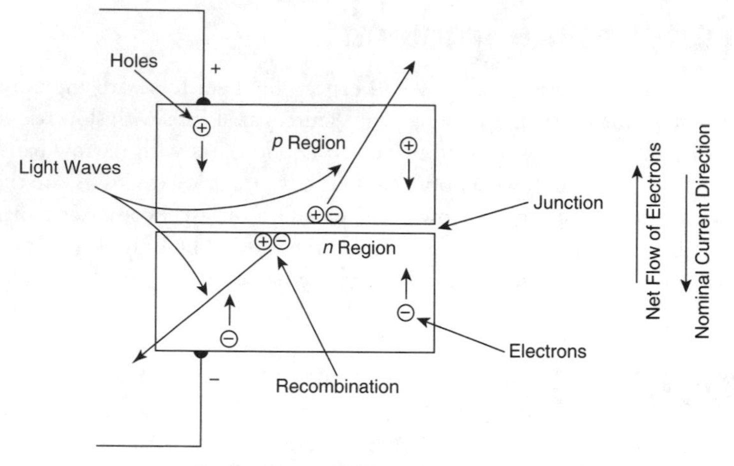

a. Current flow and light emission in an LED

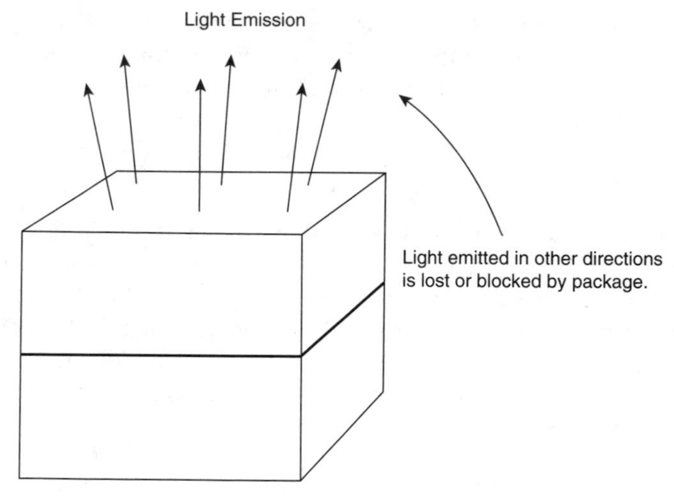

b. LED as packaged (surface-emitting)

FIGURE 9.3
LED operation.

plastics can be doped to make *p* and *n* materials, and electrical devices, including LEDs, can be made from them.

The wavelength emitted by a semiconductor depends on its internal energy levels. In a pure semiconductor at low temperature, all the electrons are bonded within the crystalline lattice. As temperature rises, some electrons in this *valence band* jump to a higher-energy *conduction band,* where they are free to move about in the crystal. The valence and conductor bands are separated by a void where no energy levels exist—the band gap that gives semiconductors many of their useful properties.

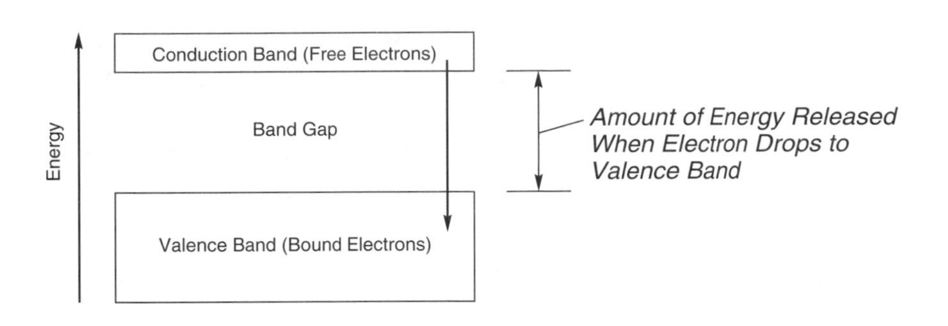

FIGURE 9.4
Semiconductor energy levels.

Conduction-band electrons leave behind a hole in the valence band, which is considered to have a positive charge. This hole can move about, as electrons from other spots in the crystalline lattice move to fill in the hole and leave behind their own hole (i.e., the hole moves from where the electron came to where the electron was). Impurity doping of semiconductors can also generate free electrons and holes. When an electron drops from the conduction level to the valence level (i.e., when it recombines with a hole), it releases the difference in energy between the two levels, as shown in Figure 9.4.

The band gap between the energy levels—and hence the amount of energy released and the wavelength emitted—depends on the composition of the junction layer of the semiconductor diode. The near-infrared LEDs used with short glass-fiber systems have active layers made of gallium aluminum arsenide (GaAlAs) or gallium arsenide (GaAs). Pure gallium-arsenide LEDs emit near 930 nm. Adding aluminum decreases the drive current requirements to improve the lifetime, and increases the energy gap so light emission is at 750 to 900 nm. Generally only the thin junction layer is made of GaAlAs, with the rest GaAs, so these devices generally are called GaAs LEDs. The usual LED wavelengths for fiber-optic applications are 820 or 850 nm. At room temperature, the typical 3-dB spectral bandwidth of an 820-nm LED is about 40 nm.

The usual LED wavelengths for glass fibers are 820 or 850 nm.

The band-gap energy also can be measured in electron volts, the amount of energy needed to move one electron through an electric field of one volt. With no bias voltage applied, this potential forms at the junction layer, so no light is generated unless a current is flowing through the LED (as you would expect). A forward bias overcomes this potential at the junction layer, allowing current to flow through the LED and generating light at the junction layer.

Other semiconductor compounds can be used to make LEDs that emit different wavelengths. Gallium arsenide phosphide (GaAsP) LEDs emitting visible red light around 650 nm are used with plastic fibers, which are most transparent in the red and transmit poorly at GaAlAs wavelengths. GaAsP LEDs are lower in performance than GaAlAs LEDs, but they cost less and are fine for short, low-speed plastic fiber links.

Visible LEDs are used with plastic fibers, which transmit poorly in the infrared.

The most important compound for high-performance fiber-optic lasers is InGaAsP, made of indium, gallium, arsenic, and phosphorus mixed so the number of indium plus gallium atoms equals the number of arsenic plus phosphorous atoms. The resulting compound is written as $In_xGa_{1-x}As_yP_{1-y}$, where x is the fraction of indium and y is the fraction of arsenic. These so-called quaternary (four-element) compounds are more complex than

InGaAsP LEDs emit at 1300 nm.

ternary (three-element) compounds such as GaAlAs but are needed to produce output at 1200 to 1700 nm. In practice, LEDs are sometimes used for short systems at 1300 nm, where many glass fibers have low chromatic dispersion, but never at 1550 nm.

Other LED characteristics depend on device geometry and internal structure. The description of LEDs so far hasn't indicated in which direction they emit light. In fact, simple LEDs emit light in all directions, as shown in Figure 9.3, and are packaged so most emission comes from their surfaces. The light is emitted in a broad cone, with intensity falling off roughly with the cosine of the angle from the normal to the semiconductor junction. (This is called a Lambertian distribution.)

More complex internal structures can concentrate output of surface-emitting LEDs in a narrower angle, by such means as confining drive current to a small region of the LED. Such designs typically require that the light emerge through the substrate, which can cause losses. One way to enhance output and make emission more directional is to etch a hole in the substrate to produce what is called a Burrus diode, after its inventor Charles A. Burrus of Bell Laboratories. A fiber can be inserted into the hole to collect light directly from the junction layer.

An edge-emitting diode emits light from its ends.

A fundamentally different configuration is the edge-emitting diode, shown in Figure 9.5. Electrical contacts cover the top and bottom of an edge emitter, so light cannot emerge there. The LED confines light in a thin, narrow stripe in the plane of the *p-n* junction. This is done by giving that stripe a higher refractive index than the surrounding material. The surrounding regions of lower refractive index create a waveguide that functions like an optical fiber, and channels light out both ends where it can be coupled into a fiber. One disadvantage is that this increases the amount of heat the LED must dissipate.

In general, the more complex the LED structure, the brighter and more tightly collimated the emitted light. Shrinking the emitting area and the region through which current passes also decreases rise time and, thus, enhances possible modulation bandwidth. Of course, as with other devices, the greater complexity comes at higher cost.

FIGURE 9.5
An edge-emitting LED.

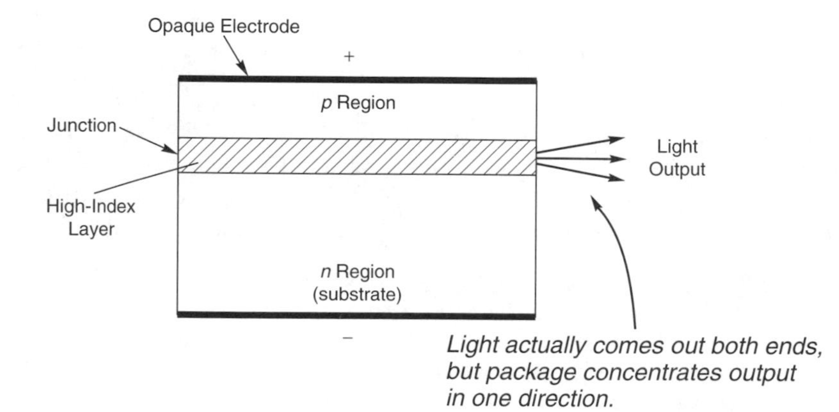

The Laser Principle

LEDs, like virtually all light sources in our everyday lives, generate light by a process called *spontaneous emission.* An atom or molecule accumulates extra energy, then quickly releases it on its own accord, without outside stimulation. Hot gas glows brilliant in the sun or bright in a flame; a hot burner on an electric range glows cherry red. If you look closely, you'll find spontaneous emission occurs on a characteristic time scale, depending on how the material stores the energy. Atoms and molecules can store only certain amounts of energy, called *energy levels.* When they release energy, they drop from a higher-energy state into one with lower energy, and the extra energy emerges as light. Spontaneous emission produces light going in all different directions at many different wavelengths, like a light bulb, a candle, or the sun. Most of these sources get their energy from heat, but LEDs get it from recombination of current carriers.

LEDs generate light by spontaneous emission.

Albert Einstein discovered that atoms and molecules also could release extra energy in another way, called *stimulated emission.* If left alone, an atom would eventually release the extra energy as spontaneous emission. However, a light wave with the right energy—matching a transition from the excited energy level to a lower one—could stimulate the atom to release the extra energy as another light wave with exactly the same energy. The new light wave also is exactly in phase with the original wave, with the same wavelength and peaks lining up exactly with the original light wave, so the two are *coherent.* Coherence is important because it organizes light, making the waves like a troop of soldiers marching in step instead of a crowd spreading in all directions as they leave a stadium.

Stimulated emission is coherent light.

Recall that you can also think of light as tiny chunks of energy, called *photons,* and you have another way to understand the process. When an atom drops from one energy level to a lower one, it normally releases a photon, which has exactly as much energy as the difference between the two states. If the atom is left alone, the new photon emerges spontaneously, but a photon with the right amount of energy can stimulate the emission of another photon with the same wavelength and phase as the first.

Stimulated emission can build up in a material. Each material has its own characteristic emission wavelengths, corresponding to *transitions* between pairs of energy levels. A single spontaneous-emission photon can stimulate the emission of another photon in the same material. These two can each stimulate emission of another photon, and so on, in a cascade of emission from a laser. The light waves are all coherent, which is important for communications.

Although stimulated emission is a general process, it can occur only in specific conditions, which is why it is extremely rare (but not impossible) in nature.

Population Inversions and Metastable States

One requirement is for more atoms to be in the higher energy state than the lower energy state that the atom is dropping to. This is called a *population inversion* because normally more atoms are in the lower energy levels. If you leave the atoms to sit long enough by

Stimulated emission requires a population inversion.

themselves, without energy input, they distribute themselves among energy levels, with more in the lower levels than in the upper ones. (This is called *thermodynamic equilibrium,* but you don't need to worry about that concept here.)

A population inversion is needed to get a laser going. The photons released by atoms in the high energy state are absorbed by atoms in the lower state. If there are too many atoms in the lower state, they'll soak up all the photons before they can find atoms in the higher state to stimulate emission from.

It also helps if the upper energy state is *metastable,* meaning that atoms tend to stay in that state unusually long before they release energy by spontaneous emission. The longer the atoms stay excited, the better the chance they can be stimulated to emit their extra energy before releasing it spontaneously.

Excitation

Semiconductor lasers are excited by a current passing through the junction layer.

Laser materials require something to excite the atoms to the high energy state they occupy in a population inversion. The usual energy sources are electric currents or light passing through the laser material.

In a semiconductor laser, the excitation comes from the current passing through the junction layer. When current carriers recombine, the free electron lingers at a higher energy level around the hole-containing atom that captured it before dropping into the hole left in the valence band and releasing its energy. This is the excited level that can be stimulated to emit light.

In an erbium-doped fiber amplifier, the excitation comes from a semiconductor laser emitting light at either 980 or 1480 nm, which excites erbium atoms, as shown in Figure 9.6. They then drop into a lower excited state, which is metastable. Light at wavelengths near 1550 nm can stimulate those excited erbium atoms to emit more light at the same wavelengths, amplifying an optical signal.

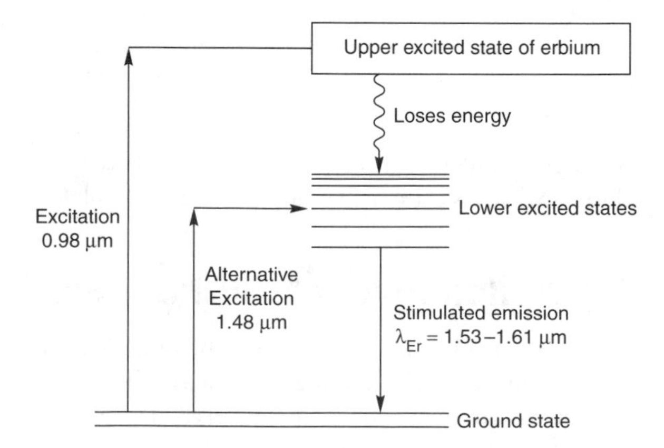

FIGURE 9.6 *Stimulated emission from erbium atoms.*

Oscillation and Amplification

The laser got its name as an acronym for *L*ight *A*mplification by the *S*timulated *E*mission of *R*adiation. However, that phrase glosses over a critical distinction between *amplification* and *laser oscillation.*

In an optical amplifier, as was shown in Figure 7.8, light makes only a single pass through the device. A weak optical signal enters the amplifier, stimulating emission from excited atoms as it passes through. The amplified signal emerges from the other end. A quantity called gain measures the increase in signal strength as it passes through the material. If the gain is 10% per centimeter in a fiber amplifier, and you started with 10 input photons, you would have 11 photons after 1 centimeter. In general you would have power

Light makes only one pass through an optical amplifier. It needs an input signal to generate an output.

$$P(L) = P_{\text{input}}\,(1 + \text{gain})^L$$

at a distance L along the fiber as long as the signals were small. If you started with 1 microwatt at the input, you would have 2.6 μW after 10 cm, 6.7 μW after 20 cm, 17.4 μW after 30 cm, and 13.8 mW after 100 cm. (In a real fiber amplifier, the output power might saturate by that level because the light would stimulate emission from all the excited atoms.)

An optical amplifier depends on an input signal to generate an output. A laser oscillator generates a beam on its own, getting a start from stimulated emission. In the simple example shown in Figure 9.7, the laser oscillator is a long, thin rod that contains excited atoms and has mirrors on both ends, which reflect most or all of the light back into the rod. If one spontaneously emitted photon goes along the length of the rod, it can stimulate emission of more photons. Suppose it's 30 centimeters long, and the output mirror transmits 10% of the light striking it and reflects the rest. An initial photon starting from one end will yield 17 photons at the other end. If 2 leak out and the other 15 are reflected back into the rod, each of the 15 will be multiplied 17 times on the way back, yielding 255 photons. The light bouncing back and forth will build up until it saturates. At that point the extra energy added to the beam on each round-trip between the mirrors will balance the light that escapes through the partly transparent mirror.

In this way, a laser oscillator generates its own beam without any external input (except the power to excite the atoms). Gain varies with wavelength, and the amplification process concentrates emission at the wavelengths where gain is strongest. In reality the laser amplifies more than one spontaneously emitted photon, and the one at the wavelength where gain is strongest is amplified the most. As you will learn later, the shape of the *laser cavity* or *resonator* also selects what wavelengths can oscillate in it.

A laser oscillator generates its own beam at a wavelength, which depends on the material and the laser cavity.

Many laser oscillators are somewhat different than the example. Often one mirror reflects all light back into the laser, while the *output mirror* allows a fraction to escape as the beam. In many semiconductor lasers, one mirror lets a little light escape for use in monitoring laser operation, while transmitting a larger fraction in a more powerful beam through the other mirror. The fraction of light transmitted depends on the type of laser. Gain per unit length also varies widely between materials. Many types of lasers exist, but only a few are used in fiber-optic systems.

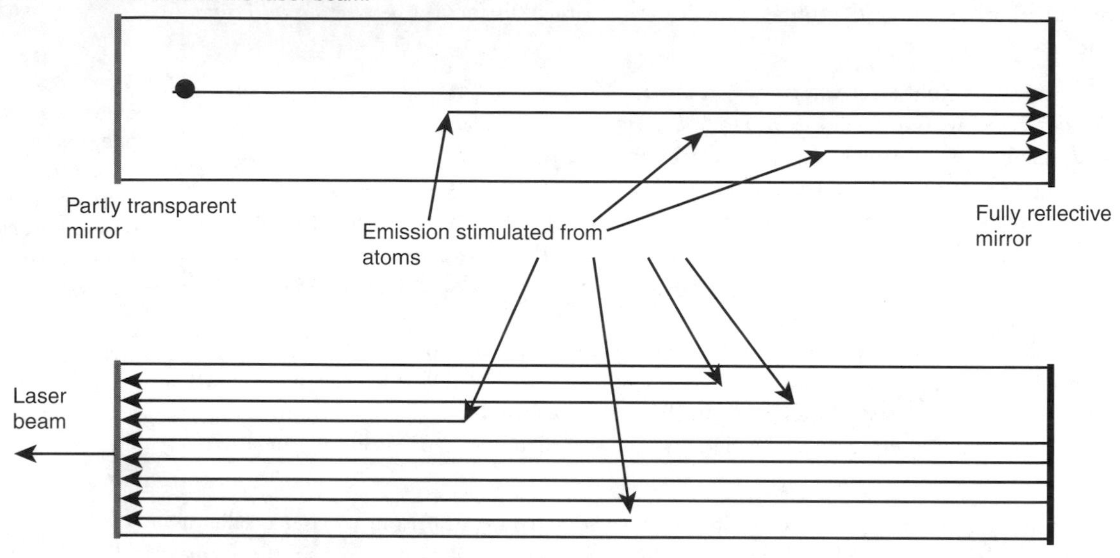

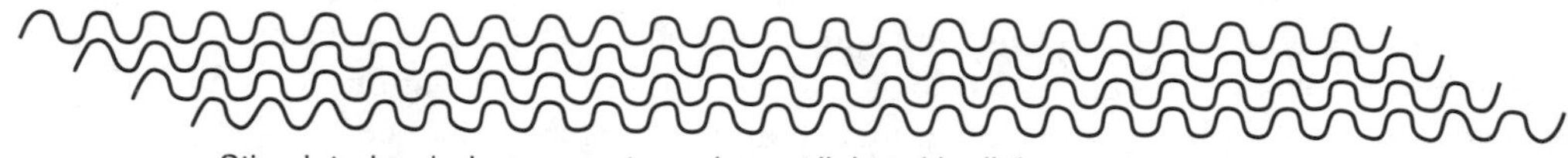

FIGURE 9.7
Laser emission.

Semiconductor Laser Sources

Semiconductor lasers resemble LEDs, but have different structures.

Semiconductor lasers are superficially like LEDs, but laser oscillation generates higher powers, confines output to a much narrower range of wavelengths, and produces somewhat more directional beams. The materials used are similar. Although some semiconductors used in LEDs are not suitable for operation as lasers, others are, and all semiconductor laser materials can work as LEDs.

The differences between lasers and LEDs arise primarily from the structure of the device that confines light and electrical current to produce laser action, and from the higher drive currents used for lasers.

Light Confinement

Efficient confinement or trapping of the stimulated emission is crucial in generating a laser beam. The more efficiently you can trap the light, the more efficient you can make your laser.

One example of light trapping is the core-cladding boundary in an optical fiber, which guides light along the core of the fiber. Conventional semiconductor lasers do something similar. The active layer at the *p-n* junction is made of a semiconductor compound with refractive index slightly higher than that of the surrounding material, as shown in Figure 9.8. Typically, the semiconductors do not differ much in composition, so they can be grown on top of each other with minimal difficulty. The junctions of different-composition semiconductors are called *heterojunctions,* and most diode lasers have them on the top and bottom of the active layer so they are called *double-heterojunction* (or double-heterostructure) lasers.

Stripe geometry in a laser confines light in a waveguide similar to the core of an optical fiber.

Looking closely at Figure 9.8, you can see that the whole junction plane is not made of the same high-index material. Only a narrow stripe is; the rest of the material in the junction plane is made of lower-index material. This is called a *stripe-geometry laser,* and the stripe serves as a rectangular waveguide similar in concept to the high-index core of an optical fiber. Typically the stripe is only a few micrometers wide and a fraction of a micrometer high and limits the laser to oscillating in a single waveguide mode. Light emerges from a spot at the edge of the wafer the same size as the waveguide, nicely matched to the core of a single-mode fiber. This extra confinement improves laser performance, and in general semiconductor lasers benefit from the tightest possible confinement of the beam within the semiconductor. (The figure does not show *electrical* confinement in the junction plane, which typically confines current flow through the narrow stripe portion of the junction.)

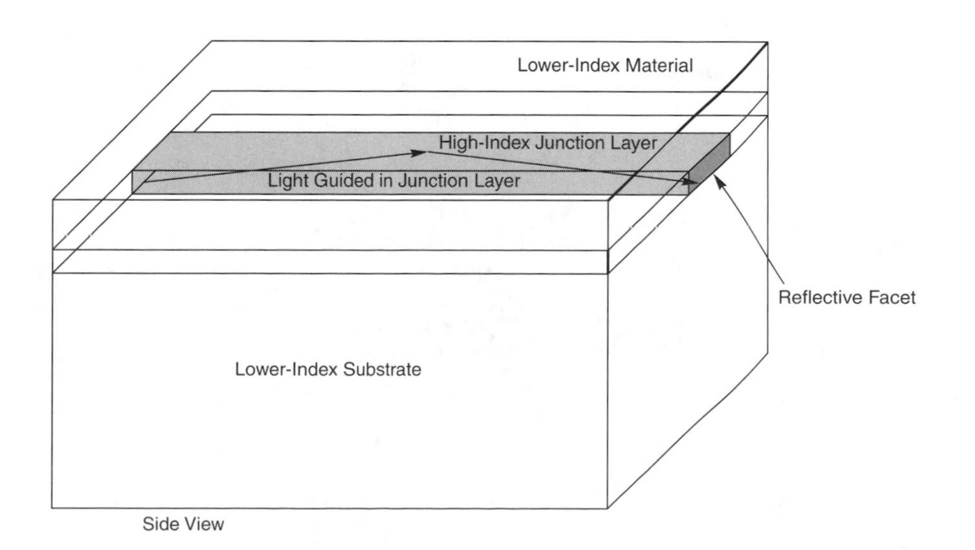

FIGURE 9.8
A double-heterojunction stripe-geometry laser.

The beam emerging from the laser is oblong in cross section. Because the emitting area is so narrow, the light waves diffract from it and spread more rapidly in the direction perpendicular to the junction (initially the thin dimension of the beam) than in the junction plane. Once you get a little way from the chip, the beam becomes a flattened cone.

A further refinement in light confinement is the use of quantum wells, extremely thin layers that further constrain where electrons and holes can recombine to emit light by differences in their internal energy states. We'll avoid the details of their operation, but you can think of quantum wells as a technique to further improve the confinement of light and energy to enhance the performance of semiconductor lasers.

Laser Beam Generation

A laser beam is formed by a resonator that reflects light back and forth through the laser medium.

Merely guiding the light along the laser stripe does not itself generate a laser beam. You must add a pair of mirrors, one at each end, to form a *resonator,* or *cavity.* Light emitted toward one mirror is reflected back through the laser medium, where it can stimulate more emission. Then it hits the other mirror and bounces back through the laser medium again. Light emitted in other directions leaks away if it is not confined. Thus stimulated emission amplifies only light aimed along the length of the laser, so it bounces back and forth between the resonator mirrors.

Figure 9.9 shows the process in more detail. In all lasers, at least one mirror must let some light escape through or around it to form the laser beam. The laser shown in Figure 9.9 is a conventional design called an *edge emitter,* because the "mirrors" are the cleaved edges of the semiconductor wafer, called *facets,* and light emerges from the edges of the wafer. The GaAs and InP compounds used in standard semiconductor lasers have high refractive indexes, so even uncoated facets reflect much of the light back into the wafer. In practice, the rear facet

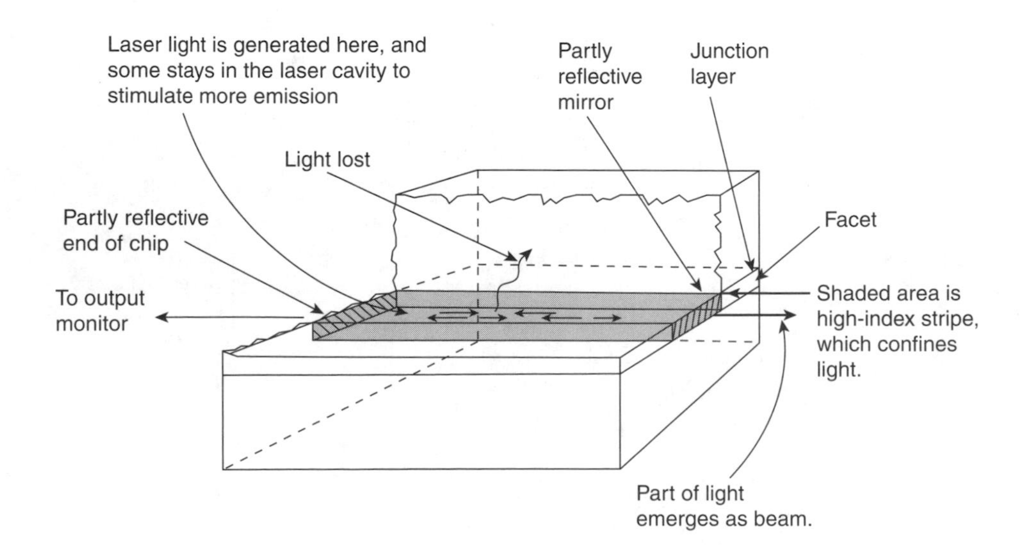

FIGURE 9.9
Operation of an edge-emitting semiconductor laser.

usually lets a small amount of light escape (which is often collected by a detector that monitors laser output power), but it may be coated to reflect most light back into the laser. In practice, the front facet is left uncoated so most light escapes, but some is reflected back into the semiconductor. Semiconductors have such strong gain that this is all the feedback needed to produce laser action.

Current Confinement

Laser structures concentrate drive current.

Laser structures also are designed to concentrate current flow through the diode. The concentration of recombining carriers at the junction or active layer increases with the current density. The likelihood of stimulated emission increases with the concentration of these excited atoms, so the higher the drive current density, the higher the carrier the concentration, and thus the higher the gain.

Designers manipulate current flow by adding insulating layers to block current flow through parts of the structure. If insulators cover all of the junction plane except the gray stripe in Figure 9.8, for example, all current will have to flow through the narrow stripe where the laser beam oscillates. This puts all the recombination in the zone where stimulated emission will contribute to the laser beam. In practice, many diode lasers have insulators in other layers, but the current confinement effect is the same.

Lasers Compared to LEDs

Both LEDs and diode lasers require a drive voltage greater than the band-gap voltage in order to generate light. Both are diodes, so they must be forward-biased, with positive bias applied to the *p*-doped material and negative bias to the *n*-type material. Operation of the two devices differs in other ways.

A diode laser must be driven above threshold current to operate as a laser.

The output power from an LED increases gradually as drive current increases from zero. However, a diode laser does not start generating a laser beam until the drive current exceeds a threshold value, as shown in Figure 9.10. When operated below that *threshold current,* the laser generates weak spontaneous emission like a rather inefficient LED; above threshold, its power and efficiency increase dramatically and the device starts operating as a laser.

The threshold is the point where the optical gain in the laser cavity exceeds the loss. As drive current increases, more carriers recombine, and are available for stimulated emission. This increases the gain within the laser resonator. Below threshold, the gain that light makes in a round trip of the laser cavity is lower than the losses it suffers from absorption and light escaping through the end mirrors. At threshold, the gain exceeds the loss, and above threshold stimulated emission increases very rapidly with drive current, as shown in Figure 9.10. Although the curve looks steep, the increment in output power does not exceed the increment in input power. Laser efficiency can be measured in two ways, either overall efficiency comparing the output power to input power, or the *slope efficiency,* which measures the extra power generated per increment in drive current. Diode lasers are much more efficient than LEDs, with slope efficiencies that can reach tens of percent.

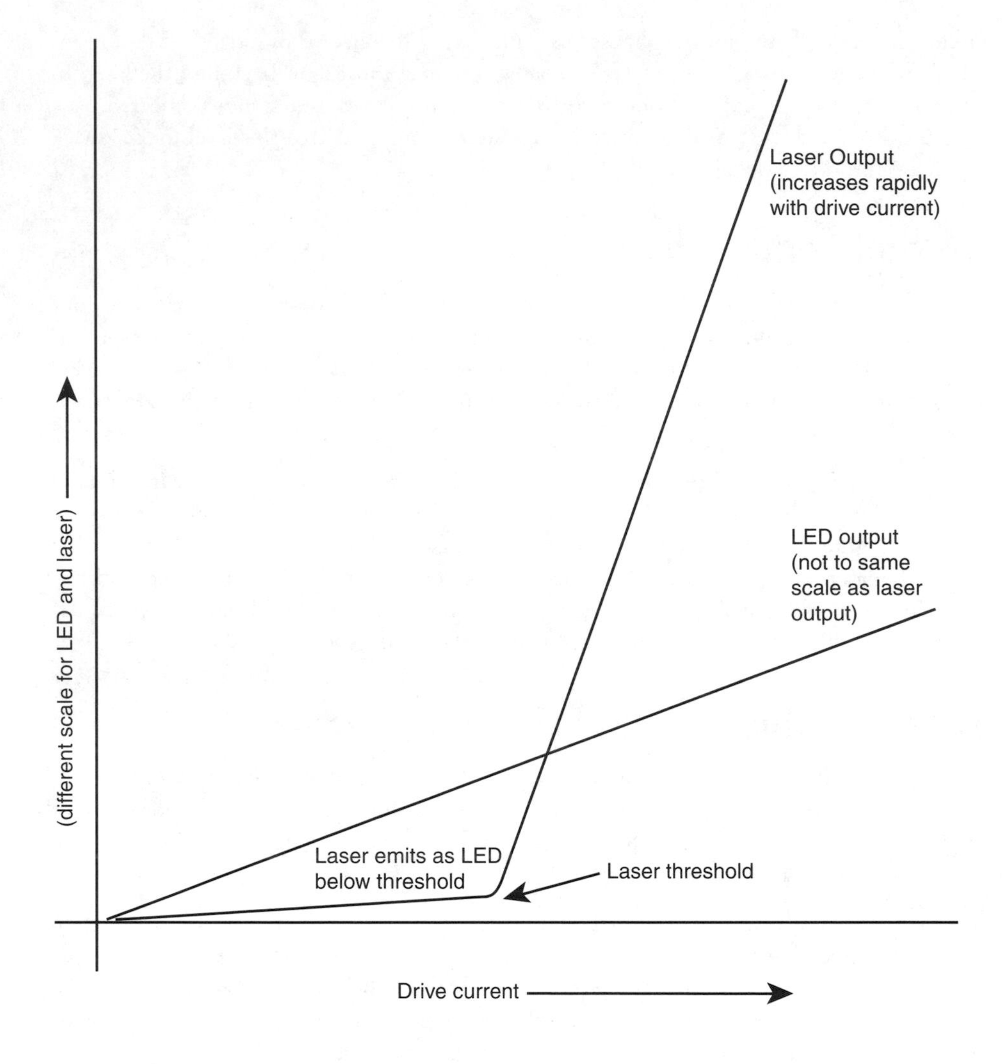

FIGURE 9.10
LED and laser power/current curves.

Diode lasers are much more efficient than LEDs.

LEDs don't have resonant cavities, because their light-emitting surfaces are made to suppress reflection, so they don't produce stimulated emission or have a threshold. That means their output increases steadily as drive current increases from zero, but the rate of increase is much less. (The LED power is not to scale in Figure 9.10.)

Important functional differences between lasers and LEDs arise from this distinction. Laser diodes are driven at higher currents per unit area than LEDs, which increases their light-generating efficiency at the cost of somewhat shorter operating lifetimes for gallium-arsenide lasers. The amplification inherent in laser action concentrates laser emission in a much narrower spectral range than LED output.

Vertical Cavity Semiconductor Lasers

An alternative design for semiconductor lasers is to have resonator mirrors above and below the active layer. This is called a *vertical-cavity surface-emitting laser* (VCSEL) because the light resonates vertically in the wafer, perpendicular to the junction, and emerges from the surface, as shown in Figure 9.11.

A VCSEL emits from its surface instead of its edge.

The mirrors in VCSELs are formed by depositing a series of layers with alternating compositions, so they selectively reflect a narrow range of wavelengths, depending on their thickness and refractive indexes. This multilayer reflector uses an effect similar to a fiber grating, but the concept of selective reflection by a stack of many thin-film layers coated on optics came before either fiber gratings or VCSELs. A VCSEL emits from a round spot on the surface of the wafer that is larger than the core of a single-mode fiber, but small enough to fit into a graded-index multimode core. Because the emitting area is circular and larger than an edge-emitting laser, a VCSEL beam is circular and does not spread out as rapidly. (The differences arise from diffraction effects.)

VCSELs have a number of attractions. They have low threshold currents and are quite efficient in converting electrical input into light. This means they consume little power and have to dissipate less heat than edge-emitting lasers, giving them a longer lifetime than edge-emitting diode lasers. They are easy to manufacture and package. Like other semicon-

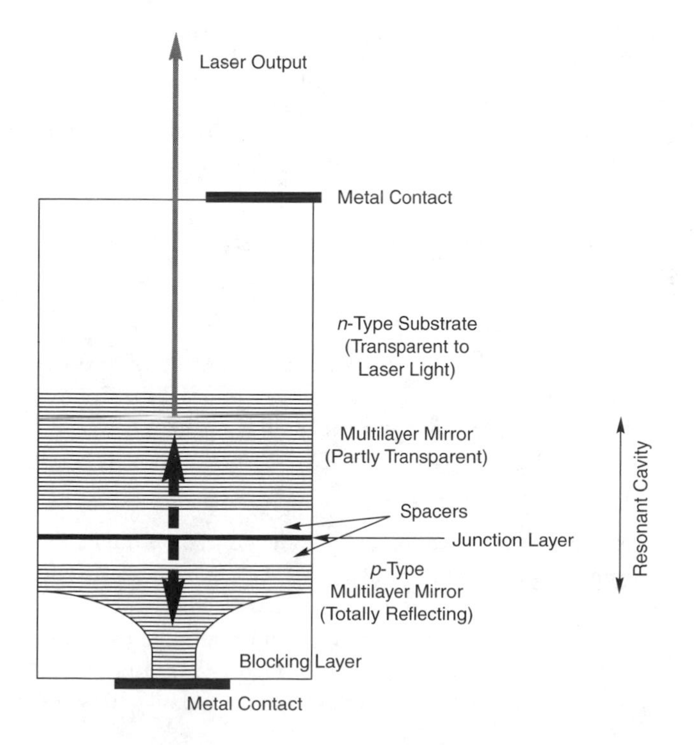

FIGURE 9.11
A vertical-cavity surface-emitting laser.

ductor lasers, they can be directly modulated at high speeds—so they can generate signals well over 1 Gbit/s.

At this writing, most VCSELs are made from GaAs semiconductor materials, which meet the requirements for making multilayer mirrors much better than InP-based compounds. The present strength of single VCSELs is as inexpensive light sources at wavelengths of 750 to 1000 nm, where they offer better lifetimes and lower costs than edge-emitting GaAs lasers and serious competition for LED sources. A crucial advantage of VCSELs is their ability to generate higher-speed signals than LEDs. This makes present VCSELs quite attractive for gigabit networks that transmit signals up to a few kilometers.

Unlike other diode lasers, VCSELs can be fabricated in two-dimensional arrays covering the surface of a wafer, which can generate separately modulated outputs that emerge from the chip surface. Such arrays of independent emitters are attractive for optical switching and signal processing, with beams going through free space as well as through fibers.

Diode Laser Wavelengths and Materials

Material composition determines the wavelengths where a diode laser has gain.

The output wavelengths of diode lasers depend both on the semiconductor compound and on the device structure. The material composition determines the range of wavelengths over which the laser has gain and can produce stimulated emission. (Composition also determines the wavelengths where LEDs can produce spontaneous emission.) The device structure selects what wavelengths in that range the laser will emit.

The diode lasers used in fiber-optic systems are made of III-V semiconductor compounds that can be fabricated on substrates of gallium arsenide or indium phosphide. The principal compounds are:

- $Ga_{(1-x)}Al_xAs$ on GaAs for 780 to 850 nm
- $In_{(1-x)}Ga_xAs_{(1-y)}P_y$ on InP for 1100 to 1700 nm
- $In_{(1-x)}Ga_xAs$ on GaAs for 980 nm

Note that some of these lasers are not used directly in fiber-optic transmitters. InGaAs lasers emitting at 980 nm and InGaAsP lasers emitting at 1480 nm are used to pump erbium-doped fiber amplifiers. In practice, almost all other InGaAsP lasers have been used for long-distance communications at 1310 and from 1525 to 1625 nm, although other wavelengths may find applications in short-distance systems.

The composition of the active layer determines the peak gain wavelength. For example, the gain of InGaAsP peaks at 1310 nm for a composition of $In_{0.73}Ga_{0.27}As_{0.58}P_{0.42}$. As you learned earlier, the process of stimulated emission amplifies the peak wavelength more than other wavelengths, narrowing the range of output wavelengths from the broad spectrum seen in an LED to the narrower range of a diode laser, shown in Figure 9.1. Other layers may have slightly different composition, and the whole structure is deposited on a substrate of either GaAs or InP, which are much easier to make in the large volumes needed for substrates.

Only certain compositions are possible in semiconductor lasers. The blend of elements in the compound affects both the band gap (which determines the output wavelength) and the spacing between atoms in the crystal. Compositions of layers in the laser must have interatomic spacing reasonably close to that of a readily available substrate material. Recent advances have eased, but not eliminated, these restrictions, by allowing some strain to accumulate between layers.

Layers in the semiconductor should have atomic spacing close to that of the substrate.

Laser Output Spectra

The spectral range of diode lasers depends on their structure. The simple edge-emitting described earlier with one mirror on each end is called a *Fabry-Perot resonator* because it's based on an optical device that Charles Fabry and Alfred Perot designed long before the laser was invented. Figure 9.12 gives a close look at the discrete lines that make up the output spectrum of a typical narrow-stripe Fabry-Perot laser. One line is strongest, but the laser oscillates simultaneously at several wavelengths. This occurs because the laser cavity allows light to resonate back and forth only if the round-trip distance between the two mirrors is an integral number (N) of wavelengths λ,

Fabry-Perot lasers have spectral bandwidths of 1 to 3 nm and emit multiple longitudinal modes.

$$2Dn = N\lambda$$

where D is the distance between the mirrors and n is the refractive index of the laser material. Each wavelength spike in Figure 9.12 corresponds to a different value of N. They span the range of wavelengths where the material gain is highest, which for diode lasers gives a typical spectral width of 1 to 3 nm.

Each spike in Figure 9.12 is a separate *longitudinal mode* of the laser, which means a resonance along the length of the laser cavity. (The modes in an optical fiber are *transverse modes*, defined by the width of the fiber; narrow-stripe diode lasers operate in a single trans-

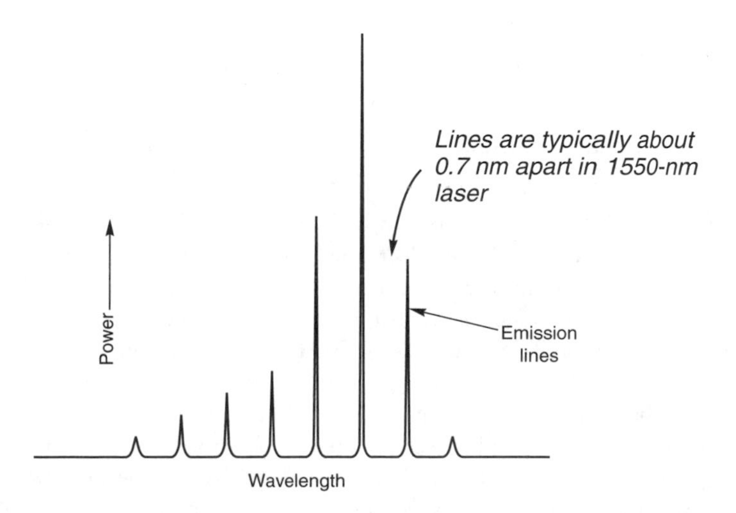

FIGURE 9.12 *Wavelengths in multiple longitudinal modes.*

verse mode.) Each of these longitudinal modes has much narrower spectral width than the entire envelope of modes emitted by the laser. The spacing between longitudinal modes depends on the cavity length and wavelength. The longer the cavity length (measured in wavelengths), the closer the modes are spaced. Edge-emitting Fabry-Perot diode lasers have short cavities, only about 0.5 m long, and their modes are about 0.6 nm apart at 1300 nm or about 0.7 nm apart at 1550 nm.

The same principles apply to VCSELs, which also have pairs of mirrors, like those for resonant cavities. However, VCSEL cavities are much shorter than edge emitters, so their longitudinal modes are separated much more widely.

Minor fluctuations during operation can make Fabry-Perot lasers "hop" between modes, shifting the emission wavelength. This is easiest to envision as the emission peak in Figure 9.12 moving from one longitudinal mode to another.

This and other problems limit Fabry-Perot lasers to systems transmitting at speeds no more than several hundred megabits per second. Optical channels must be widely spaced to prevent overlap, so lasers emitting multiple longitudinal modes can be used only in "coarse" wavelength-division multiplexing, with channel separation of at least several nanometers.

Single-Frequency Lasers

Single-frequency lasers are needed for high-speed transmission.

Distributed-feedback and distributed-Bragg reflection lasers emit only a single frequency.

For high performance, low dispersion, and closer spacing of optical channels, laser emission must be limited to a single longitudinal mode or, equivalently, to a single frequency. This has led to development of more elaborate laser resonators. Figure 9.13 shows three leading approaches.

The *distributed-feedback (DFB) laser,* in Figure 9.13(a), has a series of corrugated ridges on the semiconductor substrate, which scatter light back into the active layer. This provides feedback like the cavity mirrors on a Fabry-Perot laser, although the details of the physics are different. The *distributed Bragg reflection (DBR) laser* shown in Figure 9.13(b) works in much the same way, but the grating is etched in a region outside the zone that is pumped by electric current. In both cases, the grating ridges are spaced evenly so they scatter only a narrow range of wavelengths back into the active layer of the laser. The active layer of the laser amplifies only this selected range of wavelengths, producing very narrow spectral bandwidths at a nominal "single frequency." The wavelength depends on the line spacing in the grating and the refractive index of the semiconductor. Recall that Bragg reflection is the same effect that selects the wavelengths reflected by a fiber Bragg grating.

A different way to stabilize laser wavelength is by placing an edge-emitting semiconductor laser within an external cavity, which selects the emission wavelength. This requires coating one or both facets to suppress reflection back into the semiconductor, and adding one or two external mirrors to extend the resonator cavity beyond the laser chip. A wavelength-selective element also is added to the laser cavity. In the simple design of Figure 9.13(c), the tuning element is a diffraction grating, which serves as one external mirror, reflecting light at an angle that depends on its wavelength. (You can get the same effect by inserting a prism or some other wavelength-selective component into the laser cavity, but diffraction gratings are easier to use.) The laser chip alone emits a range of wavelengths, but when they strike the grating, most wavelengths are reflected at angles that take them

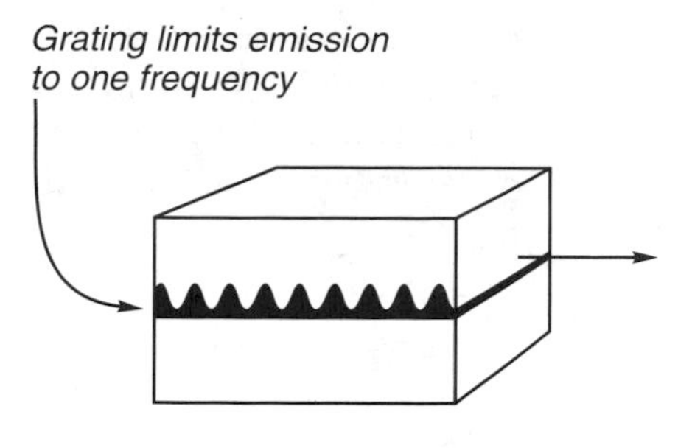

a. Distributed feedback laser.

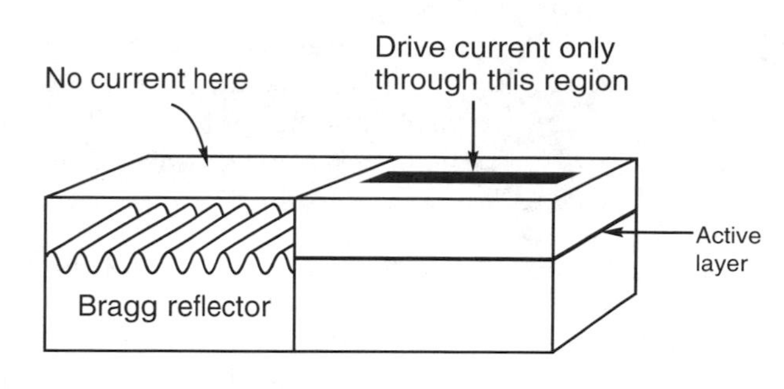

b. Distributed Bragg reflection laser.

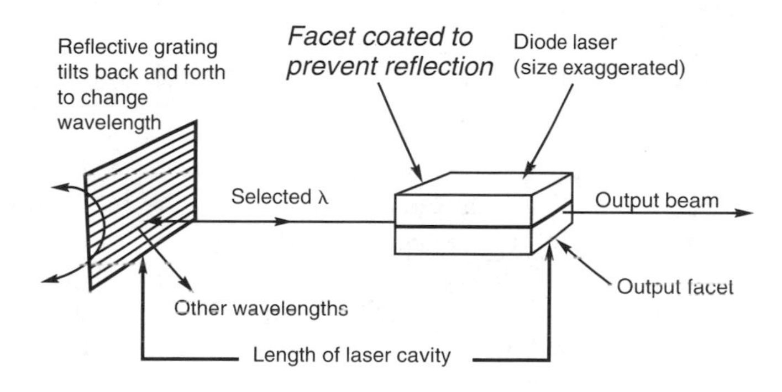

c. External cavity tunable laser.

FIGURE 9.13
Three single-frequency lasers.

away from the laser chip. Only a very narrow range of wavelengths are at the right angle to be reflected back into the laser chip for further amplification. This limits output to a single frequency.

Distributed-feedback and distributed-Bragg-reflection lasers are the types most often used to generate a single, fixed wavelength with very narrow spectral width. However, there is growing interest in lasers with output that can be tuned to emit at precise wavelengths in WDM systems.

Tunable Lasers

An external cavity laser is a good starting point for discussing *tunable lasers,* which can be changed in wavelength. You have already seen how a diffraction grating can reflect a single wavelength back into the laser chip for amplification, building up emission at a single wavelength. Turning the grating changes what wavelengths are reflected back to the laser chip, which changes the laser's output wavelength.

Tuning of the laser wavelength is important for optical networking and WDM.

As you will learn later, wavelength tunability is an attractive property for lasers used in WDM systems and optical networking, as well as for measurement instruments. Standard lasers emit only a fixed wavelength, so a system with 80 different wavelengths requires 80 different models of laser. Moreover, the service department needs spares for every one of those 80 different laser models. If a telephone company wants to install 80-channel systems, its maintenance department would need to stock every site with spares for each of the 80 wavelengths. The logistics could become a nightmare.

Tunability also can enhance the flexibility of optical networking. For example, it sometimes may be necessary to move an optical channel from one wavelength to another, because the same wavelength isn't available along its entire route. Using a tunable laser to generate the new signal would allow the wavelength to be changed without switching lasers. The laser might be tunable continuously across the spectrum, but system design would be easier if lasers were preset to emit precisely at standard wavelengths. Users would then select an optical channel, like viewers select a channel on a modern television set, without having to adjust the laser to match the desired frequency. The technology is still in its infancy, but several approaches have been developed.

Changing length of a VCSEL cavity can tune its wavelength.

We saw earlier that the resonant wavelength depends on the length of the laser cavity. Changing the cavity length has little effect on edge-emitting lasers, because their longitudinal modes are less than a nanometer apart. However, it can tune the wavelength significantly because a VCSEL cavity can be made very short, micrometers long rather than hundreds of micrometers long for edge-emitters. With a short enough cavity, only a single longitudinal mode falls within the laser's gain band. This makes possible the sort of tunable VCSEL shown in Figure 9.14. A thin, partly transparent mirror is held above the VCSEL by a movable micro-electromechanical system (MEMS) device. Vertical motion of the MEMS mirror changes the resonant wavelength in the VCSEL cavity, tuning the wavelength by over 30 nm in the laboratory.

Distributed-feedback and distributed-Bragg reflection lasers can be tuned in other ways. As you saw earlier, the wavelengths selected depend both on the grating period and on the refractive index of the material. Changing the temperature of the material can change both, by causing thermal expansion (or contraction) of the laser material as well as by directly affecting refractive index. Passing a current through a material also affects the refractive index. Generally these changes are relatively small, and allow turning over only several nanometers.

Tuning ranges can be extended to tens of nanometers by using more elaborate distributed Bragg reflectors. One example is the *sampled grating distributed Bragg reflector (SG-DBR),* which contains regions with different grating spaces that reflect a comb-like series of wavelengths. To make a tunable laser, slightly different separate sampled-grating reflectors are fabricated on each end of the active region of the laser. The laser can oscillate only at a wavelength reflected by the gratings on both ends, which can be tuned independently by changing current level or temperature. These changes shift the reflection peaks of the individual grating only slightly, but this small shift causes a much larger shift in the wavelength at which both gratings reflect to allow laser oscillation, as shown in Figure 9.15. This is sometimes called a vernier effect, because a vernier scale works in the same way to amplify the size of a small change. A related approach is the *grating-assisted coupler and sampled reflector* (GCSR), which uses a sampled grating reflector on one end and a wavelength selective coupler on the other.

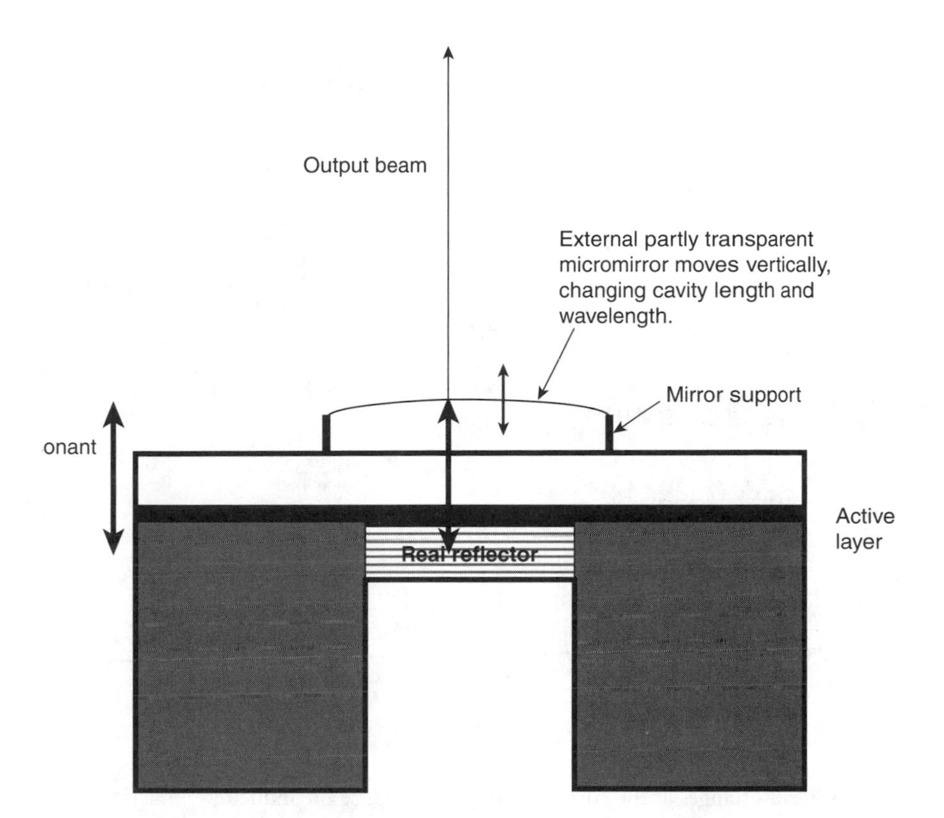

FIGURE 9.14
Tunable VCSEL relies on a moving external micromirror to change cavity length and thus wavelength.

Another approach to tuning is selecting one laser stripe from an array of several on a single semiconductor substrate. For example, if each of 12 stripes had a tuning range of 3 nm, and their center wavelengths were 3 nm apart, they could combine to cover a 36-nm range. The device could tune across one laser's 3-nm range, then switch to the next laser and tune over its range.

These approaches all have been demonstrated and are in commercial development. Other techniques are also in development. Tunable lasers do face some serious practical challenges. They must be locked to the right wavelengths for each optical channel, so they don't drift during operation. They also must be affordable, reliable, and usable in standard telecommunications facilities. The field will be interesting to watch, and is a great opportunity for innovative ideas.

Modulation and Chirp

External modulation avoids laser wavelength chirp, which causes chromatic dispersion.

Direct modulation via changing the drive current is the simplest and cheapest way to modulate a signal onto the beam emitted by a diode laser. Unfortunately, it has the undesirable side effect of shifting the laser's wavelength during the emitted pulse. The electron density

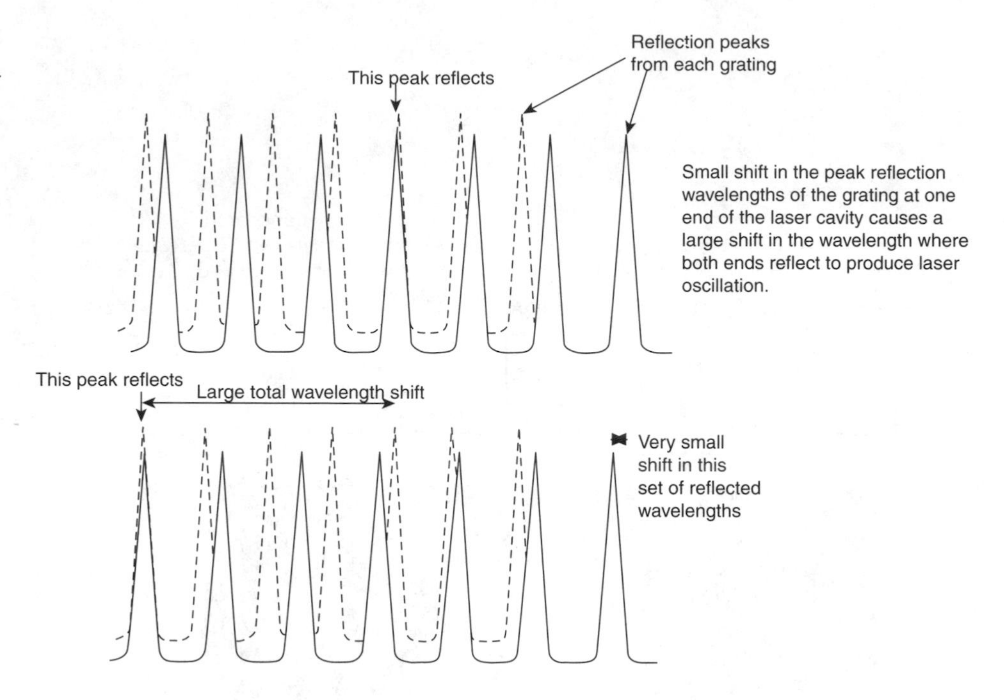

FIGURE 9.15 *Tuning of sampled-grating distributed Bragg reflector laser.*

in the semiconductor changes as the current changes, and the semiconductor's refractive index varies with the electron density. This means that modulating the current effectively changes the optical path length in the semiconductor, which equals the refractive index n times the physical distance through the semiconductor L. From the earlier equation for the resonant wavelength in an optical cavity, you can see that this means wavelength λ changes by an amount $\Delta\lambda$:

$$\Delta\lambda = \frac{2(\Delta n \times L)}{N}$$

where Δn is the change in refractive index and N is an integer, the number of wavelengths needed to make a round trip in the cavity. Although the change, called *chirp,* is small, it occurs during every laser pulse, so every pulse contains a broader range of wavelengths than it otherwise would include. The resulting dispersion can impair long-distance transmission at speeds above about 1 Gbit/s.

The cure for chirp is to drive the laser with a steady current, then modulate the steady beam externally. This is done by applying the signal to a modulator, so the fraction of the laser beam it transmits varies in proportion to the signal. You'll learn more about modulators in Chapter 16; for now all you need to know is that they modulate beam intensity but do not affect its wavelength. External modulators also are very fast, working at speeds to 40 Gbit/s. They generally are used in long-distance systems transmitting at 2.5 Gbit/s or higher.

Driving the laser source with a steady current also improves its inherent wavelength stability, because any modulation induces fluctuations in laser properties.

As mentioned earlier, wavelength is temperature-sensitive, so stabilizing the laser's operating temperature controls variations in its wavelength. Stabilization is required for high-speed transmitters.

Reliability

Reliability was a big problem with early GaAs lasers, but great improvements have been made in the technology. Nonetheless, LEDs are more reliable than edge-emitting semiconductor lasers because lasers have much higher current densities and optical power outputs. VCSELs are the most reliable lasers. InGaAsP lasers and LEDs are inherently more reliable than similar devices made from GaAs, partly because longer-wavelength photons carry less energy and partly because InGaAsP is less sensitive to temperature.

LEDs are more reliable than edge-emitting lasers; InGaAsP devices are more reliable than GaAs devices.

Operating temperature is a key factor in reliability. The threshold current of a semiconductor laser tends to increase with temperature, increasing the amount of waste heat generated within the laser. The excess heat can further increase temperature, degrading efficiency even further. Gallium arsenide is more vulnerable to this problem, which can lead to thermal runaway if laser temperature is not controlled. In addition, laser lifetime decreases as temperature increases, so accelerated aging tests are conducted at high temperatures. To control these problems, many laser transmitters are operated with active temperature stabilization, such as thermoelectric coolers. Most lasers are packaged with heat sinks, even if active cooling is not required.

An important cause of failure in semiconductor lasers is a slow decline in output power with age. To compensate for this, transmitter circuits can be designed to slowly increase drive current to maintain a constant power. In this case, failure is defined as the point at which a laser can no longer deliver the required power.

Semiconductor lasers are particularly vulnerable to damage by electrostatic discharges during handling. This problem can be overcome by care in packaging and handling, but you should be aware of its potential, and take care to ground yourself when handling lasers.

Semiconductor Optical Amplifiers

If you remove the mirrors from the ends of any laser, it can amplify light passing through it. This is true for semiconductor lasers as well as for fiber lasers, although semiconductor optical amplifiers have not gained the wide acceptance of fiber amplifiers.

A semiconductor laser without a resonant cavity can function as an optical amplifier.

A mirrorless semiconductor laser can amplify light guided through the active stripe in the junction layer. The weak light signal stimulates recombining electrons and holes to emit light at the same wavelength. The amplification per unit length—"gain" in laser terminology—is high for semiconductor lasers, so a relatively short length of semiconductor laser can amplify light many times.

I mentioned earlier that semiconductor lasers generally do not have external mirrors but rely on light reflection from the cleaved facets on the edges of the wafer. This presents a

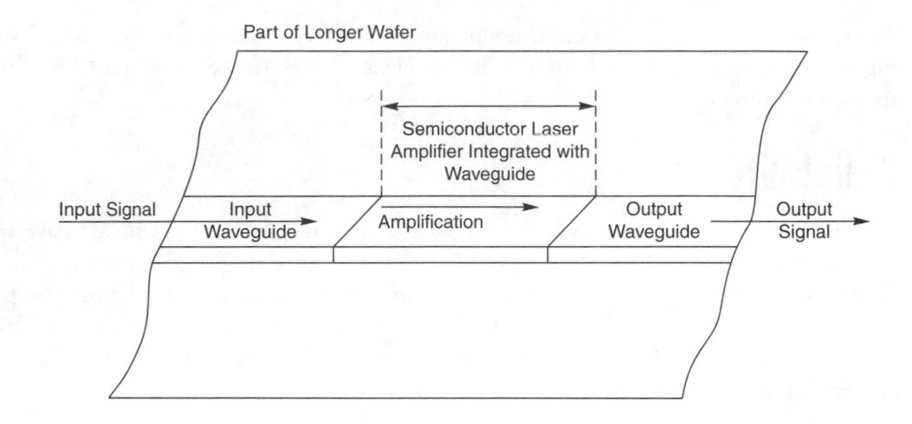

FIGURE 9.16
Semiconductor laser amplifier integrated with waveguide.

problem in designing semiconductor laser amplifiers, because such reflection could generate undesirable noise and laser oscillation that would overwhelm the amplified signal. To overcome this, the facets of semiconductor laser amplifiers used as separate devices are coated with an antireflective material. An alternative is to make the semiconductor laser amplifier part of a larger monolithic semiconductor device, as shown in Figure 9.16. In this case, a waveguide fabricated in the semiconductor layer delivers light to the amplifier stage, and another waveguide collects the light. (An isolator may be needed to prevent reflection back into the laser-amplifier stripe.) Because there are no facets at the ends of the laser, there is no reflective feedback.

Chapter 12 will describe more about the use of optical amplifiers in fiber-optic systems.

Fiber Lasers and Amplifiers

Fiber amplifiers require pump light to amplify an input signal.

Fiber lasers and amplifiers also are based on the laser principle, so they also generate light by stimulated emission. The basic principles of their operation are similar to those of diode lasers. In fiber lasers there is a resonant cavity where light oscillates to generate a beam. Fiber amplifiers lack end mirrors and a resonant cavity, and light being amplified passes through them only once. However, the details differ significantly because fiber light sources differ greatly from semiconductor diode emitters.

Energy Sources for Fiber Amplifiers

Semiconductor light sources get their energy from electric currents passing through them, releasing energy at the junction between two layers of different composition. Fiber amplifiers and lasers get their energy from external light sources. This makes them part of a larger family of *solid-state lasers* in which the light is generated by atoms embedded in a glass or crystalline host. (In the laser world, semiconductor lasers are *not* considered to be solid-state lasers because they are excited by current passing through a diode structure.)

The most important type of fiber amplifiers and lasers are based on the erbium-doped fibers described in Chapter 7. Other fiber lasers and amplifiers work on the same principles.

The light-emitting erbium atoms are mixed into the glass of the fiber core. Light from an external source travels through the fiber to excite the erbium atoms, putting them in a state where they can be stimulated to emit light at a range of wavelengths from about 1530 to 1625 nm. In fiber amplifiers the stimulation comes from a weak input signal somewhere in that wavelength range; in fiber lasers, the stimulation comes from spontaneous emission from other erbium atoms, fed back into the fiber by mirrors on the ends of the erbium-doped fiber.

The pump light must be at certain wavelengths, which excite the erbium atoms to the desired state, as was shown in Figure 9.6. The usual excitation wavelengths are 980 nm or 1480 nm. In practice, both are supplied by semiconductor lasers that illuminate the core of the fiber from one or both ends. The choice of wavelength depends on how the amplifier design is being optimized.

Erbium-doped fiber amplifiers and lasers are pumped by diode lasers emitting 980 or 1480 nm.

The fiber structure is particularly convenient for concentrating the erbium atoms and the light they generate, but it is not necessary. Other solid-state lasers are simply rods doped with a suitable light-emitting element, and pumped from the sides by light from intense lamps or semiconductor diode lasers.

Fiber Amplifiers

A fiber amplifier does not originate a signal, but instead amplifies a weak signal that enters its input end. Energy for that amplification comes from the 980- or 1480-nm pump laser, so in a sense a fiber amplifier transfers energy from the pump beam to the amplified signal.

Figure 9.17 shows one possible structure for a fiber amplifier, with the pump laser directed sending light in the direction opposite to the input signal being amplified. The pump beam also can travel in the same direction as the amplified signal. For simplicity, the figure shows only one wavelength in the input signal, but fiber amplifiers can simultaneously amplify optical channels at all wavelengths that fall within their gain band. The weak 1550-nm signal enters from a fiber on the left, and passes through an optical isolator, a device that transmits light only in one direction. The isolator blocks stray light that could return to the laser transmitter and generate noise. A separate optical filter blocks the pump beam from reaching the transmitter, while transmitting the signal beam.

Fiber amplifiers can amplify many wavelengths in their gain band at the same time.

Amplification occurs in a length of erbium-doped fiber. The pump laser steadily illuminates the doped fiber, exciting the erbium atoms so the input signal can stimulate emission. It's very important that the pump laser be operating; if the pump laser is off, the erbium atoms strongly *absorb* input signals in the 1550-nm region. Thus an amplifier with a burned-out pump laser is much worse than no amplifier at all—because it strongly absorbs the input signal.

If the pump laser is off, an erbium-fiber amplifier absorbs the input signal.

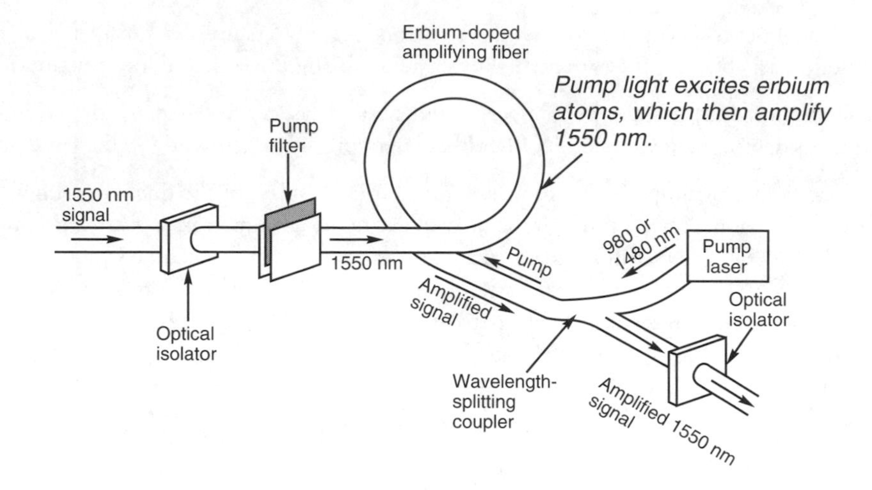

FIGURE 9.17 *Erbium-doped fiber amplifier.*

The power of the amplified signal builds up as it passes through the doped fiber. At the far end, the beam passes through a wavelength-splitting coupler that directs signals in the 1550-nm erbium band to the output fiber. Light from the pump laser enters through the other arm of the coupler. The amplified signal may pass through another optical isolator before entering the output fiber.

The erbium-amplifier C-band is 1530 to 1565 nm; the L-band is 1570 to 1625 nm.

Erbium-doped fiber amplifiers have gain across a wide range of wavelengths, so they can simultaneously amplify signals on many optical channels. One limitation is that the gain varies strongly with wavelength, as will be covered in more detail in Chapter 12. To overcome this problem, erbium-doped fiber amplifiers typically are optimized for operation in one of two bands.

The C-band, from about 1530 to 1565 nm, spans the wavelengths where gain is highest, so it typically requires only several meters of fiber. The L-band, from about 1570 to 1625 nm, has much lower gain, and typically requires around 100 meters of fiber. These bands have not been fully standardized, so they may vary among manufacturers. The band designations come from *central* and *long* wavelength bands. The gap between the two bands is intentional, to assure complete separation. The gain spectrum depends on the material used in the fiber; it differs for fibers made of fused silica and fluoride compounds, for instance.

The amplified light makes only a single pass through the fiber, from left to right in Figure 9.17. Typical amplification totals 10 to 30 dB, a factor of 10 to 1000, on each channel, although the degree of amplification may differ across the amplifier's wavelength range. Total output power—including all channels—can exceed 100 mW. As single-mode fibers, erbium-fiber amplifiers are easy to couple to communications fibers. They are not sensitive to the polarization of input light. Their wavelength range is nearly ideal because it spans tens of nanometers where optical fibers have their minimum attenuation. Their response to input signals is virtually instantaneous, giving them extremely high bandwidth. This combination of properties has made erbium-doped fiber amplifiers a spectacular success.

Fiber amplifiers are inherently analog devices that amplify input signals regardless of their speed and format. As analog devices, they can distort the signals they amplify. Fiber amplifiers reach *saturation* when they run out of energy to amplify a signal further. They can be limited by pump power, or by the need to amplify signals at many optical channels simultaneously. Noise and crosstalk are low, but as you will learn in Chapter 12, they can limit amplifier performance.

Fiber amplifiers are analog devices not sensitive to signal format. They saturate at high powers.

One important type of noise deserves mention here: *amplified spontaneous emission.* Recall that the difference between an optical amplifier and a laser is the source of the light that initially stimulates emission. In an optical amplifier, the source is a weak input signal. In a laser, the source is spontaneous emission by excited atoms. Spontaneous emission also occurs in optical amplifiers. It's normally weak, but creates random optical noise in the background. The optical amplifier can't distinguish between this background noise and the signal, so it amplifies both, creating a background noise across the entire spectral range, a bit like the low background hiss from a cassette tape recorder.

Nonerbium Fiber Amplifiers

The idea of a fiber amplifier is quite general. In principle, you can make fiber amplifiers by doping any suitable light-emitting element into the glass core of a fiber. These devices then would amplify whatever wavelengths fell in that element's band.

Materials other than erbium can be used in fiber amplifiers. Praseodymium amplifies 1310 nm; thulium amplifies 1450 to 1500 nm.

In practice, there aren't very many other good fiber amplifiers. Most solid-state laser materials either don't work well in fiber form, or don't have the right wavelengths. Neodymium is an excellent solid-state material, but its strongest wavelength is 1064 nm, outside of the useful range, and a weaker line near 1300 nm does not work well in a fiber amplifier. Ytterbium is an excellent material for high-power fiber lasers, as described below, but its 1120 nm wavelength is too short for telecommunications.

So far, the best candidates for doped fiber amplifiers are two other "rare earth" elements in the same chemical group as erbium and neodymium. Praseodymium can amplify signals in the 1310-nm region, but is far from ideal, and that band has fallen out of favor. Thulium can amplify signals at about 1450 to 1500 nm, a region sometimes called the S (for short-wavelength) band, but it is still in the early stages of development. Neither have found wide use.

Raman fiber amplifiers work on a different principle and are covered in Chapter 12.

Fiber Lasers

Add mirrors to a fiber amplifier, pump it at the right wavelength, and it can become a laser. The mirrors provide the resonant cavity needed for the laser to generate its own light, or *oscillate.* As in a diode laser, a few excited atoms release their excess energy as light, which stimulate the emission of more photons, building up a laser beam that resonates within the laser cavity.

Adding mirrors can make erbium-doped fibers into lasers.

Fiber lasers are not widely used in fiber optics, but they have found some applications, and may be used as WDM sources, tunable lasers, or where high powers are needed. A few types of fiber lasers are available, but the most important for fiber optics is the erbium fiber laser, which operates in the same wavelength range as its close cousin, the erbium-doped fiber amplifier.

A diffraction grating can tune the wavelength of a fiber laser, and generate narrow-line output.

The fiber laser may be a length of fiber with mirrors on both ends, or as a loop with couplers that deliver light from the pump laser and split off the output beam, as shown in Figure 9.18. As with external-cavity semiconductor lasers, adding a diffraction grating to the cavity of a fiber laser can limit emission to a narrow line and/or allow wavelength tuning. Other types of cavities also could make fiber lasers suitable sources for WDM systems.

A fiber laser cannot be directly modulated with an electrical signal like a diode laser. Instead it must be modulated externally. Typically the fiber laser generates a continuous beam at a narrow range of wavelength, which is modulated externally.

Fiber lasers can generate trains of pulses shorter than one picosecond.

Alternatively, fiber lasers can be designed to generate a series of very short pulses lasting less than one picosecond (10^{-12} s). The ring fiber laser shown in Figure 9.18 can be adapted for this purpose by adding an optional internal modulator to control light pulses within the laser cavity. This process, called *modelocking,* generates uniformly spaced pulses, each containing a very wide range of wavelengths. External optics can select pulses to be transmitted as a signal. It's also possible to spread out the wavelengths in each pulse, using external optics with high dispersion, then slice the pulses up to give a series of pulses at many separate wavelength channels. Applications of this concept are still in the research stages.

The highest powers available from any fiber laser come from ytterbium-doped fibers, which have generated more than 100 watts in the laboratory. Their 1120-nm wavelength does not match the needs of current fiber systems, but they can be used in industrial laser systems.

FIGURE 9.18
Erbium-fiber ring laser.

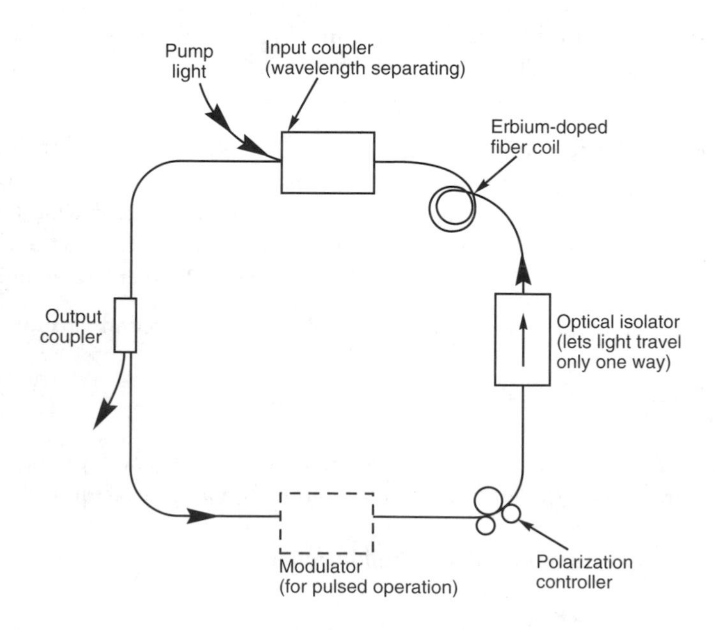

Other Solid-State Laser Sources

Other solid-state lasers also can be signal sources in fiber-optic systems. As in fiber lasers, light from an external source excites atoms in a solid material between a pair of mirrors. The laser material is a glass or crystal host doped with atoms that emits light by stimulated emission. However, the laser material is usually shaped as a rod, is much shorter than a typical fiber laser, and lacks an internal light-guiding structure.

Diode-pumped neodymium lasers can generate more than a watt near 1300 nm.

The most important of these are crystalline neodymium lasers, in which the rare earth neodymium is doped into crystals called YAG (for yttrium aluminum garnet) or YLF (for yttrium lithium fluoride). These lasers can be excited by GaAs diode lasers emitting near 800 nm. The primary neodymium line is near 1060 nm, but there are secondary lines at 1313 and 1321 nm in YLF and 1319 nm in YAG. Those fall right in the 1300-nm fiber window, but are not available from fiber lasers.

Like erbium-fiber lasers, solid-state neodymium lasers cannot be modulated directly; they require external modulators. Their big attraction is their ability to generate high power—more than a watt near 1300 nm. That's more than you want to send signals through a single length of fiber, but it can be split among many fibers to carry the same signals to many terminals, for network communications or cable television signal distribution.

What Have You Learned?

1. Wavelength, spectral width, output power, and modulation speed are key considerations in fiber-optic light sources.
2. Modulating the light source by changing the drive current is called direct modulation; it works for diode lasers and LEDs. External modulation uses a separate external device to modulate a steady beam from a light source.
3. The most common light sources for short glass fiber systems are GaAlAs LEDs or lasers operating at 820 or 850 nm. InGaAsP, lasers emit at 1200 to 1700 nm. Some InGaAsP LEDs are used at 1300 nm.
4. Red LEDs are the usual signal sources for plastic fiber systems.
5. LEDs and semiconductor lasers are both semiconductor diodes that emit light when current flowing through the diode causes electrons and holes to recombine at the junction between *p*- and *n*-doped materials. Light emission is spontaneous in an LED and stimulated in a laser.
6. The wavelength emitted by an LED or diode laser depends on the material from which it is made. Lasers emit at a narrower range of wavelengths selected by their structure.
7. Laser light is produced by the amplification of stimulated emission, with feedback from mirrors at the ends of a cavity. A population inversion is needed for laser action.
8. Lasers generate stimulated emission when the drive current is above a threshold level. They generate more power than LEDs and are more efficient.

9. Lasers are oscillators in which light resonates between mirrors to generate light; their structure determines their output. Optical amplifiers lack mirrors and feedback; they amplify weak signals that enter them.
10. A narrow stripe in the junction layer of an edge-emitting laser confines light using a waveguide effect, and generates an output beam that can couple into a single-mode fiber.
11. Vertical-cavity surface-emitting lasers (VCSELs) have mirror layers above and below the junction layer and emit light from their surfaces. Their output can be coupled easily into a multimode fiber.
12. Single-frequency lasers oscillate in a single longitudinal mode to generate the narrow-line output needed for high-speed transmission. The most common designs are distributed-feedback (DFB), distributed Bragg reflection (DBR), and external cavity lasers.
13. Tuning of laser wavelength is important for optical networking and wavelength-division multiplexing. There are several approaches.
14. Single-frequency lasers are driven with a steady current, and modulated externally to limit linewidth and dispersion effects.
15. A semiconductor laser without mirrors can serve as an optical amplifier.
16. Erbium-doped fiber amplifiers and lasers are pumped by diode lasers emitting 980 or 1480 nm.
17. Erbium-doped fiber amplifiers can amplify signals between 1530 and 1620 nm.
18. Erbium-doped fibers with mirrors on their ends can make lasers, which emit in the same range as erbium fiber amplifiers. They can be tuned in wavelength, or designed to generate a series of very short pulses.

What's Next?

Now that I have described light sources, Chapter 10 will cover fiber-optic transmitters that use these light sources.

Further Reading

Govind P. Agrawal, *Semiconductor Lasers: Past, Present and Future* (AIP Press, 1995)

Jeff Hecht, *Understanding Lasers* (IEEE Press, 1994)

C. Breck Hitz, J. J. Ewing, and Jeff Hecht, *Introduction to Laser Technology* 3rd ed. (IEEE Press, 2001)

Questions to Think About for Chapter 9

1. Below threshold a diode laser emits some light by spontaneous emission. Why does it behave like an LED?

2. A diode laser has a threshold current of 12 milliamperes. That current passes through a stripe 5 μm wide and 500 μm long. What is the current density in amperes per square centimeter?
3. The diode laser in Problem 2 emits 5 mW of light. The junction layer is 0.5 μm thick. If the light is evenly distributed across the end of the active layer, what is the optical power density in W/cm^2?
4. InGaAsP has a refractive index of about 3.5. A Fabry-Perot laser emitting multiple longitudinal modes at a nominal wavelength of 1550 nm has a cavity 500 μm long. What is the wavelength difference between two adjacent longitudinal modes?
5. Suppose the laser in Problem 4 was a VCSEL with cavity length only 10 μm, and for the time being forget about the difficulty of making InGaAsP VCSELS. Calculate the separation between two modes using the same technique.
6. A fiber amplifier is 6 meters long and produces a small-signal gain of 30 dB. Assuming that the amplifier is not saturated, and neglecting complicating factors like pump power, how long an amplifier would you need for a 20-dB gain? If the input power was 1 μW, how much does this reduce the output power in units of watts?

Quiz for Chapter 9

1. Operating wavelengths of GaAlAs LEDs and lasers are
   a. 820 and 850 nm.
   b. 665 nm.
   c. 1300 nm.
   d. 1550 nm.
   e. none of the above

2. Light emission from an LED is modulated by
   a. voltage applied across the diode.
   b. current passing through the diode.
   c. illumination of the diode.
   d. all of the above

3. Which of the following statements about the difference between semiconductor lasers and LEDs are true?
   a. Lasers emit higher power at the same drive current.
   b. Lasers emit light only if drive current is above a threshold value.
   c. Output from LEDs spreads out over a broader angle.
   d. LEDs do not have reflective end facets.
   e. All of the above

4. Laser light is produced by
   a. stimulated emission.
   b. spontaneous emission.
   c. black magic.
   d. electricity.

5. The spectral width of a Fabry-Perot semiconductor laser is about
   a. 2 nm.
   b. 30 nm.
   c. 40 nm.
   d. 850 nm.
   e. 1300 nm.

6. A distributed-feedback laser is

a. a laser that emits multiple longitudinal modes from a narrow stripe.

b. a laser with a corrugated substrate that oscillates on a single longitudinal mode.

c. a laser made of two segments that are optically coupled but electrically separated.

d. a laser that requires liquid-nitrogen cooling to operate.

7. Which of the following is an important advantage of external modulation of lasers?

a. Simpler operation

b. Does not require electrical power

c. No extra devices are needed

d. Avoids wavelength chirp that could cause dispersion

8. Which of the following is *not* an advantage of erbium-doped fiber amplifiers?

a. High gain

b. Insensitive to signal speed

c. Operation near 1300 nm

d. Broad amplification bandwidth at 1550 nm

9. What is the power source for erbium-doped fiber amplifiers?

a. Electric current passing through the fiber

b. They require no power

c. Diode pump lasers emitting at 980 or 1480 nm

d. Power is drawn from the optical signal

10. What guides light in a narrow-stripe edge-emitting laser?

a. Reflective layers on the edges of the laser wafer

b. The stripe has higher refractive index than surrounding material, so it functions as a waveguide

c. Coatings applied above and below the junction

d. Light entering it from an external optical fiber

11. Which of the following is *not* true for VCSELs?

a. VCSELs emit light from their surfaces.

b. VCSEL beams are rounder than those from edge-emitting lasers.

c. VCSELs can be made easily from GaAs or InGaAsP compounds.

d. VCSELs have low-threshold currents.

e. VCSELs have multilayer coatings as their resonator mirrors.

12. A Fabry-Perot diode laser operating at 1.3 μm has a cavity length of 500 μm and a refractive index of 3.2. How far apart are its longitudinal modes? (*Hint:* First estimate the number of waves that could fit into the cavity; then calculate the wavelengths of modes N and $N + 1$.)

a. 0.013 nm

b. 0.053 nm

c. 0.53 nm

d. 5.3 nm

e. 0.13 μm

CHAPTER

10

Transmitters

About This Chapter

Optical transmitters convert an input electronic signal into the optical form sent through optical fibers. This task involves both electronics and optics, but this chapter, like the rest of the book, concentrates on the optics. It first introduces the basic operational concepts involved in transmitters, then covers how multiplexing generates signals and modulation converts them to optical form. Finally it looks inside the box to show the functional components of transmitters.

Transmitters contain the light sources described in Chapter 9 and encode signals for the receivers covered in Chapter 11. Chapter 12 covers signal amplifiers and regenerators.

Transmitter Terminology

Strictly speaking, a fiber-optic transmitter is a device that generates an optical signal from an electronic input. An optical transmitter always contains a light source. However, the terminology can get muddled, particularly once transmitters are packaged into commercial equipment, so let's go through a few definitions before getting started.

A *transmitter* generates optical signals; a *receiver* detects them and converts them back into electronic form for equipment at the other end of the system. A *link* is the combination of a transmitter, fiber-optic cable, and a receiver used to send a signal between points. A *data link* is specifically a digital link, usually for computer data transmission, but something merely called a "link" may be digital or analog.

A *system* can be pretty much anything you want it to be. In practice, it usually means the equipment needed to generate an optical signal, transmit it, and receive it. Often this includes the electronics that process input signals to convert them into the form required to drive the light source in the transmitter. Likewise, it often includes electronics, which

A transmitter generates optical signals from an electronic input. It contains a light source.

A transceiver includes both the transmitter and receiver on that end of a system.

process the received signals. *Networks* are systems that link many points and contain many transmitters and receivers as well as cables. Later chapters will teach you more about the many different kinds of systems and networks used for telecommunications.

Two-way communications requires one transmitter and one receiver on each end of a system. It's common to package a transmitter and receiver together in a unit called a *transceiver,* which both transmits and receives signals at that end. That combination is used in many other communications systems; your telephone is a transceiver because it both transmits and receives speech.

If you're familiar with electronics, you'll recognize these terms because they come directly from electronic communications. A broadcast television station has a transmitter; your home television set is a receiver. The television transmitter puts a video signal into the right form for broadcasting; your television receiver converts that signal into a form you can watch. Remember that everything called a transmitter is not fiber optic.

Some of these definitions can be a bit hazy because they depend on packaging, and what the engineers decide to put in the boxes. Although the terms are widely accepted by engineers, sometimes marketing departments have their own ideas. Some are flat-out wrong, like short analog systems called "data links." Others are ambiguous, like "solutions," which can be anything from transmitters to systems.

Operational Considerations

A number of operational considerations shape the design and performance of a fiber-optic transmitter. These include the type of system, the type of modulation, the data rate or bandwidth, the number of optical channels, and optical power requirements.

The higher the transmission speed, the more complex the transmitter.

The higher the system performance, particularly measured as data rate, the more the transmitter has to do and, in general, the more complex it becomes. It's possible to directly modulate the output intensity of an LED by applying an electrical signal across the diode—if the signal is a simple one like analog speech or a slow stream of digital bits. The higher the performance, the more care is needed in modulation. A diode laser needs a bias current in addition to a modulation signal. A high-speed laser in a WDM system needs a thermoelectric cooler to stabilize its operating temperature and an external modulator. Very high speed circuits also require special electronic circuits able to process the high frequencies involved.

Transmitter and System Packaging

All transmitters include a light source and optical and electronic connections.

Most fiber-optic transmitters are sold as small and fairly simple packages that contain only the essentials for converting electronic signals to optical form. Even the simplest transmitters include a light source and optical and electronic connections. More complex transmitters may include circuits that put the input signal into the proper form for modulation, circuits that drive the light source with the input signal (or with a stable input to generate

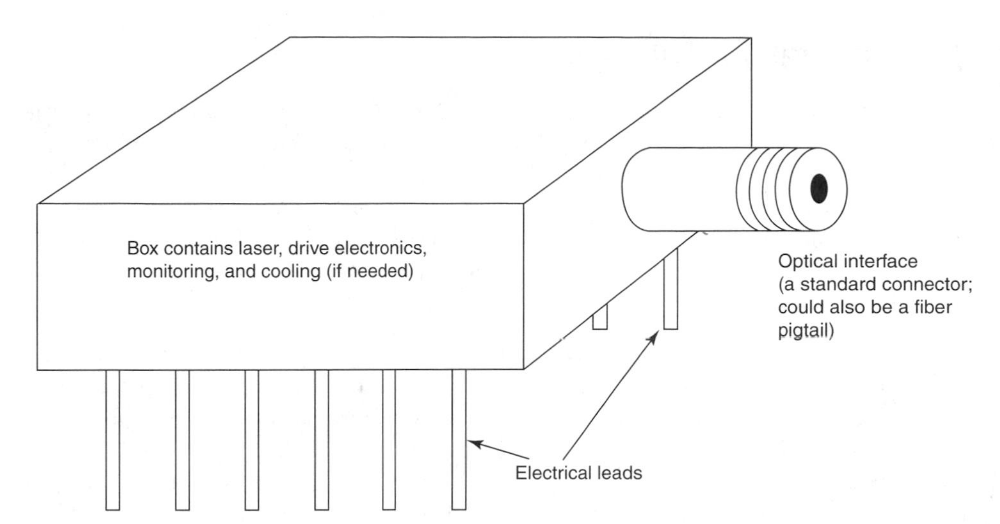

FIGURE 10.1
A simple transmitter module.

a steady output), devices to modulate and stabilize temperature, laser output monitors, and external modulators. You will learn more about these functions later in this chapter.

Transmitter modules vary considerably in design. All require some electronic interface for the input signal, and some optical interface to transfer the output signal to an optical fiber. The simplest are little more than an LED built into a connector adapter. Figure 10.1 shows a more typical module, with a laser source and drive electronics packaged in a small multi-pin module. Generally a fiber is butted against the light source inside the package, and that fiber delivers the light to the outside world. In a package equipped with a connector adapter, like the module in Figure 10.1, a jumper fiber may deliver the signal to the connector interface; in very simple transmitters, the light source may be built into the connector housing. In either case, a fiber cable mates to the connector interface. Other transmitter modules deliver output via a fiber pigtail, which can be spliced to an output fiber in a patch panel or other splice enclosure.

Often transmitter modules or boards are built into larger systems, buried inside of larger boxes except for their optical interfaces. Transmitter modules often are packaged with electronics that perform other functions, such as switching or combining electronic signals at low speeds to produce a single, higher-speed stream of data for transmission through the fiber. Those electronics perform operations before the signals are transmitted. However, they are more properly part of the communication system than part of the transmitter.

A WDM transmitter is more complex because it contains many light sources operating at different wavelengths, each with its own associated electronics. You can think of a WDM transmitter as an array of transmitters for separate optical channels, each containing a single light source driven by a separate signal at a distinct wavelength. Because our concern is the optics, we'll look at them both as individual single-wavelength transmitters, and as collective multiwavelength transmitters.

A WDM transmitter includes a module for each optical channel.

Transmitter Performance

The light source or external modulator usually limits transmitter speed.

Electronic and optical components combine to determine the performance of an optical transmitter. The light source or external modulator limits the raw speed and power. No matter what drive electronics you use, you can't make the light signals change faster than the light source (or external modulator) is capable of changing. Likewise, the electronics can't extract more power than the light source is designed to deliver—except, perhaps, in the brief interval between the time the drive power overloads the light source and the moment the light source burns out.

The output wavelength depends mainly on the light source, but also may depend on control circuits that stabilize it at an assigned value. How much stabilization is needed depends on the application. Single-channel systems generally require no wavelength stabilization, but dense-WDM systems need active stabilization to lock the laser at a standard wavelength. Because laser wavelength is temperature-sensitive, the control circuits also monitor temperature and control coolers when they are included. Cooling may also be required to limit operating temperature, because laser lifetime decreases as temperature increases.

Transmitter electronics and input signal set the modulation format and clock rate.

The transmitter electronics and the input signal set the optical signal modulation format and data rate (called *clock rate* in communications). As long as the laser or external modulator can handle the speed and power, it can transmit any format the electronics can support. Normally the transmitter electronics are designed to support standard formats.

Electronics also monitor the operation of the light source. An internal sensor may monitor output power from the rear facet of a laser, which usually transmits a small fraction of light for this purpose. Other circuits check drive-current levels, important because aging lasers need more drive current to generate the desired power level.

Analog and Digital Transmission

Optical transmitters can generate digital or analog signals.

Light sources are inherently analog devices, with output proportional to the modulating signal. An analog modulation circuit generates analog signals; a digital driver generates digital pulses.

Chapter 3 introduced the concepts of analog and digital transmission. Each have their virtues. Our eyes and ears are analog devices, so audio and video signals have to start and end in analog format. On the other hand, analog signals require much more precise reproduction and transmission than digital signals. This makes analog signals far more vulnerable to noise and distortion during transmission and processing. Digital signals can better tolerate distortion because they usually need only to detect whether a binary signal is off or on, not its shape or level. (Multilevel digital codes exist, but so far their only uses in fiber optics have been in the laboratory.) Digital electronics also are easier to design and cheaper to buy.

Digital signals are more robust than analog, but require more bandwidth to carry the same information.

Digital transmission demands faster response than analog signals carrying the same information. Recall that a single telephone voice line requires only 3000 Hz of analog bandwidth, but needs 64,000 bits per second of digital transmission capacity. This reflects the fact that digital signals contain a broader range of frequencies. You need much higher frequencies to reproduce the sharp edges of a digital pulse than the rounded edges of analog signals. The edges can be blurred as long as the receiver can tell the difference between on and off, but this still means that digital transmission requires a broader frequency range than analog transmission. That is, transmitting 1 Gbit/s requires a broader bandwidth than transmitting 1 GHz.

Although digital signals are far more common than analog in modern communication systems, some analog optical transmitters remain. As you will learn later, their most important applications are in cable television systems.

Bandwidth and Data Rate

Frequency bandwidth measures analog capacity. Bit rate measures digital capacity.

Transmission speeds are measured as bandwidth in analog systems and data rate or bit rate for digital systems. Analog bandwidth normally is defined as the frequency where the amplitude of the modulated signal drops 3 dB below the value at low frequencies, a 50% reduction in power. The digital data rate is the number of bits per second that can be transmitted with no more than a specified fraction of errors, often one incorrect bit in every trillion bits (10^{12}).

Rise time limits transmitter speed.

Both of these are quantities measured at the receiver, and don't directly measure the response or *rise time* that limits transmitters. Rise time is defined as the interval it takes light to rise from 10% to 90% of the steady-state high power level. *Fall time* is the inverse, the delay needed for the signal to drop from 90% to 10% of the maximum. If the rise and fall times are equal (they aren't always), this can be used to approximate bandwidth,

$$\text{Bandwidth (MHz)} = \frac{350}{\text{rise time (ns)}}$$

The precise relationship between bandwidth and rise time differs among light sources and transmitters.

Rise time is an important variable in selecting light sources and modulation techniques. LEDs respond more slowly to changes in drive current than lasers, with rise times from a few nanoseconds to a few hundred nanoseconds. Directly modulated lasers have rise times of a fraction of 1 ns.

Although lasers can be directly modulated at data rates of 2.5 Gbit/s, this induces a chirp in wavelength, covered in Chapter 9, which can cause excessive dispersion in long-distance systems. For that reason, external modulation is used in high-speed systems transmitting long distances. External modulators can reach data rates of 40 Gbit/s.

Multiplexing

As you've seen earlier, multiplexing is the combining of multiple signals into a single entity that can be transmitted more economically. Multiplexing takes place at or before the transmitter. There are three important types, and their handling at the transmitter differs.

FIGURE 10.2
Time-division multiplexing.

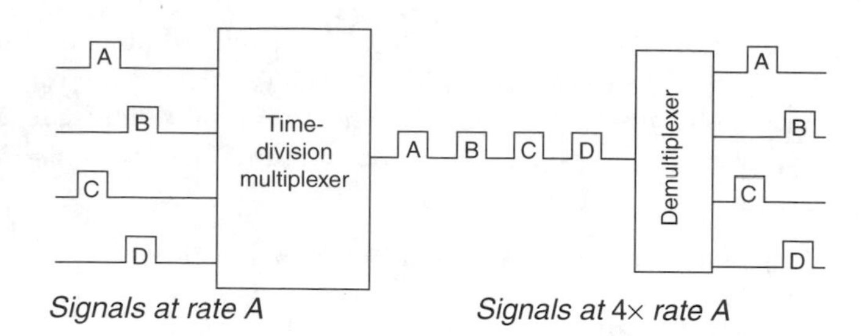

Time-Division Multiplexing

Time-division multiplexing interleaves bits from data streams to form a faster signal.

Time-division multiplexing (TDM) combines two or more digital signals by interleaving bits or bytes from separate data streams to give one faster signal, as shown in Figure 10.2. For example, 24 voice phone lines, digitized at 64,000 bits per second, can be combined into one 1.55-Mbit/s digital signal. The combined signal carries all the bits from the 24 digitized phone signals, plus extra bits that help organize the combined signals. Appendix C lists the standardized hierarchy of successively higher data rates developed for telecommunication systems.

Signals are time-division multiplexed before they reach the transmitter. The electronics that combine slower bit streams to make a signal at higher data rate may be in the same housing or rack as the transmitter. That is, incoming signals go to an electronic time-division multiplexer, which combines them into a higher speed signal, then passes that signal directly to the transmitter. The higher the speed, the more important it becomes to do the multiplexing close to the transmitter, to avoid problems transmitting the electronic signals over short distances. Care must be taken in laying out components at 10 Gbit/s, for example.

The highest time-division multiplexing rate used in current practical systems is 10 Gbit/s. Transmitters are emerging for 40 Gbit/s, but some problems remain in processing and transmitting signals at that speed.

Frequency-Division Multiplexing

Frequency-division multiplexing modulates signals onto carriers at different frequencies.

Frequency-division multiplexing is a way of combining two or more analog signals into one analog signal with a broader bandwidth. The analog signals modulate different carrier frequencies, which are then combined into a broadband signal.

This system works by assigning slots of a certain bandwidth around each carrier frequency. Modulation of the carrier spreads the signal across this range, but not into other frequency ranges. Combine the modulated carriers, and they spread across a broader range of frequencies but (at least ideally) do not overlap. This is the basis of broadcast radio and television; each station is assigned a channel based on modulating a carrier frequency. You can tune your radio or television receiver to pick out individual stations.

Cable television systems use frequency-division multiplexing to assign frequency slots to television stations. You tune to the frequencies by selecting the channel you want to watch. The main fiber-optic use of frequency-division multiplexing is in cable television distribution, but in principle it could be used elsewhere.

Like time-division multiplexing, frequency-division multiplexing is done by electronics before the signal reaches the optical transmitter.

Wavelength-Division Multiplexing

Wavelength-division multiplexing is a different matter. It transmits multiple signals through the same optical fiber at different wavelengths.

Wavelength-division multiplexing transmits signals through one fiber at multiple wavelengths.

In practice, one optical transmitter is assigned per wavelength, as shown in Figure 10.3. The input is (generally) a digital time-division multiplexed signal from an electronic multiplexer. The optical transmitter modulates that signal onto the output of a light source at a wavelength λ. Other signals modulate the outputs of other transmitters with light sources at other wavelengths.

Separate fibers deliver the output of each transmitter to an optical multiplexer, which combines them for transmission through a single output fiber. Each wavelength in the signal is a separate optical channel, which if the system is designed properly does not affect other optical channels. In this way, it works like frequency-division multiplexing for radio broadcast through the air.

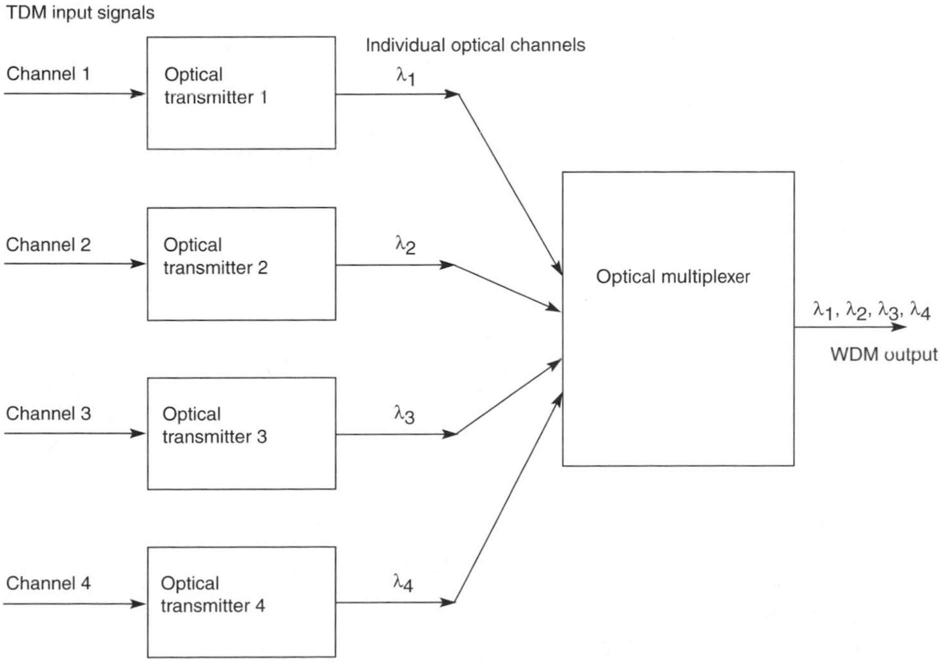

FIGURE 10.3 *Wavelength-division multiplexing at the transmitter.*

Wavelength-division multiplexing comes after the light source. Time-division multiplexing comes before the light source.

Note that wavelength-division multiplexing takes place *after* optical transmitters generate modulated optical signals at different wavelengths. Time-division and frequency-division multiplexing are both done electronically, *before* the signals modulate an optical transmitter.

Later chapters will cover other aspects of wavelength-division multiplexing in more detail, particularly its implications in overall system design and performance.

Modulation

So far, we have discussed modulation in very general terms. However, transmitter performance depends strongly on how the light source is modulated. Let's take a look at the details.

Amplitude or Intensity Modulation

Most optical carrier signals are modulated in intensity.

Most optical transmitters are modulated by signals that change the amplitude or intensity of the light they generate. Figure 10.4 shows how this works for an ideal light source that normally generates a steady coherent light wave.

The ideal light wave at the top is the *carrier signal,* like the carrier frequency of a broadcast radio or television station. *Amplitude modulation* changes the amplitude or intensity of the output light as the amplitude of the drive signal changes. This is a straightforward approach

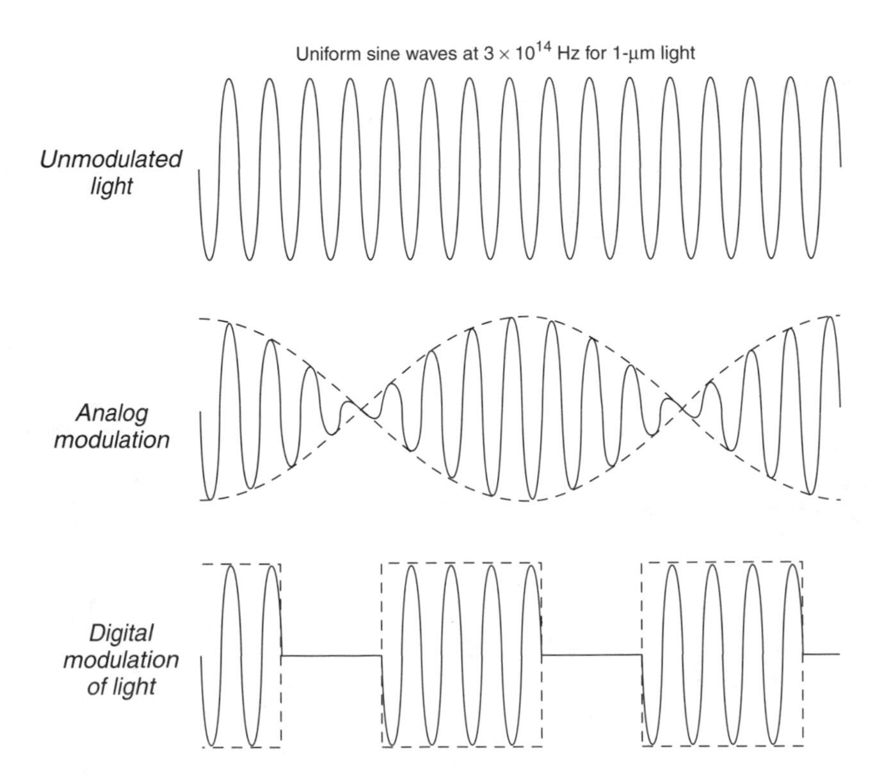

FIGURE 10.4
Amplitude modulation by digital and analog signals.

because the output power of an LED or diode laser increases with drive current. This can produce an analog signal at the middle or a digital signal at the bottom.

The figure was distorted in one important way to show the scale of the modulation. Light at 1500 nm has a frequency of about 200 THz, or 10^{14} Hz. The signals modulating the light wave are at much lower frequencies, measured in gigabits per second for a digital signal. If the figure was drawn to scale, one pulse from a 10-Gbit/s data stream would contain around 20,000 waves of light. That's impossible to draw, so the figure pretends the optical and modulation frequencies are much closer.

Amplitude modulation is standard in fiber-optic systems because it's simple and easy. Light source output naturally varies with drive current, and intensity modulators are relatively easy to build.

Direct and External Modulation

As you've seen before, semiconductor light sources can be modulated either directly or externally. Increasing the drive current increases the power emitted by either LEDs or semiconductor lasers. Diode lasers respond very quickly to the increase, but LEDs have slower response, limiting their modulation bandwidths.

Direct modulation is ideal for inexpensive transmitters, but as you learned earlier it causes an undesirable wavelength chirp, which causes excessive chromatic dispersion at high speeds. In addition, lasers develop undesirable relaxation oscillations at frequencies of a few gigahertz, which limits the maximum frequency for direct modulation. External modulators are needed at higher speeds, or when the light source cannot be directly modulated. They are covered in Chapter 16.

Diode laser sources present additional complications because of the nature of laser emission. As you learned in Chapter 9, laser emission begins after the drive current exceeds a threshold value. LEDs don't have a current threshold, so they normally are switched on and off. However, if a laser is turned completely off by shutting down the drive current, it won't start emitting again until the drive current exceeds the threshold. This poses problems at high speeds, where laser transmitters are desirable, because the laser would not start emitting light until some time after the drive current started to rise. To avoid this problem, lasers normally have two sources of drive current, which add together. One produces a steady bias current, which typically slightly exceeds laser threshold. The other is variable signal current, which actually modulates the laser output by increasing the current above the bias level, as shown in Figure 10.5.

Lasers are directly modulated by combining a bias current that turns the laser on with a signal current that modulates intensity.

This biasing arrangement leaves the laser emitting some light when it is nominally "off," but much more when it is "on." Receivers can accommodate this difference by setting proper decision thresholds for the presence or absence of pulses, as you will learn in Chapter 11.

Don't forget that the semiconductor diodes respond differently to currents and voltages. As you saw in Chapter 9, a minimum voltage has to be applied across an LED or laser to give the charges carrying current through the device enough energy to recombine at the junction level. Once that *voltage* is exceeded, current starts flowing. That current is enough to start

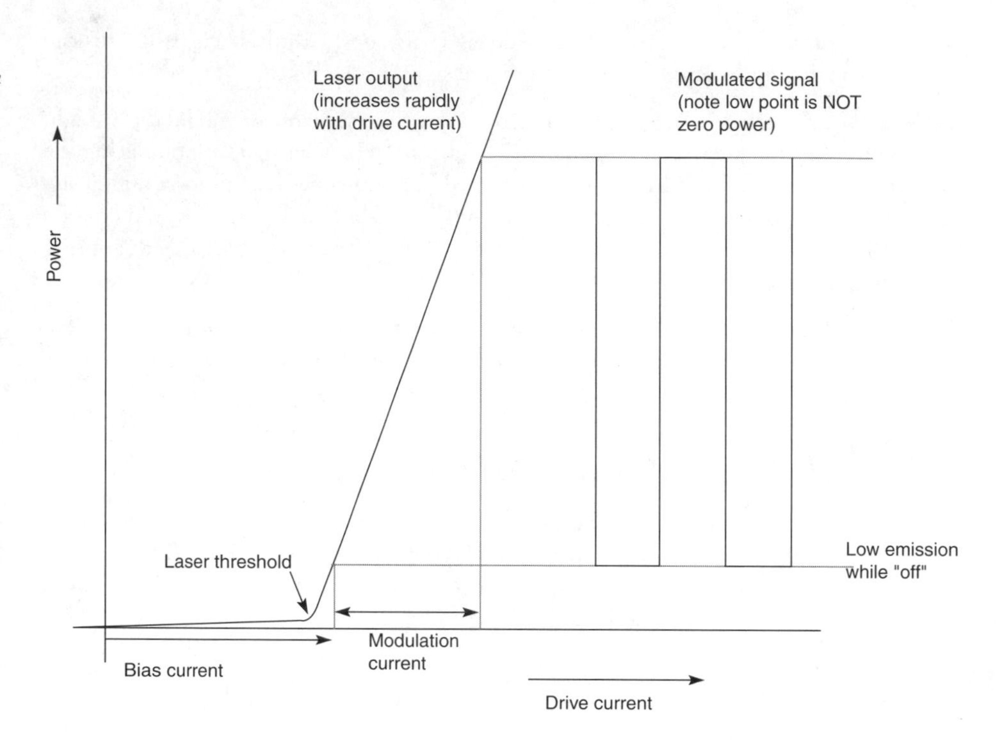

FIGURE 10.5 *Direct modulation of a laser diode.*

an LED to emit some light, but the drive current must exceed a threshold to produce stimulated emission from a diode laser.

Coherent Transmission

If you're familiar with radio, you know that amplitude modulation is not particularly sophisticated. The amplitude-modulated AM-radio band is justly notorious for its static and background noise. Modulating the frequency of the carrier signal rather than the amplitude gives the FM- (frequency-modulated) radio band much better signal quality. AM receivers pick up random spikes of amplitude noise from spark plugs and power lines, and the signal strength fades with distance from the receiver. In contrast, FM receivers do not pick up random noise spikes because they don't change the frequency of the radio signal. In addition, the strength of an FM signal depends not on its intensity but on how it changes the carrier frequency, so signals don't fade into the background; they stay strong until they start breaking up. (Other differences in AM and FM reception come from the different transmission frequencies.)

Coherent optical transmission works like an FM-band radio, but has not yet proved practical.

You might think it logical to try frequency modulation to improve optical transmission. In fact, it's been tried, but hasn't worked out very well.

Frequency modulation is one type of what optical engineers call *coherent transmission,* which works like a heterodyne (FM) radio system. The trick is to combine the incoming frequency-modulated signal with another signal kept at a constant frequency. Processing

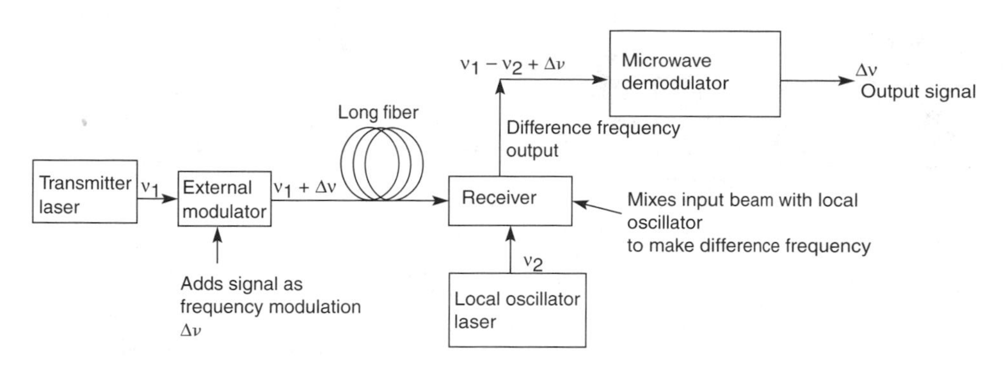

FIGURE 10.6
Coherent optical transmission.

the difference between the input signal and the constant frequency (called a local oscillator) reproduces the radio program. As shown in Figure 10.6, the optical version requires a pair of lasers, one at the transmitter and a second (the local oscillator) at the receiver. The two lasers emit slightly different frequencies, ν_1 and ν_2. The transmitter modulates the frequency of the outgoing laser beam by passing it through a suitable external modulator. Alternatively, a modulator could delay the phase of the transmitter beam, causing a phase shift. At the receiver, the incoming signal at the transmitter frequency is mixed with the local oscillator beam, to give a microwave signal at the difference between the frequencies of the two light waves. That microwave signal carries the frequency- or phase-modulated signal, which can be extracted by further processing.

Experiments have shown that coherent transmission can work, but so far it has not proven practical, and it's largely been forgotten in the explosive development of wavelength-division multiplexing and optical networking.

Single-Channel Transmitter Design

We've already covered several components of optical transmitters. To pull the whole picture together, let's consider what goes into a generalized transmitter. We'll pick a high-performance transmitter that will have all the interesting and important elements, but to prevent too many complications we will consider only one optical channel. Obviously a low-speed LED transmitter will be much simpler.

Transmitters may be packaged together with other transmitters or receivers. Many transmitters are mounted with matching receivers in a transceiver. Pairs of transceivers are attached to opposite ends of a fiber pair for two-way transmission. As channel counts grow higher, companies are packaging multiple transmitters in a single compact package.

Pairs of transceivers are attached to opposite ends of a fiber pair for two-way transmission.

Figure 10.7 shows a functional diagram of a transmitter module; we'll cover receivers in Chapter 11. The main functional elements of a typical transmitter module are:

- Housing
- Electronic interfaces
- Internal signal-processing electronics

FIGURE 10.7 *Generic optical transmitter.*

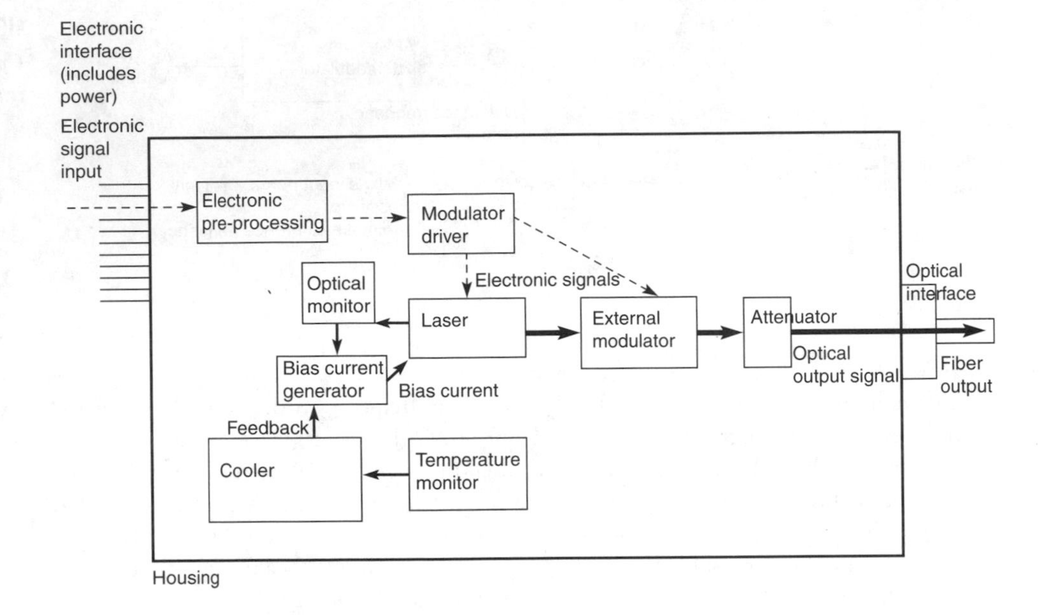

- Laser monitoring and bias circuits
- Temperature control
- Optics (including laser and external modulator)
- Optical interfaces

Some control functions can be performed by circuits outside the packaged transmitter. For example, output of the sensor that monitors laser output can be routed to external circuits that adjust bias current. Temperature data also can be routed to external controls. Here we will consider their functions rather than their locations.

Housing and Electronic Interfaces

The mechanical housing is designed to mount conveniently in other equipment.

The mechanical housing of an optical transmitter is a box designed to mount conveniently within other equipment. Often they come with a matched receiver in a transceiver. Screws, solder bonds, or other fasteners attach the housing mechanically to a printed circuit board or other substrate. More complex transmitters are packaged in cases that mount in standard equipment racks.

Typically, pins on the package provide electronic connections with other devices. They deliver power to drive the laser, the input signal, and other required signals. The transmitter also may return some information to other equipment, such as transmitter status.

Signal Processing Electronics

Internal electronics put the input signal into a form suitable for modulating the light source.

Internal electronics typically do some processing of the input signal to put it into a form suitable for modulating the light sources. One function is converting signals from the voltage variations used in electronic circuits to the current variations needed to modulate diode lasers and LEDs. Other processing may change signals to formats better suited for fiber transmission. Transmitters used in some networks may include buffers to hold input data until it can be sent, because the networks can deliver data faster than the transmitter can send the data.

Figure 10.7 shows the path of the signal from the outside connection through the electronic preprocessor, and modulator driver to either the diode laser or the external modulator. In direct-modulation systems, the modulation current is added to the bias current to drive changes in the optical output.

Laser Monitoring and Bias Circuits

A bias circuit provides the current that drives a directly modulated laser diode to a point just above threshold. This should produce a steady, low-level of power. Adding pulses of modulating current derived from the input signal produces much stronger pulses from a directly modulated laser. If a diode laser is externally modulated, the bias current generator delivers a steady current that produces the desired high output level.

In standard laser transmitters, the rear facet of the laser emits a low power level that samples the laser output. Alternatively, a coupler can split off part of the output beam. An optical monitor measures the intensity of this weak signal, and uses that information to drive a feedback circuit, which adjusts the bias current to keep output power steady. This keeps laser power from dropping with age, but does not prevent device failure.

Temperature Control

The threshold current, output power, and wavelength of a diode laser change with temperature, making temperature stabilization important in laser transmitters, particularly to achieve high performance.

Temperature stabilization is important in maintaining stable wavelength and output power.

The threshold current, I_{thresh}, increases roughly exponentially with temperature T,

$$I_{\text{thresh}}(T) = I_0 \, e^{\left(\frac{T}{T_0}\right)}$$

where I_0 is a constant and T_0 is a characteristic temperature of the laser material. The characteristic temperature of InGaAsP is about 50° to 70°K, much lower than the 120°K for GaAs, so threshold currents of InGaAsP lasers increase much faster with temperature than GaAs. The higher the threshold current, the lower the overall efficiency and the more power the laser consumes. Unless drive current is increased to compensate for the rising threshold, laser output power declines as temperature increases. Higher temperatures also tend to decrease diode laser lifetime, as does increased drive current.

The wavelength change arises from a change in the semiconductor's refractive index with temperature. The refractive index changes alter the effective length of the laser cavity (the physical length of the laser cavity times the refractive index), and thus the wavelength that resonates in the laser cavity. This means that changing temperatures make laser wavelength drift.

Cooling requirements depend on the system. Heat sinks often suffice for data rates up to hundreds of megabits. Higher-speed transmitters may need thermoelectric coolers to keep them at stable temperatures; this increases the transmitter's total power requirements, but assures proper laser operation. Keeping the laser's output power stable also improves receiver performance. Wavelength drift is not a serious issue for single-channel systems, but must be carefully controlled in wavelength-division multiplexing to avoid channel drift.

Optics and Optical Interfaces

The optical part of the transmitter starts with the laser, which generates either a steady beam or a modulated output depending on the input. Monitoring the laser's output through the rear facet controls the laser's output power through feedback to the bias-current generator.

Direct modulation adds to the laser output from a low-bias current. External modulation reduces the high output of a laser driven at a high bias.

Modulation either adds to or subtracts from the light generated by the laser source. If the laser is directly modulated, the modulation drive current increases the laser's output power above the biased level where it emits a little light. In that case, there is no external modulator. If the laser is modulated externally, the bias current drives it at a high level, and the modulator varies between fully transparent and essentially opaque. That means the external modulator *reduces* the power level in the laser beam. As with direct modulation, a small amount of power may be transmitted in the "off" state.

External modulators are very fast, and can switch signals off and on at rates to 40 Gbit/s. Their main applications are in high-end transmitters. Chapter 16 describes how they work.

An attenuator also is optional. It may be needed if the receiver is close to the transmitter, and might be overloaded by the transmitter's unattenuated power. Attenuators are most likely to be used in networks that use standard transmitters, but where the attenuation in the cable can vary over a wide range.

Wavelength Tunability

The picture will grow more complex as wavelength-tunable lasers begin to be used in transmitters. So far, most tunable lasers have been developed for laboratory use, not for commercial telecommunications. When transmitters incorporate them, they will have to add provisions for monitoring, tuning, and stabilizing the laser wavelength. Specific requirements will depend on the type of tunable lasers chosen.

Sample Transmitters

The best way to get a feeling for the internal workings of transmitters is to look at a couple of examples, deliberately simplified to aid in understanding basic concepts.

Figure 10.8 shows a very simple drive circuit for an LED. A bias voltage $+V$ is applied across a transistor, LED, and current-limiting resistor, going to ground. This provides the forward-voltage bias needed by the LED (about 1.5 V for 850-nm emitters and around 1 V

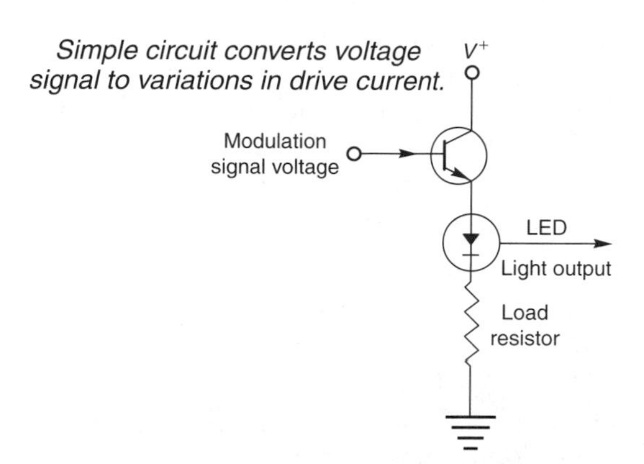

FIGURE 10.8
Simplified drive circuit for LED.

for 1300-nm LEDs). The input signal is delivered as a modulated voltage applied to the base of the transistor, which causes variations in the current passing through the LED and load resistor. The LED could be on either side of the transistor, but the circuit requires a transistor or other circuit element to modulate the drive current. The resistor limits LED current. The circuits that generate the input signal are not shown. The higher the speed and performance level, the more complex the circuit.

These transmitters can be made very compact. Figure 10.9 shows the dimensional outlines of a transceiver, which includes a 1300-nm LED transmitter able to operate at speeds to

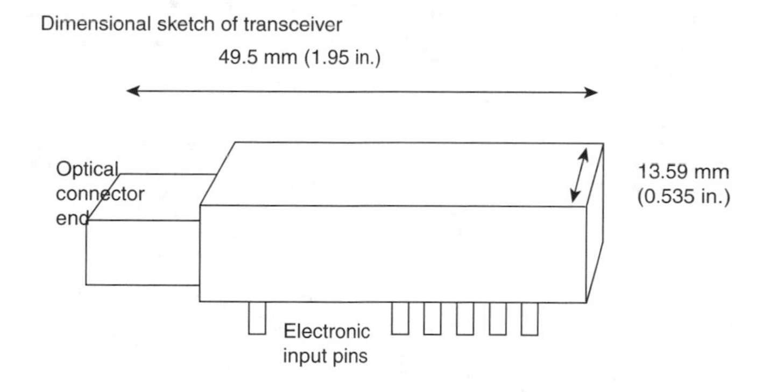

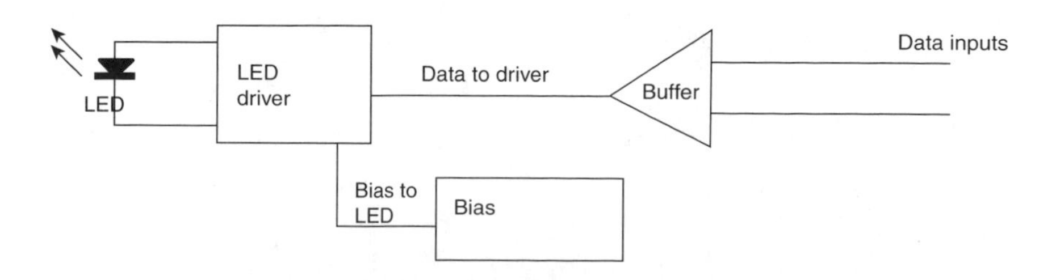

FIGURE 10.9
156-Mbit/s fiber-optic transceiver, packaged with integral small form-factor connector.

156 Mbit/s. The figure also shows a block diagram of the transmitter portion. The whole integrated package (transmitter and receiver) is just under 50 mm (2 in.) long, including a built-in connector interface at the left. (One thing to watch out for is that some mechanical specification sheets still don't label the units of dimensions. They usually include both inches and millimeters, with the metric units the higher numbers, but they may not label them. Remember what happened to Mars Climate Orbiter when a NASA contractor forgot to label units!)

Diode lasers have electrical characteristics similar to those of LEDs, but their optical operation differs. Lasers emit little light until drive current passes a threshold value, but above that threshold tend to emit higher power with more efficiency. Lasers draw much higher drive currents than LEDs, so they normally are used with smaller current-limiting resistors. Laser transmitters also may require additional components to meet the more demanding requirements of laser operation, including a sensor to monitor laser output, a temperature-sensing thermistor to monitor transmitter temperature, and a thermoelectric cooler. Typical laser transmitters are somewhat larger than the one shown in Figure 10.9, to accommodate these extra components.

Externally modulated lasers have simpler drive circuits, but the transmitters require separate electronics to drive the external modulator.

The circuit in Figure 10.10 is an example of the type used in a directly modulated laser transmitter. The drive signal is applied as a voltage to the base of a transistor, where it adds to a bias voltage and modulates drive current through the laser diode. The detector packaged with the laser monitors its output, providing feedback to an amplifier (at right), which adjusts the bias applied to the laser diode so average power remains constant. Externally modulated lasers require simpler drive circuits because they need only provide constant output power, but the transmitter then requires drivers for the external modulator.

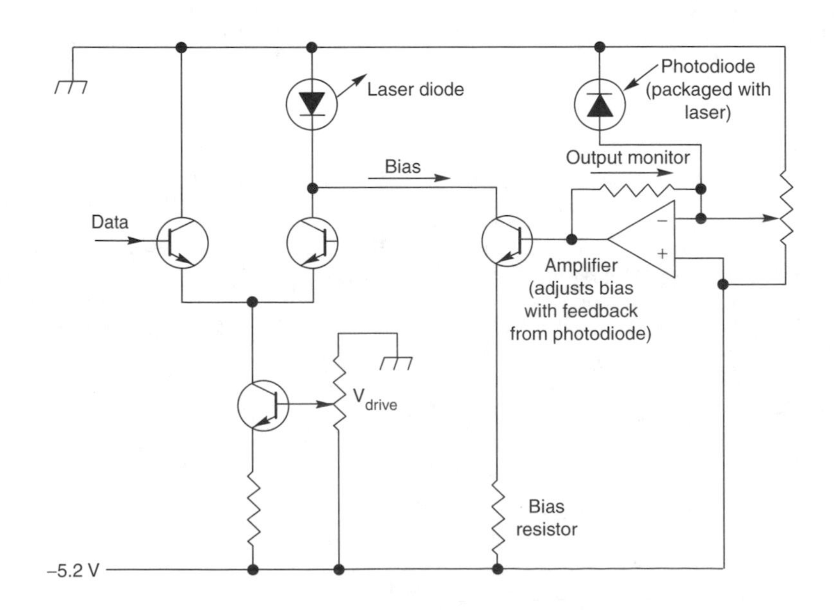

FIGURE 10.10
Laser diode drive circuitry. (Adapted with permission from a figure by Paul Shumate)

Typically the drive circuit prebiases the laser with a current close to threshold. Adding the modulated drive signal raises it above the threshold, so its starts emitting light. This approach enhances speed by avoiding the delay needed to raise the drive current above threshold. At low speeds, the prebias may be below the threshold, to avoid generating low-level emission that might confuse the receiver. At higher speeds, the laser may be prebiased to emit a little light in the off state, with addition of the modulated drive signal increasing power greatly. Prebiasing requires careful adjustments so the receiver does not mistake the low emission from the laser in its off state for a faint pulse in the on state.

As mentioned earlier, transmitters are often packaged with receivers to make transceivers, which provide two-way communications over a pair of fibers from one node to a second location. The transmitter and receiver circuits remain separate in the transceiver case, but packaging them together reduces costs and simplifies installation. Remember that these transceivers serve a different function from repeaters in which a receiver detects a weak input signal and uses it to drive a transmitter, sending a signal through a separate fiber to a *different* location.

What Have You Learned?

1. A transmitter contains a light source and generates optical signals from an electronic input. It includes optical and electronic connections.
2. A transceiver is a package that includes both the transmitter and receiver on one end of a system. It connects to a pair of fibers.
3. Transmitters range from simple to complex. The complexity increases with transmission speed.
4. Wavelength-division multiplexing requires a transmitter module for each optical channel that generates a separate wavelength.
5. Transmitter electronics and the input signal define the modulation format and the clock rate for a transmitter.
6. Optical transmitters may generate digital or analog signals. Digital signals are more robust than analog, but require more bandwidth to carry the same information.
7. Frequency bandwidth is the capacity of an analog transmitter. Data rate is the capacity of a digital transmitter. Both depend on the rise time of the transmitter.
8. Time-division multiplexing interleaves bits from data streams to form a faster signal. Frequency-division multiplexing assigns signals to different frequencies combined into one signal. Both are done electronically before the signal is fed to the optical transmitter.
9. Wavelength-division multiplexing transmits separate signals through one fiber at many different wavelengths. It requires one transmitter per optical channel. The signals are combined on the fiber after the optical transmitter.
10. Optical carrier signals are modulated in intensity.

11. Major elements of a typical transmitter are the housing, the electronic interfaces, the internal signal-processing electronics, the detector and circuits to monitor laser bias, temperature control, the laser and (optional) external modulator, and the optical interfaces.
12. Temperature stabilization is important in maintaining stable wavelength and output power from the laser.

What's Next?

Now that you have learned about fiber-optic transmitters, Chapter 11 will examine the other end of the system, the receiver.

Further Reading

HP Fiber Optic Technical Training Manual, available from *http://www.agilent.com/*

Paul W. Shumate, "Lightwave transmitters," in Stewart E. Miller and Ivan P. Kaminow, eds., *Optical Fiber Telecommunications II* (Academic Press, 1988)

Questions to Think About for Chapter 10

1. The difference between the "off" and "on" states of a laser transmitter is 20 dB. The "on" output is 1 mW. What is the "off" output at the output port of the transmitter? What is the "off" output after 30 dB of loss?
2. The diode laser used in a transmitter has a threshold current of 10 mA and normally operates at 15 mA in the "on" state. Suppose it is directly modulated with a signal without a bias current, which turns the laser on from zero current. The signal pulse has 30-ps rise time, 30-ps "on" time, and 30-ps fall time. How long is the output optical pulse, starting from the time the current crosses laser threshold?
3. Suppose you could find electronics fast enough to generate time-division multiplexed signals at 640 Gbit/s on a single optical channel. How many light waves would one bit correspond to at a wavelength of 1550 nm?
4. Time-division multiplexing requires reducing the duration of low-speed pulses and interleaving them into a combined signal. It now is done electronically. Suppose you had a way to reduce the duration of optical pulses. Can you think of a way to interleave these shortened optical pulses for time-division multiplexing? Consider a system that multiplies data rate by a factor of four. (*Hint:* Think about ways to delay pulses.)

Quiz for Chapter 10

1. Digital transmission capacity is measured as
   a. bandwidth in megahertz.
   b. rise time in microseconds.
   c. frequency of 3-dB point.
   d. number of bits transmitted per second.
2. Analog transmission capacity is measured as
   a. bandwidth in megahertz.
   b. rise time in microseconds.
   c. frequency of 3-dB point.
   d. number of bits transmitted per second.
3. If the rise time of a transmitter is 1 ns, what is its theoretical bandwidth?
   a. 1 Gbit/s
   b. 100 MHz
   c. 350 MHz
   d. 350 kHz
   e. 350 Mbit/s
4. Standard optical interfaces for transmitters are
   a. integral optical connectors or fiber pigtails.
   b. integral electronic connectors or fiber pigtails.
   c. output windows or fiber pigtails.
   d. 14-pin DIP packages.
   e. only fiber pigtails.
5. How many transmitters are needed for wavelength-division multiplexing?
   a. None; the signals are combined in the optical fiber
   b. One
   c. One per optical channel
   d. Depends on the data rate per channel
6. What provides feedback to stabilize laser intensity in a transmitter?
   a. A signal relayed from the receiver
   b. Changes in input impedance
   c. Output from the rear facet of the laser monitored by a photodiode
   d. Light scattered from the optical interface with the input fiber
   e. No feedback is required
7. What is the usual modulation method for fiber-optic transmitters?
   a. Intensity modulation
   b. Frequency modulation
   c. Wavelength modulation
   d. Voltage modulation
8. What does time-division multiplexing do?
   a. Transmits different signals at different wavelengths
   b. Shifts the frequencies of several analog signals to combine them into a single input
   c. Encrypts signals for secure transmission
   d. Interleaves several digital signals into a single data stream
9. What is the total data rate of a WDM system carrying 2.5-Gbit/s

signals at 1550, 1552, 1554, 1556, 1558, 1560, 1562, and 1564 nm?

a. 2.5 Gbit/s
b. 10 Gbit/s
c. 12.5 Gbit/s
d. 20 Gbit/s
e. 25 Gbit/s

10. With wavelength-division multiplexing, how many fibers do you need for two-way transmission of 2.5 Gbit/s signals at 1550, 1552, 1554, 1556, 1558, 1560, 1562, and 1564 nm?

a. One
b. Two
c. Four
d. Eight
e. Sixteen

11. A 1300-nm LED-based transmitter has rise time and fall time of 1.2 nanoseconds. What is the best application?

a. Transmitting 10 Gbit/s through single-mode fiber
b. Transmitting gigabit Ethernet (1 Gbit/s data rate) through graded-index multimode fiber
c. Transmitting fast Ethernet (100 Mbit/s data rate) through graded-index fiber
d. Transmitting ATM signals (155 Mbit/s data rate) through single-mode
e. Transmitting 1.5 Mbit/s through graded-index fiber

12. The LED-based transmitter in Question 11 has minimum output of −19 dBm into 62.5/125 graded-index multimode fiber. The receiver requires an input of at least −30 dBm. If fiber attenuation is 2 dB/km and other losses total 3 dB, how far away can you put the receiver?

a. 1 km
b. 3 km
c. 4 km
d. 8 km
e. 11 km

13. How is LED output modulated?

a. Externally by varying current delivered to an external modulator
b. Directly by varying drive current
c. Directly by varying voltage applied to an external modulator
d. Directly by varying LED operating temperature

14. Temperature stabilization of a laser transmitter also stabilizes what operating characteristics?

a. Laser threshold
b. Output power
c. Wavelength
d. All of the above
e. None of the above

CHAPTER

11

Receivers

About This Chapter

The receiver is as essential an element of any fiber-optic system as the optical fiber or the light source. The receiver's job is to convert the optical signal transmitted through the fiber into electronic form, which can serve as input for other devices or communication systems.

This chapter discusses the basic types of detectors and receivers, important performance considerations, and how they work. It focuses particularly on the types of detectors and to a lesser extent on the electronics.

Defining Receivers

Fiber-optic receivers detect optical signals and convert them into electronic signals usable by other equipment. The devices required to perform that task depend on the nature of the input signal. Receivers can be very simple if the input signal contains only a single optical channel, and is strong, clean, and fairly slow. More elaborate receivers are needed to deal with multiple optical channels, or to process signals that are weak, distorted, or at high frequencies.

Figure 11.1 shows major functions that may be performed at the receiving end of a system. An optical amplifier may preamplify the optical signal, increasing its amplitude so it can be detected more easily and accurately, and thus improving receiver performance. If the signal contains many optical channels, all are amplified simultaneously by roughly the same amount. (Optical amplifiers are expensive, so they are not used routinely.) A WDM signal then must go through a *wavelength-division demultiplexer,* which splits it into its composite wavelengths so each optical channel can be directed to a separate receiver. This separation is vital because the optical detectors (often called *photodetectors*), which convert signals to electronic form, are color-blind to wavelength. Although they

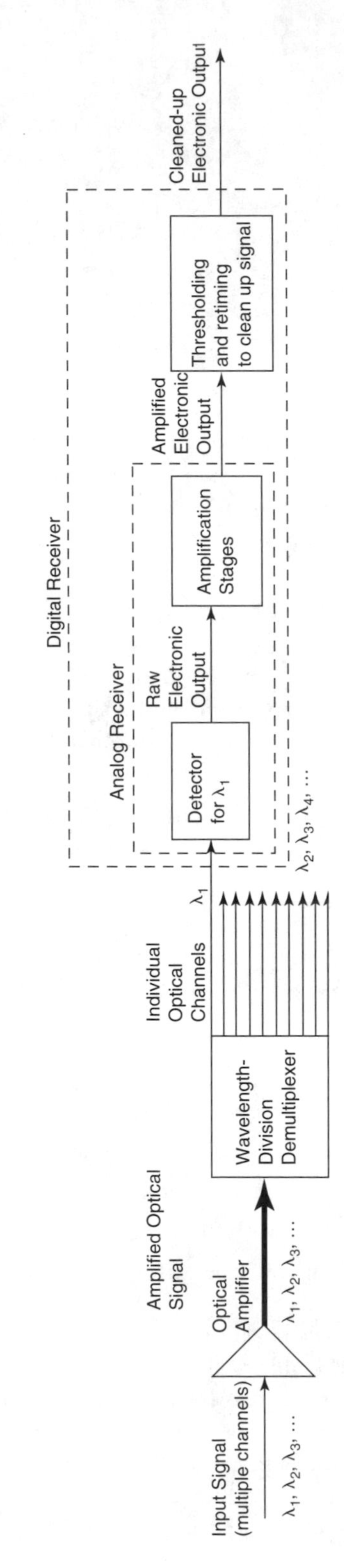

FIGURE 11.1
Major receiver functions.

do not respond equally to all wavelengths, they do not distinguish *between* wavelengths. You can think of detectors as seeing the world in shades of gray.

Most optical detectors used in fiber-optic systems generate a flow of electrons proportional to the input light, so their raw output is in the form of current. This current signal is amplified and converted to a voltage signal, a process that often requires two stages, a preamplifier and an amplifier, each optimized for different signal levels. Then the voltage signal passes through other stages that clean it up and prepare it to be sent to other devices, which appear as only one box in Figure 11.1.

Detectors are color-blind.

Every receiver does not need all these components. If fibers transmit only a single wavelength, they can deliver signals directly to a detector. A clean, powerful input signal at a low data rate may not require amplification or retiming before being delivered to other equipment, but a current usually needs to be converted to a voltage. The simplest receivers are little more than photodiodes with simple circuits attached, but high-performance receivers are quite elaborate.

It might seem logical to class all the functions shown in Figure 11.1 as part of the receiver. However, standard industry terminology evolved before the advent of optical amplifiers and wavelength-division multiplexing, which are considered strictly optical devices. Receivers are defined as strictly *opto-electronic* devices, which convert an input optical signal into electronic form, and include circuits to process that electronic signal. The standard industry approach is to treat and package receivers as distinct modules like those shown in Figure 11.2. In practice, these receivers may be integrated into packages with transmitters, with one of each forming a transceiver, as you learned in Chapter 10.

Receivers are opto-electronic devices. They generate an electric current proportional to the input light signal.

Following this industry practice, this book covers optical amplifiers separately in Chapter 12 and wavelength-division multiplexing in Chapter 15. This chapter covers receivers as opto-electronic devices. It starts with the fundamentals of detector operation, then turns to performance of the receiver and its electronic elements.

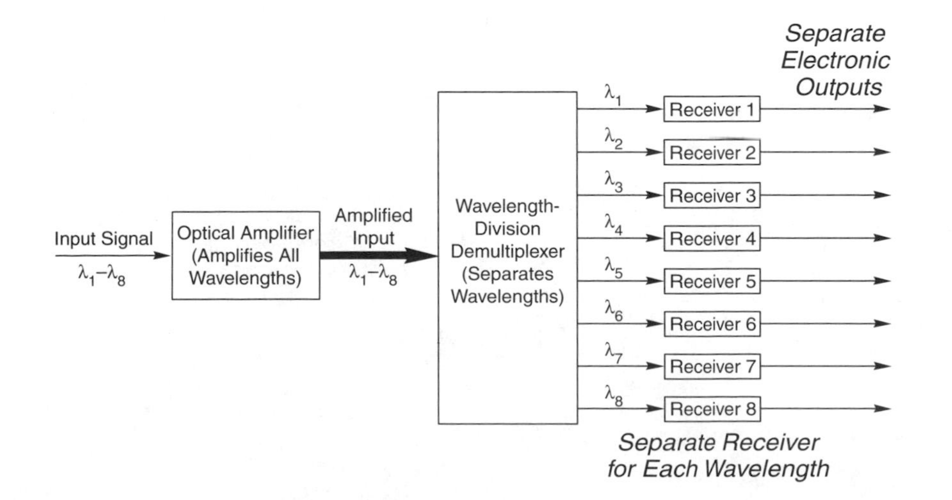

FIGURE 11.2
Splitting a WDM signal among separate receivers for each wavelength.

Detector Basics

The detectors used in fiber-optic communications are semiconductor photodiodes or photodetectors; they get their name from their ability to detect light. The simplest semiconductor detectors are solar cells in which incident light energy raises electrons from the valence band to the conduction band, generating an electric voltage. Unfortunately, such photovoltaic detectors are slow and insensitive.

Semiconductor photodiodes are reverse-biased to detect light; they produce a current proportional to the illumination level.

Photodiodes are much faster and more sensitive if electrically reverse-biased, as shown in Figure 11.3. (Recall that LEDs and lasers are forward-biased to emit light.) The reverse bias draws current-carrying electrons and holes out of the junction region, creating a depleted region, which stops current from passing through the diode. Photons with the proper energy can create electron-hole pairs in this region by raising an electron from the valence band to the conduction band, leaving a hole behind. The bias voltage causes these current carriers to drift quickly away from the junction region, so a current flows proportional to

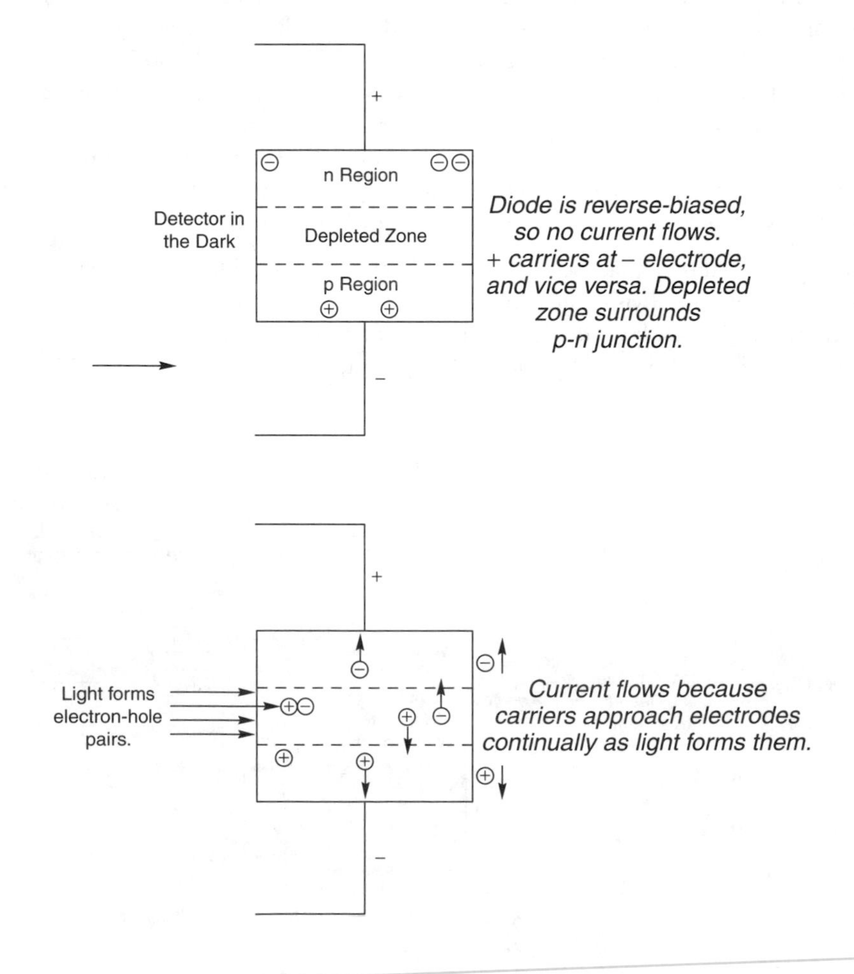

FIGURE 11.3 *Photodetector operation.*

the light illuminating the detector. Several types of detectors can be used in fiber-optic systems, as described below.

Composition and structure combine to determine the operational characteristics of photodetectors. We will start by looking at detector materials, then consider common structures. As you will see, the actual structures used are more complex than the simple example of Figure 11.3.

Detector Materials

Photodetectors can be made of silicon, germanium, gallium arsenide, indium gallium arsenide, or other semiconductors. The wavelengths at which they respond to light depend on their composition. To produce a photocurrent, photons must have enough energy to raise an electron from the valence band to the conduction band—that is, their energy must equal or exceed the bandgap energy. The need to have at least this minimum energy means that photodetector sensitivity tends to drop steeply at the long-wavelength, low-energy end of the spectrum. Other effects, such as light absorption in other parts of the device, cause the response to drop more gradually for more energetic photons at shorter wavelengths.

The wavelengths at which detectors respond depend on their composition.

Table 11.1 lists approximate spectral ranges for important detector materials used in fiber-optic systems. Note that two of the most important semiconductor materials, silicon and gallium arsenide, are not sensitive at the 1280- to 1650-nm wavelengths used in long-distance fiber-optic systems. The band gaps in InGaAs and InGaAsP depend on the material composition, so the range of operating wavelengths may vary between devices. For example, while most fiber-optic InGaAs detectors are sensitive at 900 to 1700 nm, some InGaAs detectors respond to wavelengths longer than 2200 nm.

The response of each material also varies with wavelength. Figure 11.4 shows the relative response of silicon, germanium, GaAs, and InGaAs across the range of wavelengths normally used in fiber-optic systems. Silicon and GaAs have good response at the 650-nm wavelength used in plastic fibers and the 850-nm wavelength used for short-distance transmission. Germanium has a very broad range, but response drops near 1550 nm and it suffers much higher noise levels than other materials. InGaAs has response through the entire 1250 to 1700 nm range, making it the most common material for long-wavelength

Silicon and GaAs have good response at 650 and 850 nm; InGaAs is used at 1250 to 1700 nm.

Table 11.1 Typical operating ranges of important detectors

Material	*Wavelengths (nm)*
Silicon	400–1100
Germanium	600–1600
GaAs	400–900
InGaAs	900–1700
InGaAsP	800–1600

FIGURE 11.4
Detector response curves, showing quantum efficiency relative to the peak value for the material.

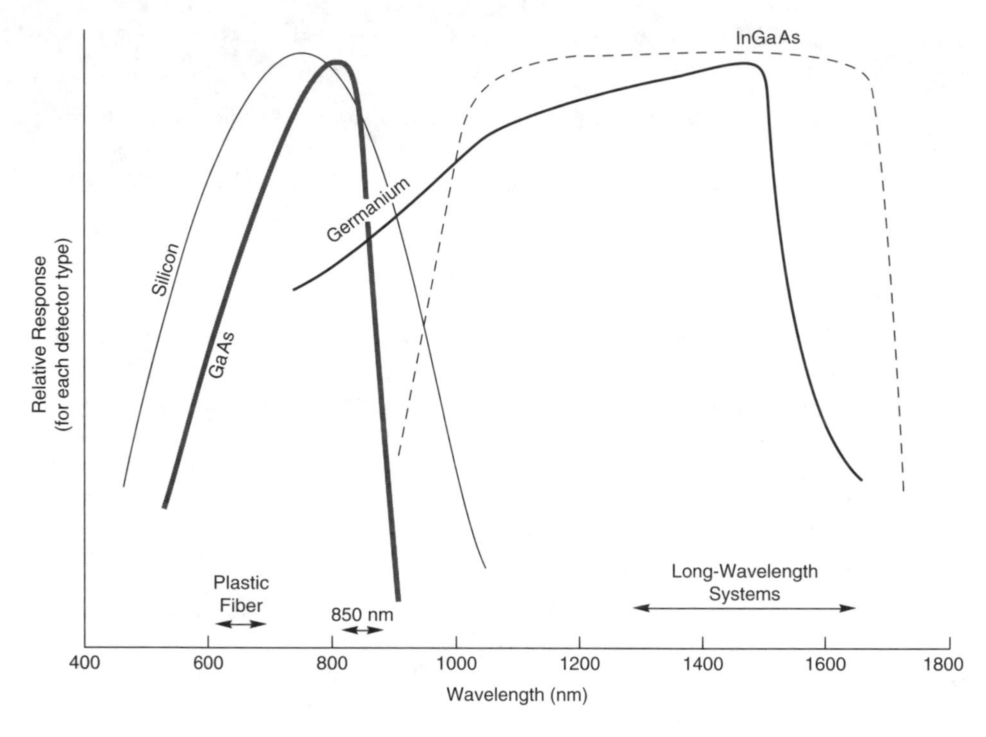

detectors. InGaAsP responds to similar wavelengths, depending on composition, but is not as widely used.

All these semiconductor materials also can be used in electronic circuits, allowing integration of detectors with amplifiers and other signal processing electronics in receivers. Detectors also can be combined with other electronic components to make hybrid receiver circuits.

pn and *pin* Photodiodes

Photodiode sensitivity is improved by sandwiching an undoped intrinsic region between the *p* and *n* regions.

You saw earlier that fiber-optic photodiodes are reverse-biased so the electrical signal is a current passing through the diode. Such a detector is said to be operating in the photoconductive mode because it produces signals by changing its effective resistance. However, it is not strictly a resistive device, because it includes a semiconductor junction, forming a *pn* photodiode. (True "photoconductive" detectors exist in which light produces current carriers that increase the conductivity of a bulk semiconductor that lacks a junction layer, but they are not used in fiber-optic systems.)

Reverse biasing draws current carriers out of the central depleted region, blocking current flow unless light frees electrons and holes to carry current. The amount of current increases with the amount of light absorbed, and the light absorption increases with the thickness of the depleted region. Depletion need not rely entirely on the bias voltage. The same effect can be obtained if a lightly doped or undoped intrinsic semiconductor region is between

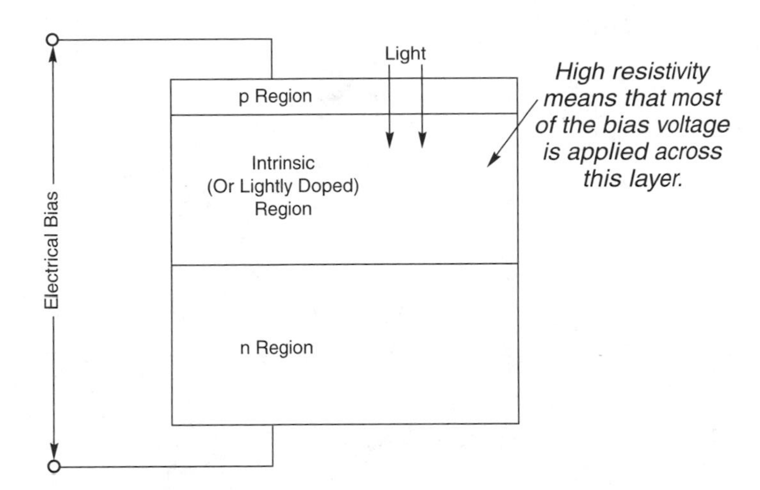

FIGURE 11.5
A simple pin *photodiode.*

the *p*- and *n*-doped regions shown in Figure 11.5. In a sense, such *pin* (*p*-intrinsic-*n*) photodiodes come predepleted because the intrinsic region lacks the impurities that can generate current carriers in the dark. This design has other practical advantages. By concentrating absorption in the intrinsic region, it avoids the noise and slow response that occur when the *p* region of ordinary *pn* photodiodes absorbs some light. The bias voltage is concentrated across the intrinsic semiconducting region because it has higher resistivity than the rest of the device, helping raise speed and reduce noise.

The speed of *pin* photodiodes is limited by variations in the time it takes electrons to pass through the device. This time spread can be reduced in two ways—by increasing the bias voltage and/or by decreasing the thickness (and width) of the intrinsic layer. Reducing intrinsic layer thickness must be traded off against detector sensitivity because this reduces the fraction of the incident light absorbed. Typical biases are 3 to 20 V, although some devices have specified maximum bias above 100 V. Typical response times range from a few nanoseconds to about 5 ps. Sensitivity of *pin* detectors is measured as amperes of current generated per watt of light. For silicon, the peak sensitivity is about 0.7 A/W at 800 nm; InGaAs has a peak of around 1 A/W near 1600 nm. An important attraction of *pin* photodiodes is a large dynamic range; their output-current characteristics can be linear over 50 dB.

pin detectors can have response times well under 1 ns and dynamic ranges of 50 dB.

The speed and sensitivity of *pin* photodiodes are more than adequate for most fiber-optic applications, and they are widely used even in high-performance systems. Sending their electrical output directly to an electronic preamplifier can boost sensitivity.

pin photodiodes are widely used because of their high speed and good sensitivity.

The designs of actual *pin* photodiodes are more complex than this simple example, particularly in fast devices, like the multigigahertz detector in Figure 11.6. Light signals at 1200 to 1600 nm pass through the antireflection coating and upper layer of InP (which are transparent at those wavelengths) and are absorbed in the intrinsic InGaAs layer. Other designs direct light through the InP substrate, which is also transparent to the signal wavelengths.

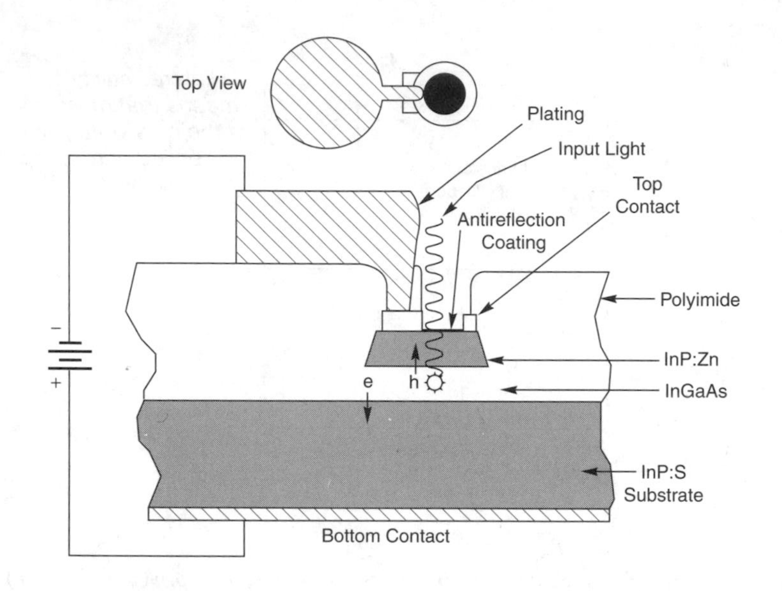

FIGURE 11.6 *Structure of a multigigahertz* pin *detector. (Copyright © 1993 Hewlett-Packard Company. Reproduced with permission.)*

Phototransistors

Phototransistors both sense light and amplify light-generated current; they are used only in low-cost, low-speed systems.

Some detectors have internal amplification, which increases their sensitivity. The simplest is the phototransistor shown in Figure 11.7. Those familiar with transistors can view it as a transistor in which light generates the base current. An alternative is to see a phototransistor as a *pn* photodiode within a transistor. In practice, light normally reaches the base through or around the emitter, which has a wide band gap so it is transparent to the wavelengths detected. Most commercial phototransistors are made of silicon; their most common uses are in inexpensive sensors, but the same devices can be adapted for low-cost, low-speed fiber-optic systems.

The generated photocurrent is amplified, like base current in a conventional transistor, giving much higher responsivity than a simple photodiode. However, this increase comes at a steep price in response time and linearity and at some cost in noise. You can see this by looking at Table 11.2, which compares several photodetectors. In practice, phototransistors are limited to systems operating below the megahertz range.

Output of a phototransistor is fed to the base of a second transistor in a *photodarlington.* This is a simple, integrated darlington amplifier, which adds a second transistor to the phototransistor to increase the responsivity. This has the trade-offs of lowering speed and increasing noise. That constrains the photodarlington's uses even more narrowly than those of the phototransistor, but it offers higher responsivity for low-cost, low-speed applications.

Avalanche photodiodes are fast detectors with internal amplification.

Avalanche Photodiodes

The *avalanche photodiode* (APD) is another photodiode that provides internal amplification. It operates at much higher speeds, making it more useful for fiber-optic communications.

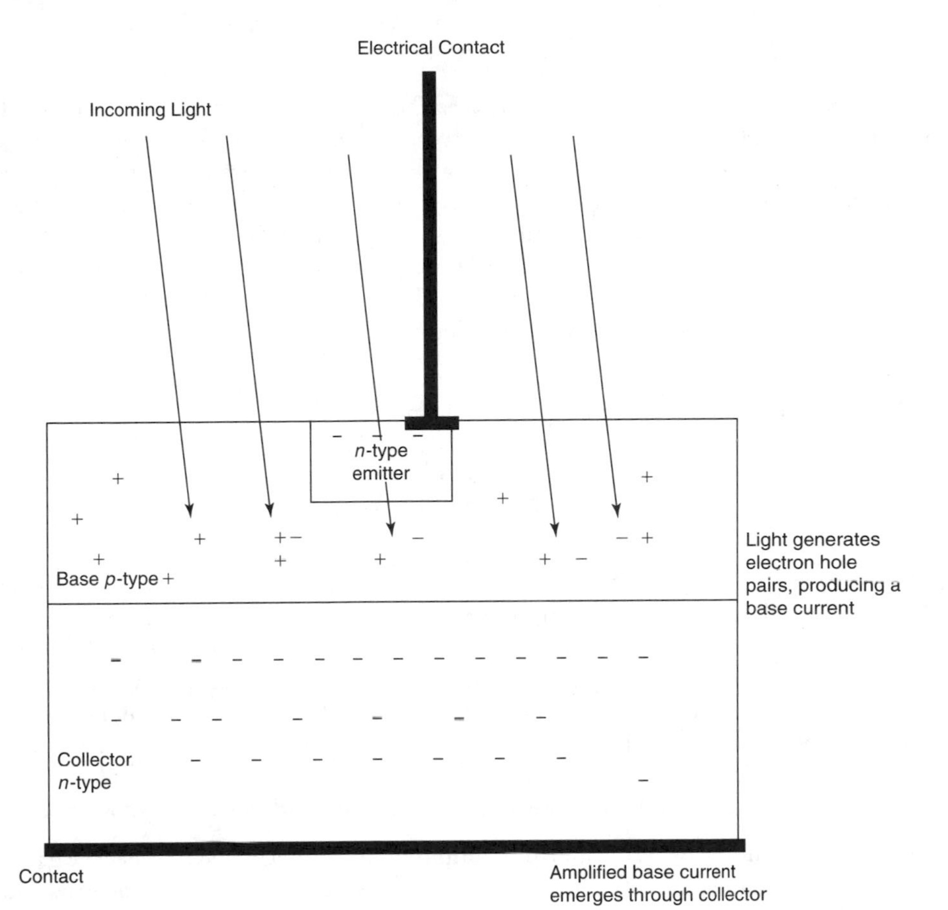

FIGURE 11.7 *Operation of a phototransistor.*

Table 11.2 Typical detector characteristics

Device	*Responsivity*	*Rise Time*	*Dark Current*
Phototransistor (Si)	18 A/W	2.5 μs	25 nA
Photodarlington (Si)	500 A/W	40 μs	100 nA
pin photodiode (Ge)	0.4 A/W	0.1–1 ns	100 nA
pin photodiode (Si)	0.5 A/W	0.1–5 ns	1–10 nA
pin photodiode (InGaAs)	0.8 A/W	0.005–5 ns	0.1–3 nA
Avalanche photodiode (Ge)	(voltage-dependent)	0.3–1 ns	400 nA (voltage-dependent)
Avalanche photodiode (Si)	10–125 A/W (voltage-dependent)	0.1–2 ns	10–250 nA (voltage-dependent)
Avalanche photodiode (InGaAs)	7–9 A/W (voltage-dependent)	0.1–0.5 ns	6–160 nA (voltage-dependent)

You can think of an APD as a two-stage device. The first stage is a conventional photodiode in which light generates current carriers. The second is an internal amplification stage based on avalanche multiplication in which a strong electric field accelerates the light-produced electrons so much that they can knock valence electrons out of the semiconductor lattice. At high voltage, the result is a near-avalanche of carriers—thus the name—that is still proportional to the amount of incident light. However, a further increase of the voltage to a point called the *breakdown voltage* V_B causes current to flow freely through the semiconductor without regard to the amount of incident light; this can damage the devices.

The multiplication factor increases as bias voltage approaches the breakdown voltage.

The factor by which each initial carrier is multiplied, called the *multiplication factor,* typically is 30 to 100. Multiplication factor M is defined as

$$M = \frac{1}{1 - \left(\frac{V}{V_B}\right)^n}$$

where V is the operating voltage, V_B is the breakdown voltage, and n is a number between 3 and 6, depending on the device characteristics. The multiplication factor can become very large as the operating voltage approaches the breakdown voltage. Care must be taken when increasing the operating voltage because reaching or exceeding the breakdown voltage can damage the device. A representative plot of multiplication factor as a fraction of breakdown voltage is shown in Figure 11.8. Note that multiplication factors of 100 require bias voltages within a few percent of breakdown. Typical APD operating voltages are 150 to 400 V in silicon but only 20 to 60 V in InGaAs, which has an inherently lower breakdown voltage.

APDs are fast, but the uneven nature of multiplication introduces noise. Avalanche gain is an average; not all photons are multiplied by the same factor. Signal power increases with

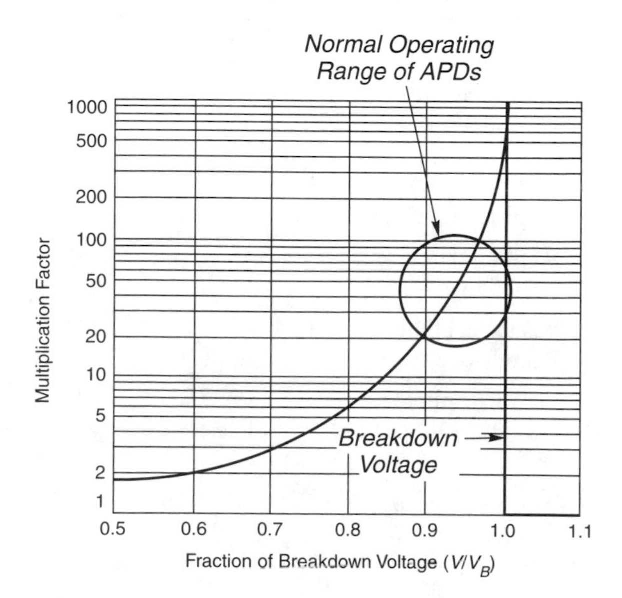

FIGURE 11.8
Increase of multiplication factor in APD.

roughly the square of the multiplication M, and at moderate values M increases faster than noise. However, as M reaches high levels, noise increases roughly as $M^{2.1}$, faster than M^2. As a result, APDs have an optimum multiplication factor to control noise; for silicon devices it is typically 30 to 100, but for InGaAs it is much lower.

Avalanche photodiodes require much higher operating voltages than the few volts normally used for *pin* photodiodes or other semiconductor electronics. The need for special circuits to provide this drive voltage, and to compensate for the temperature sensitivity of APD characteristics, makes APD receivers more complex than *pin* types. As Table 11.2 shows, APDs also suffer from higher dark voltages and cannot match the rise times of the fastest *pin* detectors.

pin-FET and Integrated Receivers

The distinctions between detectors and the receivers that contain them sometimes can be hazy. Strictly speaking, integrating a *pin* photodiode with an electronic circuit on the same substrate or a hybrid package produces a *receiver*. However, sometimes these integrated *detector-amplifiers* are simply called detectors. Some are called *pin*-FETs because they include field-effect transistor (FET) circuits along with the *pin* photodiode.

Performance Considerations

The factors that affect detector and receiver performance are complex and often interrelated. These considerations generally apply to all types of detectors and receivers, but some apply more to the detector. Remember that overall performance depends on the entire receiver system, and can be changed dramatically by replacing a detector or adding an optical preamplifier.

We can divide performance considerations into several broad categories. Four directly measure receiver performance: the strength of electrical signal generated in response to an optical input, internal noise levels, linearity, and the speed of the response. Signal coding and modulation format also play critical roles. Finally, the quality and power level of the optical input signal strongly influence performance. Let's look at each of these in turn.

Sensitivity

Sensitivity measures the response to an optical input signal as a function of its intensity, in units such as amperes (of output signal intensity) per watt (of input light). Although sensitivity may sound like a simple concept, setting a precise definition can be tricky.

Sensitivity measures how much output signal a receiver produces for a given optical input.

Detector sensitivity can be measured in two subtly different units. *Responsivity* is the ratio of electrical output from the detector to the input optical power. If the output current varies proportionally to the input, this is measured as amperes per watt (A/W). In practice, input powers usually are in the microwatt range, so responsivity is sometimes given as microamperes per microwatt (μA/μW), which is equivalent.

Quantum efficiency measures the fraction of incoming photons that generate electrons at the detector:

$$\text{Quantum efficiency} = \frac{\text{Electrons out}}{\text{Photons in}}$$

This sounds like the same thing as responsivity, but the two are not equivalent. Recall that the energy of a photon depends on the wavelength, so a 400-nm photon carries twice as much energy as an 800-nm photon. Suppose a detector generates one electron from every photon that reached it at either wavelength. In that case, the input power at 400 nm would be twice the level at 800 nm, but the output current would be the same at both wavelengths. Thus responsivity—measured relative to power—at the shorter wavelength would be only half the level at the longer wavelength.

Responsivity and quantum efficiency both depend on input wavelength and the detector material.

Both responsivity and quantum efficiency depend on the input wavelength, but their differences mean that the two curves are not interchangeable. The shape of the curve depends largely on the detector material, but the height also depends on the detector type and structure. Some curves, such as the one in Figure 11.4, show quantum efficiency at different wavelengths relative to the maximum value, rather than in absolute terms. You should always be sure of the scales you're looking at.

Typical values of responsivity from a *pin* detector alone are 0.4 to 1 A/W. InGaAs detectors typically have responsivity of 0.5 to 1 A/W in their long-wavelength range, and silicon has responsivity of 0.4 to 0.5 A/W at shorter wavelengths. (Note that according to the difference in wavelength, a silicon detector used at 800 nm should have about half the responsivity of an InGaAs detector used at 1600 nm.) Quantum efficiency of a detector with no internal amplification can approach 100%. Internal or external amplification can give much higher values by multiplying the number of electrons effectively generated by each incoming photon. Detector response also depends on operating temperature.

Receiver sensitivity is a different specification, the minimum optical input signal, usually in microwatts or dBm (decibels relative to one milliwatt) needed to operate at the required performance level. This quantity depends on the detector type as well as the receiver circuitry. For example, one receiver has sensitivity of −23 dBm with a *pin* detector, but −32 dBm with an avalanche photodiode as the detector. That reflects how internal amplification allows a receiver to process much weaker signals.

Dark Current and Noise-Equivalent Power

The electronic signal emerging from a detector includes noise as well as signal. The noise comes from many sources, including the optical input, the detector itself, and the amplification electronics. Electromagnetic interference can add noise by inducing currents in conductors in the receiver. The full complexities of noise are beyond the scope of this book, but you should understand two key concepts: dark current and noise-equivalent power.

Dark current is the noise a detector generates in the dark.

An ideal detector generates an output signal that depends only on the input light, so in the dark it should produce no signal at all. Nature isn't that kind. Any detector generates some output current when it is operated in the normal manner but receives no light at all. This *dark current* measures the electrical noise inherent within the detector, which also is present when the detector is exposed to light. It's analogous to the low level of hiss you can hear during silent intervals on an analog tape cassette. It sets a floor on the minimum detectable signal, because a signal must produce more current than the dark current in order to be detected. Dark current depends on operating temperature, bias voltage, and the type of detector.

Noise-equivalent power (NEP) is the input power needed to generate an electrical current equal to the root-mean-square noise from the detector (or receiver). This more directly measures the minimum detectable signal because it compares noise directly to optical power. NEP depends on the frequency of the modulated signal, the bandwidth over which noise is measured, the detector area, and the operating temperature. It's measured in the peculiar units of watts divided by the square root of frequency (in hertz), or $W/Hz^{1/2}$. Specified values normally are measured with a 1-kHz modulation frequency and a 1-Hz bandwidth.

Noise-equivalent power is the optical power needed to generate average detector noise.

Speed and Bandwidth

You can think of the speed of a detector in two ways. One is the time it takes to convert an input light signal into an output electronic signal. This measures how long the signal takes to pass through the detector, and has little impact on system operation because it does not change the shape of the pulses passing through. However, how fast the detector output responds to a change in the optical input changes the shape of the signal pulse, which does impact system performance. This can be measured as *rise time* or *bandwidth*.

Detector bandwidth depends on rise time.

Rise time is the time the output signal takes to rise from 10% to 90% of the final level after the input is turned on instantaneously, as shown in Figure 11.8. The *fall time* is how long the output takes to drop from 90% to 10% of the peak value after the input turns off abruptly. The two are not always symmetrical; some devices have long, slow-falling tails. Both are shown in Figure 11.8. Very slow fall times are undesirable because they can limit the detector's response speed more severely than rise time.

Rise and fall times are a measure of bandwidth, the range of frequencies that a detector can reproduce in a signal. Detectors themselves can reproduce extremely low frequencies, although receiver electronics may limit low-frequency response. The most important limitation in fiber optics is the way detector response falls off at higher frequencies. If you study electronics, you will learn that the highest frequencies are responsible for the sharp edges on a square-wave function like the input digital pulse of Figure 11.9. Lose those high frequencies, and the edges of the signal pulse become rounded, and the pulse takes longer to rise and fall, as shown in the output of Figure 11.9.

Detector bandwidth usually is defined as the frequency at which the output signal has dropped to 3 dB (50%) below the power at a low frequency. This means that only half as much signal is getting through the detector at the higher frequency. Frequencies higher than the upper bound of the bandwidth are attenuated even more. As bandwidth decreases, the higher frequencies fade away, and the pulses become more rounded.

Loss of high frequencies rounds the edges of digital pulses.

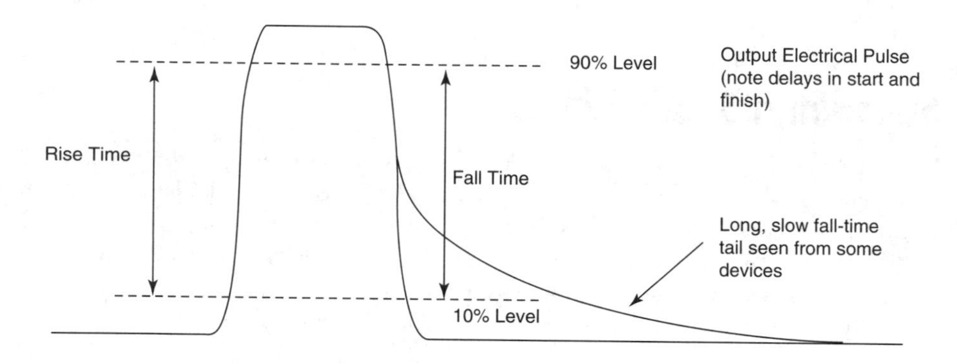

FIGURE 11.9
Rise time of a pulse.

If rise and fall times are equal, the 3-dB bandwidth can be estimated from the rise time using the formula

$$\text{Bandwidth} = \frac{0.35}{\text{Rise time}}$$

Thus a detector with 1-ns rise time has a 3-dB bandwidth of 350 MHz, while one with 10-ps rise time has a 3-dB bandwidth of 35 GHz. You can also flip the formula to estimate rise time if you know bandwidth:

$$\text{Rise time} = \frac{0.35}{\text{Bandwidth}}$$

Device geometry, material composition, electrical bias, and other factors all combine to determine bandwidth and rise time. For relatively slow devices, the rise time is proportional to the RC time constant—the photodiode capacitance multiplied by the sum of the load resistance and the diode's internal resistance. Reducing the equivalent capacitance can increase speeds. At higher speeds, two other factors limit rise time, the diffusion of current carriers and the time needed for carriers to cross the depletion region.

Linearity of Response

In an ideal detector, the output current would be a constant multiplied by the input power. This is called a *linear response,* and in practice it is available only over a limited range of input powers, called the *dynamic range.* Once the input exceeds that range, the output does not increase as rapidly, and the signal becomes distorted.

A detector has a linear response over a limited dynamic range.

As you can see in Figure 11.10, an increase of Δp in the input light produces an increase of Δi in the output current for a receiver or a detector. In the lower portion of the curve, where response is linear, $\Delta i/\Delta p$ is a constant. Once the power exceeds the detector's dynamic range, each additional photon no longer produces as many electrons in the output current, so the device starts to saturate. This happens every time the signal reaches this level, which could be every pulse in a digital system.

The result is distortion, much as when an audio speaker is driven with more power than it can handle. Exceeding the dynamic range in a digital receiver increases the bit error rate. In either case, inserting attenuators in front of the detector can reduce average signal intensity to a level within the dynamic range so the detector responds linearly. Receiver circuits also impact dynamic range, so you should consider their properties as well as those of the detector.

Detectors also need to receive a minimum level of input power, so the signal levels will exceed the background noise or dark current mentioned earlier.

Signal Coding and Modulation

Signal formats also affect detector and receiver performance. Analog modulation looks simple, because it only requires replicating the input signal. It also uses bandwidth efficiently. However, perfectly replicating the signal is difficult because detectors are not perfectly linear. Other complications include noise from the signal and the detector itself, and signal intensity. Figure 11.11 shows how distortion can affect analog signals.

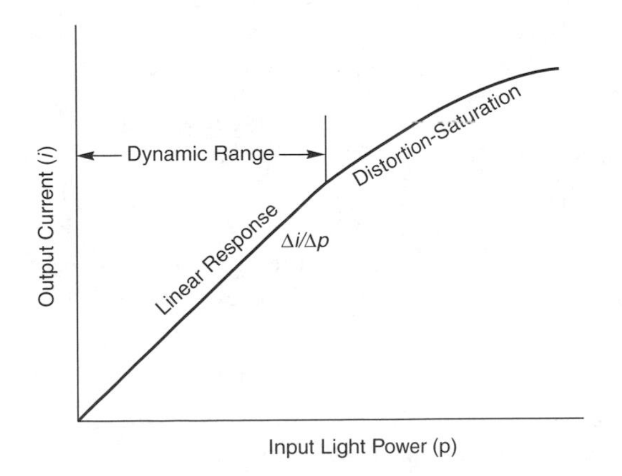

FIGURE 11.10
Receiver output as a function of input light power. Both the detector and the receiver electronics affect dynamic range.

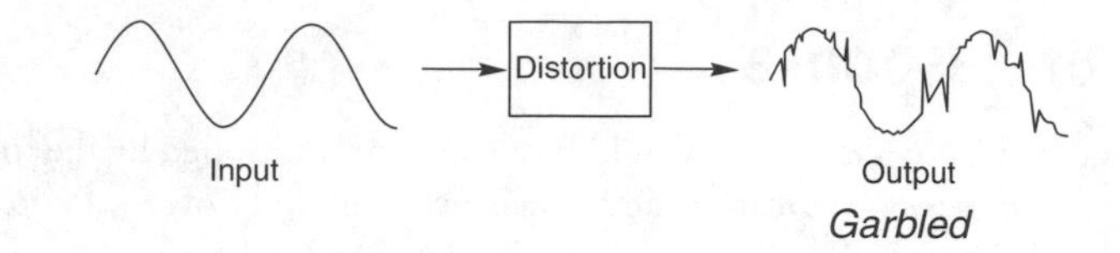

a. Analog

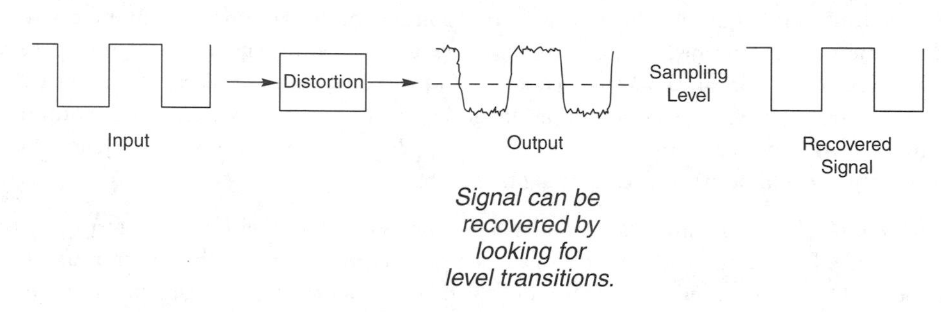

b. Digital

FIGURE 11.11 *Effects of distortion on (a) analog and (b) digital signals.*

Digital signals are more robust than analog signals, and can be extracted from noise.

Digital signals can better withstand the effects of noise and distortion than analog signals. Conventional digital signals are binary codes in which the light is either "on" or "off"—corresponding to digital 1s and 0s. This means that the receiver does not have to measure the precise signal strength, as it does to replicate an analog signal. For digital signals, it only has to know if the binary signals are "off" or "on." This can be done by looking for a minimum threshold level, shown in Figure 11.11. Any pulse that reaches this level is considered "on"; those that don't are considered "off." This actually is done by the receiver's electronic circuits, but it eases the requirements on the detectors.

Functionally, the difference between analog and digital receivers is not as great as you might think. An analog receiver contains a detector and electronics to amplify it and convert a current signal to a voltage signal. Go back to Figure 11.1 and you see these two stages also are part of a digital receiver. Even in a nominally digital receiver, these stages generate an analog output. Discrimination and retiming stages follow the amplification stage, where they regenerate a digital signal from the analog output of the first two stages.

Analog receivers must reproduce signals much more accurately than digital receivers.

The key difference is that the analog receiver has to reproduce an analog input signal very accurately or the output will be distorted, as shown in Figure 11.11(a). The analog stages of the digital receiver don't have to be that good. A digital receiver is like a sound monitor that only needs to be sensitive enough to tell if someone is talking in a room. An analog receiver is like a sound monitor that lets you understand conversations in the room. If you've ever tried to understand the announcements in an airport, rail station, or bus terminal, you know there's a big difference between sound and intelligible speech. In a sense, a digital receiver is really a cheap and dirty analog receiver with a digital back end.

It's worth noting that the "off" state is not completely off in many fiber-optic systems, because many transmitters generate a very weak optical signal for 0 and a much stronger signal for a 1. That would pose a problem if the digital signal had to be reproduced precisely. However, decision circuits can easily tell if binary signals are either in an "on" or "off" state. In that way, noise can be removed from digital signals, something that is impossible with analog signals. (You will learn more about cleaning up signals later.)

Digital transmission does not have to be binary. Some digital systems may have a series of steps in energy level, so the pulse may have, for example, four rather than two possible levels. That can carry more information, but requires telling the difference between four rather than two energy levels. Such systems have been demonstrated in the laboratory, but are not in practical use.

Signal Quality and Power

Signal-to-noise ratio measures analog signal quality.

Power and quality of the input optical signal also are important in detector response. As you just learned, digital signals can better withstand the effects of noise than analog signals. Once noise corrupts an analog signal, it is hard to remove.

Analog signal quality is measured as *signal-to-noise ratio,* usually in decibels. Normally average powers are used for both signal and noise. A 30-dB signal-to-noise ratio (or S/N ratio) means the amplitude of the signal is 1000 times higher than that of the noise. How seriously the noise level affects the signal depends on the situation, the type of signal, and who's measuring or perceiving it. If the signal is music, you'll notice a steady noise during quiet passages. A quiet conversation a few feet away might be offensive during a classical music concert, but you'd never notice it when Metallica was playing full blast. Standards have been developed for what is acceptable in analog applications such as cable television.

Bit error rate measures quality of digital signals.

Digital signal quality normally is measured as *bit error rate,* the fraction of bits received incorrectly. A normal target is 10^{-12}, or one bit in every trillion. This figure depends not only on the detector, but on the ability of the receiver electronics that process the output signal to tell the difference between 0s and 1s. A comparatively small decrease in received power can make bit error rate soar. In the plot shown in Figure 11.12, a 1-dB decrease in received power causes bit error rate to jump from 10^{-12} to 10^{-9}. All systems are not that sensitive, but some are.

Increasing signal speed can combine with detector sensitivity to limit performance at high data rates. To see how this works, consider two digital signals with the same average power. One is a series of 1 billion pulses a second; the other is a series of 10 billion pulses a second. Although the average powers are the same, that power has to be averaged over different numbers of pulses. The faster the data rate, the more pulses that average power is divided among. Each pulse in the 1 Gbit/s signal contains 10 times more energy (or 10 times more photons) than a pulse in the 10 Gbit/s signal. As the number of photons per pulse decreases, the bit error rate increases even though the average power remains constant. You can see this also in Figure 11.12, where a small increase in data rate—in this case from 500 to 600 Mbit/s—would increase bit error rate at constant power from 10^{-12} to 10^{-3}. Alternatively, the bit error rate could be kept constant by increasing received power by roughly 2 dB.

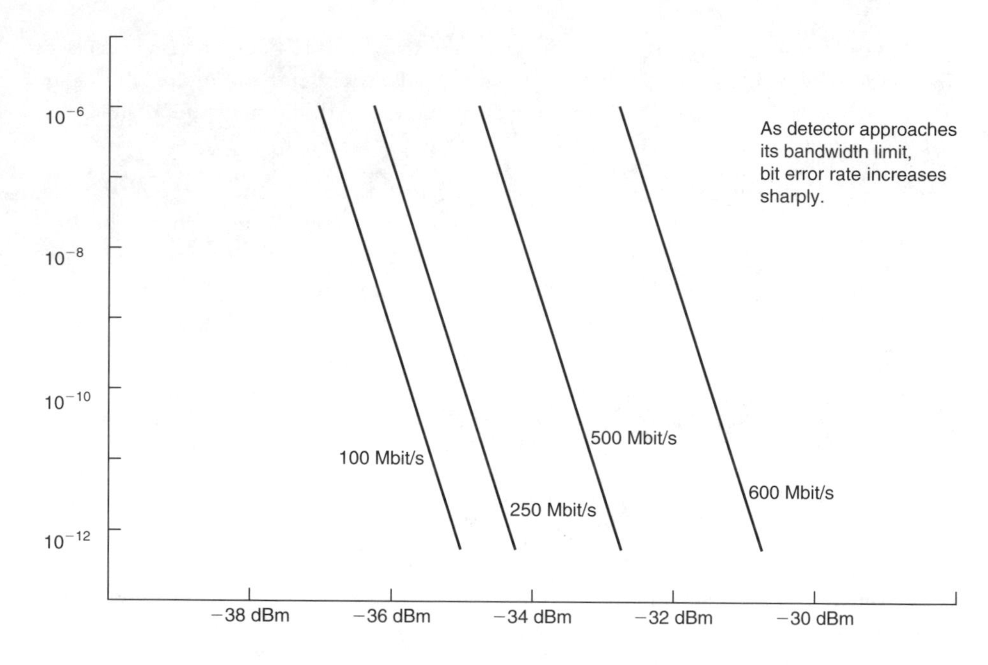

FIGURE 11.12 *Bit error rate increases as average power drops and data rate increases.*

This effect causes receiver sensitivity requirements to increase considerably as data rates per channel increase.

Electronic Functions

Detector output must be processed before other equipment can use the electronic signal.

Converting an optical signal into electrical form is only the first part of a receiver's job. The raw electrical signal generally requires some further processing before it can serve as input to a terminal device at the receiver end. Typically, photodiode signals are weak currents that require amplification and conversion to voltage. In addition, they may require such cleaning up as squaring off digital pulses, regenerating clock signals for digital transmission, or filtering out noise introduced in transmission. The major electronic functions are as follows:

1. Preamplification
2. Amplification
3. Equalization
4. Filtering
5. Discrimination
6. Timing

If you're familiar with audio or other electronics, you will recognize some of these functions. Not all are required in every receiver, and even some of those included may not be performed by separate, identifiable devices. A phototransistor, for example, both detects and amplifies. And many moderate-performance digital systems don't need special timing circuits. Nonetheless, each of these functions may appear on block diagrams such as in Figure 11.1. Their operation is described briefly below.

Preamplification and Amplification

Typical optical signals reaching a fiber-optic receiver are 1 to 10 μW and sometimes lower. If a *pin* photodiode with 0.6 to 0.8 A/W responsivity detects such signals, its output current is in the microampere range and must be amplified for most uses. In addition, most electronics require input signals as voltage, not current. Thus, detector output must be amplified and converted.

Microampere-level *pin* detector outputs must be converted to logic-level voltages.

Receivers may include one or more amplification stages. Often the first is called preamplification because it is a special low-noise amplifier designed for weak input. (An optical amplifier placed in front of a detector also may be called a preamplifier.) In some cases, as mentioned earlier, the preamplifier may be packaged with the detector. The preamplifier output often goes into an amplifier, much as the output of a tape-deck or CD preamplifier goes to an audio amplifier that can produce the power needed to drive speakers.

Equalization

Detection and amplification can distort the received signal. For example, high and low frequencies may not be amplified by the same factor. The equalization circuit evens out these differences, so the amplified signal is closer to the original. Much the same is done in analog high-fidelity equipment, where standard equalization circuits process signals from tape heads and phonograph cartridges so they more accurately represent the original music.

Filtering

Filtering helps increase the S/N ratio by selectively attenuating noise. This can be important when noise is at particular frequencies (e.g., a high-frequency hiss on analog audio tapes). It is most likely to be used in fiber optics to remove undesired frequencies close to the desired signal, such as harmonics.

Filtering blocks noise while transmitting the signal.

Discrimination

So far, the functions you've looked at are needed to reproduce the original waveform for both analog and digital receivers. However, a further stage is needed to turn a received analog signal back into a series of digital pulses—decoding and discrimination. Rectangular pulses that started with sharp turn-on and turn-off edges have been degraded into unboxy humps, as shown in Figure 11.13. Dispersion may have blurred the boundaries between pulses.

Discrimination circuits generate digital pulses from an analog input.

Pulse regeneration requires circuitry to decide if the input is on or off.

As mentioned earlier, this rounding of square pulses represents loss of high frequencies, which make up the sharp rising and falling edges of the pulse, so if they are lost, the pulses lose their square edges. The remaining low-frequency components contain most of the

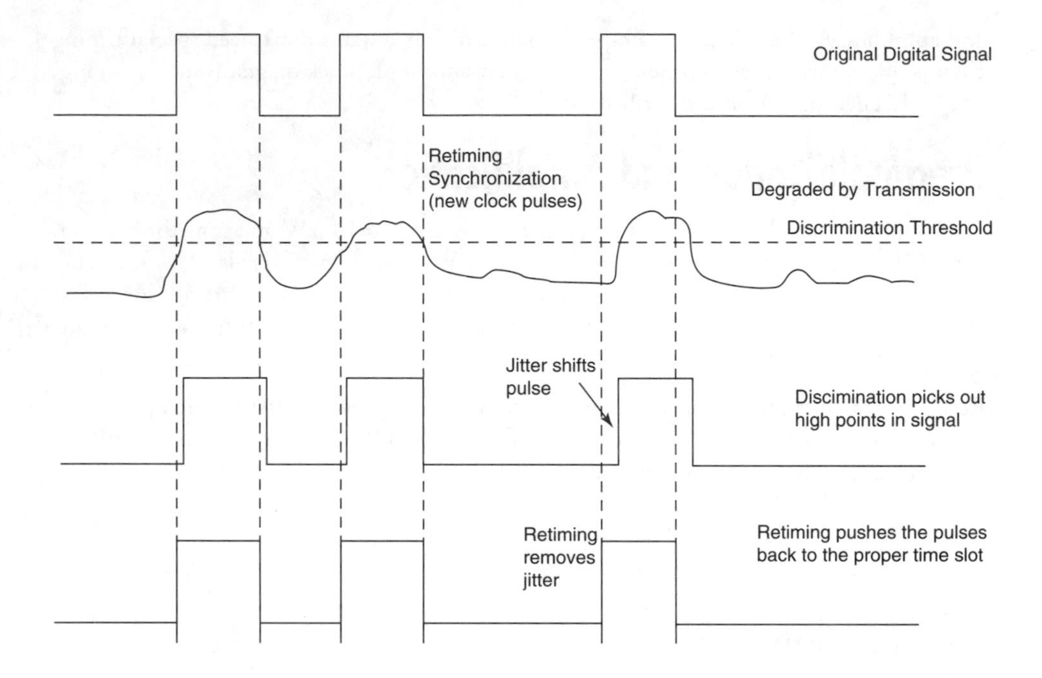

FIGURE 11.13 *Discrimination and retiming regenerate digital pulses.*

information needed, but they are not clean enough to serve as input to other electronic devices. Regeneration of clean pulses requires circuitry that decides whether or not the input is in the on or off state by comparing it to an intermediate threshold level. The decision circuit generates an "on" pulse if the power is above the threshold; otherwise it produces an "off" signal. Care must be taken in selecting this threshold level to avoid misinterpreting input; too low a threshold, for example, could turn noise spikes in the off state into signal pulses.

Timing

Timing of digital pulses often must be resynchronized.

Another essential task in many receivers, particularly in high-performance systems, is resynchronizing the signal. Digital signals are generated at a characteristic clock rate, such as once every nanosecond for a 1-Gbit/s data stream. You can see in Figure 11.13 that discrimination circuits do not necessarily spot the exact times the pulses start and end. These random errors, called *jitter,* can cause the signal to drift from the clock rate, introducing errors.

Timing synchronization recreates the clock signal (the vertical lines in Figure 11.13) and puts the regenerated pulses in the right time slots. It is an essential part of cleaning up signals at a sophisticated receiver.

Packaging Considerations

As with transmitters, packaging is important for receivers. The basic requirements are electronic, mechanical, and optical interfaces that are simple and easy to use. The main mechanical issues are mounts. Electronic interfaces must allow for input of bias voltage and amplifier power (where needed) and for output of signals in the required format. Details can vary significantly.

Optical interface requirements are simpler than for transmitters because mechanical tolerances for aligning fibers with detectors generally are looser. The active areas of detectors are larger than the cores of single-mode fibers. Larger-core multimode fibers transmit signals at slower speeds, so they normally are used only with slower detectors with larger active areas. In practice, receivers are assembled with integral fiber pigtails or connectors that collect light from the input fiber and deliver it to the detector.

In general, packaged receivers look very much like transmitters, and often the two are packaged together as a transceiver. You may have to read the labels to tell them apart. Detector modules are packaged inside receivers just as light-source modules are put inside transmitters. Internal design constraints become increasingly severe at high frequencies because of the problems inherent in high-frequency electronic transmission.

Sample Receiver Circuits

Details of receiver circuitry vary widely with the type of detector used and with the purpose of the receiver. For purposes of this book, I will show only a few simple circuits for important devices and avoid detailed circuit diagrams.

Photoconductive Photodiodes

Photoconductive photodiodes have a load resistor in series with the bias voltage.

The typical *pin* or *pn* photodiode used in a fiber-optic receiver is used in a circuit with a reverse-bias voltage applied across the photodiode and a series load resistor, such as that shown in Figure 11.14. In this mode, the photodiode is photoconductive because the photocurrent flowing is proportional to the nominal resistance of the illuminated photodiode. This simple circuit converts the photocurrent signal from a photodiode into a voltage signal.

The division of the bias voltage between the photodiode and the fixed resistor depends on illumination level. The higher the illumination of the photodiode, the more current it will

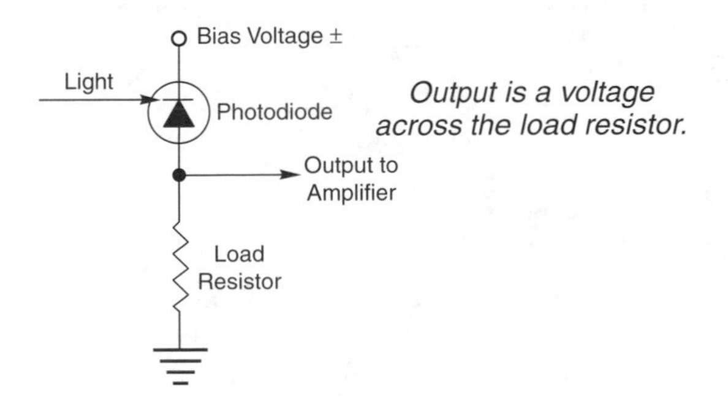

FIGURE 11.14
Basic circuit for photoconductive pin *or* pn *photodiode.*

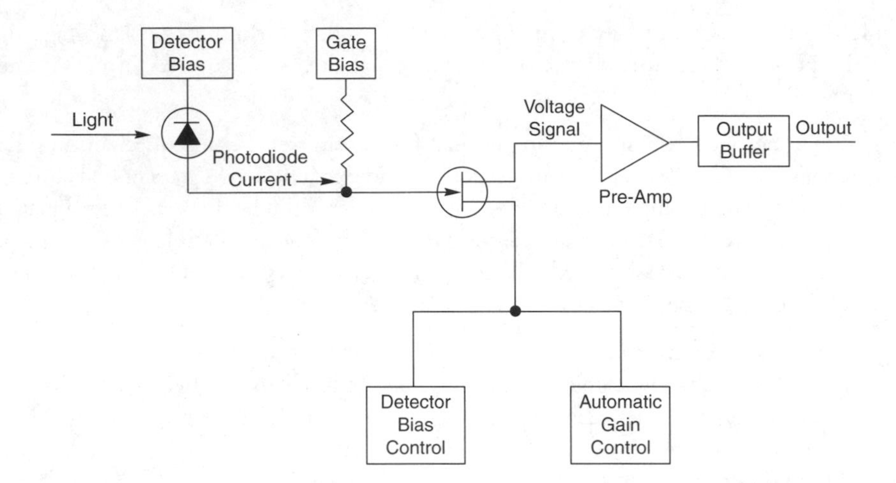

FIGURE 11.15 *Block diagram of pin-FET receiver circuit.*

conduct and, thus, the larger the voltage drop across the load resistor. In the simple circuit shown, the signal voltage is the drop across the load resistor. Most circuits are more complex, with amplification stages beyond the load resistor, as in *pin*-FET and detector-preamplifier circuits. Figure 11.15 is a block diagram of one circuit that includes automatic gain control, which can increase dynamic range by turning down the amplification factor before any other components are overloaded.

Avalanche Photodiode Circuits

The circuits used for avalanche photodiodes are conceptually similar to those used for photoconductive *pin* photodiodes. However, because of the high bias voltages required and the sensitivity of the photodiode to bias voltage, care must be taken to assure stable bias voltage. This adds to circuit complexity, as shown in the block diagram of Figure 11.16.

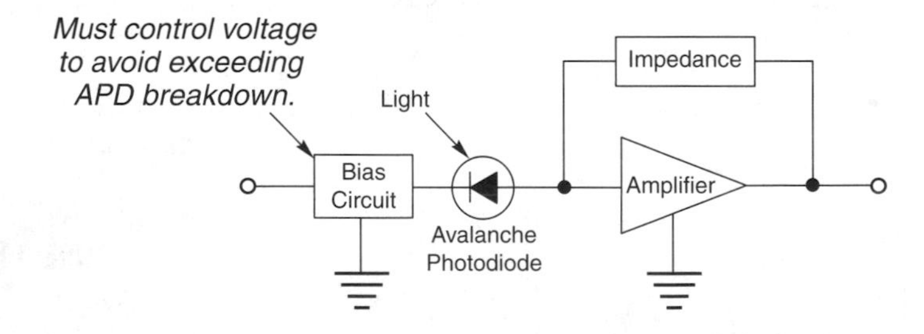

FIGURE 11.16 *Basic receiver circuit for avalanche photodiode.*

What Have You Learned?

1. A receiver detects optical signals and converts them into electronic form. It includes an optical detector, an electronic amplifier, and thresholding and retiming circuits. Any optical amplifiers and wavelength-division demultiplexing precede the receiver.

2. Detectors are color-blind, so optical channels must be separated and routed to different receivers, one per channel.
3. Most fiber-optic detectors are reverse-biased semiconductor photodiodes. Light striking the depleted region near the junction generates free electrons and holes, so a current flows through the diode and becomes the electronic output signal. Simple circuits can convert this current signal into a voltage signal.
4. The wavelengths at which detectors respond depend on their composition. Silicon is used at 400 to 1000 nm. InGaAs is used at 800 to 1700 nm.
5. *pin* photodiodes have a high-resistivity intrinsic layer between *p* and *n* layers, which improves sensitivity. They are widely used for their high speed and sensitivity.
6. Avalanche photodiodes have a high bias voltage that creates an internal cascade of electrons, which multiplies their electrical output power above that of *pin* photodiodes. They are not as fast as *pin* photodiodes and have higher noise.
7. Dark current and noise-equivalent power measure noise levels in detectors.
8. The bandwidth of a detector determines how fast its output current rises when the input signal jumps. Loss of high frequencies in the signal decreases rise time, and rounds sharp-edged signals.
9. Detectors operate best over a limited dynamic range, where their output depends linearly on the input signal. At higher powers they distort received signals.
10. Analog receivers must reproduce the shapes of input signals accurately. Digital receivers include an extra stage that detects signal transitions and retimes pulses, so they can extract digital signals from noise. This makes digital signals more robust than analog signals.
11. Signal-to-noise ratio measures the quality of analog signals. Bit error rate measures the quality of analog signals.

What's Next?

In Chapter 12, I move on to optical amplifiers and to electro-optic repeaters and regenerators, which combine the functions of receivers and transmitters.

Further Reading

S. R. Forrest, "Optical detectors for lightwave communication," pp. 569–599 in Stewart E. Miller and Ivan P. Kaminow, eds, *Optical Fiber Telecommunications II* (Academic Press, 1988)

Gerd Keiser, *Optical Fiber Communications* (McGraw-Hill, 2000), see Chapter 6 "Photodetectors" and Chapter 7 "Optical Receiver Operation."

Jim Rue and Bouchiab Nessar, "High speed avalanche photodiode optical receivers," *Fiberoptic Product News* (November 1999)

Questions to Think About for Chapter 11

1. The input signal at a receiver is −30 dBm (1 μW), which is too low for your *pin* photodiode detector. What alternatives are there to increase receiver sensitivity?
2. An input signal is −30 dBm (1 μW). If its data rate is 1 Gbit/s, how much energy does each pulse contain, remembering that a power of one watt equals one joule per second? If the signal is at 1.5 μm, how many photons does that correspond to, using the following equation?

$$E = \frac{1.989 \times 10^{-19} \text{ (joules/}\mu\text{m)}}{\lambda \text{ (}\mu\text{m)}}$$

3. If a detector has a response of 1 μA/μW and the input is −30 dBm, what is the output current? How many electrons per second does this correspond to, recalling that 1 A = 6.24×10^{18} electron charges per second? How many electrons would be contained in a 1-nanosecond pulse?
4. A silicon *pin* detector has peak sensitivity of 0.7 A/W at 800 nm. If the input signal is −20 dBm, how many electrons does a 1-ns pulse produce? Use the equations and conversion factors from Questions 2 and 3.
5. The rise time of an InGaAs *pin* photodiode is 0.005 ns. The rise time of an InGaAs avalanche photodiode at the same wavelength is 0.1 ns. All other things being equal, how much larger is the bandwidth of the *pin* photodiode? If the rise time is all that limits the bandwidth of the two devices, what are their bandwidths?
6. The breakdown voltage of a silicon APD is 100 V. What voltage should it be operated at to have an electron multiplication factor of 20?
7. The dark current in a germanium *pin* photodiode is 100 nA, compared to 1 nA in an InGaAs *pin* photodiode. Suppose that all other things are equal, and the germanium detector has a sensitivity of −15 dBm, limited by dark current. What would be the sensitivity of the InGaAs detector?

Quiz for Chapter 11

1. How many separate receivers are required for a 32-channel wavelength-division multiplexed system with an optical preamplifier that provides 15-dB gain on all channels?
   a. 1
   b. 4
   c. 8
   d. 32
   e. 100

2. When would an optical preamplifier be used at the receiver end of a system?

a. Always

b. When input power is below 1 mW

c. When the input signal contains multiple optical channels

d. When input power is below receiver sensitivity

e. Never

3. What is present in a digital receiver that is not present in an analog receiver?

a. Nothing

b. A detector

c. Thresholding and retiming circuits

d. Amplification circuits

e. Wavelength-division multiplexing

4. Photodiodes used as fiber-optic detectors normally are

a. reverse-biased.

b. thermoelectrically cooled.

c. forward-biased.

d. unbiased to generate a voltage like a solar cell.

e. none of the above

5. Silicon detectors are usable at wavelengths of

a. 800 to 900 nm.

b. 1300 nm.

c. 1550 nm.

d. all of the above

6. Which detector material is most often used in the 1550-nm window?

a. Silicon

b. InGaAs

c. GaAs

d. Germanium

e. All of the above

7. A *pin* photodiode is a

a. point-contact diode detector in which a pin makes contact with the semiconductor.

b. semiconductor detector with an undoped intrinsic region between *p* and *n* materials.

c. circuit element used in receiver amplification.

d. photovoltaic detector.

e. hybrid detector-amplifier.

8. A phototransistor

a. has an internal amplification stage based on avalanche multiplication of electrons.

b. has an external amplification stage containing a single transistor.

c. generates a photocurrent in the base of a transistor, which amplifies the signal.

d. is an ordinary transistor that generates an optical signal under bright lights.

e. is the same as a photodarlington.

9. An avalanche photodiode

a. has an internal amplification stage based on avalanche multiplication of electrons.

b. has an external amplification stage containing a single transistor.

c. generates a photocurrent in the base of a transistor, which amplifies the signal.

   d. is an ordinary transistor that generates an optical signal under bright lights.
   e. is the same as a photodarlington.

10. What type of photodetector could have a responsivity of 20 amperes per watt?
   a. Silicon *pin* photodiode
   b. Silicon avalanche photodiode
   c. InGaAs *pin* photodiode
   d. InGaAs avalanche photodiode
   e. No detectors meet this requirement

11. What bit error rate is most often specified for digital telecommunications systems?
   a. 40 dB
   b. 10^{-4}
   c. 10^{-6}
   d. 10^{-12}
   e. 10^{-18}

12. Noise equivalent power is
   a. optical input power required to generate a signal equal to the noise.
   b. noise required to equal the signal intensity.
   c. the power of the current generated when a detector is in the dark.
   d. noise present in the electrical output signal.

13. What happens when you increase the bias voltage above the breakdown voltage in an avalanche photodiode?
   a. You stabilize the output current.
   b. You stop the avalanche of electrons produced inside the semiconductor.
   c. You get in trouble because that can damage the APD.
   d. You increase signal-to-noise ratio to infinity.

14. What's the maximum value of quantum efficiency possible in a *pin* photodiode?
   a. 0
   b. 0.5
   c. 0.9
   d. 1.0
   e. 100

CHAPTER

12

Repeaters, Regenerators, and Optical Amplifiers

About This Chapter

The last three chapters have shown how transmitters and receivers send signals from one end to the other of a fiber-optic system. Sometimes the signals need a boost somewhere along the way between the transmitter and receiver. Repeaters, regenerators, and optical amplifiers give them that needed boost.

I have mentioned optical amplifiers before; they boost the strength of an optical signal by directly amplifying light. This chapter tells you more about their operation as amplifiers. Repeaters and regenerators are more closely related to transmitters and receivers because they convert the optical signal to electronic form, and may also clean up the signal like a receiver before generating a new optical signal. This chapter will explain the differences and what they mean for fiber-optic systems.

The Distance Problem

Signals fade with distance when traveling through any type of cable. The farther you go, the fainter they become, until they become too faint to detect reliably. As you saw in the last chapter, when a digital signal fades below a certain level, the bit error rate rises rapidly. Likewise, an analog signal becomes noisy or distorted, like a distant radio station.

Communication systems avoid this problem by amplifying signals along the way. As shown in Figure 12.1, a repeater or amplifier is inserted into the system at a point where

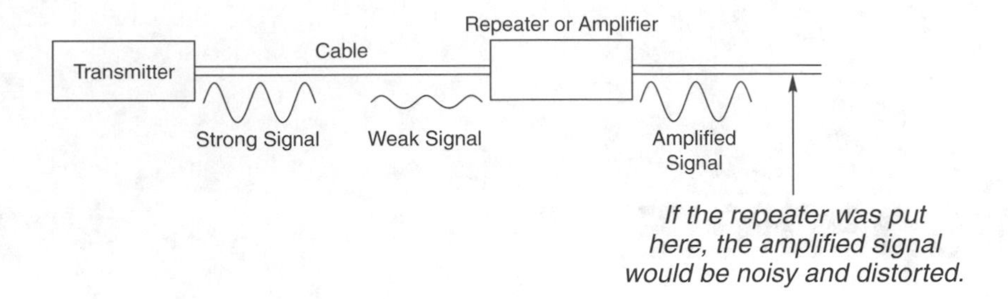

FIGURE 12.1
Amplification in a fiber-optic system.

the signal has become weak, to boost the strength of the signal so it can be transmitted through another length of cable. Many amplifiers or repeaters can be placed in sequence, each one taking a weak signal and generating a strong one. However, care must be taken to insert them at a point in the system before the signal becomes too weak, distorted, or corrupted by noise to be detected clearly. Otherwise it amplifies the noise, like turning up the volume on a noisy radio broadcast.

Fibers can transmit high-speed signals much farther than wires before amplification is needed.

Signals need amplification or regeneration if they travel far enough through either metal or fiber-optic cables. However, as you saw earlier, optical fibers attenuate high-speed signals much less than coaxial cables, so they can carry signals much farther between amplification stages. Exactly how far depends on the system design, but in general it is many times farther. For example, coaxial cables need repeaters every few thousand feet (roughly 1 km) to carry hundreds of megabits per second, while fiber systems can carry signals at the same speed for 50 to 150 km (30 to 90 mi). It's no wonder telecommunications companies prefer fibers if signals have to go far.

To be fair, it isn't always that simple because of other differences between metal cables and fibers. Optical signals do not divide the same way electrical signals do, so a fiber system in general can drive fewer terminals. Thus, if you're trying to split signals among many terminals that aren't very far away, fiber systems may need more amplification. However, the basic idea remains the same.

An optical amplifier is an analog device that amplifies optical signals using the laser principle.

A submarine "repeater" is an optical amplifier in a waterproof "repeater housing."

Optical amplifiers, repeaters, and regenerators are not interchangeable, although sometimes the terminology gets a bit sloppy. Before going further, let's take a careful look at the differences.

- An optical amplifier is a purely optical device. It uses the laser principle, described in Chapter 9, to increase the strength of an optical signal. It is an analog device, so the signal you put in is what you get out, amplified in intensity and perhaps with a dash of noise added. If the signal is noisy to start with, amplifying won't make it any better. However, optical amplifiers have a compelling simplicity. Because they leave the signal in the form of light, a single optical amplifier can simultaneously amplify dozens of optical channels in a wavelength-division-multiplexed fiber system.
- A *repeater* is an electro-optic, or opto-electronic, device that converts a weak optical signal into electronic form, then uses that electronic signal to drive a

transmitter, which repeats the input signal. You can think of it as a receiver and transmitter placed back to back, and that's pretty much how they work. In digital systems, the receiver stage typically does some processing to clean up the signal, but what is done depends on the design. The definition gets rather murky underwater. Designers of submarine cables call optical amplifiers "repeaters" because they are mounted in special waterproof "repeater housings" laid along with the cable. This usage can be confusing, particularly because submarine systems no longer use electro-optic repeaters; only optical amplifiers are laid underwater with submarine cables.

- *Regenerators* have a distinct function, to clean up a digital signal by removing noise and distortion and *regenerating* a fresh version of the signal, as you saw in Figure 11.13. They also amplify the signal in the process. Like digital receivers, regenerators have discrimination circuits that examine the time-varying signal, and decide which changes in signal strength are data bits and which are noise. Retiming circuits make sure the pulses fall into the proper time slots. The result is a series of clean digital pulses, ready for transmission through the next length of cable. Note that regenerators are explicitly digital devices, which remove noise and re-create the original digital signal.

Regenerators clean up digital signals as well as amplify them. All current regenerators are electro-optical.

In practice, the distinctions between repeaters and regenerators are hazy and somewhat historical. Before practical optical amplifiers were developed in the early 1990s, the only way to transmit signals long distances was to use repeaters to boost signal strength; their ability to regenerate signals was secondary. Once optical amplifiers came on the market, there was little need for repeaters to *amplify* signals, because fiber amplifiers usually could do the job better. However, signals still may have to be cleaned up after passing through a series of optical amplifiers, making *regeneration* the function emphasized today.

One big disadvantage of opto-electronic regeneration is that it must be done separately for each optical channel in a WDM system. Recall that detectors are color-blind, so they can't tell the difference between optical channels at 1541 and 1542 nm. Those wavelengths must be split optically, routed to separate optical receivers, then regenerated separately before being recombined for transmission through another length of fiber.

Regeneration must be done separately for each optical channel in a WDM system.

Optical amplifiers generally are preferred in modern long-distance fiber-optic systems, and they provide most of the amplification needed. Regenerators typically are used only at nodes where long-distance systems interconnect with each other, so signals must be redistributed between fibers. You'll learn more about how these systems work later; for now we'll concentrate on amplifiers and regenerators.

Types of Amplification

The differences between electro-optic and optical amplifiers deserve a closer look. Electro-optic or electronic amplification came first, in the form of repeaters. They convert the weak input optical signal into electronic form, process the signal electronically, and use that signal to drive another optical transmitter as shown in Figure 12.2(a)—like a receiver and

Electro-optic repeaters convert an optical signal into electronic form for amplification.

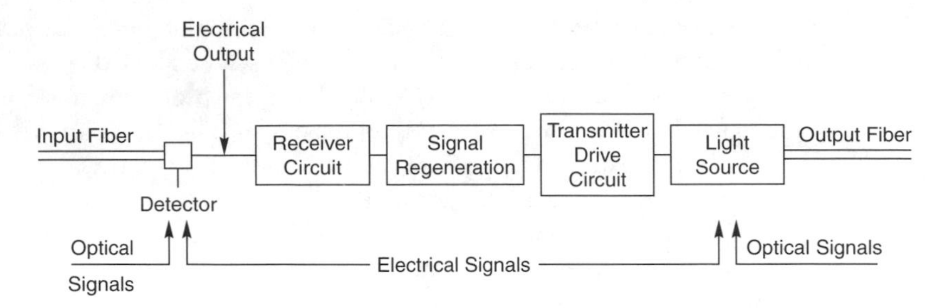

a. Electro-Optic Repeater

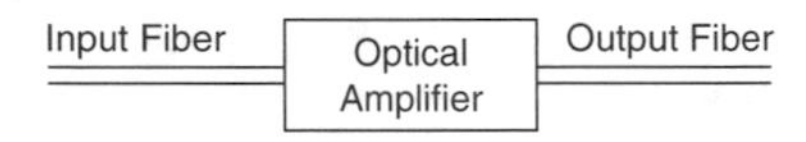

b. Optical Amplifier

FIGURE 12.2
Electro-optic and optical amplifiers.

transmitter placed back to back. Their function is very similar to electronic repeaters in copper cable or radio-frequency communication systems.

In contrast, an optical amplifier takes a weak optical input and amplifies it to generate a strong optical output signal, as shown in Figure 12.2(b). It never converts the signal into electronic form at all. (The wires attached to an optical amplifier provide electrical power to the pump laser.) Present optical amplifiers do not regenerate signals; they merely increase their amplitude, noise and all.

Optical amplifiers are simpler than electro-optic repeaters, but they cannot clean up signals.

Optical amplifiers are simple both in concept and in practice. They contain far fewer components than electro-optic repeaters and generally are far more reliable. They also offer two important operational advantages over electro-optic repeaters. A single optical amplifier can simultaneously amplify many wavelength-division-multiplexed signals carried in the same fiber, without causing them to interfere with each other. Because receivers cannot discriminate between closely spaced wavelengths, WDM signals must be demultiplexed and each one must be amplified by a separate electro-optic repeater. Optical amplifiers also are not sensitive to signal format; if you change signal speeds, you must replace electro-optic repeaters (which are designed for particular frequencies or data rates), but you often can use the same optical amplifier.

Present optical amplifiers cannot regenerate input signals. This means they cannot clean up noise or compensate for dispersion that has accumulated during fiber transmission. However, optical amplifier spacing can be limited to reduce noise, and as you learned earlier, other means can be taken to compensate for fiber dispersion.

There are two types of optical amplifiers, doped fiber amplifiers and semiconductor amplifiers, which you learned about in Chapter 9. Both work on the principle of stimulated emission that is basic to the laser, but in optical amplifiers the light being amplified comes from an outside source and makes a single pass through the medium, which lacks the mirrors present in lasers. The two families of optical amplifiers have quite distinct characteristics, so I will cover the details separately later in this chapter.

Optical amplifiers have largely replaced electro-optic repeaters in new telecommunication systems. The most widely used are erbium-doped fiber amplifiers, but other types remain in development and have some limited application. First, however, we should look at the applications for amplifiers and then learn about electro-optic repeaters and regenerators, because they still have some applications and many remain in service.

Requirements for Amplification

Optical signals may require amplification at different points in communication systems. Figure 12.3 shows some places where optical amplifiers may be used.

Optical amplification is needed in-line, after transmitters, before receivers, and after lossy components.

- *Postamplifiers* are placed immediately after a transmitter to increase strength of a signal being sent through a length of fiber. It might seem easier just to crank up the transmitter output, but that can degrade the quality of the output signal. External amplification of a lower-power transmitter output gives a cleaner signal. Postamplifiers also can generate powerful signals that can be split among many separate outputs if a single transmitter is distributing signals to many points.
- *In-line amplifiers* compensate for signal attenuation in long stretches of fiber. The goal is to amplify a weak signal sufficiently to send it through the next segment of fiber. These generally are required in long telecommunication systems but may be used in some networks where many branching points reduce transmitted power. Signals may require regeneration after a series of many amplifiers.
- *Preamplifiers* amplify a weak optical signal just before it enters a receiver, in effect increasing the sensitivity of the receiver and stretching transmission distances.

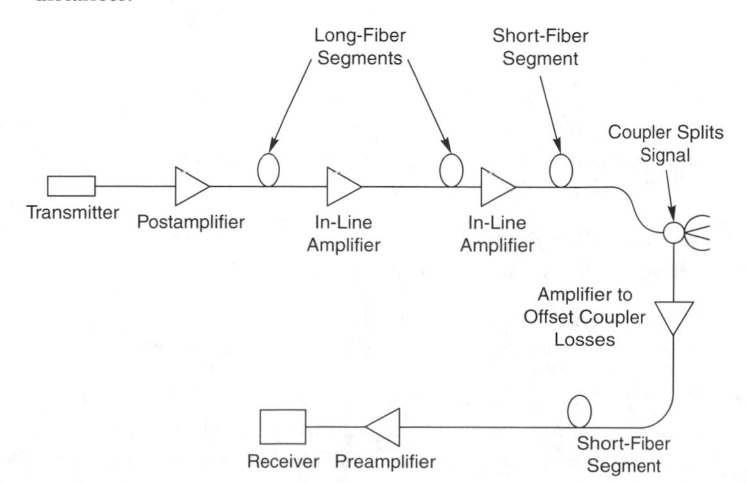

FIGURE 12.3
Roles for optical amplifiers.

- *Offsetting component losses* that otherwise would reduce signals to unacceptably low levels. Optical couplers must physically divide the signal among multiple terminals, which reduces the signal strength arriving at each one. For example, splitting a signal in half reduces each output to a level 3 dB below the input. Dividing a signal among 20 terminals reduces signal strength by 13 dB—assuming every output gets exactly $\frac{1}{20}$ of the input. Placing an optical amplifier before or after the lossy component can raise the signal strength to compensate for the loss.

Electro-Optic Repeaters and Regenerators

Electro-optic repeaters essentially link two systems end to end.

You saw earlier that an electro-optic repeater or regenerator is essentially a receiver and transmitter placed back to back in a single unit. The input end performs the usual receiver functions; the output end performs the standard transmitter functions. In the middle they amplify and typically clean up the signal. You can think of them as joining two separate fiber-optic systems together end to end.

Repeaters were widely used in long-distance systems before optical amplifiers became available, but newer systems use optical amplifiers. Most modern systems limit the use of electro-optic regeneration to major interconnection points, where signals arriving from various points are switched, reorganized, and redistributed. The incoming signals usually are converted from optical to electronic form for switching, and regeneration is a part of the process. In land systems these distribution nodes typically are several hundred kilometers apart, at major urban centers. In this case, incoming signals have gone through only a few amplifiers, typically spaced about 100 km apart. At that point the signals generally need regeneration, but receivers at the node perform the function rather than separate regenerators. Amplifiers are spaced much closer—typically about 50 km apart—in transoceanic submarine cables to avoid the need for submerged regenerators.

Some local area networks use electro-optic repeaters.

Electro-optic repeaters remain in use in some local area networks. These networks have nodes that convert optical signals into electronic form, then use the electronic signals to drive two or more transmitters, which may be optical or electronic. For example, the node may relay the input signal in optical form to the next node, but may convert it into electronic form to send to an attached terminal. You'll learn more about local area networks in Chapter 26.

Electro-optic repeaters also can make up for component losses in networks that do not operate at the 1530 to 1620 nm range of erbium-doped fiber amplifiers.

The details of repeater and regenerator operation were covered under transmitters and repeaters in Chapters 10 and 11. However, a few points deserve emphasis:

- Repeaters and regenerators are designed to operate at a specific transmission speed and format. They may contain timing circuits that generate clock signals at a specific rate, and their electronics may be optimized to operate at that rate. Circuits made to operate at 155 Mbit/s may contain components too slow to operate at 622 Mbit/s. Likewise, circuits may require specific signal formats. In short, repeaters and regenerators are not "transparent" to signal format, but optical amplifiers are.

- Regeneration can be useful in long systems where noise and pulse dispersion can accumulate to obscure the signal. Pulse dispersion is the more important problem in many cases, but dispersion compensation and careful design allow the use of optical amplifiers.
- True regeneration is usually limited to digital systems, where the digital structure of pulses makes it possible to discriminate between signals and noise. As long as the pulses can be recognized as pulses, the regenerator can produce new ones with the same timing and signal information. However, receivers have no way to remove noise from analog signals, so repeaters merely amplify it and pass it along.

Erbium-Doped Fiber Amplifiers

The most widely used optical amplifiers are erbium-doped fiber amplifiers (EDFAs). You have already learned about the basic concept of their operation as a light source in Chapter 9. They operate at wavelengths from 1530 to 1620 nm. Their success has led to interest in other types of fiber amplifiers for use at other wavelengths, which at this writing are not as well developed.

Erbium-doped fiber amplifiers are the most widely used optical amplifiers.

Function

Erbium-doped fiber amplifiers can simultaneously amplify weak light signals at wavelengths across their entire operating range. As you will see shortly, this range varies with amplifier design, but this capability is crucial for wavelength-division multiplexing. Fiber amplifiers respond very rapidly to variations in strength of the input signal, so they can amplify signals over a wide range of modulation speeds, although the response is not unlimited.

Fiber amplifiers can boost signal strength wherever necessary. In addition to being in the middle of a system, fiber amplifiers may follow the transmitter and wavelength-division multiplexer to boost output into the fiber system, or they may precede the receiver as a preamplifier.

Gain and Power Levels

Gain and output power are key measures of fiber-amplifier performance. Both are related to input power as well as to the characteristics of the amplifier. We will start by considering each wavelength channel as independent of all others, although as you will see things aren't quite that simple.

Gain and output power depend on input power as well as optical amplifier characteristics.

The *input power* is the starting point for the amplifier. As in electronic systems, it should be above a minimum value but below a maximum for the amplifier to operate properly. The input signal should be well above the background noise to provide an adequate signal-to-noise ratio. Because optical amplifiers are analog devices, they amplify whatever noise comes with the signal. Too high an input signal can cause saturation, although it should not "blow out" an optical amplifier like an audio speaker driven above its rated power.

Gain is highest for small inputs; it saturates at high inputs.

Gain is the amplification in decibels, which is a function of both input power and amplifier design. Gain is highest if the input signal is small. At higher input powers the gain can saturate because the amplifier starts running out of excited erbium atoms, which can be stimulated to amplify the signal. The output power doesn't stop increasing, because the fiber amplifier doesn't really use up its last excited erbium atom. Instead, each additional photon becomes less and less likely to stimulate emission, and thus less likely to be amplified.

This means that gain decreases as input power increases, as you can see in Figure 12.4. Increasing the input power continues to increase the output, but a small increase doesn't boost output as much as it did at a lower power level. A typical fiber amplifier has small-signal gain of about 30 dB, but the gain decreases to about 10 dB at higher powers.

Output power is the total power of the amplified signal emerging from the fiber amplifier, measured in mW or dBm. If you're measuring power in dBm and dB, the output power equals input plus gain:

$$P_{out} = P_{in} + \text{gain}$$

In WDM systems, the maximum output power is divided among the optical channels.

The maximum power is limited by the pump power as well as the amplifier structure. Typical maximum saturated output powers are 10 to 24 dBm. For a single-wavelength system, it's the power delivered on that one optical channel. For a WDM system, it's the sum of the power on all optical channels being amplified. This means that the power per channel decreases as the number of channels increases. If the maximum output is 100 mW, the amplifier can deliver 100 mW on one optical channel, 12.5 mW on each of 8 optical channels, or 2.5 mW on each of 40 optical channels.

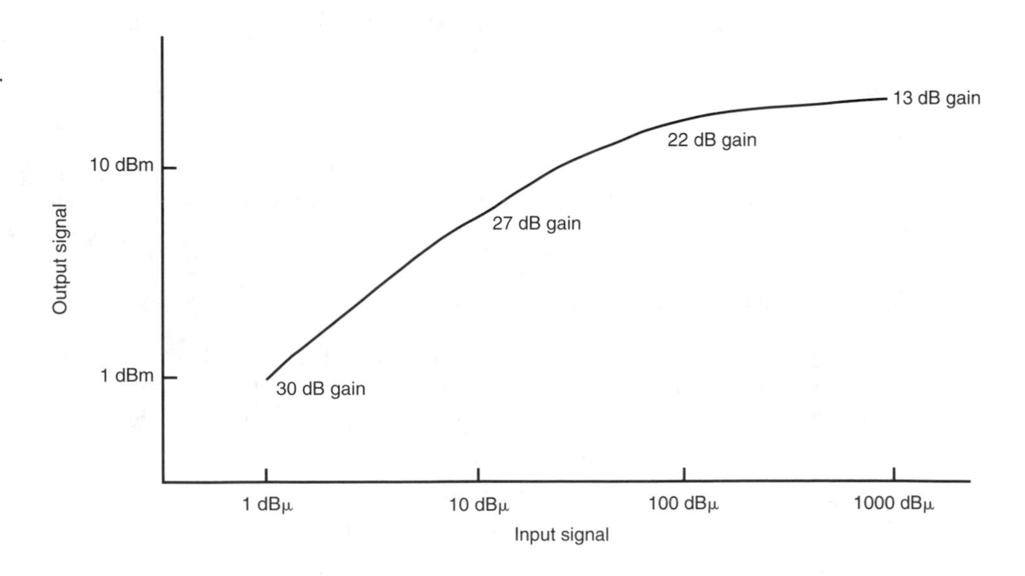

FIGURE 12.4
Saturation of fiber amplifier gain.

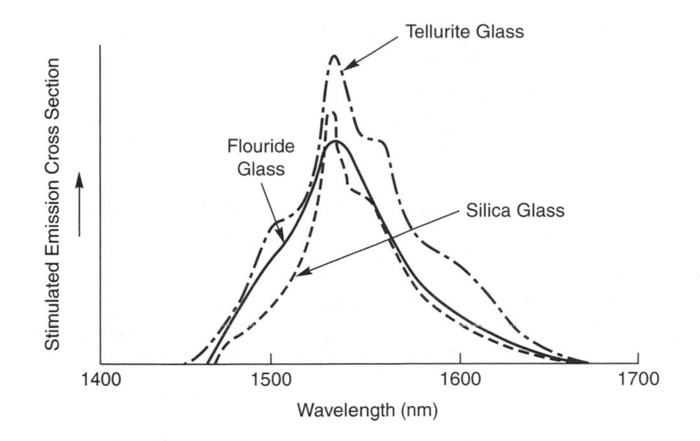

FIGURE 12.5 *Stimulated emission cross section for erbium-doped fibers of various compositions.*

Operating Wavelengths

Erbium-doped fibers can amplify light over a surprisingly wide range of wavelengths. Figure 12.5 gives an indication of this range by plotting the cross section for stimulated emission as a function of wavelengths. This cross section measures the likelihood that a photon of that wavelength can stimulate emission from an excited erbium atom. The cross section depends on the glass "host" as well as the erbium atom; it is highest for a special glass formulation containing tellurium and is somewhat lower for fluoride and silica-based glass. (The silica glass shown has extra aluminum and phosphorous to enhance erbium emission.)

You can't actually realize amplification across this entire range. Erbium atoms absorb light at the shorter wavelengths, damping possible amplification. In addition, the amplification process concentrates gain at the wavelengths where the probability of stimulated emission is highest. For relatively short lengths of fiber—a few meters—the gain is highest at 1530 to 1535 nm, as shown in Figure 12.6. This figure shows gain at various wavelengths for dif-

Small-signal gain of erbium-doped fiber amplifiers is not uniform and peaks at 1530 to 1535 nm.

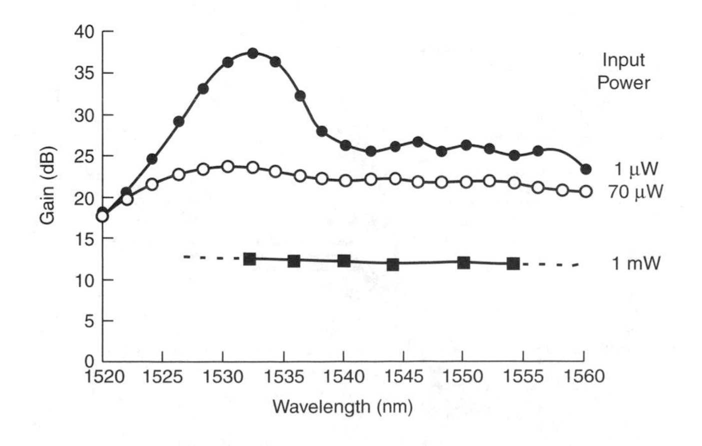

FIGURE 12.6 *Erbium-fiber amplifier gain versus wavelength at different input powers. (Courtesy of Corning Inc.)*

ferent amounts of input power. Recall that the gain is highest for small input signals. As Figure 12.6 shows, for small inputs, gain varies significantly with wavelength—by more than 10 dB from the peak between 1530 and 1535 nm to the plateau at 1540 to 1560 nm. However, for high inputs, where gain saturates, gain is more uniform across that wavelength range.

C-band erbium-fiber amplifiers have high gain at 1530 to 1565 nm; L-band amplifiers have lower gain at 1570 to 1610 nm.

Erbium-doped fiber amplifiers have lower gain at longer wavelengths, between 1565 and 1620 nm, but the stimulated emission at these wavelengths remains higher than the absorption. This produces a net gain as light travels along the fiber, so power increases along its length. Although the gain is not large, it does accumulate, so amplification is possible with a long enough fiber. Long enough in this case means 100 m or more, but the fiber can be coiled inside a case where its long length is not a problem. This opens a broader range of the spectrum for erbium-fiber amplification.

Operating requirements for the short- and long-wave portions of the erbium-fiber amplifier band conflict, so different types of amplifiers have emerged for the two operating bands. *C-band amplifiers* usually cover a range between about 1530 to 1565 nm, using several meters of erbium-doped fibers. *L-band amplifiers* are a new technology, and their wavelength range is still being established. So far most L-band amplifiers operate at 1570 to 1610 nm, but a proposed standard extends the range to 1565 to 1625 nm. Lengths are 100 m or more. (A new multicomponent glass has gain flattened across a wider range, from 1528 to 1576 nm, but it remains to be seen how much it will be used.)

Most existing optical amplifiers work in the C-band, with L-band amplifiers fairly new to the market. A 5-nm separation typically is left between the two bands, so they can be split and sent through parallel amplifiers, as shown in Figure 12.7, allowing the use of both bands simultaneously.

WDM and Optical Amplifiers

Erbium-doped fiber amplifiers were first used to amplify a single optical channel at 1550 nm, but their biggest success has been in wavelength-division-multiplexing. Unlike electro-

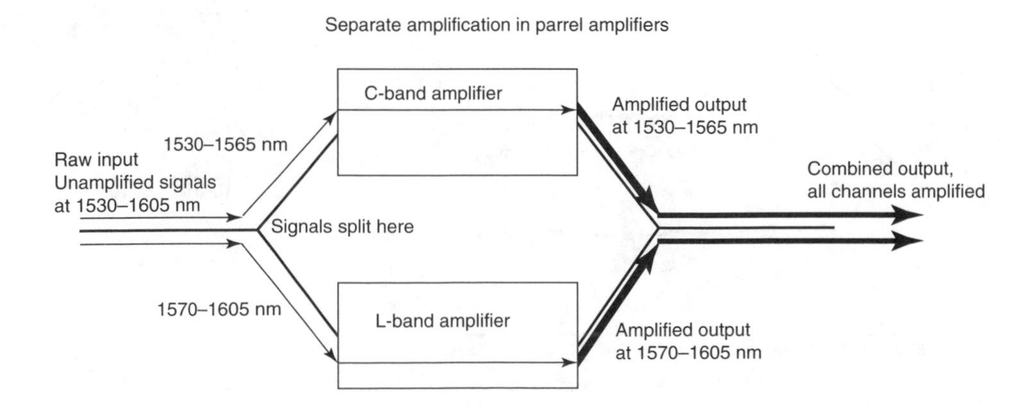

FIGURE 12.7 *High- and low-band optical amplifiers in parallel.*

optic repeaters, erbium-doped fibers can simultaneously amplify multiple wavelengths that fall within their gain bands. This has been crucial in making WDM technology practical, because it allows long-distance transmission without requiring separate repeaters for every channel carried by a fiber. Instead, a single device can amplify dozens of optical channels.

Multiwavelength operation does pose some operational complications. As mentioned earlier, all channels draw their power from the same source. The pump laser excites erbium atoms in the fiber core to states where they can amplify light at wavelengths anywhere in the erbium amplifier band. If the signal contains only one wavelength, all the erbium atoms are available to amplify that wavelength. If there are more wavelengths, the energy has to be shared among them. Thus while an erbium amplifier might be able to deliver 100 mW at one wavelength, it could supply only 2.5 mW at each of 40 wavelengths.

Fiber amplifiers must share their energy among all optical channels being amplified.

The saturation effect mentioned earlier also limits the degree of amplification possible at all wavelengths. Thus as you add more channels, the output signal at each wavelength decreases slightly as the output power saturates.

Another complication is that fiber amplifiers do not have uniform gain across their operating spectrum. As Figure 12.6 showed, small-signal gain in the C-band is highest near 1535 nm. Gain becomes more uniform across the spectrum at high-input powers, but this doesn't mean that more amplification makes the problem go away. In fact, stimulated emission tends to magnify the gain at the strongest wavelengths at the expense of weaker wavelengths. The problem is that the probability of stimulated emission at a certain wavelength increases with the strength of the signal at that wavelength. If there are twice as many input photons at one wavelength, they're twice as likely to stimulate emission than photons at another wavelength with the same gain. If in addition the gain (that is, the likelihood of stimulated emission) is lower at the weaker wavelength, the difference between the signals will increase as they pass through the amplifier. That is, the strong get even stronger. If this process isn't stopped, the weaker lines fade away and the output will contain only the strongest wavelengths.

Fiber amplifiers do not have uniform gain across their operating range.

Gain can be equalized across the spectrum in two ways. One is by adding optical filters to reduce the power on the strongest lines. The other is by adding different types of amplifiers, based on the Raman effect (described later in this chapter), which have stronger gain at wavelengths where erbium-doped fiber amplifiers have weaker gain. One or both of these techniques can equalize gain reasonably well across the spectrum, but the gain still will not be perfectly even. Later descriptions of WDM system design will cover this in more detail.

Channel spacing is an important consideration in both transmission capacity and performance. The more channels can fit through a single fiber, the higher the transmission capacity. However, increasing the channel count also multiples total power level in the fiber, which can trigger nonlinear effects that degrade transmission.

Adding channels increases capacity, but can cause side effects such as four-wave mixing.

The International Telecommunications Union has devised a standard grid of possible fiber wavelengths, defined in frequency terms. The base of the grid is 193.1 THz (193,100 GHz, or about 1552.52 nm), with nominal steps of 100 GHz (about 0.8 nm at 1550 nm) or

50 GHz (0.4 nm at 1550 nm). The grid allows for spacing at multiples of 100 GHz, and spacings of 200 GHz and 400 GHz are common.

There is a compelling organizational logic to even spacing, but it does increase the risk of noise from the nonlinear four-wave mixing described in Chapter 5. Recall that four-wave mixing is the combination of three frequencies to generate a fourth:

$$\nu_1 + \nu_2 - \nu_3 = \nu_4$$

If the optical channels are spaced an even 100 GHz apart, combining three frequencies can easily generate a fourth. For example,

$$\nu + (\nu + 300 \text{ GHz}) - (\nu - 200 \text{ GHz}) = \nu + 100 \text{ GHz}$$

This generates a signal 100 GHz above the base frequency ν, which is right on top of one standard channel. The amount of this undesirable crosstalk depends on the optical power in the interacting channels and the dispersion of the transmitting fiber.

Noise and Amplified Spontaneous Emission

As analog devices, optical amplifiers inevitably amplify any input noise that arrives with the signal. They also generate background noise by a process called amplified spontaneous emission.

Amplified spontaneous emission generates background noise in fiber amplifiers.

As you saw in Chapter 9, the light that starts simulated emission in a laser is emitted spontaneously, when an excited atom releases its excess energy without outside stimulation. A laser resonator bounces this light back and forth through the laser cavity to amplify it by stimulated emission. Fiber amplifiers lack resonator mirrors, so they don't build up a laser beam in the same way. However, spontaneous emission that occurs within the fiber can be amplified if it's guided along the fiber. This creates a background noise called *amplified spontaneous emission.*

Amplified spontaneous emission is spread across the whole operating range of a fiber amplifier, as shown in Figure 12.8. The power is much lower than at the amplified wavelengths, shown as peaks in Figure 12.8. However, it remains in the background and can be

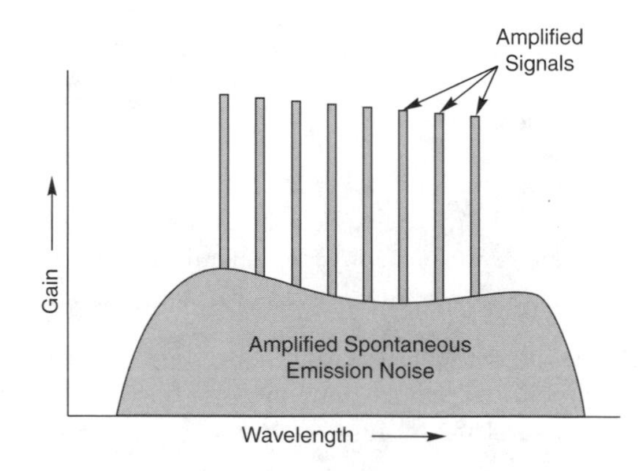

FIGURE 12.8 *Amplified spontaneous emission noise in a fiber amplifier.*

amplified in successive amplifiers. As a broadband noise, it's analogous to static in the background of an AM-radio signal.

Amplifier Spacing

Noise and gain are key factors in determining how closely repeaters must be spaced in a long-distance system. The longer the spacing, the more gain is needed to make up for the attenuation. The higher the gain, the more noise is amplified with the signal.

Amplifier spacing is closer in longer systems to keep noise from building up.

The tradeoff between amplifier cost and system performance depends on the system length. Very long systems, such as a transatlantic cable running 5000 km from the U.S. to Europe, have little tolerance for noise build-up because of their length, so fiber amplifiers typically are spaced about every 50 km, with gain a little over 10 dB. In contrast, a 600-km terrestrial system can tolerate repeater spacing of about 100 km and gain of 20 dB or more because the noise builds up over a shorter distance.

Other Doped Fiber Amplifiers

The success of the erbium-doped fiber amplifier and its compatibility with wavelength-division multiplexing pushed the telecommunications industry to shift fiber-optic transmission from 1310 to 1550 nm. However, the broad low-loss fiber window from about 1280 to 1650 nm leaves plenty of room for other fiber amplifiers that would take advantage of the large block of unused spectrum.

A proposed standard divides the spectrum from 1260 to 1675 nm into a series of fiber amplifier bands, listed in Table 12.1, but doped fiber amplifiers have been missing for all but the C- and L-bands. Thulium-doped fibers are a promising candidate for the S-band, but they remain in development.

Thulium-doped fiber amplifiers are in development for the S-band.

Praseodymium-doped fiber amplifiers have been developed for the 1310-nm region, but their performance falls short of erbium-doped fiber amplifiers, and they have generated little interest.

Table 12.1 Proposed names and ranges for optical amplification bands

Band Name	*Meaning*	*Wavelengths (nm)*	*Amplification Technology*
O band	Original	1260–1360	Praseodymium (?)
E band	Extended	1360–1460	—
S band	Short	1460–1530	Thulium-fiber (developmental)
C band	Conventional	1530–1565	Erbium-fiber
L band	Long	1565–1625	Erbium-fiber
U band	Ultra-long	1625–1675	—

Raman Amplification in Fibers

Raman amplification can extract energy from a pump beam to amplify a weak optical signal at another wavelength.

In Chapter 5, you learned about Raman scattering, which can be an undesirable nonlinear effect in a transmitting fiber. In different circumstances, the same physical phenomenon can amplify weak optical signals. It's just a matter of what beam you want.

Stimulated Raman scattering is an undesirable loss mechanism when it steals energy away from the signal wavelength and transfers it to other wavelengths. Suppose, however, that you are transmitting a strong pump beam through the fiber along with a weak signal beam. The strong pump beam excites vibrational modes in atoms in the fiber, and the weak signal beam stimulates those excited atoms to emit light at the *signal* wavelength. The result is to amplify the signal wavelength at the cost of the pump beam, which is exactly what you want in a Raman amplifier.

Like an erbium-doped fiber amplifier, a Raman fiber amplifier requires a pump beam. The process of Raman amplification can occur in ordinary silica fibers, so you don't need special doping. However, the pump energy must be considerably higher to get reasonable Raman gain. Even with high power, the Raman gain per unit length is very low, so Raman fiber amplifiers are not attractive as discrete devices. However, you can create Raman amplification in ordinary telecommunications fiber and thus distribute Raman gain along the length of a communication system. This would make the fiber itself the amplification medium. You need enough gain to offset losses in the fiber as well as other losses in the communication system.

Raman gain occurs over a fairly wide bandwidth, so it could amplify multiple wavelengths. It is not limited to as narrow a range of wavelengths, so it can amplify signals at either 1.3 or 1.55 μm, as long as you have a pump source offset by a suitable amount, to about 1.24 or 1.48 μm, respectively.

The main problem with Raman amplification is that it takes a lot of power in the pump beam—around one watt. Recent experiments with specially designed fibers have reduced this level below 700 mW, but system developers would rather use ordinary fibers. These high pump powers discourage routine use of Raman amplifiers by themselves.

Raman amplifiers have strongest gain at longer wavelengths, where erbium amplifiers have lower gain.

However, Raman amplifiers are promising for supplemental amplification and equalization when used with erbium-doped fiber amplifiers. While erbium-doped amplifiers have their highest gain at shortest wavelengths, with a peak at 1535 nm, Raman amplifiers have strongest gain at longer wavelengths. The precise gain depends on the fiber material and the pump wavelength. In silica, the peak gain is at a frequency 13 THz lower than the frequency of the pump laser, or a wavelength about 100 nm longer than the pump near 1550 nm. Gain at smaller offsets from the pump is less. If the pump is set to about 1500 nm, this means the peak is near 1600 nm, where erbium-fiber gain is low, as shown in Figure 12.9. As the figure shows, the sum of the gains of the two amplifiers is more uniform than the erbium-fiber spectrum.

In a hybrid amplifier, the Raman pump source is located at the same point as the erbium-doped fiber amplifier. As shown in Figure 12.10, a coupler directs the Ramon pump light down the input fiber, where it transfers energy to the weak input signal. The gain is highest at wavelengths where the erbium-fiber gain is low. Then the amplified signal enters the er-

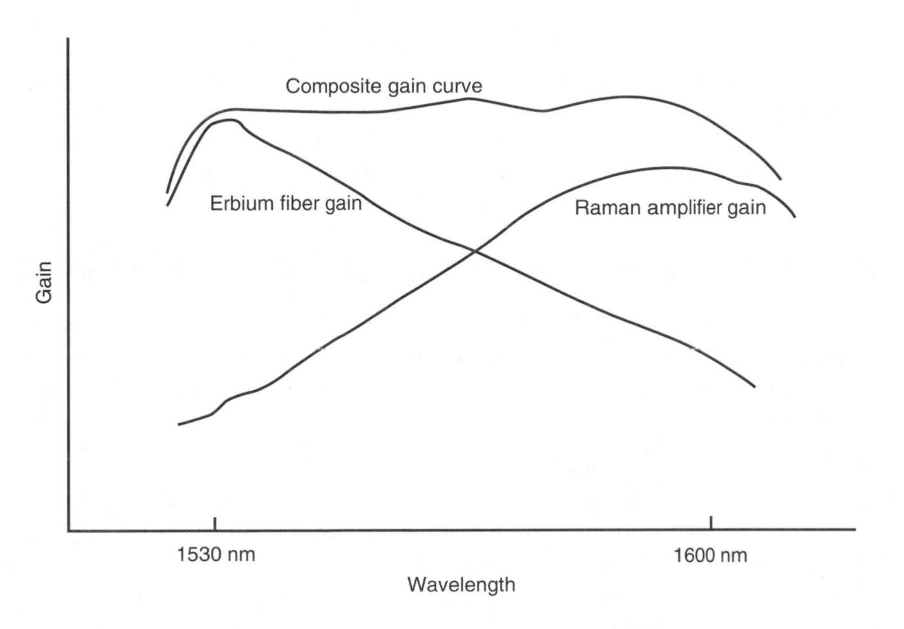

FIGURE 12.9
Raman gain equalizes spectrum of erbium amplifier to give more uniform composite gain.

bium amplifier, which has higher gain where the Raman amplification is smallest. In this way the two amplifiers add together to give more uniform gain than either one alone.

Semiconductor Optical Amplifiers

In principle, any laser can serve as an optical amplifier. Just remove the mirrors and send a signal through it, as you send a signal through a fiber amplifier. Semiconductor diode lasers are logical candidates for this approach, particularly because they are the primary light sources for most fiber-optic transmitters used in applications that require amplification. Diode lasers can amplify light over a range of wavelengths and are available for the whole

Semiconductor optical amplifiers are semiconductor lasers without reflective cavities.

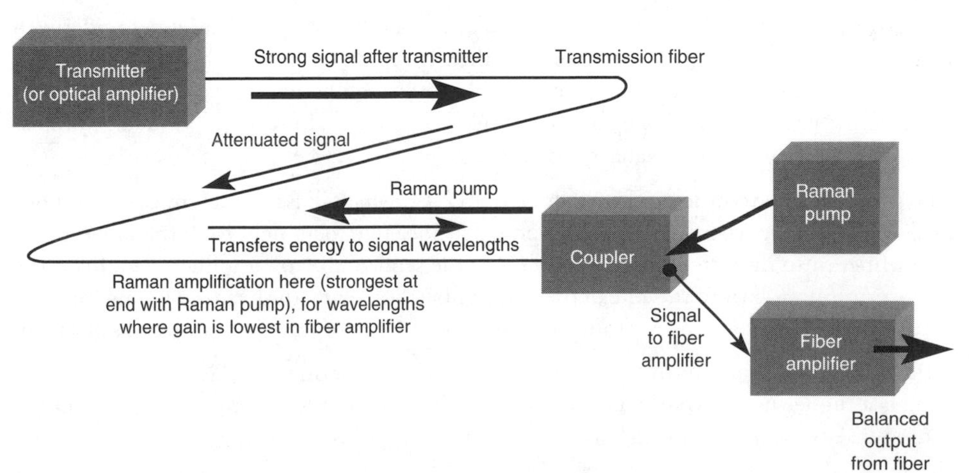

FIGURE 12.10
Layout of hybrid Raman/erbium amplifier.

region from 1250 to 1675 nm. They have very high gain per unit length, so compact devices can provide the required amplification. They can be integrated on a semiconductor substrate with other optical components, with planar waveguides transporting the light between components like the cores of optical fibers. They also can switch and control optical signals and convert them to other forms. However, they are not as well developed as erbium-doped fiber amplifiers.

Characteristics of Semiconductor Optical Amplifiers

Semiconductor optical amplifiers can switch signals off and on and be integrated with other optical components.

Semiconductor optical amplifiers share some operating characteristics with other optical amplifiers. They have a characteristic gain that is high for small input signal levels but that saturates at high powers. They also have a peak output power and can amplify light across a range of wavelengths.

A crucial difference comes from their mode of operation. As in semiconductor lasers, stimulated emission comes from carrier recombination at the junction layer. This recombination occurs only when a current is flowing through the device. In a semiconductor optical amplifier, this current can be modulated, turning the amplifier off and on. When the amplifier is off, it absorbs the input signal, so nothing gets through. When the amplifier is on, it generates an amplified output signal. Thus a semiconductor optical amplifier can modulate the signal as well as amplify it. (Erbium-doped fiber amplifiers also block light when the pump laser is off, but it's impractical to modulate them by turning the pump laser off and on.)

A second crucial difference is structural. Fiber amplifiers are fibers, discrete devices that are physically separate from transmission fibers, but which can easily be coupled to other fibers. Semiconductor optical amplifiers are planar devices—thin, flat layers like the light-emitting stripes in semiconductor lasers. As such they integrate well with other planar devices and the planar waveguides mentioned briefly in Chapter 6, making it possible to combine them with other components on a monolithic substrate like an electronic integrated circuit. (Optical integration isn't as easy as electronic integration, but that's another matter.)

Limitations of Semiconductor Optical Amplifiers

Light is hard to transfer from an optical fiber to a semiconductor optical amplifier.

The structural difference between a fiber and a semiconductor optical amplifier underlies a major drawback of the semiconductor amplifier. It's easy to transfer light from a fiber to a fiber, or from a planar waveguide to another planar waveguide. It's not very hard to transfer light from a laser stripe into the core of a single-mode fiber. However, it's difficult to transfer light from a fiber into a planar waveguide.

The problem is the geometry, shown in Figure 12.11, which illustrates a semiconductor optical amplifier placed between a pair of fibers. The idea is to focus light from the single-mode input fiber onto the active stripe, amplify it in the semiconductor amplifier, then focus the intense output beam into the core of the output fiber. Squeezing the beam emerging from the 9-μm core of a single-mode fiber into an active stripe 1 μm or less thick is a serious problem.

Noise levels are higher in semiconductor amplifiers than fiber amplifiers.

Other problems center on the operating features of semiconductor optical amplifiers. One issue is a higher noise level than erbium-doped fiber amplifiers, an important problem because noise accumulates through a series of optical amplifiers.

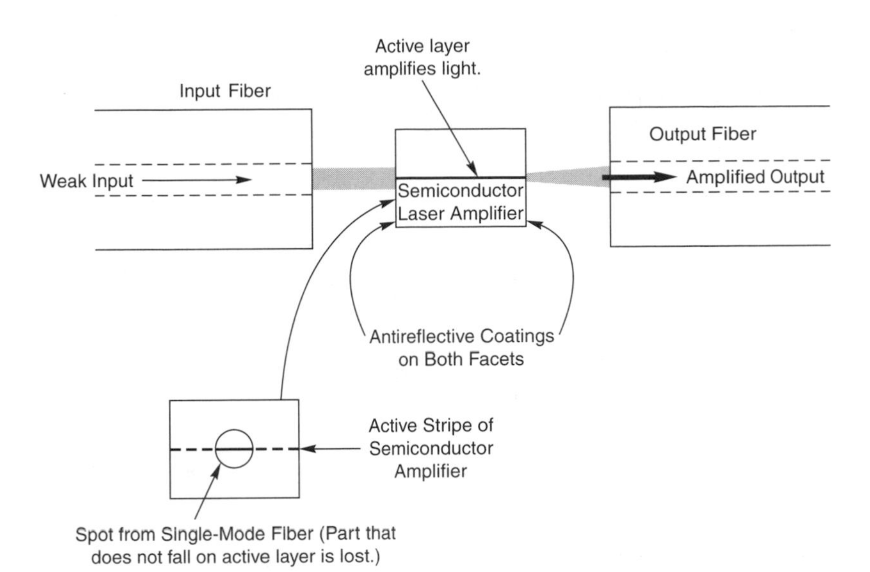

FIGURE 12.11 *Semiconductor laser amplifier.*

Semiconductor optical amplifiers can respond very quickly to changes in the input signal, but this is a mixed blessing. The response is so fast that the output changes as intensity of analog input signals changes—and the gain changes with it as a function of input power. Signal gain might be 30 dB when signal intensity is low but only 20 dB when intensity is high, leading to serious distortion of analog signals.

A more subtle problem is light reflection from the ends of the laser cavity. The high refractive index of semiconductor materials makes it difficult to completely suppress reflection from the facets at the edges of the wafer. Such reflections can introduce instabilities and noise into an optical amplifier; semiconductor optical amplifiers are particularly vulnerable to this effect because of their high gain.

An additional problem is that semiconductor optical amplifiers are sensitive to the polarization of input light, so they amplify light of different polarizations by different amounts. Standard fibers do not control the polarization of light they transmit, so uncontrolled fluctuations in polarization—normally not an issue with fiber-optic systems—can affect the amount of amplification, so the gain depends on an uncontrollable factor. Engineers don't like that sort of thing, because it can introduce noise.

Developers have made progress in solving many of these problems, at least for some applications. But they remain obstacles for others. These effects are prime reasons why semiconductor optical amplifiers have not seriously challenged erbium-doped fiber amplifiers, as in-line amplifiers in telecommunication systems. However, semiconductor optical amplifiers have strengths that offset their weaknesses for other applications.

Semiconductor optical amplifiers can be integrated with other planar optical devices on a wafer.

Integrated Semiconductor Optical Amplifiers

You can make planar semiconductor waveguides and other components from the same materials as semiconductor lasers and semiconductor optical amplifiers. This makes it possible

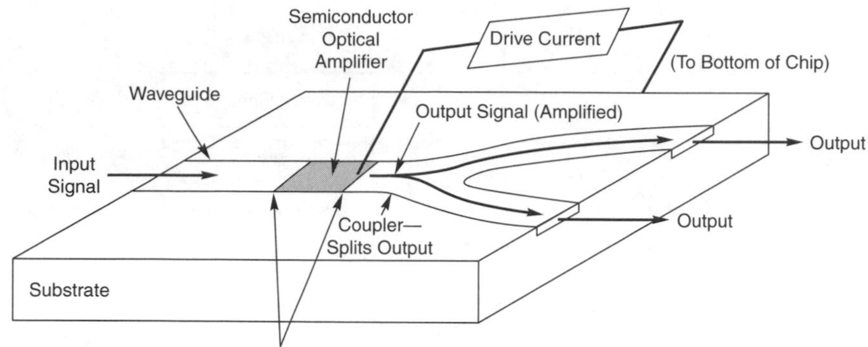

FIGURE 12.12
Integrated semiconductor optical amplifier.

to integrate various components on the same substrate, guiding light between them along passive planar waveguides.

The integration of a semiconductor optical amplifier with other planar waveguide devices is shown in Figure 12.12. The optical amplifier differs from the passive parts of the waveguide in two ways. The amplifier region is doped to have a junction layer in the plane of the waveguide, while the rest of the waveguide may not have a *pn* junction. In addition, a voltage is applied across the semiconductor amplifier, causing a current to flow and to produce recombination at the junction layer. When the weak input signal passes through the amplifier zone, it stimulates emission and is amplified. Integrating the optical amplifier with the waveguide prevents reflections at the ends of the amplifier zone. It also avoids coupling losses if the light is already in the waveguide.

For convenience, Figure 12.12 shows only a simple waveguide delivering the weak optical signal to the amplifier and a coupler that divides the output between two outputs on the right.

However, you can add more components if you have the real estate on the wafer and if they're compatible with semiconductor waveguide technology. Usable components include switches, couplers, and modulators, although semiconductor optical amplifiers themselves can serve some of these functions.

Figure 12.13 shows how this capability might be used in practice. You need to multiplex the outputs of six diode lasers emitting at different wavelengths and then distribute the combined WDM signal to six locations. A simple approach is to mix the signals in a single device called a *star coupler,* shown at the middle, but the coupler and dividing the signals among the six identical outputs causes a 15-dB loss, leaving only a weak signal at each output. In this example, an initial 1-mW output from each output drops 15-dB as it passes through the coupler and then is amplified another 15 dB by the semiconductor optical amplifier, so each output includes 1 mW of light at each wavelength.

Switching, Modulation, and Signal Control

Semiconductor optical amplifiers can modulate and switch signals.

Semiconductor optical amplifiers respond to changes in drive current as quickly as diode lasers. This means that you can modulate the gain and hence the output signal by changing

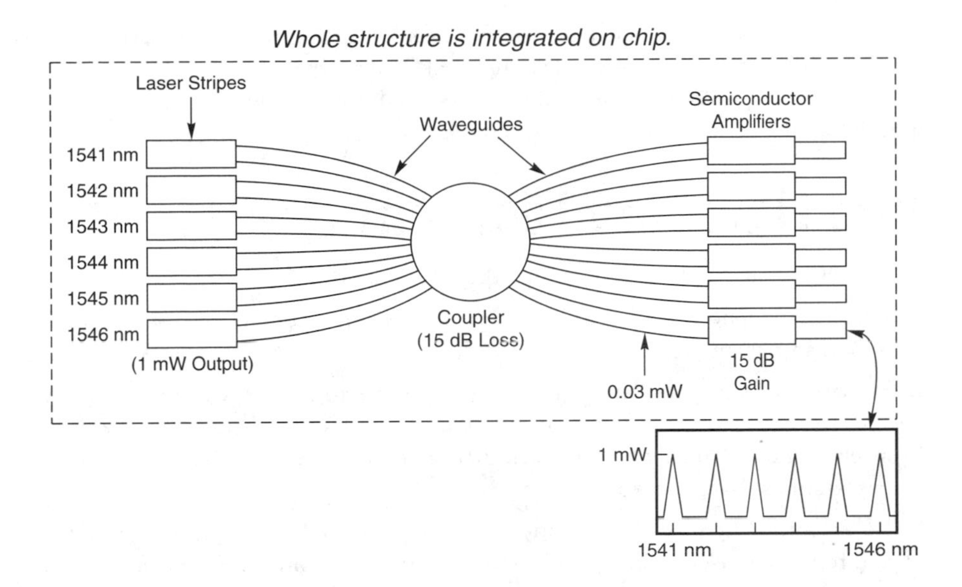

FIGURE 12.13 *Semiconductor amplifier integrated with laser array and waveguide coupler.*

the drive current passing through the optical amplifier. This is a degree of signal control impractical with erbium-doped fiber amplifiers.

You can use this modulation capability to switch signals between a pair of waveguides, as shown in Figure 12.14. The input signal is divided between two waveguides, each leading to an optical amplifier. You switch signals between the two outputs by turning one semiconductor amplifier on and the other off. More refined schemes are also possible. You will learn more about modulation in Chapter 16.

Regeneration in Optical Amplifiers

The current generation of optical amplifiers cannot regenerate signals. However, researchers are working on ways to adapt the technology for regeneration and have per-

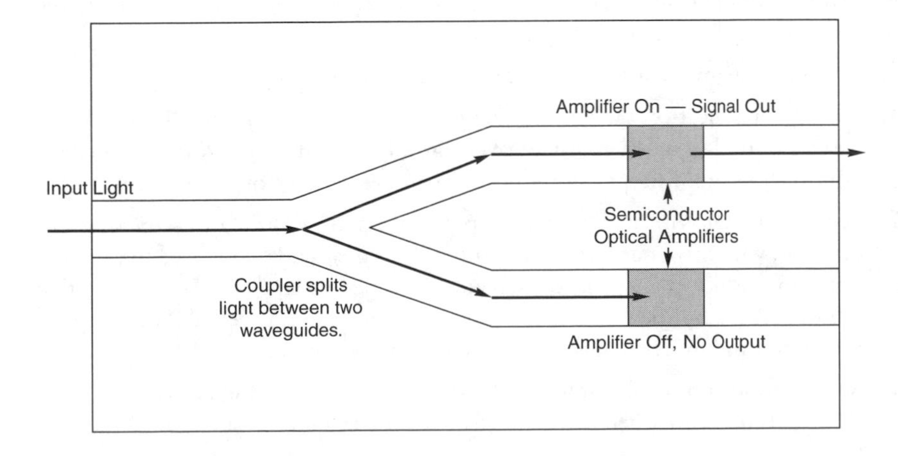

FIGURE 12.14 *Semiconductor optical amplifiers used as switches.*

formed some interesting laboratory demonstrations. The goal is to develop devices that—like electro-optic regenerators—can clean up weak input signals, regenerating sharp, correctly timed digital pulses to eliminate dispersion and distortion effects. At this writing, it's hard to say what will prove the best approach.

What Have You Learned?

1. Signals require amplification because they fade with distance.
2. Amplification increases the strength of a signal. Regeneration cleans up a signal as well as amplifying it.
3. Electro-optic repeaters convert a weak optical signal to electronic form, amplify it, and use the electronic output to drive another optical transmitter. They consist of a receiver and transmitter back to back. Repeaters *per se* are rarely used today except in certain local-area networks.
4. Regenerators resemble electro-optic repeaters, but include a stage that regenerates and retimes the electronic signal before driving the transmitter. Regeneration typically is done at the ends of systems or at switching nodes where signals are rearranged and routed to their destinations.
5. Optical amplifiers directly increase the strength of an optical signal, using the laser principle of stimulated emission to amplify the optical signal. They are insensitive to data rate or signal format.
6. Erbium-doped fiber amplifiers are the most important optical amplifiers in practical use. C-band amplifiers operate at 1530 to 1565 nm; L-band amplifiers now operate at 1570 to 1610 nm.
7. Gain and output power of fiber amplifiers depend on input power as well as characteristics of the amplifier. Gain saturates at high input powers.
8. Erbium-fiber amplifiers can simultaneously amplify many optical channels in their operating band, but the total output power on all channels is limited.
9. Gain is not uniform across the wavelength range of erbium-fiber amplifiers, and must be equalized in WDM systems.
10. Amplified spontaneous emission can cause noise in erbium-fiber amplifiers.
11. Raman amplification can extract energy from a powerful pump beam to amplify a weaker optical signal at another wavelength. Raman amplifiers can be added to erbium-fiber amplifiers to smooth total gain over a range of wavelengths.
12. Semiconductor optical amplifiers amplify light passing through a semiconductor junction with drive current passing through it; their ends are coated to prevent reflection, which makes a laser oscillate. Their performance does not match that of erbium-fiber amplifiers, but they are available at more wavelengths and can be integrated with other semiconductor waveguide components.
13. Semiconductor optical amplifiers can serve as modulators and switches because their gain turns off when the drive current is turned off.

What's Next?

Chapter 13 moves on to the connectors and splices that bridge the gaps between optical fibers, and connect them to transmitters, receivers, and other components.

Further Reading

P.C. Becker, et al., *Erbium Fiber Amplifiers: Fundamentals and Technology* (Academic Press, 1999)

International Engineering Consortium, "Raman amplification design in wavelength division multiplexing (WDM) systems tutorial," http://www.iec.org/tutorials/raman

Yan Sun, et al., "Optical Fiber Amplifiers for WDM Optical Networks," *Bell Labs Technical Journal 4,* pp. 187–206 (Jan–Mar 1999)

Questions to Think About for Chapter 12

1. An erbium-fiber amplifier has small-signal gain of 30 dB. If it is operated in that high-gain mode, how far can signals travel between amplifiers if fiber loss is 0.25 dB/km? Neglect all other losses.
2. An erbium-fiber amplifier is operated with higher total input power, so its gain is only 12 dB. What amplifier spacing is needed in the same type of fiber? Neglect all other losses.
3. Signals require regeneration after passing through five of the high-gain amplifiers in Question 1, but not until they have passed through 100 of the low-gain amplifiers in Question 2. What are the total spans between repeaters for the two systems?
4. An erbium-fiber amplifier can generate a maximum all-line output of 20 dBm. The input on each of 40 optical channels in its operating range is −20 dBm. If the maximum all-line output can be divided equally among all channels (an unrealistically optimistic assumption), what is the highest possible gain?
5. Gain in a C-band erbium-fiber amplifier varies 4 dB across the range from 1530 to 1565 nm. If you don't use any equalization, and the output power after passing through a series of amplifiers can vary no more than 25 dB, what is the longest series of amplifiers you can use? Assume the variation is the same for each amplifier in the series.
6. Equalization reduces the range of gain in a C-band erbium-fiber amplifier to 0.5 dB. Making the same assumptions, how many amplifiers can the signal pass through?

7. You want to install large-effective-area fiber in part of a system transmitting 40 optical channels to reduce nonlinear interactions between the wavelengths. Where should you install it and why?

Quiz for Chapter 12

1. Amplifiers are needed
   a. to overcome the threshold for driving an optical fiber.
   b. to compensate for fiber attenuation.
   c. only with copper-wire systems.
   d. to convert optical signals into electronic form.
2. What is the difference between amplification and regeneration?
   a. Regeneration retimes and cleans up the signal as well as amplifying it.
   b. Regeneration does not increase signal power.
   c. There is no difference.
   d. Regeneration is done optically; amplification is electronic.
3. What can optical amplifiers do that electro-optic repeaters cannot?
   a. Compensate for fiber dispersion
   b. Retime signals
   c. Operate at a wide range of signal speeds without adjustment
   d. Convert signal wavelengths
4. What can electro-optical repeaters do that optical amplifiers cannot?
   a. Compensate for fiber dispersion
   b. Retime signals
   c. Operate at a wide range of signal speeds without adjustment
   d. Both a and b
   e. None of the above
5. Erbium-doped fiber amplifiers operate at which of the following wavelengths?
   a. 1530 to 1620 nm
   b. 1280 to 1330 nm
   c. 750 to 900 nm
   d. At all important fiber windows
   e. Only at exactly 1550 nm
6. How many different wavelengths can you transmit using a fiber amplifier with operating range 1540 to 1565 nm if your signals are spaced at the 100-GHz spacing recommended by the International Telecommunications Union? (Remember the speed of light is 299,792,458 m/s.)
   a. 8
   b. 16
   c. 25
   d. 31
   e. 32

7. The gain of an erbium-doped fiber amplifier is 5 dB higher at 1535 nm than at the other wavelengths between 1540 and 1560 nm that it transmits. What do you need to do to equalize gain?

a. Nothing, gain will saturate eventually.

b. Add a filter that attenuates 1535-nm light by 5 dB and transmits the other wavelengths without loss.

c. Add a filter that attenuates all wavelengths but 1535 by 5 dB.

d. Replace the amplifier; it's defective.

8. Fiber amplifiers and semiconductor optical amplifiers both increase signal strength by

a. spontaneous emission of light at the signal wavelength.

b. stimulated emission of light at the signal wavelength.

c. Raman amplification of the signal light.

d. converting the light into electrical form and amplifying the current.

e. They share no common mechanism.

9. How does a semiconductor optical amplifier differ from a semiconductor laser?

a. Only a laser can generate stimulated emission.

b. Only an amplifier can generate stimulated emission.

c. An amplifier does not require an electric drive current.

d. An amplifier has nonreflective ends.

e. There is no difference.

10. How does a semiconductor optical amplifier differ from a fiber amplifier?

a. A semiconductor amplifier can modulate the light it amplifies more easily.

b. A semiconductor amplifier can be integrated on a wafer with other planar optical components.

c. Gain per unit length is higher in a semiconductor amplifier.

d. Semiconductor amplifiers are not widely used as in-line amplifiers.

e. All of the above

11. How can Raman amplification supplement an erbium-doped fiber amplifier?

a. Raman amplification has higher gain.

b. Raman amplification has lower noise.

c. Raman amplification can equalize gain across the erbium-fiber operating range.

d. Raman amplification has gain outside the erbium-fiber operating range.

e. It sounds fancier so the supplier can make more money by putting one in the same box and pretending it does something.

12. Which of the following could not be used to extend the transmission range of a system operating at 1300 nm?

a. An electro-optic repeater

b. An electro-optic regenerator

c. A semiconductor optical amplifier

d. An erbium-doped fiber amplifier

CHAPTER

13

Connectors and Splices

About This Chapter

Connectors and splices link the ends of two fibers both optically and mechanically. The two are not interchangeable. A *connector* is mounted on the end of a cable or optical device so it can be attached to other cables or devices. Like electrical connectors, fiber-optic connectors can be plugged and unplugged. In contrast, splices are permanent junctions between a pair of fiber ends. The cables attached to your television and stereo have connectors on the end so they can be plugged into other components. Splices are the optical equivalent of permanent solder joints.

This chapter starts by explaining their applications and their common operating principles, then describes connector properties and types, and splicing. Chapter 14 will cover couplers, which in fiber optics are quite different from connectors and splices.

Applications of Connectors and Splices

Connectors and splices both make optical and mechanical connections between a pair of fibers. Their job is to transfer light efficiently and hold the fibers together. They differ in how they do that job, and as a result they have different places in fiber-optic systems. You can think of connectors as being designed for connections that may have to change, while splices are used for permanent connections.

Electrical connectors are common in modular electronic, audio, or telephone equipment, although you may think of them as plugs and jacks. Their purpose is to connect two devices electrically and mechanically, such as a cable and a stereo receiver. A plug on the cable goes into a socket in the back of the receiver, making electrical contact and holding the cable in place. Both the electrical and mechanical junctions are important.

If the cable falls out, it can't carry signals; if the electrical connection is bad, the mechanical connection doesn't do any good. (You'll understand the problem all too well if you've ever tried to find an intermittent fault in electronic connectors.)

Connectors make temporary connections among equipment that may need to be rearranged.

Fiber-optic connectors are intended to do the same job, but the signal being transmitted is light through an optical fiber, not electricity through a wire. That's an important difference because, as you learned in Chapter 4, the way light is guided through a fiber is fundamentally different from the way current travels in a wire. Electrons can follow a convoluted path through electrical conductors (wires) if the wires make good electric contact somewhere. However, fiber cores must be precisely aligned with each other to transfer optical signals—just how precisely you'll see later.

Electrical connectors are used for audio equipment and telephones because the connections are not supposed to be permanent. You use fiber connectors for the same reason. For permanent connections, you splice or solder wires, and you splice optical fibers. Permanent connections have some advantages, including better mechanical stability and—especially for fiber optics—lower signal loss. However, those advantages come at a cost in flexibility; you don't want to cut apart a splice each time you move a computer terminal or telephone.

Splices and connectors are used in different places.

Fiber-optic connectors and splices are far from interchangeable. Connectors are normally used at the ends of systems to join cables to transmitters and receivers. Connectors are used in patch panels where outdoor cables enter a building and have their junctions with cables that distribute signals within the building. They are used where configurations may need to be changed, such as at telecommunication closets, equipment rooms, and telecommunication outlets. Examples include the following:

- Interfaces between devices and local area networks
- Connections with short intrabuilding data links
- Patch panels where signals are routed in a building
- The point where a telecommunication system enters a building
- Connections between networks and terminal equipment
- Temporary connections between remote mobile video cameras and recording equipment or temporary studios

Splices are used where junctions are permanent or where the lower loss of splices is critical. For example, long cable runs are spliced because the cable segments should never need to be disconnected. Splice loss is lower, and splices are smaller and fit into cables better. Splices generally are stronger, and with the right equipment are easier to install in the field.

In practice, this means that you usually put connectors on the ends of cables, and splices in the middle. Broken cables are repaired by splicing the fibers and mending the cable in the field. Connectors can be installed in the field or in the factory, but factory installation is easier. Sometimes the two techniques are used together to speed installations. For example, a cable may be connected to equipment in a building by splicing a fiber from the cable to a fiber pigtail attached to a factory-mounted connector.

The distinctions between connectors and splices are not always as sharp as they might seem. Some common fiber connectors resemble connectors used for metal coaxial cables, and most are obviously different from common splices, which weld, fuse, or glue two fiber ends together. However, between these extremes are such hybrids as demountable splices, which nominally bond fibers together permanently, but can be removed if desired.

Distinctions between splices and connectors are not always sharp.

The same basic considerations apply to transferring light between fibers whether in connectors or splices. We will cover them first for connectors, then recall these principles when we discuss splices.

Fiber-to-Fiber Attenuation

The key optical parameter of fiber connectors and splices is the attenuation, the fraction of the signal light that is lost. Loss is measured in decibels for a mated pair of connectors or a complete splice, that is, for light going from one fiber to the other. Light actually passes through a pair of connectors, but the loss of one connector is no more meaningful than the loss of one end of a pair of fibers that has been spliced; the signal doesn't go anywhere in either case.

The most important optical characteristic of connectors and splices is loss.

Typical attenuation is a fraction of a decibel for connectors and around 0.1 dB for splices. Manufacturers specify losses for specific fiber types; as you will learn later in this chapter, mismatched fibers can have much higher loss. The rest of this section is based on the assumption that a splice or connector joins the ends of two fibers. Connectors also can be mounted on transmitters, receivers, and other components, but although details differ, the principles are the same. This section concentrates on loss mechanisms that are important for connectors. The same principles apply to splices, but some effects are more important for one type of fiber connection than the other.

Connector and splice losses are caused by several factors, which are easier to isolate in theory than in practice. These factors stem from the way light is guided in fibers. The major ones are as follows:

- Overlap of fiber cores
- Alignment of fiber axes
- Fiber numerical aperture
- Fiber spacing
- Reflection at fiber ends

These factors interact to some degree. One—overlap of fiber cores—is really the sum of many different effects, including variation in core diameter, concentricity of the core within the cladding, eccentricity of the core, and lateral alignment of the two fibers.

Overlap of Fiber Cores

Offset of fiber cores by 10% of their diameter can cause a 0.6-dB loss.

To see how core overlap affects loss, look at Figure 13.1, where the end of one fiber is offset from the end of the other. For simplicity, assume that light is distributed uniformly in the

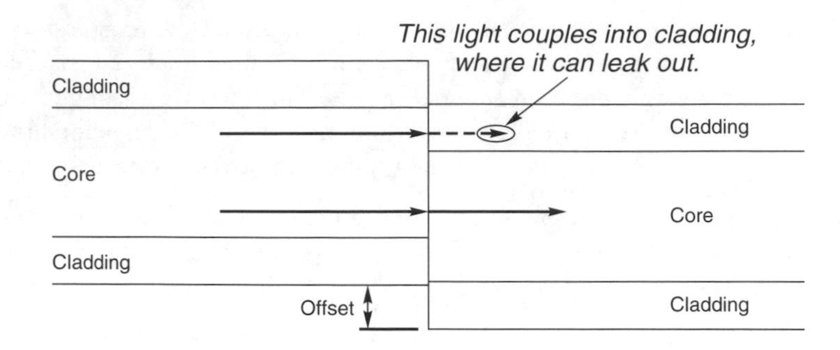

FIGURE 13.1
Offset fibers can cause loss.

cores of identical fibers and that the two fiber ends are next to each other and are otherwise well aligned. The loss then equals the fraction of the input-fiber core area that does not overlap with that of the output fiber. If the offset is 10% of the core diameter, the excess loss is about 0.6 dB.

Mismatches of emitting and collecting areas also occur if core diameters differ. Suppose that the fibers were perfectly aligned but that the 50-μm nominal fiber core diameter varied within tolerance of ±3 μm, as specified on a typical commercial graded-index fiber. With simple geometry, you can calculate the loss for going from a fiber with core diameter d_1 to one with core diameter d_2. (You also can use radius if you want—the factor of two differences from diameter cancels out—but usually core diameter is what's specified.) The relative difference in area is

$$\text{Loss} = \frac{(d_1^2 - d_2^2)}{d_1^2}$$

For the worst case, going from a fiber with a 53-μm core to one with a 47-μm core, the difference is a factor of 0.21. If light was distributed uniformly through the core, that fraction of the light, about 1 dB, would be lost. Fortunately things are rarely that bad, because light normally is concentrated toward the center of the core, and core diameter rarely varies as much as the maximum allowed by the specifications.

The same principles apply for single-mode fiber, but in that case the critical dimension is mode-field diameter, which is typically slightly larger than core diameter. The formula for relative loss is the same, but the tolerances are much tighter because single-mode fibers have much smaller cores. For example, a nonzero dispersion-shifted fiber has mode-field diameter of 8.4 ± 0.5 μm at 1550 run. Although the diameter tolerance is very small, going from the largest fiber that meets these tolerances to the smallest can be costly in loss:

$$\text{Loss} = \frac{(8.9^2 - 7.9^2)}{(8.9)^2}$$

The result is essentially the same as for the maximum variation in core diameter of 50-μm fiber, 0.21, or 1 dB. As with multimode fiber, things are rarely this bad in practice. A

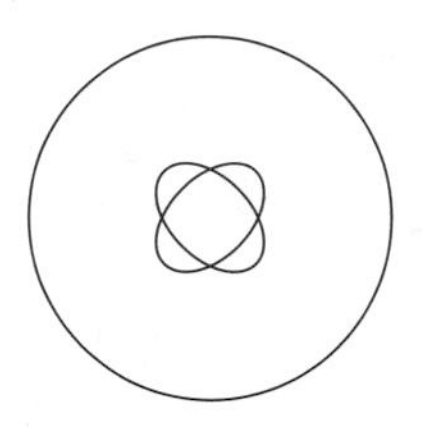

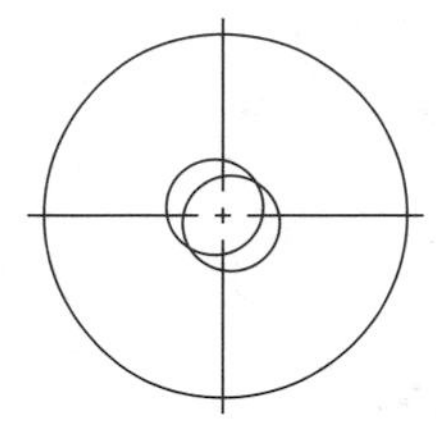

FIGURE 13.2
Losses arise when cores are elliptical or off center.

single-mode beam is most intense at the center of the core, and most specified single-mode connector losses are 0.1 to 0.5 dB.

You can get into serious trouble if the fiber types are mismatched, so signals go from a multimode fiber into a single-mode fiber. This makes it vital for you to know the fiber type. Going from a 62.5-μm graded-index fiber to a single-mode fiber with a 9-μm core results in 97.9% of the light being lost, a 17-dB loss. Even going between 62.5- and 50-μm multimode fibers causes a 1.9-dB loss, 36% of the light.

Mismatches in area also can arise from other factors. The fiber core may be slightly elliptical, or the core might be slightly off center in the fiber. Figure 13.2 shows these problems in exaggerated scale. Variations in cladding or coating dimensions can throw off alignment in connections that hold the fiber in position by gripping its outside.

Alignment of Fiber Axes

As you learned earlier, light must be directed straight along the fiber axis to be guided through a fiber. This is very different from electronic connections, which only need to make the attached wires touch each other for electrons to pass between them. This makes alignment of fiber axes critical to low optical connection loss.

Angular misalignment of fiber ends can cause significant losses.

Figure 13.3 shows how losses from fiber alignment increase with the angle between the two fibers. As the angle θ between the fibers increases, light from the first fiber enters the second fiber at a larger angle to the axis. Although a light ray passing directly along the axis may still fall within the fiber's acceptance angle, other rays can leak out. Losses are worst

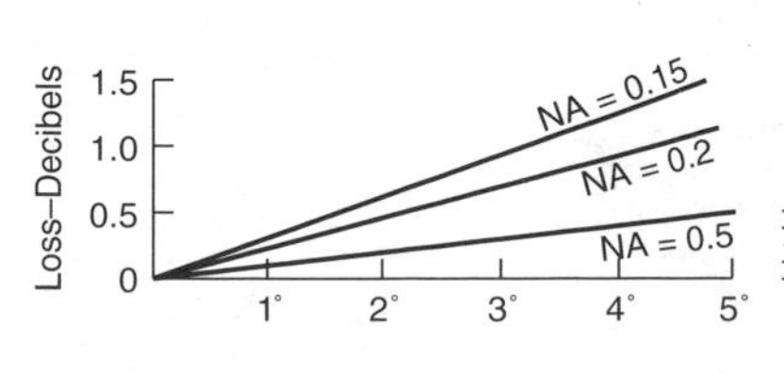

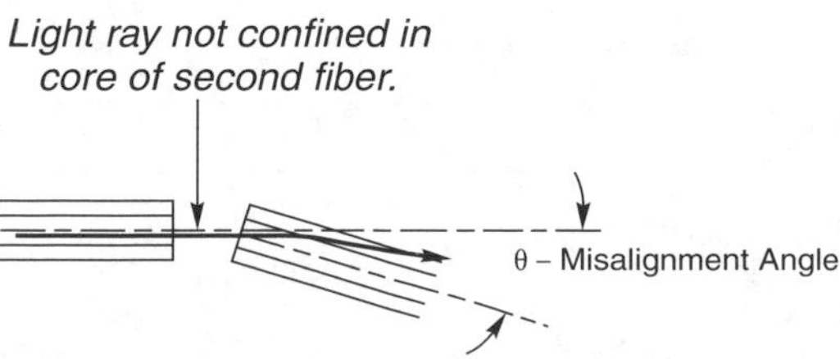

FIGURE 13.3
Misaligned fiber axes cause losses.

for fibers with small numerical apertures, while fibers with larger NAs can collect light entering over a wide range of angles. A good connection should align the two fibers very closely.

Fiber Numerical Aperture

Differences in NA can contribute to connection losses.

Differences in NA between fibers also contribute to connection losses. If the fiber receiving the light has a smaller NA than the one delivering the light, some light will enter it in modes that are not confined in the core. That light will quickly leak out of the fiber, as shown in Figure 13.4. In this case, the loss can be defined with a simple formula:

$$\text{Loss (dB)} = 10 \log_{10} \left(\frac{NA_2}{NA_1} \right)^2$$

where NA_2 is the numerical aperture of the fiber receiving the signal and NA_1 is the NA of the fiber from which light is transmitted. The NA must be the measured value for the segment of fiber used (which for multimode fibers is a function of length, light sources, and other factors), rather than the theoretical NA. Note also that there is no NA-related loss if the fiber receiving the light has a larger NA than the transmitting fiber.

Spacing Between Fibers

Numerical aperture also influences the loss caused by separation of fiber ends. Light exits a fiber in a cone, with the spreading angle—like the acceptance angle—dependent on numerical aperture. The more the cone of light spreads out, the less light the other fiber can collect, as shown in Figure 13.5. In this case transfer losses increase with numerical aperture because the larger the NA of the output fiber, the faster the light spreads out. The formula

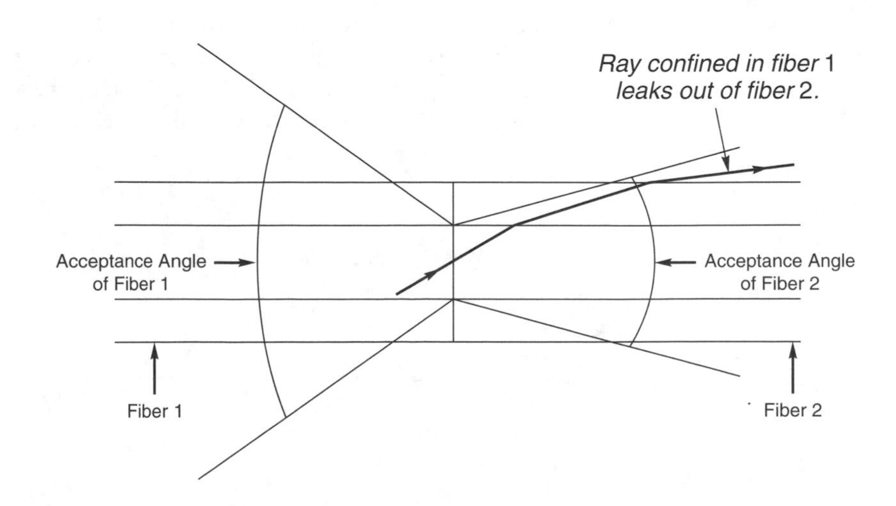

FIGURE 13.4
Mating fibers with different NAs can cause losses.

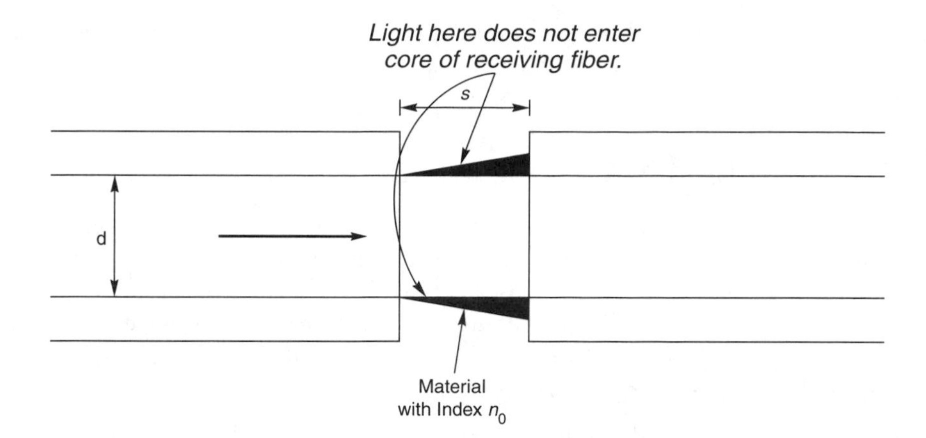

FIGURE 13.5
End-separation loss.

for the end-separation loss is rather involved, even assuming the transmitting and receiving fibers are identical:

$$\text{Loss (dB)} = 10 \log_{10}\left(\frac{d/2}{\frac{d}{2} + S \tan\,(\arcsin)\left(\frac{\text{NA}}{n_0}\right)}\right)$$

where d is core diameter, S is the fiber spacing, NA is the numerical aperture, and n_0 is the refractive index of the material between the two fibers. Figure 13.6 shows a plot of the loss for

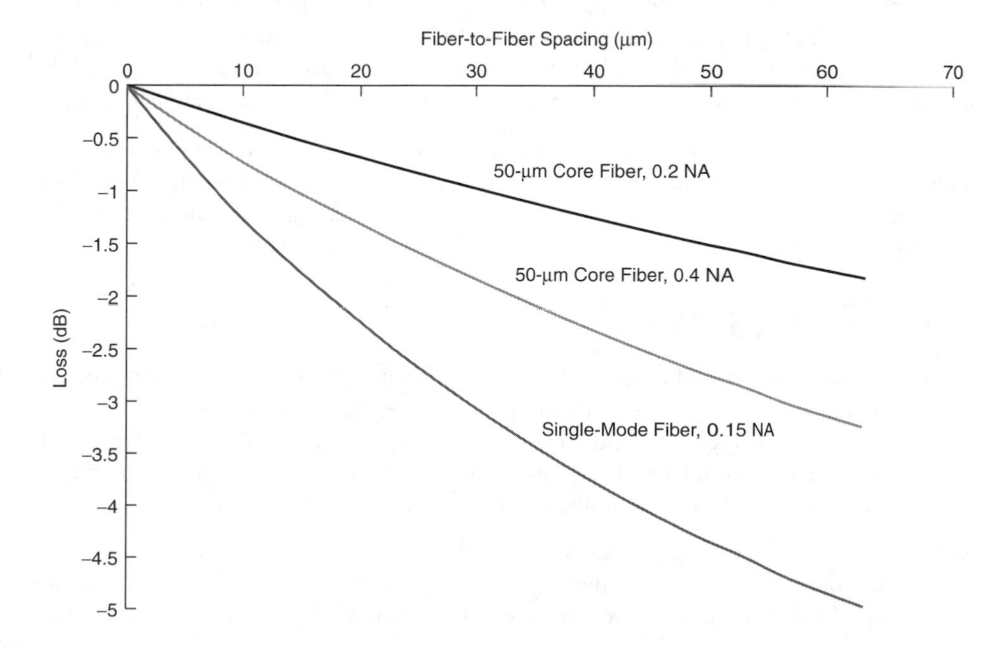

FIGURE 13.6
Loss caused by fiber spacing for three types of fiber, neglecting reflection loss.

three different fibers, two with 50-μm cores and NAs of 0.2 and 0.4 and one single-mode fiber with 0.15 NA; the material between the fibers is air, which has a refractive index almost equal to 1.

End Reflection Loss

Fresnel reflection causes a 0.32 dB loss if there is a gap between fiber ends.

Leaving a space between fibers also causes end-reflection loss from a process called *Fresnel reflection,* which occurs whenever light passes between two materials with different refractive indexes. This loss occurs for all transparent optical materials, even ordinary window glass. If you look from a lighted room out into the dark, the reflections you see on the window come from Fresnel reflection. If you look carefully, you'll note they come from both the front and the back surfaces of the glass panes.

Fresnel reflection loss depends on the difference in refractive index between the two materials. For light going from the core of a fiber with refractive index n_{fiber} into another material with refractive index n, this fraction of light reflected R is

$$R = \left(\frac{n_{\text{fiber}} - n}{n_{\text{fiber}} + n}\right)^2$$

This fraction increases with the index difference. For light going between glass and air, it is about 3.4% (0.15 dB) if the glass has a refractive index of 1.45 and 4% (0.18 dB) if the refractive index is 1.5. You can get the result directly in dB by using

$$\text{loss (dB)/surface} = -10 \log\left(1 - \left(\frac{n_{\text{fiber}} - n}{n_{\text{fiber}} + n}\right)^2\right)$$

These numbers give the loss per glass–air interface. The light suffers the same reflection in going from air back into glass, and there are two reflections in going through an interfiber gap from glass to air to glass. The total loss these two reflections cause as light goes through the gap is 0.30 to 0.36 dB, depending on the core index, with a typical value of about 0.32 dB.

To reduce this loss, the fiber ends can be butted together (carefully, to avoid damage), an approach used in most connectors. Other alternatives are coating the fiber ends with layers of antireflective materials (which have a refractive index between glass and air), or filling the gap with a transparent gel with refractive index close to that of the fiber.

Other End Losses

One dust particle can block most light transfer between a pair of single-mode fibers.

So far I have assumed that the fiber ends are cut and polished perfectly clean, and precisely perpendicular to the fiber axis. We all know things are never *that* good. The core of a single-mode fiber is less than 10 μm in diameter, so one dust particle could block virtually all light from the core. The fiber ends may not be perfect. Repeated mating and unmating of a connector can scratch the ends of fibers installed so they contact one another.

With all these loss mechanisms, it is no wonder why early developers were very worried about fiber-optic connections. (They developed multimode fiber because they did not think single-mode connections were feasible.) Tremendous progress has been made, but connection losses still can be significant in designing fiber-optic systems, as you will learn in Chapter 21. Typical losses of good single- or multimode connectors are 0.1 to 0.5 dB, while those of good splices typically are 0.1 dB or less.

The installation of a connector or splice is a critical variable, so mated fibers need to be tested. Repeated matings and unmatings can change attenuation by damaging fiber ends. Generally manufacturers specify a typical connector loss, along with changes they expect to be caused by mechanical and environmental factors during use.

Internal Reflections

Attenuation is not the only potentially harmful thing that can happen within a connector. Back-reflections from fiber ends can affect the performance of laser light sources. As mentioned in Chapter 9, the operation of a semiconductor laser depends on light reflected from the front and rear facets back into the laser cavity. Reflection of other light from a fiber junction back into the laser can cause noise. The stray light can stimulate emission within the laser cavity, generating a spurious signal—in other words, noise.

Back-reflections from fiber ends can affect laser light sources.

The best way to prevent such noise is to stop light from being reflected back toward the laser transmitter. As mentioned above, fiber ends can be butted together to avoid creating a reflective air gap. Another approach is to fill any gap between the fibers with a fluid or gel that has the same reflective index as the glass. This works better for permanent splices than for connectors, because opening a "wet" connector could contaminate the fluid or gel.

Another approach is to cleave and polish the ends of the fiber at an angle slightly off the perpendicular, so any light reflected back along the fiber is not guided all the way to the transmitter. You can think of this as taking advantage of the losses in coupling light into tilted fibers, since this reflection essentially tilts the light at an angle to the fiber. The big difficulty is making sure the tilted ends are aligned properly to avoid high losses, which requires rotating the fiber ends until they match.

Devices called optical isolators have been developed to suppress back-reflection in a different way, by transmitting light in only one direction. You'll learn more about them in Chapter 15.

Analog systems are the most vulnerable to reflection noise, so it is a particular concern for analog cable-television systems. Some cable-TV systems use angled-fiber connectors to control the problem, but they require careful installation.

Mechanical Considerations in Connectors

So far I have concentrated on optical characteristics of fiber connections. Mechanical considerations also are important, and they differ markedly for connectors and splices. Thus we will shift from connections in general to connectors in particular. Important considerations range from size, shape, and ease of use to mechanical integrity.

Mechanical considerations are important for fiber connectors.

Virtually all fiber connectors are designed well enough to stay in place under normal conditions. Ideally they should withstand physical stress applied during their use, from the normal forces in mating and unmating to the sudden stress applied by a person tripping over a cable. Connectors also must prevent dirt and moisture from contaminating the optical interface.

Ease of Use and Size

Size and shape determine a connector's ease of use and compatibility with other equipment.

The size and shape of a connector determine its ease of use and compatibility with communications equipment. Connector design has evolved greatly over the years, as engineers have refined requirements for usability. Many early connectors were screw-on designs, adapted from coaxial cable connectors. The industry has tended away from twist-on designs to connectors that snap into place, with a latch that holds them in place until released. Like the familiar snap-in jack on modern telephones, this design is very easy to use. It also is adaptable for duplex (two-fiber) connectors simply by clamping a pair of single-fiber connectors together. Many of these designs have structures that guide the connector into place only if it is inserted in the right orientation—vital in duplex connectors. Some are designed to help technicians insert connectors "blindly" into sockets they can't see, like when you're reaching behind a cabinet and plugging a connector into its socket by feel.

Density of connections has become an important consideration as the complexity of fiber systems has increased. The first generation of connectors were comparable in size to coaxial cable connectors, about 9 to 10 mm (0.35 to 0.4 in.) wide. A new generation of small form-factor connectors are designed to fit two (or more) fiber connections into the same space as an older single-fiber connector. As you will learn later in this chapter, some are single-fiber connectors, and some hold two or more fibers. The newer designs also can snap into place in tight confines when connectors are closely packed on equipment.

Durability

Typical fiber connectors are specified for 500 to 1000 matings. Most can be torn from cable ends by a sharp tug.

Durability is a concern with any kind of connector. Repeated mating and unmating of fiber connectors can wear mechanical components, introduce dirt into the optics, strain the fiber and other cable components, and damage exposed fiber ends. Typical connectors for indoor use are specified for 500 to 1000 mating cycles, which should be adequate for most uses. Few types of equipment are connected and disconnected daily. Specifications typically call for attenuation to change no more than 0.2 dB over that lifetime.

Connectors are attached to cables by forming mechanical and/or epoxy bonds to the fiber, cable sheath, and strength members. (Usually the fiber is epoxied, and the other bonds are crimped.) That physical connection is adequate for normal wear and tear but not for sudden sharp forces, such as those produced when someone trips over an indoor cable. That sharp tug can detach a cable from a mounted connector, because the bond between connector and fiber is the weakest point. The same is true for electrical cords, and the best way to address the problem is to be careful with the cables.

Because sharp bends can increase losses and damage fibers, care should be taken to avoid sharp kinks in cables at the connector (e.g., where a cable mates with a connector on a patch panel). Fibers are particularly vulnerable if they have been nicked during connector installation. Care should also be taken to be certain that fiber ends do not protrude from the ends of connectors. If fiber ends hit each other or other objects, they can easily be damaged, increasing attenuation.

Environmental Considerations

Most fiber-optic connectors are designed for use indoors, protected from environmental extremes. Keeping them free from contaminants is even more important than it is for electrical connectors. Dirt or dust on fiber ends or within the connector can scatter or absorb light, causing excessive connector loss and poor system performance. This makes it unwise to leave fiber-optic connectors open to the air, even indoors. Many connectors and patch panels come with protective caps for use when they are not mated. These caps are the sort of things that are easily lost, but they should not be.

Fiber ends must be kept free of contaminants to avoid excess losses.

Special hermetically sealed connectors are required for outdoor use. As you might expect, those designed for military field use are by far the most durable. Military field connectors are bulky and expensive, but when sealed they can be left on the ground, exposed to mud and moisture. They are designed to operate even after having one end stuck in mud and wiped out with a rag! Normally, nonmilitary users will avoid outdoor connectors or house them in enclosures that are sealed against dirt and moisture.

Connector Structures

I have talked about fiber connectors in fairly general terms so far for an important practical reason—many different types have been developed. This was a logical response to the difficult technical problems faced in developing durable, low-loss optical connectors. However, order has emerged from the chaos as users insisted on standard types to simplify their logistics.

Most connectors in common use today have some common elements, shown in generic form in Figure 13.7. The fiber is mounted in a long, thin cylinder called a *ferrule,* with a hole sized to match the fiber cladding diameter. The ferrule centers and aligns the fiber and protects it from mechanical damage. The end of the fiber is at the end of the ferrule, where it typically is polished smooth. The ferrule is mounted in the connector body, which is attached to the cable structure. A strain-relief boot shields the junction of the connector body and the cable.

Connectors have common elements, including ferrules, bodies, and strain-relief boots.

Most standard fiber connectors lack the male–female polarity common in electronic connectors. Instead, fiber connectors usually mate in adapters (often called *coupling receptacles,* or *sleeves*) that fit between two fiber connectors. Similar adapters are mounted on devices such as transmitters and receivers, to mate with fiber connectors. Although this approach

Most fiber connectors are mated through coupling receptacles between connector pairs.

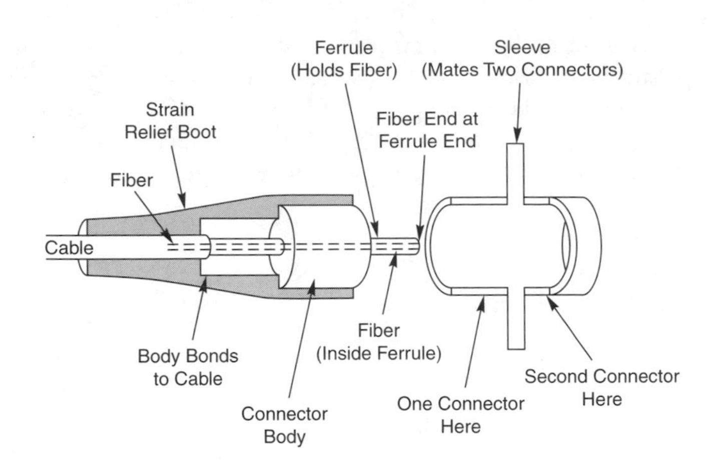

FIGURE 13.7
A simplified generic fiber connector with coupling receptacle or adapter.

requires the use of separate adapters, which must be kept in stock to link cables, it otherwise reduces inventory requirements. The adapters can also be designed to mate one type of connector to another.

Ferrules are typically made of metal or ceramic, but some are made of plastic. The protective plastic coating is stripped from the fiber before it is inserted in the ferrule. The hole through the ferrule must be large enough to fit the clad fiber and tight enough to hold it in a fixed position. Standard bore diameters are $126 + 1/-0$ μm for single-mode connectors and $127 + 2/-0$ μm for multimode connectors, but some manufacturers supply a range of sizes (e.g., 124, 125, 126, and 127 μm) to accommodate the natural variation in fiber diameter. Adhesive is typically put in the hole before the fiber is pushed in to hold the fiber in place. The fiber end may be pushed slightly past the end of the ferrule and then polished to a smooth face.

The ferrule may be slipped inside another hollow cylinder (also called a sleeve) before it is mounted in the connector body. The body, typically made of metal or plastic, includes one or more pieces that are assembled to hold the cable and fiber in place. Details of assembly vary among connectors; cable bonding is usually to strength members and the jacket. The end of the ferrule protrudes beyond the connector body to slip into the mating receptacle. A strain-relief boot is slipped over the cable end of the connector to protect the cable-connector junction.

Many connectors are mounted on transmitters, receivers, junction boxes, or other devices. The interface is the same as for standard connectors or adapters, but the mounting is different. In transmitters, the light source may deliver light to the connector through a fiber pigtail or be mounted directly at the interface. In receivers, the interface may deliver light directly to the detector or to a fiber pigtail that carries it to the detector. In a junction box or patch panel, you are likely to find an array of connector adapters, ready for you to plug an input cable in one side and an output cable in the other, as shown in Figure 13.8. Fiber connector adapters also can be mounted in wall outlets, like telephone wiring. As in junction boxes, what you see are the adapters.

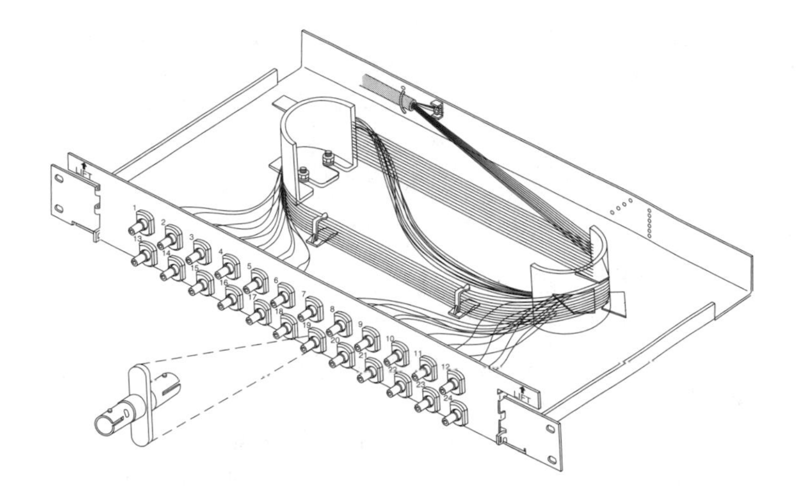

FIGURE 13.8
Connector panel. (Courtesy of Corning Cable Systems, Hickory, N.C.)

Connector Installation

Fiber connectors may be installed in the factory or in the field.

Fiber users face important trade-offs in deciding where and how to install connectors. The tight tolerances needed for low loss are easier to reproduce in a factory environment. However, field installation gives much more flexibility in meeting system requirements and allows on-the-spot repairs. Each approach has its advantages, and connector manufacturers have taken some steps to offer users the best of both worlds.

A big advantage of factory installation is that it's the cable supplier's responsibility to do it right. Trained technicians mount and test the connectors, with all the equipment they need in a controlled environment. Generally, they can mass-produce standard lengths of connectorized cable economically. That's fine for short jumper cables used in patch panels, but it's more difficult to supply the many different lengths needed for intrabuilding cable.

An intermediate step is to supply cable segments with factory-mounted connectors on one end and fiber pigtails on the other. These pigtails can be spliced to cables in the field, using mechanical or fusion splices described later in this chapter. This is a quick and easy approach using splicing equipment that many field technicians already have. Factory polishing makes connector losses low, and many types of connectors can be used. However, it requires additional splicing hardware and can add to costs. It works best for many-fiber loose-tube cables.

Field installation of the complete connector enhances flexibility and has low consumable costs. Labor costs may be low, depending on the location, but installation results depend on both the skill of the technician and the forgivingness of the connector design. It generally takes more time and skill on the part of the technician than splicing a premounted connector, and it requires some special tools. It works best for tight-buffered cables. Some manufacturers supply field connectorization kits with some of the most sensitive alignments already done.

Connector manufacturers continue efforts to simplify connector installation. One example is a field-installable connector with a built-in mechanical splice. This allows the incoming fiber to be spliced to a factory-polished and -installed fiber stub only an inch long—an extremely short pigtail.

Connecting Single- and Multifiber Cables

Connector installation is simplest for single-fiber cables. However, most cables contain two or more fibers, complicating matters.

The simplest case is the duplex connector, connecting a cable with two fibers, one transmitting in each direction. It is often made from two single-fiber connectors arranged side by side in a single housing. This is not quite enough for most practical applications, because the connected fibers must be matched so they are transmitting signals in the same direction, called *fiber polarity*. To ensure that the proper fibers are connected, most duplex connectors are keyed, so they can be inserted in only one way, preventing the wrong fibers from being mated.

Multifiber cables may be broken out at each end into many single-fiber connectors.

Multifiber cables are often broken out into separate fibers at patch panels or junction boxes, with single connectors on each fiber, as shown in Figure 13.8. This is often the simplest approach for multifiber cables, especially where they come into buildings and must be split to distribute signals to different areas. It is also the usual approach for multifiber cables within buildings.

An alternative is multifiber connectors, which simultaneously connect many fibers, greatly simplifying installation and reducing space requirements. The basic idea is to arrange a set of fibers in a fixed format and to mate them to the corresponding fibers from a different cable arranged in the same format. Such connectors require precision alignment to avoid the loss problems I described earlier and tend to have somewhat higher loss than single-fiber connectors. They are most often used with ribbon cables.

Standard Connector Types

Many connector designs have been standardized by the IEC and other organizations.

During the 1980s, almost every manufacturer of fiber-optic connectors seemed to have its own designs. Some remain in production, but much of the industry has shifted to standardized connector types, with details specified by standards organizations such as the Telecommunications Industry Association, the International Electrotechnical Commission, and the Electronic Industries Association. Standards groups, in turn, have developed standards for more than two dozen connector types, most of them widely used types and some of them new types developed for emerging needs.

I can't hope to cover the whole variety of connectors in any detail; that's best done by consulting catalogs and product specifications. However, I will discuss a few examples of important types used for single- and multimode glass fibers. Other types of connectors may be used for plastic fibers and large-core fibers. I divide them loosely into families, which sometimes overlap: single-fiber connectors that snap or twist in place, polarizing connectors, multifiber connectors, and small form-factor connectors.

Snap-in Single-Fiber Connectors (SC)

Figure 13.9 shows a widely used snap-in connector, the SC connector developed by Nippon Telegraph and Telephone of Japan. Like most fiber connectors, it is built around a cylindrical 2.5-mm ferrule that holds the fiber, and it mates with an interconnection adapter or coupling receptacle. Pushing the connector latches it into place, without any need to turn it in a tight space, so a simple tug will not unplug it. It has a rectangular 9-by-7.9 mm cross section that allows high packing density on patch panels and makes it easy to package in a polarized duplex form that assures the fibers are matched to the proper fibers in the mated connector.

The SC is a widely used snap-in single-fiber connector.

Twist-on Single-Fiber Connectors (ST and FC)

Figure 13.10 shows a widely used twist-on connector, the ST connector long used in data communications. It may look familiar because it is one of several fiber connectors that evolved from designs originally used for copper coaxial cables. Like the SC, it is built around a 2.5-mm cylindrical ferrule and mates with an interconnection adapter or coupling receptacle. However, it has a round cross section and is latched into place by twisting it to engage a spring-loaded bayonet socket.

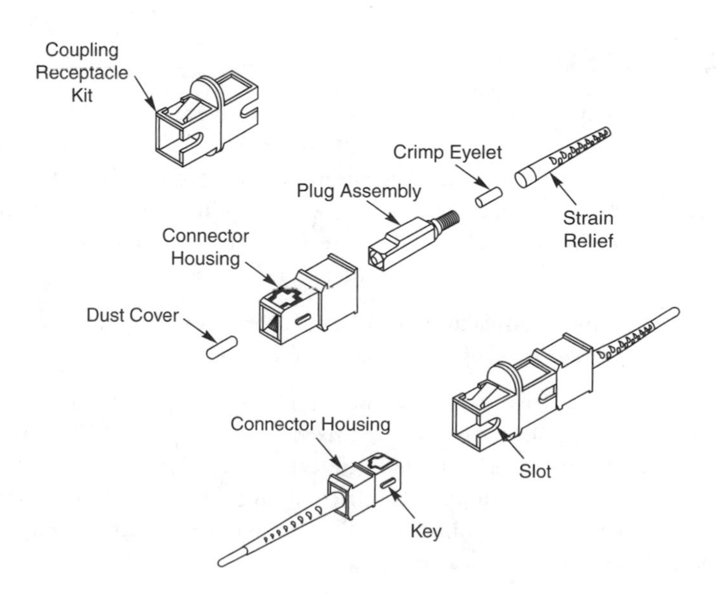

FIGURE 13.9
SC connector, expanded and assembled. (Courtesy of AMP Inc.)

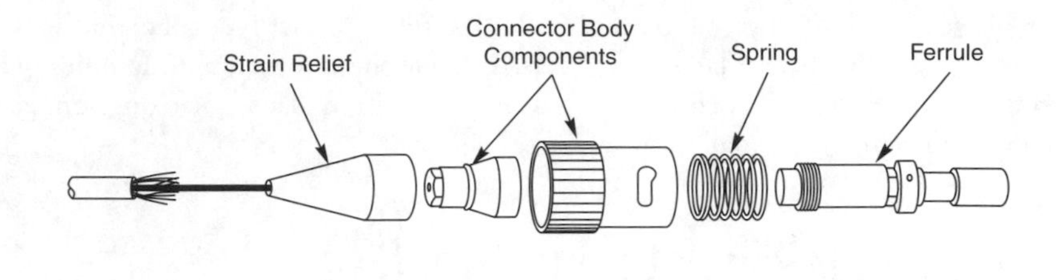

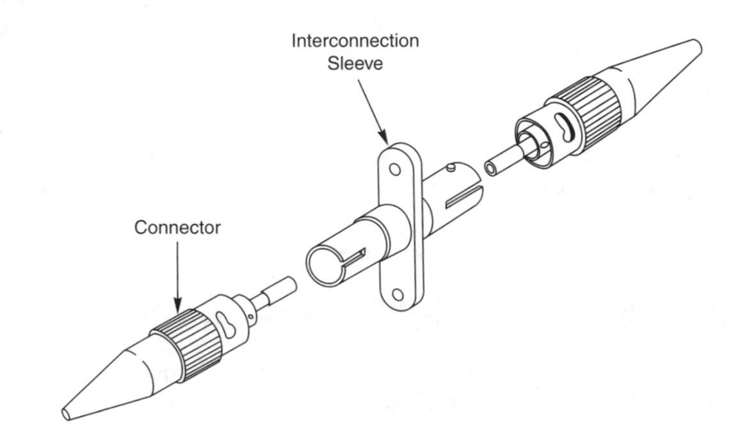

FIGURE 13.10
ST connector, expanded and assembled. (Courtesy of Corning Cable Systems, Hickory, N.C.)

Another design for a twist-on connector is the FC (sometimes called FC-PC). Its structure is similar to that of the ST, but it is threaded and screws in place rather than twisting to latch. One drawback of such twist-on connectors is that they generally cannot be mounted in pairs as a duplex connector.

Duplex Connectors

Duplex connectors are keyed to mate in only one orientation.

Duplex connectors include a pair of fibers and generally have an internal key so they can be mated in only one orientation. *Polarizing* the connector in this way is important because most systems use separate fibers to carry signals in each direction, so it matters which fibers are connected. Attach the connector the wrong way and you have one transmitter sending signals to the other transmitter while the two receivers stare at each other through a dark fiber, each waiting in vain for the other to send a signal.

One simple type of duplex connector is a pair of SC connectors, mounted side by side in a single case. This takes advantage of their plug-in-lock design.

Other duplex connectors have been developed for specific types of networks, as part of comprehensive standards. One example is the fixed shroud duplex (FSD) connector specified by the Fiber Distributed Data Interface (FDDI) standard, shown in Figure 13.11. Another is the retractable shroud duplex (RSD) connector developed by IBM for local-area networks. Both superficially resemble long, flat wall plugs for electric lamps but are designed to meet fiber-optic network standards.

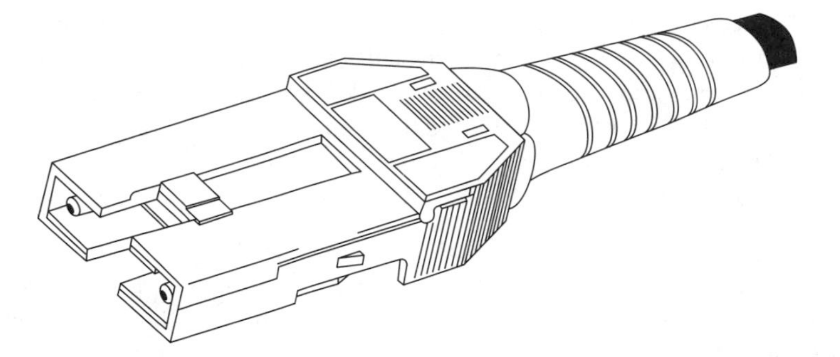

FIGURE 13.11 *Fixed shroud duplex (FSD) connector for FDDI network. (Courtesy of Corning Cable Systems, Hickory, N.C.)*

Polarizing Connectors

In addition to duplex connectors, which are polarized so they can be mated in only one orientation, special connectors are made for use with single fibers that transmit polarized light. Their role is to orient polarizing fibers so the orientation of polarized light is the same in both input and output fibers.

MT Multifiber Connectors

A family of multifiber connectors is built around the MT ferrule, which aligns up to a dozen fibers parallel to each other, as shown in Figure 13.12. As you may suspect from this arrangement, the MT connector was developed for use with multifiber ribbon cable. Coatings are removed before the fibers are mounted in the ferrule, leaving 125-μm fibers mounted on 250-μm centers. The ferrules also include a pair of 0.7-mm holes, running parallel to the fibers on the outer sides of the ferrule. These holes accommodate precision metal guide pins, which align mated ferrules with tight tolerances so the fibers match up properly.

The MT ferrule can align up to a dozen fibers in a multifiber connector.

Pairs of MT ferrules can be mated together with guide pins and held in place with metal clips. Alternatively, the MT ferrules can be mounted within connector bodies, which mate together, usually with an adapter, while the guide pins align the ferrules precisely with each other. Many of these designs have male–female polarity.

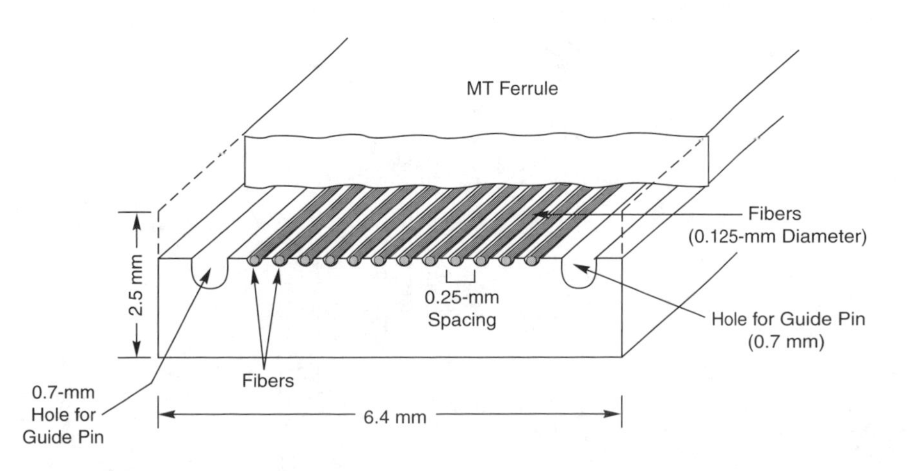

FIGURE 13.12 *MT ferrule holds a dozen fibers in parallel grooves.*

Depending on details of the design, the guide pins may be supplied separately for insertion when the connectors are mated or may be installed permanently in one ferrule (typically one that is permanently mounted in a case rather than on a cable). It is the guide pins that assure the precise alignment of fiber ends. The connector bodies provide the mechanical force holding the ferrules in place; typically they are spring-loaded and snap into position, like SC connectors. There are a variety of connectors built around MT ferrules.

One such design is the rectangular-format MPO connector, shown in Figure 13.13. A pair of connectors mate in an adapter, connecting up to 12 fibers. The male connector (shown in the figure) has the guide pins; the female connector does not. Some companies have modified designs with minor differences, such as allowing two plugs to mate with each other without an adapter between them. These connectors are intended for installations that require many fiber connections.

Trying to align many fibers at once stresses mechanical tolerances, so typical losses of multifiber connectors can be higher than those of single-fiber connectors, up to about 1 dB. (Duplex connectors typically have about the same loss as single-fiber connectors.) However, multifiber connectors greatly reduce installation costs for multifiber systems.

Small Form-Factor Connectors

Small form-factor connectors fit into tighter spaces.

A number of small form-factor connectors have been developed in recent years to fill the demand for devices that can fit into tight spaces and allow denser packing of connections. Some are miniaturized versions of older connectors, built around a 1.25-mm ferrule rather than the 2.5-mm ferrule used in SC-type connectors. Others are based on smaller versions of MT-type ferrules, or other designs. Most have a push-and-latch design that adapts easily

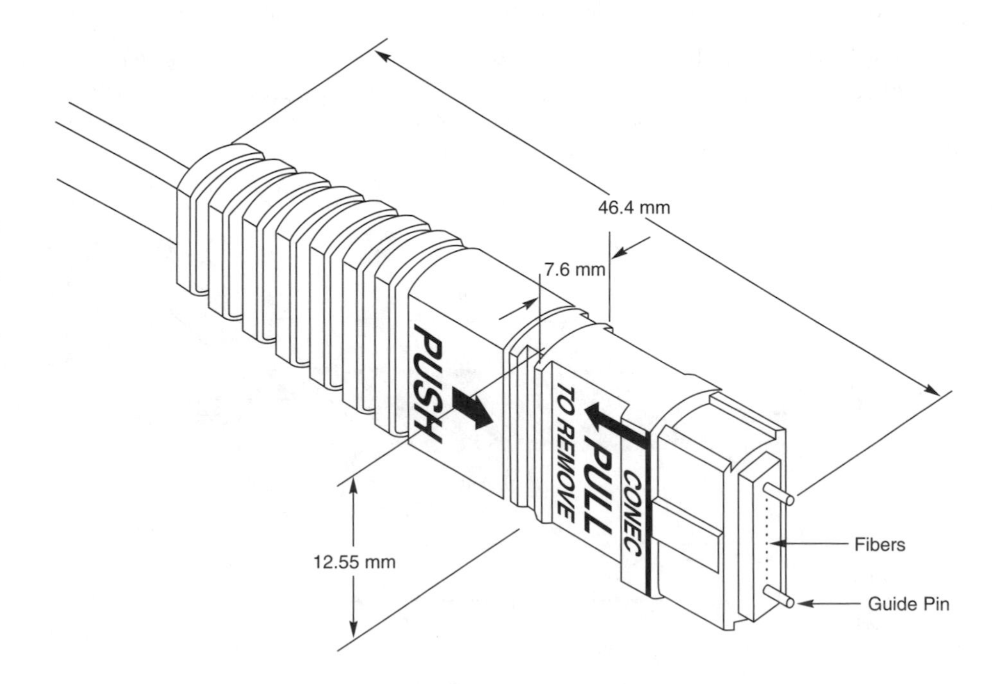

FIGURE 13.13
Male MPO connector assembly (Courtesy of US Conec.)

to duplex connectors. Figure 13.14 shows a sampling. Typical dimensions are 5-mm square for the plug end of the connector, with ferrule in the middle, and 10-by-13 mm for a duplex adapter.

The MT-RJ connector shown in Figure 13.14(a) is derived from the MT design, with a miniature two-fiber ferrule and an overall size about the same as a standard telephone jack (an RJ45 connector). The pins protruding from the MT-RJ connector are used for alignment; the two fibers are between them, and don't show on this scale. The MT-RJ is designed to be used in intrabuilding communication systems like interior telephone wiring, and can fit into the same slot in a wall plate as the socket for an RJ45 telephone jack. MT-RJ connectors are designed as plugs and jacks, like RJ45 telephone connectors. Adapters can be used with some designs, but are not required for all.

The LC connector in Figure 13.14(b) borrows features from both electronic telephone jacks and the SC fiber connector. Externally, it resembles a standard RJ45 telephone jack. Internally, it resembles a miniature version of the SC, with a 1.25-mm ceramic ferrule holding the fiber inside a plastic case. Like the SC, it mates with other connectors in an adapter. The LX.5 connector also uses a 1.25-mm ferrule in a miniature SC-like plastic case, but the connector has an integral end cap that covers the ferrule end when it is out of the adapter; plugging it into the adapter opens the end cap automatically.

The MU connector is also a miniature version of the SC, but unlike the LC retains the push–pull external latching mechanism of the SC. It also uses a 1.25-mm ferrule. Other small form-factor connectors have been designed around 1.25-mm ferrules, which differ in details such as latching mechanism and assembly procedures. One small form-factor connector, the Fiber-Jack, is built around a 2.5-mm ferrule.

An alternative structure, used in the VF-45 connector in Figure 13.14(c), is to insert the connector into the socket at a 45° angle, pressing a pair of fibers into V grooves inside a plastic adapter. Pressure and the 45° angle holds the fibers in place in the grooves. This approach avoids the need for expensive ferrules, reducing connector costs.

Splicing and Its Applications

Splices weld, glue, or otherwise bond together the ends of two optical fibers in a connection that is intended to stay connected. "Temporary" splices may be made in special cases, including emergency repairs to broken cables and testing during installation or renovation of a cable system. Splices may be made during installation or repair.

Splices make permanent bonds between fibers. Applications include joining lengths of cable outside buildings and emergency repairs.

Splices generally have lower loss and better mechanical integrity than connectors, while connectors make system configuration much more flexible. Table 13.1 compares their features. Typically, splices join lengths of cable outside buildings, and connectors terminate cables inside buildings. Splices may be hidden inside lengths of cable, or housed in special splice boxes; connectors are typically attached to equipment or patch panels at cable interfaces.

It may seem strange to list "permanent" as an advantage of splices and "nonpermanent" as an advantage of connectors. However, each has its advantages. A splice to fix a broken

FIGURE 13.14
Small form-factor connectors. (a) MT-RJ connector. (Courtesy of Corning Cable Systems) b) Duplex LC connector.(Courtesy of Agere Systems.) (c) VF-45 connector. (Courtesy of 3M Telecom Systems Division)

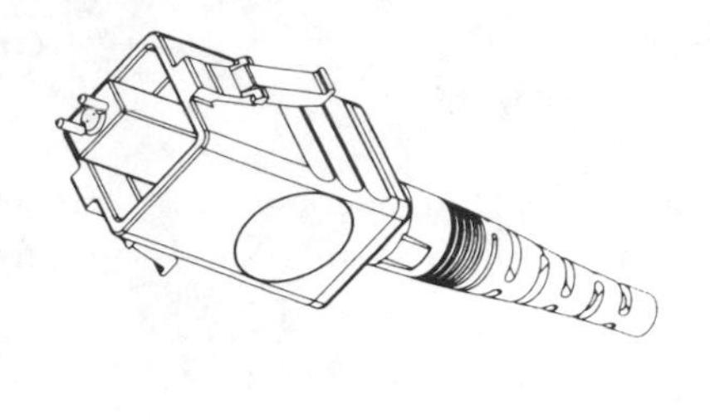

a

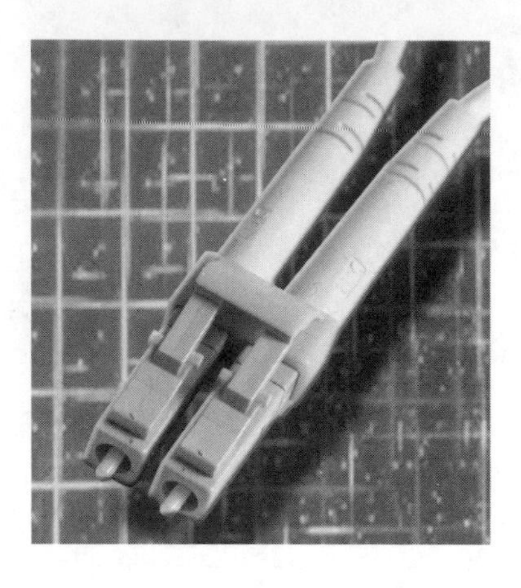

b

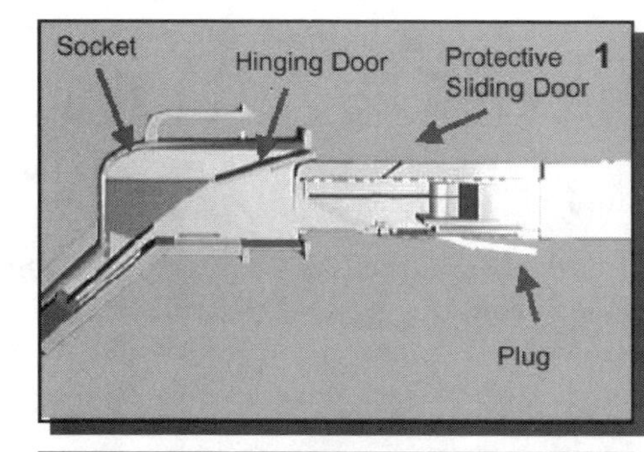

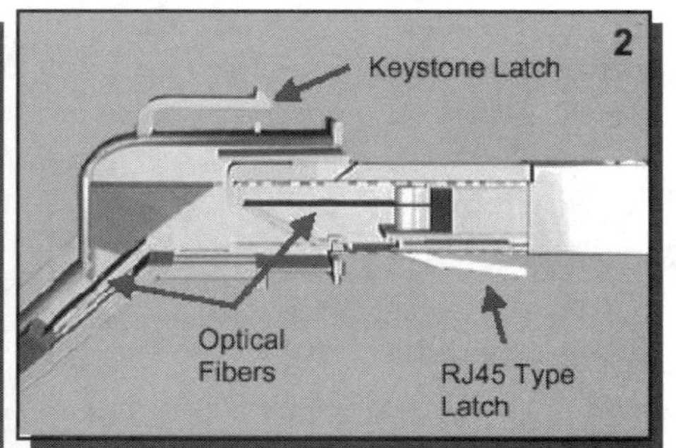

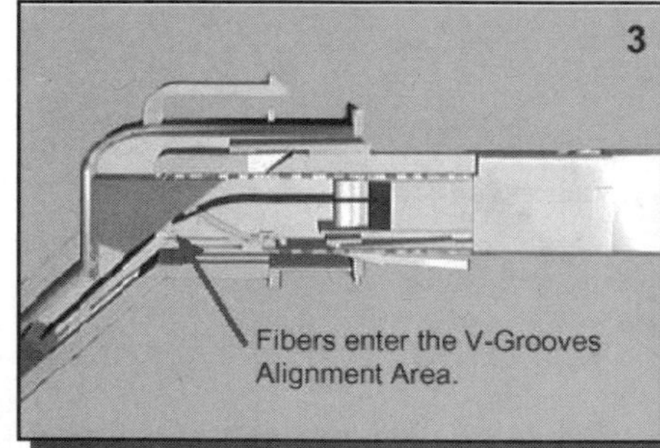

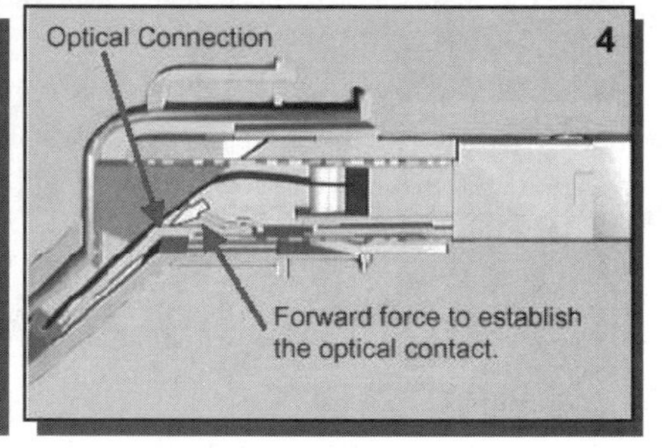

c

underground cable should be permanent, but you don't want to attach cables permanently to indoor terminals that may be moved or replaced.

Splices have lower attenuation than connectors.

The lower attenuation of splices is important for installing systems that span tens to thousands of kilometers. Bare fiber comes in lengths to about 25 km, but most cables are too bulky to fit that much on a manageable spool. In practice, outdoor cables are spliced in the field at least every several kilometers, or more often depending on the configuration.

Table 13.1 Comparison of connector and splice advantages

Connectors	*Splices*
Nonpermanent	Permanent
Simple to use once mounted	Lower attenuation
Factory installable on cables	Lower back-reflection
Allow easy reconfiguration	Easier to seal hermetically
Provide standard interfaces	Usually less expensive per splice
	More compact

The physical characteristics of splices are important in many outdoor applications. The spliced cables must withstand hostile outdoor environments, so the splices are housed in protective enclosures. (Fibers spliced during cable manufacture are protected by the cable structure that surrounds them.) Generally outdoor enclosures are sealed to protect against moisture and temperature extremes, but can be re-opened if repairs or changes are needed—like their electronic counterparts on copper cables.

Splicing Issues and Performance

Three main concerns in splicing are the optical characteristics of the finished splice, its physical durability, and the ease of splicing.

Attenuation and Optical Characteristics

The same factors that contribute to loss in connectors can cause splice loss, although differences between the two processes mean that some mechanisms are less important for splices than for connectors.

Splices bond the two fiber ends together by melting (fusing) them, gluing them, or mechanically holding them in a tight structure. This tends to align the fibers with tighter tolerance than in a connector, giving lower attenuation. As long as these processes bond the two fiber ends together with no intervening air space, they reduce or eliminate fiber-spacing loss and minimize back-reflection. High back-reflection is a sign of a bad splice.

Splices align fibers more accurately than connectors, so they have lower attenuation.

Differences between the fibers being spliced cause *intrinsic losses.* Mechanisms include variations in the size and shape of the fiber core, core eccentricity or offset, and differences in refractive-index profile. The inevitable manufacturing tolerances cause slight variations even in nominally identical fibers. These mechanisms are the same as those affecting connector loss, which you learned earlier.

Extrinsic losses arise from the nature of the splice itself. They depend on alignment of the fiber ends, quality of end preparation, refractive-index matching between ends, contamination, end spacing, waveguide imperfections at the junction, and angular misalignment of bonded fibers. Again, the same mechanisms affect connector loss, although their impacts may differ.

Good splices can have loss near 0.05 dB.

Typically intrinsic and extrinsic losses are comparable in magnitude for well-made splices. Fortunately, the total splice loss often is less than the arithmetic sum of the two types. Total loss can be very low, near 0.05 dB, in properly made splices, but imperfect junctions can suffer from high loss. A single 10-μm dust particle in the wrong place can block the core of a single-mode fiber. With proper tools and procedures, attenuation is comparable for splices of single- and multimode fibers. (Measurement anomalies can make some splices seem to be "gainers," but this effect is not real.)

Back-reflection normally is very low in good splices. High back-reflection or attenuation is a sign of a defective splice.

Specified values for splice loss assume the fibers are correctly matched. As in connectors, mismatched fibers can cause significant losses. For single-mode fibers, the most important mismatches are in mode-field diameter. This may be inevitable when different fibers are being spliced together for dispersion management. Natural variations in fiber characteristics can raise loss above the average values of 0.05 to 0.1 dB. The worst type of mismatch is splicing multimode fiber to single-mode with light going into the single-mode fiber, which can cause attenuation of nearly 20 dB. Such mistakes can happen because virtually all telecommunication fibers have the same 125-μm core diameter. Color-coding in cables identifies individual fibers, but does not distinguish different types.

Strength

Fibers are more vulnerable to damage at splices. Stripping the plastic coating can damage the fiber surface.

If you pull a spliced copper wire, you expect the splice to fail long before the wire itself fails. Optical fibers likewise are more vulnerable at splices, with the specific mechanisms depending on the splice type.

Stripping coatings from fibers can damage them before splicing, causing microcracks that later become points of failure. This is particularly likely for mechanical stripping. In fusion splicing, contaminants can weaken the melted zone, while thermal cycling from heating and cooling can weaken adjacent parts of the fiber. In practice, fusion splices tend to fail near the splice interface, but not exactly at the junction point.

Typically both mechanical and fusion splices are protected by coatings, claddings, and/or jackets that bond to the fibers and protect them from mechanical and environmental stresses.

Ease of Splicing

Splices often are made in the field, making the ease of splicing a critical concern. This has led to the development of special equipment, which I'll describe in more detail below.

Types of Splicing

There are two basic approaches to fiber splicing: fusion and mechanical. Fusion splicing melts the ends of two fibers together so they fuse, like welding metal. Mechanical splicing holds two fiber ends together without welding them, using a mechanical clamp and/or glue. Each approach has its distinct advantages. Fusion splicers are expensive, but they require almost no consumable costs, and fusion splices have better optical characteristics. Mechanical splicing requires less equipment (and no costly fusion splicer), but consumable costs per splice are much higher.

Fusion Splicing

Fusion splicing welds fiber ends together.

Fusion splicing is performed by butting the tips of two fibers together and heating them so they melt together. This is normally done with a fusion splicer, which mechanically aligns the two fiber ends, then applies a spark across the tips to fuse them together. Typical splicers also include instruments to test splice quality and optics to help the technician align the fibers for splicing. Typical splice losses are 0.05 to 0.2 dB, with more than half below 0.1 dB. The basic arrangement of a fiber splicer is shown in Figure 13.15.

Individual fiber splicers are designed differently, but all have the common goal of producing good splices reliably. Many are automated to assist the operator. They are expensive instruments, with prices starting at thousands of dollars and reaching tens of thousands of dollars for the most sophisticated models. Major differences center on the degree of automation and the amount of instrumentation included. Most models share the following key elements and functions:

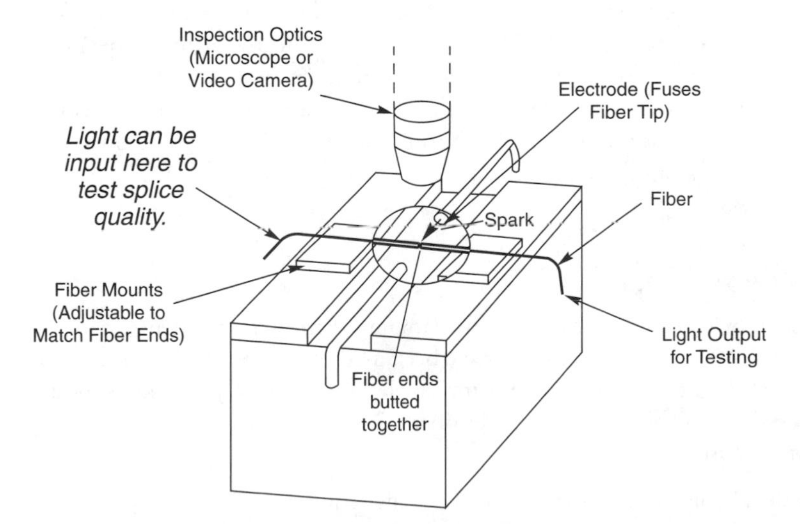

FIGURE 13.15 *Key components of a fiber splicer.*

- A fusion welder, typically an electric arc, with electrode spacing and timing of the arc adjustable by the user. The discharge heats the fiber junction. Portable versions are operated by batteries that carry enough charge for a few hundred splices before recharging. Factory versions operate from power lines or batteries.
- Mechanisms for mechanically aligning fibers with respect to the arc and each other. These include mounts that hold the fibers in place, as well as adjust their position. More expensive splicers automate alignment and measurement functions.
- A video camera or microscope (generally a binocular model) with magnification of 50 power or more so the operator can see the fibers while aligning them.
- Instruments to check optical power transmitted through the fibers both before and after splicing. Typically, light is coupled into a bent portion of the fiber on one side of the splice and coupled out of a bent portion on the other side. With proper calibration, this can measure the excess loss caused by the splice. (This may be missing from inexpensive field splicers.)

Before fusion splicing, plastic coatings must be removed from the fiber, and the end must be cleaved perpendicular to the fiber axis.

Fusion splicing involves a series of steps. First, the fiber must be exposed by cutting open the cable. Then the protective plastic coating or jacket must be stripped from a few millimeters to a few centimeters of fiber at the ends to be spliced. The fiber ends must be cleaved to produce faces that are within 1° to 3° of being perpendicular to the fiber axis. The ends must be kept clean until they are fused.

The next step is alignment of the fibers, which may be done manually or automatically. After preliminary alignment, the ends may be "prefused" for about a second with a moderate arc that cleans their ends and rounds their edges. These ends are then pushed together, allowing power transmission to be tested to see how accurately they are aligned. After results are satisfactory, the arc is fired to weld the two fiber ends together. Care must be taken to ensure proper timing of the arc so the fiber ends are heated to the right temperature. After the joint cools, it can be recoated with a plastic material to protect against environmental degradation. The spliced area can also be enclosed in a plastic jacket. The entire splice assembly is then enclosed mechanically for protection, which in turn is mounted in a splice enclosure. The case around the individual splice provides strain relief.

Mechanical Splicing

Mechanical splicing gives higher losses but requires simpler equipment than fusion splicing.

Mechanical splices join two fiber ends either by clamping them within a structure or by gluing them together. A variety of approaches have been used in the past, and many are still in use. The extremely tight tolerances in splicing single-mode fiber often require special equipment not needed for splicing multimode fiber. Those extra requirements typically make single-mode splicing more expensive.

In general, mechanical splicing requires less costly capital equipment but has higher consumable costs than fusion splicing. This can tilt the balance toward mechanical splicing for organizations that don't perform much splicing, or for emergency on-site repair kits. Mechanical splices tend to have slightly higher loss than fusion splices, but the difference is not dramatic. Back-reflections can occur in mechanical splices, but they can be reduced by using epoxy to connect the fibers, or by inserting into the splice a fluid or gel with a refractive index close to that of glass. This index-matching gel suppresses the reflections that can occur at a glass–air interface. There are several types.

The *capillary splice* relies on inserting two fiber ends into a thin capillary tube, as shown in Figure 13.16. The plastic coating is stripped from the fiber to expose the cladding, which is inserted into a tube with an inner diameter that matches the outer diameter of the clad fiber. The two fiber ends are then pushed into the capillary until they meet (often with index-matching gel inserted to reduce reflections). Compression or friction usually holds the fiber in place, although epoxy may also be used.

A capillary splice holds two fiber ends in a thin tube.

Alignment of the fiber ends depends on mechanical alignment of the outside of the fibers. The result is a simple splice that is easy to install and can compensate for differences in the outer diameters of fibers. However, it is not designed to compensate for other differences between the fibers being joined.

The *rotary* or *polished-ferrule* splice is a more elaborate type that can compensate for subtle differences in the fibers being spliced. As with other splices, the plastic coating is first removed from the fiber. Then each fiber end is inserted into a separate ferrule, and its end is cleaved and polished to a smooth surface. The two polished ferrules then are mated within a jacket or tube, and rotated relative to each other while splice loss is monitored. The ferrules are fixed in place at the angle where splice loss is at a minimum. Although this technique is more complex and time-consuming than capillary splicing, it offers a more precise way of mating fibers. Its sensitivity to rotation of the fiber around its axis makes it suitable for splicing polarization-sensitive fibers.

Fibers also can be spliced by butting them together in V-shaped grooves, as shown for fiber ribbons in Figure 13.17. (Recall that MT-family connectors and the VF-45 connector also butt fibers together in V grooves.) The fibers are placed in opposite ends of the same groove, and are pushed together until they contact. Then a separate matching plate is

V-groove splices are valuable for multifiber ribbon cables.

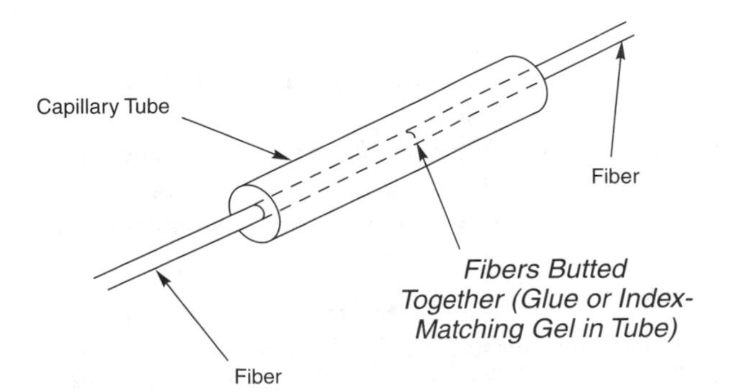

FIGURE 13.16
Capillary splice joins two fibers.

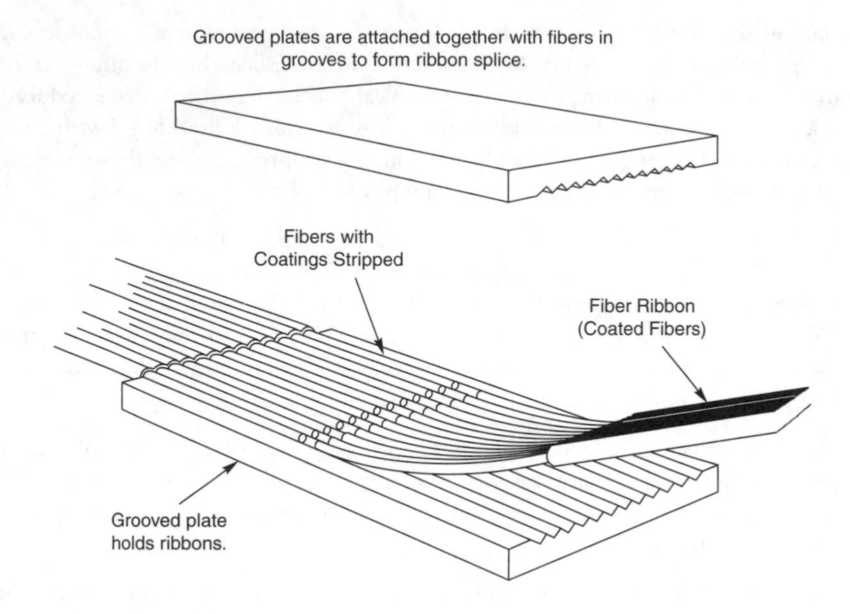

FIGURE 13.17 *Mass-splicing of 12-fiber ribbon in V-grooved plate.*

applied on top. The fiber ends can be inserted into separate grooved plates, which can have covers applied and the ends polished before they are mated with another plate, as in MT-family connectors. The *V-groove splice* is particularly useful in multifiber splicing of ribbon cables, where each parallel fiber slips into a separate groove. Special splicers are sold for this purpose.

The *elastomeric splice* has an internal structure similar to a single-fiber V-groove splice, but the plates are made of a flexible plastic held in a sleeve, and the groove is tapered toward the center. The plates are assembled in a sleeve before the fibers are installed. First an index-matching gel or epoxy is inserted into the hole, then one fiber is inserted until it reaches about halfway through the splice. Then the second fiber is inserted from the other end until it pushes against the first. This type of splice is useful in field kits for emergency fiber repairs by technicians with little fiber experience, giving typical loss of 0.25 dB, adequate for such repairs.

Splicing Requirements

Commercial splicing equipment is designed to serve a variety of needs and be used in a variety of environments. In practice, fusion splicing normally is done by technicians who work primarily with fiber, whether installing new cables or repairing existing ones. Telephone companies are likely to have vans equipped with specialized fiber equipment for these purposes. Fusion splices tend to be used mostly for cables with long outdoor runs, where loss is a major concern.

Mechanical splices are more likely to be used by nonspecialists to repair shorter cables indoors, where final loss of the cable is less important than fixing it promptly. For example, the technician responsible for maintaining a corporate local-area network may use mechanical splices to patch a damaged indoor cable, or to make a few splices needed to reconfigure the system. The cost per splice is higher than fusion splicing, but the overall cost is much lower because of the high cost of a fusion splicer.

Splice Housings

Fiber-optic splices require protection from the environment, whether they are indoors or outdoors. Splice enclosures help organize spliced fibers in multifiber cables, and also protect splices from strain and contamination.

Splice housings typically contain a rack such as the one shown in Figure 13.18, which contains an array of individual splices. This rack is mounted inside a case that provides environmental protection. Individual fibers broken out from a cable lead to and from the splices. To provide a safety margin in case further splices are needed, an excess length of fiber is left in the splice case. Like splice enclosures for telephone wires, fiber-optic splice cases are placed in strategic locations where splices are necessary (e.g., in manholes, on utility poles, or at points where fiber cables enter buildings).

Splice housings organize splices in multifiber cables.

Fiber splice enclosures should be designed to

- Hold cable strength member tightly
- Block entrance of water
- Provide redundant seals in case one level fails
- Electrically bond and ground any metal elements in the cable (e.g., strength members and armor)
- Be re-enterable if the splice must be changed or repaired
- Organize splices and fibers so they can be readily identified
- Provide room for initial splicing and future modifications
- Leave large enough bend radii for fibers and cables to avoid losses and physical damage

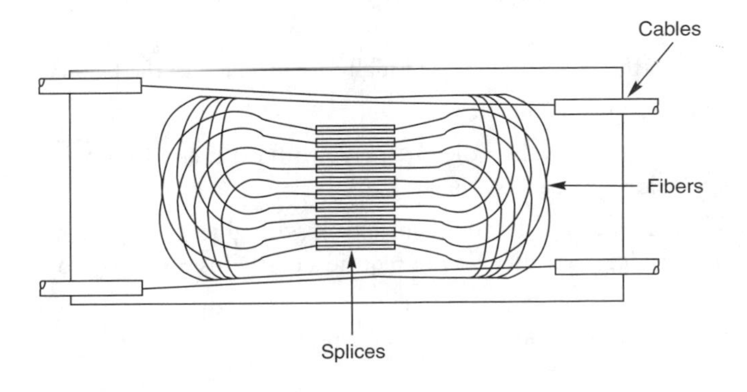

FIGURE 13.18 *Splices arrayed inside housing.*

What Have You Learned?

1. Connectors hold fiber ends together in a temporary connection. They are used where equipment may need to be rearranged.
2. Splices permanently attach and align fiber ends. They are used for permanent junctions while installing or repairing fiber.
3. The most important optical specification of connectors and splices is attenuation, the loss in transferring a signal between fibers.
4. Causes of loss include mismatch of fiber cores, misalignment of fiber axes, differences in numerical aperture, spacing between fibers, reflection at fiber ends, and dirt in fiber junctions. Alignment tolerances are tighter for small-core single-mode fibers than for larger-core multimode fibers.
5. Fresnel reflection causes a 0.32-dB loss if there is an air gap between fiber ends. To avoid this, fibers are butted together in connectors or spliced together mechanically.
6. Back-reflection is an important parameter because it can cause noise if the light returns to laser transmitters. Butting fiber ends together, fusion splicing, or filling the junction between fibers can reduce this reflection. So can angled connectors.
7. Mechanical properties of connectors are important. They should withstand hundreds of matings, keep fiber ends clean, and hold cables in place.
8. Most connectors contain cylindrical ferrules that hold the fiber inside a connector body. Most fiber-optic connectors lack male–female polarity and mate through interconnection adapters or coupling receptacles.
9. Many types of connectors have been marketed and remain available, but only a few are in wide use.
10. Duplex connectors are used for pairs of fibers, one carrying signals each way. Multifiber cables may be attached to multifiber connectors or broken out to individual connectors.
11. Standard-form connectors include the snap-in SC and the twist-on ST and FC.
12. Small form-factor connectors are about half the size of standard-form connectors. They include the LC, MT family, MT-RJ, MU, Fiber-Jack, and VF-45. MT-family and MT-RJ connectors house two or more fibers; the others often are used in duplex form.
13. Splices are normally made in the field; connectors usually are installed in the factory.
14. Splices have lower attenuation than connectors.
15. Fusion splices melt two fiber ends together; they are made with expensive fusion splicers. Typical loss of a fusion splice is less than 0.1 dB.
16. Mechanical splices hold fiber ends together mechanically or with glue. Losses are slightly higher than fusion splices, but they do not require a costly fusion splicer to install.
17. Splices are mounted in indoor or outdoor enclosures for protection against stress and the environment.

What's Next?

In Chapter 14, I will look at fiber-optic couplers, which join three or more fiber ends, and other passive optical components. Chapter 15 will cover the optics used in wavelength-division multiplexing.

Further Reading

Bob Chomycz, *Fiber Optic Installer's Field Manual* (McGraw Hill, 2000), see Chapter 11, "Splicing and Termination."

Hassaun Jones-Bey, "Connector pace accelerates to meet telecomm demand," *Laser Focus World,* September 1999, pp. 137–139.

Gerd Keiser, *Optical Fiber Communications* (McGraw Hill, 2000), see Chapter 5, "Power Launching and Coupling," for general discussion of light transfer.

Kathleen Richards, "SFF connector battle is far from over," *Lightwave,* October 1999, pp. 43–46.

Note that manufacturers often have detailed information on their own connectors.

Questions to Think About for Chapter 13

1. What is the loss caused by core-diameter mismatch when going from a single-mode step-index fiber with 9-μm core to a graded-index multimode fiber with 50-μm core?
2. You are transferring light from a 62.5/125-μm graded-index fiber with $NA = 0.275$ to a single-mode step-index fiber with 9-μm core and $NA = 0.13$. You're going to lose a lot of light from the core-size mismatch. How much loss comes from the *NA* mismatch? How much from the area mismatch, assuming even light distribution?
3. You saw earlier that Fresnel reflection loss for an air gap between a pair of fibers is 0.32 dB. Recall that the loss depends on the difference between the refractive indexes of the material in the gap and the glass. If you have water with refractive index of 1.33 in the gap, what is the Fresnel loss?
4. If two step-index fibers with core radius a are offset a distance d from each other, as shown in Figure 13.1, the area of the two cores that overlap is

$$A_{\text{overlap}} = 2a^2 \arccos\frac{d}{2a} - d\left(a^2 - \frac{d^2}{4}\right)^{0.5}$$

Suppose your connector makes a 1-μm error in aligning the otherwise identical cores of two step-index single-mode fibers with 9-μm cores. How much loss does that cause?

5. Using the formula of Question 4, go back and estimate how precisely the same step-index single-mode fibers would have to be aligned to have offset loss of only 0.3 dB. (*Hint:* You can try the formula for different values of offset if you program it into a computer spreadsheet.)
6. Why can't two twist-on connectors be assembled into a unit as a duplex connector?
7. A major telephone carrier puts you in charge of field repairs for a major urban center. You need to outfit a special truck for skilled technicians to use in repairing breaks in overhead and buried cables. What type of splicer do you buy?
8. A large retail company hires you to manage its data-transmission networks. The company has many regional offices in separate cities, each with fiber running to a dozen desks. You want to supply every office with a repair kit in case someone trips over a cable. What type of splice equipment do you buy?

Quiz for Chapter 13

1. Connectors
   a. permanently join two fiber ends.
   b. make temporary connections between two fiber ends or devices.
   c. transmit light in only one direction.
   d. merge signals coming from many devices.
2. Index-matching gel in a connector
   a. holds the fibers in place.
   b. keeps dirt out of the space between fiber ends.
   c. prevents reflections at fiber ends.
   d. eliminates effects of numerical aperture mismatch.
   e. all of the above
3. What is the excess loss caused by the mismatch in core diameters when a connector transmits light from a 62.5/125 multimode fiber into a 50/125 multimode fiber?
   a. 0 dB
   b. 0.19 dB
   c. 0.8 dB
   d. 1.9 dB
   e. 12.5 dB
4. The largest excess loss probably will occur in which case?
   a. Transfer of light from a single-mode to a multimode fiber
   b. An air gap of 2 μm is left between identical fibers
   c. A 20-μm soot particle is spliced near the core of a pair of single-mode fibers
   d. A 2° angle is left between a pair of fibers when they are spliced
   e. Index-matching gel is left out of a mechanical splice

5. How many matings and unmatings is a typical fiber-optic connector rated to survive?

a. 100

b. 1000

c. 10,000

d. 100,000

e. 1 million

6. Ferrules do what in a fiber-optic connector?

a. Relieve strain on the cable

b. Allow adjustment of attenuation

c. Hold the fiber precisely in place

d. Prevent back-reflection

7. How does an SC connector attach mechanically to an adapter or patch panel?

a. It pushes straight in and snaps into place

b. With a special tool

c. It must be screwed into place

d. It twists with a bayonet-type latch

e. Only with duct tape

8. Which of the following mechanisms is *not* used in small form-factor connectors?

a. Some latch in place like telephone jacks

b. Some twist in place like coaxial-cable connectors

c. Some use 1.25-μm miniature ferrules

d. Some use V-grooves for alignment

e. Some mount a pair of fibers in a single ferrule

9. Splices

a. permanently join two fiber ends.

b. make temporary connections between two fiber ends or devices.

c. transmit light in only one direction.

d. merge signals coming from many devices.

10. Typical splice loss is around

a. 0.01 dB.

b. 0.1 dB.

c. 0.5 dB.

d. 0.8 dB.

e. 1.0 dB.

11. What would happen if fibers with identical outer diameters but different core diameters were spliced together?

a. The splice would fail mechanically

b. Loss would be high in both directions

c. Loss would be high going from the large-core fiber to the small-core fiber, and low in the opposite direction

d. Loss would be high going from the small-core fiber to the large-core fiber, and low in the opposite direction

e. Impossible to predict

12. Splice housings are important because they

a. reduce splice attenuation.

b. protect splices from physical and environmental stresses.

c. prevent hydrogen from escaping from splices.
d. allow measurement of splice attenuation.
e. contain light sources.

13. Identify the connector type that is *not* a small form-factor type.

a. MT-RJ
b. VF-45
c. LC
d. ST
e. Fiber-Jack

14. Which connector mounts two fibers in the same ferrule?

a. MT-RJ
b. VF-45
c. LC
d. ST
e. Fiber-Jack

CHAPTER 14

Couplers and Other Passive Components

About This Chapter

A variety of components manipulate optical signals in fiber-optic systems. They fall into two broad categories, passive components that require no outside power supply, and active components that draw external power. This chapter covers couplers and other passive components that are not involved in wavelength-division multiplexing. Chapter 15 covers WDM optics including couplers, and Chapter 16 covers switches, modulators, and other active components.

Couplers split input signals into two or more outputs, or combine two or more inputs into one output. As you will learn, optical coupling is more complex than its electronic counterpart because of the nature of optical signals. This chapter will explain how optical couplers work and describe the technologies used. It also will cover other important passive components not intended specifically for WDM applications, including attenuators and optical isolators.

Coupler Concepts and Applications

Connectors and splices join two fiber ends together. That's fine for sending signals between two devices, but many applications require connecting more than two devices. Connecting three or more points is a job for a coupler. (*Taps* usually connect three points, by taking part of a signal passing through a communication line.) Couplers differ from switches in that couplers make unchanging connections, but switches can alter the connections. (Switches are covered in Chapter 16.)

Couplers are used in many places that you may not notice. For example, you need a coupler to connect both a telephone and an answering machine to the same telephone

Couplers connect three or more points.

line. It may plug into the wall and give you two adjacent sockets, one for the phone and the other for the answering machine. It may be attached to the back of the answering machine, so you see two sockets, one for the line to the wall, the other for the line to the phone. Or it may be inside a single unit that contains both a phone and answering machine, dividing the signals between the two units in the same box. But the coupler must be there.

Couplers and taps are easy to make for electronic equipment. Electric current flows as long as you have physical contact between conductors; you don't have to line them up carefully, as you must with fiber cores. In addition, electronic signals usually are in the form of voltage. If you hook 1, 2, or 20 identical resistors across an *ideal* voltage source, each will see the same voltage signal as shown in Figure 14.1. Of course, that's not exactly what happens in a telecommunication system. Putting more resistors in parallel across the signal source lowers the total resistance, so the voltage across the load will drop, depending on transmission-line resistance and other source characteristics. However, if the system is carefully designed, many loads in parallel will all have voltages close to what they would see individually. Thus electronic coupling can be as simple as hooking up wires to a signal source.

FIGURE 14.1
Optical and electronic couplers.

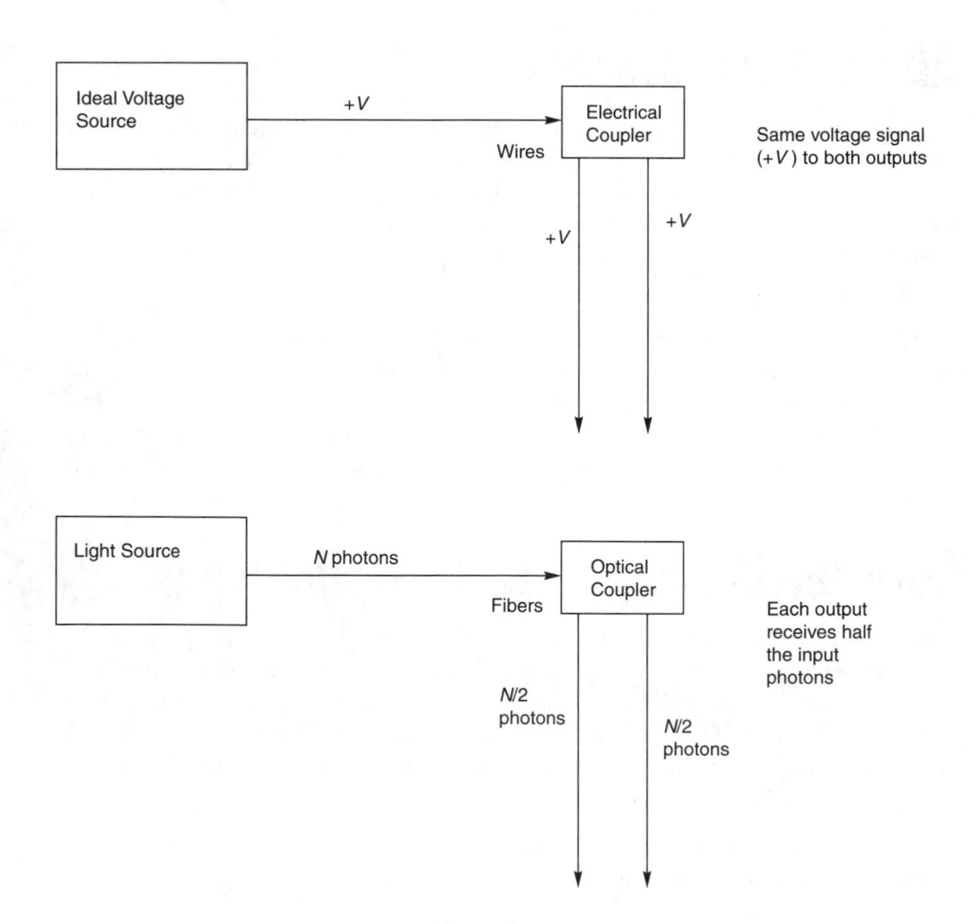

Optical signals are transmitted and coupled differently than electrical signals. First, you have to direct light into fiber cores, not merely make physical contact anywhere on the conductor. In addition, the nature of the optical signal is different. An optical signal is not a potential, like an electrical voltage, but a flow of signal carriers (photons), similar in some ways to an electrical current. Unlike a current, an optical signal does not flow *through* a receiver on its way to ground. The optical signal *stops* in the detector, which absorbs the light. That means you cannot put multiple fiber-optic receivers in series optically, because the first would absorb all the signal (except in certain special circumstances). If you want to divide an optical signal between two or more output ports, they must be in parallel. However, because the signal is not a potential, you cannot send the whole signal to all the ports. You must divide it between them, so no terminal receives a signal as strong as the input. Divide an optical signal equally between two terminals, and each gets half, as shown in Figure 14.1.

An optical signal must be divided among output ports, reducing its strength.

The need to divide an optical signal limits the number of terminals that can be connected to a passive coupler, which merely splits up the input signal. With more ports, less signal reaches each one if the signal is divided equally, as shown in Table 14.1. Each doubling of the number of outputs reduces signal strength by 3 dB. Add too many outputs, and the signal grows too weak to detect reliably. The maximum number of ports depends on receiver sensitivity and other elements of system design.

The loss shown in Table 14.1 is the best case possible, assuming all the input light emerges from one of the outputs. In practice, things aren't that good, particularly if the coupler has

Table 14.1 Loss from splitting signals equally in passive couplers with no excess loss

Number of Output Ports	*Fraction of Input in Each Output*	*Loss in dB*
2	0.5	3.01
4	0.25	6.02
5	0.2	6.99
8	0.125	9.03
10	0.1	10
15	0.067	11.76
20	0.05	13.01
25	0.04	13.98
50	0.02	16.99
100	0.01	20
200	0.005	23.01
400	0.0025	26.02
1000	0.001	30

many output ports. You can divide a signal in half efficiently, but dividing it into 50 or 100 equal parts is harder; some inevitably gets lost, reducing output signal strength. The difference between input signal and the sum of all the outputs (P_1 to P_n) is called *excess loss.*

$$\text{Excess loss (dB)} = -10 \log \left(\frac{(P_1 + P_2 + \cdots + P_n)}{P_{\text{input}}} \right)$$

Note that Table 14.1 assumes the input signal is divided equally among the output ports. This does not have to be the case; you can design couplers that send 90% to one output and 10% to a second. It also assumes that the couplers are passive devices, which draw no input power and merely divide the input power among output ports. Most couplers fall into that category, but there are a few exceptions, which I will explain later.

Coupler choice and design depend on applications.

Couplers have many different applications, so many types have been developed, which divide signals in different ways. Dividing signals between two outputs is the simplest example. As shown in Figure 14.2, couplers may need to split off a small fraction of the signal for each of several terminals, deliver identical signals to many different terminals, or direct different wavelengths to different places. Each of these applications requires a different type of coupler.

In the local-area network of Figure 14.2(a), you need to direct a small fraction of the optical signal from the server to each terminal. Thus you might use a coupler that transmits 90% of the signal through the network and diverts only 10% to each terminal. (This turns out to be rather inefficient, because coupler losses accumulate around the ring, so real local-area networks usually use different architectures.)

For cable-television distribution, shown in Figure 14.2(b), you want to divide the same signal into many equal portions, one for each subscriber. For this application, you use a simple 1-to-*n* splitter, which divides the signal equally among *n* outputs.

For wavelength-division demultiplexing, shown in Figure 14.2(c), you want to separate several wavelengths carried in the same fiber and distribute them to different places. Although you can consider this as a type of coupler, wavelength-division multiplexing has become so important it is covered separately in Chapter 15.

These examples highlight some functions couplers can serve in fiber-optic systems. They can combine signals from different sources, separate signals carried at different wavelengths and route them to different destinations, or split signals among two or more receivers. Note that the couplers do not change the signals; they merely combine, divide, or separate them.

Coupler Characteristics

Several optical characteristics determine the use and function of couplers. The most important of these are:

- Number of input and output ports
- Signal attenuation and splitting
- Directionality of light transmission (which way the light goes through the coupler)

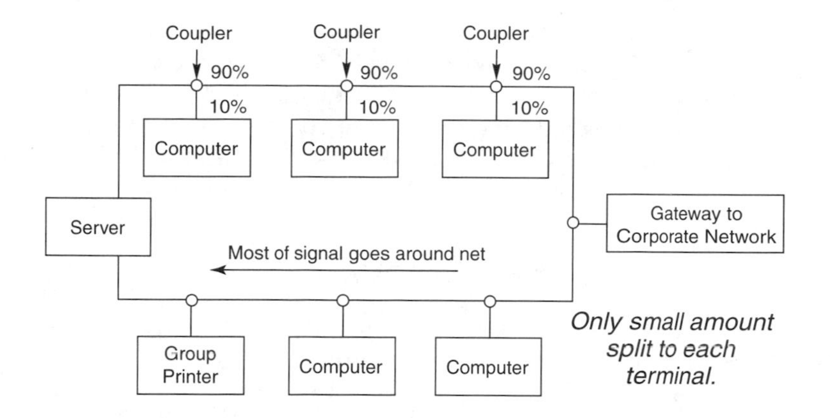

a. Local-Area Network

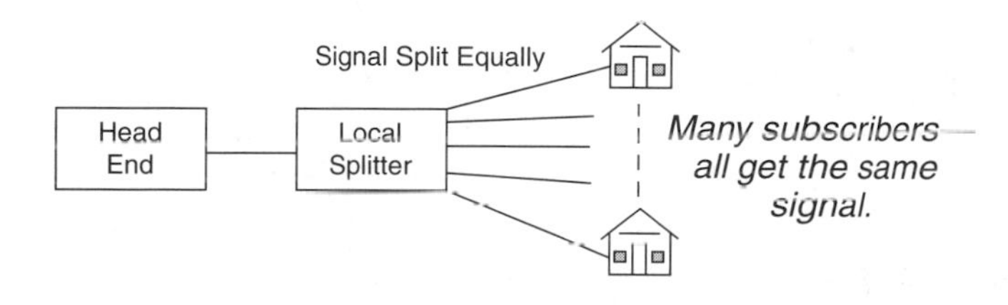

b. Cable Television

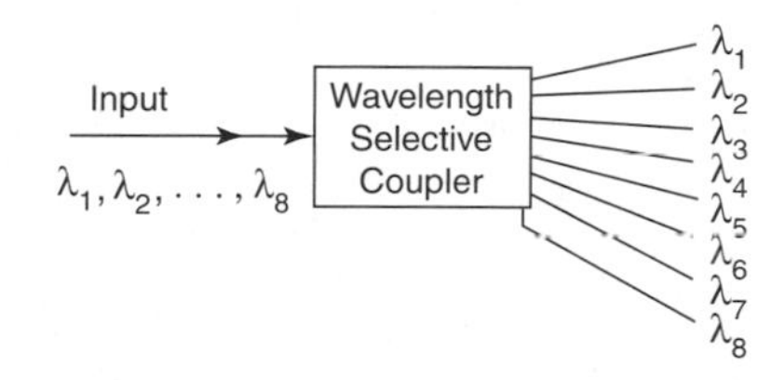

c. Wavelength-Division Demultiplexing

FIGURE 14.2
Different coupler applications: (a) tapping signals in a local-area network; (b) delivering cable television to many subscribers; and (c) splitting wavelengths.

- Wavelength selectivity
- Type of transmission: single- or multimode
- Polarization sensitivity and polarization-dependent loss

Number of Ports

Various applications require different numbers of input and output ports. For example, to split one input between two outputs, you need a 1-by-2 (1 input, 2 output) coupler. To divide one signal between 20 outputs, you need a 1-by-20 coupler. And if you want to combine 10 inputs and distribute the combined signal to 10 outputs, you need a 10-by-10 coupler.

The number of input and output ports depends on the application.

Outputs are usually, but not always, distinct from inputs.

In most cases, the inputs are distinct from the outputs, but in some cases they are not. For example, a 1-by-2 coupler implicitly is dividing the signal from one input port between a pair of different output ports. However, it is not automatically clear if a 10-by-10 coupler has 10 terminals that serve as both inputs and outputs, or 10 inputs and 10 distinct outputs (a total of 20 terminals). Usually the input and output ports are distinct, but not always, depending on the technology chosen, which I will cover later.

Signal Splitting and Attenuation

The number of ports alone does not tell how the signal is being divided among them. Most couplers divide signals equally among all output ports, but some divide the light unequally. For example, a coupler that follows an optical amplifier may split off 1% of the output signal to an optical performance monitor, which verifies that all the expected optical channels are present in the output. Another example is unequal division of a signal between fibers delivering signals to receivers at different distances from the coupler.

Both signal splitting and excess loss contribute to port-to-port attenuation. As you saw in Table 14.1, a 3-dB loss is inevitable when splitting an optical input signal equally between two outputs. A coupler that delivers 90% of the input to one output and 10% to the other has loss of 0.46 dB on the 90% port and 10 dB on the 10% port. Excess loss, as mentioned earlier, is essentially the light wasted within the coupler, which does not reach one of the outputs. Generally excess loss is small, but it is not safe to ignore. Specified port-to-port losses include both splitting losses and excess loss.

Port-to-port attenuation also may be specified for *undesired* light, such as signals going the wrong way through a directional coupler, or wavelengths that are supposed to go out other output ports. These values essentially measure noise.

Directionality

Transmission depends on which way light goes in a directional coupler.

Many couplers are sensitive to the direction of light passing through them. They're not exactly one-way streets, but they tend to keep light going in the same direction. Figure 14.3(a) shows an example of one such directional coupler, where incoming light branches between two diverging outputs, like a waveguide that splits. Light that enters the left port is split between the two outputs on the right side of the figure. However, if light enters the upper-right port, virtually all of it will go to the left port because of the coupler geometry. (A small fraction of light will reach the lower right port, but the loss will be high.) It's as if the waveguides were grooves and you were rolling marbles down them—once in a while the marble might bounce out the lower-right port, but most times it would go to the right.

Other couplers show little or no sensitivity to light direction. Figure 14.3(b) is one example, where light entering from the three ports on the left is reflected from a mirror on the right, which scatters light to all three ports. It doesn't matter which port the light enters; reflected light emerges from all three—including the input port. Such a coupler mixes the light from all inputs and delivers it to all outputs.

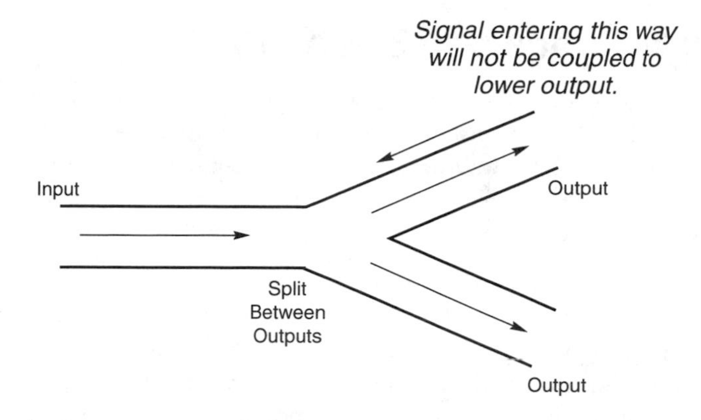

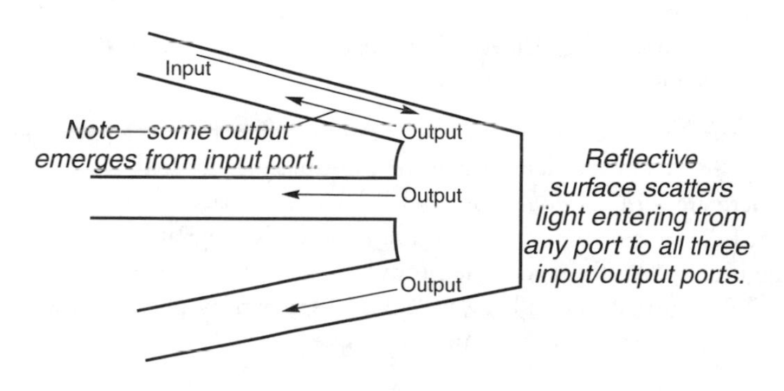

FIGURE 14.3
Directional and nondirectional couplers.

Most directional couplers are really *bidirectional* devices; that is, they can transmit light in either direction, but the light keeps going in that direction. The 1-by-2 coupler in Figure 14.3(a) is such a device. If you direct light through the upper-right port, it will keep going in the same direction and emerge through the single port at the left. Note the difference from a *nondirectional* coupler. In a bidirectional coupler, light going in a port on either side emerges only from the ports on the other. In a nondirectional coupler, light going in any one port emerges from all the ports, including the input. There isn't much demand for truly nondirectional couplers, but they can be made if you need them.

Generally, directionality or bidirectionality is an advantage in couplers, because it sends the signal in the direction you want it. Light headed in the wrong direction—back toward the transmitter in an input fiber, for example—can cause problems such as generating noise in laser transmitters.

Directionality or suppression of reflection back toward the source generally is measured in decibels. If a 1-mW (0-dBm) signal goes through a coupler with 50-dB directionality, only 0.01 μW (−50 dBm) will go in the wrong direction.

Wavelength Range and Selectivity

The mechanisms that divide or combine light in couplers typically depend in some way on the wavelength of light being transmitted. In some cases the dependence may be quite small, so the coupling ratio varies little between, say, 1200 and 1650 nm. In others the variation can be quite strong, directing wavelengths just a nanometer apart out different ports. These extremes represent two different classes of couplers.

Properties of wavelength-insensitive couplers vary little with wavelength.

Wavelength-insensitive couplers change their transmission little over their intended operating range. They are used in applications where light of all wavelengths is supposed to be treated equally. For example, a wavelength-insensitive coupler would be used in a system where the transmitter wavelength is not specified precisely, but the coupler has to split the signal the same way over the entire range of possible operating wavelengths. Such couplers also may be used to divide multiwavelength signals so all outputs contain all optical channels. One example would be splitting off a small fraction of optical amplifier output to an optical performance monitor to verify transmission on all channels between 1530 and 1565 nm. Performance of these couplers is specified as the same within a certain tolerance over a range of operating wavelengths.

Wavelength-selective couplers intentionally send light of different wavelengths in different directions. They are used in wavelength-division multiplexing, covered in Chapter 15. As you will learn, there are various types. Some separate widely spaced wavelengths, such as the 980-nm light from pump lasers and 1550-nm band optical signals in optical amplifiers, as shown in Figure 14.4. Others separate optical channels that are spaced closely in wavelength. The devices usually are called wavelength-division multiplexers (or demultiplexers), but strictly speaking they are special-purpose couplers.

Other Transmission Sensitivities

As you will learn later in this chapter, several technologies can be used for couplers. Some of these coupler designs are limited to either single- or multimode fiber. These should be clearly identified.

Light transmission in some couplers is a function of the polarization of light, which can cause an effect called *polarization-dependent loss*. That is, the transmission of vertically polarized light differs from that of horizontally polarized light. This can be a problem because

FIGURE 14.4
Wavelength-selective coupler.

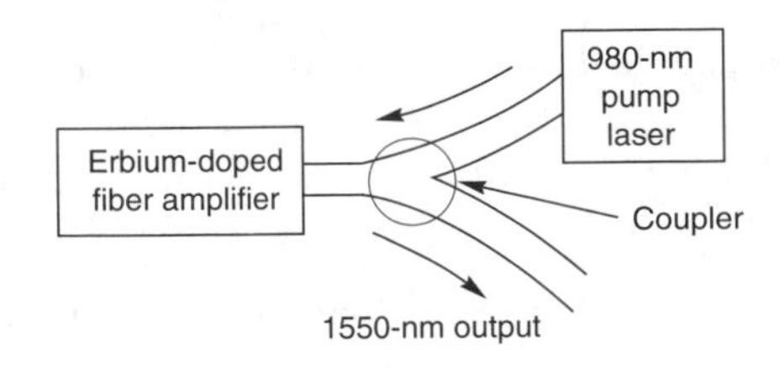

Coupler transmits 980-nm pump through top port and 1550-nm amplifier output through bottom port

most fiber-optic systems do not constrain polarization. If the coupler loss depends on polarization, random variations in input polarization can essentially modulate the light output, inducing noise and degrading signal quality. The larger the polarization-dependent loss, the more significant this noise becomes.

Coupler Types and Technologies

Couplers may look superficially similar, but there are several distinct families and underlying technologies. The names vary to some extent among manufacturers. The configurations or types reflect the function of the coupler. Various technologies can be used to meet different performance requirements.

Coupler Configurations

Coupler configurations define the function, often by the number of input and output ports. The four main types are shown in Figure 14.5 and described next:

T and Y couplers, sometimes called *taps,* are three-port devices, which split one input between two outputs. They may divide the signal equally between the two outputs or split it in some other ratio. Some T couplers are analogous to electrical taps that take part of a signal from a passing cable and relay it to a terminal; they are often shown as one fiber coming off a cable in a T configuration, as in Figure 14.5. Others have a Y-shaped geometry, with two outputs branching at an angle from one input, and are called Y couplers. The directional coupler in Figure 14.3(a) is an example of a Y coupler. They are often—but not always—directional.

T couplers are three-port devices.

Tree, or 1-to-n, couplers generally take a single input signal and split it among multiple outputs, as shown in Figure 14.5. Some have a pair of inputs that are each divided among multiple outputs. Some may combine multiple inputs to one or two outputs, so they actually are combiners. They are generally directional.

Star couplers get their name from the geometry used to show their operation in diagrams such as in Figure 14.5, a central mixing element with fibers radiating outward like a star. They have multiple inputs and outputs, often (but not always) equal in number. There are two basic types. One type is directional, mixing signals from all input fibers and distributing them among all outputs, like the upper star coupler in Figure 14.5, often made by fusing fibers together. These are bidirectional devices because they also can transmit light in the opposite direction. A second is nondirectional, instead taking inputs from all fibers and distributing them among all fibers—both input and output—as with the lower star coupler shown in Figure 14.5.

Wavelength-selective couplers distribute signals according to their wavelengths. Their main uses are to route WDM signals to their proper destinations and to separate wavelengths transmitted for different purposes through the same fiber, such as separating the light pumping an optical amplifier from the amplified signal. Wavelength-selective couplers are supposed to block other wavelengths from reaching the wrong destination. Chapter 15 covers them in detail.

Wavelength-selective couplers distribute signals according to their wavelength.

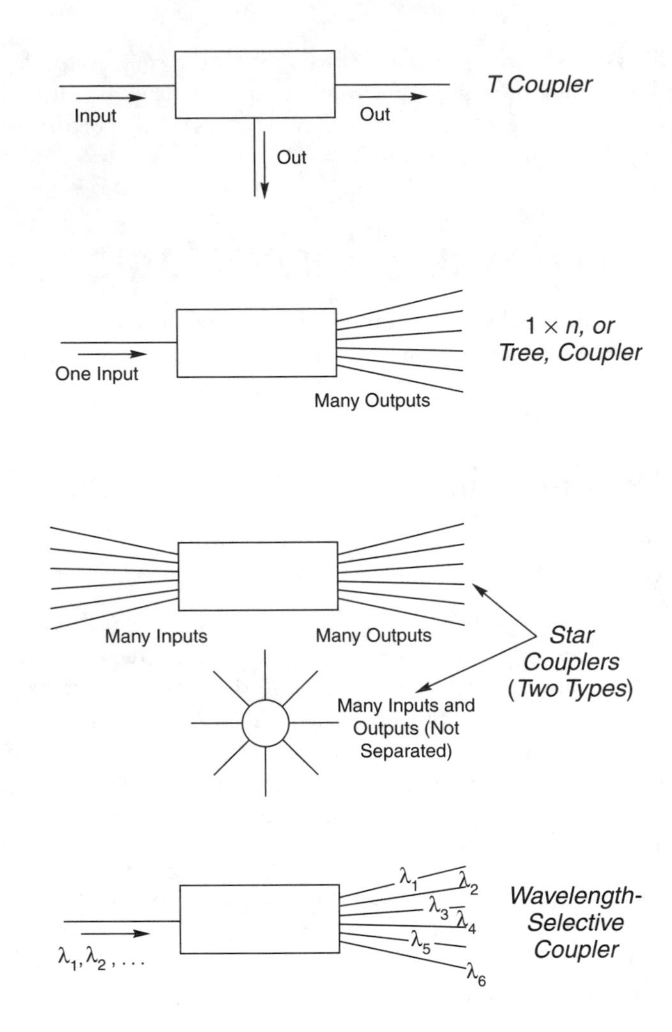

FIGURE 14.5
Important coupler types.

Bulk and Micro Optics

Micro optics are tiny versions of conventional lenses and other optical components, shrunk in size to work with fibers.

In the world of fiber optics, *bulk optics* are conventional lenses, mirrors, and diffraction gratings, the sort of things you can hold in your hands. Bulk optics do not have to be large; they may be made quite small to match the dimensions of optical fibers and light sources. Such *micro optics* may be tiny, but they are still based on the same optical principles as larger bulk optics, so I will cover them together.

Bulk optics were the basis of many early types of couplers, and they still work Figure 14.6 shows a simple example, the use of a device called a *beamsplitter* to split one input signal into two outputs. Like a one-way mirror, the beamsplitter transmits some light that hits it and reflects the rest. Collect the light from the two outputs in fibers, and you have a T coupler.

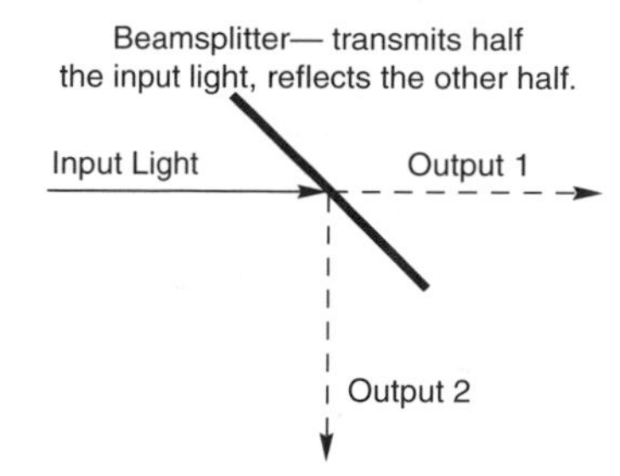

FIGURE 14.6
Bulk optical coupler: A beamsplitter divides a signal in half.

Bulk optical couplers often include lenses that expand, collimate, or focus light. The simple coupler of Figure 14.6 generally works better if a lens expands the light emerging from a fiber and focuses it onto a large area of the beamsplitter. This function is *collimation,* and such optics are called *collimators.* Then additional lenses focus the output beams into output fibers. Standard lenses with curved surfaces may be used; generally they are tiny, to match the sizes of fibers.

Alternatively, gradient-index (GRIN) lenses (or rods) may be used. These are rods or fibers in which the refractive index of the glass changes either with distance along the rod or with distance from the axis. The refractive-index gradient makes GRIN lenses focus light in a way functionally equivalent to ordinary lenses, but GRIN lenses are smaller and easier to adapt to fiber systems.

GRIN lenses are rods or fibers with refractive index graded so they act like ordinary lenses.

Another application of bulk optics is the use of a diffraction grating to separate wavelengths. A diffraction grating is an array of closely spaced parallel grooves, which act together to scatter light at an angle that depends on its wavelength, generating a rainbow of colors.

Fused-Fiber Couplers

Normally you can't transfer light between fibers just by touching them together. The light-guiding cores are covered by claddings that keep light from leaking out. If you want to couple light between fibers, you have to transfer it between the cores. That means you have to remove the claddings from one side of each fiber so the cores can touch. That is the basis of fused-fiber couplers, made by melting together fibers, usually with claddings removed partly or totally from one side, as shown in Figure 14.7. Often the fibers are twisted together to improve light transfer. Fused-fiber couplers sometimes are called *biconic* couplers, which should not be confused with biconic connectors, an early type in only limited use today. They are the most common technology used to make couplers.

Fused-fiber couplers are the most widely used type.

Although Figure 14.7 shows the cores merged, they don't have to merge completely in the middle zone. A phenomenon called evanescent-wave coupling lets light pass through a thin zone of lower refractive index—say, a few micrometers of remaining cladding—between the higher-index fiber cores. This is possible because some light actually travels in the inner portion of the cladding, as you learned in Chapter 4.

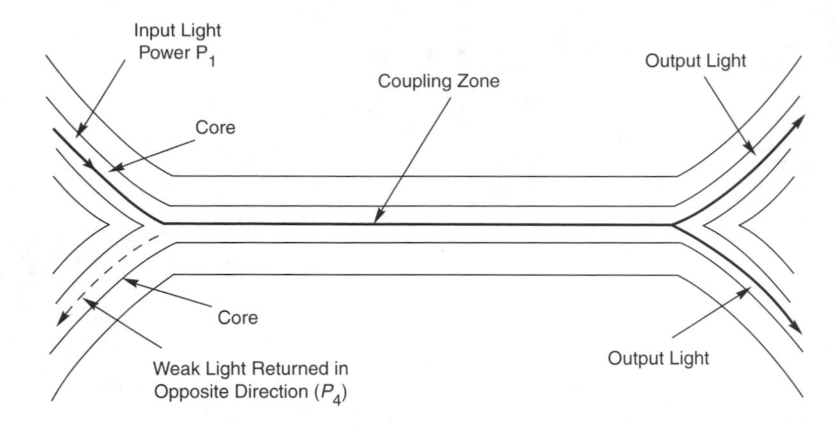

FIGURE 14.7
A 2 × 2 fused-fiber coupler.

Fusing two fibers produces a 2 × 2 coupler with two inputs and two outputs. In practice, these are often turned into 1 × 2 couplers by cutting one fiber end inside the case. This design is inherently directional, although it is bidirectional in the sense that light can go through it in either direction. If light enters the fiber end at upper left in Figure 14.7, the only way light can reach the fiber end at lower left is by reflection or scattering. Directivity is measured by comparing the input power, P_1, to the power reflected back through the other fiber end on the input side, P_4:

$$\text{Directivity (dB)} = -10 \log \left(\frac{P_4}{P_1} \right)$$

For a typical fused-fiber coupler, the directivity is 40 to 45 dB.

The details of fused-fiber coupler operation depend on whether the fibers are multimode or single-mode. In multimode couplers, the higher-order modes leak into the cladding and into the core of the other fiber; the degree of coupling depends on the length of the coupling zone, and does *not* depend on wavelength. In single-mode fibers, light transfers between the two cores in a resonant interaction that varies with length. If all the light enters in one fiber, it gradually transfers completely to the other, then transfers back as it travels farther, shifting back and forth cyclically. The distance over which the cycling takes place depends on the coupler design and the wavelength, as Chapter 15 will describe in more detail.

The fused-fiber coupler design can be extended to multiple fibers using the same basic principles. The important change is adding more fibers, so signals from all input fibers mix in the coupling zone and emerge out of all the output fibers. This approach can be used to make star couplers with many distinct inputs and outputs. Such multifiber fused couplers are bidirectional.

Planar Waveguide Couplers

As you learned earlier, optical fibers are not the only type of optical waveguides. Like fibers, planar waveguides confine light in a region of high refractive index surrounded by material with a lower refractive index. The planar waveguide may be a thin strip embedded in the surface of a flat substrate, as you saw in Figure 6.11, or it could be a strip deposited on the top of a flat substrate. Air and the substrate combine to serve the function of the cladding in a fiber.

Waveguide patterns are written using the same techniques that write circuit patterns onto semiconductor wafers. The simplest types of waveguides are the straight lines you have seen earlier, but much more complex patterns can be formed. These include waveguides that branch or merge, making them the planar waveguide equivalents of fused-fiber couplers.

Simple waveguide couplers are branched planar waveguides.

A simple type of waveguide coupler is a Y-shaped structure that divides one input waveguide into two outputs, as shown in Figure 14.8. The actual split angle is much smaller than shown. If the output waveguides split at equal angles, the light divides equally between them. This approach can be extended to more than two outputs by putting more Y couplers on the outputs of the first, to divide each output further. It also is possible to have the waveguide split into more than two outputs, but that structure may not divide the power evenly.

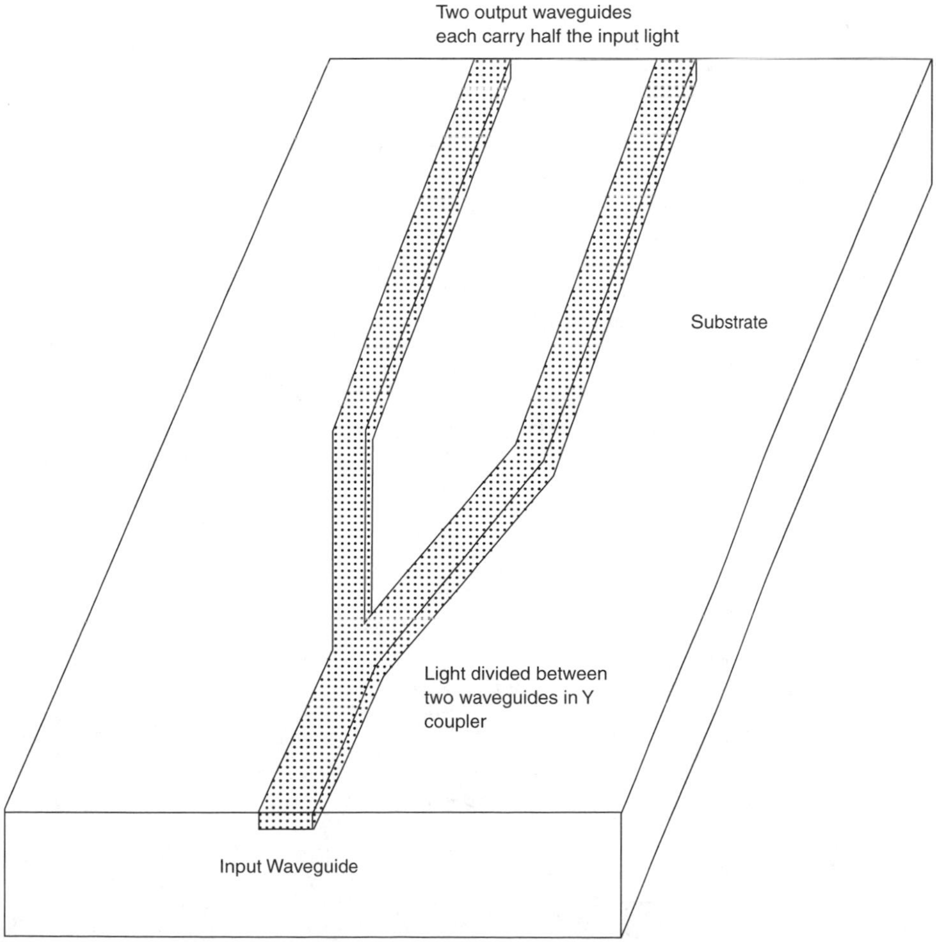

FIGURE 14.8
Planar waveguide splits in two, so light divides equally between arms of the Y coupler.

Light also can be transferred between two closely spaced waveguides on the same substrate. Like the cores of optical fibers, the high-index zones of planar waveguides allow some light to leak into the surrounding lower-index material as evanescent waves. If a second high-index waveguide is close to the first, these evanescent waves can leak into it. This light transfer is the basis of the *evanescent-wave coupler,* shown in Figure 14.9.

Evanescent-wave couplers depend on light leakage between two closely spaced waveguides.

An evanescent-wave coupler gradually transfers light between the two waveguides, along the region where the waveguides are closely spaced. In Figure 14.9, light enters in the top waveguide, and gradually transfers to the lower waveguides. This continues until all light shifts from the upper to the lower waveguide at a point called the *transfer length,* which depends on the optical characteristics of the waveguide. At this point, all the light is in the lower waveguide, and the process reverses, with the light starting to shift back from the lower waveguide to the upper one. Thus the distribution of light energy between the two waveguides oscillates back and forth between them with distance, as shown in the lower part of Figure 14.9. This oscillation stops at the end of the coupling region, determining the final distribution of light. This is the same process that occurs in single-mode fused-fiber couplers.

Designers select lengths and optical properties of the two parallel waveguides to give the desired distribution of light (e.g., 50/50 or 75/25). In practice some light is lost within the waveguide and in transferring between the two waveguides.

Surface waveguides can be fabricated in complex patterns on a variety of materials. When they are made on the same substrate with many other devices, they are often called *integrated optics,* but then they usually contain active devices such as lasers, switches, or modulators. Chapter 16 covers such devices.

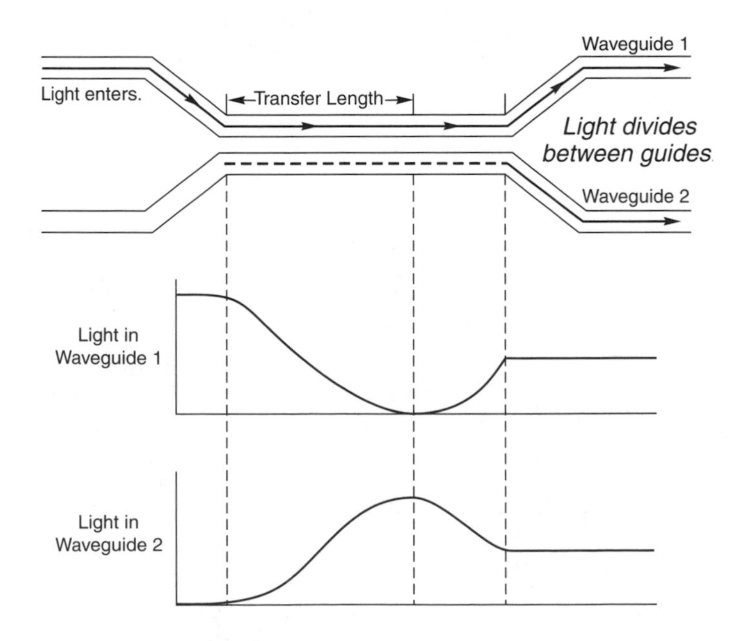

FIGURE 14.9
Light transfer between two evanescently coupled waveguides.

Planar waveguides can be made in a variety of materials. They include glass, semiconductors such as gallium arsenide, and materials such as lithium niobate with properties that change when an electric field is applied to them.

One important practical issue with planar waveguides is transferring light between them and fibers. The thin, flat profile of a planar waveguide is a poor match for the cylindrical core of a single-mode fiber and presents some problems for the larger cores of multimode fibers. You need lenses and other optics to couple light efficiently from a fiber into a waveguide or from a waveguide into a fiber. In practice, this has been a major limitation on the use of planar waveguide components.

Active Couplers

Devices called *active couplers* also look to the user as if they split signals from fiber-optic transmission lines, but if you look closely, they work quite differently. An active coupler is essentially a dedicated repeater in which the signal from a receiver drives two (or more) transmitters, which can generate optical and/or electronic output signals. This means that active couplers are not passive devices. However, they do function as couplers, so they are mentioned here rather than in Chapter 12 on repeaters.

An active coupler is a repeater with two or more outputs.

Active couplers are mostly used in local-area networks. For example, a fiber that runs to a network node may drive a receiver that generates two electronic outputs. One goes to an optical transmitter, which generates an optical signal to send through the next length of fiber in the network. The other is transmitted in electronic form to the terminal attached to that network node. Other configurations also are possible.

Attenuators

As you learned in Chapter 11, too much light can overload a receiver. Attenuators reduce light intensity, by transmitting only a fraction of the input light. They are needed when a transmitter could deliver too much light, such as when it is too close to the receiver.

Attenuators reduce light intensity uniformly across the spectrum.

An attenuator is a type of optical filter, which should affect light of all wavelengths transmitted by the system equally. Attenuators are like sunglasses, which protect your eyes from being dazzled by bright lights. Fiber-optic attenuators generally absorb the extra light energy, which is too little to heat the attenuator noticeably. They should not reflect the unwanted light, because in a fiber-optic system it could return through the input fiber to cause noise in a laser transmitter.

Most attenuators have fixed values that are specified in decibels. For example, a 5-dB attenuator should reduce intensity of the output by 5 dB. Attenuators designed for general optics use may have attenuation specified as the percent of light transmitted *(T)* or as *optical density.* Optical density is defined as:

$$\text{Optical Density} = \log_{10}\left(\frac{1}{T}\right)$$

This should look familiar, because it's close to the formula for attenuation in decibels, without the factor of 10. You can think of optical density as 0.1 times attenuation in dB, so a filter with optical density of 2 has a 20-dB loss.

Variable attenuators also are available, but they usually are precision instruments used in making measurements.

If you're familiar with electronics, it may be tempting to think of an attenuator as an optical counterpart of a resistor. This is not a good general analogy. An attenuator does limit the flow of light like a resistor limits current flow—but resistors also serve other circuit functions, such as providing voltage drops, and controlling circuit loads. The only job of attenuators in a fiber-optic system is to get rid of excess light.

It's important to distinguish between attenuators and other types of optical filters. Attenuators should have the same effect on all wavelengths used in the fiber system. That is, if the attenuator reduces intensity at one wavelength by 3 dB, it should do the same at all wavelengths. Other types of filters typically do not affect all wavelengths in the same way. For example, a filter might transmit light in the 1530 to 1565 erbium-amplifier band, but have 50 dB attenuation in the 980-nm pump band. In fiber-optic systems, the term *filter* is used for filters in which light transmission varies significantly with wavelength. Wavelength-selective filters are key components for wavelength-division multiplexing, covered in Chapter 15.

Optical Isolators

Optical isolators transmit light only in one direction.

Optical isolators are devices that transmit light only in one direction. They play an important role in fiber-optic systems by stopping back-reflection and scattered light from reaching sensitive components, particularly lasers. You can think of them as optical one-way streets with their own traffic cops or as the optical equivalent of an electronic rectifier (which conducts current only in one direction).

The inside workings of optical isolators depend on polarization. An optical isolator includes a pair of linear polarizers, oriented so the planes in which they polarize light are 45° apart. Between them is a device called a Faraday rotator, which rotates the plane of polarization of light by 45°. As shown in Figure 14.10, what happens to the light passing through an optical isolator depends on the direction it is going.

First consider light going from left to right. The input light is unpolarized, but the first polarizer transmits only vertically polarized light. Then the Faraday rotator twists the plane of polarization 45° to the right. The second polarizer transmits light if its plane of polarization is 45° to the right of vertical, which in this case is all the light that passed through the first polarizer and the Faraday rotator. The signal goes through unimpeded except for a 3-dB loss because half of the input signal is blocked by the input polarizer.

Now consider light going in the opposite direction, from right to left. The polarizer on the right transmits only light polarized at 45° to the vertical. The Faraday rotator turns it another 45° to the right, so the plane of polarization is horizontal. That horizontally polarized light is blocked by the polarizer on the left, which transmits only light polarized vertically.

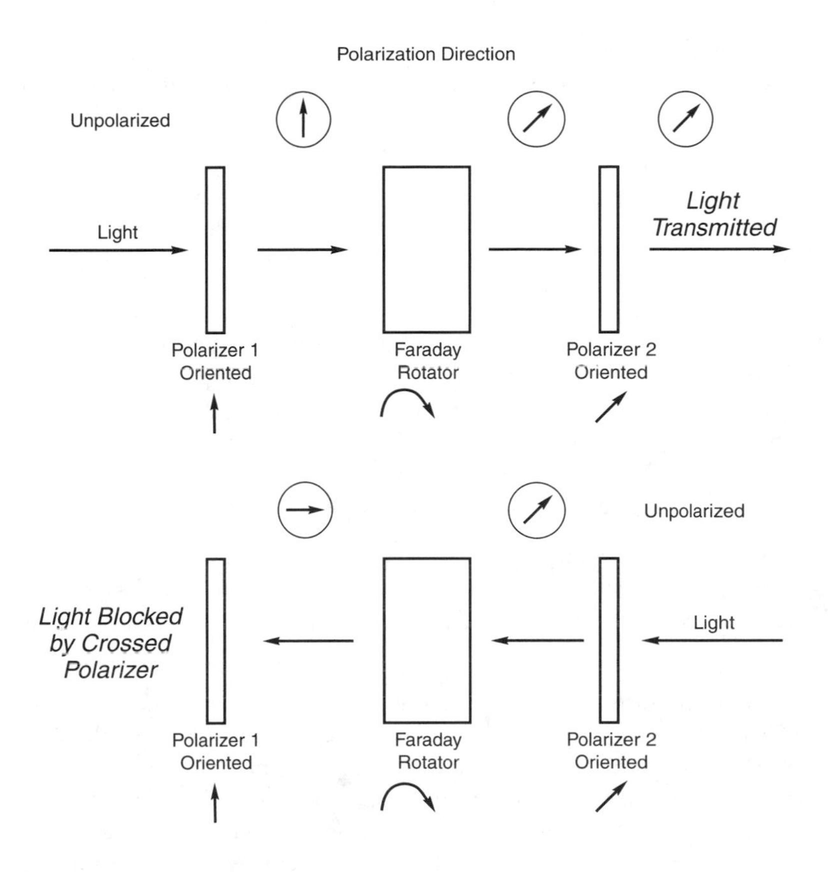

FIGURE 14.10
An optical isolator transmits light in only one direction.

Light blocking is never perfect, but with good polarizers an optical isolator can attenuate light headed in the wrong direction by 40 dB or more, protecting lasers from noise that might affect their performance.

One drawback of this simple design is the 3-dB loss arising from the input polarizer, which blocks the half of the input signal that is not vertically polarized. More refined designs can avoid this by separating the input signal into two beams, one vertically polarized, the other horizontally polarized. The vertically polarized light goes through the isolator as shown. The horizontally polarized light can either be rotated 90° in polarization, then passed through a similar optical isolator, or passed through an optical isolator that transmits only horizontally polarized light. This design is more complicated and requires more components, but avoids the 3-dB input loss.

Optical Circulators

An optical circulator sends light in one direction through a series of ports.

The optical circulator is a cousin of the optical isolator in both its function and design. Its function is to serve as a one-way street for light passing through a series of ports, so light that enters in port 1 must go to port 2, and any light entering at port 2 goes to port 3, and so on. Like the optical isolator, it uses polarization to do its job.

One way to make an optical circulator is with a pair of optical isolators. One can be inserted between port 1 and port 2, blocking light going backwards from port 2. A second can be inserted between ports 2 and 3, blocking light trying to go back from port 3 to port 2. However, these designs lose the blocked light.

A more elegant and efficient design is shown in Figure 14.11, which is assembled from three types of components, polarization-sensitive beam displacers, Faraday rotators, and devices called *waveplates,* which rotate polarization in a *different* way than Faraday rotators. Each requires a bit of explanation before you try to follow the light through the figure.

The beam displacers are made from strongly birefringent materials that bend light of different polarizations in slightly different directions. The input light is unpolarized, but on entering the crystal it breaks down into two beams, one polarized vertically and one polarized horizontally. In the diagram, the vertically polarized beam is bent upwards, and the horizontally polarized beam goes straight through. The vertically polarized beam is bent up no matter which way the light is traveling, to the right or left in the diagram.

The Faraday rotator always rotates polarization by 45°, whether the beam is going forward or backward through it. This means that a beam going back and forth through it will have its polarization rotated a total of 90°. Recall that this is an essential feature that makes the optical isolator work.

The waveplate works differently. When light goes through in one direction, it rotates the polarization 45°. When light goes through in the other direction, it rotates the polarization −45°, or 45° in the other direction. That description may make it sound strange, but this means that when light makes a round trip through a waveplate, its polarization is first rotated 45° in one direction, then rotated back to the original direction on the return trip, for an overall rotation of zero. That is, a beam that makes a round trip through a waveplate has the same polarization it started with.

Now go back to Figure 14.11 and trace the paths of the two polarizations from port 1 to port 2. The vertically polarized input is deflected up, then rotated +45° by the waveplate and another +45° by the Faraday rotator, a total of 90°, making it horizontally polarized so it passes straight through the second beam displacer. Then it is rotated −45° by the waveplate and +45° by the Faraday rotator, a net change of zero, so it remains horizontally polarized through the third displacer and out port 2. The horizontally polarized input, in contrast, goes straight through the first displacer, and is rotated −45° by the waveplate and +45° by the Faraday rotator, a net of zero. It then goes straight through the second displacer. At the second rotator, it is rotated +45° by the waveplate and +45° by the Faraday rotator, a total of 90°, which makes it vertically polarized so the displacer bends it upwards—and aims it out Port 2, where it is supposed to go.

The tricky part is following the path from port 2 to port 3 (right to left in Figure 14.11). The beam displacer splits the two polarizations so they pass through the second rotator. This time the top (horizontally polarized) beam is rotated +45° by the waveplate and +45° by the Faraday rotator, changing its polarization to vertical. The middle beam displacer bends the beam upward, and it emerges from the upper side of the middle beam displacer on its way to port 3. In this case, it is rotated −45° by the waveplate and +45° by

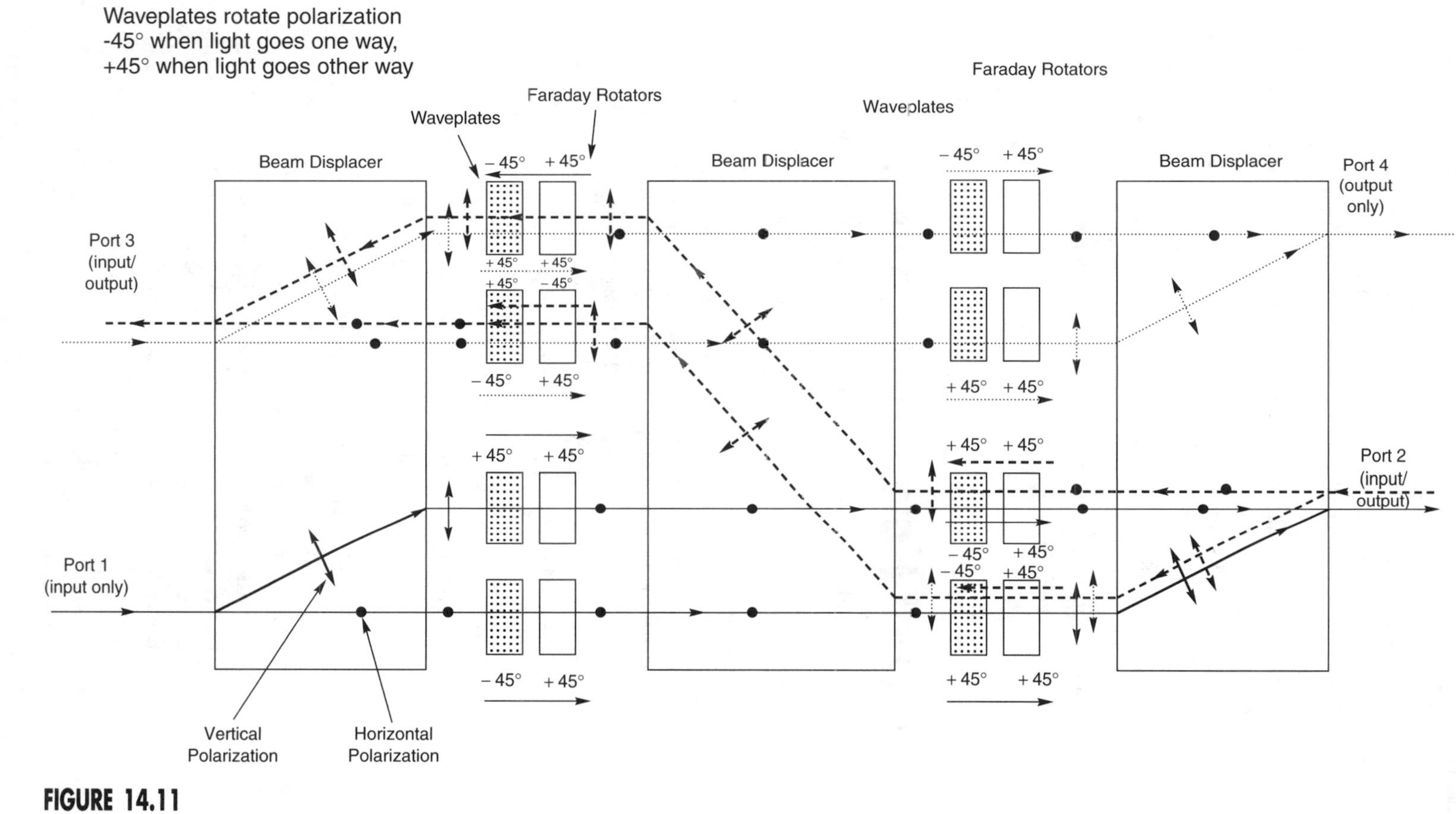

FIGURE 14.11

Optical circulator directs light from port 1–2, 2–3, and 3–4, without allowing it to go backwards.

the Faraday rotator, so it remains vertically polarized, and is deflected downward to port 3. The bottom (vertically polarized) light from port 2 is deflected downward, where the waveplate rotates it −45° and the Faraday rotator rotates it +45°, leaving it vertically polarized as it enters the middle beam displacer. It's bent upward, and arrives at the lower position on its way to port 3. Here the polarizer rotates it +45° and the Faraday rotator rotates it +45°, changing it to horizontally polarized light that goes straight through the beam displacer to port 3.

Each level of the optical circulator is identical, so the steps can be repeated as long as you want. The crucial tricks are separating the polarizations, routing them through different components, and taking advantage of the different ways Faraday rotators and waveplates rotate polarization.

What Have You Learned?

1. Couplers connect three or more fibers or ports. Dividing an optical signal among two or more ports reduces its strength because it divides the photons in the signal.
2. Several different types of couplers are used; their design depends on the application.
3. Direction is important in couplers. Most couplers are directional in the sense they transmit signals from one or more inputs to one or more outputs, with little light going from one input to another. Most designs also are bidirectional, in the sense that the input and output ports could be reversed to change light coupling.
4. T or Y couplers, or taps, are three-port devices. Tree, or 1-to-*n,* couplers divide one input among *n* output ports. Star couplers have multiple inputs and outputs. Outputs are usually distinct from inputs.
5. Wavelength sensitivity is important in couplers. It is desirable for wavelength-division multiplexing, but not for most other applications.
6. Many couplers are made from bulk or micro optics, such as beamsplitters.
7. GRIN lenses are rods or fibers with refractive index graded so they refract light like ordinary lenses.
8. Fused-fiber couplers transfer light between the cores of two fibers melted together. Single-mode fused-fiber couplers work differently than multimode versions. Multifiber fused-fiber couplers are possible.
9. There are two types of planar waveguide couplers. Some simply divide light between two waveguides branching in a Y from a single-input guide. Others rely on evanescent-wave coupling to transfer light between two parallel waveguides. Evanescent-wave couplers are sensitive to wavelength.
10. Active couplers are repeaters with two or more outputs.
11. Attenuators block light uniformly across a range of wavelengths to reduce signal strength at the receiver.

12. Optical isolators transmit light in only one direction. They rely on polarizing optics and Faraday rotators.
13. Optical circulators route light through a series of ports, feeding output from one port to the next, and taking input from the second port and routing it to a third. They rely on birefringent crystals, Faraday rotators, and other polarization rotators.

What's Next?

Chapter 15 covers wavelength-selective optics used for wavelength-division multiplexing.

Further Reading

Morris Hoover, "New coupler applications in today's telephony networks," *Lightwave,* Vol. 17, March 2000.

Luc B. Jeunhomme, *Single-Mode Fiber Optics: Principles and Applications* (Marcel Dekker, 1990), see Chapter 6, "Passive Components."

Questions to Think About for Chapter 14

1. Suppose your input signal is −10 dBm and your receivers require a signal of at least 0 dBμ. You want to distribute signals to as many terminals as possible. If there is 3 dB of fiber loss between you and each receiver, how many terminals can you deliver signals to? How much coupler loss does this correspond to on each channel? Assume you can buy a star coupler with as many ports as you want, which has no excess loss.
2. Suppose that all star couplers available for the system described in Question 1 have excess loss of 3 dB. How many terminals can you reach with these couplers, and what is the total loss per channel?
3. An alternative design is to cascade a series of 3-dB T couplers. The first splits the signal in half, then each output has its own 3-dB coupler, dividing that output in half, yielding one-quarter of the original output. Adding more layers further divides the signal. Suppose you can get as many 3-dB couplers as you want and each one has no excess loss. How many terminals can you divide signals among in Question 1?
4. A local-area network includes 90/10 couplers, which split 10% of the input signal and deliver it to a local terminal. Suppose you have 10 of them in series and the input power is −10 dBm. What is the power delivered by the last coupler out each of its ports?

5. If your receiver requires 1 μW of power, how many more 90/10 couplers could you have in series before the 10% side does not deliver enough power for reliable operation? Assume the same −10 dBm input as in Question 4.
6. An optical amplifier delivers 0.5 mW/channel on each of 32 channels. You want to monitor its performance by diverting a small portion of its output to an optical performance monitor that requires 1 dBμ input per optical channel. What fraction of the output power do you need to divert to the performance monitor?
7. Neglecting excess internal losses, what is the difference in attenuation between the following two optical circulators? The first uses the simple optical isolators of Figure 14.10—one oriented from port 1 to 2 and the second from port 2 to 3, with a 50/50 T coupler splitting input signals from port 2 between two routes. The other is the more complex optical circulator of Figure 14.11.

Quiz for Chapter 14

1. You have a coupler that divides an input signal equally among 16 outputs. It has no excess loss. If the input signal is −10 dBm, what is the output at any one port?
   a. −12 dBm
   b. −20 dBm
   c. −22 dBm
   d. −26 dBm
   e. −30 dBm
2. A 1 × 20 coupler has output signals of −30 dBm at every port if the input signal is −10 dBm. What is its excess loss?
   a. 0 dB
   b. 1 dB
   c. 2 dB
   d. 4.2 dB
   e. 7 dB
3. A coupler splits an input signal between two ports with a 90/10 ratio. If the input signal is −20 dBm and the coupler has no excess loss, what is the output at the port receiving the smaller signal?
   a. −21 dBm
   b. −29 dBm
   c. −30 dBm
   d. −31 dBm
   e. −110 dBm
4. What type of coupler could distribute identical signals to 20 different terminals?
   a. T coupler
   b. Tree coupler
   c. Star coupler
   d. M × N coupler
   e. Wavelength-selective coupler
5. What type of coupler divides one input signal between two output channels?
   a. T coupler
   b. Tree coupler
   c. Star coupler
   d. M × N coupler
   e. Wavelength-selective coupler

6. You find a coupler with four ports and no label on it. You measure attenuation from port 1 to the other three ports. The values are: −40 dB to port 2, 3 dB to port 3, 3 dB to port 4. What type of coupler do you have?

a. Star coupler with three unequal outputs

b. Tree coupler with three unequal outputs

c. A directional 2-by-2 coupler with two inputs and two outputs

d. A nondirectional T coupler

e. A broken coupler

7. A Y coupler that equally divides light between two outputs has a 3-dB loss on each channel. What is the right explanation?

a. The 3-dB figure is excess loss.

b. Half the photons that enter the coupler go out each output, corresponding to a 3-dB loss on each channel.

c. The coupler polarizes the light going out each port, causing 3-dB loss.

d. Every coupler has at least 3-dB loss no matter how it divides the input light.

e. The optics are dirty, causing loss of half the input light.

8. Evanescent waves cause light energy to transfer between channels in what type of coupler?

a. Planar-waveguide coupler

b. Single-mode fused-fiber coupler

c. Bulk optical coupler

d. Multimode fused-fiber coupler

e. a and b

f. b and d

9. An attenuator

a. is a filter that blocks one wavelength but transmits others.

b. polarizes input light, causing loss of the other polarization.

c. reduces light intensity evenly across a range of wavelengths.

d. selectively blocks photons produced by spontaneous emission.

10. How many polarizers does a simple optical isolator use to block transmission of light in the wrong direction?

a. None

b. 1

c. 2

d. 3

e. 4

CHAPTER

15

Wavelength-Division Multiplexing Optics

About This Chapter

Wavelength-division multiplexing (WDM) multiplies the signal capacity of fiber-optic systems, but it also places stringent requirements on system optics. Optical components must be able to combine and separate optical channels at very closely spaced wavelengths. Specific requirements vary with system design, and several technologies are available.

This chapter will first outline the diverse requirements for separating wavelengths in fiber-optic systems, then cover the technologies developed for various types of wavelength separation. Some of these technologies are closely related to the couplers described in the last chapter, while others were developed specifically for WDM. This chapter concentrates on passive technologies to separate wavelengths. Chapter 16 covers switching and modulation technologies, many of which are used together with WDM to create optical networks. The boundaries between active and passive technologies can be hazy, and sometimes the two overlap.

WDM Requirements

The best-known application of wavelength-division multiplexing is the transmission of many optical channels through a single fiber at different wavelengths. This requires optics to combine signals generated by separate transmitters at one end, and to separate the optical channels for processing in separate optical receivers at the other end. Packing channels closely together increases capacity, but it also makes the job of separating them more difficult.

WDM requires optics to combine optical channels at one end and to separate them at the other.

The same issues of combining and separating wavelengths also arise in other areas of fiber-optic systems. In erbium-doped fiber amplifiers, the pump laser beam must be separated from the amplified signals. Some systems transmit signals through the same fiber in two directions using two different wavelengths, which also must be separated. These technologies also are considered wavelength-division multiplexing. First I will describe requirements for WDM systems; then I will cover the technologies used to separate wavelengths.

WDM Systems

Multiplexers combine optical channels. Demultiplexer separate them.

Figure 15.1 shows a simple example of a WDM system. Optical signals at eight different wavelengths from separate transmitters are combined in a *multiplexer* at the left. The signals travel through the same fiber from the multiplexer to an *add-drop multiplexer* in the middle, which routes one wavelength, λ_4, to a point at the bottom and picks up another signal at the same wavelength, called λ_{4*}, to show it is a different signal.

At the right side, the eight signals are split in a *demultiplexer* and routed to separate receivers, one for each wavelength. As you learned in Chapter 11, receivers are color-blind in the sense that they respond in the same way to all wavelengths they can detect. This means that the signals must be completely separated, because any stray light at a different wavelength will show up as noise in the receiver output. For example, if some light from the signal at λ_5 reached Receiver 6, the receiver would think it belonged in channel 6, and it would interfere with the actual λ_6 signal.

Demultiplexers must separate optical channels completely, with low crosstalk.

You can think of the multiplexer and demultiplexer as mirror images of the same device. The multiplexer takes separate wavelengths and combines them, and the demultiplexer takes combined wavelengths and separates them. Key operating considerations differ between the two. Multiplexers should have low insertion loss and avoid scattering light back to any of the transmitters. Demultiplexers must reliably separate the optical channels, with

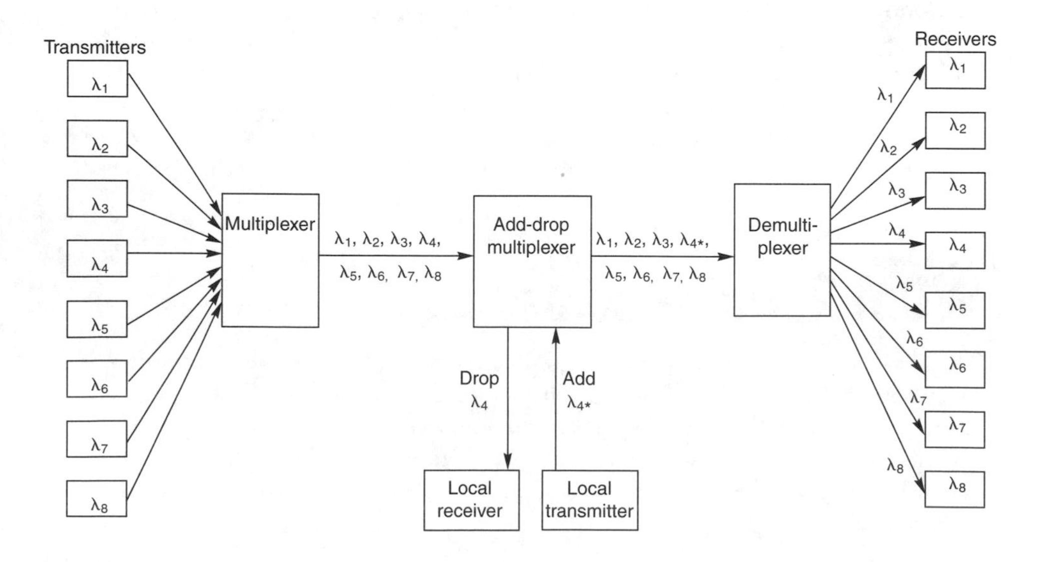

FIGURE 15.1
A WDM system.

low leakage of light from one optical channel into an adjacent channel. In practice, the two devices can be used as mirror images of each other, although sometimes the multiplexers may have wider channel spacing than the demultiplexers to reduce insertion loss.

The add-drop multiplexer serves a different function, picking out one or more wavelengths from a combined signal so they can be "dropped" at a location part way along a system. It also can "add" signals transmitted from the midpoint station onto empty channels. In Figure 15.1, it is adding a signal λ_{4*} to replace the signal that has been dropped at the λ_4 wavelength.

As the technology evolves, WDM systems are incorporating more complex functions. The optical channels on a fiber may be separated and switched, using optical switches I will describe in Chapter 16. Manipulating signals as optical channels may require wavelength conversion, a concept also described in Chapter 16. Some systems may use wavelength routing, a concept involving passive optics described in this chapter, although it is equivalent to the switching devices covered in Chapter 16.

Channel Density

Channel density is a key variable in WDM systems. The optics provide uniformly spaced slots for optical channels, although the system may not use all those slots. The spacing of the slots determines the density. The International Telecommunications Union grid has specified one set of standard center frequencies separated by 100 GHz, which corresponds to about 0.8 nm in the erbium-fiber amplifier band. Some commercial systems have a 50-GHz grid spacing. Developers are talking about spacings of 25 GHz and even 12.5 GHz, but it is not clear when or if that technology will become commercial.

Typically, spacings of 200 GHz or less are called *dense wavelength-division multiplexing*, or *DWDM*. This allows many more channels to be squeezed into the system, as shown in Figure 15.2, which compares optical channels in a system with 40 channels separated by 100 GHz and 4 channels separated by 1000 GHz, a spacing of about 8 nm. Such wide spacings often are called *wide wavelength-division multiplexing*, or *WWDM*. Channel spacings as wide as 20 to 25 nm are used in some systems, particularly those operating in the 850 or 1300 nm bands. A few systems with even wider spacing have been demonstrated at visible wavelengths with plastic fibers, but are not in wide use.

Channel spacings of 200 GHz or less are dense wavelength-division multiplexing (DWDM).

The wider the spacing, the easier and less expensive it is to make multiplexing and demultiplexing optics. On the other hand, system bandwidth may limit the total range of usable wavelengths—for example, to the 1530- to 1565-nm range of C-band erbium-fiber amplifiers. That limits the number of usable channels.

Optical Channel Separation

Demultiplexing optics need to separate optical channels cleanly, providing a high degree of isolation between them. Isolation should be 20 to 40 dB, so the input on adjacent channels is reduced 20 to 40 dB below the input on the desired channel.

Demultiplexing optics should transmit only the selected optical channel.

The best way to see how this works is to consider the transmission of a demultiplexer as a function of wavelength for one output port—that is, for an individual optical channel.

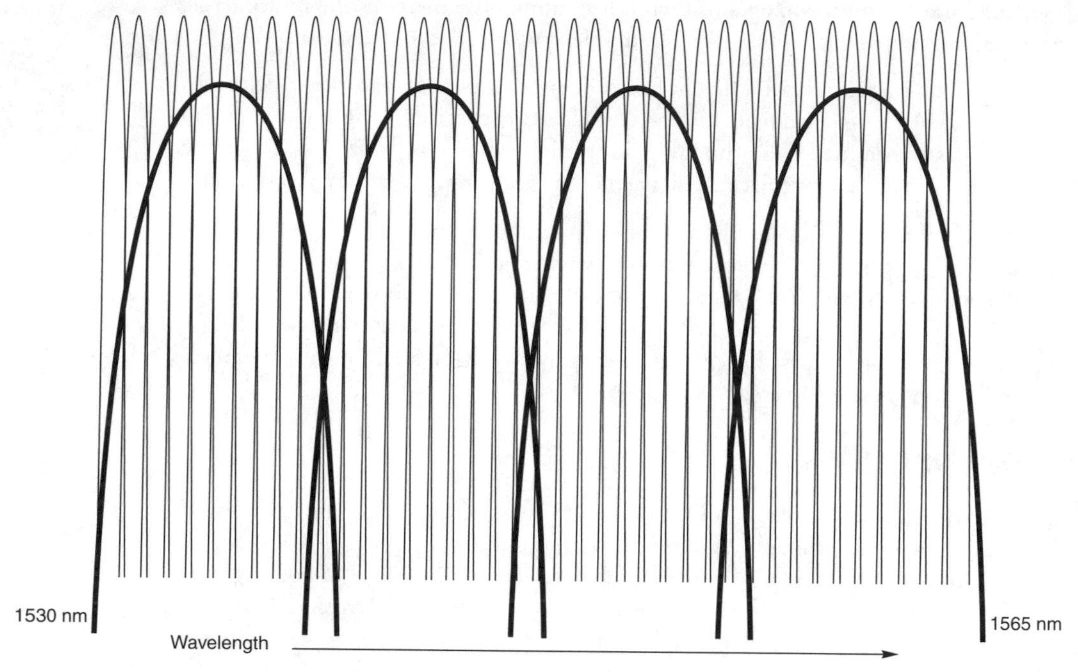

FIGURE 15.2
Channel spacing in WDM systems.

Ideally, the port should transmit all the light at the center of the optical channel, and no light outside it. As usual, reality is a compromise. Normally the peak transmission at the center wavelength is not 100%, with a typical loss 3 to 5 dB. The transmission drops off steeply at longer and shorter wavelengths, as shown in the curves in Figure 15.2.

Normally the channels are equally spaced, so the points where the curves intersect match the widths of the channels or, equivalently, the channel separation. In Figure 15.2, this is 100 GHz for the tightly packed DWDM channels, and 1000 GHz for the loosely packed WWDM channels. The actual transmission curves depend on the technology used, and normally spread out more at the bottom than in these idealized curves.

Crosstalk occurs when optical channels overlap.

The points where the curves overlap is where crosstalk can occur. The amount of crosstalk also depends on the range of wavelengths emitted by the transmitter on each channel, not shown here. In general, the transmitter emission also drops sharply away from the peak wavelength, leaving an effective wavelength gap between adjacent optical channels. The range of wavelengths emitted by the transmitter depends on the modulation scheme as well as on the light source itself.

Examples like in Figure 15.2 simplify channel spacing in two ways. In actual systems, some of the channels may not be "populated" with transmitters and receivers; that is, the slot is available, but not in use at the moment. In addition, some designs may intentionally leave some slots open to avoid possible crosstalk. The gap normally left between C- and L-band erbium-fiber amplifiers is one example. Standard C-band amplifiers stop at 1565 nm, and L-band amplifiers don't start until 1570 nm, leaving a 5-nm gap. Some WDM systems leave similar gaps between blocks of wavelengths, perhaps dropping one or two 100-GHz slots between groups of eight optical channels.

Add-Drop Multiplexers

A full demultiplexer separates all the optical channels in a fiber, but in many cases you may want to separate only one or two channels from a larger number of channels. This is the job of an add-drop multiplexer, like the one shown in the middle of Figure 15.1.

An add-drop multiplexer diverts one or more optical channels, and may add new signals in their place.

To pick off one optical channel, you pass the light through an optical device that treats the selected channel differently from other channels, such as a wavelength-selective filter. If it reflects the selected channel, it should transmit all other wavelengths; likewise, if it transmits the selected channel, it should reflect all other wavelengths in the system. Figure 15.3 shows how an add-drop multiplexer reflects a selected channel while transmitting other channels, so it can drop one optical channel at a local node while transmitting all other channels. Other optics may be used to add a signal at the same wavelength to replace the dropped channel.

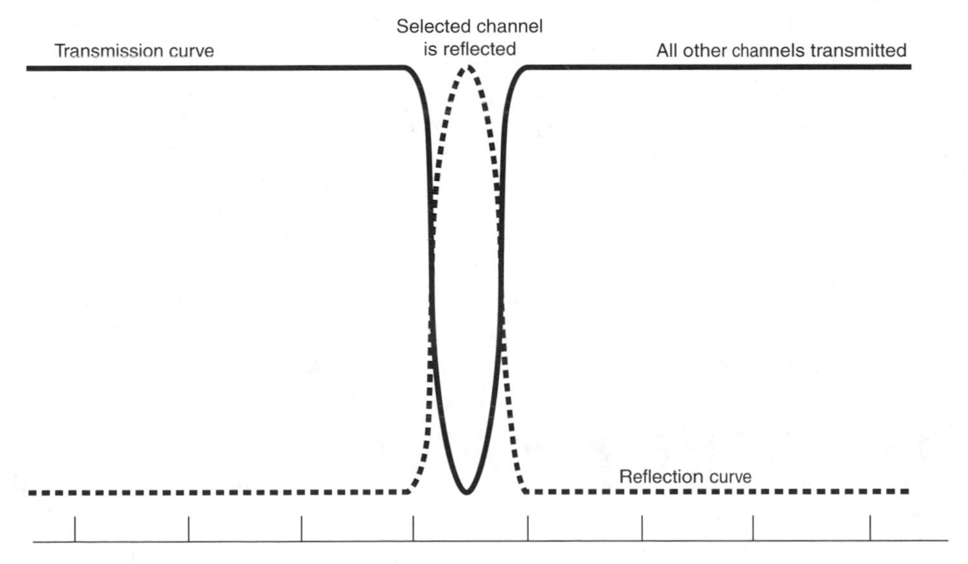

FIGURE 15.3
Add-drop multiplexer reflects one channel and transmits others.

Note that such an add-drop multiplexer is a fixed device that always adds and drops the *same* wavelength at the same point. Changing the selected wavelengths requires changing the optics or adding a switch that can make the change.

Wavelength Routing

A wavelength router directs different wavelengths to different points.

Another type of demultiplexing is called *wavelength routing.* A wavelength router is essentially an optical demultiplexer that directs different wavelengths to different points instead of to different receivers. Optical signals thus are routed to particular destinations according to their wavelength. (As you will learn in later chapters, this sort of routing is not the same as the routing used in Internet transmission.)

To envision wavelength routing, take a fresh look at Figure 15.1, but in your mind replace receivers 1 through 8 with destinations 1 through 8. The destinations could be different locations in the system, or simply different cables at an optical node. Wavelength routing is another way of organizing an optical network.

Coarse Wavelength Separation

So far, we've concentrated on wavelength-division multiplexing that separates wavelengths that are relatively closely spaced, even for the "wide" WDM system of Figure 15.2. In these systems, all the wavelengths are signals in the same general transmission band, which need to be separated. In some other cases, the wavelengths that must be separated are in different transmission bands.

Wavelength-selective couplers separate pump light from amplified signal in erbium-fiber amplifiers.

One case is in erbium-fiber amplifiers. As you learned earlier, erbium-fiber amplifiers are pumped by 980- or 1480-nm light from diode lasers. Both that pump light and the signal being amplified must pass through the erbium-doped fiber, but elsewhere the pump light must be separated from the signals being amplified at wavelengths of 1530 to 1620 nm. This requires wavelength-selective couplers on the ends, which transmit the amplified signals to the output fiber, but block the pump light or direct it out another port. A typical coupler would separate light in the two bands, with the pump light going in one direction (to and from the pump laser), while the signal would go in another (to the output fiber).

Some systems transmit signals on only two optical channels widely separated in wavelength, one at 1300 nm and the other near 1550 nm. These systems also use wavelength selective couplers to transmit 1300 nm to one output and 1550 nm to the other. Some of these systems are bidirectional, with 1300 nm transmitted in one direction and 1550 nm in the other. Many of these systems are older ones, installed in the past.

Optical Filters and WDM

The optical devices most often used to selectively transmit certain wavelengths are called *filters.* The term covers a broad range of devices, including the attenuators described in Chapter 14, and you should understand what they are and how they work. Filters play important roles in WDM systems, although other technologies also may be used.

Sunglasses are a familiar type of optical filter, and like the filters used for WDM, sunglasses come in many varieties. Ordinary gray-green sunglasses are simple attenuators that block a uniform fraction of the light across the spectrum, and don't obviously change the colors of the world. Polarizing sunglasses transmit light of only one polarization, blocking the other polarization. The world doesn't look obviously different through polarizing sunglasses unless you look at certain parts of the sky or surfaces that look unusually bright or dark. Colored sunglasses and some photographic filters make the world look colored because they block other shades. Thus blue or red sunglasses make other objects seem to be those colors.

In the world of optics, "filter" often is a broad term applied to components that filter out part of the incident light and transmit the rest. Many types, such as photographic filters and most sunglasses, absorb the light they don't transmit. The only places such absorbing filters are used in fiber-optic systems are where it's important to absorb undesired light, such as in attenuators and optical isolators. In WDM systems, the wavelengths that are not transmitted through the filter normally are reflected so they can go elsewhere in the system. Such filters are like mirror shades and one-way mirrors, which reflect most incident light, but transmit enough for you to see through them (if you're looking into a brighter area).

WDM filters transmit selected wavelengths and reflect others.

The term "filter" is used a little differently in WDM. Typically it means one specific type of filter, the *interference filter,* which I describe below. Other types of WDM optics such as fiber Bragg grating *act* like filters, in the sense that they block some light, but they are not considered quite the same. I'll use interference filters as a way to explain the operation of WDM optics, then cover other types of wavelength-selective optics.

Interference Filters

Interference filters are made by depositing a series of thin layers of two materials with different refractive index on a flat piece of glass. Alternating layers are deposited of each material. Typically the materials are insulators or dielectrics, which do not conduct electricity, so these filters are sometimes called *dielectric filters.*

The layers in an interference filter selectively transmit a narrow range of wavelengths, and reflect other light.

The difference in refractive index between the two layers causes reflection at each surface. (The basic phenomenon is the same as the Fresnel reflection that occurs in connectors with an air gap between the fibers.) The more identical pairs of alternating layers, the more the reflection builds up—at most wavelengths.

Light is transmitted only at certain wavelengths, which are selected by the optical characteristics of the layers. As in a resonant laser cavity, the light wave has to make a round trip between the layers in an integral number of wavelengths. Light waves at these wavelengths are in phase with each other, so they add constructively in the transmitted beam. Transmitted wavelengths λ are given by the formula

$$N\lambda = 2nD \cos \theta$$

where N an integer, n is the refractive index of the layer, D is the layer thickness, and θ is the angle the incident light makes to the normal. Figure 15.4 shows how an interference filter transmits light with one wavelength per pair of layers (or one-half wave per layer), and

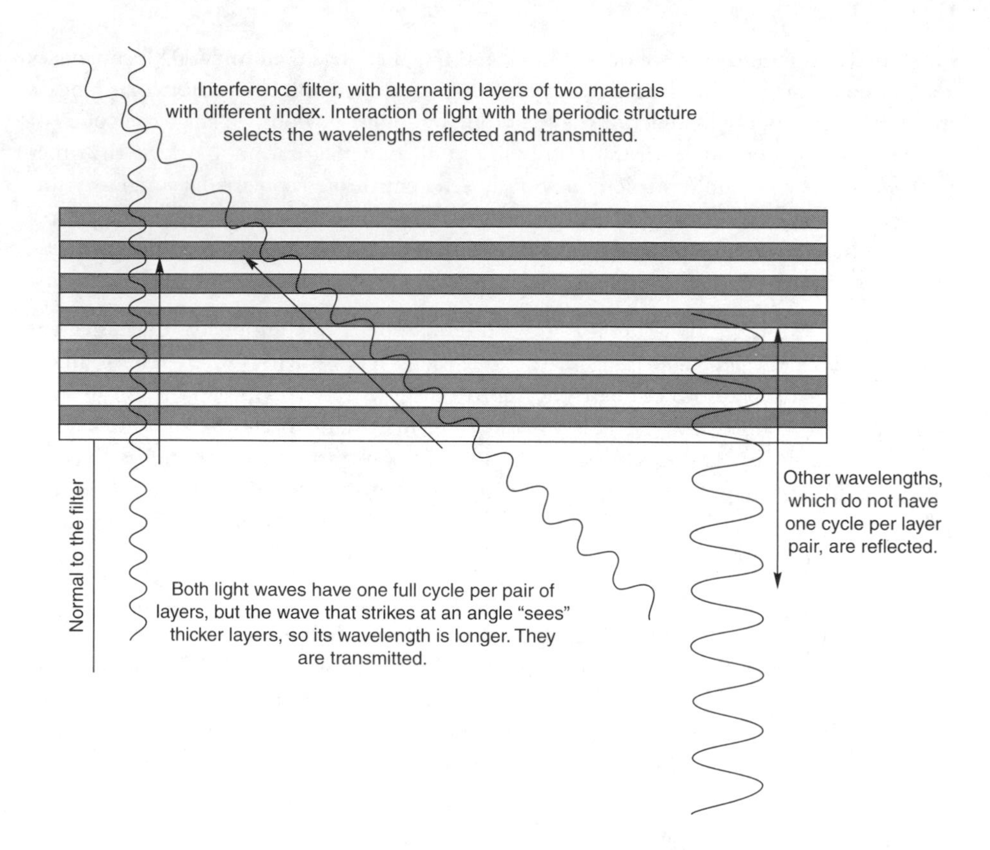

FIGURE 15.4 *Wavelength selection in an interference filter.*

reflects other wavelengths. Note that the wavelength transmitted depends on layer thickness, refractive index, and angle of incidence on the filter.

From an optical standpoint, the transmitted wavelengths are in phase and interfering constructively, so the waves add in intensity. Waves at other wavelengths are out of phase, so they interfere destructively, canceling their amplitude in the transmitted beam. Instead they are reflected.

From the user's standpoint, these effects are analogous to the wavelength selection effects of fiber gratings covered in Chapter 7. However, there is an important difference. Fiber gratings selectively *reflect* a narrow range of wavelengths, while interference filters selectively *transmit* a narrow range of wavelengths. This becomes important in designing demultiplexers, as you will learn later in this chapter.

The precise selection of wavelengths transmitted and reflected, and the shape of the reflection and transmission curves, depend on details of filter design including thicknesses and compositions of the layers, and the numbers of layers in the "stack." Normally the more layers, the finer the resolution, and the narrower the range of wavelengths selected.

The design of interference filters is a well-developed and highly specialized art, and it can produce carefully controlled results. With the right choice of material compositions and layer thicknesses, engineers can coat thin glass plates with interference filters that strongly reflect one wavelength while transmitting almost all the light at nearby wavelengths, so they are widely used in optical multiplexers and demultiplexers.

Although interference filters are considered "bulk optics" because they are discrete components, the filters used in WDM optics are quite small, typically a few millimeters across.

Line, Band, and Cutoff Filters

Interference filters can be made with various transmission characteristics by adjusting their composition, the thickness of layers, and the number of layers. Three types of filters important in WDM optics are the *line filter,* the *band filter,* and the *cutoff filter;* Figure 15.5 shows their transmission characteristics.

Line and band filters either reflect or transmit light in a selected range of wavelengths. If the range of wavelengths is narrow, they are called *line* filters; an example would be a filter to pick out one 100-GHz optical channel. Filters that select a broader range of wavelengths are called *band* filters; an example would be a filter that selects a 10-nm chunk of the erbium-fiber amplifier band. Line-rejection or band-rejection filters reflect the selected band while transmitting nearby wavelengths; line-pass or band-pass filters transmit the selected wavelengths while reflecting adjacent wavelengths.

Line and band filters select a range of wavelengths.

Note that you can arrange filters in various ways so one type can serve different functions. For example, a line-pass filter normally transmits one wavelength while reflecting other light. However, you can use a line-pass filter as a line-rejection filter simply by collecting the *reflected* light rather than the transmitted light.

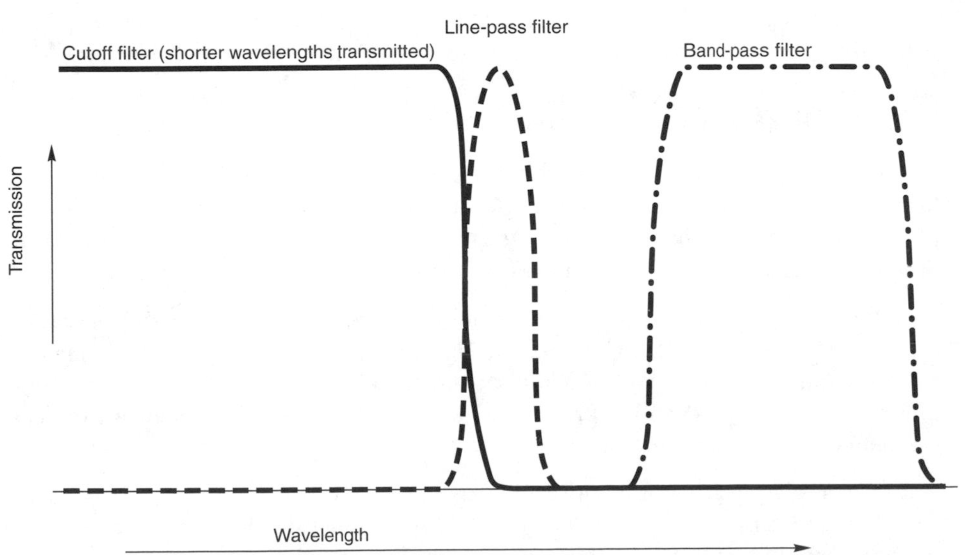

FIGURE 15.5 *Transmission of line, band, and cutoff filters.*

Cutoff filters make a sharp transmission between transmitting and reflecting at a certain wavelength.

Cutoff filters are designed to make a sharp transition between transmitting and reflecting at a certain wavelength. For example, a filter designed to separate optical channels directed to C- and L-band erbium-fiber amplifiers would have a cutoff wavelength at 1567 nm, with shorter wavelengths reflected to the C-band amplifier and longer wavelengths transmitted to the L-band amplifier. The cutoff filter in Figure 15.5 transmits short wavelengths but blocks long ones.

It's important to realize that these filters are designed for use in a specific range of wavelengths. Outside of the specified band, their transmission may vary. Thus you can't count on a cutoff filter that reflects wavelengths shorter than 1567 nm to also reflect light at 1300 nm—unless you've checked its properties at that wavelength.

Equalizing Filters

Equalizing filters compensate for uneven gain of optical amplifiers at different wavelengths.

In Chapter 12, you learned that fiber amplifiers do not amplify all wavelengths equally because their gain varies with wavelength. This can cause problems after a series of amplifiers because signals at the stronger wavelengths will overwhelm those at weaker wavelengths.

Equalizing filters are designed to compensate for this difference by attenuating the wavelengths that would otherwise be amplified most strongly. As shown in Figure 15.6, the filter transmission is intended to offset the gain. The higher the gain, the lower the transmission; the lower the gain, the larger the fraction of the light transmitted. For example, if amplifier gain is 2 dB higher at 1535 nm than at 1550 nm, the filter should transmit 2 dB more light at 1550 nm. The filter could be put before or after the amplifier, but the idea is the same—after passing through both, the output should be uniform across the amplifier's gain band, as shown at the bottom of Figure 15.6.

Equalization becomes more important the longer the system is. You can live with 0.5-dB gain differential in each of a series of five amplifiers, because that builds up to a modest 2.5 dB difference. But that's not acceptable in a submarine cable with 100 amplifiers, where the difference would become so high that the weaker channels would be lost.

Fixed and Tunable Filters

Some filters can be tuned to select different wavelengths.

The standard interference filters described above always transmit light in the same way as long as light strikes them at the same angle. Such fixed optical filters are fine for many applications, but tunable filters also are attractive for use in instruments or systems that require adjustment. A few different approaches are possible.

One simple approach is to tilt an interference filter, because the wavelength it selects depends on the angle at which light strikes it. Another is to use a prism or diffraction grating to spread out a spectrum and pick a narrow range of wavelengths from that spectrum. However, neither approach has proved able to meet the stringent requirements of dense wavelength-division multiplexing.

A more complex approach is to move an interference filter that is made so its optical characteristics vary along its length. This approach can meet high-resolution requirements, but requires special filters and mechanical movement of the filter.

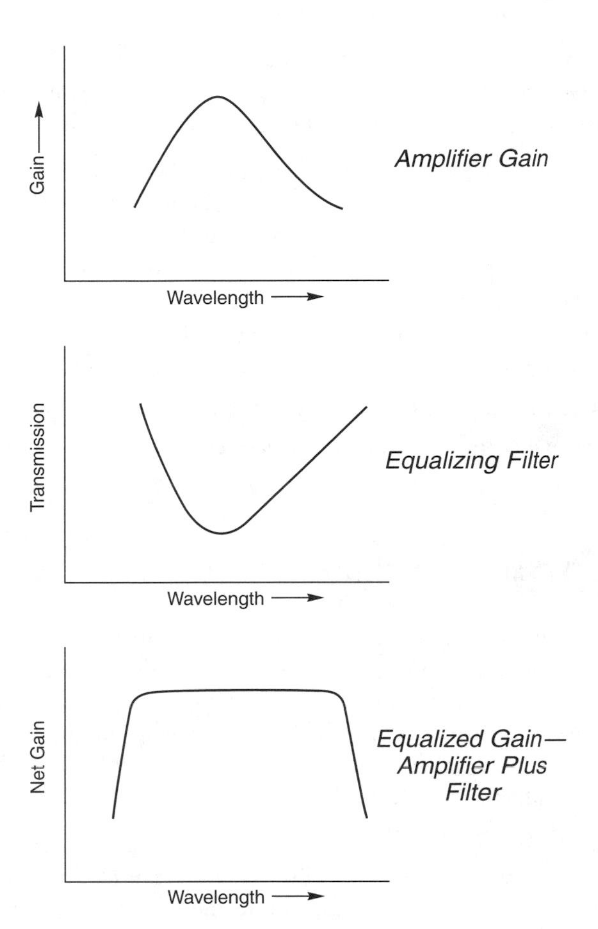

FIGURE 15.6
Equalizing filter balances uneven gain in a fiber amplifier.

A more common approach is the Fabry-Perot interferometer, which is essentially an optical cavity similar to those used as laser resonators, but without the laser medium inside. It consists of two partially transparent mirrors aligned parallel to each other, so light bounces back and forth between them. Normally air fills the space between them. Light bounces back and forth many times once it enters the cavity, so interference effects select wavelengths that resonate in the cavity. That is, an integral number *(N)* of wavelengths λ equals a round-trip distance in the cavity ($2L$):

A Fabry-Perot interferometer can be a tunable filter.

$$2L = \frac{N\lambda}{n}$$

where n is the refractive index of the material between the mirrors, a quantity needed to account for the difference between the wavelength in empty space and the material.

The cavity transmits light at wavelengths that match this resonant condition, like an interference filter. In fact, the Fabry-Perot interferometer is just a simple version of an interference filter, with a single cavity instead of a stack of layers. Normally the Fabry-Perot cavity is short, so the spacing between wavelengths is large. Adjusting the cavity length changes the wavelength selected, tuning the filter. You adjust the length either by moving the mirrors or by tilting them so light follows a longer path between the mirrors.

Acoustic waves can create density waves in glass; changing the sound frequency adjusts the wavelength selected.

Another common approach is the acousto-optic filter, where acoustic waves travel through a transparent material such as glass. The atomic vibrations produced by the acoustic waves create regions of higher and lower density within the glass. The denser regions have higher refractive index, creating a multilayer structure in the glass. As in an interference filter or fiber grating, these regular high-index zones selectively scatter light of certain wavelengths selected by the spacing between them. Tuning the acoustic frequency changes the grating spacing, and hence the selected wavelength—making a tunable filter.

The big advantage of tunable filters is obviously their tunability. Their disadvantages are much greater cost and complexity than fixed wavelength filters.

WDM Technologies

Several technologies are competing for WDM applications.

Interference filters and other technologies can be used to separate and combine wavelengths in WDM systems. Several approaches are now competing for WDM applications, creating a real technology "horse race." Some technologies appear to have advantages for certain types of WDM systems, but the field is still evolving, and no single approach dominates. Although these technologies work in different ways, they can achieve the common goal of optical multiplexing and demultiplexing.

You've already learned about one important WDM technology, interference filters, which selectively transmit certain wavelengths. Fiber Bragg gratings work similarly by reflecting specific wavelengths. WDM applications require the use of many interference filters or fiber gratings, with each one picking off an individual wavelength or group of wavelengths. The details differ between the two, as I will explain below.

Diffraction gratings spread out a spectrum of wavelengths that pass through them, so other optics can pick off individual wavelengths. Waveguide arrays also spread out a spectrum, but pick off individual wavelengths internally.

Fused-fiber couplers and Mach-Zehnder interferometers treat wavelengths yet another way, directing alternating optical channels out a pair of ports, effectively interleaving wavelengths. We'll see how this works below.

Interference Filters for WDM

Using interference filters for WDM requires taking light out of the fiber and passing it through a set of filters that sorts the light out by wavelength. Typically a lens collimates or focuses the light emerging from the input fiber, which then passes through one or more filters. When the demultiplexing is finished, separate lenses collect the separated optical channels and focus them into individual output fibers.

A narrow-line interference filter typically transmits a single optical channel while reflecting other wavelengths. Several interference filters can be cascaded to pick off a series of six wavelengths, as shown in Figure 15.7. The first filter transmits channel λ_1 while reflecting all other channels. The remaining channels hit the second filter, which transmits channel λ_2 while reflecting the four remaining channels. In this arrangement you need $n - 1$ filters to isolate n optical channels.

Cascaded interference filters can pick off one wavelength at a time for demultiplexing.

The concept is simple and straightforward, but interference filters are not perfect. Although they reflect *virtually* all the incident light at other wavelengths, some is lost, and these losses add up after a series of reflections. Picking off one wavelength at a time works fine for 8 channels, but the losses could grow excessive if you have 16 channels.

To prevent such losses, optical signals can be divided into groups of channels, which are then split up individually. Figure 15.8 shows such a system built from high- and low-pass filters plus 8-channel demultiplexers that pick off one channel at a time, as in Figure 15.7. In this scheme, incoming light first hits a high-pass filter, which reflects all light with wavelength less than λ_{17}. The shorter wavelengths are diverted to a low-pass filter, which reflects light with wavelengths longer than λ_9. Each of those sets of 8 channels is directed to an 8-channel demultiplexer. Wavelengths from λ_{17} to λ_{40} are routed to another low-pass filter, which reflects all light with wavelengths greater than λ_{24}. Channels λ_{17} to λ_{24} then go to an 8-channel demultiplexer, while the longer wavelengths pass are sent to another long-pass filter, which splits them into 8-channel groups for demultiplexing.

Filters can divide optical channels into groups, then separate the groups into individual channels.

This approach does not reduce the *total* number of filters needed, but it does reduce the number of filters any optical channel is going to encounter before reaching a receiver. The upper limit for the configuration shown in Figure 15.8 is 10.

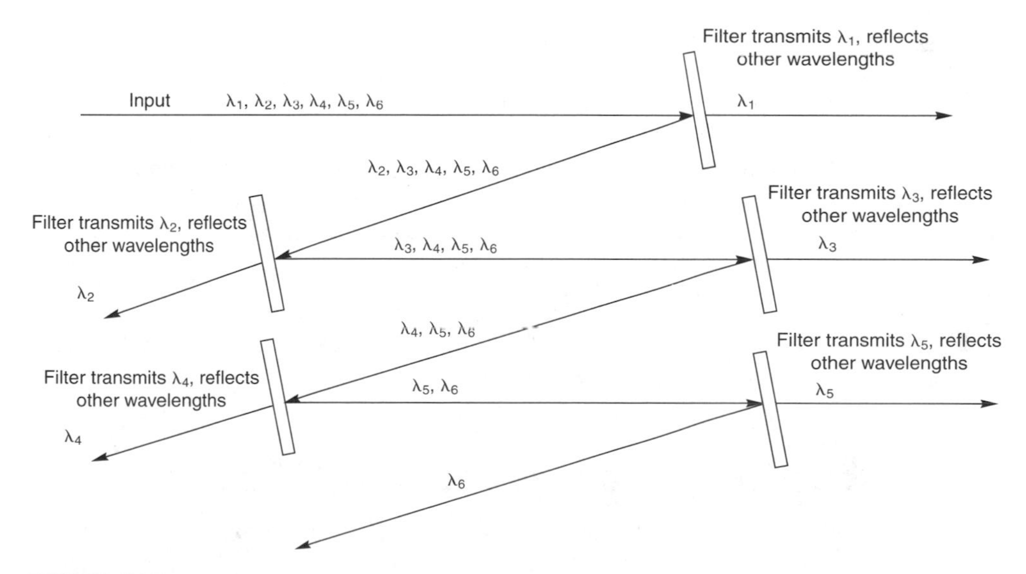

FIGURE 15.7
Interference filter WDM picks off one wavelength at a time.

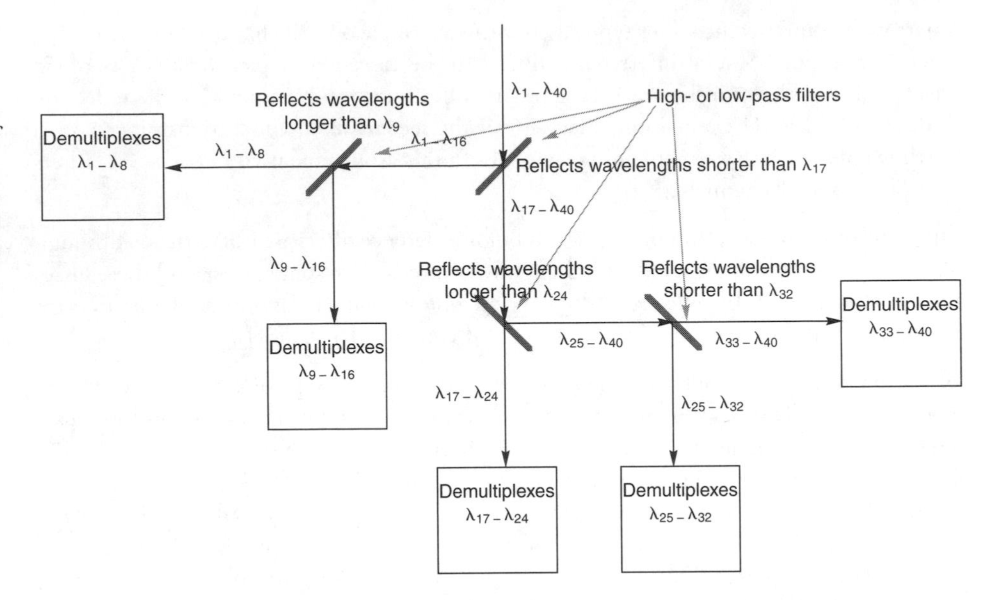

FIGURE 15.8 *Demultiplexing 40 channels by separating blocks of channels.*

The arrangement shown in Figure 15.8 has another important advantage. It's modular, so you don't need to start with all 40 channels. You could start with channels λ_{17} to λ_{24}, then add the high- and low-pass filters to split off other wavelengths as you needed to add the extra channels. This is the way telephone companies like to work, adding capacity only as they need it. They save money in the short term, and in the long term prices for WDM equipment are likely to decrease.

Filter WDMs can be upgraded modularly, but require about as many filters as the system has channels.

Interference filters are widely used for WDM, and it's worth reviewing their advantages. First, the underlying technology is well developed. Interference filters have been around for many years, although the extremely narrow-line filters used in DWDM systems were only developed very recently. Filters can be made very small—a few millimeters across—a good match for fiber-optic systems. They have good performance and can be assembled in modular units, so users can upgrade their systems several channels at a time, instead of jumping from 1 to 40 channels. On the down side, you need roughly as many filters as you have optical channels—adding to costs, complexity, and optical losses.

Remember also that interference filters do not always have to separate every single wavelength out of an optical signal. A single optical filter could transmit a single optical channel in an add-drop multiplexer, with the remaining channels reflected and collected for transmission through the rest of the system.

Fiber Bragg Gratings and Optical Circulators

Fiber gratings reflect the selected wavelength and transmit other wavelengths.

Fiber Bragg gratings can be grouped together in ways similar to interference filters, but they have some significant functional differences. Generally they reflect a single selected wavelength and transmit the rest, as shown earlier in Figure 7.5. Recall that interference filters generally *transmit* the selected wavelength.

Bragg gratings also are fibers, which is a mixed blessing. As fibers, Bragg gratings are easy to couple to other optical fibers. This makes it easy to get the transmitted light out of the Bragg grating, but the reflected light presents a problem. Tilting an interference filter directs the reflected light away from the source of the input light. However, light reflected inside the fiber grating is reflected straight back toward the input fiber. This reflected light has to be separated.

Separating that light is the job of a device inconspicuously called an *optical circulator* in Figure 7.5. Functionally it's a cousin of the optical isolator in the sense of being an optical one-way street, but it's considerably more complex inside, as you saw in Figure 14.11.

Fiber gratings must be used with optical circulators, which route signals in only one direction.

Figure 15.9 takes a closer look at the function of a typical optical circulator used with a fiber grating. It's a three-port device that allows light to travel in only one direction—from port 1 to port 2, then from port 2 to port 3. This means that any light reflected back from the fiber grating at port 2 is directed not back to port 1, but on to port 3. In Figure 15.9, the reflected light is the optical channel at λ_8. If there are more ports, the circulator must keep light going only in one direction, as if it was going around an optical traffic circle, as shown in Figure 15.9. The optical circulator shown in Figure 14.11 can be extended to serve a series of ports, although making a complete circle and delivering signals back to the first port is tricky.

In the usual implementation of an optical circulator for use with a fiber grating, port 1 is only an input, port 2 is both an input and output, and port 3 is only an output. Such iso-

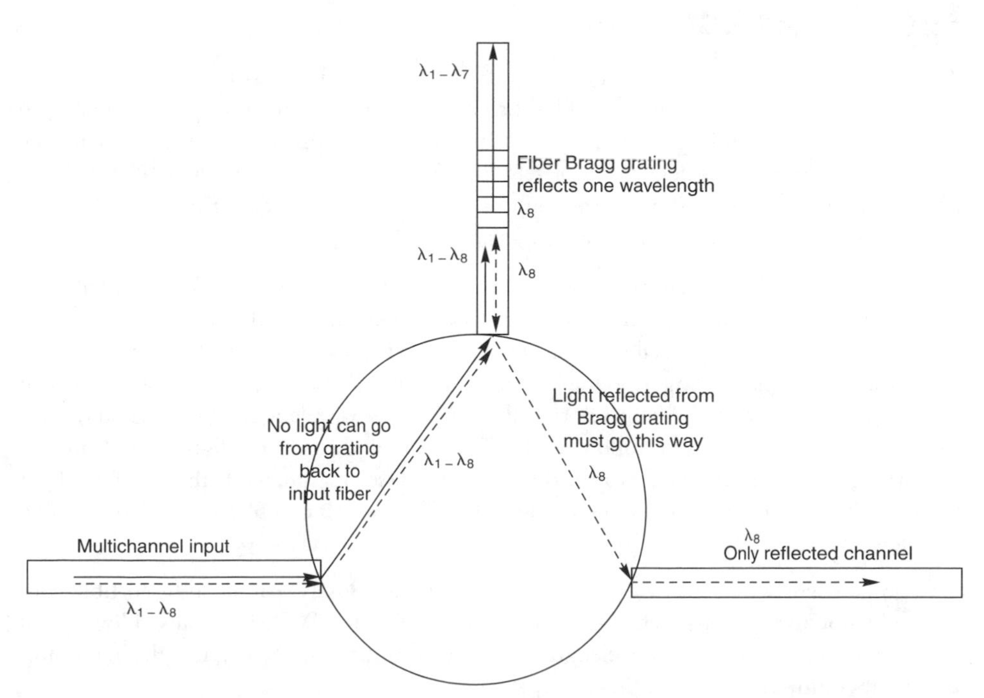

FIGURE 15.9
Optical circulator directs reflected light to next port, so no light goes backwards.

lators can be implemented in various ways. The design of Figure 14.11 is fairly complex but has low loss; other designs are possible, such as using optical isolators or directional couplers. In practice, the need for optical circulators is often considered a potential drawback of using fiber gratings.

Fiber gratings also can separate wavelengths in other ways. Two or more gratings can be fabricated in series along the fiber to reflect multiple wavelengths. For example, in Figure 15.9, a grating that reflected λ_4 could be added above the one reflecting λ_8. Signals at λ_4 would pass right through the grating reflecting λ_8, in both directions—when they come from the optical circulator, and after the grating reflects them back.

Earlier, Figure 7.6 showed how a fiber grating could serve as an add-drop multiplexer. The grating would reflect signals at the wavelength being dropped, while other optical channels would pass straight through. Because the fiber grating can reflect signals at the dropped channel in both directions, it can serve as the drop as well as the add, when a second optical circulator is used on the other end. In this case, the added signal is directed into the grating through the second optical circulator. The grating then reflects the added optical channel, adding it to the other optical channels it transmits, and the optical circulator directs the combined output of the fiber grating to the output port.

In principle, groups of fiber gratings could be assembled into arrays that served the same logical functions as the array of interference filters in Figure 15.7. However, the cost and complexity of optical circulators discourages such arrangements.

Fused-Fiber Couplers

Fused-fiber couplers can separate wavelengths by directing them out different ports.

The fused-fiber couplers described in Chapter 14 are inherently sensitive to wavelength. As in waveguide couplers, the amount of light transferred between the fused fibers depends on the length of the coupling region, as measured in wavelengths. Over some characteristic distance, the light is transferred completely from one output to the other. This distance is longer when measured in shorter wavelengths, because more of them fit into the same distance, opening a way to separate wavelengths.

You can use this effect to separate wavelengths, as shown in Figure 15.10. Light initially enters the top of the two fused fibers. Gradually, the light shifts to the bottom fiber. If the fused region is long enough, all the light transfers into the lower fiber, and the process starts over again, this time shifting from the bottom to the top. The degree of shifting depends on how many wavelengths the light has travelled, so shorter wavelengths shift back and forth first, with longer wavelengths following. In Figure 15.10, the 980-nm light shifts from top to bottom fiber and back to the top at the end of the fused region, while the 1550-nm light has shifted only from the top to the bottom fiber. This process completely separates the two wavelengths.

The process works best for two wavelengths that are not closely spaced, so fused-fiber couplers are not used for separating optical channels in dense-WDM systems. However, it works fine for widely spaced wavelengths, like the pump and signal wavelengths in erbium-doped fiber amplifiers used in this example.

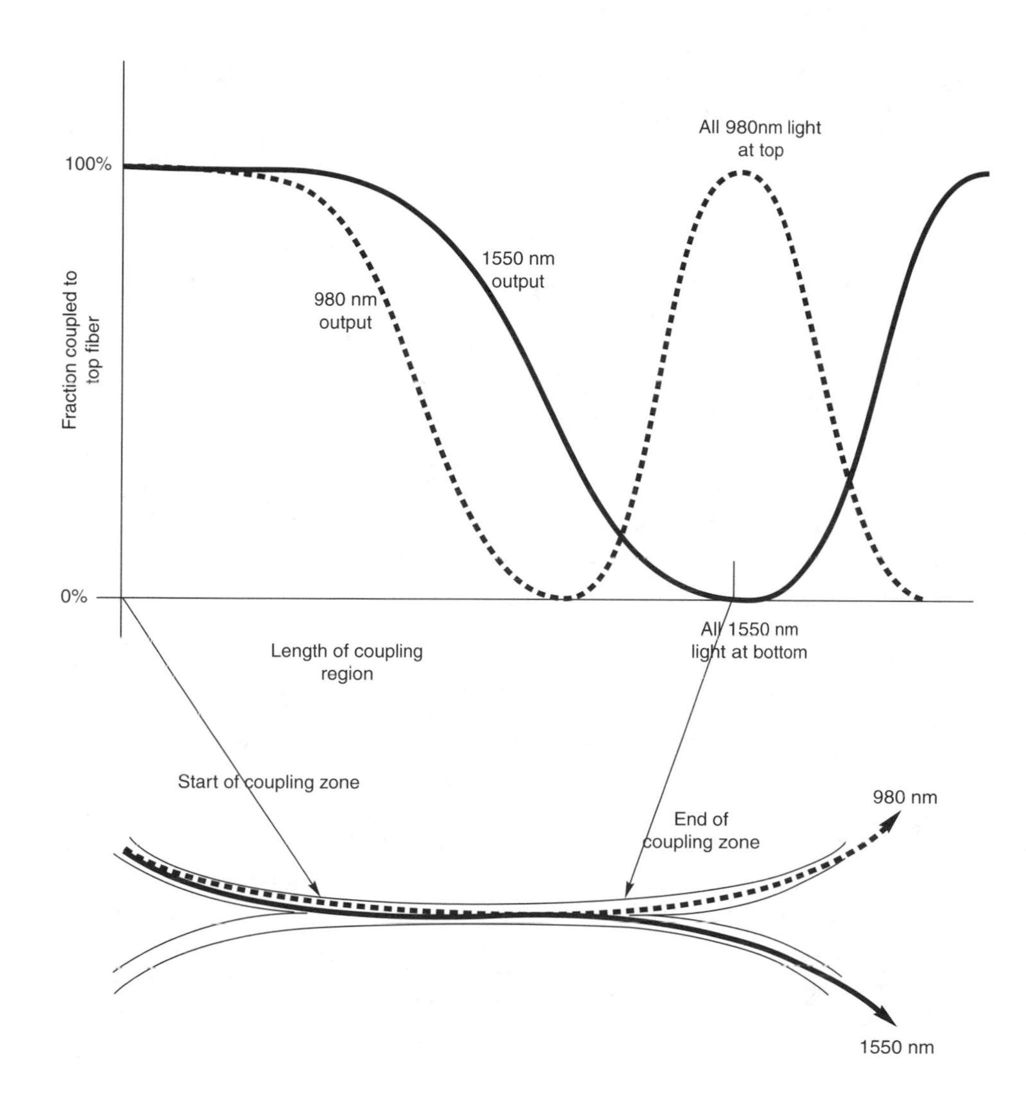

FIGURE 15.10
Fused-fiber coupler splits two wavelengths.

Mach-Zehnder Interferometers

Fused-fiber couplers are essential components in another type of device for dense wavelength-division multiplexing called a *fused-fiber Mach-Zehnder interferometer,* or more simply an *interleaver.* Unlike other multiplexers and demultiplexers, these devices split groups of evenly spaced optical channels into sets of odd and even channels, by using the interference of light in a fiber structure. To see how they work, we'll start with a look at the concept of a Mach-Zehnder interferometer, named after the physicists who invented it.

Mach-Zehnder interferometers interleave wavelengths, separating odd and even optical channels.

Interferometers pass light waves along two different paths, causing interference between the waves. As you learned earlier, coherent light waves can add or subtract their amplitude, producing constructive or destructive interference. In a Mach-Zehnder interferometer, a device called a *beamsplitter* splits an input beam into two parts, which pass along different routes, then are combined in a second beamsplitter. Figure 15.11 shows how this can be done with fused-fiber couplers serving as the beamsplitters. Input enters through one fused-fiber coupler, where it is divided between two fibers that form the arms. Light in the two arms recombines in a second fused-fiber coupler.

Incrementing frequency of an optical channel by $\Delta\nu$ shifts output between arms of the interleaver.

The relative phase of the light emerging from the interferometer arms determines its distribution between the two outputs of the output coupler. This phase depends on wavelength as well as the length of the arms. As the wavelength changes, the distribution of light between the two arms changes. In the case shown in Figure 15.11, at λ_1 all the light emerges from the top arm, but at λ_2 all the light emerges from the bottom arm. Every time the wavelength changes by that increment, the output light shifts between arms of the second fused-fiber coupler.

The increment in wavelength that shifts the output depends on the difference in effective length ΔL between the two interferometer arms. If we assume the two arms have the same refractive index *n,* this equals

$$\Delta L = \left(2n\left[\frac{1}{\lambda_1} - \frac{1}{\lambda_2}\right]\right)^{-1} = \frac{c}{2n\Delta\nu}$$

(The two arms could have different refractive indexes, which would complicate things.) Note that it's simpler to express the difference in terms of the change in *frequency,* $\Delta\nu$, rather than a change in wavelength. In fact, the increment in spacing of optical channels is uniform only if the spacing is measured in terms of a change in frequency $\Delta\nu$ rather than a change in wavelength. Thus the signals are at frequencies ν_1, $\nu_1 + \Delta\nu$, $\nu_1 + 2\Delta\nu$, $\nu_1 + 3\Delta\nu$, and so on. The difference between taking increments in terms of wavelength and in terms of

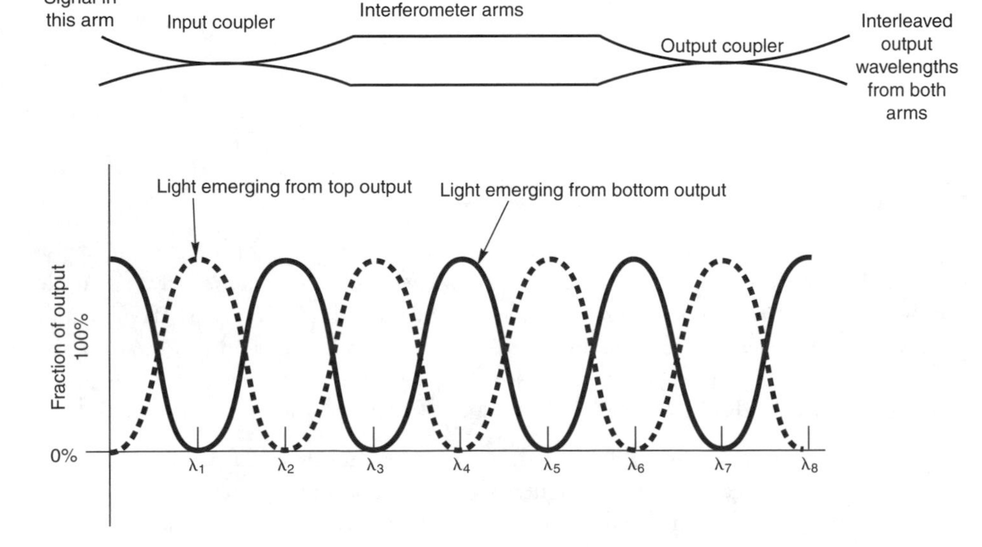

FIGURE 15.11 *Fused-fiber coupler Mach-Zehnder interferometer interleaves wavelengths.*

frequency is small, but can be significant if you're designing systems. I label optical channels by their wavelength for convenience, but remember that actual channel spacing is uniform in frequency, *not* wavelength.

Interleavers essentially split odd and even optical channels. Thus in Figure 15.11, signals at λ_1, λ_3, λ_5, and λ_7 emerge entirely from the top output, while signals at λ_2, λ_4, λ_6, and λ_8 emerge entirely from the bottom output. This makes a fused-fiber Mach-Zehnder interferometer an effective wavelength interleaver that can demultiplex a set of uniformly spaced optical channels. The interleaver also can work backwards, shuffling odd and even optical channels together.

A single interleaver does not completely demultiplex the signals. In the example of Figure 15.11, you still have λ_1, λ_3, λ_5, and λ_7 in the top output, and λ_2, λ_4, λ_6, and λ_8 in the bottom output.

Interleavers separate or combine signals in several stages.

Those channels also must be separated, but you can do that by repeating the process, as shown in Figure 15.12. In this case, the first interleaver has the finest resolution—100 GHz (about 0.8 nm) in the example. The next interleaver needs to split the remaining channels, so it splits channels twice as far apart—200 GHz. The final interleaving demultiplexer must split the two channels left in each output fiber, so it needs 400-GHz spacing (about 3.2 nm).

Note that Mach-Zehnder interferometers can be built with planar-waveguide technology as well as with fused-fiber couplers. So far, the main application of planar waveguides has been in the multi-arm arrayed waveguide devices described below, but two-arm Mach-Zehnder interleavers also can be made using planar waveguides.

As you can see, interleaving is a different process than picking one signal off at a time with an interference filter or fiber Bragg grating. Demultiplexing eight channels requires the same number of components—seven—whether you use interference filters or interleavers.

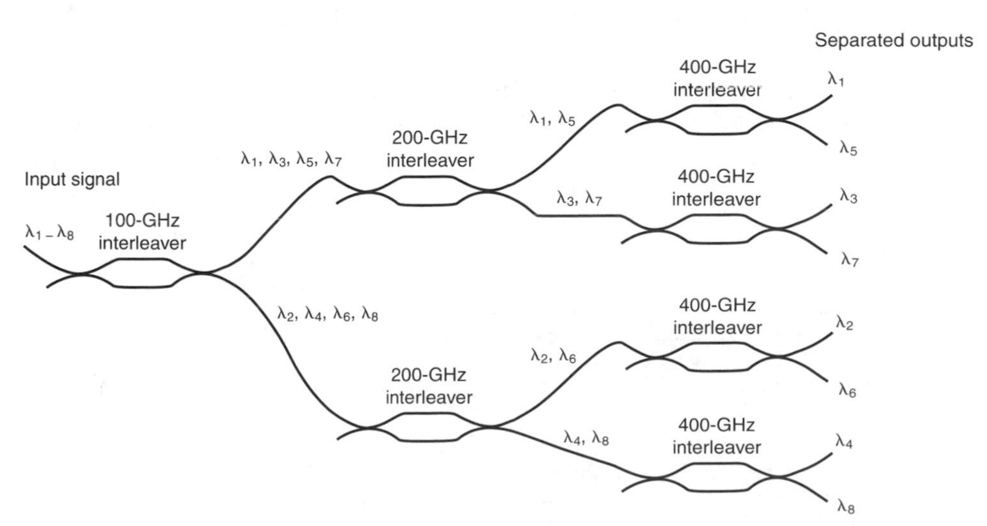

FIGURE 15.12
Interleavers demultiplex optical channels in stages.

However, with interleavers all signals pass through three components; with the interference filter demultiplexer shown in Figure 15.7, the final two channels have to pass through all seven filters, so they experience more loss than the other channels.

The differences between interleavers and other demultiplexers are important to remember. As you will see later, these differences are vital in designing hybrid WDM systems that take advantage of the strengths of two (or more) different approaches.

Bulk Diffraction Gratings

Diffraction gratings separate wavelengths by spreading out a spectrum.

I mentioned earlier that a diffraction grating—a series of parallel grooves or lines—diffracts light in a way that spreads out a spectrum. Interference between light waves scatters different wavelengths from the grating at different angles. The gory details of the optical physics aren't important here; what matters is that the wavelengths spread out, like a rainbow. You can see the same rainbow effect in a CD if you tilt it back and forth while looking at light reflected from it. The pits that store data on the CD are arranged in grooves that wind in a tight spiral around the disk, forming sets of parallel spots that act like a diffraction grating.

You can use the same effect to separate wavelengths, with suitable optics to focus the input light, collect the reflected light, and focus it into the output fibers.

Figure 15.13 shows a grating demultiplexer. This device uses a gradient-index (GRIN) rod lens in which the refractive index varies through a block of solid glass, producing the same focusing effect as a standard lens. (A GRIN lens is easier to align than a standard lens for this application.)

Input on three optical channels, λ_1, λ_2 and λ_3, enters through the bottom fiber. The GRIN rod focuses the input light onto a diffraction grating at the back end of the rod, which is set at an angle to reflect the light at the right angle. The grating diffracts each wavelength at a different angle, and the GRIN rod focuses each wavelength onto an output fiber. When everything is properly aligned (which, of course, is a big part of the job), λ_1 emerges from the top output fiber, λ_2 emerges from the middle output fiber, and λ_3 emerges from the lowest output fiber.

Such simple bulk diffraction grating demultiplexers work well for separating a few wavelengths that are widely spaced, but they don't give high channel isolation between closely spaced wavelengths. However, the way that diffraction gratings spread out a continuous spectrum of wavelengths is an advantage for measurement instruments. If you want to

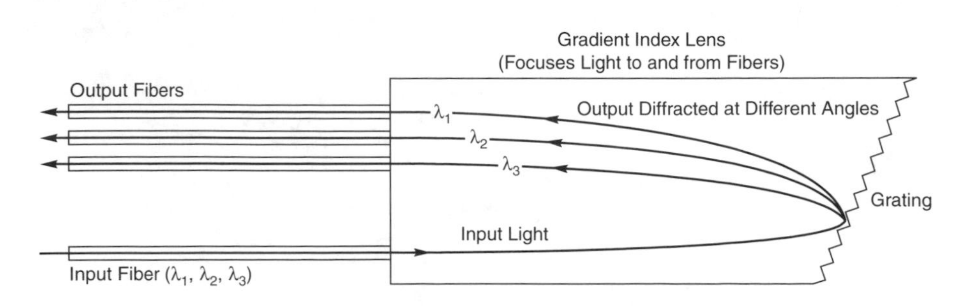

FIGURE 15.13
A grating coupler with GRIN rod separates three wavelengths.

measure the distribution of power as a function of wavelength, you usually want resolution finer than you need to look at a single channel. Scanning a continuous spectrum gives better resolution, and a diffraction grating makes this spectrum available. Thus measurement instruments are likely to use diffraction gratings to spread out the spectra that they measure. (The same is true for optical performance monitors, which measure the distribution of optical power across the spectrum in communication systems, to check that all channels are operating properly.)

Special diffraction gratings called *echelle gratings* offer higher resolution than ordinary gratings, which makes them potentially attractive for use in DWDM. That technology and other grating multiplexers are still in development.

Waveguide Array

Arrays of curved planar waveguides can separate many optical channels at once.

Another way to separate waveguides is to pass them through an array of planar waveguides, as shown in Figure 15.14. The multichannel input signal enters through the input coupler at lower left, which distributes signals to an array of several curved waveguides that run from the input coupler to an identical output coupler. Like the arms of a Mach-Zehnder interferometer, the waveguides differ in length, producing interference effects where their outputs mix in the output coupler.

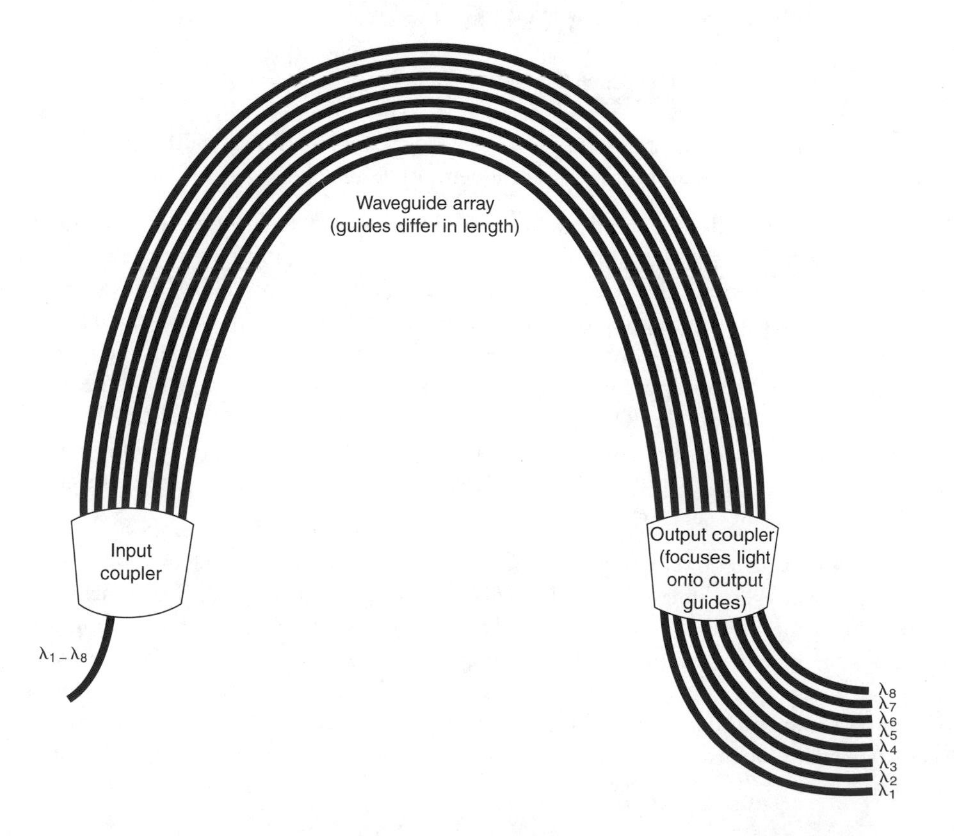

FIGURE 15.14
Array waveguide demultiplexer.

As in a Mach-Zehnder interferometer, interference effects cause each wavelength to emerge from the array at a different angle. In a sense, the array spreads out a spectrum like a diffraction grating. However, the design of the output coupler and the output waveguides focuses the light into the discrete output waveguides at lower right. There's one output guide for each optical channel. The wavelength spacing depends on the difference in optical path lengths among the curved waveguides, but you don't need to worry about those details.

What matters from a functional standpoint is that the waveguide array is a monolithic device that can separate many optical channels at once. Typical numbers are 16 to 40 outputs, with spacing at 50, 100, or (for low channel counts) 200 GHz. The input coupling loss is higher than many other optical demultiplexers at several decibels, but a single waveguide array (or *arrayed waveguide*) can separate up to 40 optical channels.

Although waveguide arrays are expensive, you only need one to separate up to 40 channels. This makes them much simpler for systems with high channel counts, where the cost *per channel* can be lower than when using larger numbers of demultiplexing components that cost less individually.

Building Multiplexers and Demultiplexers

Multiple technologies can be integrated in a single multiplexer or demultiplexer.

So far you've learned the general principles of WDM technologies. A few other considerations go into building actual multiplexers and demultiplexers.

You should realize that these technologies are building blocks. Actual systems may integrate two or more WDM technologies into a single multiplexer or demultiplexer. These hybrid WDMs can take advantage of the best features of the different technolgoies. For example, you could use a Mach-Zehnder interleaver as the first stage, to break up an 16-channel signal with 100-GHz spacing into two 8-channel signals with 200 GHz spacing. Then interference filters with 200-GHz resolution could break up the 8-channel signals. This could cut costs significantly because 200-GHz filters are much less expensive than 100-GHz filters. Similarly, two interleaving stages—one at 50 GHz, the other at 100 GHz—could break down a 32-channel signal with 50 GHz spacing to four separate 8-channel signals with 200-GHz channel spacing.

In talking about general principles, it's easiest to assume that channels are uniformly spaced, and that every slot is filled. This does not happen in general. The spacing of channel slots generally is uniform, but the channels may be grouped into blocks, and not all of the slots may be filled. For example, one or two channel slots may be left between 8-channel blocks in a system like the one shown in Figure 15.8, if the band-pass filters do not have sharp enough cutoffs. In practice, a gap of about 5 nm normally is left between the erbium-fiber amplifier C-band at 1530 to 1565 nm and the L-band at 1570 to 1620 nm.

The ability to add more channels at reasonable cost is a major practical concern. Telecommunications carriers want room to increase their transmission capacity, but they don't want to pay for large amounts of equipment they can't use immediately. For example, they might like to build the 40-channel system shown in Figure 15.8 in increments of eight channels at a time. This has to be traded off against the possibly lower overall cost of installing a larger system all at once.

Finally, remember that every channel slot does not have to be used immediately. A prudent company may initially use only two slots in a 16-channel system, waiting to buy more expensive transmitters and receivers until they're needed. Like installing spare fibers in cables, the extra capacity costs little until it's needed. It's important to keep this in mind when you hear about systems with large numbers of channel slots; in general, all the channels are not likely to be populated except on the busiest routes. As when you install new electrical wiring in a house, you want to have capacity not just for the equipment you already own, but for what you expect to add in coming years.

What Have You Learned?

1. WDM optics combine optical channels at the input end of a system and separate them after transmission through a fiber. Multiplexers combine channels; demultiplexers separate them.
2. An add-drop multiplexer goes in the middle of a system. It can both drop existing channels at an intermediate point and add new channels to a fiber carrying WDM signals. The added signal can replace the dropped channel.
3. Channel density depends on spacing between channels. Standard spacings for dense-WDM spacings are 200, 100, and 50 GHz. Channel spacing may be as wide as 25 nm for "wide" WDM.
4. Wavelength routers direct different wavelengths to different points.
5. Separating pump wavelengths from the outputs of fiber amplifiers also requires wavelength-division multiplexing.
6. WDM filters transmit selected wavelengths and reflect others.
7. An interference filter uses multiple thin layers to selectively transmit a narrow range of wavelengths; others are reflected. Interference filters can select a very narrow range of wavelengths.
8. Cutoff filters make a sharp transmission between transmitting and reflecting at a certain wavelength.
9. Equalizing filters compensate for the unequal gain of optical amplifiers across their operating ranges.
10. Most filters have fixed wavelength response, but some can be tuned to transmit different wavelengths. They include acousto-optic filters and Fabry-Perot interferometers.

11. Interference filters can select closely spaced optical channels. A cascaded series of interference filters can pick off one wavelength at a time to demultiplex optical channels. Each filter transmits one channel and reflects the rest.
12. Interference filters can select groups of optical channels as well as individual channels.
13. Fiber Bragg gratings reflect the selected wavelength and transmit other wavelengths. They must be used together with optical circulators for demultiplexing, but have high resolution in selecting optical channels.
14. Fused fiber couplers can separate wavelengths by directing them out different ports, but their resolution is limited.
15. Mach-Zehnder interferometers interleave wavelengths, directing alternating channels out of each of two inputs. They are sometimes called interleavers, and can separate closely spaced optical channels. Interleavers separate channels in a series of stages.
16. Bulk diffraction gratings separate wavelengths by spreading out a spectrum. They are often used for measurement instruments.
17. A single waveguide array can separate 16 to 40 closely spaced optical channels.

What's Next

Chapter 16 will cover optical switches, optical modulators, and other active elements in optical networks.

Further Reading

M. S. Borella, J. P. Jue, D. Banerjee, B. Ramamurthy, and B. Mukherjee, "Optical Components for WDM Lightwave Networks," *Proceedings of the IEEE,* Vol. 85, No. 8, pp. 1274–1307, August 1997.

Kenneth A. McGreer, "Arrayed waveguide gratings for wavelength routing," *IEEE Communications Magazine,* Vol. 36, No. 12, December 1998, pp. 62–68.

J. J. Pan and Y. Shi, "Combining gratings and filters reduces WDM channel spacing," *Laser Focus World,* September 1998.

Questions to Think About for Chapter 15

1. An erbium-doped fiber amplifier can transmit signals at wavelengths between 1530 and 1565 nm. How many optical channels can you fit in this range with 200 GHz spacing? How many channels with 100 GHz spacing?

2. A transatlantic fiber-optic cable contains 100 optical amplifiers. It needs equalizing filters to balance the gain of the erbium-fiber amplifiers across their operating ranges. If the receivers used on the system have a dynamic range of 20 dB, how closely do the equalizing filters have to balance gain? Assume all filters and amplifiers are identical.
3. A typical interference filter for demultiplexing 100-GHz channels has 0.5-dB loss on the reflected channels and 2.0-dB loss on the transmitted channels. How much loss does the signal suffer on the first channel picked off (λ_1) in Figure 15.7? What is the loss for the last channel of eight channels picked off (λ_8) in a similar arrangement? What channel in an 8-channel system suffers the highest total loss?
4. A typical fiber Bragg grating has 99.9% reflection (0.0043-dB loss) and 0.2-dB loss for transmitted wavelengths. Assume loss of 1 dB in the optical circulator. What is the loss for the first channel of eight channels picked off in a cascaded series of fiber Bragg gratings? What are losses for the seventh and eighth channels?
5. What should the difference in path lengths be in a Mach-Zehnder interferometer designed to interleave optical channels separated by 50 GHz? By 200 GHz? Assume the refractive index of the material is 1.5 and is uniform for both arms.
6. You want to separate 16 optical channels that are uniformly spaced 50 GHz apart with optical interleavers. How many interleavers do you need? How many interleavers does each optical channel pass through?
7. A 40-channel arrayed waveguide demultiplexer has average loss of 8 dB for each channel processed. How does this compare to the highest loss of the 40-channel interference-filter demultiplexer shown in Figure 15.8? You can use the results from Question 3 to give you the loss for the 8-channel demultiplexing boxes. What's the minimum loss?

Quiz for Chapter 15

1. What is the broadest channel spacing that is considered "dense" WDM?
   a. 400 GHz
   b. 200 GHz
   c. 100 GHz
   d. 50 GHz
   e. 0.8 nm
2. What does an add-drop multiplexer do?
   a. Converts all optical signals in a fiber to electronic form
   b. Amplifies optical signals after attenuation has reduced signal strength below 1 μW per optical channel
   c. Adds and drops optical channels at intermediate locations without interfering with other signals on the fiber
   d. Adds and drops optical channels at an intermediate point while regenerating other channels on the fiber

e. Switches signals between different wavelengths at an intermediate point in the system

3. What selects the wavelengths transmitted by an interference filter?

a. The thickness and composition of layers deposited on glass

b. The composition of the glass plate on which it is deposited

c. Coloring dyes added to the layers in the interference filter

d. Parallel ridges formed in the uppermost layer

e. Only the refractive index of the surface layer

4. You want to reflect light at wavelengths longer than 1567 nm and transmit light at shorter wavelengths. What type of filter do you want?

a. Color filter

b. Cutoff filter

c. Band-pass filter

d. Line filter

e. Attenuation filter

5. What type of filter is tunable in wavelength?

a. Interference filter

b. Cutoff filter

c. Band-pass filter

d. Fabry-Perot interferometer

e. Line filter

6. An acousto-optic filter is

a. a type of cutoff filter.

b. a tunable filter.

c. an interference filter.

d. a color filter.

e. impossible to build.

7. Interference filters

a. reflect the selected wavelength and absorb other wavelengths.

b. reflect the selected wavelength and transmit other light.

c. transmit the selected wavelength and reflect other wavelengths.

d. transmit the selected wavelength and absorb other light.

8. What type of WDM system requires an optical circulator?

a. Interference filters

b. Fiber Bragg gratings

c. Mach-Zehnder interferometers or interleavers

d. Bulk diffraction gratings

e. Tunable optical filters

9. Fiber Bragg gratings

a. reflect the selected wavelengths and absorb other wavelengths.

b. reflect the selected wavelength and transmit other light.

c. transmit the selected wavelength and reflect other wavelengths.

d. transmit the selected wavelength and absorb other light.

10. What type of technology is used in an interleaver?

a. Cutoff filter

b. Interference filter

c. Fiber Bragg grating

d. Mach-Zehnder interferometer

e. Fabry-Perot interferometer

11. How many interference filters do you need to make an 8-channel demultiplexer that picks off one channel at a time?

a. 4

b. 6

c. 7

d. 8

e. 9

12. How many fiber gratings do you need to make an 8-channel demultiplexer that picks off one channel at a time?

a. 4

b. 6

c. 7

d. 8

e. 9

CHAPTER 16

Optical Switches, Modulators, and Other Active Components

About This Chapter

The evolution of fiber-optic systems into optical networks makes the ability to modify optical signals increasingly important. Signals may be modulated in intensity or other properties, switched between fibers, converted from one wavelength to another, or modified in other ways. The devices that do this are called *active components,* because they act upon the signal.

You already have learned about two important families of active components, light sources and optical amplifiers, covered in Chapters 9 and 12, respectively. In this chapter you will learn about optical switches, modulators, and wavelength converters, which modify the light signals from these sources.

Defining Active Components

A bit of explanation can clarify the useful distinction between active and passive components. The development of new components and concepts has shifted the definitions over time, as is typical in fast-moving fields. The original definition of active component was something that required external power to act upon a signal. In contrast, *passive components* drew no power from outside. By this definition, a laser or optical amplifier is obviously an active component, and an optical fiber or attenuator is obviously passive.

Active components do something that changes a signal; typically they draw external power.

Another way to separate active and passive components is to consider how they act on a signal. In this view, a passive component always performs the same operation on a signal, such as splitting it into constant proportions, or attenuating it by a fixed amount. In contrast, an active device can modify its effect on the signal—for example, modulating its intensity or switching it between fibers. Thus, active devices control the actions of the system. This definition has the advantage of focusing on how the device affects the signal, but has its own limitations. If strictly applied, it would consider a laser as a passive device if it generated a steady beam that was modulated externally, because the laser does not change the signal. Note also that an active component does not have to actually change the signal; like an emergency switch that can divert an optical signal in case of failure, it may be a *potentially* active component.

Modulators and switches are active components.

In practice, it's best to regard active components as those that either draw power from outside or can modify the signal in changing ways. The most important active components in current optical systems—optical switches and modulators—do both. Some outside power is needed to change the signal intensity or redirect the light, and both modulation and switching change over time.

Some haziness in terms is inevitable because signals can be changed in different ways. Recall that Chapter 14 mentioned active couplers, which combine a receiver with a pair of transmitters to split signals between two outputs. The function is the same as a passive coupler, but it's implemented with active components. Likewise, photochromic sunglasses modulate light transmission by turning dark in intense sunlight, although this effect is much too slow for use in communications.

It's best not to worry too much about drawing the finest possible line between passive and active components. As new technologies emerge, the distinction is likely to continue evolving. The important part is to remember that, in general, active devices do something that changes a signal in fundamental ways and passive devices merely attenuate, combine, or separate signals.

Modulators and Modulation

Light must be modulated to transmit a signal. As you learned in Chapters 9 and 10, the simplest way to modulate light intensity is internally, by changing the drive current passing through an LED or laser light source. Unfortunately, this approach runs into a number of limitations as speeds increase above about 1 Gbit/s. The maximum data rate, average output power, and the extinction ratio between "on" and "off" power all are limited. In addition, direct modulation can distort analog signals and shift the output wavelength during the pulse, an effect called "chirp," which you learned about earlier. Finally, laser output power and wavelength are easiest to stabilize if the laser emits light steadily.

External modulation is attractive for high-performance systems.

These limitations are most serious in high-performance systems. Output power and extinction ratio limit receiver performance and spacing between optical amplifiers. Modulation speed limits the transmission capacity. Nonlinearities can distort analog signals, important in cable-television systems. Wavelength chirp increases the dispersion penalty. In short, if you want high performance, you generate a steady laser beam and modulate it with an external device called a *modulator.*

An optical modulator changes how much light it transmits in response to an external control signal. Many types have been developed for other applications, but fiber-optic systems are particularly demanding because they require modulation at gigabit rates, much faster than most modulation mechanisms. For example, liquid crystal devices cannot respond fast enough for fiber-optic modulators, but are fine for laptop computer displays, which operate much slower.

Modern fiber-optic systems use two main families of modulators. Electro-optic modulators rely on changes in the way certain planar waveguides carry light. Electro-absorption modulators are semiconductor diodes that in their internal structure resemble lasers, but are switched between states that transmit and absorb light. We will look at them separately.

Electro-Optic Modulators

Electro-optic modulation depends on what is called the *electro-optic effect,* a change in the refractive index of certain materials when an electric field is applied to them. The change affects light passing through the material virtually instantaneously. The velocity of light in a material is the speed of light in a vacuum divided by the refractive index, so increasing the refractive index slows down the light; reducing it speeds up the light. The change is proportional to the voltage applied to the material.

Electro-optic modulators rely on changes in refractive index caused by an electric field.

When you look at a waveguide, you measure the effect of this change in refractive index as a shift in the phase of the light waves compared to what the phase would have been without the applied voltage. A shift of a half a wavelength—180°—would leave the shifted light completely out of phase with the unshifted wave. The phase shift normally is measured by this comparison:

$$\text{Phase shift } (\Delta\Phi) = 180° \times \frac{V}{V_{180}}$$

where V is the voltage applied to the modulator and V_{180} is the voltage needed to shift the phase a half-wavelength, or 180°.

Merely delaying the light modulates its phase but not its intensity. To modulate the intensity, an electro-optic modulator splits the input light equally between a pair of parallel waveguides. In the example shown in Figure 16.1, a modulated voltage is applied to one waveguide, but not to the other. This modulates intensity of the light where the two waveguides merge at the right. If the waveguides are equal lengths and the voltage is zero, the light waves are in phase when they combine, so the waves add constructively, producing a signal. The light is "on." However, if you apply the voltage needed to delay the signal by 180°, the light in the two waveguides is out of phase when they merge. The two waves interfere destructively, canceling each other out, and the output intensity is nominally zero.

Delaying the light phase causes interference effects that modulate the output intensity.

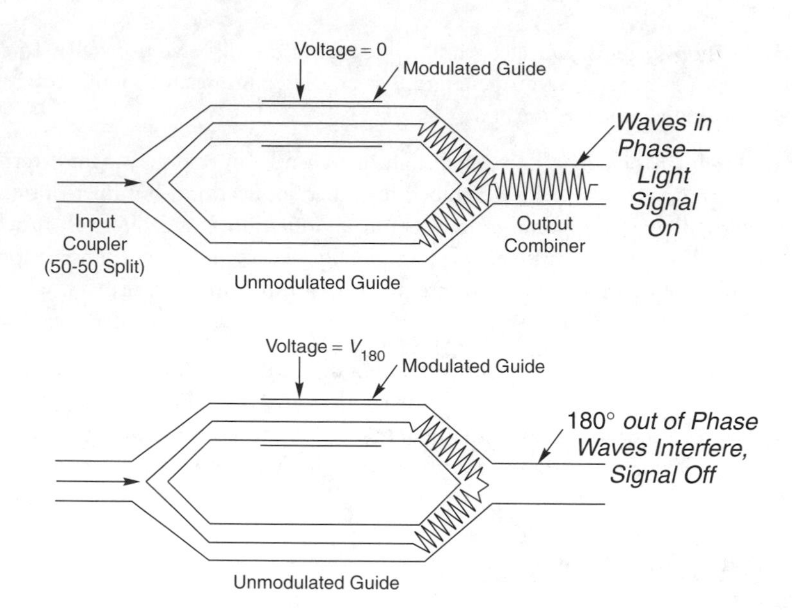

FIGURE 16.1 *Simple electro-optic waveguide modulator.*

In practice, a little light remains when the signal is nominally zero, a quantity measured by the extinction ratio, which compares the output power P_{on} in the on state with that in the off state P_{off}.

$$\text{Extinction ratio (dB)} = -10 \log \left(\frac{P_{off}}{P_{on}} \right)$$

The same approach works for analog modulation, but in this case you adjust the voltage so the delay varies continuously between 0° and 180°. The result is a continuous variation in output intensity shown in Figure 16.2. (Note that the peak power is lower than the laser output by an amount that equals the insertion loss of the modulator, even when there is no voltage applied.)

Actual electro-optic modulators are more complex. Often voltages are applied across *both* waveguides, but with the opposite polarities, so the voltage delays the phase of one wave while speeding the phase of the other. In this case, a voltage of $+V_{180}/2$ is applied to one waveguide, and $-V_{180}/2$ is applied to the other, giving the same modulation with lower voltage. Typically the voltage signal applied to each channel is the sum of two signals, one a bias that sets the operating level, the other the modulating signal. For example, the bias may set the modulator to normally transmit a certain average power, with the variations in the modulation voltage changing the transmitted power above and below that level.

A further complication is that refractive index can vary with the polarization of light. In glass and many other materials, the refractive index is nearly identical for light of different polarizations, but in other materials it varies significantly with the orientation of the polarization relative to the crystal axes. Materials in which the refractive index differs significantly for vertically and horizontally polarized light are called *birefringent.*

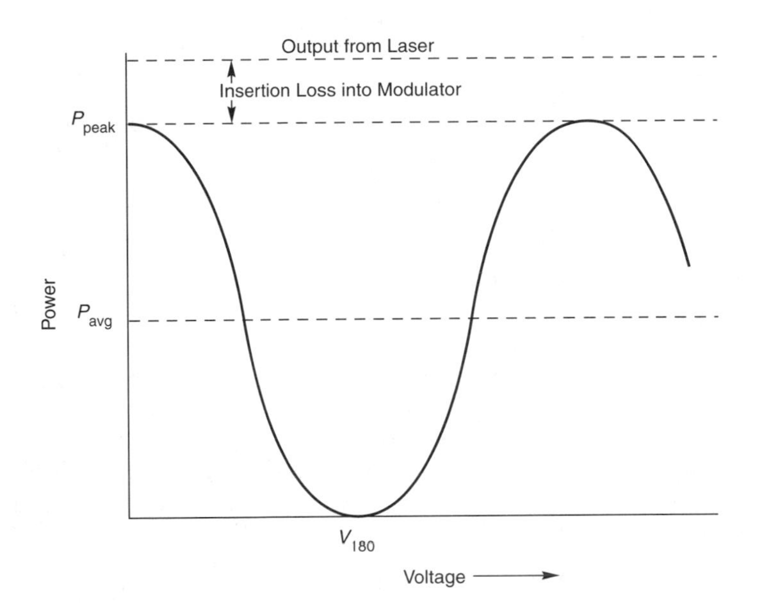

FIGURE 16.2
Variation in modulator output with voltage.

Modulators are affected because the magnitude of the electro-optic effect also depends on polarization, so an applied voltage does not change the refractive index the same amount for vertically and horizontally polarized light. Thus an electric field that delays vertically polarized light by 180° may delay horizontally polarized light only 120°, so interference would not cancel out the horizontally polarized component of the light. One way to avoid this problem is by using a polarizing filter to block the undesired polarization before the light reaches an electro-optic modulator, so only one polarization is transmitted.

In theory, electro-optic modulators can be made of any material that displays the electro-optic effect and is transparent at the signal wavelength. In practice, the only material in wide use at 1.3 and 1.55 μm is lithium niobate ($LiNbO_3$). Waveguides are made by diffusing titanium or hydrogen into the lithium niobate, raising the refractive index of a narrow stripe that forms a waveguide. One process raises the refractive index for one polarization but depresses it for the other, so only one polarization stays in the guide, while the other diffuses into the substrate. The goal is to eliminate the need to polarize light before sending it through the modulator.

Electro-optic modulators are made of lithium niobate.

Lithium niobate modulators are widely used today. The technology is well developed and they can be modulated at rates to 40 Gbit/s for digital transmission. As waveguide devices, they can be integrated with some other optical devices. However, they cannot be integrated with light sources because lithium niobate is not a light emitter.

An electro-absorption modulator is a semiconductor diode that is reverse-biased so modulation makes it absorb rather than emit light.

Electro-Absorption Semiconductor Modulators

An electro-absorption semiconductor modulator is a waveguide device based on different principles than the electro-optic modulator. The electro-absorption modulator has a structure similar to that of an edge-emitting semiconductor laser, and the two can be integrated

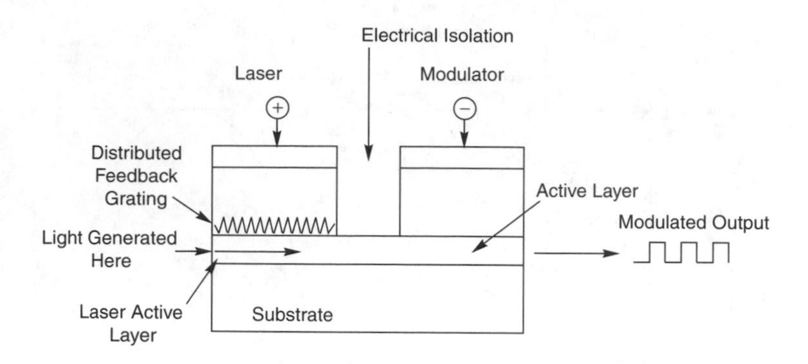

FIGURE 16.3 *An electro-absorption modulator integrated with a semiconductor laser.*

on the same chip, as shown in Figure 16.3. In this arrangement, the laser and modulator share an active layer, so light generated within the laser stripe is coupled directly to the modulator waveguide.

Despite their common structure, an electro-absorption modulator operates in a quite different mode than a semiconductor laser. The laser is forward-biased so current flows through it, causing current carriers to recombine and generate light. The modulator is operated with a reverse bias, like a *pin* photodetector. When the modulator is unbiased, no current flows and it is transparent to light at the laser wavelength. However, when the bias voltage is applied, the laser light can produce electron-hole pairs that are pulled in opposite directions by the bias voltage, causing a net absorption at the laser wavelength. Increasing the bias increases the absorption, blocking the beam.

The laser and modulator are electrically isolated from each other. A steady current drives the laser, so it generates a steady optical output. The input signal drives the modulator. For zero applied voltage, the optical output is at its highest level. Applying a higher voltage to the modulator increases light absorption. (Note that this means that high voltages generate no light output.)

Although the laser and modulator sections of the integrated structure have similar structures, they are not identical. The structure of the active layer also differs between the two. The laser may include a distributed-feedback grating in the cavity or a distributed Bragg grating in the waveguide. Thicknesses of the active layers differ, as does the doping that differentiates between the laser and modulator sections. Nonetheless, the two devices can be fabricated on the same substrate, forming a single, integrated light source and modulator.

Like electro-optic modulators, electro-absorption modulators are polarization sensitive, although integrating them on the laser chip simplifies packaging. They are made from InGaAsP semiconductors, so they can readily match laser wavelengths.

Variable Attenuators

Other types of modulators are available, and are used for other optical applications. However, they generally cannot change their light transmission fast enough to modulate signals at gigabit speeds. These devices essentially serve as variable attenuators or slow off–on modulators (which are functionally equivalent to off–on switches). Some use the same principles as the optical switches described below.

Switching in Optical Networks

Earlier you learned that the telecommunication network can be seen as an array of pipes and switches. The switches are the key difference between the old fiber-optic systems that merely piped signals from point to point, and the emerging optical network. Optical switches allow an optical network to process and direct light signals, as well as pipe them from place to place. This greatly enhances the functionality of an optical network, and makes optical switches very important.

Optical switches allow an optical network to process signals optically.

Switches serve various functions in telecommunications, and you should understand a bit about these applications before looking at specific switches. The most familiar type of switching is directing signals from point to point. The telephone network and the Internet do this in different ways, but the end user does not see a big difference without looking closely. Users merely know that they dial a phone number or enter an Internet address.

Other switching functions are less obvious to users. Telecommunications companies install switches to route signals around failed equipment, so a single failure won't knock out an entire telecommunication system. This function is called *protection switching,* and it relies on having backup routes. One common approach is to connect several cities in a ring that has extra capacity reserved. A cable break triggers the protection switch to divert signals that would have gone through the broken fibers to pass through the reserve fibers, as shown in Figure 16.4.

Telecommunication companies are making increasing use of switches to change the services they provide to customers, a process called *provisioning.* Traditionally, carriers sent technicians to physically connect lines to customers, but now that the network is changing faster, it is more economical to install switches and make the changes remotely. A similar function is dynamically changing the transmission capacity of parts of the network to meet special needs. For example, a carrier might reconfigure the network around a sports stadium for a World Series or Superbowl. Optical switches are rarely used for *circuit switching,* making temporary connections such as those needed for a telephone call, because fibers handle much larger blocks of information.

Provisioning changes the services delivered over telecommunication lines.

Optical switching is still a young technology, and both hardware and applications are evolving. Most nonprotection switching is still electronic, but optical switching is coming, because it can manage higher-capacity transmission. Let's take a more careful look at the functional requirements for various types of switches.

Protection Switching

Protection switching is a simple but vital function. In case of a cable break or equipment failure, the network must redirect signals along a different path that will bypass the failure.

Protection switching sends signals through a backup fiber.

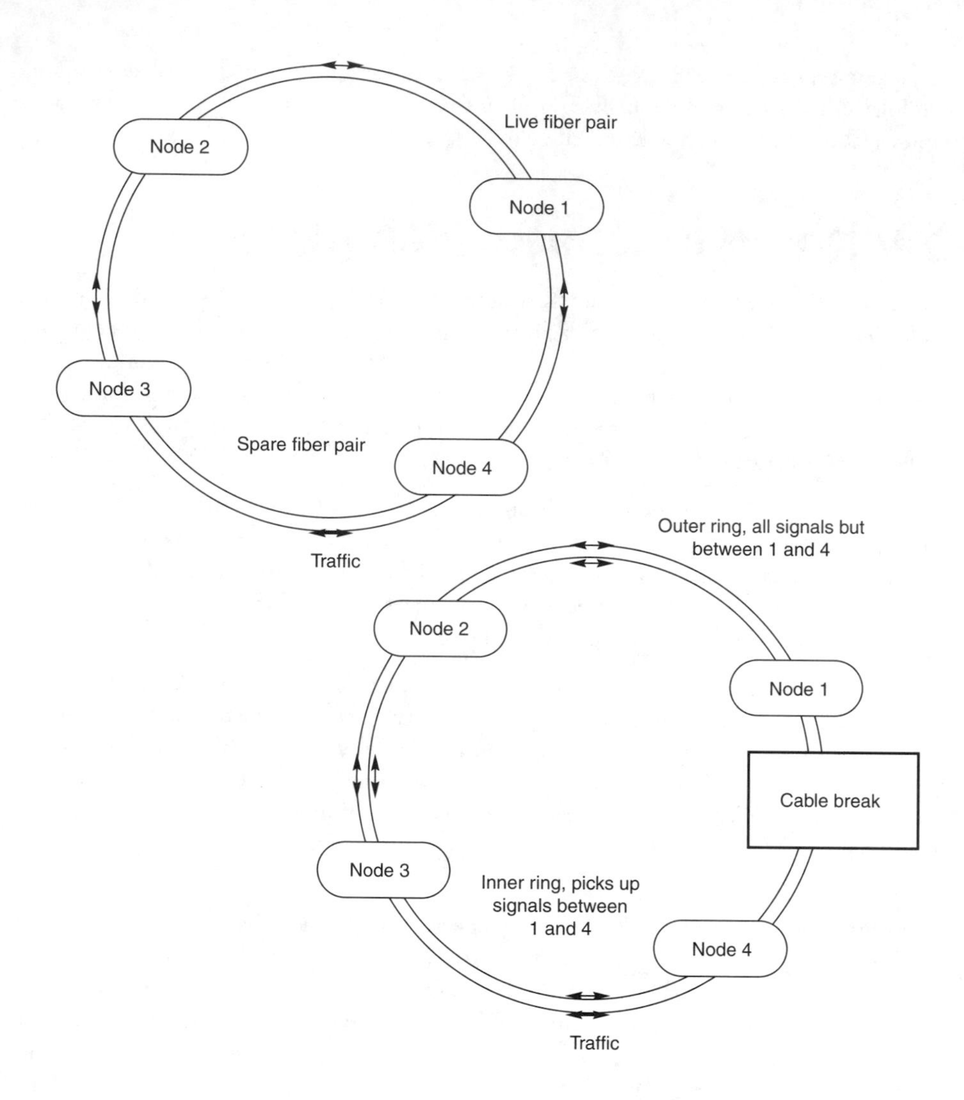

FIGURE 16.4
Protection switching. In case of a fiber break, switches at nodes 1 and 4 redirect traffic between those nodes over the spare fiber pair.

Ideally this should be done automatically in a small fraction of a second, so no telephone connections over the broken link are likely to be dropped. Some data will be lost, but protocols exist to retransmit the data.

Signals can be switched to a backup route at the transmitter output or at a node somewhere along the system. The equipment required is relatively simple. Something must detect the failure and command the switch to divert its signals along the backup fiber. Standard protection switches have two possible outputs, the main fiber and the backup. Their job is to sit and wait. Technicians may test them, but they're not used regularly and repeatedly. If there is a failure, the switch typically will be reset after repairs. This technology is common, and is part of an important telephone-industry standard called SONET, which you will learn about in Chapter 20.

Remote Provisioning and Reconfiguration

Provisioning is the changing of network configuration to alter the services delivered to customers, or to provide new services. Traditionally it has been done manually, but now there is much interest in *remote provisioning.* This is a robotic equivalent of sending a technician to a remote site to rearrange cables. The goal is to save money and help telecommunications companies manage their networks more efficiently. Remote provisioning is like going to your basement to set switches when you want to move a phone in your home. That analogy shows both the appeal and the difficulty—switching in new phone lines would be easier than stringing new wires, but you would have to install extra equipment to make it work.

Provisioning changes network configurations to deliver new services.

Provisioning schedules are comparatively leisurely. If you want a new phone line, it doesn't have to be switched on in seconds. However, the operation usually is more complex than protection switching. Remote provisioning is still rare, but it's being designed into new equipment.

Cross-Connects and Circuit Switching

Directing signals among many possible users is a more complex task than protection switching. Figure 16.5 shows the basic idea. Signals must be directed from any of N possible inputs to any of M possible outputs. Switches that perform this task are called *cross-connects* or *switching fabrics.* They perform the same function as the old-fashioned telephone operator who sat at a switchboard plugging pairs of wires into sockets that led to different telephone lines. Monstrous banks of electro-magnetic switches did the same task a generation ago; now special-purpose electronic computers lie at the heart of telephone switching offices.

Cross-connects make connections among multiple inputs and outputs.

Optical cross-connects are just beginning to appear in the core of the telecommunication network, where they can transfer high-speed optical signals among input and output fibers. So far, most optical cross-connects can handle only a limited number of inputs and

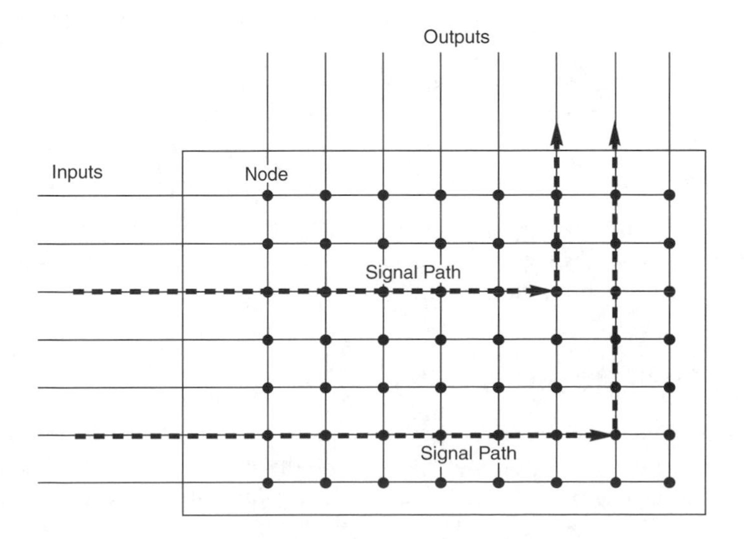

FIGURE 16.5
Optical cross-connect.

outputs, such as 8 × 8 switches, with 8 inputs and 8 separate outputs. Optical cross-connects have been demonstrated with up to 1000 inputs and outputs in the laboratory, and commercial versions are in development.

Although it is simplest to think of optical cross-connects as giant-scale versions of switchboards, that is not a very accurate view. Nobody today makes the equivalent of phone calls that stream 10 Gbit/s between two end terminals. Optical cross-connects are mainly used for load management, to deliver capacity where it is required. They may be used more for making connections as transmission requirements grow. Today that function is mostly done by converting high-speed optical signals to electronic form and processing them through electronic switches.

WDM and Optical Switching

WDM channels may be switched together or separately.

So far, we have not considered how many optical channels are transmitted through each fiber. Wavelength-division multiplexing is an important issue in optical switching because different applications require different treatment of optical channels. Some applications require switching all optical channels carried by a fiber in the same way; others require the optical channels be separated and switched independently.

All optical channels carried by the fiber need to be redirected for protection switching; the standard approach is to simultaneously switch them all to a backup fiber. Switching of all optical channels also may be needed when directing large volumes of traffic; for example, transmitting a large volume of traffic through a series of major switching nodes in a long-distance network. Network managers may organize transmission so one fiber carries signals from New York to Cleveland, which are then switched to another fiber for transmission from Cleveland to Chicago. A separate fiber from New York may carry signals that the Cleveland switch directs to Detroit. Simultaneously switching multiple channels in the same fiber simplifies switch operation.

On the other hand, traffic management often requires redirecting optical channels that arrive through the same fiber into several different fibers. For example, another fiber may carry signals from New York for distribution to other cities in Ohio; that is, one wavelength may go to Toledo, another to Akron, a third to Columbus, and a fourth to Cincinnati. In that case, a demultiplexer would separate the WDM signals from New York, then the switch in Cleveland would process them separately.

Redirecting individual wavelengths requires first separating the wavelengths. Depending on the configuration, this may require either isolating one wavelength with an add-drop multiplexer, or completely demultiplexing all the wavelengths.

Wavelength Routers

A wavelength router separates signals by wavelength.

One type of WDM switch deserves special mention—the *wavelength router.* Essentially a wavelength router is a special-purpose demultiplexer that directs optical channels to different destinations, depending on their destination. You can think of it as a conventional WDM

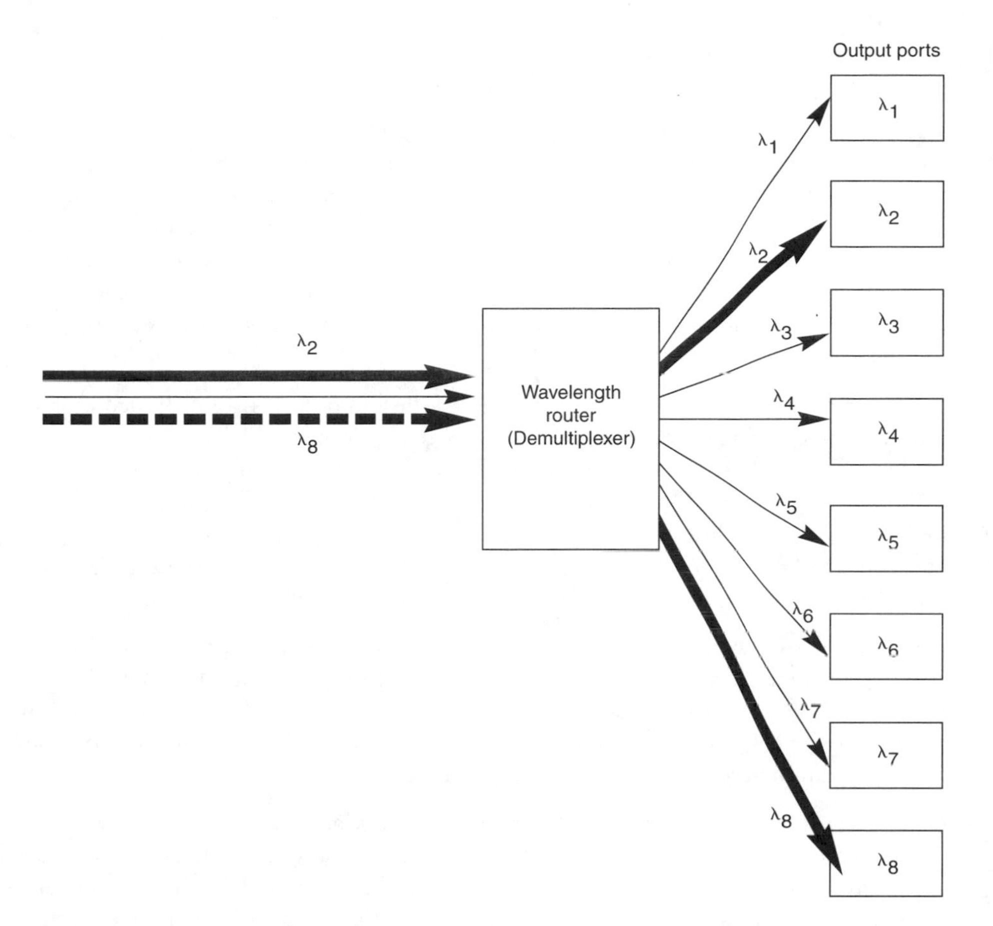

FIGURE 16.6
Wavelength router directs input signals by their wavelength.

demultiplexer with fixed output ports. It gets its name from the fact that it routes input signals to their destinations based on their wavelength, as shown in Figure 16.6. Any input signals at λ_2 is routed to port 2, while any input signal at λ_8 is routed to port 8. If you switch the wavelength, you switch the output port. Although present applications are limited by the need for wavelength converters, wavelength routers can provide a distinct function.

Note that wavelength routers are distinct from Internet routers, as described below.

Switches and Routers

Switches connect circuits. Routers direct data packets based on their headers.

The difference between switches and routers is an important one in telecommunications, but is easy for newcomers to misunderstand. Although both switches and routers direct signals, they do so in different ways and operate on different kinds of signals.

So far this section has concentrated on switches and switching. Originally, switches made physical connections between electrical circuits, like a wall switch connects a light fixture to an electric power line, turning on the light. Old-fashioned electro-mechanical switches made physical connections between the wires running from your telephone and the wires running to your neighbor's phone.

Circuit-switched systems reserve dedicated channels.

Today, most switching is electronic, with solid-state circuits making connections. Once calls are digitized, your call does not have a whole wire (or fiber) to itself, but it does have a fixed time slot in the series of pulses being transmitted. Engineers still call this connection a *circuit* (or sometimes a *virtual circuit*), although it is not a set of wires dedicated to your conversation. Such circuit-switched systems reserve a guaranteed capacity for each call. It's functionally the same as having your own dedicated pair of wires during your entire conversation, always available whether or not you are talking. Your entire conversation follows the same route.

An alternative approach is called *packet switching*. Instead of holding a dedicated channel open for you all the time, you share the system with many other users. The signals you send are divided into data packets, with headers added to indicate their destination. Devices called *routers* read the headers, then decide where to send the packet based on that information and network conditions at the moment. You can think of them as drivers of parcel delivery trucks who read the label (the header) at your door, then decide the best route to take the package to its destination. The Internet is the most familiar example of packet switching.

Note that there are important functional differences between switches and routers. Switches set up a circuit and leave it alone as long as it's carrying signals. When the connection is finished, the switch hangs up and waits for another call. Switches don't pay any attention to the content of the call beyond the initial information needed to make the connection, and monitoring to see that the line is still in use.

Routers have a more complex job. They must read the headers of each and every packet, then direct it to one of many other routers part-way to the packet's destination. The packet is likely to go through a series of routers. Each router in sequence reads the header and sends the packet closer to its destination. Like mail sorters, routers may bundle together packets that are going in the same direction, to be sorted and redistributed at their destination. In addition to reading the headers, the routers monitor network conditions to establish the best routes for sending data packets.

It's important to remember that circuit switching and packet routing are different operations, with distinct requirements and hardware. Electronics can do both. So far, optical circuit switches are available, but there are no true optical routers on the market that read headers on packets of data in optical form and route the light signals to their proper destinations. True optical routers are in the research and development stage.

Transparent versus Opaque Switches

Transparent optical switches let light go straight through; opaque switches do not.

Optical switches can be divided into two broad categories: *transparent* and *opaque*. The names imply the key difference. Optical signals go straight through a transparent switch without being converted into any other form. One example is a mirror that moves back and forth, directing incident light into one of two possible outputs. The same optical signal that enters the switch is reflected from the mirror, and goes out one of the two possible outputs.

In an opaque switch, the signal is converted into some other form before switching. A simple example is a switch that converts the optical signals into electronic form, processes them electronically, then sends the output signals in one of two possible directions. This sort of switch is considered opaque because the light signal does not go straight through it. Even though both input and output signals are in the form of light, the light is converted into some other form in between.

Free-space Optical Switching

Unlike electrical signals, optical signals can travel freely through the air or empty space. This means that optical switches can have internal gaps, unlike electrical switches that must have continuous physical connections so electrons can flow through them.

Signals may pass through free space inside an optical switch.

Free-space optical switching simply means sending signals between points through empty space instead of through optical fibers. For example, a mirror might be tilted to one of several positions, each one aiming a beam striking the mirror in a different direction. The "MEMS" switches described below are good examples.

Optical Switching Technologies

Several technologies can be used for optical switching, and more are in development. The essential idea is to move the beam from one point to another. This may be done mechanically by moving an optical component, or in other ways that shift or deflect light without moving parts. The technologies differ in how fast they can redirect a beam, and in how many different directions they can point it. Some can switch a beam between two directions, while others can aim it over a range of angles.

Opto-Mechanical Switches

Opto-mechanical switches redirect signals by moving fibers or optical components so they transfer light into different fibers. (They are considered distinct from the micro-electromechanical or MEMS switches considered below.) Figure 16.7 shows a simple example. The input signal comes through the fiber on the left. A mechanical slider moves that fiber

Opto-mechanical switches move fibers or optics to redirect signals.

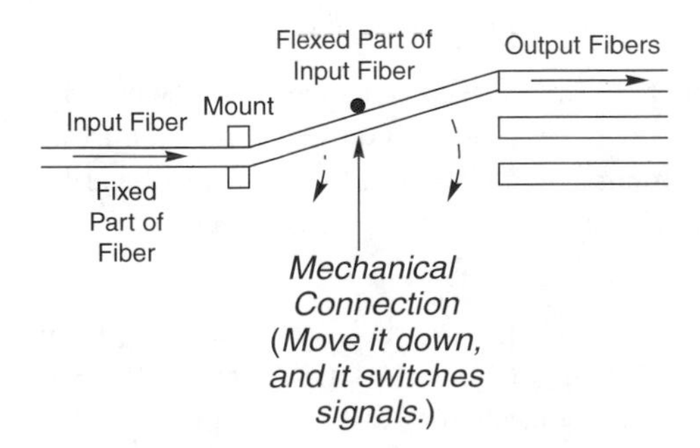

FIGURE 16.7
An opto-mechanical switch.

up and down, latching into one of three position. Each position directs light from the input fiber into a different output fiber. In this design, the slider flexes a short length of the input fiber.

Many other designs are possible. Instead of moving a fiber, an opto-mechanical switch could move a mirror or lens to focus light into different fibers. The switch could be toggled mechanically or electronically. With precise optics, you can make an optical cross-connect that focuses light from one of several input fibers onto one of several output fibers. Collimating and collecting optics can focus the beam into the core of the output fiber.

The common element of all opto-mechanical switches is that their operation involves mechanical motion of an optical component. Precise motion is important because fiber alignment tolerances are tight, although large collecting optics can ease requirements. Although opto-mechanical switches are simple in concept, they are far more demanding in practice than for electrical switches. Another disadvantage is that telecommunication companies generally prefer to avoid moving parts in our solid-state age.

Nonetheless, opto-mechanical switches have come into wide use because they are the simplest and cheapest optical switches available. They are used mainly for protection switching and other applications where it is vital to be able to switch signals when necessary, but where you hope it isn't necessary very often. They also are used in some instruments.

MEMS Switches

MEMS switches redirect light using tiny moving micromirrors.

Micro-Electro-Mechanical Systems (MEMS) are tiny mechanical structures made by a series of steps involving deposition and etching of a substrate material. These structures can be used as optical switches, which deflect light by moving back and forth. Although this could be called opto-mechanical, in practice MEMS switches are considered as a separate class of devices.

MEMS technology is adapted from the photolithographic methods of making integrated electronic circuits. Doping and deposition build up a series of patterned layers on a semiconductor substrate—in practice, on silicon. Then some of the material is etched away to leave mirrors supported by posts, as shown in Figure 16.8. Circuits deposited on layers below the suspended mirrors can carry currents, which generate electromagnetic forces that can pull on the mirrors, tilting them. The tilting mirrors scan reflected light across space and can direct beams to output ports. They require about 10 volts to activate, can switch position in microseconds, and can operate for hundreds of millions of cycles.

Originally developed for use in displays, optical MEMS have been adapted for switching. Arrays of mirrors are fabricated on silicon substrates. With careful design, complex mirrors can be made to tilt back and forth in two dimensions, so they could scan both vertically and horizontally. Figure 16.9 shows an example of such a mirror, encircled by a pair of rings that can tilt it in two dimensions.

Such tilting mirror structures can scan over a range of angles, so they can collect light from many distinct input ports and direct it to any of many distinct outputs. This gives tremendous flexibility, but it makes accuracy essential. If the mirrors drift from their assigned positions, the output can go to the wrong port.

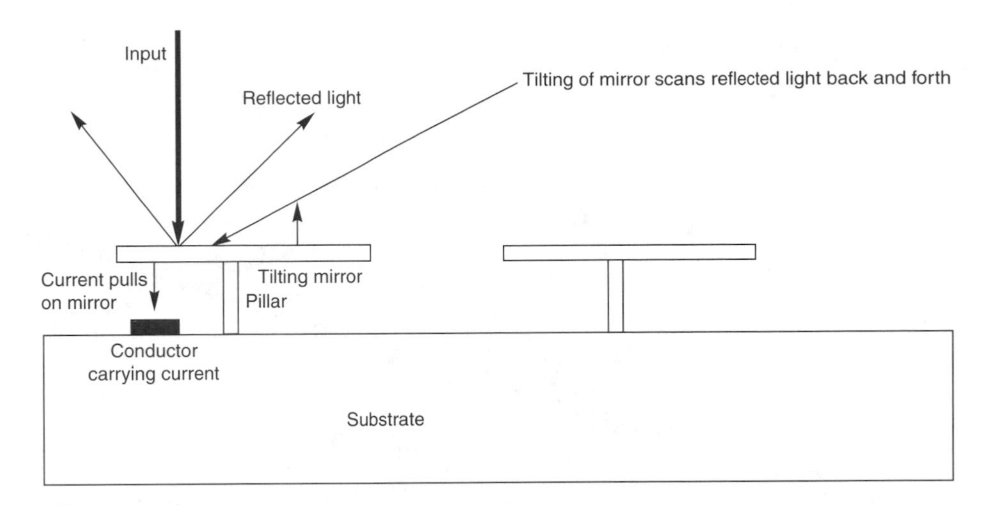

FIGURE 16.8
MEMS mirrors tilt back and forth.

An alternative design moves mirrors between two distinct positions, where they latch in place. You can think of them as being either in the "up" or "down" position. If they are down, light goes through their position unchanged. If the mirror is up, it reflects light in a single alternative direction. Because the mirror latches into place, it always reflects light in the same direction, easing the need for alignment.

Some MEMS switches latch into two distinct positions.

Latching structures are somewhat more complex than tilting mirrors, but moving between fixed positions is attractive for some applications. Advocates sometimes call these latching mirrors "digital" MEMS because they have only two positions, the equivalent of "off" and "on." This is a careful choice of words, because it implies tilting mirrors are "analog" and thus imprecise and obsolete.

MEMS technology is developing rapidly, and appears very promising for optical switching. Nonetheless, some important questions remain to be answered, particularly how long the moving parts can retain the precision required for accurate switching.

Bubble Switches

Bubbles moving back and forth in liquid guides can switch light by total internal reflections.

Bubbles that move back and forth in liquid guides also can serve as an optical switch. In these devices, two grids of planar optical waveguides cross each other on a substrate, as

FIGURE 16.9
Two-axis tilting mirror. The center mirror pivots on two axes defined by the two surrounding rings. (Courtesy of Lucent Technologies)

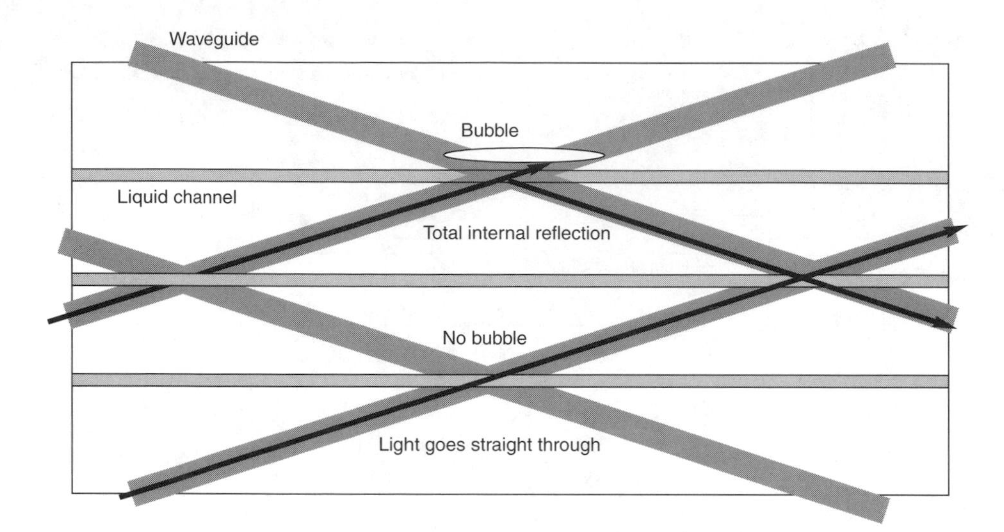

FIGURE 16.10
Bubble switch.

shown in Figure 16.10. These waveguides have a higher refractive index than the substrate, so they guide light across the flat device. Channels containing a liquid with the same refractive index as the waveguides cross through the junction points of the waveguides.

As long as liquid fills the channel at the junction point, the light sees a continuous waveguide and goes straight through the junction. This changes when a bubble moves into the junction point. The refractive index in the bubble is much lower than that in the liquid, and the waveguides cross at a sharp angle, which is beyond the critical angle for total internal reflection. When the bubble is in place, it causes total internal reflection, diverting the light down the other waveguide, as shown in Figure 16.10.

The same techniques used to control ink-jet printers can move bubbles back and forth in the channels, which are sealed to keep the liquid from escaping. The bubbles can be formed by vaporizing small amounts of the liquid, with expansion regions left in the liquid channels to allow for changes in volume. Bubble switches have no mechanical moving parts, although the liquid does move. The technology is considered promising but still young.

Electro-Optical Switches

Voltages applied to planar waveguide channels switch signals in electro-optical switches.

The electro-optic waveguide technology used for the electro-optic modulators described earlier in this chapter also can be used to make a solid-state optical switch with no solid or liquid moving parts. To make a switch, the single input waveguide is replaced by a pair of input guides that meet in a 2 × 2 coupler connected to the active section. Then the single output of the two parallel guides in the active section is replaced by a 2 × 2 coupler splitting the signal between a pair of output guides. Figure 16.11 shows the idea.

As in the modulator, operation depends on applying a voltage to one or both of the parallel electro-optic guides in the active section. This changes the refractive index, delaying the

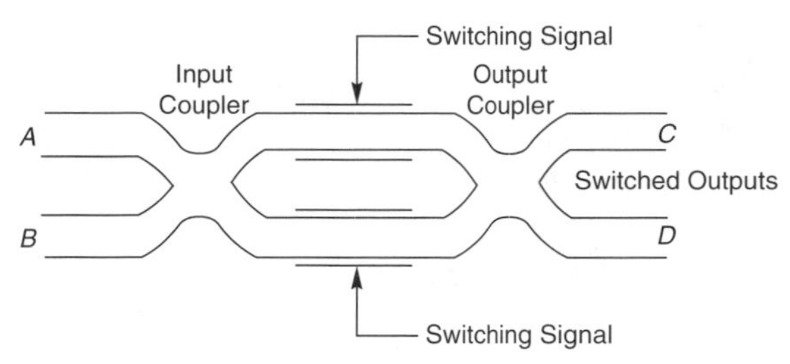

Switching Signal	Connections
V_1	A–C, B–D
V_2	A–D, B–C

FIGURE 16.11
A 2 × 2 electro-optic switch.

phase of light in one waveguide relative to the other. Changing the relative phase of the output signal from the two guides by 180° switches it from one output port to the other. This can switch a single input from one output channel to the other. If separate input signals are entering the top and bottom ports, a 180° phase shift can swap the two between different outputs, as shown in Figure 16.11. If you merely want to switch a signal off and on, you can switch it off by directing the light to a port that goes nowhere. The result is a solid-state switch with no moving parts and a very quick response time. The only changes are in the drive voltage. Normally electro-optical switches are made of lithium niobate, because applying a voltage across the waveguide causes a large change in refractive index.

This approach works well for switching one or two ports, but more complex configurations are more difficult. The simplest way to build more complex switches is by cascading a series of waveguide switches. For example, you could make a 1 × 4 waveguide switch by cascading three 1 × 2 switches, with the two outputs of the first switch providing the inputs for the second pair of switches. This cascade would have a total of four outputs.

Thermo-Optic Switches

Thermo-optic switches rely on thermal changes in refractive index.

Interference switches are based on changing the refractive index to shift the relative phases of light emerging from the two arms of the device, which may be parallel waveguides on a single substrate or discrete components. The refractive index can be changed by techniques other than changing the voltage applied to the arms. In thermo-optic switches, changes in temperature affect the refractive index, shifting the relative phase of the light waves so light emerges from one port or the other. Different materials are used, but the interference principle is the same as in the electro-optic waveguide switch.

Thermo-optic switches are slower than electro-optic switches, but they are simpler and are also solid-state devices without moving parts. Their millisecond response times are adequate for most types of optical switching, and they are used more widely in practical systems than electro-optic switches.

Liquid-Crystal Switches

Liquid-crystal switches work by changing light polarization.

Another switching technology, long used in optical displays, is liquid crystals. Liquid crystals get their name because their large molecules tend to orient themselves in the liquid phase, although they do not form a fixed lattice like a solid crystal. This molecular alignment can polarize transmitted light. The types of liquid crystals used in displays have another important property—applying an electric field can change their orientation, and thus change their effect on the polarization of transmitted light.

For displays or switches, liquid crystals are sandwiched in a thin layer between two parallel glass plates with electrodes applying a voltage across the liquid. The voltage switches between two states, typically one that rotates the polarization, and one that leaves the polarization unchanged. Adding a polarizer makes the device function as a switch or display.

For example, suppose a vertical polarizer is put on the top of a liquid crystal device, so light passes through the polarizer before entering the liquid crystal. Applying a voltage then rotates polarization 45° as the light passes through the liquid crystal layer when it is reflected by the rear surface, and rotated another 45° as it passes back through the liquid crystal. Thus the light exiting the liquid crystal has rotated 90° to be horizontally polarized, and is blocked by the polarizer. Those regions would look dark on a liquid crystal display.

The switch shown in Figure 16.12 works in a similar way. Input light is separated into its two polarizations and reflected off a liquid crystal plate. In this case, the liquid crystal rotates polarization 45° when the voltage is off, but does not affect polarization when it is on. As a result, the reflected light has different polarizations when the voltage is off or on. Polarizing optics deflect that light in different directions, depending on its polarization, so the light emerges from different ports depending on the voltage applied to the liquid crystal.

Other Types of Switches

Other interactions between light and materials can be used as the basis of optical switches. The basic requirement is for an effect that can be switched off and on to deflect light in one of two (or more) directions.

One type in development is an *electroholographic optical switch,* based on a crystal that can record a hologram as a pattern of electric charge distributed through the material. Visible light records the holographic pattern in the crystal when the switch is made. The recorded hologram has no effect on light passing through the crystal when no electric field is applied. However, applying a voltage affects the optical properties of the crystal, causing the hologram to deflect infrared light at 1300 or 1550 nm—switching it in a different direction.

Another example is the *acousto-optic switch,* based on creating an acoustic wave pattern in a solid. The acoustic wave sets up regions where density is higher and lower than in the bulk material, which can deflect light beams passing through the solid at various angles. Changing the acoustic signals switches the optical signals.

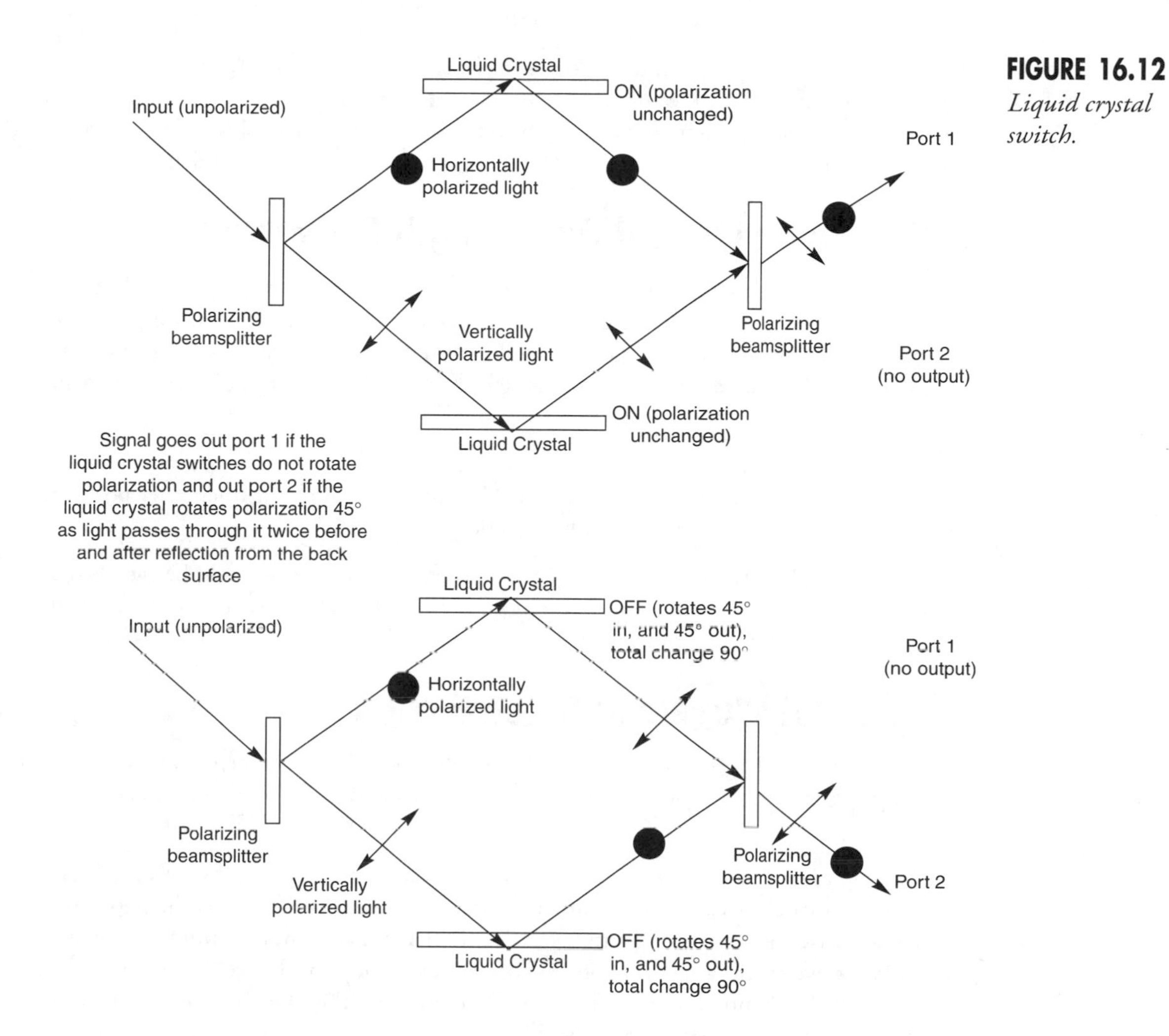

FIGURE 16.12 *Liquid crystal switch.*

Wavelength Conversion

As optical networks develop, they are likely to require devices that shift optical signals from one wavelength to another. The need is likely to arise because you can't be sure the same wavelength will be available along the entire route a signal has to travel. For example, an optical channel may be available at 1540 nm from Chicago to Indianapolis, but the only channel available from Indianapolis to Cincinnati may be at 1542 nm. Sending a signal from Chicago to Cincinnati then would require converting the wavelength at Indianapolis.

Optical networking may require shifting optical signals from one wavelength to another.

Wavelength conversion is a way to enhance the flexibility of an optical network. Essentially, it allows a signal to be shifted between lanes of the optical information highway.

At present, neither the requirements nor the technology for wavelength conversion are well developed. It might seem ideal to be able to dynamically shift from one input wavelength to any other output wavelength, but it is not yet clear how important this will be in optical networking, nor how expensive it would be to have output tunable in wavelength.

Opto-Electro-Optical Wavelength Conversion

Wavelengths can be converted by driving a transmitter at the desired wavelength.

Today, the simplest way to change wavelengths is to convert the optical signal to electronic form, then use the electronic output to drive a transmitter at the desired output wavelength. This *opto-electro-optical (OEO) wavelength converter* is really a special-purpose repeater, which generates a signal at a wavelength different than the input. It could be made tunable by using a tunable laser in the transmitter.

This approach is effectively being used today when signals are switched at terminal nodes. When a terminal switch regenerates an optical channel to redirect it to another port, the switch may drive a transmitter at another wavelength.

OEO wavelength converters also could be used separately, so signals could be switched to different wavelengths without going through a switch. These devices are likely to be the first generation of wavelength converters.

All-Optical Wavelength Conversion

Purely optical wavelength converters could avoid the complications inherent in the special-purpose repeaters used for opto-electro-optical conversion. Different approaches are under study.

The leading approach at present is cross-gain modulation of a semiconductor optical amplifier. In this case, a weak signal beam at the input wavelength modulates the population of current carriers in a semiconductor optical amplifier as it amplifies a continuous signal at the output wavelength. The variations in current carrier density change the gain and thus the power at the output wavelength, effectively converting the signal to the output wavelength. The concept is promising but still in development.

An alternative is to use nonlinear effects such as stimulated Raman scattering or four-wave mixing to shift the signal to the desired wavelength. For example, a fiber could be pumped with a strong pump beam at frequency ν_1 while carrying a signal at ν_2, and four-wave mixing could produce a signal at the frequency $2\nu_1 - \nu_2$. So far efficient conversion remains a problem.

Integrated Optics

Integrated optics combine multiple functions on the same monolithic device.

Earlier, you learned about passive planar waveguide devices that guide light in ways similar to optical fibers but are rectangular in cross-section. Planar waveguides can be integrated with other planar components to make *integrated optics* that combine multiple functions in the same monolithic device. The concept has been around since the late 1960s, when it was proposed as an optical counterpart to integrated electronic circuits. However, optical integration of more than a few components has only begun to emerge from the laboratory.

Some planar waveguide components already are in wide use, including the electro-optic modulators and switches described earlier in this chapter. Edge-emitting semiconductor lasers and semiconductor optical amplifiers, as well as electro-absorption modulators, are in a sense planar waveguide components, because their active layers function as planar waveguides. They also are widely available.

The new technology is combining these waveguide components on a single substrate with other components. The array waveguide demultiplexer in Figure 15.14 is an integrated-optic device. So is the combination of an electro-absorption modulator and a semiconductor laser in Figure 16.3. Another example is the combination of four semiconductor laser stripes, tuned to emit at different wavelengths, with a waveguide coupler and a semiconductor optical amplifier, as shown in Figure 16.13. This device functions as a tunable laser with a broad tuning range. Only one laser stripe emits at a time; its wavelength can be adjusted by temperature tuning, although only over a limited range. The combination of four gives a broader tuning range; more laser stripes can be added if desired. Their output is mixed in a coupler and directed to a semiconductor optical amplifier, which compensates for loss in the coupler to generate a stronger output signal.

Some integrated optic devices are purely passive, such as the array waveguide demultiplexer, which contains no active components. Two broad classes of integrated optics do incorporate active devices. One class is based on semiconductors and includes semiconductor lasers, electro-absorption modulators, and semiconductor optical amplifiers. The other is based on lithium niobate and includes electro-optic modulators and switches, based on changes in the refractive index of lithium niobate caused by electric fields.

Integrated optics have not been as successful as their electronic counterparts. One reason is that government and industry have invested far more money in developing electronic technology over a long period of years. Another is that interactions involving light are not as strong as those involving electrons. You can make transistors into microscopic dots on an

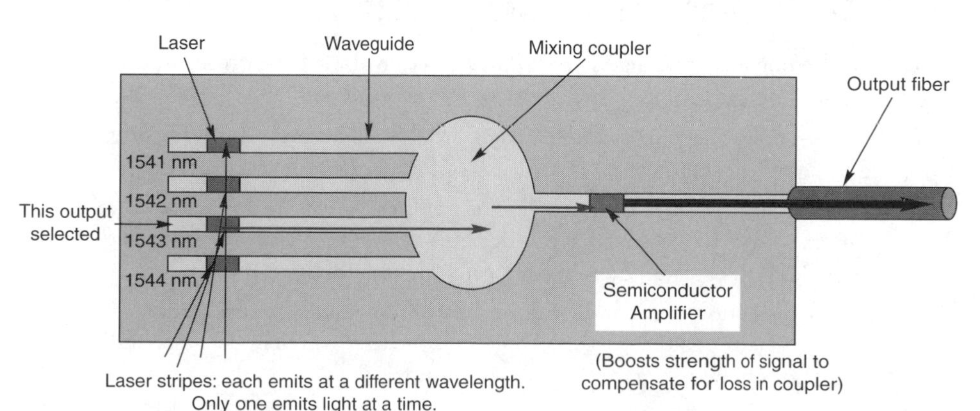

FIGURE 16.13
Four laser stripes integrated as tunable laser.

integrated electronic circuit because electromagnetic fields affect the paths of electrons over very short distances. However, integrated optic modulators and switches typically must be a few centimeters long to change the direction of light. In addition, planar waveguides have much higher loss than optical fibers, so losses accumulate in long, integrated optic components.

You also may encounter a separate class of devices sometimes called *integrated opto-electronics.* As the name suggests, they integrate electronic components as well as optical components on the same substrate. Typical examples include integrating detectors with amplification circuits, and drive circuits with laser or LED light sources. In some cases these are not true monolithic circuits, but hybrid circuits that bond together two chips made of different semiconductors—one for the laser or detector, and the other for the electronics.

Optically Controlled Modulation and Switching

All-optical control is attractive but hard to achieve.

So far, I have said very little about direct optical control of active devices. Electronics control electronic logic and switching devices, but light generally does not control present optical modulators and switches. Instead, electronic circuits remain in control of practical optical components.

Intense research is under way on all-optical modulation and switching. This is attractive for many applications, because it would avoid the need to convert optical signals into electronic form to control optical operations. Progress has been made in some areas, but other types of optical control are more difficult. It's hard to see how light could directly control the operation of MEMS tilting mirrors, for example. For now, most work remains in the realm of research.

What Have You Learned?

1. Active components change signals. They typically draw power from an external source. Modulators and switches are important examples.
2. In high-performance transmitters, an external modulator produces the output signal by modulating the intensity of a continuous beam from the laser source.
3. Electro-optic modulators rely on electric fields to change the refractive index of lithium niobate in a planar waveguide. This delays the phase of light in one of two parallel guides, causing interference, which modulates light intensity.
4. Electro-absorption semiconductor modulators are reverse-biased diodes. Applying a drive current causes them to absorb light at the laser wavelength. Laser and modulator can be integrated on the same chip.
5. Optical networks require optical switches to direct and process signals.

6. Protection switching sends signals through a backup fiber in case of failure. Provisioning changes the services delivered over telecommunication lines.
7. Cross-connects can connect any of multiple inputs to any of multiple outputs. A telephone switch is a good example.
8. The WDM channels carried in a single fiber may be switched separately or collectively.
9. A wavelength router directs input signals according to their wavelength. The same wavelength always goes to the same destination.
10. Switches connect circuits or reserved channels. Routers direct data packets based on their headers. The standard telephone network is circuit-switched. Internet data is transmitted by packet switching and directed by routers.
11. Opto-mechanical switches move fibers or bulk optics to redirect signals.
12. MEMS are micro-electro-mechanical systems made by etching tiny mechanical structures from a semiconductor. They include tiny micromirrors, which can be moved to switch optical signals. MEMS mirrors may tilt over a continuous range, or latch into distinct positions.
13. Bubble switches direct signals by moving bubbles back and forth in liquid guides to the junction points of planar waveguides. The bubbles redirect signals by total internal reflection.
14. Electro-optical and thermo-optical switches change the refractive index of planar waveguides, causing interference, which switches the signal between a pair of output ports.
15. Liquid crystal switches affect the polarization of light; they are used together with polarizers to switch light.
16. Wavelength converters are needed for WDM systems. An opto-electro-optical converter is a special-purpose repeater that drives a transmitter at a different wavelength. Other wavelength converters rely on nonlinear effects, or on modulating the gain of a semiconductor optical amplifier.
17. Integrated optics combine multiple functions on the same monolithic device.

What's Next?

Chapter 17 describes the units and techniques used for optical measurements.

Further Reading

Jaafar M. H. Elmirghani, and Hussein T. Mouftah, "All-optical wavelength conversion: Technologies and applications in DWDM networks," *IEEE Communications,* Vol. 38, No. 3, pp. 86–92 (March 2000)

Andrew Leuzinger, "Liquid-crystal technology implementation for optical switching," *Integrated Communications Design* (November 2000, pp. 30–32)

K. Okamoto, *Fundamentals of Optical Waveguides* (Academic Press, February 2000)

Alice White, "Planar waveguide integrates WDM devices," *Laser Focus World* (October 1999, pp. 117–121)

Questions to Think About for Chapter 16

1. An external modulator with a 20-dB extinction ratio modulates the output of a 1-mW laser. The signal then passes through 20 km of fiber with loss of 0.5 dB/km. Neglecting other losses, what are the powers at the detector when the light is off and when it is on?
2. What are the power levels in Question 1 if the external modulator has insertion loss of 3 dB?
3. What are the power levels for Question 1 if the external modulator has 3-dB insertion loss and an extinction ratio of 10 dB?
4. One important issue in switch design is the number of elements required to switch the signals. Suppose you have a simple cross-connect switch such as the one shown in Figure 16.4, with a single switch element at each node, which either transmits or reflects the beam. If you have 10 inputs and 10 outputs, how many switching elements do you need? What if you have 100 inputs and 100 outputs?
5. A tilting-mirror switch can reflect the light input from a single input port to any of N output ports. With this design, how many switching elements do you need for a 10×10 switch? A 100×100 switch? Assume the tilting mirror can point the beam at as many output ports as needed.
6. How does the bubble switch shown in Figure 16.9 scale? What number of bubble-waveguide intersections do you need for an $N \times N$ optical cross-connect?

Quiz for Chapter 16

1. What phase shift do you need to cause destructive interference between two coherent beams of light, canceling them out?
   a. 0°
   b. 45°
   c. 90°
   d. 180°
   e. 360°
2. What material is used in electro-optic modulators and switches?
   a. Lithium niobate
   b. Gallium arsenide

   c. Indium phosphide
   d. Silica on silicon
   e. Any of the above

3. How should an electro-absorption modulator be biased to block light transmission?
   a. No bias is required; it is normally opaque
   b. Reverse bias
   c. Forward bias
   d. It must be biased in the same direction as the integrated laser light source

4. Telecommunications customers use provisioning switches for
   a. back-up during repairs of a failed transmission line.
   b. making temporary circuit connections to direct telephone calls.
   c. changing services provided to customers.
   d. routing data packets over the Internet.

5. Operation of an optical cross-connect is analogous to
   a. fuses that block electrical power transmission if a circuit overloads.
   b. a telephone switchboard that makes connections between callers.
   c. a fleet of trucks delivering parcels over the best available routes.
   d. municipal water services that pipe water to all homes and businesses.

6. A wavelength router
   a. directs incoming signals to outputs according to their wavelengths.
   b. is an Internet router able to process WDM signals at multiple wavelengths.
   c. is an Internet router that can process optical signals at only one wavelength.
   d. is an optical cross-connect that converts optical signals to different wavelengths.

7. The difference between switches and routers is
   a. switches are mechanical and routers are electronic.
   b. switches are optical and routers are mechanical.
   c. switches connect circuits and routers direct packets.
   d. switches direct packets and routers reserve channels.
   e. just a difference in marketing buzzwords.

8. What kind of switch converts an optical signal to electronic form, then uses the electronic signal to drive another optical transmitter?
   a. Transparent
   b. Opaque
   c. Opto-mechanical
   d. Electro-optical
   e. Bubble

9. An opto-mechanical switch
   a. uses light to mechanically move an electrical switch.
   b. mechanically moves a fiber, mirror, or lens to redirect optical signals.

c. mechanically moves an electronic switch to redirect optical signals.
d. uses light to mechanically move an optical switch.
e. none of the above

10. What type of optical switch can direct light across a range of angles to any of the many possible output ports?
a. An electro-optical switch
b. A pop-up MEMS switch that latches in one of two positions
c. A tilting-mirror MEMS switch
d. A bubble switch
e. A liquid-crystal switch

11. How could you assemble a 1 × 8 electro-optical switch?
a. By dividing one input waveguide into eight optical waveguides
b. By moving one input fiber to connect with one of eight output fibers
c. By tilting a mirror to one of eight possible positions directing light to different outputs
d. By cascading a series of seven 2 × 2 switches, with the two outputs of the first going to inputs of two separate second-stage switches, and the four outputs of those switches going to inputs of four final-stage switches
e. It's impossible.

12. What is the key difference between a MEMS switch and an opto-mechanical type?
a. The MEMS switch has no moving parts
b. Opto-mechanical switches have only two possible outputs
c. MEMS switches are miniature monolithic devices with movable elements
d. None of the above

13. Which type of switch operates by changing the polarization of light?
a. MEMS
b. Opto-mechanical
c. Bubble
d. Liquid-crystal
e. Electro-optical

14. Which of the following is not an application of optical switches?
a. Protection switching around a damaged fiber
b. Changing transmission lines to serve a new customer
c. Redirecting signals at the terminal point of a cable
d. Balance transmission load among several possible routes for a telecommunications carrier
e. Routing Internet packets

CHAPTER

17

Fiber-Optic Measurements

About This Chapter

Fiber-optic technology has its own distinct set of measurements, based largely on a mixture of optical and electronic techniques. This chapter covers the important optical aspects of fiber-optic measurements. It covers measurement units, the quantities measured, and the types of measurements performed, pointing out differences between optical and electronic measurements.

The emphasis here is on basic optical concepts that you need to know when working with fiber optics. The next chapter covers fiber-optic test equipment, along with its use in troubleshooting fiber-optic systems.

Basics of Optical Power Measurement

Most important fiber-optic measurements involve light, in the same way that important electronic measurements involve electric fields and currents. There are some exceptions, such as the length and diameter of optical fibers and cables, the sizes of other components, and the electrical characteristics of transmitter and receiver components. Because this is a book about fiber optics, I will mention such measurements only in passing. However, I will talk about measuring some things other than light, because you cannot qualify the properties of optical fibers if you consider only light. For example, to measure the dispersion of light pulses traveling through an optical fiber, you must observe how light intensity varies as a function of time, which requires measuring time as well as light.

When you're working with light, you need to know what can be measured. The most obvious quantity is optical power, which like electrical voltage is a fundamental measuring

Fiber-optic measurements involve light and other quantities, such as the variation of light with time.

stick. However, power alone is rarely enough; it usually must be measured as a function of other things, such as time, position, and wavelength. Wavelength itself is important because optical properties of optical components, materials, light sources, and detectors all depend on wavelength. Other quantities that are sometimes important are the phase and polarization of the light wave. You need to learn a little more about these concepts before getting into more detail on measurement types and procedures.

Optical Power and Energy

Specialized terminology makes fine distinctions about quantities related to optical power.

People have an intuitive feeling for the idea of optical power (measured in watts) as the intensity of light. However, a closer look shows that optical power and light intensity are rather complex quantities and that you need to be careful what you talk about. Table 17.1 lists the most important quantities, which are described in more detail later.

Each photon or quantum of light carries a characteristic *energy,* as you learned in Chapter 2. Energy is often denoted by E, but the symbol Q is often used in optics. The amount is a function of the wavelength or frequency of the electromagnetic wave. *Photon energy* is easiest to express as the frequency of the wave (ν) times Planck's constant h, which equals 6.63×10^{-34} joule-second, or 4.14×10^{-15} electron-volt-second.

$$\text{Energy} = h\nu$$

When working in wavelength units, the formula for photon energy (in joules) is

$$\text{Energy } (J) = \frac{hc}{\lambda}$$

Table 17.1 Measurable quantities related to optical power

Quantity and Symbol	*Meaning*	*Units*
Energy (Q or E)	Amount of light energy	joules
Optical power (P or ϕ)	Flow of light energy past a point at a particular time (dQ/dt)	watts
Intensity (I)	Power per unit solid angle	watts per steradian
Irradiance (E)	Power incident per unit area	W/cm^2
Radiance (L)	Power per unit solid angle per unit projected area	$W/steradian\text{-}m^2$
Average power	Power averaged over time	watts
Peak power	Peak power in a pulse	watts

where c is the speed of light (approximately 300,000 km/s) and λ the wavelength in meters. If the wavelength is expressed in micrometers, the formula for photon energy (in electron volts) becomes

$$\text{Energy (eV)} = \frac{1.2406}{\lambda(\text{in } \mu\text{m})}$$

Photon energy is the energy carried by a single photon. Normally a pulse of light contains many photons, and the total pulse energy carried by the pulse is the sum of all the photon energies. You can think of energy as the total number of photons that arrives at a destination. However, only knowing the total energy does not tell you if the energy arrived in a single pulse lasting a tiny fraction of a second, or in a slow but steady trickle that took all day.

Power (P or ϕ) measures the rate of energy transfer per unit time. You can think of it as the rate at which photons (or, equivalently, electromagnetic waves) carry energy through the system or arrive at their destination. The rate of energy transfer can vary with time, so power is a function of time. Mathematically, power P is expressed as a rate of change or derivative with respect to energy (Q):

$$\text{Power} = \frac{d(\text{energy})}{d(\text{time})}$$

or

$$P = \frac{dQ}{dt}$$

Sometimes called *radiant flux,* optical power is measured in watts. One *watt* is defined as the flow of one joule of energy per second. Watts are units used to measure the transfer rates of all types of energy, including electrical energy or heat. That is, a watt of light delivers the same amount of energy per second as a watt of electricity. (Note, however, that the power ratings of light bulbs in watts measure how much electrical power they use, not the amount of light they radiate, which is much lower.)

Optical and Electrical Power

Both optical and electrical power measure more fundamental quantities. For light and other types of electromagnetic radiation, the power is proportional to the square of the amplitude of the electromagnetic wave, shown as A in Figure 17.1, as well as to the number of photons received per second. The wave amplitude measures the strength of the electrical field in the wave.

Optical power is proportional to the square of the light wave amplitude.

Electrical power is usually given as the product of the voltage (V) times the current (I):

$$P = VI$$

or, power (in watts) = volts × amperes. The relationship can take other forms if you use Ohm's law, $V = IR$ (voltage = current × resistance):

$$\text{Power} = \frac{V^2}{R} = I^2 R$$

FIGURE 17.1
Properties of an electromagnetic wave.

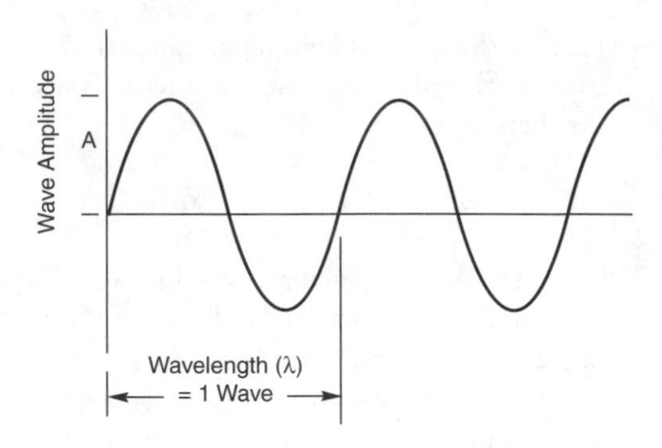

Recall that voltage across a resistance is the strength of the electric field, and you can see that electrical power looks like optical power. It's easy to measure the voltage or current in electronics, but it's not easy to measure the amplitude of the electric field in optics. Thus in electronics you may measure the voltage and current and multiply them to get power, but in optics you measure power directly.

A closer comparison of optical power and electronic power shows more about their differences and similarities. The energy carried by an electron depends on the voltage or electric field that accelerates it. Earlier, I mentioned the electron volt as a unit of energy. One *electron volt* is the energy an electron carries after it is accelerated through a potential of one volt. The total power is thus the number of electrons passing through a point times the voltage that accelerated them.

Energy per photon depends on the photon wavelength.

Each photon has a characteristic energy, which depends on its wavelength or frequency. If the light is at a steady level, the amplitude of the light wave measures the number of photons per unit time. Thus the total energy delivered by the light is the energy per photon times the number of photons (the wave amplitude).

This makes the two types of power look the same, and that stands to reason. Electrical power is the energy per unit time delivered by electrons, where the electron energy depends on the voltage, that accelerated the electrons. Optical power is the energy per unit time delivered by photons, each of which has a fixed energy that depends on its wavelength. The total power measures the rate at which these photons are arriving.

There is a complication to this picture. Normally, a constant voltage accelerates all electrons to the same energy. However, all photons arriving at a given point may not have the same energy unless they all have the same wavelength. Lasers can deliver monochromatic light, with all photons having almost the same energy, but optical measurements were developed long before lasers, when there was no easy way to account for differences in photon energy. Instead of trying to count photons, optical power measurements usually average out the differences and give the results in watts.

Peak and Average Power

Power is an instantaneous measurement of the flux of energy at a given moment. This means that it can vary with time. In general, two types of power are measured in optical systems: *peak power* and *average power.*

Power is an instantaneous measurement; it varies with time.

The peak power is the highest power level reached in an optical pulse, as shown in Figure 17.2. This power may not be sustained long. In a fiber-optic system, this is the highest level reached while a signal is being transmitted.

The average power measures the average power received over a comparatively long period, often a second. In a communication system, this is the average over many pulses and quiet intervals. For a digital fiber-optic system in which the transmitter is sending "on" pulses half the time (a 50% duty cycle), the average power is half the peak power because the power is either fully on or fully off. In the example of Figure 17.2, the average power is less than half the peak power because the power is lower during most of the pulse, and because the interval between pulses is slightly longer than a single pulse.

Peak power is the highest level in an optical pulse; average power is the average over an interval.

The average power of an ideal digital transmitter also depends on the modulation scheme. Some modulation patterns and data streams do not keep the transmitter emitting light half the time.

In practice, fiber-optic measurements average power levels over many pulses to give average power rather than peak power. However, transmitter output may be specified as peak power, so it pays to check.

Pulse energy is another measurement that can be valuable. An example is trying to calculate how many photons arrive per pulse in high-speed systems, because that number drops with data rate. Suppose the average power in two signals is 10 μW. If one signal carries 2.5 Gbit/s and the second carries 10 Gbit/s, the pulses in the faster system will be only one-fourth the duration, and during that interval they will deliver only one-fourth the energy carried by a pulse lasting four times as long. If you delve deeply into communication theory, you will find that the ultimate limits on communications often are stated as the minimum pulse energy needed to deliver a bit of information.

The ultimate limits on communications come from pulse energy.

Pulse energy measures the total energy received during a pulse, as shown in the shaded area in Figure 17.2. If the power is uniform during the length of the pulse, as in a

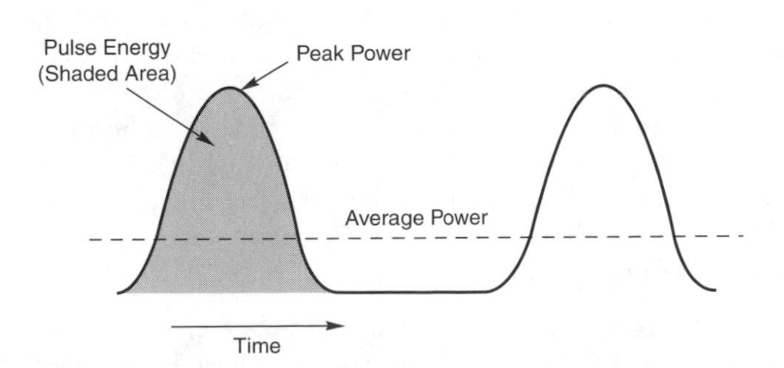

FIGURE 17.2
Peak and average power, and total pulse energy.

series of square digital pulses, the pulse energy Q is the product of the power P times the time t:

$$Q = P \times t$$

If the instantaneous power varies over the length of the pulse, you need to integrate power over the pulse duration, which is mathematically expressed as:

$$Q = \int P(t)\,dt$$

As long as the power remains level during the pulse, multiplication works fine.

Optical Power Measurement Quirks

The definition of decibels looks different for power and voltage.

Before I go deeper into measuring various forms of optical power, I'll warn you about a few potentially confusing measurement quirks. In electrical measurements, the decibel power ratio is often defined in terms of voltage or current. These are in the form

$$\text{Power ratio (dB)} = 20 \log\left(\frac{V_1}{V_2}\right) = 20 \log\left(\frac{I_1}{I_2}\right)$$

where V and I are voltage and current, respectively.

Fiber-optic measurements usually give a different-looking equation in terms of powers P_1 and P_2:

$$\text{Power ratio (dB)} = 10 \log\left(\frac{P_1}{P_2}\right)$$

Why the different factor preceding the log of the power ratio? Because electrical power is proportional to the square of voltage or current. If you measure the ratio of voltage or current, you have to square it to get the power ratio, which is the same as multiplying the log of the ratio by 2. You don't have to do that if you measure power directly, either optically or electrically. Electrical measurements are usually in voltage or current, but optical measurements are in power, so it may seem that the difference is between optical and electrical. However, the real difference is between measuring power directly or indirectly. Both formulas are correct, but be careful to use the proper one.

Optical power can be measured in decibels relative to 1 mW (dBm) or 1 μW (dBμ).

A second potentially confusing point is measurement of optical power in some peculiar-seeming units. Normally, power is measured in watts or one of the metric subdivisions of the watt—milliwatts, microwatts, or nanowatts. Sometimes, however, it is convenient to measure power in decibels to simplify calculations of power level using attenuation measured in decibels. The decibel is a dimensionless ratio, so it can't measure power directly. However, power can be measured in decibels relative to a defined power level. In fiber optics, the usual choices are decibels relative to 1 mW (dBm) or to 1 μW (dBμ). Negative numbers mean powers below the reference level; positive numbers mean higher powers. Thus, +10 dBm means 10 mW, but −10 dBm means 0.1 mW. Zero means there is no difference from the reference level, so 0 dBm is 1 mW.

Such measurements come in very handy in describing system design. Suppose, for instance, that you start with a 1-mW source, lose 3 dB coupling its output into a fiber, lose another

10 dB in the fiber, and lose 1 dB in each of three connectors. You can calculate that simply by converting 1 mW to 0 dBm and subtracting the losses:

Initial power	0. dBm
Fiber coupling loss	–3. dB
Fiber loss	–10. dB
Connector loss	–3. dB
Final Power	–16. dB

Convert the −16 dBm back to power, and you find that the signal is 0.025 mW; however, that often isn't necessary because many specifications are given in dBm. This ease of calculation and comparison is a major virtue of the decibel-based units.

Types of Power Measurement

As Table 17.1 indicates, optical power can be measured not just by itself but also in terms of its distribution angle or space. In many cases (e.g., measuring the brightness of illumination), it is important to know not just total power but also power per unit area. The main concern of fiber-optic measurements is with total power (in the fiber or emerging from it) or power as a function of time, but you should be aware of other light-measurement units to make sure you know what you're measuring.

LIGHT DETECTORS

Light detectors measure total power incident on their active (light-sensitive) areas—a value often given on data sheets. Fortunately, the light-carrying cores of most fibers are smaller than the active areas of most detectors. As long as the fiber is close enough to the detector, and the detector's active area is large enough, virtually all the light will reach the active region and generate an electrical output signal.

The response of light detectors depends on wavelength. As you learned in Chapter 11, silicon detectors respond strongly at 650 and 850 nm but not at the 1300 to 1700 nm wavelengths used in long-distance systems. On the other hand, InGaAs detectors respond strongly at 1300 to 1700 nm but not to the shorter wavelengths. In addition, detector response is not perfectly uniform across their entire operating region. You have to consider the wavelength response of detectors to obtain accurate measurements.

Recall also that detectors cannot distinguish between different wavelengths within their operating regions. If eight WDM channels all reach the same detector, its electrical output will measure their total power, not the power of one channel.

In addition, individual detectors give linear response over only a limited range. Powers in fiber-optic systems can range from over 100 mW near powerful transmitters used to drive many terminals to below 1 μW at the receiver ends of other systems. Special detectors are needed for accurate measurements at the high end of the power range.

IRRADIANCE AND INTENSITY

Irradiance (E) is power per unit area. Intensity (I) is power per unit solid angle.

Things are more complicated when measuring optical power distributed over a large area; all the power may not be collected by the detector. Then another parameter becomes important, *irradiance* (denoted E in optics), the power density per unit area (e.g., watts per square centimeter). You cannot assume that irradiance is evenly distributed over a given area unless the light source meets certain conditions (e.g., that it is a distant point source such as the sun and that the entire area is at the same angle relative to the source). Total power (P) from a light source is the irradiance (E) collected over area (A). This can be expressed as an integral:

$$P = \int E\, dA$$

over the entire illuminated surface. If the irradiance (E) is uniform over the entire area, this becomes

$$P = EA$$

where A is the area.

The use of the symbol E for irradiance is a potential source of confusion because E is often used to denote energy. Strictly speaking, the symbol for optical energy was supposed to be Q, but E is now used more widely. Irradiance is rarely measured in fiber optics, but you can always spot it by looking for units of power per unit area.

The term *intensity* (I) also has a specific meaning in light measurement—the power per unit solid angle (steradian), with the light source defined as being at the center of the solid angle. This measures how rapidly light is spreading out from the source.

Irradiance and intensity are often confused, and the power per unit area is often called intensity. This mistake is understandable because both units measure the distribution of power, one over a surface area, the other over a range of angles. The easiest way to tell is to look at the units; if someone measures "intensity" in watts per square centimeter, they're really talking about irradiance. There are fewer situations where power is measured per steradian.

Optical fibers confine light within the small fiber core, so most measurements concern total power. However, irradiance and intensity may be important when measuring the concentration of power inside a fiber core, or when beams are directed through free space by optical switches.

Radiometry measures light at all wavelengths. Photometry measures only light visible to the human eye.

It also is important to remember that, in general, light is *not* distributed uniformly across space within fiber cores or laser beams, and in some cases that light distribution can be important.

RADIOMETRY AND PHOTOMETRY

Specialists in optical measurement often divide the field into two areas, *radiometry* and *photometry.* The terms are often used interchangeably, but they are not synonyms. Photometry is the science of measuring light visible to the human eye, at wavelengths of 400 to 700 nm. Light not visible to the human eye doesn't count in photometric measurements. Radiometry measures light at all wavelengths, whether or not the human eye can sense it. Fiber-optic measurements must be made in radiometric units, with watts measuring optical power, because most fiber-optic signals are outside the visible range.

Photometric units measure power visible to the human eye in *lumens,* which is the unit used to measure the output of light bulbs. Lumens weigh the contributions of different wavelengths according to the eye's sensitivity. Thus light at 550 nm, where the eye is most sensitive, counts more on a photometric scale than wavelengths where the eye is less sensitive, such as 450 or 650 nm. Photometric units ignore light at 850, 1300, or 1550 nm, so you should never use them for fiber-optic measurements. You can use a radiometer-photometer, which is an instrument calibrated in both units, but a true photometer won't work. (You may sometimes find instruments called "photometers," which do measure in radiometric units, but this is because some people in the industry are sloppy with their terminology.)

So far I've talked about ideal photometers and radiometers, but the real world doesn't work quite that nicely. Detectors do not have uniform response across the electromagnetic spectrum. As shown in Figure 17.3, their response varies markedly with wavelength and is limited in range. This makes it essential to calibrate radiometers to account for detector response at the wavelengths being measured.

Fiber-optic power meters are calibrated for wavelengths used in fiber systems.

In practice, fiber-optic power meters are calibrated specifically for major wavelengths used in fiber systems, at 650, 850, 1300, and 1550 nm. (This means that in the strict sense they are not true radiometers because they work only at certain wavelengths.) You should also recognize that fiber-optic power meters, radiometers, and photometers all measure average power. Their response times are much slower than signal speeds, and their displays cannot track instantaneous power fluctuations.

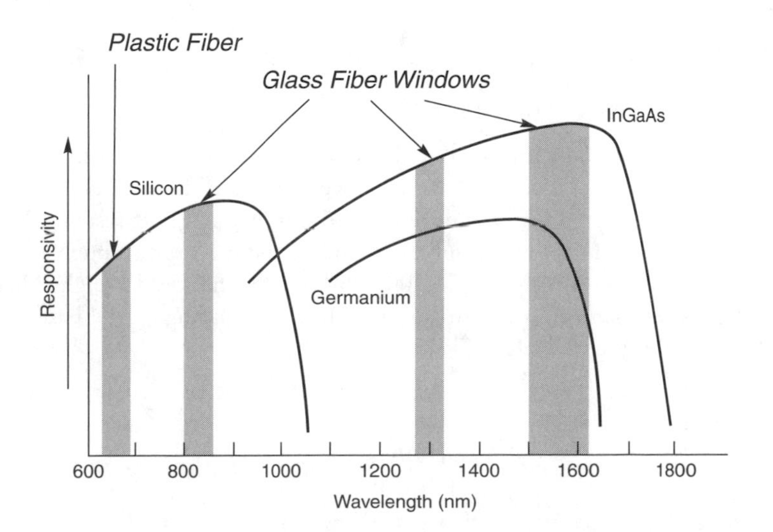

FIGURE 17.3
Detector response at different wavelengths.

Wavelength and Frequency Measurements

Wavelength is critically important in WDM systems.

Wavelength-measurement requirements vary widely, depending on the application. Precise knowledge of source wavelengths is critical in dense WDM systems, where the transmission channels are closely spaced and must be matched to the transmission of demultiplexing components. Knowledge of the spectral response of system components also is vital in WDM systems, particularly for filters used in demultiplexing signals. On the other hand, wavelength need not be known precisely in systems carrying only one wavelength.

Wavelength and Frequency Precision

So far, I have usually described wavelengths in nice round numbers, such as 1550 nm. That's common in optics; engineers who work with light think in terms of wavelength. However, wavelength is not as fundamental a characteristic of a light wave as its frequency. The wavelength depends on the refractive index of the medium transmitting the light; the frequency is constant. This is why standard channels and spacing for WDM systems are specified in terms of frequency.

Earlier, you learned that the wavelength in vacuum equals the speed of light divided by frequency, ν:

$$\lambda = \frac{c}{\nu}$$

However, this equation holds only in vacuum. When the light is passing through a medium with refractive index n, the equation becomes

$$\lambda = \frac{c}{n\nu}$$

which means the wavelength decreases by a factor $1/n$.

Precise measurements and calculations are vital with WDM systems.

I have used round numbers in much of this book because they're usually good enough. Why punch 10 digits into your calculator when you can learn the same concept by punching only 2 or 3? Those approximations don't work for dense-WDM systems. You have to use exact numbers or you get into trouble. To understand why, run through a set of calculations first using the approximation of 300,000 km/s for the speed of light; then use the real value. Let's calculate the vacuum wavelength corresponding to the base of the ITU standard for WDM systems, 193.1 THz.

Using round numbers,

$$\lambda = \frac{3 \times 10^8}{(193.1 \times 10^{12})} = 1553.60 \text{ nm}$$

Using the exact value for *c*, the wavelength is

$$\lambda = \frac{2.997925 \times 10^8}{(193.1 \times 10^{12})} = 1552.52 \text{ nm}$$

The difference is less than 0.1%, but that's enough to shift the wavelength by more than one whole 100-GHz frequency slot in the ITU standard. In short, the wavelength tolerances in dense-WDM systems are too tight to get away with approximations. You have to be precise.

Because of the importance of precision, frequency units may be used in measurements rather than the more familiar wavelength units. You should be ready to convert between the two when necessary, always using the precise formulas.

If you've been watching carefully, you will note that the wavelengths given so far are for light in a vacuum, where the refractive index is exactly 1. Why don't we use the wavelengths in air, which has a refractive index of 1.000273? It's primarily a matter of convention and simplicity. Physicists have long used vacuum wavelengths, and adjusted them slightly—when necessary—for transmission through air. If you wanted to convert frequency to wavelength in air, you would have to add a factor of *n* to all your equations, which could introduce errors. In addition, light often goes through other media, such as the glass in an optical fiber.

The wavelengths assigned to optical channels are the values in vacuum, not in air.

From a physical standpoint, frequency is a more fundamental quantity. A light wave with a frequency of 193.1 THz oscillates at the same frequency in a vacuum, air or glass. Yet the wavelengths differ because the refractive index differs among the three media. This is a major reason that the standards for optical channels are stated in frequency rather than wavelength.

Ways of Measuring Wavelength

It is not easy to measure wavelength precisely; it takes sophisticated instruments and carefully controlled conditions. Precise measurements became essential as WDM systems packed channels close together, making it critical to separate and identify optical channels precisely. A few basic concepts are critical to understanding these measurements and the specific instruments covered in the next chapter.

Wavelength measurements can be absolute or relative.

Wavelength measurements can be absolute or relative. *Absolute measurements* tell you the precise wavelength (or frequency) of a light source. *Relative measurements* tell you the difference between the wavelengths of two light sources, often in frequency units. In general, these comparisons are easier to make.

In practice, absolute measurements are made by comparing the wavelength with some well-defined standards of length or frequency. One way is to monitor changes in interference in two arms of an interferometer as the length of one is changed; another is to measure the different in frequency between the unknown source and a standard, such as a laser with precisely defined wavelength.

The accuracy of both absolute and relative measurements depends critically on calibration of the measurement, accuracy of the standards, and the comparison process. For example, precise measurements require accounting for the fact that air has a refractive index of 1.000273 at room temperature near 1550 nm. Although that number is only slightly higher than the refractive index of a vacuum, the small wavelength shift corresponds to a frequency difference of about 50 GHz in the 1550 nm window. That's significant for DWDM systems.

Linewidth Measurements

In addition to measuring the central wavelength of a laser source, you often need to measure the range of wavelengths in the signal, called the *linewidth* or *spectral width.* Where fiber dispersion is an issue or where a WDM system carries closely spaced wavelengths, the linewidth should be small and often is measured in frequency units—for example, 150 MHz for a temperature-stabilized DFB laser emitting continuously. On this scale, frequency units are more convenient than wavelength; at 1550 nm, 100 GHz is about 0.8 nm, so 150 MHz is about 0.0012 nm. (External modulation adds to that linewidth.)

At lower speeds, where dispersion is not a critical concern, such as where a simple diode laser is modulated directly, the linewidth is much larger and is generally measured in wavelength units. In this case, wavelength units are more convenient.

Spectral Response Measurements

Spectral response measures how systems and components respond to different wavelengths.

In addition to knowing the wavelength of the transmitter, you need to know how a fiber-optic system and its components respond to different wavelengths. This is called *spectral response.* For most components, the most important response is loss or attenuation as a function of wavelength. In the case of filters, multiplexers, and demultiplexers, you need to know how light is divided as a function of wavelength. That is, you need to know how much light is routed in different directions at various wavelengths. For optical amplifiers, the important feature is gain as a function of wavelength.

Spectral response measurements require a properly calibrated light source that emits a suitably narrow range of wavelengths. Figure 17.4 illustrates the problem by comparing two light sources with the transmission of a fiber Bragg grating that selectively reflects at 1555 nm for wavelength-division demultiplexing. A narrow-line source such as a tunable laser can accurately measure the response of the fiber grating, but a broadband source cannot, because its light contains a range of wavelengths much broader than the range of wavelengths the grating reflects.

Phase and Interference Measurements

Phase measures a light wave's progress in its oscillation cycle.

Like other types of waves, electromagnetic waves have a property called *phase.* Recall that a light wave is made up of an electric field and a magnetic field, each of which periodically rise and fall in intensity. Figure 17.5 shows how the amplitude of a light wave rises and falls in a regular cycle. The amplitude follows the pattern of a sine wave, rising to a positive peak, declining through zero to a negative peak, then rising again through zero to a positive

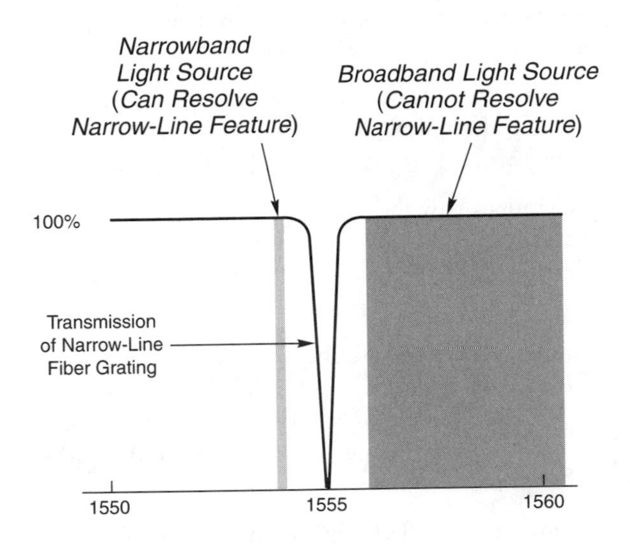

FIGURE 17.4
Only a narrowband source can measure transmission of a narrow-line demultiplexer.

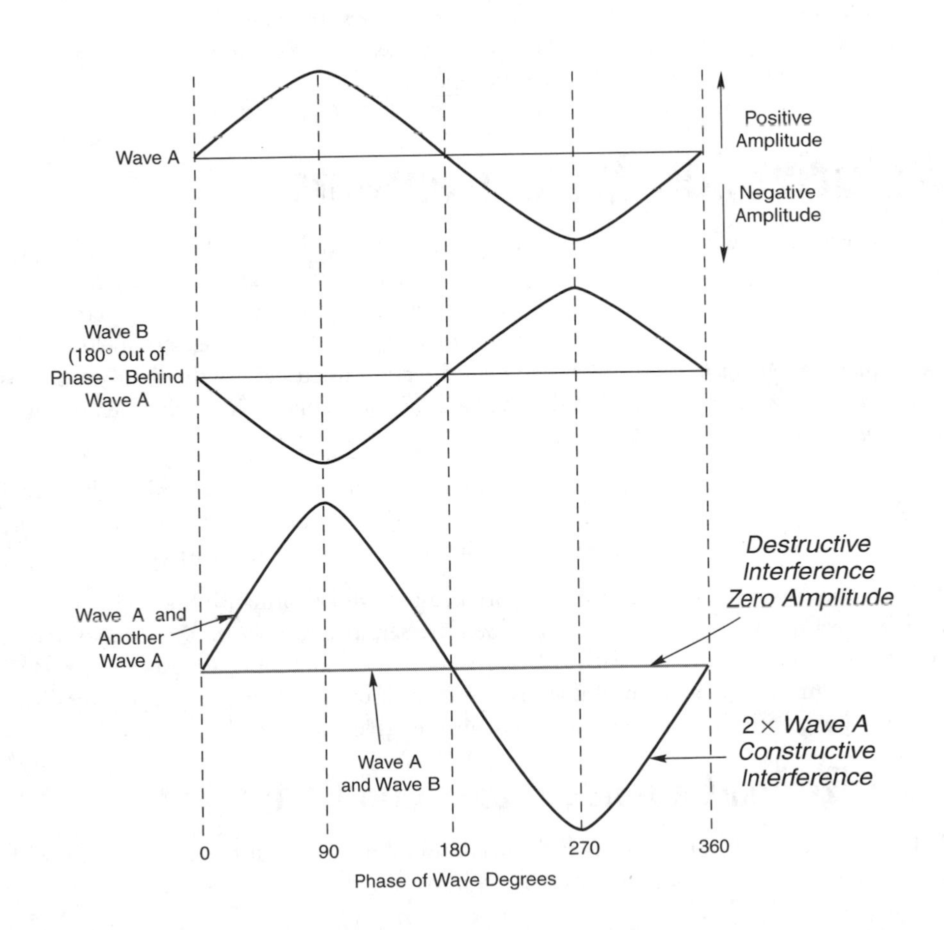

FIGURE 17.5
Phase and interference of light waves.

peak. One wavelength is one complete cycle, from zero through the positive and negative peaks back to the zero point where the amplitude again is increasing.

Phase is measured relative to other waves, and measurements are made by comparing progress through the sine-wave cycle. Phase shifts typically are measured in degrees (between 0° and 360°), or in radians (between 0 and 2π).

The phase shift between two waves determines how their amplitudes add or interfere. As shown in Figure 17.5, the combined amplitude of two light waves is the sum of their instantaneous amplitudes. If the peaks line up, their amplitudes add, but if the peaks of one wave line up with the valleys of the other, they cancel.

Knowing the phase shift is particularly important because the operation of many fiber-optic devices depends on phase shifts. Recall that the electro-optic modulators described in Chapter 16 combined light from two parallel waveguides. Modulating the refractive index of the arms changed the phase of light waves passing through the two arms, so they either combined to generate light or canceled to produce no output.

Phase measurements also are based on interference effects. Two beams are combined and their combined intensity is measured, often as conditions are changed to cause phase shifts. Note that this requires the beams to be nominally identical except for their phase, with the same wavelength and comparable intensities.

Polarization Measurements

Polarization is the alignment of the electric fields that make up light waves.

You saw earlier that light waves are made up of electric and magnetic fields oscillating perpendicular to each other and to the direction the light wave is traveling. The polarization direction is defined as the alignment of the electric field, which automatically sets the direction of the perpendicular magnetic field. Light waves with their electric fields in the same plane are *linearly polarized.* If the field direction rotates along the light wave, it is *elliptically* or *circularly polarized.* And if the fields are not aligned with each other, the light is *unpolarized.*

Polarization is measured by passing light through a polarizer, which transmits light only if its electric field is aligned in a particular direction. The fraction of light transmitted by the polarizer indicates the degree of polarization in the direction of the polarizer.

Polarization has taken on increasing importance in high-performance fiber-optic systems and in certain fiber-optic sensors. Some fibers are sensitive to polarization. In addition, a number of components are sensitive to polarization, including some optical isolators, couplers, waveguides, and some multiplexers and demultiplexers. This makes it important to measure polarization dependance of such quantities as loss, gain, and dispersion.

Polarization-dependent loss is the maximum difference in loss for different polarizations.

Polarization-Dependent Loss and Gain

Polarization-dependent loss measures the maximum difference in attenuation for light with various degrees of polarization. For example, if loss is 3 dB when transferring horizontally polarized light from a fiber to a waveguide but 6 dB for vertically polarized light at the same junction, the polarization-dependent loss is 3 dB (assuming that those are the minimum and maximum losses). This is measured by adding polarization analyzers to conventional loss measurement instruments.

Polarization-dependent gain is the inverse; it measures the maximum difference in gain for light of various polarization states. This is an important quantity for semiconductor optical amplifiers.

Polarization-dependent loss and gain are important because they can introduce noise into fiber-optic systems. As you learned earlier, most single-mode fibers are circularly symmetric and do not discriminate between the two orthogonal polarization modes. Although these fibers carry both polarizations, the relative light intensities in the two modes fluctuates randomly. Normally those random fluctuations even out over the length of the system. However, components that preferentially block one polarization can translate those random fluctuations into intensity noise, so it's important to measure the polarization response of components.

Polarization-Mode Dispersion

In Chapter 5, you learned how polarization-mode dispersion can limit high-speed fiber transmission. Measurements are particularly important because polarization mode dispersion is sensitive to environmental conditions and was not well characterized for older fibers. Special instruments are built for this purpose.

Time and Bandwidth Measurements

As you learned in Chapter 5, system bandwidth is limited by the spreading or dispersion of pulses in the fiber, transmitter, and receiver. This means that time and bandwidth measurements in fiber-optic systems are related. You can think of them as different ways to measure the information transmission capacity of the system. Time measurements directly measure how fast the system can respond to a pulse. Bandwidth depends on this time response, but it also can be measured directly.

Bandwidth and time measurements both indicate system transmission capacity.

Pulse Timing

Figure 17.6 shows the key parameters in measuring pulse timing.

- *Rise time* is the time the signal takes to rise from 10% to 90% of the peak power.
- *Pulse duration* normally is the time from when the signal reaches half its maximum strength to when it drops below that value at the end of a pulse. This is called *full width at half maximum,* abbreviated FWHM.
- *Fall time* is the interval the signal takes to drop from 90% to 10% of peak power.

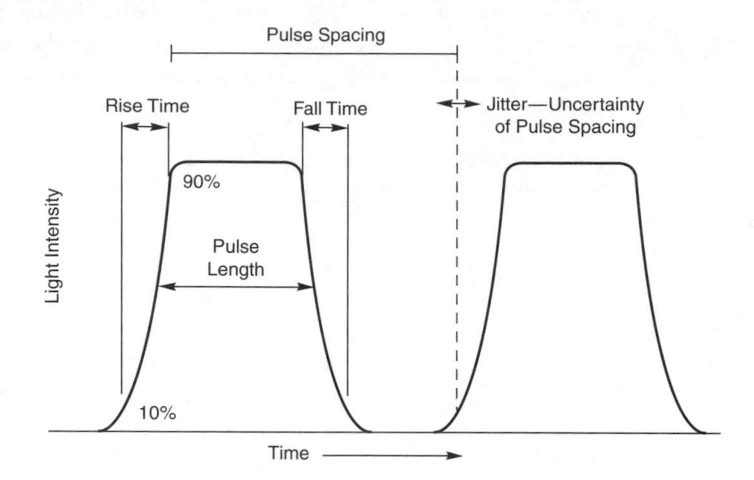

FIGURE 17.6
Pulse timing.

- *Pulse spacing* or *pulse interval* is the interval between the start of one pulse and the point where the next should start. If the signal is on for 1 ns and there is a 1-ns delay before the next pulse can start, the pulse spacing is 2 ns. The pulse spacing generally means the interval between transmitting one data bit and transmitting the next, whether the bits are "0s" or "1s."
- *Repetition rate* is the number of pulses or data bits transmitted per second, which in practice is the pulse spacing divided into 1:

$$\text{Repetition rate} = \frac{1}{\text{pulse spacing}}$$

  Thus if the pulse spacing is 1 ns, the repetition rate is 1 Gbit/s.
- *Jitter* is the uncertainty in the timing of pulses, typically measured from the point at which they should start.

Measurements are made by feeding the optical signal to a detector, which generates an electronic output that instruments measure. This means that time response measured for the light pulse also includes the time response of the detector and the instrument. This effect can be significant at high data rates and must be considered in making fast measurements.

Repetition rates in even the slowest fiber-optic systems are extremely fast on a human scales—a million or more pulses per second—so you cannot see signal-level variations in real time. They are recorded on an oscilloscope or other display, which allows you to see events that are too fast for your eyes to perceive.

Bandwidth and Data Rate

Bandwidth measures the highest analog frequency a system can transmit.

Bandwidth and data rate are separate quantities with distinct meanings. Both depend on pulse timing, but can be measured by themselves. *Bandwidth* is an analog measurement of the highest frequency the system can transmit; *data rate* or *bit rate* is a digital measurement of the number of bits per second actually transmitted. Both deserve a bit more explanation.

An analog transmission system can carry signals at a range of frequencies. That is, it responds to a certain range, analogous to the nominal 20 to 20,000 Hz response of the human ear. Normally, the bandwidth of fiber-optic systems is limited only at high frequencies. If you measure the signal received as a function of frequency, it drops at high frequencies, as shown in Figure 17.7. Typically the bandwidth is defined as the point where response has dropped by one-half or 3 dB—about 5 GHz in Figure 17.7.

As you can see from Figure 17.7, the system does transmit higher frequencies, but it attenuates them more strongly, so their intensity at the output is much weaker. Thus signals at 5.5 GHz are attenuated about 3 dB more than those at 5 GHz, and 6 dB more than those at 3 GHz. In practice, the higher attenuation means you can't count on the signals reaching the receiver in usable form. The usability of an analog signal is measured by the signal-to-noise ratio, described later in this chapter.

The speed limit of digital systems is measured as a *data rate* or *bit rate,* that is, how many bits can be transmitted through the system per second. The data rate response is not as easy to plot as the frequency response. In practice, the quality of a digital signal is measured as a *bit error rate,* also described later in this chapter. The poorer the signal quality, the more bits are received incorrectly. The maximum data rate is the highest transmission speed that meets error-rate specifications.

Data rate measures the speed of digital systems.

Bandwidth and data rate are related, but not equivalent. Communication theory shows that any signal—digital or analog—can be considered as the sum of signals at many different frequencies. The ideal square pulses of digital signals are made up of a series of analog signals at

Limited bandwidth rounds the corners of digital pulses.

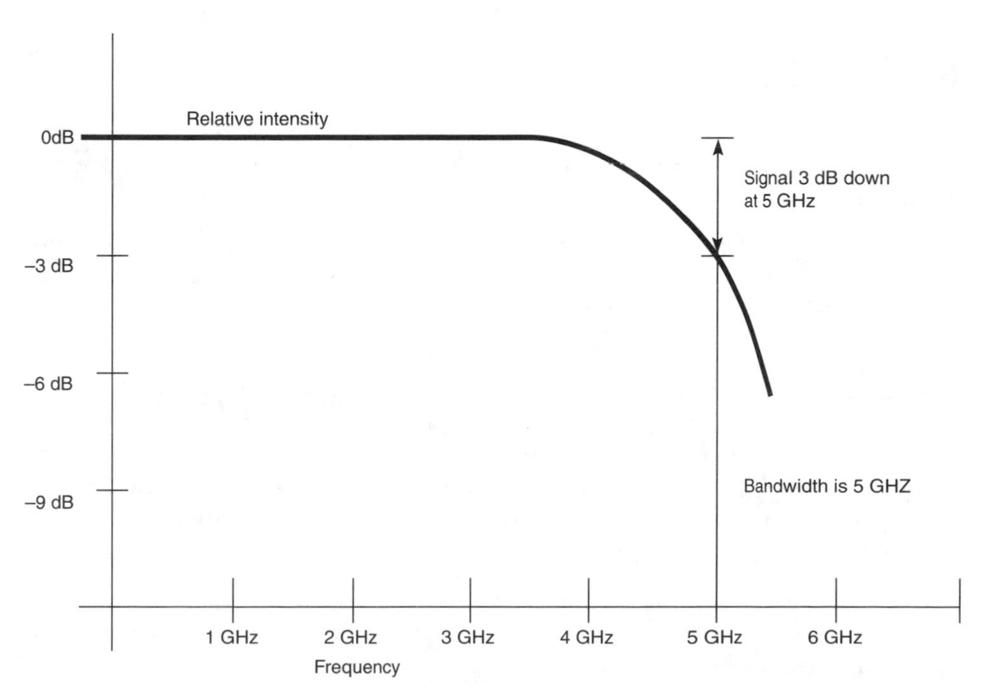

FIGURE 17.7
Frequency response of an analog fiber-optic system.

multiples of the square-wave frequency, that is, the bit rate. The higher frequencies give the square waves their sharp corners, so limited bandwidth rounds the corners of digital pulses.

Signal Quality Measurements

Transmission quality of telecommunication systems can be assessed directly by measuring how well the output signal reproduces the input. Different concepts are used for analog and digital systems.

Signal-to-Noise Ratio

Signal-to-noise ratio measures analog transmission quality.

The usual way to assess the quality of an analog communication system is to measure the ratio of signal power to noise. The higher the *signal-to-noise ratio* (often written S/N), the higher the quality of the signal. What is an acceptable signal-to-noise ratio depends on the application and the user. The background hiss on analog audio cassette tapes is a good example; you notice it more in quiet passages of classical music in a quiet room than you would listening to loud rock music in a speeding car with the windows down. Users of analog transmission systems may set standards that define acceptable signal-to-noise levels. That is done by the cable television industry, the main user of analog fiber-optic systems.

Signal-to-noise ratio also can assess the performance of individual analog components. Optical amplifiers are important examples, where the ratio of the output signal to the background noise determines performance. The concept is fairly intuitive; the more signal and the less noise, the better. Typically signal-to-noise ratios are measured in decibels.

Bit Error Rate

The fraction of incorrect bits is the bit error rate.

Bit error rate (or ratio) measurements compare digital input and output signals to assess what fraction of the bits are received incorrectly. They offer a quantitative measurement of signal quality.

In practice, a special instrument generates a randomized bit pattern, which is transmitted through the system. The total number of bits transmitted are counted. So are errors that occur when the signal bit interpreted by the receiver does not match the transmitted signal. The more wrong bits, the worse the transmission quality.

As you would expect, the bit error rate increases as received power drops, as well as when the system approaches other performance limits such as maximum data rate. The increase in error rate is quite steep, and can be more than a factor of 100 when the input signal drops by 1 dB, if the system is operating near its performance limits. Other factors may set a minimum bit error rate when input power is adequate, and excess power can cause errors by overloading the receiver.

Users set standards for acceptable bit error rates. A typical target for telecommunications and data transmission is 10^{-12} (one error in a trillion bits). Higher error rates may be acceptable in other applications, such as video transmission.

Eye-Pattern Analysis

One popular way to assess performance of digital fiber-optic links is to superimpose a series of pulses on an oscilloscope display. This is called *eye-pattern analysis* because it produces the eye-shaped pattern shown in Figure 17.8.

An eye pattern superimposes waveforms of successive bits to assess signal quality.

The oscilloscope traces each received pulse on the screen. If there was no noise, each trace would follow exactly the same line, overlaying other pulses. Adding noise to the signal causes the intensity to vary randomly during the signal pulse, blurring the trace vertically. Likewise, jitter that varies the time when pulses arrive at the receiver spreads the lines horizontally.

What the eye pattern really measures is the repeatability of pulses reaching the instrument. The better the transmission quality and the more uniform the received pulses, the more open the eye will appear. If the eye starts to close—leaving less clear space in the center—it indicates that transmission errors are likely because it's becoming hard to tell the high points of the signal (the top of the eye) from the low points (the bottom of the eye).

The eye pattern measures pulse repeatability.

Careful interpretation of the eye pattern can yield important information on fiber-link performance. Some important points for interpreting eye patterns include:

- Height of the central eye opening measures noise margin in the received signal.
- Width of the signal band at the corner of the eye measures the jitter.
- Thickness of the signal line at the top and bottom of the eye is proportional to noise and distortion in the receiver output.
- Transitions between the top and bottom of the eye show the rise and fall times of the signal.

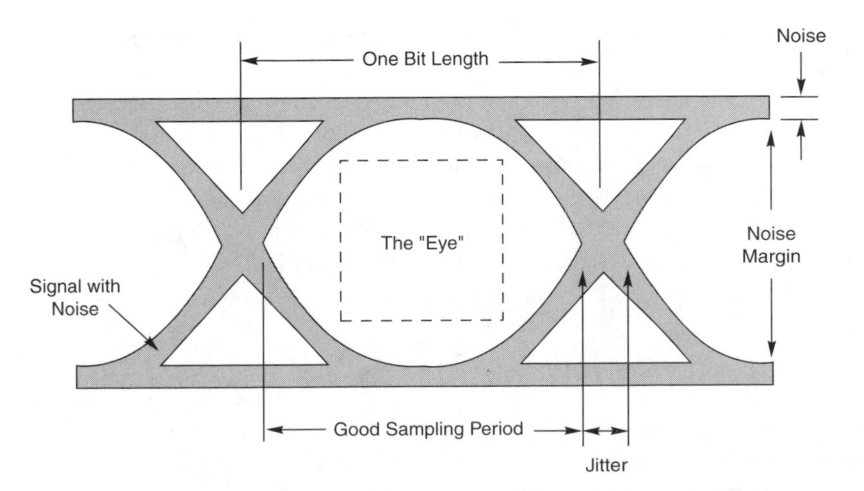

FIGURE 17.8
An eye pattern.

Fiber-Specific Measurements

So far I have described two broad classes of measurements: optical quantities that apply to fiber optics and communications system performance.

Another broad class of measurements is specific to optical fibers and fiber-optic systems. It's impossible to cover them comprehensively in a general introduction to fiber optics, so I will concentrate on a few key fiber parameters, such as attenuation and power in fibers. I will begin with general measurement concepts, then turn to important fiber-optic procedures. Chapter 18 describes important types of test equipment and outlines simple troubleshooting procedures.

Calibration

Calibration is essential for verifying the accuracy of measurements.

Calibration is an essential element of any measurement procedure. You calibrate the bathroom scale when you check that it reads zero before you stand on it to weigh yourself. When you make electrical measurements, you check that the current meter reads zero when the current probes are not in contact with the circuit.

For more precise measurements, you want to make sure the readings are accurate at more points on the scale. To double-check your bathroom scale, you could place a hundred-pound weight on it. To check a current meter, you could connect it to a standard current source designed to deliver one milliampere.

To be really exacting, you might want to go a step further and check the source of your calibrations. Instead of borrowing weights from your neighbor's barbells to test your bathroom scale, you might borrow a set of weights from the local university, which have been compared against precise standard masses. Then you could compare the university's weights against the barbells and find, for example, that the barbell weights weigh 97.5 pounds instead of the 100 pounds stated on the box.

This sort of calibration and comparison is done for precision measurements. Check a good set of test instruments, and you will find dates on which they were calibrated, and you may find that the calibrations are "traceable" to an organization like the National Institute of Standards and Technology, which provides measurement standards in the United States. You don't need this sort of exacting precision for every fiber-optic measurement, but it's essential for precision work. For example, in a short data link you might only need to know optical power to within 20%. However, in a state-of-the-art DWDM system with 50-GHz channel spacing, you may need to measure wavelength of individual light sources with accuracy of 0.01% to make sure channels are adequately spaced.

Fiber Measurement Standards

Measurement standards assure that results are compatible.

Serious fiber-optic measurements often refer to cryptic-seeming codes, such as EIA/TIA-455A. These codes identify standards written by industry and professional organizations, which specify how to make the measurements so the results are comparable to those made by other groups.

Standards often go into excruciating details in specifying the techniques and equipment used for particular measurements. However, those details can be as important in getting the right answer as not leaning against the sink can while standing on your bathroom scale. The essential point is to make sure that everyone's measurements are comparable and repeatable. Fiber specialists have found out the hard way that results can differ depending on whether the labs are or are not air-conditioned.

The details of fiber standards are beyond the scope of this book. Table 17.2 lists major standards organizations and Web sites where you can find more information.

Measurement Assumptions

One reason that standards describe procedures in so much detail is to limit the number of assumptions made in making measurements. We inevitably make implicit and explicit assumptions, and sometimes those assumptions can lead to the wrong results.

If you check that your bathroom scale read zero when nothing is on it, and 100 pounds with calibrated weights, you still make an assumption when you weigh yourself. You assume that the rest of the scale is accurate, and that if you go 50 steps up from 100 pounds, the scale is accurately measuring you at 150 pounds. However, it's possible that the scale might be a few pounds off, so what reads 150 pounds is actually 145 or 155 pounds.

We also make many assumptions in fiber optics, and some of them also may be wrong. It's reasonable to assume that light is distributed the same way along the length of a single-mode fiber, but not in a multimode fiber. Light may have to travel through a kilometer of graded-index fiber before distributing itself evenly among all the possible transmission modes. Differences between the real and the assumed mode distribution can affect measurements of the

Table 17.2 Standards organizations

ANSI	American National Standards Institute	*www.ansi.org*
ASTM	American Society for Testing & Materials	*www.astm.org*
Bellcore	Bellcore (now Telcordia Technologies)	*www.telcordia.com*
EIA	Electronics Industries Alliance	*www.eia.org*
ICEA	Insulated Cable Engineers Association	*www.icea.net*
IEC	International Electrotechnical Commission	*www.iec.ch*
IEEE	Institute of Electrical and Electronics Engineers	*www.ieee.org*
ITU	International Telecommunications Union	*www.itu.int*
NEC	National Electrical Code (administered by National Fire Protection Association)	*www.nfpa.org*
NIST	National Institute for Standards and Technology	*www.nist.gov*
TIA	Telecommunications Industry Association	*www.tiaonline.org*
UL	Underwriters Laboratories	*www.ul.com*

fiber's light-acceptance angle and numerical aperture. Likewise, we might assume that connector loss is identical in both directions, but be fooled because the core of one fiber is slightly smaller than the other, so the loss is 0.3-dB higher in one direction than in the other.

Careful adherence to standards can avoid pitfalls, which can be subtle in more sophisticated measurements.

Fiber Continuity

Fiber continuity checks can verify system function. The simplest test is to see if light can pass through the fiber.

A major concern in installing and maintaining fiber-optic cables is system continuity. If something has gone wrong with the system, you need to check if the cable can transmit signals. If it can, you know you have another problem. If it can't, you need to find the break or discontinuity. In some cases, the break may be obvious—a cable snapped by a falling tree limb or a hole dug by a careless contractor. However, such damage is not always obvious, and the cable route may not be readily accessible.

Early fiber technicians developed a quick-and-dirty test of fiber continuity. One shined a flashlight into the fiber, and a second on the other end looked to see if any light emerged. That is far from ideal because flashlight beams do not couple efficiently into optical fibers. It also requires people at each end of the fiber—one to send the light and the other to look for the transmitted light—and those people must be able to communicate with each other.

Now specialized instruments can do a much better job, often without requiring people on both ends. You will learn more about them in Chapter 18. Optical time-domain reflectometers (OTDRs) and optical fault indicators send pulses of light down the fiber and look for reflections that indicate a fault. Optical test sets measure power transmission. Fiber identifiers can tell if exposed fibers are carrying signals (they work by bending the fiber and observing light that leaks out at the bend). Visible fault identifiers send visible red light through the fiber, and visual inspections show if any is leaking out.

Optical Power

Measurements of optical power require knowing the wavelength and duty cycle.

Optical power is the quantity most often measured in fiber-optic systems. I described the basic principles earlier. The power may be output from a transmitter or optical amplifier, power emerging from a length of optical fiber, or power in some part of a system. The wavelength must be known so that the detector can be calibrated for that wavelength. Duty cycle—the fraction of the time the light source is on—should also be known to interpret properly measurements of average power. The usual assumption is 50% (half on, half off) for digital modulation, but under certain circumstances that may be far off (e.g., if a series of 1s is being transmitted in NRZ code).

Normally, power is measured where the light emerges from a light source or fiber. Fiber-optic power meters collect the light from the fiber through an optical connector, which directs the light to a detector. Electronics process the detector output and drive a digital display that shows the power level in linear units (nanowatts to milliwatts) or in dB referenced to either 1 mW or 1 μW. Measurement ranges are automatically switched across the dynamic range, which is typically a factor of one million. Typical measurement accuracy is ±5%.

Measuring optical power stops the beam, because it's absorbed by the detector. If you want to sample the power level in a transmitted signal, you need a coupler to divert a calibrated fraction of the light to a detector and transmit the rest.

It is important to keep input power within the dynamic range of the power meter. Excessive powers won't be measured correctly, and in extreme cases, exposure to excess power can damage some detectors.

Attenuation

Attenuation is the most important property of passive optical components, because it determines what part of an optical signal is lost within the component and how much passes through. It is always a function of wavelength, although the wavelength sensitivity varies widely. In fibers, the variation with wavelength is significant; in some other components, it is negligibly small.

Optical test sets measure cable loss.

As you learned earlier, attenuation is measured by comparing input and output powers. The standard way to measure cable loss with an optical test set—which includes a light source and transmitter—is shown in Figure 17.9. First the light source is connected to the power meter through a short launch cable, and the power is adjusted to a convenient level (−10 dBm in this case). Then a short receive cable is added between the launch cable and the power meter; a power change no more than 0.5 dB verifies the receive cable is good. The meter is again adjusted to the desired level. Then the cable to be tested is connected between launch and receive cables and the power it transmits is read (−14.2 dBm in this case). The difference, 4.2 dBm, is taken as the total attenuation of the cable being tested, including fiber, connectors, and splices. For more precise measurements, the loss should be measured in both directions through the test cable, because connector attenuation may differ slightly in the two directions.

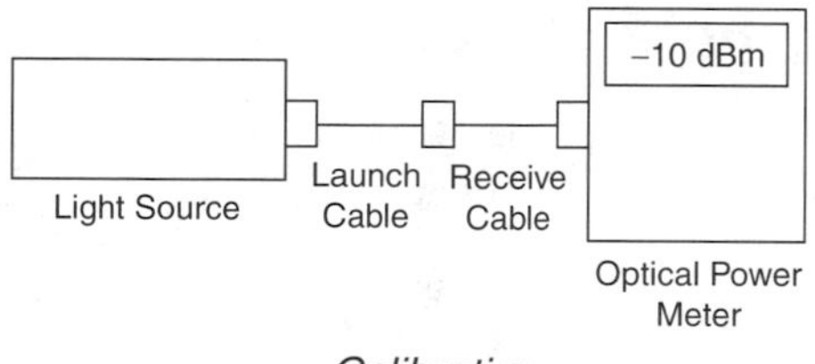

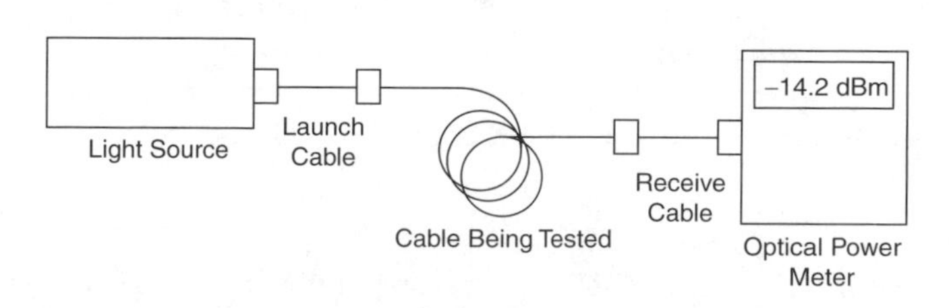

FIGURE 17.9
Cable loss measurement

The same principles can be used to measure the attenuation of other components, such as couplers, or of segments of cable installed in a system. If cable ends are located at different places, the tests can be performed by technicians working at both ends, one with a light source and the other with a power meter, or by temporarily installing a "loop-back" cable to send the signal back to the origination point through a second fiber. Inevitably, small losses are measured less accurately than large ones.

Optical time domain reflectometers can also measure attenuation, as described in Chapter 18.

Precise measurements of fiber attenuation rely on cutting back fibers to compare power emerging from short and long lengths.

This simple comparison technique is adequate for most purposes, but it does not precisely measure pure fiber loss, because it includes loss within the connectors at each end. More precise measurements of fiber loss alone require the cut-back technique. First, power transmission is measured through the desired length of fiber. Then the fiber is cut to a short length (about a meter) and the power emerging from that segment is measured with the same light source and power meter. Taking the ratio of those power measurements eliminates input coupling losses (which occur in both measurements), while leaving the intrinsic fiber transmission loss (which is present only in the long-fiber measurement).

The cut-back method is more accurate for single-mode fibers than for multimode fibers, because of the way mode distribution changes along the fiber. Accurate measurement of long-distance attenuation of multimode fibers requires use of a mode filter to remove the higher-order modes that gradually leak out of the fiber. However, this won't accurately measure the loss of short multimode fibers, which depends on propagation of the high-order modes.

One special problem with single-mode fibers is that light can propagate short distances in the cladding, throwing off measurement results by systematically underestimating input coupling losses. To measure true single-mode transmission and coupling, fiber lengths should be at least 20 or 30 m.

Mode-Field and Core Diameter

Mode field diameter is the region occupied by light in a single-mode fiber.

As you learned earlier, fiber core diameter can vary because of manufacturing tolerances. In addition, mode-field diameter—the diameter of the region occupied by light propagating in a single-mode fiber—is somewhat larger than the core diameter. These quantities can be measured.

Practical interest in the mode-field and core diameters depends on the distribution of light, and measurements are, therefore, based on light distribution. One approach is to scan across the end of the fiber with another fiber of known small core diameter, observing variations in light power collected by the scanning fiber. Other approaches rely on observing the spatial distribution of light near to or far from the fiber—the near-field and far-field intensity patterns. Those distributions of optical power can be used to calculate the core diameter.

A related quantity important for both single- and multimode fibers is the refractive-index profile, the change in refractive index with distance from the center of the fiber. This also is measured by examining the light distribution across the fiber.

Numerical Aperture and Acceptance Angle

The numerical aperture measures how light is collected by an optical fiber and how it spreads out after leaving the fiber. It measures angles, but not directly in degrees or radians. Although NA is widely used to characterize fiber, it isn't NA that is measured, but the fiber acceptance angle, from which NA can be deduced.

Numerical aperture is not measured directly; it is calculated from the acceptance angle.

Numerical aperture and acceptance angle are most important for multimode fibers. As mentioned earlier, measured numerical aperture depends on how far light has traveled through the fiber, because high-order modes gradually leak out as light passes through a fiber. The measured numerical aperture can be larger for shorter fibers, which carry a larger complement of high-order modes, than it will be for long fiber segments. Measurements are made by observing the spread of light emerging from the fiber.

Cutoff Wavelength

Cutoff wavelength, the wavelength at which the fiber begins to carry a second waveguide mode, is an important feature of single-mode fibers. The measured effective cutoff wavelength differs slightly from the theoretical cutoff wavelength calculated from the core diameter and refractive-index profile. As with core and mode-field diameter, cutoff wavelength is a laboratory rather than a field measurement.

Cutoff wavelength is measured by observing where stripping out the second-order mode causes an increase in loss.

Normally, the cutoff wavelength is measured by arranging the fiber in a test bed that bends the fiber a standard amount. Fiber attenuation as a function of wavelength is measured twice. First, the fiber is bent in a manner that causes the second-order mode to leak out almost completely. Second, the fiber is arranged so it transmits both first- and second-order modes. These two measurements are compared, giving a curve such as the one in Figure 17.10, which shows excess loss as a function of wavelength. In this case, λ_c is the effective cut-off wavelength, which is defined as the wavelength above which second-order

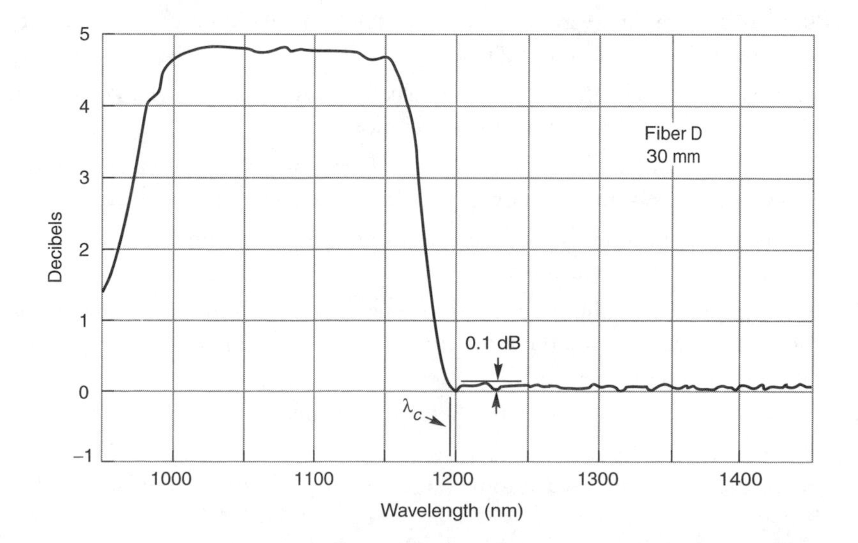

FIGURE 17.10 *Measurement of effective cutoff wavelength (Courtesy of Douglas Franzen, National Institute of Standards and Technology)*

mode power is at least a certain amount below the power in the fundamental mode. The measurement finds this value by locating the point where excess loss caused by stripping out the second-order mode is no more than 0.1 dB.

What Have You Learned?

1. Each photon has a characteristic energy, defined as Planck's constant (h) times the frequency (ν). Photon energy also equals hc/λ, where c is the speed of light and λ is the wavelength.
2. Optical power measures the transfer of light energy, and is defined as the change in energy with time. Power is proportional to the number of photons passing a given point per unit time.
3. Optical power is proportional to the square of the light wave amplitude.
4. Optical power can be measured in decibels relative to a power level. The units dBm are powers relative to 1 mW; dBμ measures power relative to 1 μW.
5. Peak power is the highest level in an optical pulse; average power is the average over a longer interval, typically a second or more.
6. Pulse energy can limit communications because detectors must receive a minimum amount of energy to recognize a pulse.
7. Fiber-optic power meters are calibrated for wavelengths used in fiber systems; they measure average power.
8. Accurate conversions between wavelength and frequency are critical in WDM systems, where the channels are based on *frequencies,* not wavelengths. Always use the exact value for the speed of light in a vacuum in conversions, 2.997925×10^8 m/s, and remember that wavelengths conventionally used are those in vacuum, not in air or glass.
9. Wavelengths can be measured absolutely, or relative to another wavelength.
10. Phase measures a light wave's progress in its 360° oscillation cycle. It is measured relative to the phase of other light waves.
11. Polarization measures the alignment of electric fields in light waves. Polarization-dependent loss can affect system performance.
12. Bandwidth measures the highest analog frequency a system can transmit. The speed of digital systems is measured by the maximum data rate that can be transmitted.
13. Signal-to-noise ratio measures analog transmission quality. Bit error rate measures the quality of digital transmission.
14. An eye pattern superimposes waveforms of successive bits to assess signal quality. The more "open" the eye, the more similar the successive pulses are, and the better the transmission quality.
15. Measurement standards assure that results are compatible. Calibration verifies measurement accuracy.
16. Precise procedures are needed to measure fiber or cable attenuation accurately.

What's Next?

Chapter 18 covers test equipment and troubleshooting techniques.

Further Reading

Dennis Derickson, ed., *Fiber Optic Test and Measurement* (Prentice Hall PTR, Upper Saddle River, NJ, 1998)

Edward F. Zalewski, "Radiometry and Photometry," section 24 in Michael Bass, ed., *Handbook of Optics,* 2nd ed., Vol. 2, (McGraw-Hill, New York, 1995)

Catalogs of test equipment manufacturers typically include tutorials on fiber-optic test and measurement. Two good ones are:

Agilent Technologies, *Lightwave Test and Measurement Catalog* (see *www.agilent.com*)

Exfo Electro-Optical Engineering, *Lightwave Test & Measurement Reference Guide* (see *www.exfo.com*)

Questions to Think About for Chapter 17

1. A 1-Gbit/s signal has an average power of 1 μW at the receiver. What is the average energy in each pulse?

2. A 10-Gbit/s signal has an average power of 1 μW at the receiver. What is the average energy in each pulse?

3. The wavelength being transmitted in Questions 1 and 2 is 1550 nm. How many photons does each pulse contain?

4. Suppose these systems were operating at a wavelength of 850 nm. How many photons would be in each pulse at average power of 1 μW and data rates of 1 and 10 Gbit/s? Assuming that noise levels are constant relative to photons per bit, how would that affect signal-to-noise ratios?

5. The air pressure at the Keck Telescope on the top of Mauna Kea in Hawaii, 4.2 km above sea level, is about 60% of that at sea level. The refractive index of air is 1.000273 at standard temperature and pressure. Assume that the refractive index of air is proportional to density. What are the wavelengths of light with a frequency of 193.1 THz in a vacuum, in air at sea level (standard temperature and pressure), and at the top of Mauna Kea? Be sure to use the exact value of the speed of light, 299,792.5 km/s. How does this compare with the shift in wavelength between optical channels at 50 GHz spacing at the same frequency?

6. What is a frequency difference of 100 GHz equivalent to at the 850 nm wavelength of gallium arsenide lasers?

Quiz for Chapter 17

1. Optical power is

a. light intensity per square centimeter.

b. the flow of light energy past a point.

c. a unique form of energy.

d. a constant quantity for each light source.

2. What measures power per unit area?

a. Irradiance

b. Intensity

c. Average power

d. Radiant flux

e. Energy

3. A digitally modulated light source is on 25% of the time and off 75% of the time. Its rise and fall times are instantaneous. If its average power is 0.2 mW, what is the peak power?

a. 0.2 mW

b. 0.4 mW

c. 0.8 mW

d. 1.0 mW

e. Impossible to calculate with the information given

4. The light source in Problem 3 is left on for 10 s. How much energy does it deliver over that period?

a. 0.2 mW

b. 0.2 mJ

c. 0.8 mJ

d. 2 mJ

e. 8 mJ

5. Light input to a 10-km long fiber is 1 mW. Light output at the end of the fiber is 0.5 mW. What is the fiber attenuation in dB/km?

a. 0.3 dB/km

b. 0.5 dB/km

c. 1 dB/km

d. 3 dB/km

e. 5 dB/km

6. Optical power is proportional to the

a. square of the optical intensity.

b. square of the optical energy.

c. wavelength times the speed of light.

d. square of the voltage applied to the detector.

e. square of the electric-field amplitude.

7. An old meter with its labels worn off measures the output of a 1550-nm laser transmitter at zero lumens even when you turn it to peak sensitivity. What's wrong?

a. The laser is burned out.

b. The meter is reading in radiometric units, which only measure power at 1300 nm.

c. The meter is calibrated in photometric units, which only measure light visible to the eye.

d. You are using a dead optical energy meter.

e. You are using an electrical power meter.

8. Optical channels are spaced 50 GHz apart in a DWDM system. What wavelength difference does this correspond to at 1550 nm?
   a. 0.2 nm
   b. 0.4 nm
   c. 0.5 nm
   d. 0.8 nm
   e. 50 nm
9. A continuous 1-mW beam delivers light at 193.1 THz for one second. About how many photons is this equivalent to? Use a value of 6.626×10^{-34} J/Hz for Planck's constant *h*.
   a. 10^9 photons
   b. 193.1×10^9 photons
   c. 7.82×10^{12} photons
   d. 193.1×10^{12} photons
   e. 7.82×10^{15} photons
10. Pulse duration is
    a. the interval between the time the rising pulse reaches half its maximum height to the time the falling pulse drops below that height.
    b. the interval between successive peaks of the pulse.
    c. the time it takes the pulse to rise from 10% to 90% of its maximum value.
    d. half the cycle of a periodic sine wave.
    e. the time from the start of one pulse to the start of the next.
11. System bandwidth is measured as the
    a. number of bits per second transmitted with no errors.
    b. number of bits per second transmitted with a bit error rate of 10^{-12}.
    c. maximum frequency transmitted with no decline in power from lower frequencies.
    d. frequency at which power has dropped 3 dB from the power at lower frequencies.
    e. wavelength at which power has dropped 3 dB from the power at lower wavelengths.
12. What is the standard measure for transmission quality in digital systems?
    a. Signal-to-noise ratio
    b. Bit error rate
    c. Attenuation from transmitter to receiver
    d. 3-dB bandwidth
    e. Pulse interval
13. Jitter measures
    a. rise time of a digital pulse.
    b. duration of a digital pulse.
    c. shaking of your test instruments caused by people walking through your lab.
    d. uncertainty in bit error rate.
    e. uncertainty in pulse timing.
14. What does an open eye pattern indicate?
    a. Good-quality transmission because a series of digital pulses are nearly identical

b. Good-quality transmission because an analog carrier signal is at peak intensity
c. That a signal of at least one milliwatt is reaching the instrument
d. That the fiber is broken so it cannot transmit noise
e. That you have managed to stay awake through the whole chapter

15. What do you need to measure accurately the loss of a length of cable with connectors on each end?

a. An optical power meter
b. A light source and optical power meter
c. A light source, optical power meter, and launch cable
d. A light source, an optical power meter, a launch cable, and a receive cable
e. A light source, an optical power meter, a launch cable, a receive cable, and a bit error rate test set

CHAPTER

18

Troubleshooting and Test Equipment

About This Chapter

Fiber-optic troubleshooting usually involves checking for damaged or incorrectly installed equipment, or for components that have failed or malfunctioned in some way. Sometimes the problems are simple, but usually they require specialized test equipment to analyze and identify the problem.

This chapter opens by summarizing common problems likely to be encountered in fiber-optic systems. Then it describes important test equipment and its operation, drawing on measurement concepts you learned in Chapter 17. Finally it reviews some simple procedures to track down problems. The techniques covered are specific to fiber optics, not general to communications. That means you will learn ways to spot a broken cable, but not how to diagnose the hardware or software in an electronic switching system.

Fiber-Optic Troubleshooting

The goal of troubleshooting is to diagnose and correct problems. This chapter, like the rest of this book, focuses on fiber-optic equipment, but that still leaves a wide range of potential problems. Fiber-optic systems generally are reliable, but they are not perfect. As Murphy said, "If anything can go wrong, it will."

Our first thought when something fails is usually of a dramatic problem. When your telephone service goes out, you suspect a wire has snapped somewhere. Likewise, it's logical to suspect that a broken cable is behind the sudden failure of a fiber transmission line. Often it is, with culprits ranging from careless backhoe operators to gophers. But

Not all failures have obvious causes like cable breaks.

cable breaks are not always obvious; construction workers may not notice if their equipment doesn't expose broken cable ends, and gophers don't drag the gnawed cables out to show you. You may need special equipment to spot the damaged cable. In fact, it's much more efficient to use test equipment to spot damaged cable from a convenient point than to drive along a cable route looking for a suspicious construction site.

In practice, many other things can cause complete failures. Fibers may be bent too tightly at junction boxes and snap in response to a small strain. Strain, crushing, or contamination may have damaged the fibers inside a cable without visible impact on the cable itself. Dirt may have gotten into a connector and blocked light transmission. A laser may have failed in the transmitter, or a component may have failed in the receiver power supply. WDM components may have gradually drifted out of tolerance until signals no longer fall at their assigned wavelengths.

Many failures occur when something changes, so a good starting point is to ask if users changed the system in any way just before it failed.

Some failures are complete; others merely degrade transmission.

All problems are not complete failures. Noise on your phone line can make it impossible to use a dial-up modem, transmit faxes, or carry on a conversation, but you may still get a dial tone. Likewise, a defective connector or an electronic malfunction in the transmitter or receiver might make a fiber line noisy, or attenuate the signal and increase the bit error rate (or, in analog systems, decrease the signal-to-noise ratio).

Fiber-optic troubleshooting also involves installing new services or changing existing ones. Not all installations involve new cable. Generally telecommunications companies lay more fibers than they initially use. The extra fibers are spares for future expansion, called "dark fibers." Similarly, wavelength-division multiplexing systems are installed with slots for additional optical channels that are not populated with transmitters until the capacity is needed for expansion.

You can't assume these systems are ready to handle the intended traffic. You may need to measure fiber properties such as attenuation, chromatic dispersion and polarization-mode dispersion, to verify it can transmit signals at the required speed and over the intended distance. In some cases, you may need to adjust transmitter power or add (or remove) attenuators to make sure the proper power levels reach the receiver.

You also may need to verify that the fibers go to the proper destination. One common cause of installation failures is that the fibers don't make the proper connections. Someone may have plugged a connector into the wrong socket, and the fiber may go to a dead line.

In short, troubleshooting is a complex task, and the more complex the system, the more complex it is to troubleshoot. Although manual inspection can spot some problems, test equipment generally can help you diagnose and locate problems faster and more accurately. Thus you will spend most of this chapter learning about test equipment and how it can help you. First, however, you should consider what types of measurements you need for what jobs.

Testing, Installation, and Troubleshooting

We can divided measurement and troubleshooting tasks into three broad classes which sometimes overlap.

Testing, installation, and troubleshooting are distinct but related tasks.

- *Testing* evaluates equipment in the laboratory, factory, or field. Research engineers test new systems they have assembled on the laboratory bench to see how well they operate. Quality control technicians test newly built equipment before shipping it to customers. Field engineers test equipment in the field to check that it meets specifications, both after initial installation as part of routine maintenance.
- *Installation* requires testing to verify that equipment works as intended. This may be done both before and after actual installation. If you're installing cable in a difficult site, you may want to spend a few minutes checking that it arrived with all the fibers intact before spending a week laying it. Before adding a transmitter at a new wavelength to a WDM system, you may verify that the system will transmit that wavelength properly. In some cases, you may need to trace the route of a fiber through a system, as you would do in installing an extra home phone line.
- *Troubleshooting* occurs after problems happen with a fiber-optic system, when you try to find the source of the problem and then repair it. Unlike testing and installation, you know there is a problem, but you have to find it. Troubleshooting can require checking many points of possible failure.

The overlaps are many. Installation becomes troubleshooting when you try to turn on the system and it doesn't work. The problem may be a fiber connector plugged into the wrong socket, or it may arise from the interaction of two or more components that are slightly out of specification.

Typically, similar test equipment is used in testing, installation, and troubleshooting. You are likely to use different versions of an instrument on a factory floor and in an all-wheel drive van loaded with troubleshooting equipment for field work, but the measurement principles are the same.

Measurement and Signal Transmission

One important practical consideration in all measurements is whether or not they interrupt communication signals.

Some measurements can be made without interrupting traffic.

Interrupting traffic is not a major concern if the system is not operating because it is being installed or has failed. Interruptions *are* an issue if a system is partly operational, such as when new wavelengths are being added to a WDM system. In that case, you need to be certain that tests on one wavelength channel do not disrupt the operation of others. Avoiding interruption is particularly important for preventative maintenance and testing, when your goal is to verify correct system operation.

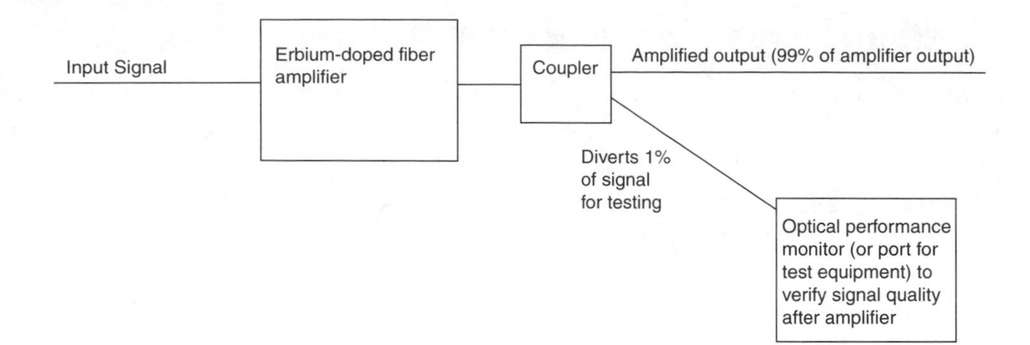

FIGURE 18.1 *Splitting off a small fraction of light for measurement or optical performance monitoring.*

One way to avoid interruptions is to make measurements with a wavelength not used in the fiber system. Fiber transmission can be tested with a signal at 1625 nm, which is outside the operating range of normal fiber systems. This is called *out-of-band testing,* and can spot problems such as damaged fibers or bad connectors without disrupting service. However, it obviously can't verify operation of transmitters at other wavelengths. Another approach is to include taps along the system, which divert a small fraction of light to test equipment or an optical performance monitor, as shown in Figure 18.1.

Test and Measurement Instruments

Catalogs from major equipment makers list many types of equipment designed for fiber-optic test and measurement. Trying to cover them all would take a book in itself, so I will focus on the equipment you are most likely to encounter, particularly in general field service, installation and operation. You may well encounter other equipment in manufacturing or in research and development, but much of it is specialized for such tasks as evaluating performance of erbium-doped fiber amplifiers.

Power, attenuation, wavelength, and signal quality are key measurements for troubleshooting.

The basic measurements are those you learned in the last chapter, including power, energy, attenuation, wavelength, and signal quality. Some instruments directly measure these quantities and tell you the results. Others process and interpret the raw data to give results you need to interpret system performance.

We will concentrate on a few major areas (some of which may be measured together, such as optical power and wavelength in an optical spectrum analyzer):

- Fiber continuity and attenuation (usually measured by comparing power levels)
- Optical power
- Wavelength
- Signal quality
- Polarization

Different instruments measure these quantities differently, and may be used for different applications. During an installation, you may need to check continuity between transmitter and receiver; other times you may want to know attenuation of the fiber between them.

Optical Power Meters

Optical power meters are among the simplest optical measurement instruments. As shown in Figure 18.2, they include a fiber connector, a calibrated detector or detectors, electronics to amplify and process the signal, and a digital display. Most are compact and portable, with autoranging digital readouts that show power on either decibel or watt scales. They are invaluable tools that can be adapted for many measurements.

Optical power meters are calibrated for specific wavelengths.

Power meters typically are calibrated for use at the standard fiber windows at 850, 1300, and 1550 nm. Many also are calibrated for the 660 and 780 nm wavelengths used in some short-distance systems. Some are calibrated for a wider range of wavelengths, particularly for use in WDM systems. Some require switching between separate detectors for long and short wavelengths; others use germanium detectors that can span the range from 780 to 1550 nm.

It is important to take care to check wavelength when taking measurements because detector response and calibration factors differ between wavelengths. Most instruments can store measurements, and many come with computer interfaces.

Light Sources for Testing

To measure attenuation and other characteristics of fiber systems, you need a light source as well as an optical power meter. A variety of fiber-optic test sources are available, designed for different applications, and you must match the source to the measurement task. In simple light sources the wavelength is fixed, but in more sophisticated (and expensive) instruments the output wavelength is tunable with various degrees of precision.

Test sources supply light at fiber system wavelengths.

Test sources deliver a continuous, calibrated power level for measurements of power and loss. They also may have a separate output mode that modulates signals at an audio frequency to aid in identifying the fiber under test. Some have two outputs. Important types include:

- *Broadband sources* normally emit a broad range of near-infrared wavelengths; a familiar example is a tungsten lamp. Typically the entire range of wavelengths is passed through whatever is being measured, with output power measured by a wavelength-selective instrument (an *optical spectrum analyzer,* described below). Although the total amount of light is high, the amount of light per unit wavelength is relatively low. Table 18.1 summarizes the key features of the most important types, tungsten lamps, edge-emitting LEDs, and fiber amplifiers operating with no input signal so they generate only amplified spontaneous emission. They generally are not used in the field.

Broadband light sources include tungsten lamps, fiber amplifiers, and edge-emitting LEDs.

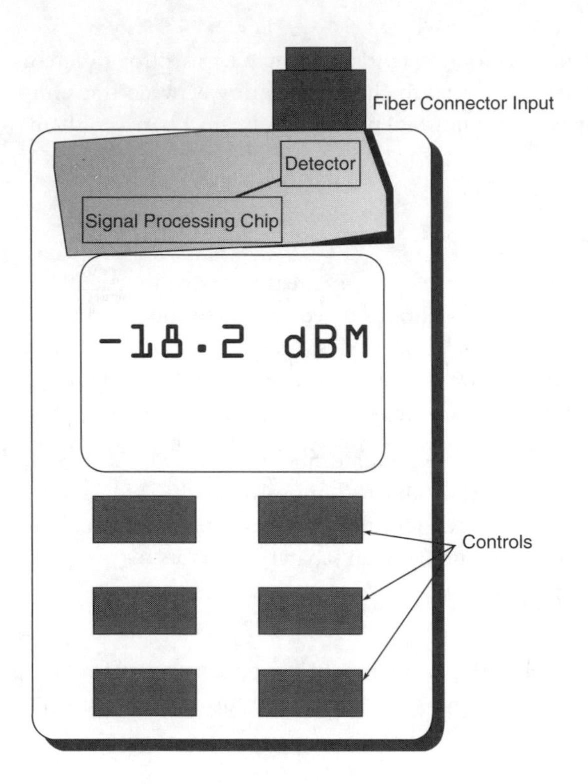

FIGURE 18.2
Handheld optical power meter.

Table 18.1 Broadband sources and their characteristics

	Tungsten Lamp	*Edge-emitting LED*	*Amplified Spontaneous Emission from Erbium-fiber Amplifier*
Wavelength range (nm)	Broadband across infrared	Depends on composition	1500–1600 nm
Spectral width	Whole near-infrared	50–100 nm per LED, multiple LEDs can be used	Peak amplitude 30–40 nm
Total power into single-mode fiber	1 μW	100 μW	1 to 10 mW
Peak power per nm into single-mode fiber	−63 dBm/nm	−25 dBm/nm	−10 dBm/nm

- *Monochromators* are laboratory sources long used in nonfiber measurements. They differ from other types in that their broadband emission is tuned internally by selecting a narrow range of wavelengths with a prism or diffraction grating. Only that narrow range is used for measurements.

- *LED sources* are also used in small portable field instruments. Typically these have center wavelengths of 850, 1300, or 1550 nm and spectral bandwidths of 50 to 100 nm for testing the major glass-fiber windows. They launch microwatts to tens of microwatts into a fiber.

- *Fixed diode laser sources* normally operate at the 850, 1300, and 1550 nm fiber windows. They typically have linewidth of a few nanometers, and deliver more than 100 mW into a fiber. Like LED-based light sources, they normally are used in small, portable instruments designed for field use. Their advantage is higher power levels, allowing measurements through longer lengths of fiber.

- *Tunable laser sources* are lasers that are tuned to change their output wavelength, as described in Chapter 9. Individual lasers have limited tuning ranges, but several may be combined to give a wider range of wavelengths. Tunable lasers can have very narrow spectral widths, and can deliver about a milliwatt in that power range. They generate very precise wavelengths for tests of DWDM components and systems.

Typically, test sources emit continuous beams for measurements, but their output can be modulated for special purposes and to identify the signal or the transmitting fiber.

Field test sources normally come with a selection of connectors and adapters to allow their use with a variety of equipment.

Optical Loss Test Sets

An *optical loss test set* combines a light source with an optical power meter calibrated to work with it. The amount of power emitted by the light source is known, so the power meter measurement indicates how much the received signal has been attenuated. Optical loss test sets also can measure attenuation by comparing power levels with and without the component being tested, as you saw for a cable in Figure 17.9. Optical loss test sets are offered as distinct instruments, but you can think of them as a light source packaged with a power meter.

An optical loss test set includes a light source and power meter calibrated to work together.

Optical return loss test sets are different in that they measure light reflected back toward the source rather than transmitted through the system. Reflections are very important because they can cause noise in edge-emitting semiconductor lasers, degrading system operation. Reflection measurements, sometimes called *reflectometry*, can check for potential problems.

Fiber-Optic Talk Sets

Many types of measurements require two technicians working at different locations to test a system running between the two points. Fiber-optic talk sets were developed to allow them to communicate with each other and coordinate their tasks. Similar equipment is used on copper telephone wires. Fiber-optic talk sets include a simple transmitter and receiver that can send and receive voice signals through optical fibers. In a sense, they turn any available fiber into a telephone line. They generally have headsets attached.

Typical talk sets also can generate a 2-kHz signal for fiber identification. Multifunction talk sets also can generate a continuous signal to measure fiber attenuation.

Visual Fault Locators

A visual fault locator spots faults by sending red light down a fiber.

A visual fault locator is a hand-held troubleshooting instrument that sends red light from a semiconductor laser down a fiber to check for faults such as cracked fibers or defective splices. The visible light travels along the core until it reaches a fault, where it leaks out. Light leaking through the fault can be seen through plastic coatings and jackets under suitable illumination. Infrared light in the signal leaks out at the same point, but your eyes can't see it.

Attenuation of glass fibers is much higher at the 630 to 670 nm wavelengths of red light than in the 1300 to 1650 nm transmission window, but the red light can still travel up to 5 km through standard fibers. Note that the fibers must be exposed to use visual fault location effectively. If the red light leaks out inside a thick cable wrapped in black plastic, you can't see it. The technique is particularly valuable in equipment bays and other places inside buildings where fibers are exposed.

Shining a flashlight beam down a multimode fiber can serve the same function, and has long been used to trace fiber continuity as well. However, the flashlight couples little light into a single-mode fiber.

Live Fiber Detectors

A little signal light leaks out when fibers are bent.

The ability to locate fibers carrying live traffic also is important in troubleshooting. This isn't easy because, as you learned earlier, light passing through a fiber does not generate electromagnetic fields or other external signs of its passage. However, traffic can be detected by bending the fiber, as shown in Figure 18.3. The bend causes light in the core to exceed the critical angle for total internal reflection when it hits the core-cladding boundary, so it leaks out.

Detection of live traffic can be invaluable, but the technique requires care. Sensitive detectors are needed because only a small fraction of the signal leaks out. (Recall that you can't see the invisible infrared signal wavelength.) Bending the fiber too tightly can weaken or break it. Thus you use special instruments that can sense the signal wavelengths after they pass through plastic fiber coating and jackets. Sensitivity of the technique depends on the type of fiber.

Optical Time-Domain Reflectometers

The optical time-domain reflectometer (OTDR) is a powerful and versatile instrument for fiber-optic measurements. It's essentially an optical radar that shoots a short light pulse down the fiber. Glass atoms scatter a small fraction of the light back toward the instrument because of the Rayleigh scattering that you learned about in Chapter 5. Irregularities such as splices, connectors, and defects in the fiber reflect and scatter additional light.

An OTDR is an optical radar that analyzes fiber properties.

The OTDR plots intensity of the light returned to the instrument as a function of time, as shown in Figure 18.4. The plot contains two major components. Rayleigh scattering from the glass produces the background that gradually declines in intensity with distance from

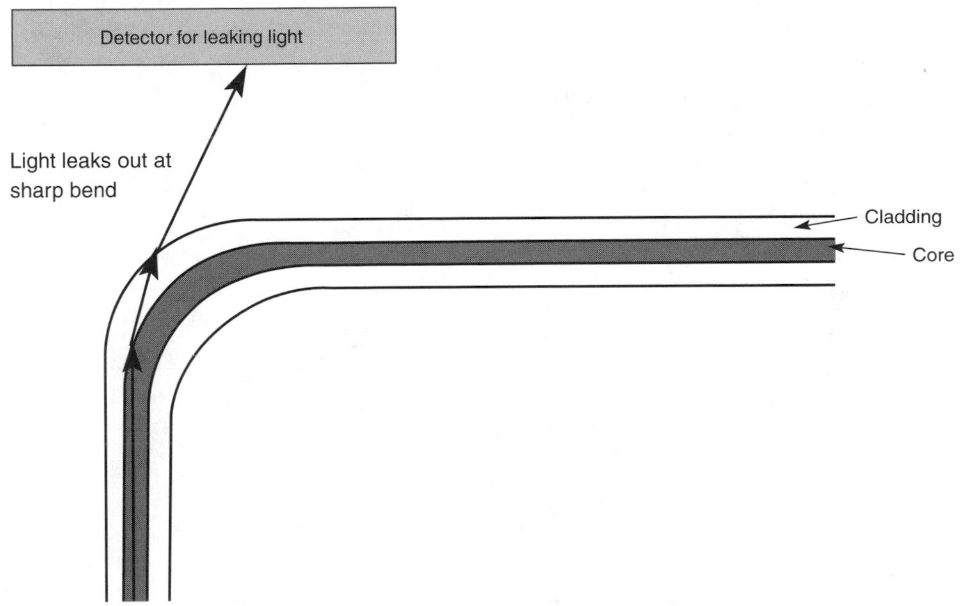

FIGURE 18.3
Live fiber detector: If light leaks out at a bend, a signal is passing through the fiber.

the instrument. If you looked through an infinitely long perfect fiber, you would see only this gradually declining line. The slope of the decline measures fiber attenuation. The peaks are reflections from discontinuities, analogous to the blips on a radar screen that mark objects spotted by radar.

OTDRs cannot measure the near end of the fiber well.

The distance scale at the bottom is calculated from the time light takes to return. Although the optical pulse is very short and the receiver responds very quickly, the signal received from the part of the fiber closest to the instrument is not useful. This "dead zone" ranges from a few meters to perhaps 20 meters.

The largest peak in Figure 18.4 is reflection from the end of the fiber. The next largest peak is reflection from a connector. Look carefully and you can see that the signal just after the connector is slightly lower than it was before the connector; this drop is the connector loss. Mechanical splices also reflect some light back to the instrument and have a measurable loss, but both loss and reflection are smaller in this example. Major flaws like a broken fiber resemble the end of the fiber.

The plot also shows a fusion splice, which has a small loss but does not reflect light back to the OTDR. Overly sharp fiber bends, which allow light to leak from the fiber core, look similar on OTDR plots.

FIGURE 18.4
Features of an OTDR plot.

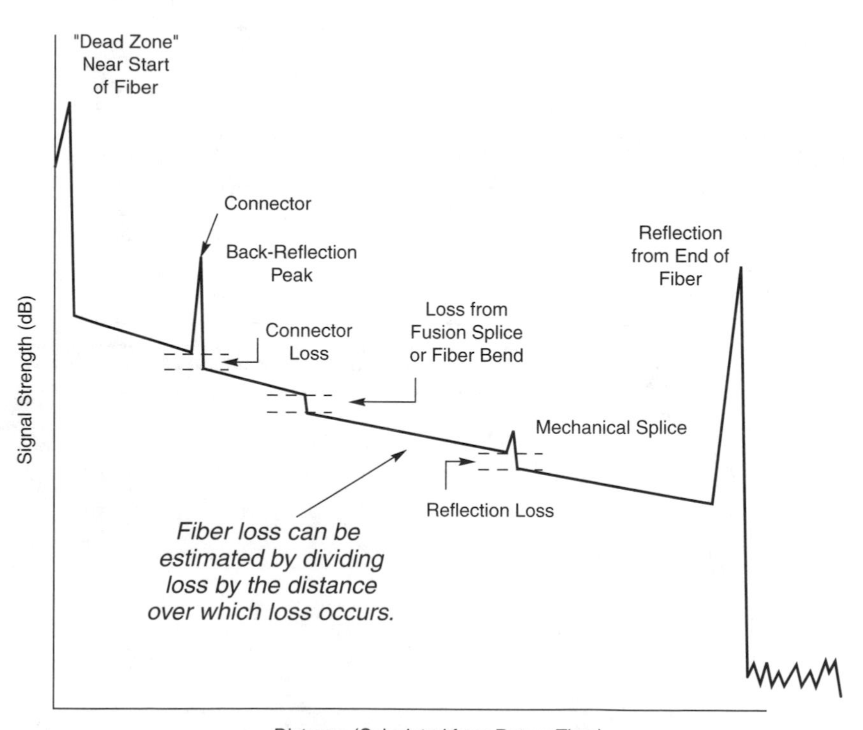

Early OTDRs required the operator to interpret results by reading data plotted on the screen. Modern instruments analyze the data internally, giving readings for quantities such as fiber attenuation, length, and the locations of reflections. They also tabulate the peaks on the curve as "events," listing their location, reflection, and loss, and interpreting their nature. Figure 18.5 shows an actual OTDR display. Options allow you to expand the scale to study suspect parts of the cable in more details. The instruments also include options for printouts or computer output for further analysis.

The OTDR's major attraction is its convenience and ability to spot cable faults remotely at distances to tens of kilometers by sending test signals from one end of a cable segment. This allows a technician to locate a damaged fiber in a long cable without driving along the whole cable route. This instrument can locate a break to within a matter of meters, except in the dead zone, pinpointing the location to investigate. OTDRs are widely used in field measurements, and generally built for field use.

OTDRs can spot cable flaws remotely up to tens of kilometers away.

It is important to realize that OTDR loss measurements are not as accurate as direct measurements of attenuation. The fraction of light scattered back toward the instrument may vary with the type of fiber. Some splices between different fibers appear to be "gainers," with the excess backscatter in the more distant part of the cable making it look as if the signal power has increased.

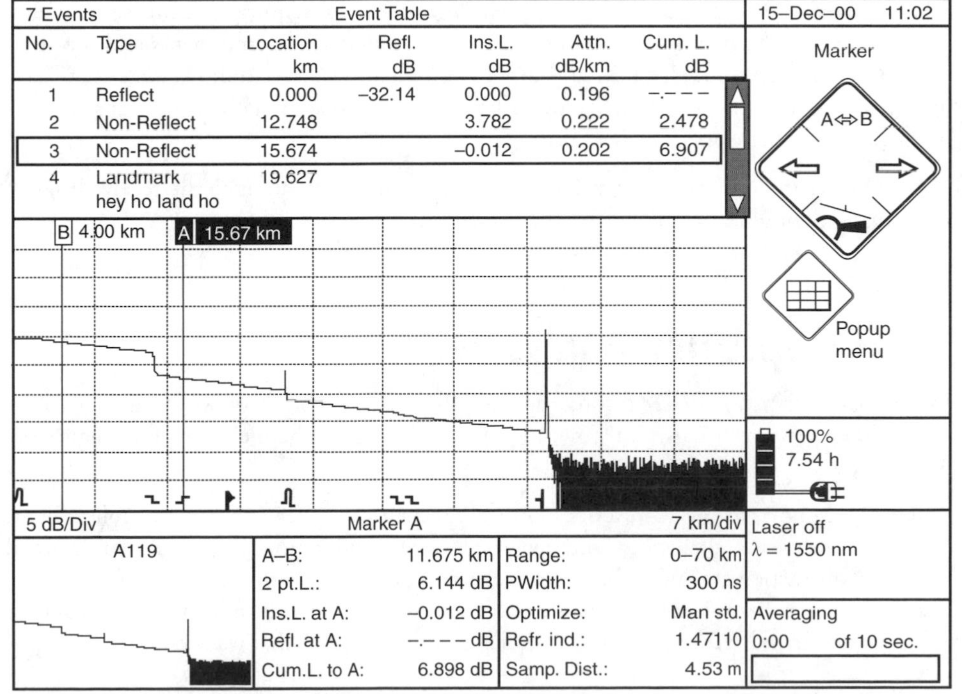

FIGURE 18.5

OTDR display includes log of locations checked, potential anomalies, and loss between points on cable. Spike at right is end of the fiber; reflections beyond that point are noise. (Courtesy of Agilent Technologies)

Oscilloscopes, Analyzers, and Eye Patterns

Oscilloscopes and analyzers perform eye-pattern analysis.

As you learned in Chapter 17, eye-pattern analysis is based on superimposing a series of received signal pulses on top of each other to show how precisely they replicate each other and verify a clear distinction between on and off states. Originally that was done by aligning pulses on an oscilloscope, and this can still be done where necessary.

Modern oscilloscopes or "communications analyzers" are programmed to perform functions automatically, such as eye-pattern analysis. Optional plug-in modules can provide other functions. Normally oscilloscopes are test instruments used in the laboratory or factory, but they also can be used at terminal points and switching centers to diagnose system performance when needed.

Special Test Sets

Some fiber parameters such as chromatic dispersion and polarization-mode dispersion require complex measurements. They are best performed by test sets designed to measure the raw parameters and process the data internally, yielding the desired measurement results. Typically they are used in the lab, factory, or switching center.

Chromatic dispersion testers are large and complex systems that make measurements and calculate results.

Polarization mode dispersion and *polarization dependent loss* require measurements with a *polarization analyzer,* that separates light into its two polarized components and measures their transmission through fibers and other components. Polarization characteristics have become important for high-performance fiber systems, so some polarization analysis systems have been developed for field use.

Bit error rate testers transmit a random bit sequence and compare it to the received signal, measuring the number of bits incorrectly received.

Optical Spectrum Analyzers

Optical spectrum analyzers measure power as a function of wavelength.

An *optical spectrum analyzer* records optical power as a function of wavelength by scanning the spectrum. It can plot power levels on all WDM channels for performance assessments.

Figure 18.6 shows the basic idea of an optical spectrum analyzer. Input optics collect light, generally from an optical fiber, and focuses it onto a diffraction grating. As you learned earlier, a diffraction grating spreads out a spectrum of wavelengths. This spectrum is then focused onto a flat surface, where a narrow aperture or slit transmits a narrow band of wavelengths to a detector. The detector measures the power at that wavelength.

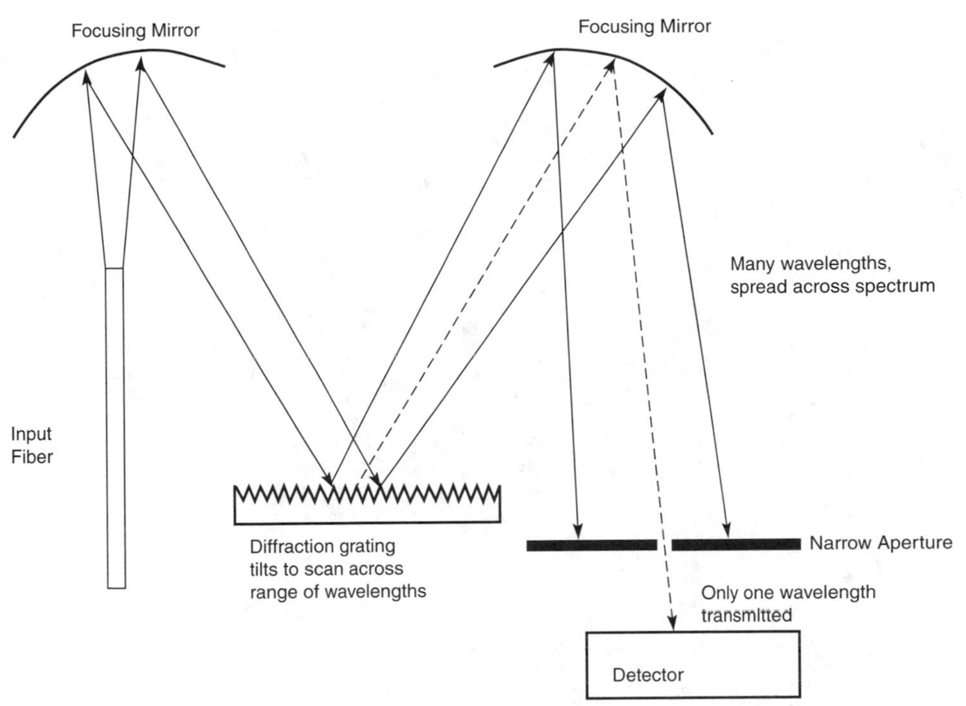

FIGURE 18.6
Optical spectrum analyzer.

In this design, tilting the diffraction grating moves the spectrum, so a different wavelength passes through the slit. Tilting the grating slowly and continuously scans the entire spectrum across the slit. With proper calibration, the power measured at a certain time is correlated with the wavelength passing through the slit at that instant. This makes it possible to plot power against wavelength, and assess the power levels on different optical channels in a WDM system, as shown in Figure 18.7.

Other designs are possible. The slit and detector can move instead of the grating, scanning across the plane where the spectrum is spread out. Alternatively, the spectrum can be spread across an array of detectors, with each one detecting a separate wavelength, an approach used in optical performance monitors, described below.

Note that optical spectrum analyzers measure power over long intervals of time relative to bits, so they show average power on each optical channel.

Optical spectrum analyzers average power over long intervals.

Optical spectrum analyzers began as high-performance laboratory instruments. As wavelength-division multiplexing spread through the telecommunication network, optical spectrum analyzers were adapted for use in the field and in switching centers. They are important instruments in verifying the proper function of WDM systems and in troubleshooting if problems should arise.

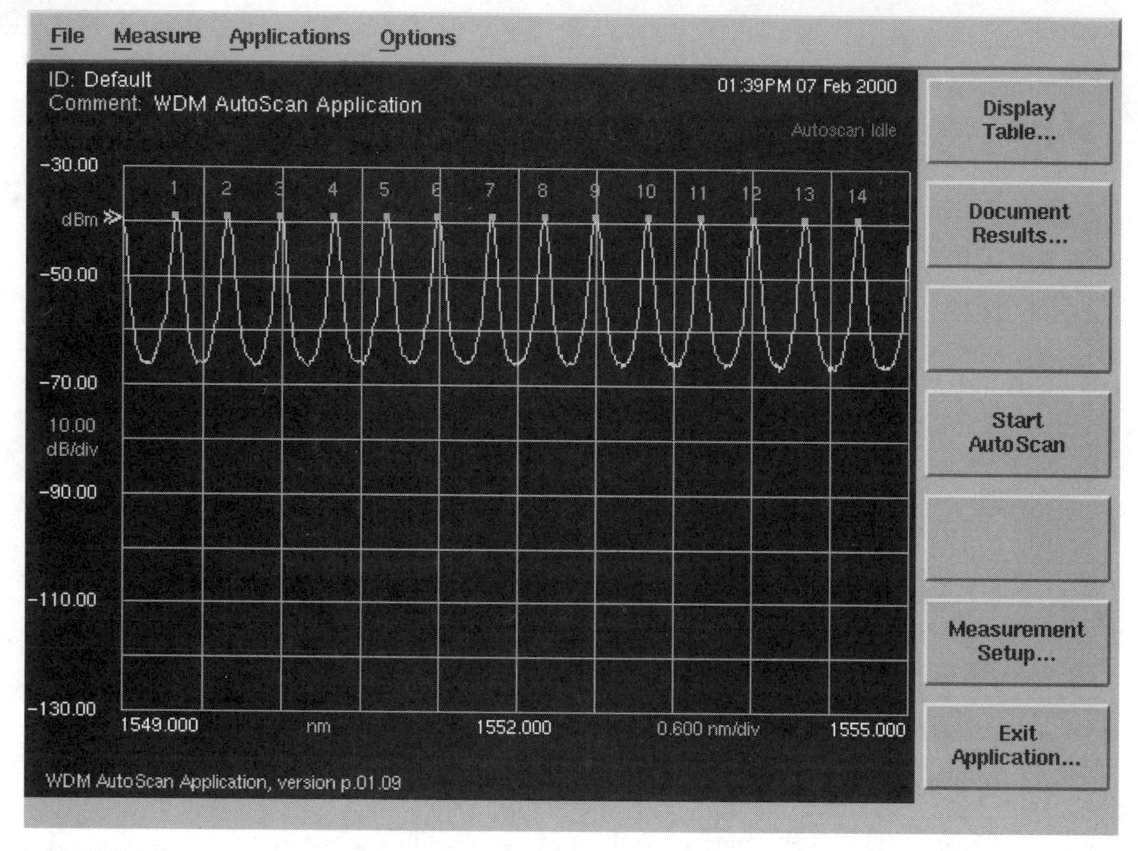

FIGURE 18.7

Optical spectrum analyzer measures wavelength, power, and signal-to-noise ratio of channels in a WDM system. A graphic plot (above) displays the channels; a numerical display (right) gives measurements. (Courtesy of Agilent Technologies)

Optical Performance Monitors

Optical performance monitors are simple versions of spectrum analyzers.

An *optical performance monitor* is a simple version of an optical spectrum analyzer that is permanently installed as part of a WDM system. As shown in Figure 18.8, an optical performance monitor spreads a spectrum across an array of photodetectors. The position of the detector element measures wavelength; the power received by that element indicates the power at that wavelength. Output of the optical performance monitor goes to electronic systems that verify that the signals are at the proper wavelengths and of adequate intensity.

Optical performance monitors generally are part of monitoring systems installed as part of WDM systems to assure their proper operation.

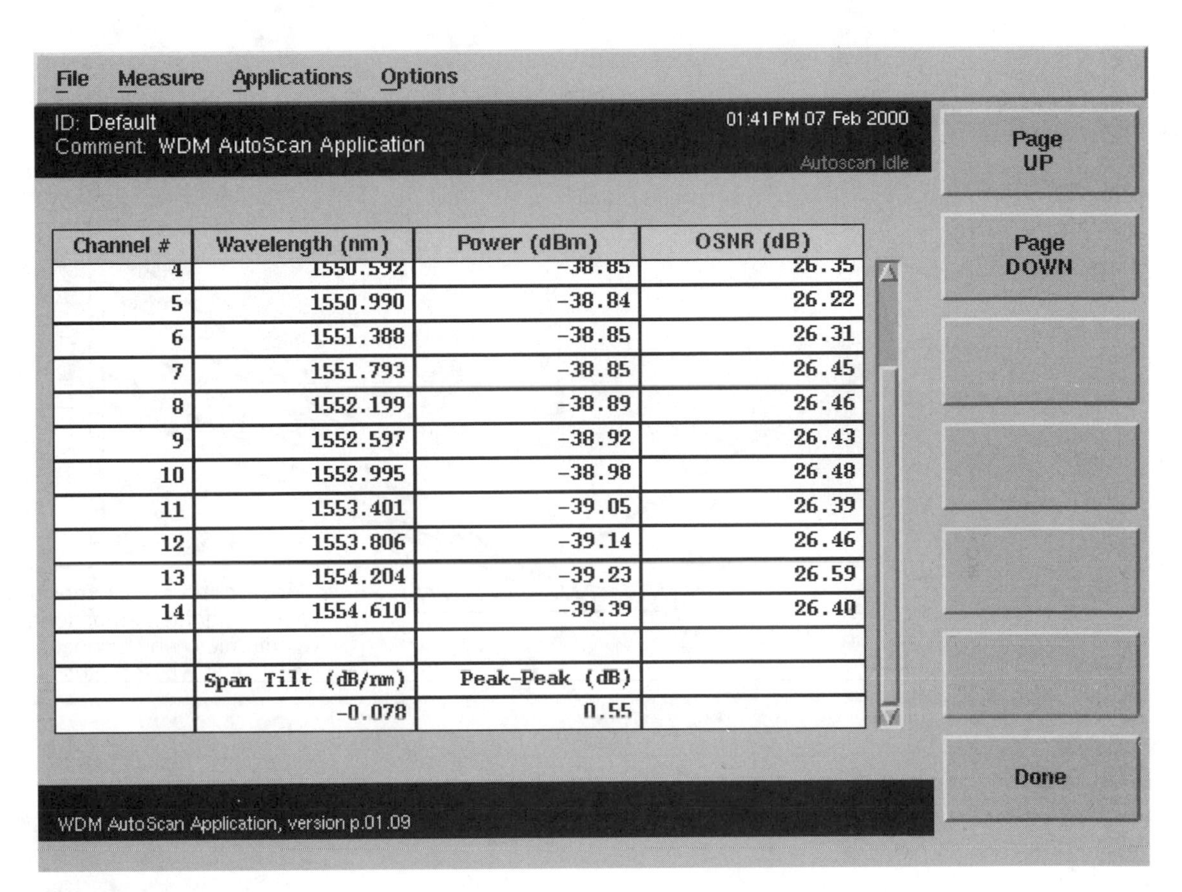

File Measure Applications Options

ID: Default
Comment: WDM AutoScan Application

01:41PM 07 Feb 2000

Autoscan Idle

Channel #	Wavelength (nm)	Power (dBm)	OSNR (dB)
4	1550.592	-38.85	26.35
5	1550.990	-38.84	26.22
6	1551.388	-38.85	26.31
7	1551.793	-38.85	26.45
8	1552.199	-38.89	26.46
9	1552.597	-38.92	26.43
10	1552.995	-38.98	26.48
11	1553.401	-39.05	26.39
12	1553.806	-39.14	26.46
13	1554.204	-39.23	26.59
14	1554.610	-39.39	26.40
	Span Tilt (dB/nm)	Peak-Peak (dB)	
	-0.078	0.55	

Page UP

Page DOWN

Done

WDM AutoScan Application, version p.01.09

FIGURE 18.7
(Continued)

Wavelength Meters

Accurate wavelength measurements are essential for building and operating WDM systems. Optical spectrum analyzers give approximate values, but wavelength meters can give more precise values.

Wavelength meters are built around a device called the Michelson interferometer. It consists of a beamsplitter and a pair of mirrors, as shown in Figure 18.9. The beamsplitter divides input light into equal portions, with one-half transmitted through the beamsplitter, and the other half reflected to the other *arm* of the interferometer.

A wavelength meter is based on the Michelson interferometer.

In Figure 18.9, the lower arm is a fixed length; a mirror at the end reflects light back toward the beamsplitter. Half the reflected light passes through the beamsplitter to the detector at top; the other half is reflected back toward the light source. The mirror at the end of the

FIGURE 18.8
Optical performance monitor.

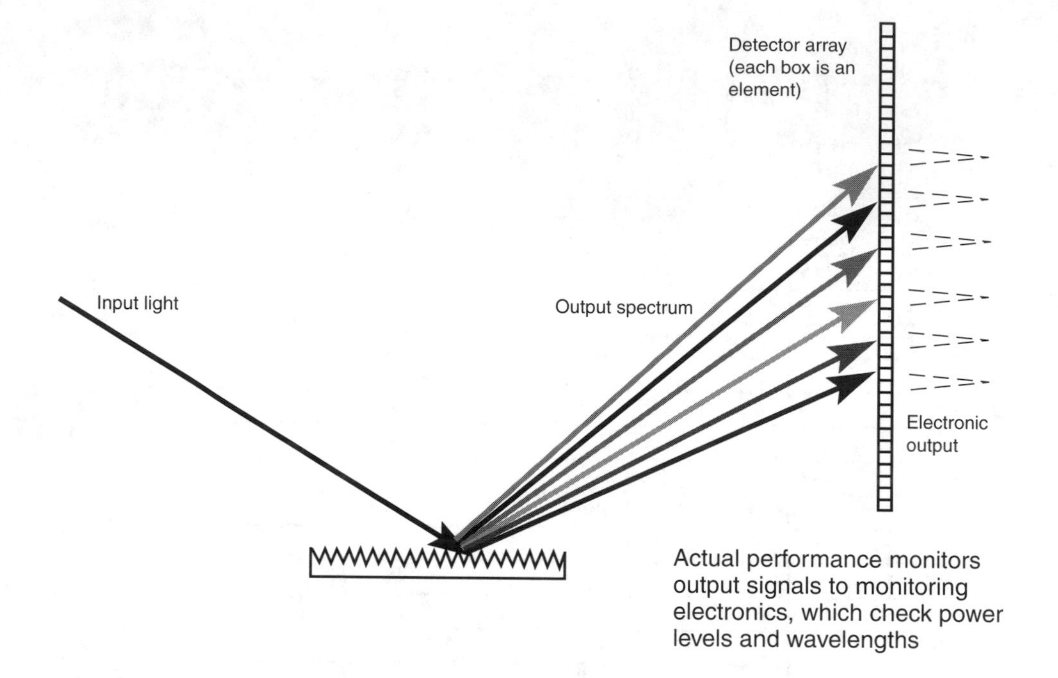

FIGURE 18.9
Wavelength meter counts interference fringes as mirror moves over a known distance.

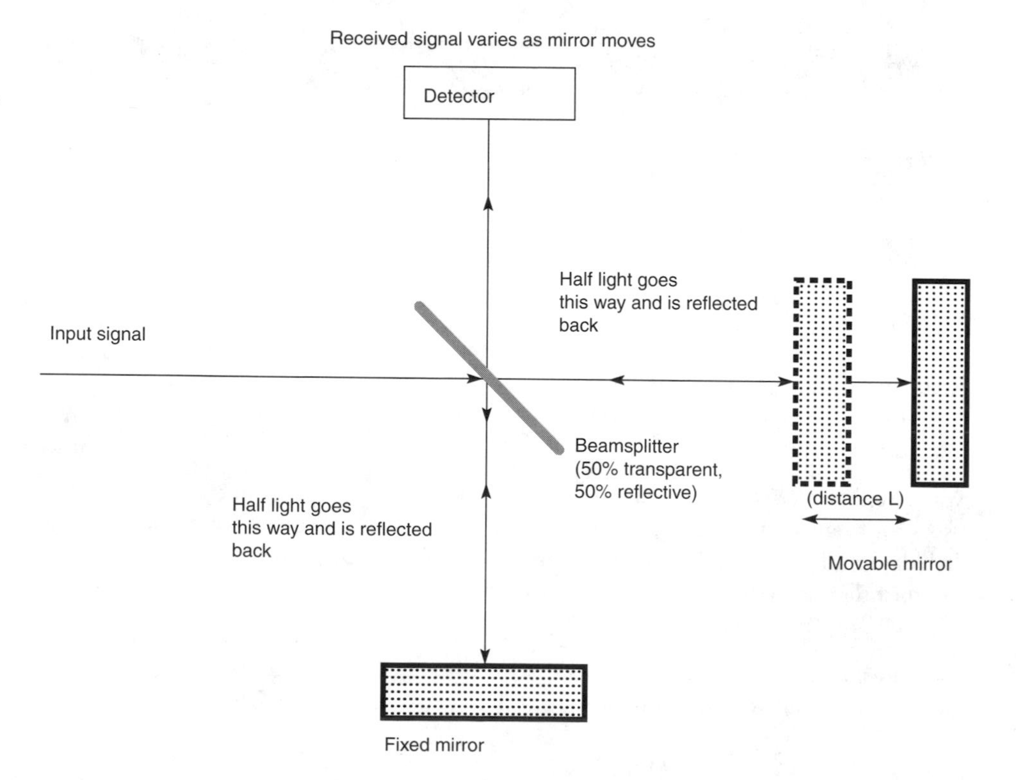

horizontal arm is moved back and forth, so its length varies. That mirror also reflects light back toward the beamsplitter, with half the light passing through and the other half reflected toward the detector.

The amplitudes of the beams reaching the detector from each arm are identical. Because the light came from the same source, the waves from the two arms are nominally identical, so they can interfere constructively or destructively. Suppose you initially adjust the moving mirror so the light amplitudes cancel out at the detector. Then you move the mirror a known distance, counting each time the light reaching the detector goes through a light-dark-light cycle. If you know the distance L precisely and find that the light goes through N cycles as the mirror moves that distance, you can calculate wavelength from the equation

$$\lambda = \frac{2L}{N}$$

(The factor of 2 comes from the fact that the light makes a round trip through the arm, so the light travels twice the distance of the arm.)

Precise measurements of wavelength require accounting for the refractive index of dry air, 1.000273 at 1550 nm, and standard temperature and pressure. Extremely precise measurements require considering air pressure and humidity as well, which changes the refractive index of air slightly.

Wavelength meters give digital measurements of both wavelength and power at that wavelength for use in the factory or at operating sites. They can be calibrated by measuring known wavelengths.

Troubleshooting Procedures

Fiber-optic systems are far too diverse to give a single, all-purpose guide to troubleshooting. Because this book concentrates on principles of the technology, this section covers basic concepts rather than details. Specific procedures depend on the nature of the system, and often require looking at the electronic parts of the network to isolate the problems.

Troubleshooting is a systematic way of performing tests to isolate problems. You may perform it in different ways depending on the tools you have at hand and your starting location. If you are testing a faulty point-to-point transmission line, your first task is to isolate whether the fault is in the transmitter, the cable, the receiver, or an attached connector. Figure 18.10 gives two alternative approaches for a single technician, starting at either the transmitter or receiver end. The basic idea is to test what you have easy access to first, before traveling to a remote site. A pair of technicians working at each end and able to communicate with each other would follow a different procedure.

Troubleshooting systematically analyzes problems.

The OTDR's ability to spot problems remotely makes it invaluable for troubleshooting point-to-point systems from a single location, although the distance measurements may

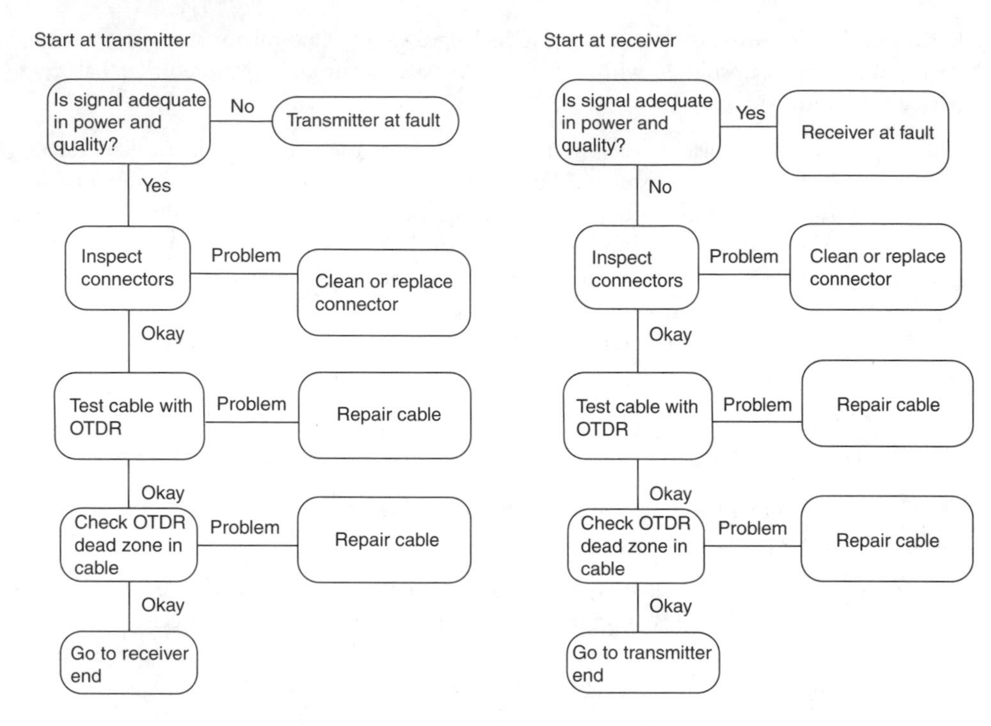

FIGURE 18.10 *Troubleshooting procedures for point-to-point link.*

not be exact. Cables usually include "storage loops"—extra lengths of cable that allow room to replace damaged cable. Nonetheless, if you can pinpoint a break in a 100-km cable to the nearest 20 m, you have an excellent head start in finding the damage.

Other equipment also assists in diagnosing faults in cable systems. Table 18.2 lists some typical cable problems, their causes, and ways to identify and fix them.

Table 18.2 Typical problems in point-to-point fiber systems

Problem	*Origin*	*Diagnostic Equipment*	*Repair*
Excess connector loss	Dirt or damage	Inspection microscope	Clean, reinstall, replace
Excess loss localized in fiber or cable	Excess bending	Visual fault locator or OTDR	Straighten fiber or cable
Excess splice loss	Aging, stress, or contamination	OTDR or visual fault locator	Reinstall
Fiber break	Physical damage to cable	OTDR, visual fault locator	Splice or replace cable

Office local-area networks offer other types of problems in equipment such as patch panels and entry boxes. Cables in office buildings also face different menaces than outdoor cables. People can snap fibers by tripping over loose cables or tugging on equipment attached to the cable. Breaks are most likely near connectors. Cables hidden in walls or suspended ceilings can be cut accidentally. The more equipment connected, the more things can go wrong.

The majority of cable runs in local-area networks generally are too short for OTDRs to be useful, but LANs usually cover small areas, making on-site service easier. Table 18.3 lists a

Table 18.3 Local-area network troubleshooting

Equipment to Check	*Tasks*
Does not interrupt network operation	
Patch cords and panels	Trace cabling to verify proper connections are made; correct if needed.
Patch cables and fibers	Check bend radius; correct if too small.
Transmission cables	Inspect for damage or tight bends. Check for signs of recent construction near cable.
Powered equipment	Check power is on and monitors show normal operation. Verify configuration.
Cable connections	Verify connections are secure.
Outdoor plant	Inspect for signs of damage or evidence of recent construction or other disruption.
May interrupt network operation	
Cable connections	Wiggle connectors to hunt for intermittent connections.
Interrupts network operation	
Patch cables and cable connections	Disconnect, test cables, and inspect and clean connectors at all termination points.
Transmission cables	Measure attenuation and compare to specified values.
Long cables	Test with OTDR and compare to records, looking for changes.
All cables with laser transmitters	Check for back reflection toward transmitter, which could disrupt laser operation.
Transmitters	Measure output power level and compare to records and specifications.
Receivers	Measure power arriving through fiber and compare to records. Verify it equals transmitter output minus cable loss.

number of steps you can take to check system function. They are divided into tests that don't disrupt system operation and those that interrupt service.

Failure versus System Degradation

Failures may be total or merely degrade system operation.

One important way of sorting problems is to determine whether they result in permanent and total system failure or merely degrade system operation. (Degradation includes those maddening intermittent problems that often go away temporarily when you try to diagnose them.) Always check for any changes in system or terminal configuration. If the network "worked fine until I moved my workstation to clean the desk," check for damage to the attached cable or dirt in the connector. If a user just installed new communications software, testing the receiver and fiber output may show it's not a hardware problem at all.

Typically—but not always—total failures mean some key component has failed or power is down to an essential part of the system. Any obvious changes in the environment are the leading suspects. If a construction crew is digging a hole in the street where your cable runs and that part of the system goes down, they're your number one suspect. Likewise, check for recent changes in junction boxes; somebody may have replaced a cable incorrectly and plugged the connector into the wrong adapter.

Intermittent problems suggest loose connectors or partial damage to connectors. A simple check is to wiggle cables and connectors while monitoring for changes in operation.

Degraded operation, such as reduced data transmission speed, may indicate noise in the system. You can rule out a cable break, but not damage to the fiber or bends that cause large light losses. The transmitter or receiver may be generating excess noise, or producing weak signals. Dirt in a connector may be the problem. System degradation also may be a warning of imminent failure of a component or subsystem. Generally such degraded operation is likely to require test equipment to track down the problem.

WDM Troubleshooting

WDM troubleshooting requires special techniques and instruments.

Other challenges arise in systems transmitting multiple wavelengths through the same fiber. So far we've assumed that each fiber carries only one signal. WDM systems carry multiple signals, which means more potential problems to track down.

A total failure of a WDM system—where all channels are out and no light comes through the system—can be treated much like total failure of a single-channel system. No individual laser failure would knock out an entire WDM link, but a power failure could. Cable breaks also could disable an entire WDM system. An obvious first test is to see if power is reaching the receiver end.

Table 18.4 lists some potential failure modes and what parameters to check first as the most likely causes. Optical spectrum analyzers can check the power level on each optical channel. Optical performance monitors can verify operation of optical amplifiers. Wavelength meters may be needed for precise measurements of optical wavelengths.

Table 18.4 WDM system failures and likely causes

Type of Failure	*Check First*
All channels down, no power reaching receiver	Cable break, power failure at transmitter site.
All channels down, power reaching receiver	Optical channels at proper wavelengths using optical spectrum analyzer. Verify power at receiver, performance of demultiplexer. Measure noise levels.
One channel down; all others operating	Failure of transmitter or receiver on that channel. Failure of multiplexer or demultiplexer on that channel. Noise on that channel.
Some channels down, others operating	Drift of some optical channels from assigned wavelengths. Performance of optical amplifiers. Noise on affected channels. Problems in multiplexer or demultiplexer. Look for patterns in failures.
Degradation of one channel only	Degradation of transmitter or receiver on that channel, possibly including wavelength drift. Misalignment of that channel in multiplexer or demultiplexer. Background noise.
Degradation of multiple channels	Problems in multiplexer or demultiplexer. Problems in optical amplifiers. Broad background noise (e.g., amplified spontaneous emission).
Degradation of all channels	Entire multiplexer or demultiplexer. Problems in optical amplifier. Damage to cable or connectors causing noise.

What Have You Learned?

1. The goal of troubleshooting is to diagnose and correct problems.
2. Some failures are complete; others merely degrade transmission. Not all failures have obvious causes.
3. Testing evaluates equipment in the laboratory, field, or factory. Installation requires testing to verify equipment works as intended. Troubleshooting responds to problems.
4. Some instruments directly measure power, attenuation, wavelength, and signal quality; others interpret raw data to analyze system performance.

5. Optical power meters are calibrated for the standard wavelengths used in fiber-optic systems.
6. Test sources include LEDs, fixed diode lasers, broadband sources, and tunable lasers, which operate at fiber system wavelengths.
7. An optical loss test set includes a light source and power meter calibrated to work together to measure attenuation. Optical return loss test sets measure light reflected back to the source.
8. A visual fault locator spots faults by sending red light down a fiber, so you can look for scattered red light.
9. Bending a fiber allows a little signal light to leak out; detecting this light is a simple test for live fibers.
10. An optical time-domain reflectometer (OTDR) analyzes fiber properties by sending a short light pulse down a fiber and measuring light scattered back to the instrument. It allows a technician at one end of a cable to spot distant flaws.
11. Oscilloscopes and communications analyzers perform eye-pattern analysis.
12. Optical spectrum analyzers spread a multiwavelength signal into a spectrum of light and measure the power at each wavelength. They are valuable for testing WDM systems.
13. A wavelength meter uses a Michelson interferometer to measure the wavelength of light.
14. Preferred troubleshooting techniques depend on the tools you have at hand, your starting location, and the type of system you are analyzing.
15. One way to classify problems is to determine whether they cause permanent and total system failures or merely degrade operation.
16. WDM troubleshooting requires special instruments such as optical spectrum analyzers and wavelength meters.
17. Some tests can be performed without disrupting system operation; others interrupt service.

What's Next?

In Chapter 19, we will turn to the basic concepts behind fiber-optic communication systems and optical networking.

Further Reading

Bob Chomycz, *Fiber Optic Installer's Field Manual* (McGraw-Hill, 2000)

Dennis Derickson, ed., *Fiber Optic Test and Measurement* (Prentice Hall, 1998)

Jim Hayes, editor, *Fiber Optics Technician's Manual* (DelMar Publishers, 1996), see Larry Johnson, "Fiber optic restoration" (Chapter 16), and Jim Hayes "Fiber optic testing," (Chapter 17).

Catalogs:

Agilent Technologies *Lightwave Test and Measurement Catalog (www.agilent.com)*

Exfo Electro-Optical Engineering *2000 Lightwave Test & Measurement Reference Guide (www.exfo.com)*

Questions to Think About for Chapter 18

1. Devise a troubleshooting procedure for testing your ability to connect to the Internet using a personal computer, an external modem, and a standard telephone line.
2. Which is more likely to fail: an old fiber-optic cable left undisturbed in an underground duct, or a new office cable that was plugged into a new computer?
3. What is the advantage of measuring fiber attenuation at 1625 nm?
4. You want to test a WDM system by transmitting signals through it one wavelength at a time at the same power generated by the standard transmitters. What sort of light source should you use?
5. You test a 50-km cable with an OTDR and find a sharp peak at 25 km and no signal returning from greater distances. What does this tell you about the cable?
6. When the moving arm in a wavelength meter moves 7 mm, the instrument counts 9,000 fringes. What's the difference between the wavelength in vacuum and the wavelength in air ($n = 1.000273$)?

Quiz for Chapter 18

1. A standard optical power meter would *not* be calibrated for measurements at
   a. 850 nm.
   b. 1300 nm.
   c. 1400 nm.
   d. 1550 nm.
2. An optical loss test set includes a(n)
   a. wavelength meter and power meter.
   b. optical spectrum analyzer and power meter calibrated across the same range of wavelengths.
   c. power meter and a length of fiber calibrated to work together.
   d. light source and a power meter calibrated to work together.

e. power meter, light source, wavelength meter, and optical spectrum analyzer.

3. A visual fault indicator does what?

a. Shines red light down the core of a fiber to make visible any flawed points where light leaks from flaws

b. Illuminates the plastic coating of a fiber with red light to spot any uneven spots on the surface

c. Shines light through the hollow zone of a loose tube cable to search for any fiber fragments in the cable

d. Illuminates the outside of a fiber with ultraviolet light to cause fluorescence where light leaks from the fiber

e. None of the above

4. You can test a fiber to see if it's carrying an optical signal by

a. pointing the end at a white piece of paper and looking for fluorescence.

b. scraping away the cladding and monitoring for light leaking out.

c. bending the fiber and monitoring with an infrared sensor for light leaking out.

d. removing the plastic coating and looking for light in the fiber.

e. all of the above

5. You use an optical time-domain reflectometer to analyze a fiber with an 10-μm air gap at a connector. What would you expect the OTDR to show at the point where the connector is installed?

a. Nothing

b. A strong reflection accompanied by a loss

c. A strong reflection accompanied by a gain

d. A flat region across the gap

e. A sharp drop in scattering in the air gap, followed by higher scattering in the glass

6. What type of test equipment can best identify a fiber that has been moved and connected to the wrong point in a patch panel?

a. An OTDR

b. An optical loss test set

c. An optical power meter

d. An optical spectrum analyzer

e. Manual inspection and comparison with records

7. What type of instrument can display an eye pattern?

a. Oscilloscope

b. Optical loss test set

c. Optical power meter

d. OTDR

e. Optical spectrum analyzer

8. An optical spectrum analyzer can record

a. the spectrum of a light source.

b. all the wavelengths transmitted by a WDM system.

c. the optical channels amplified in an erbium-doped fiber amplifier.

d. the wavelengths transmitted by an optical multiplexer.

e. all of the above

9. A wavelength meter is based on what optical system?

a. Mach-Zehnder interferometer

b. Optical spectrum analyzer

c. Michelson interferometer

d. Fabry-Perot interferometer

e. Ross-Perot interferometer

10. How would you locate a break in an aerial cable?

a. Drive along the line and look for fallen cables

b. Measure cable loss with an optical return loss meter

c. Measure the cable with an OTDR

d. Measure power level at the receiver

e. Listen to police radio for reports of drunk drivers hitting utility poles

11. The bit error rate of the fiber-optic system connecting your building to the Internet has reached 10^{-3}. What possibility can you rule out when you start troubleshooting?

a. A backhoe broke the cable

b. The laser transmitter has gotten too warm and drifted off wavelength

c. Dirt in a connector

d. Moisture has contaminated an outdoor splice

e. A kink in the cable at a junction box

12. An optical spectrum analyzer shows that your WDM system delivers no signal at all at one wavelength but other channels are working fine. Which of the following could have caused the problem?

a. A gopher gnawed one fiber in the cable

b. The laser for that channel failed

c. Dirt has gotten into a connector

d. The cable is kinked at the junction box

e. Failure of an optical amplifier

CHAPTER

19

System and Optical Networking Concepts

About This Chapter

The fiber-optic components described in earlier chapters are assembled into systems to provide communication services. This chapter takes a closer look at basic system concepts that were introduced in Chapter 3. To understand telecommunications systems and the emerging optical network, you need to learn both the specifics of how fiber-optic systems transmit signals and the tasks these systems perform. This chapter is the first of several that will teach you about optical networks and the services they provide.

An Evolving Network

We live at a time when telecommunications is changing very fast. The technology used to transmit signals, the nature of telecommunications services, and the structure of the industry all are shifting rapidly. This chapter and those that follow concentrate on teaching you the basic principles of fiber-optic systems. Don't be surprised if the details change. The terminology is evolving even faster than the hardware.

I will start with an overview of key areas, then turn to some specific concepts that are important buildings blocks of fiber-optic communications.

Telecommunication Systems

Chapter 3 introduced the concept of telecommunication systems. Telecommunication networks deliver voice, video, and data services around the globe. Extensive interconnections among global, national, and local systems allow you to send electronic mail to Japan, a fax to Africa, or to phone someone in England.

The tele-communications network is evolving rapidly.

Telecommunication networks are global.

The term *telecommunications* does not specify what technology delivers signals. Traditional electronic telecommunications relied on radio waves transmitted through the air or electronic signals transmitted through copper wires. Much of the global telecommunications network remains electronic, mostly on the edges where transmission demands are modest. Copper wires carry signals to telephones and fax machines in your home and office, to your personal computer, and to many television sets. Radio signals deliver broadcast radio and television, and send voice and data to mobile phones.

Fiber-optic equipment is common in the heart of the network. Generally the more signals that need to be carried, the more likely they will travel through fiber. National long-distance networks, submarine cables, and "trunk" cables between local telephone switching offices are examples.

You need to understand telecommunications in general to understand the role of fiber optics. This book can't cover all of the details, but it can explain how and where fiber optics are used. This chapter concentrates on the building blocks of telecommunication systems, as well as fundamental concepts that affect fiber systems. Later chapters cover standards, fiber system design concepts, and specific types of fiber systems.

Optical Networking

Optical networks manage signals in the form of light.

Optical networking is becoming an integral part of telecommunications. Earlier you saw that a telecommunications system can be seen as "pipes," which transmit signals, and "switches," which organize and direct them. Early fiber-optic links were merely pipes, which transmitted signals from one switch to another. Optical networks add switching, so that signals can be organized and managed as light instead of purely electronically. The details are still being worked out.

Today, optical networking is done mostly in long-distance backbone systems. Optical networking technology is spreading to other parts of the telecommunication system that carry high-speed signals, but most of the network—especially the slower outer portions—remains electronic.

All-optical networks process signals entirely in optical form, without converting them to electronic form. In practice, only limited parts of the fiber-optic network are all-optical because most switching remains electronic. Thus many optical networks are not *all* optical.

Changing Technology and Business

Both technology and business are changing tele-communications.

Changing businesses and technologies are driving changes in telecommunication systems.

Just 20 years ago, the structure of the American telephone network was determined largely by the monolithic Bell System, formally known as the American Telephone and Telegraph Corporation. For most Americans, the giant AT&T was "the telephone company." A single backbone network linked regional nodes, which in turn fed signals to and from local networks. The network resembled a centrally designed system of veins and arteries that carried information.

Now long-distance communications is largely separate from local telephone service, and many companies offer local or long-distance phone service (although few offer both). The network has become more diverse and complicated. Several companies operate their own long-distance backbone systems. Nonetheless, much of the earlier structure remains behind the scenes, and most "competitive" local phone companies deliver service to homes by leasing wires from the "Baby Bells" formed by the first breakup of AT&T. Corporate changes cause other complications. (Mergers have changed my local phone company from Nynex to Bell Atlantic to Verizon, but the same wires run to my house.)

Technological changes are reshaping the telecommunications network as well. Currently a hot debate rages between advocates of "traditional" telephone service on dedicated circuits, and converting voice calls to data packets like those carried on the Internet. We'll talk more about that idea in Chapter 25.

Telecommunication Network Structure

Today's global telecommunication network carries a mixture of voice, video, and data signals. Its structure is far more complex and fluid than in the days when it carried only long-distance telephone calls for national telephone monopolies. Let's look briefly at the overall topology before turning to the structural elements that build up the network.

Nodes and Connection Points

Telecommunication networks consist of links running between nodes, as shown in Figure 19.1. Typically there are multiple routes between pairs of nodes, both to handle traffic flow and provide backup in case of cable failures. Switches or routers at the nodes direct signals to the next node, or distribute them locally.

Telecommunications networks are made of links between nodes.

Each node is a point that serves many users—it could be a neighborhood distribution center, a telephone switching center in a small town, or a major switching center in a big city. These nodes, in turn, distribute signals farther, to other nodes or to individual customers. Small nodes connect to larger ones and customers; larger nodes connect to smaller ones and to high-capacity backbone systems. The traffic is highest on the backbone routes between major population centers. Although individuals make more local calls, they are spread through many local cables, while long-distance calls are carried by a few high-capacity cables.

Types of Services

We generally think of telecommunications as including voice, video, and data signals. Thanks to the Internet, digital data accounts for the largest share of traffic volume in the United States.

Telecommunications includes voice, video, and data signals.

These services are distinct and have different transmission requirements, although they may travel over the same lines. Voice requires little bandwidth per channel but is highly

FIGURE 19.1
Network connects nodes.

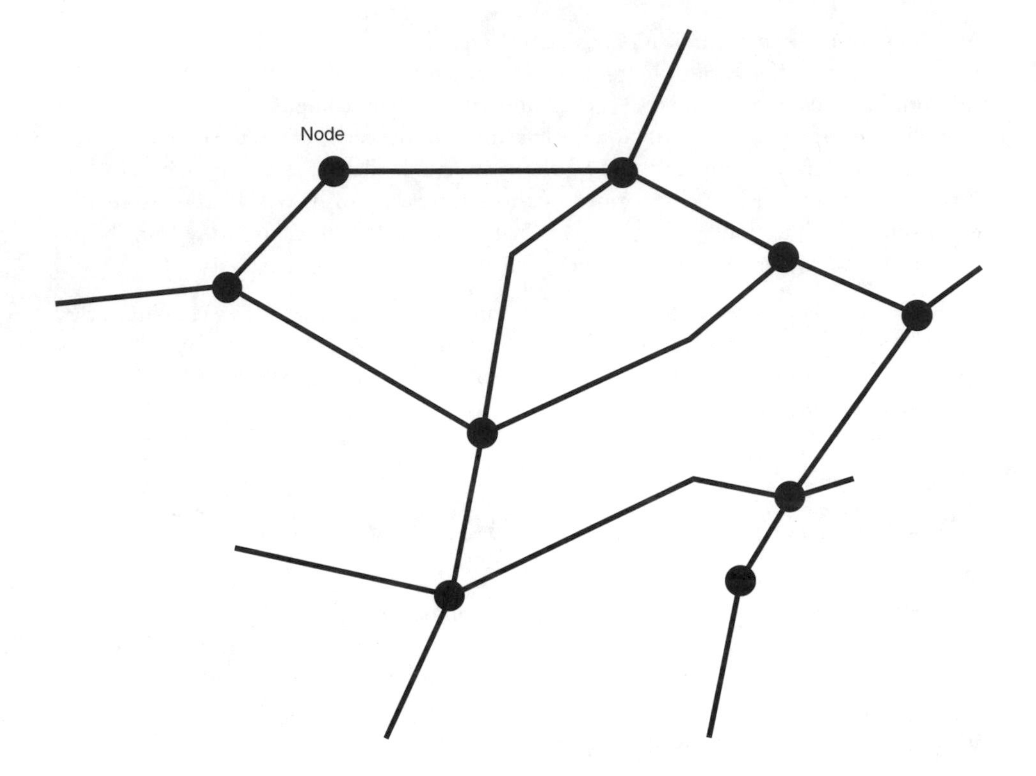

sensitive to delays, so "voice channels" or "switched circuits" normally are reserved for telephone calls. Video requires much more capacity, and also is sensitive to delays; in practice, it often is transmitted separately. In contrast, Internet transmission varies widely with time, but can tolerate delays, leading to the design of the modern Internet.

In practice, voice, video, and data are transmitted partly on dedicated networks and partly on a single shared network. Dedicated networks typically are fine-tuned for one application, but it's expensive to build new networks, so new services are often added to existing networks, like dial-up modems and facsimile transmission were added to the voice telephone network.

Circuit and Packet Switching

The crucial operating difference between voice telephone transmission and the Internet is how signals are packaged.

The telephone system is circuit-switched. The Internet is packet-switched.

Since the telephone was invented in the nineteenth century, voice calls have been made over dedicated circuits. Originally calls went over pairs of wires strung between phones; now they are assigned a reserved slot in a stream of other data. This reserved capacity means that once a connection is made, the phone responds the instant you start talking, and the person at the other end hears you immediately.

The Internet was designed around packet switching, which groups data together into packets for transmission. Each packet has an address header, which is read by network elements called *routers,* that directs the packet to its destination. Packets can fill the distribution pipeline much more efficiently because they don't have to reserve particular capacity, and they can be delayed. That's fine for electronic mail or for downloading pages from the Web, but annoying when it makes streaming video flow unevenly, or conversations over the Internet have puzzling timing.

Circuit- and packet-switched signals usually are separated. Packet-switched signals can be repackaged to go over a circuit-switched line, but with a cost in efficiency. Internet-style packet transmission generally is cheaper per unit capacity than circuit switching, but it can induce delays that pose problems for voice and video traffic. That fundamental issue is at the base of the debate about using Internet formats for voice transmission.

Standards, Protocols, and Layering

Telecommunication systems follow standards that enable different parts of the network to talk with each other. We take them for granted as long as they work, and they play a vital role in modern systems. If the telephone network in Seattle cannot interpret the sequence of bits transmitted by a telephone network in London, it won't matter if both callers speak English.

You will learn more about standards in Chapter 20, but a few definitions are important now.

Standards, or *protocols,* are conventions for encoding data in particular formats that can be interpreted by other equipment using the same protocol or standard. The *Internet Protocol (IP)* is used for Internet transmission. *SONET* and *Asynchronous Transfer Mode (ATM)* are used for both voice and data, although they were designed with circuit-switched networks in mind.

Standards define shared formats used to exchange data.

Dedicated logic circuits generally can translate digitized signals from one protocol to another. However, the translation circuits add to costs, and the results may not be efficient. For example, Internet Protocol data takes more room when repackaged for transmission in SONET format.

The telecommunications network is now so complex that it has multiple *layers* of standards. Each layer is intended to perform certain functions, while "hiding" functions performed on other layers. You can compare this to software programs serving different functions on a personal computer. A word processor generates a file in one format. If you attach that word-processing document to an e-mail message, your e-mail software will transmit the message and document as one or more data packets. The Internet may manipulate the data packets further before they arrive at their destination, where another computer decodes the data packets and converts them back into an e-mail message and an attached document in the original format. If all goes well, all you see is software that transmits and receives the e-mail message and attached file. You shouldn't have to worry about the details any more than you should have to worry about digitizing your voice for transmission over long-distance phone lines.

Transmission Topologies

The basic building blocks of a telecommunication network are systems that link two or more points. Interconnect these small system building blocks, and you have a larger-scale network.

You can group these system building blocks into a few basic categories. Here we will concentrate on the fiber-optic versions.

- **Point-to-point systems,** which simply carry signals back and forth between two points.
- **Point-to-multipoint, or broadcast, systems,** which distribute identical signals from a central facility to multiple terminals that may (or may not) be able to send signals back.
- **Networked systems,** which transmit signals among many terminal points somehow linked to each other.
- **Switched systems,** which make temporary connections between pairs of terminals or subscribers attached to the system.

Figure 19.2 shows each of these building blocks. You can find variations on most of these concepts, such as broadcast systems that include networking or switched elements. Large systems are assembled from these elements to meet particular needs. For example, an office building may have a hierarchy of networks, with local-area networks in each department, each local-area network connected to a node in a network that serves the whole floor, and each of those floor-wide networks, in turn, linked to a building-wide network, which is linked to the Internet. Similarly, the telephone network employs levels of switched systems, which direct calls first to large cities, then to districts within the cities, to neighborhoods, and finally to individual homes.

Let's look briefly at each of these building blocks and their uses.

Point to Point

Point-to-point transmission links pairs of terminals.

Point-to-point transmission is the simplest fiber-optic system. It provides two-way communication between a pair of terminals that are permanently linked together, with a transmitter and receiver on both ends.

Conceptually, the distance between terminals doesn't matter. The two could be on opposite sides of the room or on opposite sides of the ocean. If the link is too long for the transmitter to send signals through the entire length of fiber, optical amplifiers or repeaters can be added to boost signal strength. Examples of point-to-point links range from a cable linking a personal computer and a dedicated printer to a transatlantic submarine cable.

If you look closely enough, you can break other fiber-optic systems into point-to-point links. That reflects the reality that any fiber system has a transmitter on one end and a receiver on the other, although couplers in between may split the signals among multiple fibers. It also reflects the fact that point-to-point fiber links generally are easy to build.

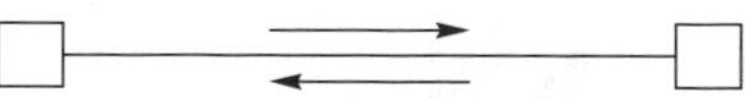

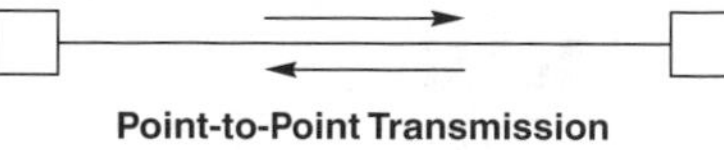

Point-to-Point Transmission

Main Terminal

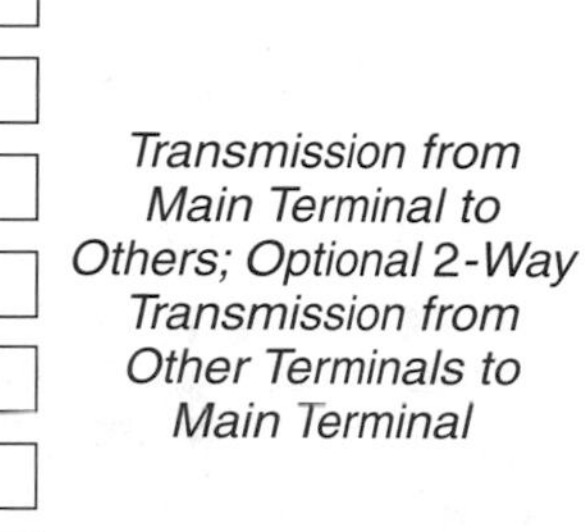

Point-to-Multipoint Transmission

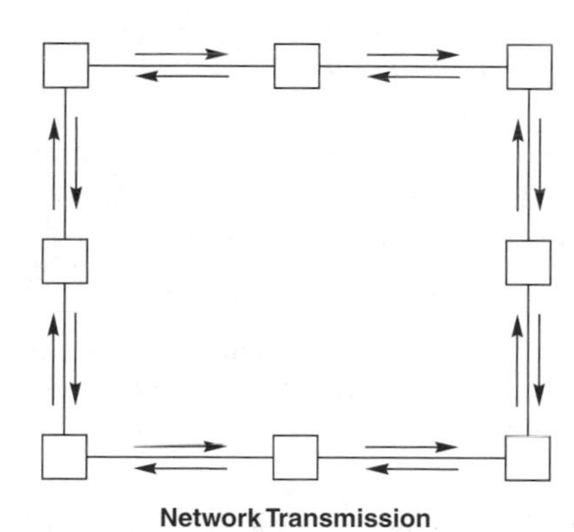

2-Way Transmission between Any Pair of Terminals

Network Transmission

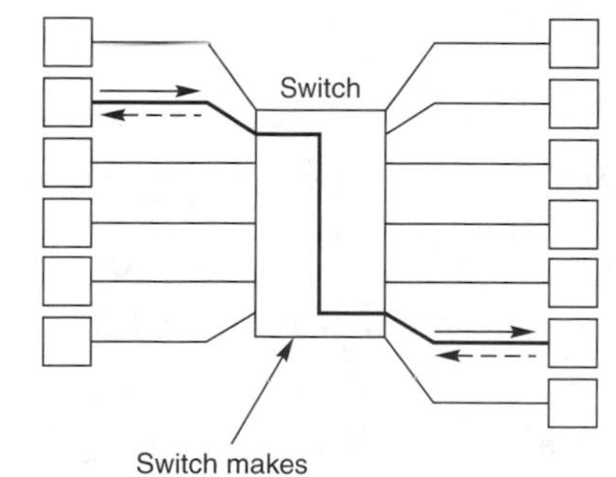

Switched Transmission

FIGURE 19.2
Types of transmission.

Point to Multipoint (Broadcast)

Point-to-multipoint transmission uses one transmitter to serve many terminals.

Another family of systems sends the signal from one transmitter to many terminals. This is sometimes called broadcasting, because it is analogous to the way a radio or television transmitter broadcasts signals through the air to many receivers. In a fiber-optic system, the terminals may or may not return signals to the central transmitter. If there is a return signal it is often at a lower speed than the broadcast transmission.

Because they serve many terminals, point-to-multipoint transmitters generally send higher-power signals than those in point-to-point systems. The basic design can vary considerably, as shown in Figure 19.3. A tree or star coupler can split the signal from one transmitter to drive many terminals. Or the split signals can drive optical amplifiers that amplify the signal from the main transmitter and send it to terminals (or another stage of optical amplifiers in a multilevel system).

Like a point-to-point transmission system, a point-to-multipoint system is fixed, with permanent connections between transmitters and receivers. Typically, point-to-point systems include multiple levels of signal distribution. For example, the head end of a cable-television system sends signals to local distribution nodes, which in turn send signals to neighborhood nodes, which distribute signals to individual homes. Typically these signals send relatively few signals "upstream" from home terminals to the head end where signals originate. A *pure* point-to-multipoint system distributes identical signals to all terminals, but modern cable-television systems can distribute some unique signals to individual homes.

The main transmitter in a point-to-multipoint system is more "important" than the terminals it serves. Even in a two-way system, it sends most information handled by the system, and is essentially an information provider (whatever you think of the offerings of your local cable system). Individual terminals provide little or no information, and they can go only to the main transmitter; they generally cannot communicate directly with each other. If the main transmitter fails, a point-to-multipoint system is off the air.

Networks

Networks allow many terminals to talk with each other.

A network system differs fundamentally from a point-to-multipoint system because all terminals are treated (more or less) equally. All terminals can both send and receive signals, and all can send signals to any other terminal on the network. Figure 19.4 shows a few simple variations. Small networks are often called local-area networks; larger ones may be called metropolitan-area or wide-area networks.

Terminals can be connected to the network in various ways. One approach is to put them all around a ring or loop. Signals may be split or tapped from the loop to serve each terminal, or they may pass through each terminal where they can be modified. Because fiber couplers have high attenuation, some networks are built as a collection of point-to-point links that are regenerated at each terminal. That approach is used in a few standard fiber networks, including FDDI. Another design is the star network, shown in Figure 19.4, where signals to and from each terminal pass through a central point, which is either a passive coupler that divides input light or an active coupler that receives and retransmits the signal.

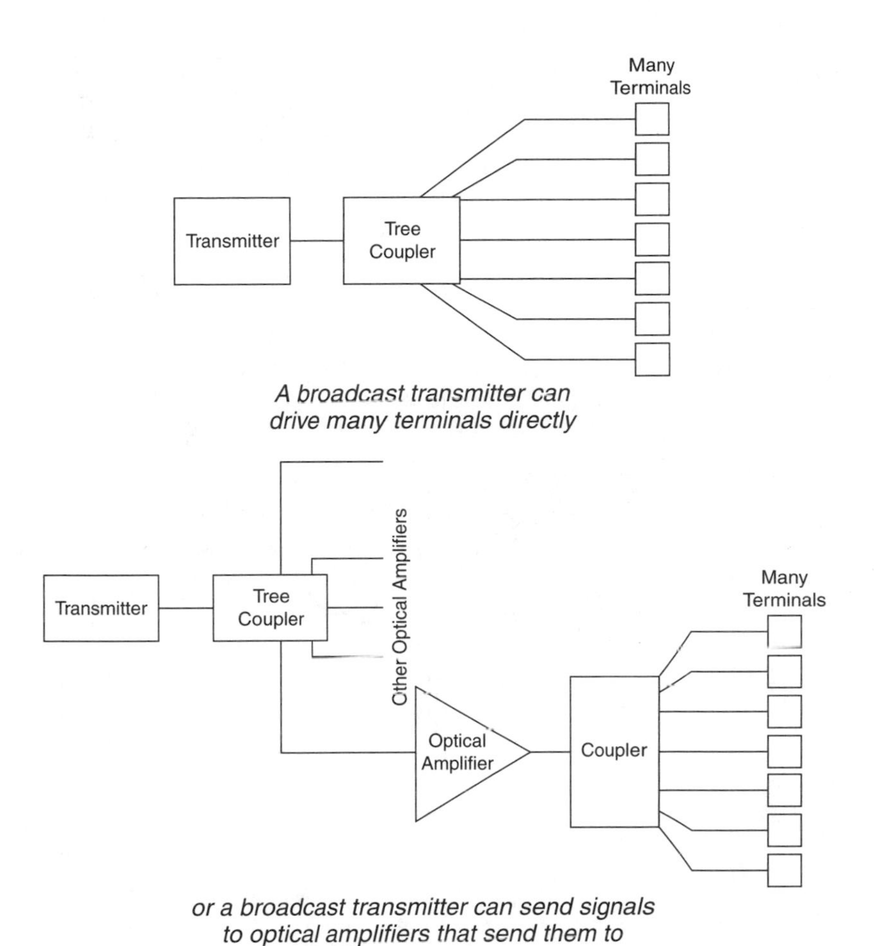

FIGURE 19.3
Point-to-multipoint transmission.

This approach is most often used with Ethernet and other networks where one node connects many terminals to the network backbone.

An alternative is the *mesh* network, also shown in Figure 19.4, where interconnections do *not* organize terminals in a ring or star configuration, but instead form a mesh-like grid. In the example, each node has links to at least three other nodes. This creates multiple routes between nodes, a robust architecture.

A mesh network lacks a highly organized geometry.

As you can see, a mesh lacks a highly organized geometry. You can't just direct signals from one node to the next because typically there is no single *next* node. Switches or routers must be programmed where they should direct signals.

Networks can be linked to make a network of networks.

Networks can be linked together with other networks to make a network of networks, as shown in Figure 19.5. In this example, each small-scale (department) network interfaces with a larger-scale (floor-wide) network, which in turn interfaces with an even larger

FIGURE 19.4
Network transmission.

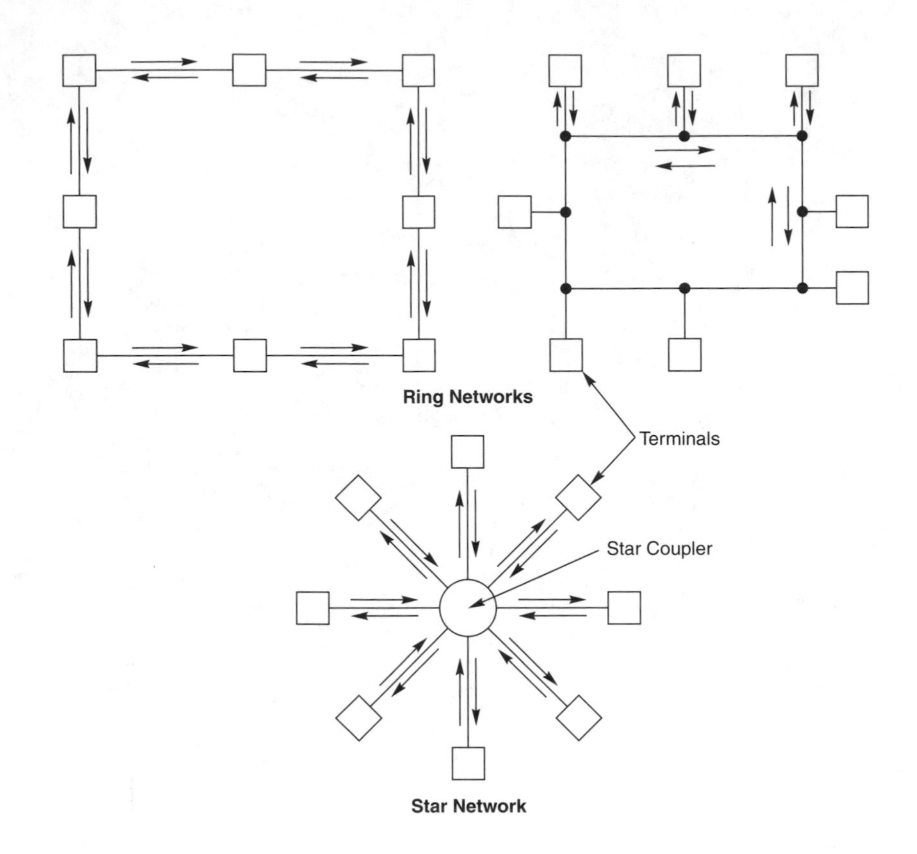

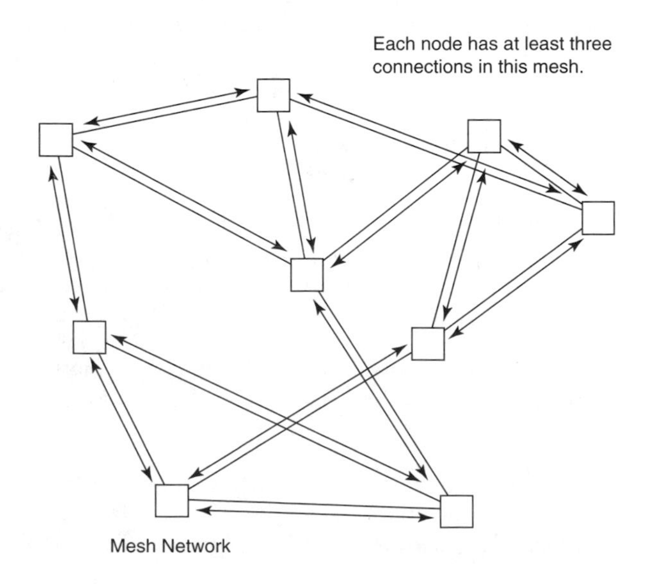

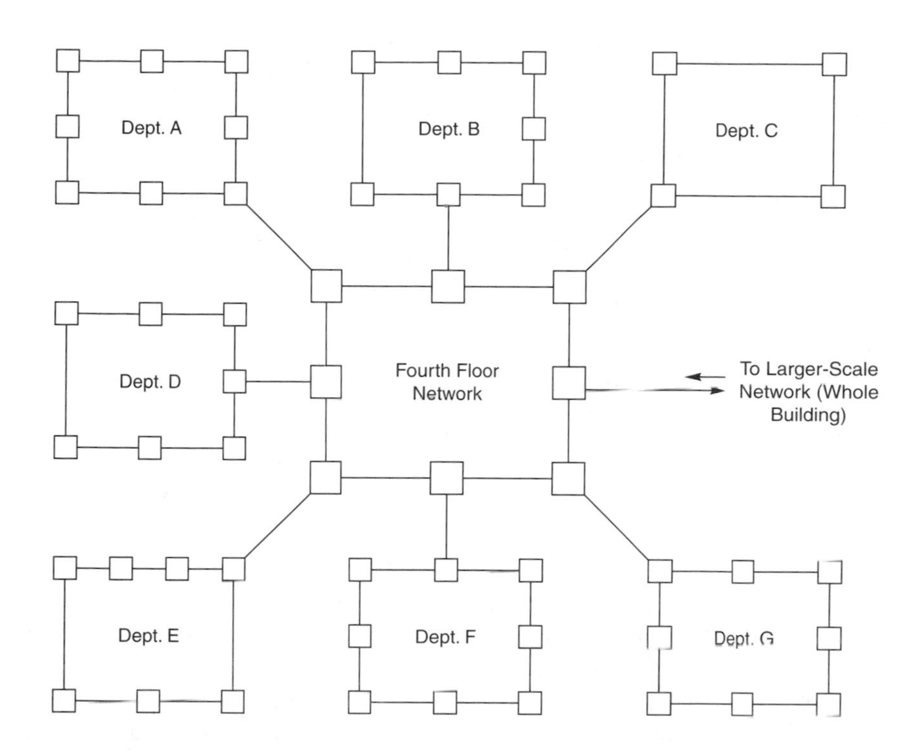

FIGURE 19.5
A network hierarchy, with small networks interconnected to make a large network.

(building-wide) network. Scaling this network-of-networks approach even further leads to the Internet.

As with point-to-point and point-to-multipoint systems, networks have permanent connections to each node and, except in mesh networks, the routing of signals can be changed only by rearranging cables. This means that the same terminals always talk to each other unless the configuration is changed. In practice, network connections usually go through patch panels with connectors that allow attaching and removing terminals. A typical example is a local-area network (LAN) for personal computers in an office. The terminals do not have to be identical, but do require a common protocol to talk with one another.

Switched Transmission

Switching allows temporary connections between pairs of terminals.

Adding switches to a communication network makes it more flexible. Switches allow any pair of terminals to send and receive signals directly to and from each other. The connections are inherently temporary, so each terminal can talk—at different times—to any other terminal, as shown in Figure 19.2. Depending on the system design, multiple terminals may be linked together at once. The telephone network is the standard example.

Switching increases system complexity, but adds tremendous power by making temporary connections to send signals between any pair of terminals linked to the switch. You can assemble switches in series, so each one directs signals at a different level. This allows the global telephone network to send calls around the world. To give a simplified example, one switch might direct a long-distance call to your state, another to the city where you live, a third to your part of town, a fourth to your block, and a fifth to your home. In practice, several of these switches may be in the same place—typically those serving your part of town, your block, and your home all are installed in the local telephone-company switching office.

Most switching in the telephone network now is done electronically. In effect, the fiber-optic elements of the telephone network are point-to-point systems linking electronic switches. That is changing as optical switching technology matures and optical networking spreads, as you learned in Chapter 16.

Switching and Routing

Switching and routing are different operations.

Switching and routing are different operations in telecommunication systems, although they serve similar functions in directing signals to their destinations. A detailed comparison of the two is beyond the scope of this book, but I can give you a general idea of the distinction between them.

Switches are relatively simple-minded hardware. A switch makes connections that create a circuit linking two points, using simple rules based on information like the area code of a telephone number. The connections may be physical connections that carry electrons from the power lines through a lamp in your home or direct light through optical fibers to a receiver somewhere in the telephone system. Switches in the telephone system create a temporary path from your phone to your grandmother's when you call her and maintain that path until you finish your conversation. As long as the phones are off the hook, the switches hold that path open even if no one is talking.

Routers read headers on each data packet to direct signals.

Routers are more complex devices used for packet switching. Each data packet is given a header that specifies the destination for the block of data it carries. The router reads this address and determines the best route for the data packet to follow, based on information such as traffic conditions on the network mesh. It then sends the packet to another router at a node nearer to the destination. That router repeats the process, sending the packet to another node, and so on until it reaches its destination.

You can think of a router as a clerk in an opto-electronic parcel delivery service. The clerk reads the electronic label, decides where a packet should go next, then puts the packet on an optical conveyor belt to the next node. Another clerk at the next stop repeats the procedure. Each decision is based on conditions at the time, so two successive data packets may not travel the same route, and may not arrive at the same time, but they should reach the same destination. In that way, it's very much like the mail. (It's also like the mail in that packets can get lost without the sender realizing it.)

Unlike switches, routers do not create temporary circuits. Nor do routers maintain an open pathway for additional data packets. What routers offer is the ability to pack data much more efficiently onto transmission lines than switches can, as shown in Figure 19.6. A switch holds the line open whether or not you're talking, leaving lots of open slots. Routers can pack data signals more tightly because they don't have to reserve slots.

Routers are more complex than switches, and need more processing power to examine the address on every data packet passing through. Switches merely read the initial address and

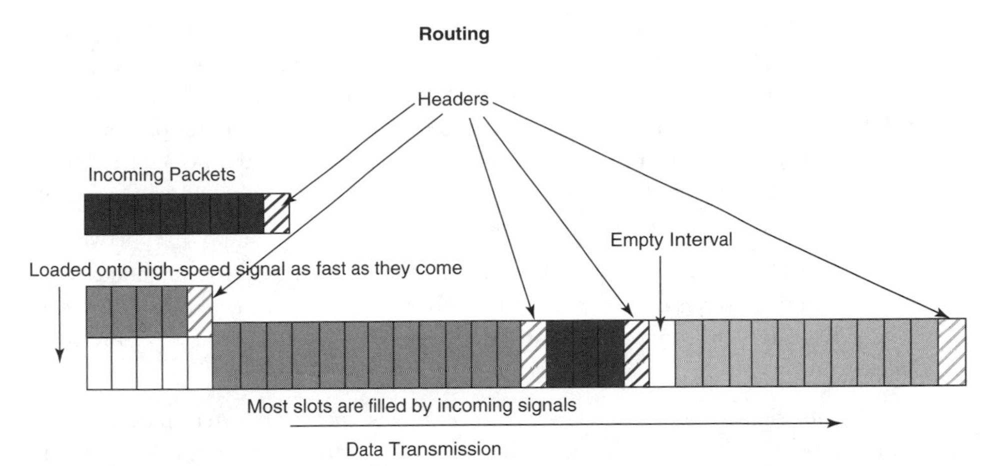

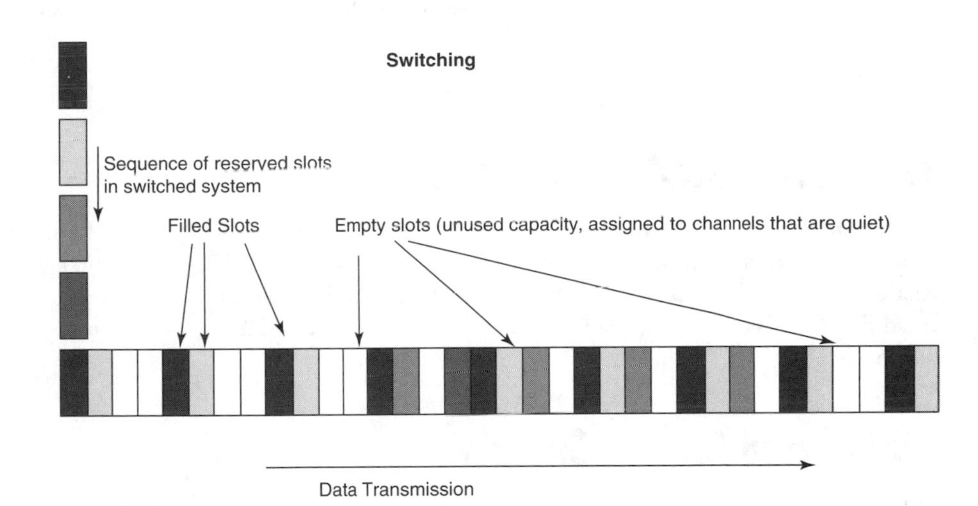

FIGURE 19.6
Routers use transmission capacity more effectively by packing signals closely together. Switches leave slots open because they reserve capacity.

make a circuit connection once per call. Switches are faster and cheaper because they are much simpler, but they also are less flexible than routers. Each technology has its advantages.

Broadcasting, Networking, and Switching

The distinction between broadcasting, networking, and switching is important because it determines how signals are distributed to terminals attached to the system.

Broadcasting, networking, and switching distribute signals differently.

Broadcasting sends the same signals to everybody. You and your neighbors receive the same radio and television broadcasts, if you have the right antennas and receivers. Likewise, you and your neighbors receive the same cable television channels, although the neighborhood sports fanatic may pay extra for a decoder that unscrambles the premium sports channels.

Many forms of networking also distribute the same signals to or through all nodes, as in the ring networks of Figure 19.4. However, each terminal is designed to collect and decode only the signals directed to it. Thus if your computer is on a local-area network, all the traffic on the network may pass through your node, but your computer will see only messages directed to you. Old-fashioned party-line telephones worked on similar principles. Several homes shared a single phone line, but each home had a different ring, and you were only supposed to pick up the phone when you heard your ring.

Routed networks like the Internet pass data packets through a series of routing nodes. Each packet goes through a series of terminals, but not through the entire network. Those terminals read the headers, but only the terminal designated as the destination gets to read the data packet.

Switching sends signals only between a pair of terminals, so the other terminals never receive that signal. (The signal does pass through nodes or switches, but not to other terminal devices.) Unlike a network, you get a dedicated private line, although only a temporary one. With modern private-line phones, every home on the block has a separate phone line, so your neighbors can't tie up your line or hear your calls unless the wires get crossed.

Transmission Formats

Signals are transmitted through optical fibers by modulating light with a signal. Different systems use different modulation techniques, depending on application requirements. Analog and digital signals can each be modulated in different ways. We will first review the differences between analog and digital signals and then consider variations on the two approaches.

Digital and Analog Signals

Digital signals are used more widely than analog signals.

In Chapter 3, you learned that signals could be transmitted as continuous analog variations in intensity (or some other parameter) or as a series of digital pulses. Analog transmission was used for many years and remains common in cable-television systems and a few other applications. It can pack more information into less bandwidth than digital signals but is much more vulnerable to noise and distortion, and analog signals cannot be manipulated as easy as digital signals.

Digital transmission has grown much more popular for most telecommunications applications. It requires simpler electronics and can encode any form of information, making it possible to merge digital data streams from many different sources and transmit them in one combined signal. Although your home telephone line carries analog signals, the telephone network converts them to digital form for easier switching and processing. Thanks to the low noise of digital telephony, a transatlantic telephone call sounds no different than one across the street.

Digital and analog signals are ways of representing information. An analog signal varies in a continuous way, like the variations in air pressure we sense as sound. A digital signal encodes these variations as numbers, which can be used to reconstruct the original vibrations our ears hear as sound.

Analog and digital are signal types; modulation is the transmission format for those signals.

Some specific concepts are used to describe signal format, which you should take time to understand.

A *carrier* is the nominal frequency or wavelength at which a signal is transmitted. It is a pure frequency generated by an oscillator or light source, before a signal is added. When a radio station says it has a frequency of 102.5 MHz, that means the carrier frequency. The carrier frequency is much higher than the signal frequency or bit rate.

Modulation is the process of adding a signal to a carrier wave for transmission. Modulation changes the amplitude, frequency, or phase of the carrier wave. The transmitter modulates the carrier wave; the receiver extracts the signal from the carrier.

Coding is the representation of a digital signal—that is, the pattern of 1s and 0s, which transmits the information in the signal. Different coding schemes represent the same series of bits in different ways.

Let's look at some important examples.

Amplitude Modulation

Virtually all fiber-optic systems use amplitude modulation in which the light intensity varies in proportion to the instantaneous signal strength. You can see the basic idea if you look back at Figure 10.4. The signal varies much more slowly than the light waves (you couldn't see the light waves if they were drawn to true scale). The stronger the signal, the more light from the transmitter.

Analog and digital fiber-optic systems use amplitude modulation.

Amplitude-modulated analog signals vary continuously in strength, in direct proportion to the continuously changing input signal. The signal format depends on the source. You may have an analog voice signal as the source, but most analog fiber-optic systems carry cable television as described in Chapter 27.

Digital Coding Schemes

Nominally, amplitude-modulated digital signals are a series of bits; however, there are several distinct coding techniques, as shown in Figure 19.7. Each major scheme has its own distinct characteristics:

Digital bit strings can be coded in several different ways.

- *NRZ (no-return-to-zero) coding*—Signal level is low for a 0 bit and high for a 1 bit and does not return to 0 between successive 1 bits.
- *RZ (return-to-zero) coding*—Signal level during the first half of a bit interval is low for a 0 bit and high for a 1 bit. Then it returns to 0 for either a 0 or 1 in the second half of the bit interval.

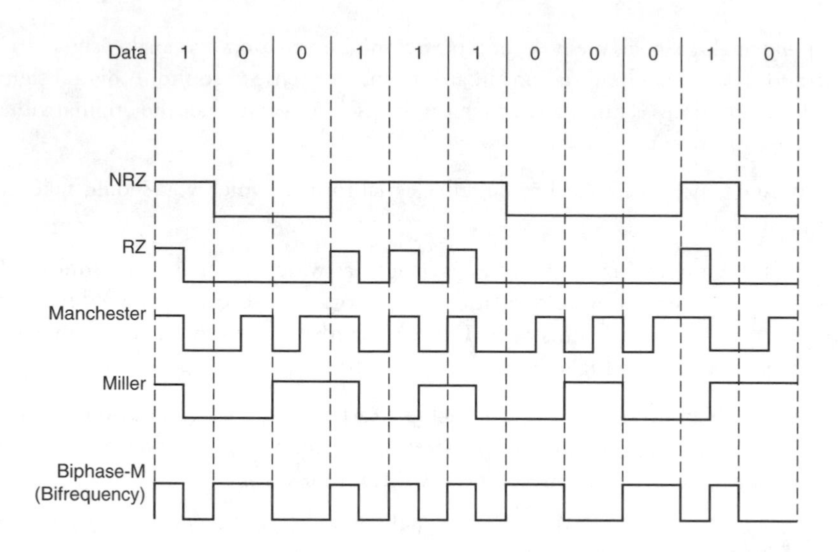

FIGURE 19.7
Digital data codes.

- *Manchester coding*—Signal level always changes in the middle of a bit interval. For a 0 bit, the signal starts out low and changes to high. For a 1 bit, the signal starts out high and changes to low. This means that the signal level changes at the end of a bit interval only when two successive bits are identical (e.g., between two 0s).
- *Miller coding*—For a 1 bit, the signal changes in the middle of a bit interval but not at the beginning or end. For a 0 bit, the signal level remains constant through a bit interval, changing at the end of it if followed by another 0 but not if it is followed by a 1.
- *Biphase-M or bifrequency coding*—For a 0 bit, the signal level changes at the start of an interval. For a 1 bit, the signal level changes at the start and at the middle of a bit interval.

Each type of coding has its own advantages and disadvantages. NRZ coding is simple to implement, and requires no more than one transition per bit, easing bandwidth requirements. However, long intervals of 0s or 1s can produce a steady signal level, making it easy to lose clock timing. RZ coding has more transitions during "on" bits, but can produce long intervals of 0 signal during a sequence of 0s. Codes that always have transitions during bits, such as Manchester and biphase coding, require higher bandwidth but generate their own clock signal to aid in timing. Other types of digital coding also are possible, but the details fall outside the scope of this book.

Error Correction Codes

Normally, eight bits of data represent a byte for computer data transmission. Adding an extra bit to each byte can verify data accuracy. Adding two or more extra bits allows error correction, which improves system performance and reduces power requirements at the receiver.

In computer data storage, *parity bits* are extra bits added to detect the presence of an error. For example, one bit can be added to an 8-bit byte so the number of "1s" always is even or odd. If a system expecting an odd-parity signal detects a set of 9 bits including four 1s, it knows one of those bits is incorrect, although it can't tell which one.

Adding two or more bits per byte gives enough extra information to identify which bit is incorrect, so the error can be corrected at the receiver. Use of an *error-correction code* adds to system overhead, but the improvements in bit error rate and transmission distance can be worthwhile.

Soliton Transmission

Soliton transmission is an alternative to the usual methods of transmitting a series of amplitude-modulated digital pulses through a fiber-optic system. Solitons are pulses that rise and fall in intensity in a specific pattern, which allows them to regenerate themselves as they travel along a fiber.

Solitons are self-regenerating pulses that do not change shape along a fiber.

You can think of solitons as doing a delicate balancing act between two competing effects that degrade the transmission of other pulses. One is chromatic dispersion, which stretches out pulses carrying a range of wavelengths as they travel along a fiber. The other is self-phase modulation, which spreads out the range of wavelengths as pulses pass through an optical fiber. The details depend on some rather abstruse mathematics that you don't want to worry about, but for the proper pulse shape the two types of stretching offset each other. This means that the shape of the pulse—how its intensity changes with time—is not affected by chromatic dispersion. Solitons can exist in places other than fibers—the first were waves spotted in nineteenth-century canals—but their only use in communications is for fiber-optic transmission.

Soliton pulses have proved surprisingly robust in optical fibers. They suffer attenuation, but optical amplifiers can compensate for the loss, allowing long transmission distances. The input pulses don't have to match the ideal soliton shape exactly, because fiber transmission gives them the proper soliton shape. A laser source can generate a series of soliton pulses, and an external modulator can generate a signal by blocking certain pulses, as shown in Figure 19.8. Thus the presence of a pulse can mean a 1, and the absence can indicate a 0. Note that solitons must be spaced a certain distance apart to keep them from interfering with each other and that their return to 0 at the end of the pulse makes this signal RZ coded.

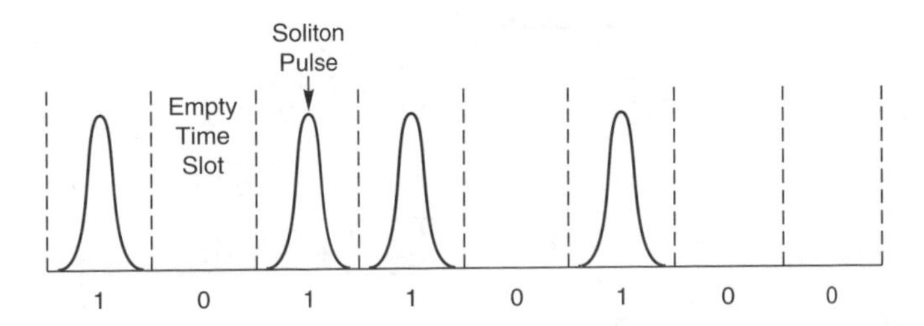

FIGURE 19.8
Soliton pulses modulated with a digital signal.

Soliton pulses can be transmitted at high rates, but their broad wavelength spread limits how closely soliton channels can be packed for WDM. Soliton systems are in development, but not in wide use.

Frequency Modulation

You may be familiar with the idea of frequency modulation from radio and television. Instead of modulating the amplitude or strength of the carrier, frequency modulation changes the frequency. Amplitude modulation was developed first for radio and is simpler to implement, but you can hear the difference in performance when you switch between AM and FM bands on your radio. Frequency modulation works better for broadcast signals because it does not pick up noise caused by random fading of signal amplitude. It's widely used to broadcast analog video signals.

The digital counterpart of frequency modulation is called *frequency-shift keying* (FSK). It shifts the frequency a detectable amount when the signal shifts from on to off. For example, the transmitter might jump between 998.3 MHz for off and 998.4 MHz for on.

Coherent fiber-optic systems have been demonstrated but are not in practical use.

Frequency modulation is more difficult to implement optically than it is electronically, but systems have been demonstrated. They rely on the principle of *coherent,* or heterodyne, transmission, which is adapted from heterodyne radio transmission. Figure 19.9 shows the basic idea. The laser transmitter emits a frequency ν_1, which is modulated by the signal. At the receiver, that light is mixed with light from a second laser at a nearby frequency ν_0, giving an intermediate frequency signal at the difference frequency, $\nu_1 - \nu_0$ (or, strictly speaking, the absolute value of that difference). The receiver electronics extract an output signal from that intermediate frequency signal.

The major attraction of coherent detection is that it avoids noise encountered in direct detection, making the receiver sensitive to fainter signals and allowing more loss between transmitter and receiver. It also allows the use of frequency-shift modulation, which won't work with direct detection. The greater sensitivity could stretch transmission distance or allow higher data rates. However, advances in coherent transmission and frequency modulation have not kept up with the rapid progress in WDM systems, which are now considered much more practical.

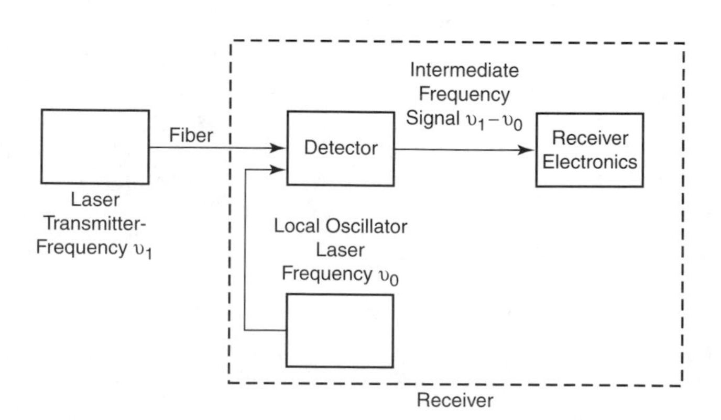

FIGURE 19.9
Coherent transmission.

Phase-Shift Modulation

Another approach to modulation is to vary the phase of the transmitted wave continuously in proportion to the signal. In principle, you can extract the phase-shift information by combining the transmitted wave with another identical wave in a system such as the coherent transmitter of Figure 19.9. The digital counterpart of phase modulation is phase-shift keying, which alters the phase shift by a fixed amount between on and off states.

The idea sounds good and has been demonstrated in the laboratory. However, it has not yet proven practical.

Transmission Capacity

Transmission capacity is a key figure of merit, which measures the amount of information a communication system can carry. In fiber-optic systems it usually is specified for a single fiber. Two-way communication usually requires a pair of fibers, one to carry signals in each direction. (In rare cases a single fiber carries signals simultaneously in both directions.) Capacity may also be given for an entire cable, particularly for submarine cables. Total cable capacity is the sum of the capacities of all the fiber pairs.

Transmission capacity is the amount of information a fiber can carry.

The usual measurement of capacity for analog systems is the bandwidth in megahertz or gigahertz. For digital transmission, capacity is measured in megabits, gigabits, or terabits per second—that is, how fast the system can transmit bits.

Capacity is increased by multiplexing, the combination of many signals into one. As you learned in Chapter 3, you can multiplex electronic signals in time or frequency, and multiplex optical signals by wavelength. Let's take a closer look at each of these approaches, and see how they combine to determine capacity.

Electronic and Time-Division Multiplexing

Electronic multiplexing began long before optical fibers were first used in telecommunications. Electronic equipment combines two or more separate input signals into a single output signal. That combined signal is transmitted through a communication system and then "demultiplexed" to break it into its original components. Multiplexing takes different forms in digital and analog systems.

Electronic multiplexers combine two or more signals to produce a signal that drives a fiber-optic transmitter.

Digital systems use time-division multiplexing (TDM), which combines several input signals into a single bit stream, as shown in Figure 19.10. In the example shown, four separate 1.5-Mbit/s inputs feed into a multiplexer. The multiplexer combines the signals, selecting first one pulse from input 1, then a pulse from input 2, and so on, in sequence. Essentially, the multiplexer shuffles the pulses together and retimes them because the lower-speed pulses are too long to stuff into a faster stream of bits. At the other end of the system, a demultiplexer sorts the bits out, putting bit 1 into channel 1, bit 2 into channel 2, and so forth. Interleaving also can be done byte by byte or in larger chunks.

Time-division multiplexers generate a single bit stream.

FIGURE 19.10
Time-division multiplexing combines digital signals.

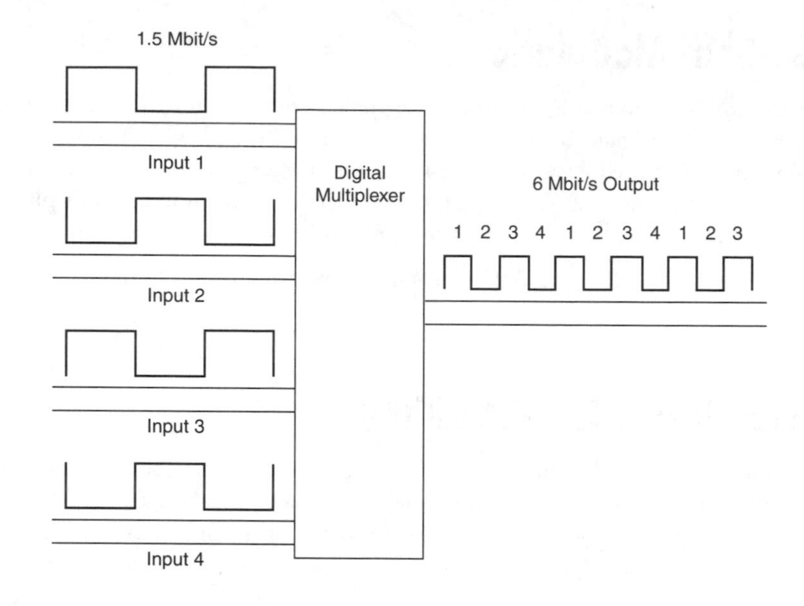

Time-division multiplexing can shuffle incoming data bits together so perfectly because all the input data signals must arrive at the same rate. That is, you can only make a 6 Mbit/s output signal by interleaving four 1.5 Mbit/s data streams. A time-division multiplexer could not combine signals at 3, 2, and 1 Mbit/s to yield one 6 Mbit/s data stream, although the total input and output rates match. You would need to combine those signals in some other way.

As you will learn in Chapter 20, time-division multiplexing works with a fixed hierarchy of data rates. Several signals at one rate are merged to make one signal at a higher rate, and several signals at that rate are merged to make one at an even faster rate. This is an inherent limitation of time-division multiplexing, but not all fast digital signals are assembled in this way.

Analog multiplexing modulates multiple carrier frequencies.

Analog multiplexing works differently, by modulating carrier signals at multiple higher frequencies with the analog input signals. This is often called *frequency-division multiplexing* because it multiplexes by carrier frequency. Radio stations share the broadcast spectrum in the same way, transmitting at different frequencies so they do not interfere with each other. In fiber-optic systems, the higher-frequency signals are combined to generate one analog signal spanning the entire frequency range.

In the example shown in Figure 19.11, each input signal covers 0 to 6 MHz. The first signal modulates a 100-MHz carrier, generating signals with frequencies of 100 to 106 MHz. The second modulates a 120-MHz carrier, generating signals from 120 to 126 MHz. The third and fourth signals modulate carriers at 140 and 160 MHz, respectively, to generate signals at 140 to 146 and 160 to 166 MHz. Bandpass filters at the receiver separate the frequencies and the original 6-MHz signals are regenerated. This example leaves empty space between channels, but where spectrum is scarce, as in television broadcasting, the channels would be packed closely together.

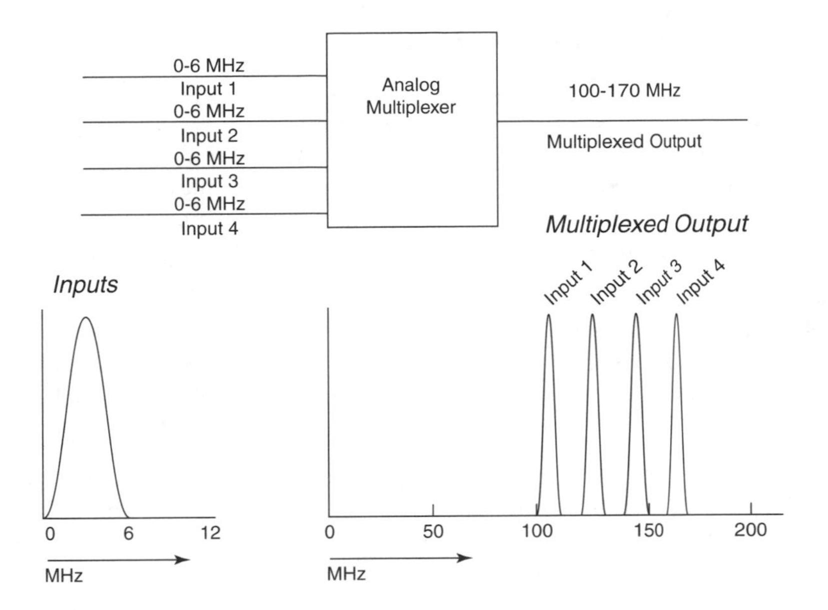

FIGURE 19.11
Analog frequency-division multiplexing.

Both analog and digital electronic multiplexers generate composite signals at the combined bandwidth or data rate. Digital signals can be further multiplexed to higher data rates, but this is rarely done for analog signals. The maximum data rate for a single channel depends on the transmitter, receiver, and fiber capacity.

Multiplexing in Packet Switching

High-speed signals can be generated directly, without TDM.

You don't need to interleave input data streams by time-division multiplexing to generate a high-speed digital signal. You can build up high-speed digital signals by assembling them directly from data bundled in packets, as shown in Figure 19.12. The packets do not have

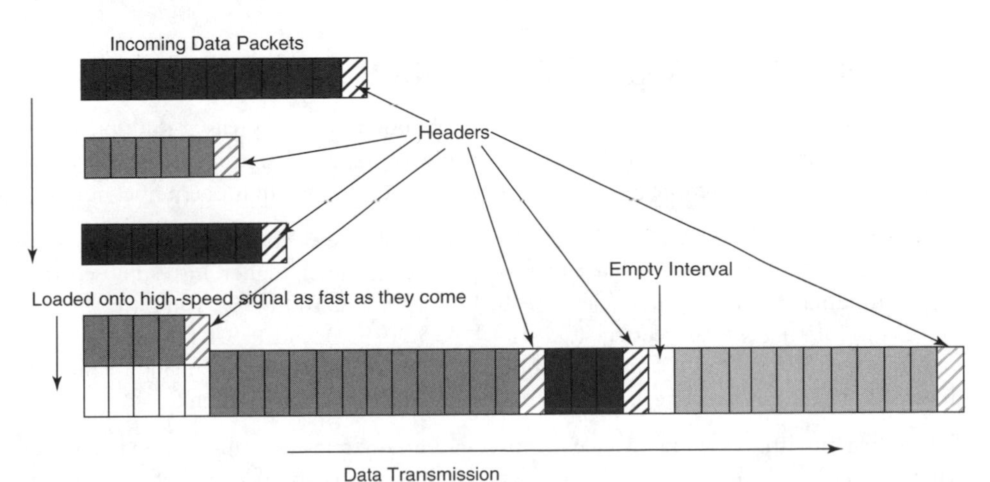

FIGURE 19.12
High-speed signal assembled from data packets.

to arrive at a particular rate or contain a particular number of bits, although they must have headers that specify their destination.

The data rate is set by the clock speed of the transmitter, which generates a specified number of pulses per second. Data packets are lined up for transmission in the sequence of their arrival. The transmitter sends one packet at a time, starting with the header, which contains address information, and proceeds through the rest of the bits in sequence. If another packet is waiting, the transmitter starts sending it; otherwise it sends blank intervals.

It's like lining people up to get onto a moving staircase where each step holds one person. The people line up at the base, and one gets onto each passing step. As long as people are in line, every step is filled. Steps only go empty if no one is left in line. To make the analogy better, you could imagine tour groups lining up, with a single leader acting as the "header" of each group.

Statistical multiplexing combines multiple packet streams that may have different data rates.

Multiple packet streams are combined by a technique called *statistical multiplexing,* rather than by interleaving bits. Incoming data packets on each channel accumulate in separate storage buffers. The multiplexer takes packets from each buffer in turn, keeping track of the traffic on each channel, then allocates more time to the busiest channels.

Statistical multiplexing is used for transmitting Internet Protocol (IP) traffic. Like time-division multiplexing, it combines data streams from multiple sources into a single faster signal. However, it does not require a rigid hierarchy of incoming data rates. Because statistical multiplexers average bursty traffic over many channels, the total capacity of the input channels may be higher than their output capacity. That is, a statistical multiplexer might have 10 inputs delivering up to 100 Mbit/s, but one output able to transmit only 600 Mbit/s. That design can work as long as the average input is below the peak capacity. For example, if each input channel averages only 20 Mbit/s, the combined inputs should be safely below 600 Mbit/s most of the time. (This sort of averaging is common in telephone networks, which don't have output connections for every input line because most lines are used only a small fraction of the time. Problems only arise when everyone makes long-distance calls on Mother's Day.)

Wavelength-Division Multiplexing

WDM transmits signals at multiple wavelengths through one fiber.

Wavelength-division multiplexing is the transmission of different signals at multiple wavelengths through the same optical fiber. You've already learned a little about WDM, and you're going to be hearing a lot more because it's one of the hottest topics in modern fiber optics.

Conceptually, wavelength-division multiplexing is much like electronic frequency-division multiplexing. The difference is that optical frequencies are much higher. Instead of separate signals modulating radio frequencies at 100, 120, 140, and 160 MHz, you have separate signals modulating optical frequencies of 193.1, 193.2, 193.3, and 193.4 THz, as shown in Figure 19.13. The optical channels function like separate frequencies in analog frequency-division multiplexing, except that the signal modulating the carrier is digital.

WDM began with very wide wavelength spacing. Early systems typically carried two wavelengths, 1310 and 1550 nm, doubling the transmission capacity of a single fiber. The ad-

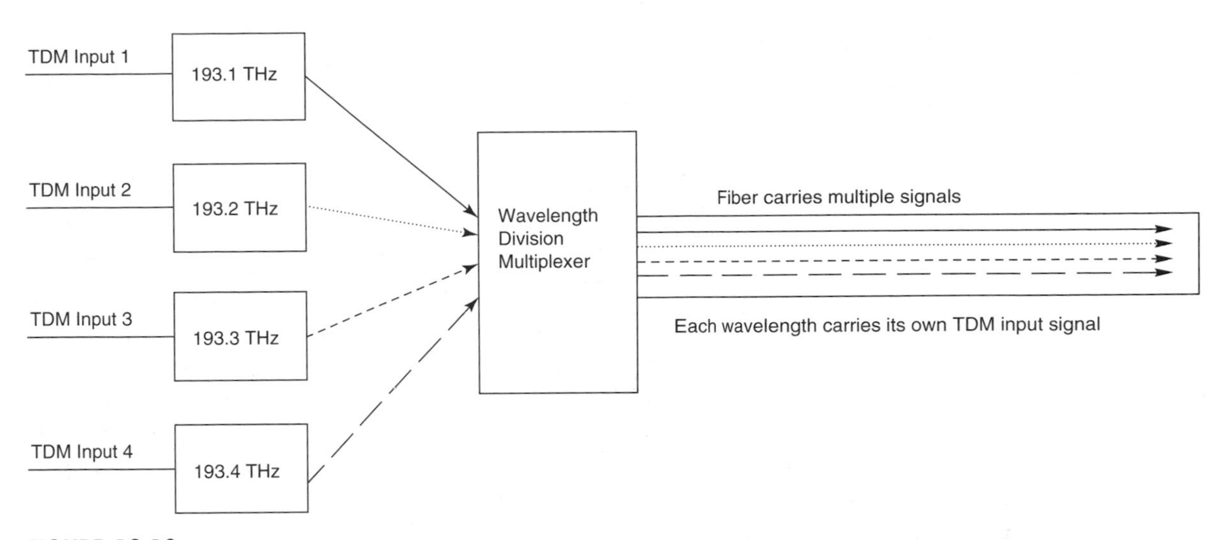

FIGURE 19.13
Wavelength-division multiplexing.

vent of erbium-doped fiber amplifiers and the huge demand for bandwidth led to the development of techniques for packing wavelengths more closely together. As you will learn later, the degree of packing can vary. Common spacings are 1000, 400, 200, 100, and 50 GHz in frequency units, corresponding to 8, 3.2, 1.6, 0.8, and 0.4 nm in the 1550 nm band. When channels are spaced at 200 GHz or less, they are often called *dense wavelength-division multiplexing,* or *DWDM.*

The closer WDM channels are spaced, the higher the transmission capacity.

The more closely the channels are spaced, the higher the transmission capacity of the system. The packing density is limited by the need to keep the wavelengths separated. In general, the higher the modulation speed, the broader the spectral width of the optical signal, and the more room needed between wavelength channels. Thus 2.5 Gbit/s signals can be packed together more tightly than 10 Gbit/s signals. This creates tradeoffs between time-division and wavelength-division multiplexing, as you will learn in Chapter 22.

The total capacity of the fiber is limited by the range over which wavelength-division multiplexing is possible. In amplified systems, the limit is set by the operating range of the optical amplifiers. More channels can be added at other wavelengths in systems that do not require amplifiers.

Total capacities can reach staggering levels. Manufacturers have offered commercial systems at total data rates to about 1.6 Tbit/s—160 channels at 10 Gbit/s each. Research laboratories have reached peak data rates of several terabits per second, but only over limited distances.

So far, real-world systems do not reach those upper limits. Precision narrow-band transmitters are extremely expensive. Narrow-line WDM optics also are expensive. Typically systems are installed with wavelength slots available for many optical channels, but with only a few populated. Telecommunications companies reserve the extra channel slots for future expansion. Adding the extra capacity costs relatively little now, and will avoid the high costs of laying new cable when—as everyone expects—the demand for capacity multiplies and it pays to install extra transmitters and receivers.

Transmission Distance

A secondary figure of merit is the distance a fiber-optic system can transmit signals. Distance can be surprisingly hard to pin down because it can be expressed in different ways.

You might not think distance is very important because it's possible to reach most populated points on the Earth via voice, fax, or e-mail. Yet those signals generally are relayed between communication systems. To reach New Zealand from the United States, you have to route calls through long-distance lines on land to the landing points of fiber-optic cables that go under the Pacific Ocean. To reach Antarctica, you rely on satellites passing over the continent, or radio links from New Zealand.

Even if we narrow our focus to a fiber-optic system, we can look at distance in two ways. One is the end-to-end distance between terminal points or regenerators, where the signal is broken up into pieces to be directed elsewhere, or regenerated for transmission through another fiber-optic link. The second is the spacing between optical amplifiers. Systems shorter than about 100 km generally don't need amplifiers, and it's possible to stretch amplifier spacing to a few hundred kilometers. However, longer systems need amplifiers.

Amplifier spacing depends largely on fiber attenuation.

Different effects cause the two limits. Amplifier spacing depends primarily on fiber attenuation, with transmitter power and receiver sensitivity also important. It marks the point where signal strength has declined to a level that requires amplification.

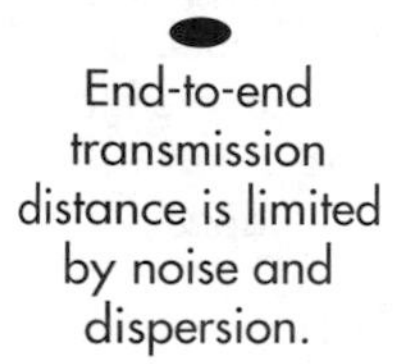

Regenerator spacing or end-to-end distance depends on degradation of the signal quality, primarily due to noise and dispersion. The dispersion accumulates from fibers; noise accumulates from optical amplifiers. This is a longer distance than between amplifiers, and with careful management can be stretched to thousands of kilometers.

As you will learn, there are tradeoffs between the two distances. The farther apart the amplifiers, the more noise they add to the signal, so the shorter the end-to-end distance. That's why repeater spacing is much closer in cables crossing an ocean than in one running several hundred kilometers on land.

Optical Networking

As I mentioned at the start of this chapter, optical networking is a new concept that is evolving rapidly. It's an outgrowth of wavelength-division multiplexing and optical switching, which promises to enhance the power and flexibility of telecommunications networks.

The tremendous growth in telecommunications traffic presents problems as well as opportunities. Managing the sheer volume of traffic is one problem. Other difficulties include breaking that volume into more manageable chunks, and structuring transmission systems to meet user needs.

Time-division multiplexing is a powerful technology but not a flexible one. It works well only with a limited set of fixed data rates. You often have to break a high-speed TDM signal down into its component parts to extract one of the slower signals that was merged into the fast data stream.

Wavelength-division multiplexing is more flexible because WDM channels are entirely independent. Many optical channels share the same fiber, but they don't have to share the same format. One can carry Gigabit Ethernet, another a 2.5 Gbit/s stream of Internet data, a third can carry a digitized high-definition television signal, and a fourth can carry a 2.5-Gbit/s stream of TDM data. All you need are transmitters, receivers, fibers, and optical amplifiers that can handle the raw data rate required.

WDM is more flexible than TDM because WDM channels are independent of each other.

Optical switching makes it possible to pick one or more of these signals out of a fiber without disturbing the others. An add-drop multiplexing switch can do that in the middle of a transmission line. At the end of a transmission line, a demultiplexer can separate and redirect the optical channels. Some might go to local customers, and others could be routed to several different cities.

Together, wavelength-division multiplexing and optical switching give telecommunications carriers much more control over their network than would be possible with time-division multiplexing at extremely high speeds. Carriers call this *granularity*, meaning that the traffic can be managed in smaller chunks than otherwise possible. The more granular the structure, the better they can meet customer needs and the more efficiently they can manage their networks.

Carriers also hope to automate their new optical networks much more than existing electronic switching systems. They want technicians to change services delivered to customers from a remote control center without having to visit the switch to change equipment manually. They want to be able to shift transmission loads around busy spots or damaged equipment. These capabilities aren't inherently optical, but will come with the new generation of optical networking systems.

The final shape of the optical network is far from determined. The types of services used are evolving as well as the technology. Internet, voice, and video services all have different requirements, and these shift as the services themselves evolve. Several years ago, most Internet traffic was text electronic mail and file transfers; now graphics-intensive Web sites account for the bulk of Internet traffic. If and when streaming video becomes widespread, it will impose its own requirements on Internet traffic. Meanwhile, optical technology is also evolving, as component engineers learn to build new components, and system engineers learn how the new components can best meet their needs. It's clear the next several years will see major advances, but it's far from clear how they will affect network architecture and system operation.

Cost and Reliability

New telecommunication technology also has to meet stringent real-world requirements for cost and reliability. A cutting-edge system that transmits 5 Tbit/s through a single fiber may be hot stuff in the lab. However, no carrier will use it if it's down half the time, or costs five

Cost and reliability are crucial in real-world systems.

times more than an alternative system. Research labs can achieve spectacular results with a small army of specialists fine-tuning delicate equipment, but a phone company doesn't want to send a senior engineer with a Ph.D. up a telephone pole every morning to adjust an optical interface unit.

Soaring demand is driving rapid expansion of communications capacity, but it's not clear how fast demand is rising. The industry has changed so much in the last decade that no single carrier can give solid estimates of traffic growth, and it's not even clear what is an accurate measure of traffic. Rough estimates are that Internet traffic is roughly doubling each year, and that its volume probably exceeds that of slower-growing telephone traffic. Carriers need more capacity to keep up, or develop new business, driving the rapid expansion of fiber-optic systems.

Reliability is as essential as overall capacity. This affects the overall topology of a communication network as well as the design of specific equipment. Most high-capacity systems connect major nodes to two or more other nodes, so a single failure won't knock out service. If branch cables provide service, a single failure can disable an entire branch, as shown in Figure 19.14. Ring or mesh networks are more robust because connections still remain after a single failure, so traffic can be re-routed to all terminals.

Telecommunication carriers typically test new equipment before deciding to buy large quantities. *Field tests* assess both performance and reliability in an actual operating environment. Transmitters, receivers, and switches usually go into the same buildings as existing equipment; cables run through standard ducts or are hung from ordinary poles. The insides of buildings may seem benign environments, but some new equipment may demand stringent temperature controls or draw large amounts of power.

Outdoor equipment must withstand much harsher environments. Overhead cables and outdoor boxes bake in the summer sun and freeze in winter cold. Water seeps into under-

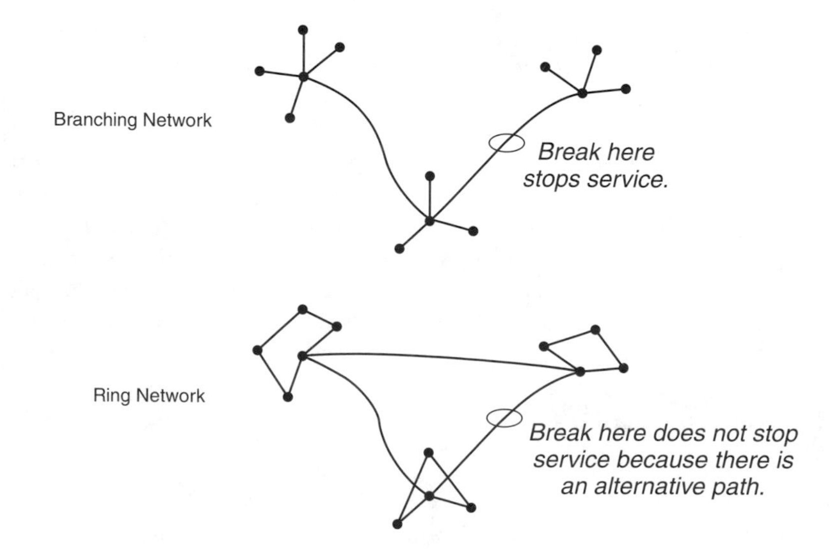

FIGURE 19.14
Branch and ring topologies for connecting network nodes.

ground ducts and manholes. Well-made cables are robust, but other equipment can pose problems, as can old copper cables.

Construction expenses are a crucial issue. Digging up a street can cost many times more than buying the cable to install underneath it, especially in downtown areas. Rebuilding 7.5 miles of highways in downtown Boston will cost at least $14 billion by the time the "Big Dig" is finished at the end of 2004. The job includes installing 5,000 miles of fiber-optic cable and 200,000 miles of copper cable. That's why carriers try to avoid new construction. Carriers first used wavelength-division multiplexing to multiply the capacity of existing fibers and avoid laying new ones. When they do dig up the streets, they typically install extra ducts for future expansion, and lay cables with many more fibers than they need immediately. The extra "dark" fibers are cheap insurance against future construction costs.

Room for future expansion is a common thread in many system designs. Although system manufacturers offer WDM systems that can carry 40, 80, or more channels, typically no more than a few wavelengths are turned on when the system is installed. Companies wait to install the rest of the transmitters and receivers until they need the extra capacity. They expect they will benefit from steady reductions in hardware costs and continuing improvements in performance.

What Have You Learned?

1. The telecommunications network is evolving rapidly, and changing in response to changes in business and technology. The network extends around the globe.
2. Optical networking is the management of signals in the form of light, usually in the heart of the network.
3. Telecommunications networks are made of links between nodes. They carry voice, video, and data signals, which have different transmission requirements.
4. The telephone system is circuit-switched. The Internet is packet-switched.
5. Standards or protocols allow different equipment to exchange signals in a common format.
6. Point-to-point transmission links pairs of terminals permanently. Point-to-point multipoint or broadcast transmission sends signals from one transmitter to many terminals.
7. Networks link many terminals with each other in various configurations.
8. Packet switching uses routers to deliver signals in a network; the routers read data headers on the packets to determine their destination. The Internet uses packet switching.
9. Circuit switching establishes temporary connections between pairs of terminals; it is used for telephone systems.
10. Both analog and digital fiber-optic systems use amplitude modulation.
11. There are several different digital coding schemes.

12. Transmission capacity measures the bandwidth or data rate of a communication system.
13. Time-division multiplexing is the interleaving of incoming data from several incoming bit streams to generate a composite stream of data. All input is at the same rate, and the output is a multiple of the input.
14. Frequency-division multiplexing uses low-frequency analog input signals to modulate multiple higher carrier frequencies.
15. Statistical multiplexing combines multiple packet streams, which may have different data rates to generate one higher-speed digital output. It is used for Internet transmission.
16. Wavelength-division multiplexing transmits separate signals at different wavelengths through a single fiber.
17. Transmission distance may be measured by spacing between amplifiers or regenerators. Amplifier spacing depends on power; regenerator spacing depends on noise and dispersion.
18. Optical networking allows signals to be managed in smaller chunks. WDM gives more flexibility than TDM because the optical channels are independent.
19. Cost and reliability are crucial issues in real-world systems.

What's Next?

In Chapter 20, you will learn about the standards developed for fiber-optic systems.

Further Reading

Roger L. Freeman, *Fundamentals of Telecommunications* (Wiley-Interscience, 1999)

Gary M. Miller, *Modern Electronic Communications,* 6th ed., (Prentice Hall, 1999)

Jean Walrand and Pravin Varaiya, *High-Performance Computer Networks,* 2nd ed. (Morgan-Kaufmann, 2000)

Questions to Think About for Chapter 19

1. Internet routers read headers on data packets, and use that information to direct the packets toward the proper destination. What type of network architecture would you expect to be used to connect routers in the Internet backbone? Why?
2. Can you use headers to control the flow of data packets around a ring network in which signals pass through all the terminals anyway?

3. An electronic time-division multiplexer generates signals at 1 Gbit/s. If it has eight inputs, what are their data rates?

4. How would an optical time-division multiplexer work?

5. How is wavelength-division multiplexing analogous to frequency-division multiplexing?

6. What are the prime limitations on amplifier spacing and regenerator spacing? How do the two affect each other?

Quiz for Chapter 19

1. What type of telecommunications is packet-switched?

a. Cable television

b. Telephones

c. The Internet

d. All

e. None

2. What type of telecommunications is circuit-switched?

a. Cable television

b. Telephones

c. The Internet

d. All

e. None

3. Optical networking is

a. a way to switch and organize signals optically.

b. any transmission of signals in optical form.

c. a communication system that uses only light.

d. a futuristic concept unlikely to be implemented.

e. a trademarked marketing term.

4. A router does which of the following?

a. Makes circuit-switched connections between terminals

b. Broadcasts signals to many points

c. Optically directs light signals to their destinations

d. Reads packet headers and directs signals to their destinations

e. Is equivalent to a switch

5. A switch does which of the following?

a. Makes circuit-switched connections among terminals

b. Broadcasts signals to many points

c. Converts packet headers to circuit-switching directions

d. Reads packet headers and directs signals to their destinations

e. Is equivalent to a router

6. Amplitude modulation is used for

a. digital transmission of a single 2.5 Gbit/s optical channel over fiber.

b. analog transmission of cable-television signals.

c. AM radio broadcasting.
d. optical transmission of Internet data.
e. all of the above

7. What is the proper name for digital coding in which a strong signal means a 1 and a low or zero signal means a 0?
    a. No return to zero (NRZ)
    b. Return to zero (RZ)
    c. Manchester coding
    d. Frequency-division multiplexing
    e. Phase modulation
8. Interleaving incoming bit streams to produce a faster output signal is called
    a. packet switching.
    b. frequency-division multiplexing.
    c. time-division multiplexing.
    d. statistical multiplexing.
    e. wavelength-division multiplexing.
9. Simultaneously transmitting separate signals through an optical fiber at different wavelengths is called
    a. packet switching.
    b. frequency-division multiplexing.
    c. time-division multiplexing.
    d. statistical multiplexing.
    e. wavelength-division multiplexing.
10. What type of multiplexing requires all incoming signals to be at the same data rate?
    a. packet switching
    b. frequency-division multiplexing
    c. time-division multiplexing
    d. statistical multiplexing
    e. wavelength-division multiplexing
11. Transmission capacity of an optical fiber is the
    a. total amount of information the fiber can transmit.
    b. distance between amplifiers.
    c. number of wavelengths the fiber can transmit, regardless of data rate.
    d. distance from end to end.
    e. data rate that can be transmitted at 1550 nm.
12. A fiber-optic system can transmit 2.5 Gbit/s on each of 40 optical channels, with an amplifier spacing of 100 km. The company that operates the system has installed transmitters and receivers at only 4 wavelengths. What is the data rate of the installed system?
    a. 2.5 Gbit/s
    b. 10 Gbit/s
    c. 40 Gbit/s
    d. 100 Gbit/s

CHAPTER

20

Fiber System Standards

About This Chapter

For two people to communicate, they must speak the same language. Communication systems likewise work only if transmitters and receivers attached to them speak the same language. Communication engineers have devised standards to assure that equipment from different companies will be able to interface properly.

This chapter will introduce you to the system-level standards most important for fiber-optic systems. Some are specific to fiber optics; others also cover other communications technologies. The topic of standards is complex and continuously evolving, so I will not go into much depth, especially for standards with little direct impact on fiber-optic systems. However, you should at least learn to recognize the most important standards and their functions.

Why Standards Are Needed

As you learned earlier, signals can be transmitted in a variety of ways, with different types of digital or analog coding. However, those differences only scratch the surface of the immense potential for variations. You can think of those physical differences as being similar to the distinctions among the media you use to communicate with other people—speech, the written word, sign language, pictures, and so on.

There are many other levels of variations in signal formats, just as people speak many different languages or computer programs store data in different formats. Unlike human languages, signal formats are designed by engineers to transmit signals efficiently and economically. Their choices depend on the types of signals being carried, the distances and types of terminals involved, and the hardware and software they have available. The results can vary widely with factors such as time and network scale.

These differences become a problem when you want networks to connect to each other or when you want to combine two or more generations of equipment, such as existing telephones with new digital switches and transmission lines. Then you need common languages and ways of translating signals between formats. That's when you need standards.

Standards establish the languages spoken by communications systems as well as the medium they use to transmit signals. Engineers from various organizations work on committees that develop the standards, usually sponsored by industry groups and organizations responsible for standards, such as the International Telecommunications Union, the American National Standards Institute, and the Telecommunications Industry Association. The standards they write are intended to make sure that communications systems can understand each other.

The importance of standards increases with the scale of the system, the degree of interconnection, and the variety of services it carries. For a simple fiber-optic link connecting two points, all that is really needed is for the transmitters and receivers at the two terminals to speak the same language. If you want to make that link part of a network, it has to speak a language the other terminals understand.

Some formats are proprietary, meaning they were devised by one company (or a group of companies) for use on its own equipment. Typically those standards are optimized for that equipment. Standard formats, on the other hand, are used by many companies making similar equipment, so you can attach Company A's terminal to Company B's transmitter and send signals to Company C's receiver. You can think of standards as agreements by everyone at a technical meeting to speak English or by everyone sending electronic mail to send messages in text-only format.

Standards have evolved considerably over the years, changing with both the marketplace and the technology. In the 1970s, AT&T was effectively America's telephone monopoly, so it set the standards for telecommunication systems. Since the 1984 breakup of AT&T, industry groups have come to set the standards. Many standards have become international, so you can make phone calls to Brazil, send faxes to India, and dispatch e-mail to Indonesia.

Changing standards have accommodated changes in industry practice. In the 1960s, telephone lines carried only analog voice telephone conversations. By the 1970s, the telephone network started to convert to digital voice transmission between switching centers. In the 1980s, the telephone network started to handle more computer data and video transmission, plus fax signals. Today, the high-speed lines operated by long-distance carriers are digital data highways that transmit a wide variety of signals, all digitized into a common form that can be reconverted to other formats at the receiving end. Standards make this multipurpose system possible.

Standards are crucial in an open, deregulated market.

Standards are crucial to the function of an open, deregulated market for equipment that must interconnect. You need to be sure you can plug any phone you buy today into the telephone jack in your wall and use it with any local or long-distance carrier. Most standards take into account the existence of old equipment and can accommodate much of it.

You can use your digital PCS cell phone to call your grandmother on the heavy black dial phone she has used since 1952. Neither you nor your grandmother should notice the automatic electronic conversion between the two formats. You should remember, however, that some new standards do not accommodate old equipment, such as standards for digital television transmission, which make no effort to talk with the "old" analog set you bought brand new in 2000.

Families of Standards

Families of standards have been developed to meet specific transmission requirements. Typically this means sending voice, video, or data signals in different environments. For example, there are standards for how to digitize voice telephone calls, and how to interleave the data streams from individual conversations to give higher and higher data rates. Other standards specify formats for data transmission over certain types of computer networks.

Families of standards exist for voice, data, and video systems.

Standards for voice, video, and data signals evolved separately, and remain somewhat distinct. One family of well-established standards is based on circuit switching of signals originally derived from voice telephone networks. This family of standards covers a wide range of capacities, from single voice circuits to high-speed digital lines carrying the equivalent of hundreds of thousands of phone conversations. They also can carry digitized video and computer data, although they are not optimized for those applications.

Data transmission standards likewise were optimized to handle computer data. Today most rely on some form of packet switching. The Internet Protocol is a flexible standard that can interface with local area networks using a variety of standards such as Ethernet. The common feature of these standards is that they are designed to handle data flow at uneven or "bursty" rates from computers rather than the steady rates generated by voice channels.

Distinct sets of standards also exist for analog and digital video, but they are not well-integrated, and standards for converting between the two formats are not well defined.

Interface standards integrate these diverse signals for transmission on a single telecommunications network. Telephone-based standards merge computer data into the flow of digitized voice signals on a network largely optimized for circuit switching. SONET, the *Synchronous Optical Network,* is the leading example in North America. Alternative data-based standards seek to merge voice signals into the larger flow of data on the Internet, based on Internet Protocol. Many industry analysts expect voice to merge into the flow of Internet data in the long term, but telephone companies worry that call quality could suffer. (Judging from personal experience, call quality *does* suffer when transmitted over the Internet with current technology.)

SONET was developed for circuit-switched voice signals.

Like voice signals, video signals require a connection that seems continuous to the user, although in theory the connection may seem continuous without being circuit switched. Video differs both in its requirement for much higher transmission bandwidth and in the fact that digital techniques can compress video data streams much more effectively than voice signals.

Most video transmission traditionally has been over separate media, such as over satellite links to broadcasting stations, so video standards are not well integrated. Many cable television systems have been modified to supply homes with data and voice service, but they separate those signals from video for long-distance transmission. Efforts to convert video from analog to digital format pose further complications. For that reason, this chapter will concentrate mostly on voice and data standards.

Layers of Standards

Modern standards are structured in layers that serve different functions.

Modern standards have been developed as a series of *layers,* each of which serves a different function. Essentially each layer provides a set of interfaces for certain users, concealing deeper layers that they don't need to worry about. The layering structure comes from the *Open System Interconnection (OSI)* model developed by the International Organization for Standardization.

Figure 20.1 shows the voice layer of the telephone network. The top view is what you see when making a phone call. You may be calling across the country, and your voice may be converted into bits and shuffled with other bits from other telephone calls, but you don't have to worry about those details. They're all in the "cloud" of the telecommunication system at the bottom, which you can think of as a black box that sends the signals. When you make a call, you see a system behave like the view at top, where a pair of dedicated wires go from your phone to the phone you're calling. The same thing happens if your computer sends a fax.

The services you see are the top of a stack of layers.

Look deeper, and you can see the other layers shown in Figure 20.2. The top layer specifies the services seen by customers, such as voice, fax, and data transmission. A lower layer specifies the data format for digital interchange of information—essentially the packaging of data bits representing these signals, which may be used in switching or directing signals. An even lower layer specifies how the signals are arranged into a data stream transmitted at one wavelength through an optical fiber, or transported physically on metal cables or via radio links. For fiber transmission, many of these physical channels can be stacked together at different wavelengths.

The services shown at the top layer have their own standard formats, which may cover more than one service. Standard analog voice telephone lines can carry facsimile signals and data from dial-up computer modems as well. The faxes and modems generate analog signals (which you can hear as tones on the phone line) that carry digital data over a line that the telephone company treats like a phone call or "voice channel." Many companies transmit digital data signals directly to the telephone system, using services such as leased lines; data also can be linked to the Internet.

The next layer is the adaptation layer, which converts these signals into a standard format for data interchange. The adapters convert these signals into digital form and then assemble the raw bits into a standard format used by whatever organization is carrying the data.

The third layer down shows the two most common interchange formats. ATM stands for *asynchronous transfer mode,* widely used in the telephone industry. IP is the *Internet Protocol,*

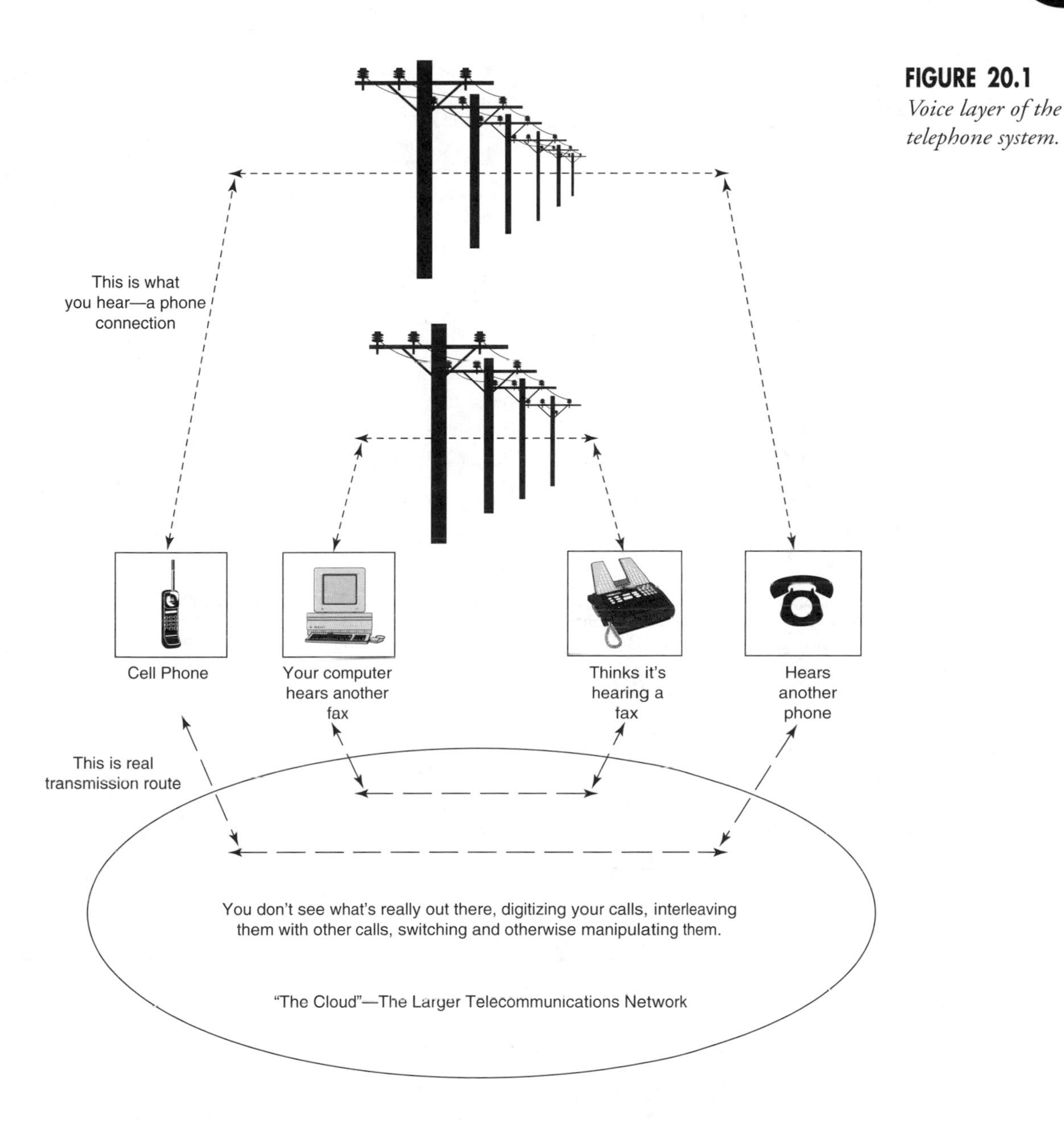

FIGURE 20.1
Voice layer of the telephone system.

generally used for Internet data transmission. These are ways of packaging bits that I will describe later in this chapter. Note that they say nothing about the transmission medium. They are simply logical packages of bits, such as the 8-bit *bytes* used in a computer. All the signals generated by the various services can be intermingled here; the data packages include labels for sorting them out at the other end. (ATM and IP formats do not serve identical purposes, and sometimes signals in IP format are repackaged into ATM format, although this is not shown in Figure 20.2. As I warned you, this can get complicated.)

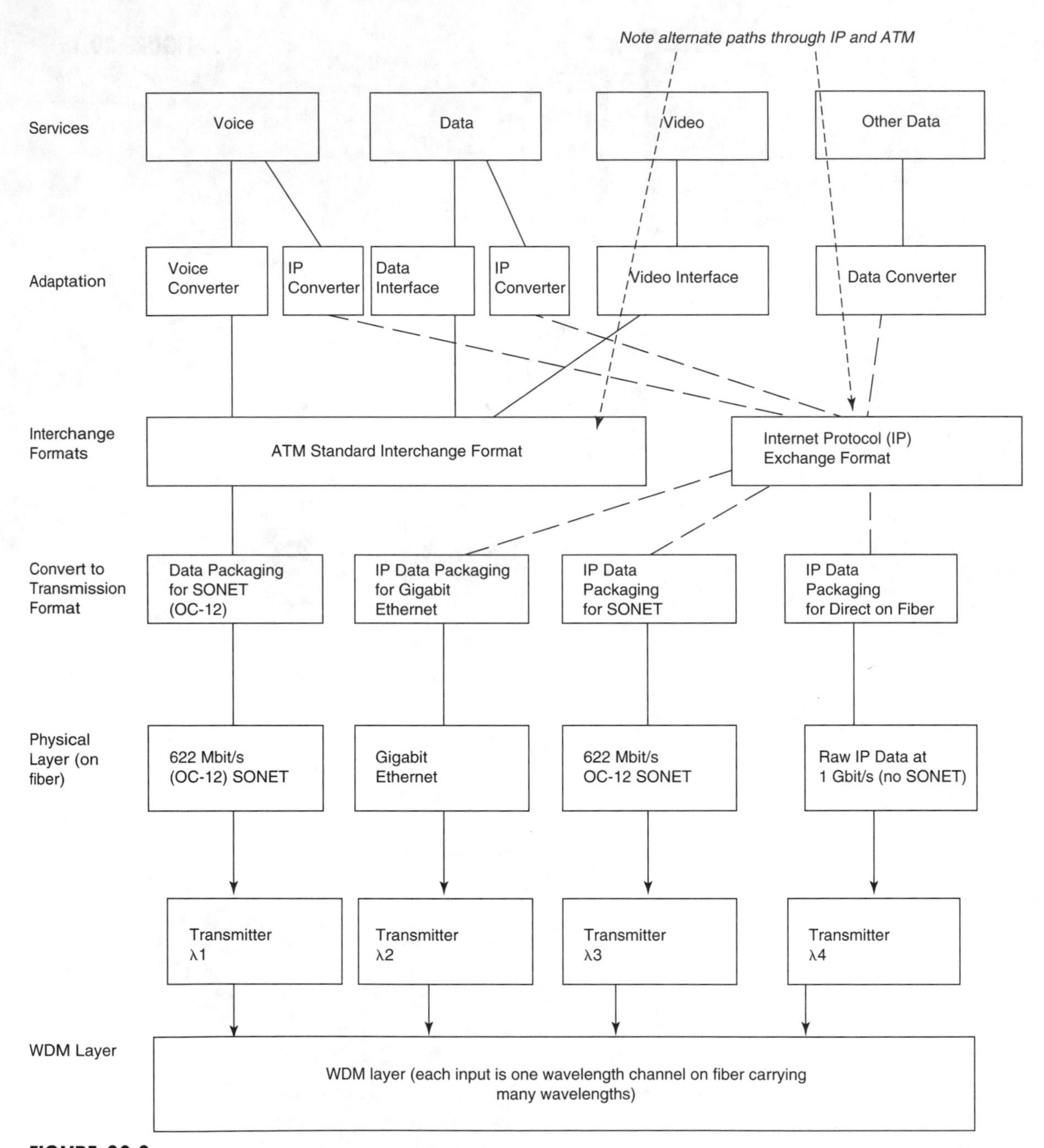

FIGURE 20.2
Standard layers for communications services, data interchange and transmission.

That raw interchange format data must be repackaged into forms that match the transmission requirements of particular media. This is the fourth layer down in Figure 20.2, a second set of conversions that generate signals in the proper form for transmission.

The *physical layer* in Figure 20.2 shows data formats developed for transmitting a series of bits physically through the transmission medium. The figure shows three possible formats: SONET at 622 Mbit/s, Gigabit Ethernet, and raw IP data at 1 Gbit/s.

Each of these channels in the physical layer may correspond to one optical channel in a fiber carrying many WDM channels, as shown at the bottom of Figure 20.2. Each physical layer channel drives a transmitter at a separate wavelength, which emits light transmitted through a single fiber in the WDM layer. In this case, the *WDM layer* corresponds to the long-distance backbone of the telecommunications network. Signals reaching the other end go back up through the same series of layers, and are converted back to their original form.

You'll note Figure 20.2 shows different paths that signals can follow through two interchange formats. In the long-distance telephone network, signals generally are converted into the ATM format for interchange. The Internet converts data signals to IP format. Normally ATM signals feed into SONET systems; IP may feed into ATM, directly into SONET, into Gigabit Ethernet, or into a transmitter it drives directly with IP signals. SONET signals carrying different interchange formats must be decoded in different ways.

Interchange formats are needed when signals pass through two or more networks with different formats. If signals travel only within a single network, it's often easier to leave them in the native format. For example, a data signal sent from one terminal to another on a local-area network need not be converted into IP format because it isn't leaving that network.

The layered structure of standards may seem confusing, but it can simplify design and development. For example, engineers designing a telephone don't need to worry how the signal will be digitized; they need to know only the format of the signal on the wire leaving the phone. Likewise, engineers designing the digitizing electronics need only know the format of the delivered input signal and the format required for the output signal. Standardizing the layer interfaces divides the job of system engineering into reasonable chunks. Thus the engineer generating a SONET signal need not worry if the SONET signal will be wavelength-division multiplexed; all that matters is that the signal is in SONET form.

Trying to make sense of all these formats and protocols would take a book in itself, so I'll concentrate here on the standards most important for fiber optics. The goal is not to make you an expert, but to help you understand enough to get along.

Transmission Format Concepts

What we call a transmission format is a way of coding and packaging information. Although some analog systems remain important, I will concentrate on digital systems to explain basic concepts.

Transmission formats are ways of coding and packaging information.

In addition to transmitting the signal you are sending, the system must transmit routing and decoding information. That means that the transmitted data must identify its destination and how it should be interpreted (e.g., as voice, video, or a particular kind of data). This information can be encoded in various ways, usually as *headers,* which precede the data, depending on the type of system. As the signals pass from one network or layer to another, the headers used in the first layer or network can become part of the data block transmitted by the next. Networks may also transmit other *overhead* information, such as priority of a transmission or check bits to verify the signal was received correctly.

Networks developed for different purposes have taken different approaches in transmitting signals. We can view them in two different ways—the choice of data-packaging methods and the task of linking networks.

Time-Division Multiplexing and Packet Switching

Time-division multiplexing guarantees transmission capacity.

The two fundamental approaches to organizing digital signals for transmission are *time-division multiplexing* and *packet switching.* As you learned earlier, time-division multiplexing allocates fixed slots for each incoming data stream, as shown in Figure 20.3. In this example, each of three input channels has its own slot reserved in the data stream. The high-speed data of input A always fills its slot, the slower data of input B only fills part of its slot some of the time, and the system keeps a slot for input C even if it isn't sending any data.

Time-division multiplexing is used for voice telephone service. It has the advantage of guaranteeing adequate transmission capacity, as you expect when you make a connection on telephone lines. However, it guarantees capacity at the cost of efficiency. When channel B has

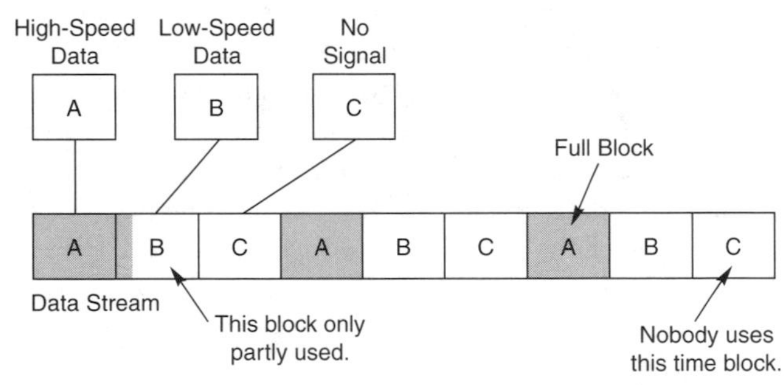

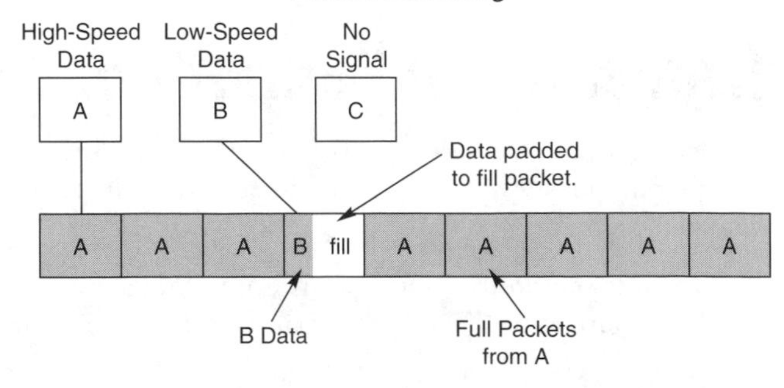

FIGURE 20.3
Time-division multiplexing compared with packet switching.

only a little to say, it still gets a whole voice channel worth of slots. When you're stuck on hold on line C, with no signal going out, the slots reserved for you go unused.

A more efficient way to use a limited capacity is to break the input signals into blocks of data called packets, and then to drop them into slots as they become available. As shown at the bottom of Figure 20.3, this approach fills the limited number of slots available. The system asks each input channel if it has a packet to send. A always has one, so each time its turn comes it sends a packet. B and C get slots if they need them, but if not, A can use the slots. Priorities may have to be assigned to keep A from hogging all the slots.

Packet switching uses capacity more efficiently.

The two systems direct signals in different ways. In time-division multiplexing, the system assigns a regular slot in the continuing data stream, based on input information, such as the phone number you dial. That slot is assigned all through the route between the two phones, and the system knows that as long as the call continues, bits from that call are contained in the same slot. If the signal starts in byte 15 in a series of 32, it will be in that position as long as the call continues. When the call ends, the slot can be assigned to another call. Thus the destination information comes from the signal's place in the bit stream, and headers don't have to be transmitted repeatedly after the connection is made.

Instead of using addresses to assign slots, packet switching attaches a header containing address information to every data packet. Routers at nodes in the system read these headers to see where to transmit the packets. It's not the position that directs the data, it's the header attached to it. Including a header does add overhead to the data transfer, but in practice the increase in efficiency from reducing the number of empty time slots more than offsets the extra overhead books used for the headers.

Packet-switched signals include address headers.

Packet switching was developed for data transmission, where input is inherently "bursty"—high at some times and slow or zero at others. The flexibility of allocating packets works well for this purpose and pays big dividends in efficiency. The downside of packet switching is that the data-flow rate varies with time. That variation can be annoying when it makes the Internet appear to be the World Wide Wait, but you can live with the delays for data transmission.

On the other hand, delays can cause serious problems with voice or video channels, which need to be guaranteed transmission at a constant speed. Time-division multiplexing better fits their needs, but it is still not efficient. A compromise is to develop protocols that give voice and video channels priority so they are routed at the proper rate, while other signals with lower priority can be delayed if necessary. This is done in some packet-switching systems, and is included in a new version of the Internet Protocol that is not yet widely used.

Internetworking

Another way to look at the telecommunications network is as an interconnection of networks. What we call the Internet evolved from systems developed to make "internetwork" connections among computer networks at large organizations.

This viewpoint is based on the assumption that individual networks already exist, with their own transmission standards. The function of the Internet or internetworking protocols is to make connections among these networks. Thus the Internet has transmission protocols, which differ from those developed for telephone systems.

Internetworking protocols are essentially standards for the interchange of data among networks that can have different architecture. Local-area networks typically carry signals at moderate speeds—tens to hundreds of megabits per second are common. Internetworking links typically operate at somewhat higher speeds, particularly as a larger fraction of traffic within a local network is headed to destinations outside the network. They also usually span greater distances.

If you step back and forget their different origins, the Internet and the long-distance telephone/telecommunications network resemble each other in many ways. However, specific standards differ, reflecting the important differences between the Internet and the telephone system.

Interchange Standards

Interchange standards such as asynchronous transfer mode (ATM) and the Internet Protocol (IP) are built around packet switching. Their function is to package information into packets that carry routing and other necessary information. Both were developed to make efficient use of network resources, but they differ because their designers did not share the same priorities or concerns and the two systems initially carried different types of traffic.

Both asynchronous transfer mode (ATM) and the Internet Protocol (IP) use packet-switching techniques to package digital signals into packets for efficient transmission. However, the two interchange formats were designed for different purposes and thus offer different features. Developed largely for telephone applications, ATM assigns priority codes to packets from different sources, allowing it to create "virtual circuits" for high-priority voice and video signals. These virtual circuits behave much like switched circuits and time slots in time-division multiplexed systems. IP behaves more like standard packet-switched systems.

Asynchronous Transfer Mode (ATM)

ATM is a packet-switched format that creates "virtual circuits."

Asynchronous transfer mode (ATM) was developed by the telecommunications industry to meet requirements for transmitting high-priority voice and video signals as well as lower-priority data. In ATM, the packets are fixed-length *cells* containing 53 bytes. The first 5 bytes are a header, which identifies the data and its destination, while the remaining 48 bytes contain the user data, as shown in Figure 20.4.

ATM functions as an interchange and data-packaging format, the third layer in Figure 20.2. It takes 48 bytes of user data at a time, and adds a header, which provides address destination and specifies the service priority. The top grade of service is constant bit rate, which simulates circuit-switched voice signals. Other grades promise various types of average bit rate, or low-priority service that essentially sends signals when the system gets around to it.

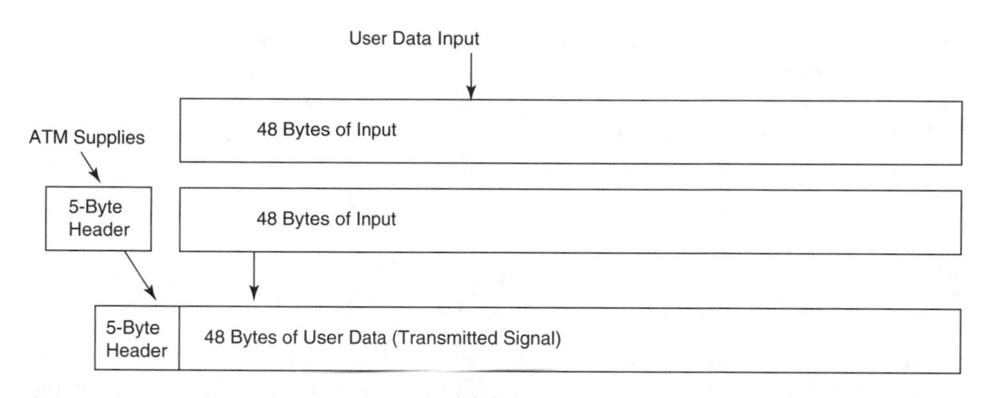

FIGURE 20.4
Asynchronous transfer mode cells.

As a packet-switching system, ATM can intermingle high-speed video signals with low-speed voice signals and data. Transmission priorities depend on the type of signal, so high-bandwidth and delay-sensitive video signals have higher priority than computer data. The short cell structure was chosen for maximum throughput for the mix of voice, data, and video signals expected when the standard was developed.

The packet-switching aspect of ATM should not be evident to users, who should receive signals at the same rate they would with a dedicated line or time slot. Its function is to help the service provider make the most efficient possible use of resources—that is, not spend as much on hardware as would be needed with dedicated lines. In short, ATM is a resource-allocation system that lets cheaper packet-switched circuits act like dedicated phone lines.

Internet Protocol (IP)

The Internet also uses packet switching, but the Internet Protocol (IP) specifies different ways of breaking data into packets. Instead of fixed-length cells, it generates variable-length packets, with the header information indicating the frame length as well as the address.

Internet Protocol (IP) is a packet-switching system with variable-length frames, developed for the Internet.

The IP standard evolved over a number of years and was specifically designed for connecting networks that transmit digital data. Because of that origin, the widely used version 4 (IPv4) does not assign a priority code to each packet. Instead, nodes attempt to transmit every packet on a best-effort. If too many packets show up, the node eventually discards packets that haven't been transmitted after a certain time. That can leave your Web browser sitting doing nothing, as it waits for a packet that will never come. A new version of the protocol, IPv6, does assign priority codes to packets, but it is not yet widely implemented, so you can't count on it being available.

IP transmission has spread rapidly with the growth of the Internet, and works well for data transmission. It also can be used to process voice and video, but with IPv4 they may suffer noticeable and annoying transmission delays, particularly for signals transmitted any distance. Developers hope that IPv6 will alleviate these problems.

Many suppliers of telecommunications systems are pushing IP systems for transmitting voice as well as data. They argue that ATM equipment is an unnecessary expense, and some advocate signals through fiber directly in IP format, as shown at the bottom right of Figure 20.2, eliminating SONET as well. Telephone companies are reluctant to make that change because they worry it will degrade voice quality, and it remains to be seen what will happen.

Fiber-Transmission Standards

The telecommunications and networking industries have developed a variety of standards for data transmission over fiber. These standards structure digital transmission, specifying signal transmission rates and how signals are structured to carry information, such as their destinations.

Unlike interchange standards, these standards usually assume a particular transmission medium. Some are oriented toward long-distance transmission, others to networking or connecting networks, and a few seek to cover both domains. I will describe a sampling of important standards, but you should remember that new ones are always being proposed, and some old ones are slowly fading from sight. Later chapters will give more details on their operation.

Digital Telephone Hierarchy

The digital telephone hierarchy is a series of time-division multiplexed rates.

Back when AT&T was *the* telephone company for most of the United States, it devised a set of standards for the then-new idea of digital telephone transmission. This digital telephone hierarchy remains in use at speeds of 45 Mbit/s or less. This standard is sometimes called the *plesiochronous digital hierarchy,* or PDH, because it uses many independent clocks for timing. The International Telecommunications Union devised a similar—but not identical—standard that is used in much of the rest of the world. You'll learn more about these standards later in Chapter 23 on the global telecommunications network; Appendix C lists them.

The starting point for the digital telephone hierarchy is electronics that converts a standard analog voice telephone signal into a digital signal at 64,000 bit/s. Other circuits interleave the bit streams from 24 digitized phone lines into a single sequence of 1.5 Mbit/s, called the DS1 rate, which is transmitted through T1 lines. That, in turn, feeds into systems operating at successively higher T2 and T3 (or DS2 and DS3) rates. The top of that original hierarchy, the T4 (DS4) rate of 274 Mbit/s, is not used today.

The digital telephone hierarchy is a sequence of time-division multiplexing steps, each one combining multiple signals at slower speeds and essentially interleaving them into fixed slots. As you learned earlier, these fixed slots guarantee the capacity to carry voice telephone signals, at the cost of leaving some capacity unused. The data rates are at speeds that were convenient for telephone transmission when the systems were designed, but they don't necessarily meet the needs of other users, such as video transmission.

Another constraint of the digital hierarchy is that you have to step back down the ladder to extract a lower-speed signal. That is, you must break a high-speed signal into its component parts to extract one of the components. In addition, the format omits some control features that would aid in operation of a modern network.

SONET/SDH

The *Synchronous Optical Network* (SONET) is a standard designed to make fiber transmission more efficient than the old digital telephone hierarchy. SONET is the North American standard; its international counterpart is the *Synchronous Digital Hierarchy* (SDH). SONET and SDH were designed to use ATM as an interchange format. They differ only slightly, to accommodate differences between telephone standards at lower rates.

SONET/SDH is a hierarchy of digital rates for fiber links.

SONET is a time-domain multiplexing system that organizes transmitted data into 810-byte blocks called *frames,* which include headers containing information on signal routing and destination as well as blocks of data from the signal itself. The headers are generated and inserted as the system packages input signals into SONET frames. The frames can be switched individually without breaking the signal up into component parts.

Developed to carry mixed traffic over fiber, SONET/SDH explicitly defines a series of transmission speeds. The SONET base rate (OC-1) is 51.84 Mbit/s, which with overhead accommodates the widely used T3 rate of the North American digital telephone hierarchy. The next step is the 155.52-Mbit/s OC-3 rate, nominally produced by interleaving frames from three OC-1 signals; this is also the base level of SDH. Beyond that are OC-12 at 622 Mbit/s, OC-48 at 2.5 Gbit/s, OC-192 at 10 Gbit/s, and OC-768 at 40 Gbit/s. Appendix C tabulates these transmission rates.

The frame structure in SONET explicitly allocates a certain number of slots per second for input signals. Although the frames may look like packets in their structure, they are organized as a circuit-switched TDM system, in the sense that they guarantee transmission capacity. They also can handle packet-switched data, but they handle it in a circuit-switched way. That is, they create a virtual circuit to transmit the signal.

SONET frames guarantee capacity like circuit-switched systems.

The SONET/SDH standards do more than specify data rates and frame sizes. They also specify that the network be arranged in a ring topology as shown in Figure 20.5, rather than in the older hub-and-spoke or branching system. A SONET ring includes a complete set of redundant fibers, so that if a cable is broken, the SONET hardware will detect the failure and automatically re-route signals through the backup fibers. This allows the system to continue operating when a cable fails, but requires a duplicate set of fibers to guarantee access.

Fibre Channel

Fibre Channel is a standard developed for transmission at speeds of 12.5, 25, 50, 100, 200, and 400 Mbit/s. It adds 2 error-correction bits to each 8-bit byte, so this corresponds to nominal data rates of 133, 266, 531, 1062, 2124, and 4125 Mbit/s. The European-style spelling of *fibre* in a deliberate choice, which the developers say symbolizes the use of metal

Fibre Channel transmits variable-length frames at rates to 4 Gbit/s.

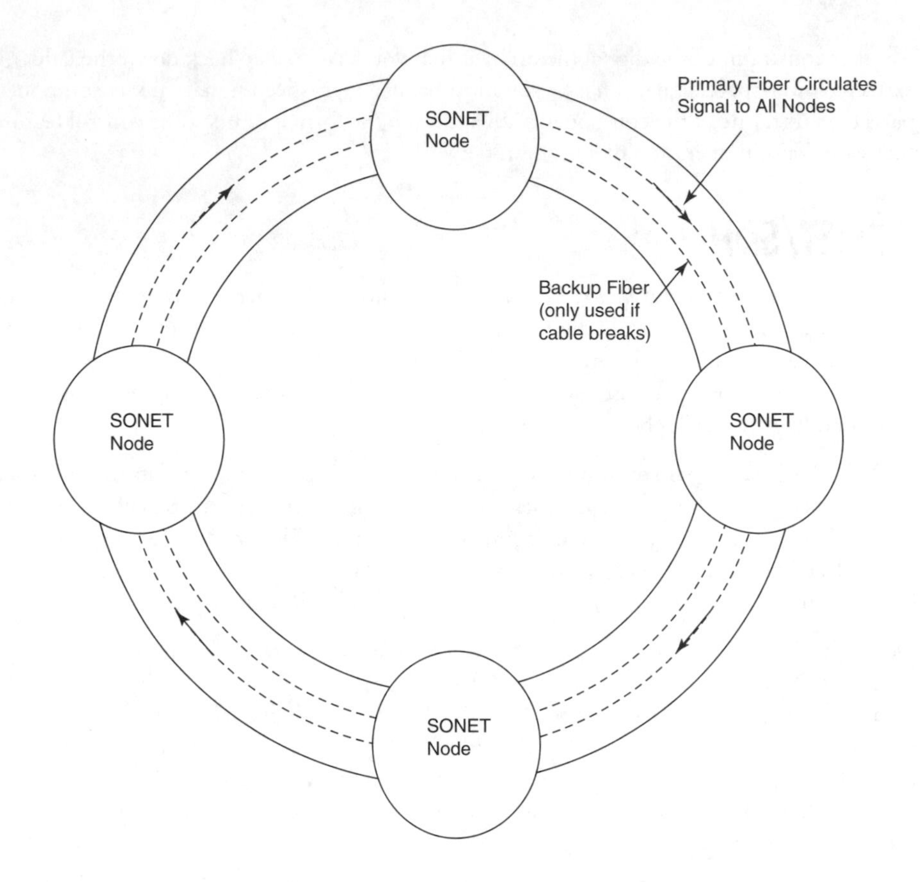

FIGURE 20.5A
Intact SONET ring links all nodes through one fiber.

cables in addition to single-mode and graded-index fibers. Fibre Channel transmits data in variable-length frames.

Fibre Channel operates on three layers, one that groups data in frames, a second that encodes and decodes the frames, and a third that provides physical transportation of signals over various media. You can think of these as the interchange and physical layers in Figure 20.2. It can handle point-to-point transmission between a pair of devices (such as a high-speed disk array and a computer), transmission around a loop, or transmission through a switched network. Unlike SONET, it was developed primarily for short-distance transmission in a computer system, network, building, or campus situation. Sometimes called a storage-area network, it is often used for high-speed data transfer between computers and peripherals.

Ethernet Protocols

Ethernet has been extended to 1-Gbit/s and 10-Gbit/s rates.

Ethernet has become a widely used standard for data transmission and local-area networks. Originally developed for 10 Mbit/s transmission, it has since been expanded to include *Fast Ethernet* at 100 Mbit/s and *Gigabit Ethernet* at 1 Gbit/s. Developers are now finalizing *10-Gigabit Ethernet* for transmission at 10 Gbit/s. The transmission format adds two error-

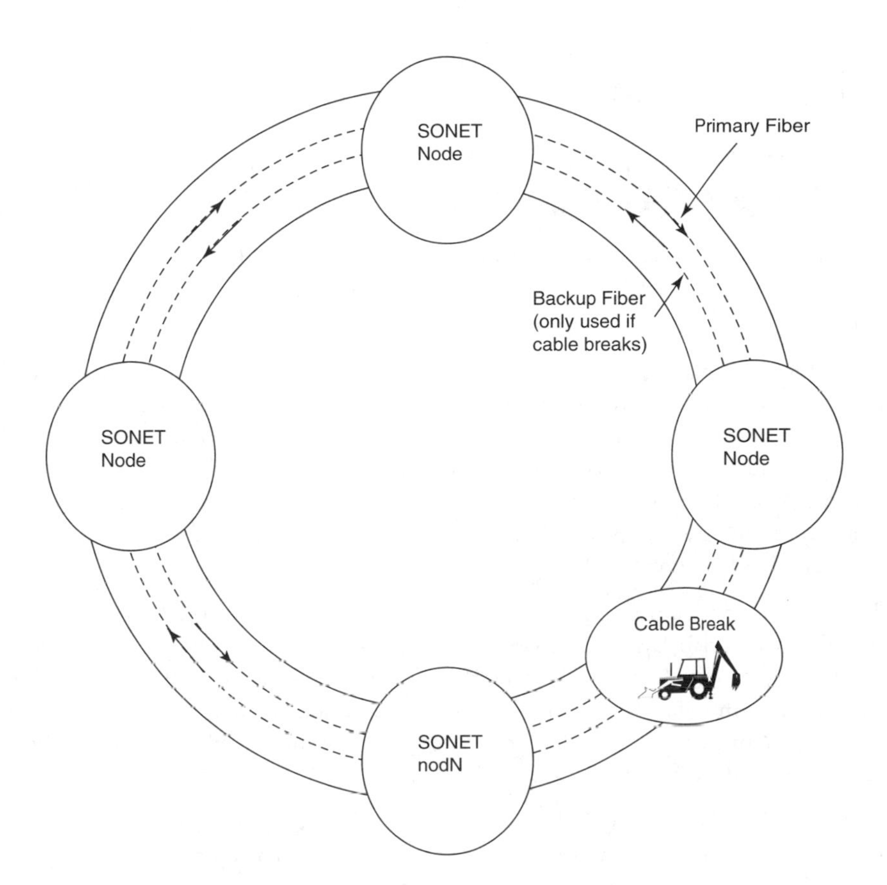

FIGURE 20.5B *Broken SONET ring still links all nodes through primary and backup fibers.*

correcting bits to each 8-bit byte, but the numbers do not count the correction bits. Thus Gigabit Ethernet actually transmits 1.25 Gbit/s including overhead bits.

Ethernet is a widely used format that uses inexpensive electronics and interfaces with other standard formats including IP. Ethernet and Fast Ethernet are widely used in computer networks. Gigabit Ethernet is spreading for applications that require higher transmission speeds, and 10-Gigabit Ethernet is likely to follow. Single-mode fiber can transmit Gigabit Ethernet up to 5 km. The high transmission speed and low equipment costs have led to much interest in Gigabit Ethernet for fiber-to-the-home or neighborhood systems. You will learn more about Ethernet transmission in Chapters 25 and 26.

Fiber Distributed Data Interface (FDDI)

The *Fiber Distributed Data Interface (FDDI)* is a network standard covering transmission at 100 Mbit/s. It can serve up to 500 nodes on a dual-ring network with up to 2 km between nodes. Originally developed for graded-index multimode fiber, the standard has been expanded to cover single-mode fibers and copper wires. The wire version is sometimes called CDDI, for Copper Distributed Data Interface.

Initially designed as a backbone connecting slower local-area networks, FDDI is now used in local area networks, which are covered in Chapter 26.

Video Standards

Traditional video standards cover analog transmission.

Video transmission standards are in a state of flux. Traditional video broadcasting and cable television signals are in analog formats. An effort is underway to shift to digital transmission for *high-definition television* (HDTV), but that change has been very slow.

Three different formats exist for analog television. North America and Japan use NTSC, with 525-line displays transmitted at the rate of 60 interlaced half-frames per second. Most of the rest of the world uses two other formats, PAL and SECAM, which transmit 625-line images at the rate of 50 interlaced half-frames per second. The critical difference in frame rates comes from the difference in the frequency of AC power lines, 60 Hz in North America and Japan, but 50 Hz in most other countries. These analog video formats require 6 MHz of bandwidth per video channel.

Each broadcast television station modulates a carrier at an assigned radio frequency with the 6 MHz video signal. Cable networks do the same, but they may assign different frequencies to different channels. This results in both broadcast and cable systems being frequency-division multiplexed, with different video channels modulating carriers signals at different frequencies. Cable television systems are the main application of analog fiber optics, as you will see in Chapter 27.

Digital television standards are still being finalized.

Digital television (DTV) standards are being adopted around the world, with a main goal being a higher resolution than existing analog sets. Digital data compression can squeeze a raw digital high-definition video data rate of about 1.5 Gbit/s down to 19.2 Mbit/s, which can be squeezed into existing 6-MHz video channels. Alternatively, a single 6-MHz video channel could carry two or more digitized signals with lower resolution. However, many details remain to be worked out.

The transition from analog to digital formats is likely to be very messy. Digital standards are not fully set, particularly for cable television, but current schedules call for phasing out analog broadcasts starting as early as 2006. You will learn more about these issues in Chapter 27.

Optical Networking Standards

Optical channel standards are in development.

The advent of wavelength-division multiplexing and the development of optical switching technology is creating a need for optical networking standards. Some have been proposed, but many are still in development.

One important standard has been set, a wavelength "grid" specified by the International Telecommunications Union. It specifies the wavelength grid in frequency units from a base of 193.1 THz, with offsets of 100 GHz or 50 GHz, listed in Appendix D.

Other standards are under consideration. A proposal from ITU adds three optical networking layers to the bottom of the layers of telecommunications standards described ear-

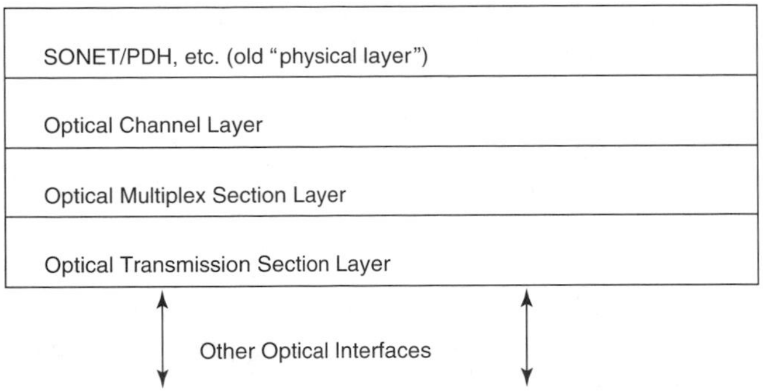

SONET/PDH, etc. (old "physical layer")

Optical Channel Layer

Optical Multiplex Section Layer

Optical Transmission Section Layer

Other Optical Interfaces

FIGURE 20.6
Optical networking layers.

lier. As shown in Figure 20.6, the optical channel layer would be below the SONET/SDH layer, and would cover end-to-end transport of one wavelength in a WDM system. Below that is the optical multiplex section layer, which covers simultaneous processing of optical channels at multiple wavelengths. The bottom layer, the optical transmission section layer, covers transmission of multiwavelength signals through a fiber. The standard leaves room for other optical interfaces with the optical transport layer.

Data transmitted through the optical channel will carry an "optical wrapper" of overhead data. The actual data transmitted would be sandwiched between a block of operations and maintenance information, and a block of data for forward error correction.

An optical wrapper will carry overhead data.

Many details have yet to be worked out. One issue is to what extent transmission should be "optically transparent" or "optically opaque." The critical difference is whether optical signals are switched and transmitted only as light (a transparent system), or whether they would be converted into electronic form for some types of processing, making transmission opaque. Current proposals envision some parts of the network being transparent, but other parts having electronic interfaces.

Other protocols are being developed for protecting channels and restoring service in case of an outage on either the single-wavelength level or the whole-fiber level. (Although errant backhoes and broken cables are the symbols of failure, as optical channel counts increase other types of failures grow more likely.) A long list of other standards are also in development for tasks such as transferring signals between carriers and directing signals through optical cross-connects.

What Have You Learned?

1. Standards specify coding techniques so different systems can understand each other. The importance of standards increases with the scale of the system.
2. Families of standards exist for voice, data, and video systems.
3. Modern standards are developed for layers that perform different functions in a telecommunication system.

4. Time-division multiplexing combines digital signals for high-speed long-distance transmission by interleaving input channels at lower data rates. It guarantees slots for service.
5. Packet-switching merges data streams by statistical multiplexing of incoming packets. This uses resources more efficiently because it does not leave empty slots, but it does not guarantee service.
6. Packet headers carry information on how to route packets; they are followed by data.
7. Asynchronous Transfer Mode (ATM) is an interchange standard that packs data into cells. It assigns priorities to cells, depending on the type of signal they carry, so voice signals get high priority.
8. Internet Protocol is an internetworking standard for packet-switching.
9. The digital telephone hierarchy is a series of time-division multiplexed digital data rates to 45 Mbit/s, originally developed in the 1960s but still in use.
10. SONET/SDH specifies transmission of digital data in frames at an ordered series of rates starting at 51.84 Mbit/s through optical fibers. It specifies a ring topology to guarantee that service continues if a cable is broken.
11. Fibre Channel carries data rates of 133 to 4126 Mbit/s through fibers or metal cables. Applications include linking computers to storage systems, point-to-point transmission, networks, and connections between networks.
12. The Ethernet family includes standard Ethernet at 10 Mbit/s, Fast Ethernet at 100 Mbit/s, Gigabit Ethernet at 1 Gbit/s, and the new 10-Gigabit Ethernet at 10 Gbit/s.
13. Most present video transmission is in analog form using the NTSC, PAL, and SECAM standards. Standards for digital television broadcast have been developed, but many details remain to be worked out.
14. Optical networking standards specify wavelength channels. They may also cover optical channels, optical multiplexing, and optical transmission of WDM signals.

What's Next?

In Chapter 21, you will learn the basic elements of designing optical systems transmitting at a single wavelength. Chapter 22 will cover optical networking design.

Further Reading

John C. Bellamy, *Digital Telephone,* 3rd ed., (Wiley-Interscience, 2000)

Roger L. Freeman, *Fundamentals of Telecommunications* (Wiley-Interscience, 1999)

Jean Walrand and Pravin Varaiya, *High Performance Communication Networks* (Morgan Kaufmann, 2000).

Web Resources

ATM: *http://www.atmforum.com/*

FDDI: *http://www.ineng.com/univercd/cc/td/doc/cisintwk/ito_doc/fddi.htm*

Fibre Channel: *http://www.fibrechannel.com*

Gigabit Ethernet: *http://www.gigabit-ethernet.org/*

Questions to Think About for Chapter 20

1. Follow the voice signals in Figure 20.2 through the layers in the diagram. What function does Internet Protocol serve?
2. Many makers of telecommunications equipment are proposing to transmit IP signals directly on fiber, without going through ATM or SONET coding. What advantages might this have?
3. What difference between ATM and IPv4 formats is most important for voice transmission?
4. A major advantage of packet switching is that it can combine signals that arrive at uneven rates to use transmission capacity efficiently. Suppose you have four packet-switched input signals, which can arrive at peak rates of 1 Gbit/s. However, on average the packets account for only about 20% of the peak capacity. If all goes well, can you squeeze those four input channels through a 1 Gbit/s output?
5. You can pack 24 voice channels on one T1 carrier, four T1 carriers into a T2 channel, and seven T2 carriers into a T3 signal. How many voice channels can a T3 signal carry?
6. How many voice channels can an OC-192 signal carry, assuming an OC-1 carrier transmits the equivalent of one T3 carrier?

Quiz for Chapter 20

1. Which of the following are *not* defined by telecommunications industry standards?
   a. Data transmission formats on optical fiber
   b. Transmission speeds in digital telecommunications
   c. Interchange formats for signals sent to other countries
   d. Data transmission in local-area networks
   e. Monthly telephone service charges
2. What kind of a standard is Asynchronous Transfer Mode (ATM)?
   a. Data-interchange format
   b. Fiber transmission
   c. Analog television
   d. Time-division multiplexing
   e. Financial transfer for banking

3. Which of the following is a SONET data rate?
   a. T3, 45 Mbit/s
   b. 100 Mbit/s
   c. OC-3, 155 Mbit/s
   d. 1 Gbit/s
   e. None of the above
4. How does packet-switching combine signals from different sources?
   a. Assigns each one a different wavelength
   b. Packages them into a series of packets, with headers indicating destinations
   c. Assigns each one a different time slot in a sequence of bits
   d. Transmits them simultaneously at different frequencies
   e. None of the above
5. Packet-switching has what advantage over time-division multiplexing?
   a. Packets always reach their destination on time
   b. Packets use available transmission lines more efficiently, avoiding empty time slots
   c. Packet-switching assures line availability
   d. Packets avoid signal interruptions
   e. The two cannot be used for the same type of communications
6. Which of the following are interchange formats? (More than one answer is possible.)
   a. ATM
   b. FDDI
   c. Fibre Channel
   d. Gigabit Ethernet
   e. Internet Protocol (IP)
   f. SONET
7. Which of the following are fiber-transmission formats? (More than one answer is possible.)
   a. ATM
   b. FDDI
   c. Fibre Channel
   d. Gigabit Ethernet
   e. Internet Protocol (IP)
   f. SONET
8. Which transmission format is primarily used for long-distance telecommunications?
   a. SONET
   b. FDDI
   c. Fibre Channel
   d. Gigabit Ethernet
   e. Ethernet
9. How fast are data transmitted at the OC-192 rate?
   a. 51.84 Mbit/s
   b. 100 Mbit/s
   c. 155 Mbit/s
   d. 1 Gbit/s
   e. 10 Gbit/s
10. Which of the following is an analog transmission standard?
   a. FDDI
   b. Fibre Channel
   c. NTSC
   d. HDTV
   e. SONET

CHAPTER

21

Single-Channel System Design

About This Chapter

Now that you have learned the ideas behind fiber-optic communication systems, it's time to look at how they are designed. Design is a big topic, so it is split into two chapters. This chapter covers design of single-channel systems to meet loss and bandwidth requirements. Chapter 22 covers design considerations for wavelength-division multiplexing and optical networking.

Loss budgets and transmission capacity, or bandwidth, are crucial in both single- and multi-channel systems, but the basic principles are the same for both. You calculate loss budget to be sure that enough signal reaches the receiver to give adequate performance. Likewise, you must calculate pulse dispersion, or bandwidth, to be sure the system can transmit signals at the proper speed. Some simple guidelines will give you rough assessments. In the real world, you also must consider cost-effectiveness, and make trade-offs among various approaches to find the best performance at the most reasonable cost. Single-channel design techniques can be applied to each channel in a multiwavelength system.

These two chapters will not prepare you for heavy-duty system design. However, they will give you the basic understanding of design concepts and technical trade-offs you need to assess fiber-optic systems.

Variables

Design of a fiber-optic system is a balancing act. You start with a set of performance requirements, such as sending 2.5 Gbit/s through a 5-km cable. You add some subsidiary goals, sometimes explicitly, sometimes implicitly. For example, you may demand cost as low as possible, less than another alternative, or no more than a given amount. Your system might need an error rate of no more than 10^{-15} and should operate without interruption for at least 5 years.

Design of fiber-optic systems requires balancing many cost and performance goals.

You must look at each goal carefully to decide how much it is worth. Suppose, for instance, you decide that your system absolutely must operate 100% of the time. You're willing to pay premium prices for transmitters, receivers, and super-duper heavily armored absolutely gopher-proof cable. But how far should you go? If that is an absolute must because of national security and you have unlimited quantities of money, you might buy up the entire right of way, install the cable in ducts embedded in a meter of concrete, and post guards armed with tanks and bazookas to make sure no one comes near the cable with a backhoe. If its purpose is just to keep two corporate computers linked together, you might be satisfied with laying a redundant gopher-proof cable along a second route different enough from the first that no single accident would knock out both.

Many variables enter into system design.

That somewhat facetious example indicates how many variables can enter into system design. In this chapter, I will concentrate on the major goals of achieving specified transmission distance and data rate at reasonable cost in the simple case of a fiber carrying only one optical channel. Many design variables can enter into the equation, directly or indirectly. Among them are the following:

- Light source output power (into fiber)
- Coupling losses
- Spectral linewidth of the light source
- Response time of the light source and transmitter
- Signal coding
- Splice and connector loss
- Type of fiber (single- or multimode)
- Fiber attenuation and dispersion
- Fiber core diameter
- Fiber NA
- Operating wavelength
- Optical amplifiers
- Direct versus indirect modulation of transmitter
- Switching requirements
- Receiver sensitivity
- Bit error rate or signal-to-noise ratio
- Receiver bandwidth
- System configuration
- Number of splices, couplers, and connectors

- Type of couplers
- Costs
- Upgradability of design

Many of these variables are interrelated. For example, fiber attenuation and dispersion depend on operating wavelength as well as the fiber type. Coupling losses depend on factors such as fiber NA and core diameter. Some interrelationships limit the choices available. For example, the need to achieve low fiber loss may require operation at 1300 or 1550 nm, and the need for optical amplification may dictate 1550 nm.

Some variables may not give you as many degrees of freedom as you might wish. For example, you may need to interconnect several computer terminals. You have enough flexibility to pick any of the possible layouts in Figure 21.1, but you still have to connect all the terminals, and that requires enough optical power to drive them all.

The type of system dictates the features you consider. If your goal is to connect computer terminals on a single floor of an office building, coupling loss will be more important than fiber attenuation. If your goal is to span the Pacific Ocean, fiber attenuation, dispersion, and optical amplifiers will be your main concerns. Designers of a transpacific cable must carefully consider how to achieve maximum transmission capacity, but the office-network designer must try to minimize coupling losses.

Often you can reach similar performance goals in more than one way. As you will learn in Chapter 22, a typical example is whether to transmit one signal at 10 Gbit/s or four WDM signals at 2.5 Gbit/s. Both choices can deliver a total of 10 Gbit/s, so the choice must depend on other factors, such as the cost, expected reliability, and potential for future upgrades.

Real fiber-optic system design is inherently a complex task, like trying to solve many simultaneous equations in algebra. The best way to understand the concepts is to look at them one at a time, before you worry about how they interact. For that reason, the examples that follow are kept simple, without worrying about the complex trade-offs that affect real design decisions.

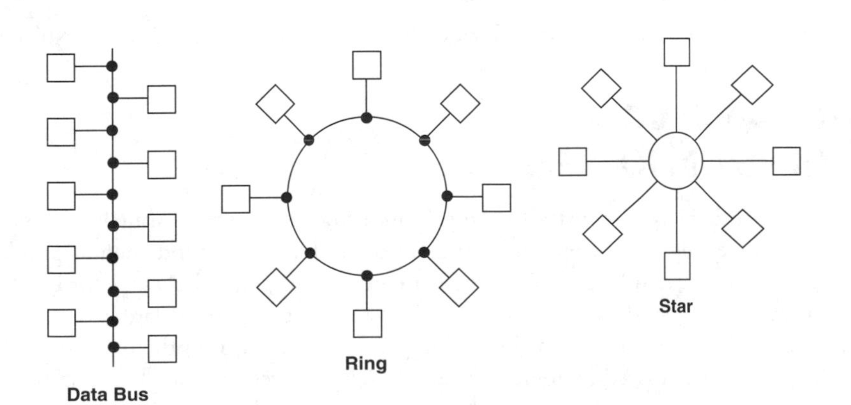

FIGURE 21.1
Three ways to interconnect computer terminals.

Power Budgeting

Power budgeting verifies that enough light reaches the receiver for proper system operation.

Power budgeting is much like making sure you have enough money to pay your bills. In this case, you need enough light to cover all optical transmission losses and to deliver enough light to the receiver to achieve the desired signal-to-noise ratio or bit error rate. That design should leave some extra margin above the receiver's minimum requirements to allow for system aging, fluctuations, and repairs, such as splicing a broken cable. However, it should not deliver so much power that it overloads the receiver.

One note of warning: be sure you know what power is specified where. You can lose 3 dB if the transmitter manufacturer specifies output as peak power but the receiver manufacturer specifies average power.

In simplest form, the power budget is

The difference between transmitter output and receiver input equals the sum of system losses and margin.

$$\text{Power}_{\text{transmitter}} - \text{total loss} + \text{amplification} = \text{margin} + \text{receiver sensitivity}$$

when the arithmetic is done in decibels or related units such as dBm. The simplicity of these calculations is a main reason for using decibel units.

Remember that optical amplification can offset loss in the system budget. Optical amplifiers are expensive, but that high cost is justified in some cases. You wouldn't buy a $3000 optical amplifier so you could replace a $100 laser source with a $10 LED, but you would if you could avoid spending $10,000 on an electro-optic regenerator.

All losses in the system must be considered. These include

- Loss in transferring light from source into fiber
- Connector loss
- Splice loss
- Coupler loss
- Fiber loss
- Fiber-to-receiver coupling loss

Some of these losses have been covered in detail earlier, but others deserve more explanation.

Light Collection

MATCHING LEDs TO FIBERS

Significant losses can occur in coupling light sources to fibers.

Typically, little light is lost in transferring from a fiber to a receiver, but large losses can occur in transferring light from the source into the fiber. The fundamental problem is matching the source emitting area to the fiber core. This is particularly true for LEDs with large emitting areas, as shown in Figure 21.2. If the emitting area is larger than the fiber core, some emitted light is lost in the cladding and dissipated from the fiber. In addition, some light rays are emitted at angles outside the fiber's acceptance angle. The smaller the fiber core and the numerical aperture, the more severe these losses. LEDs couple light effi-

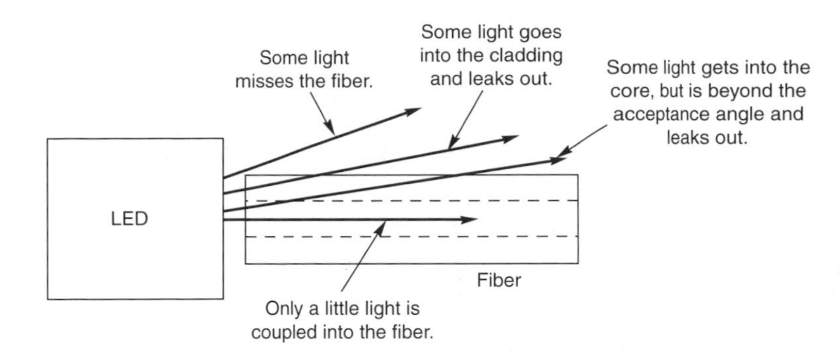

FIGURE 21.2
Light losses in transferring LED output into a fiber.

ciently into large-core step-index fibers, but losses are much larger for LEDs when coupled to 62.5/125 multimode graded-index fibers. LED output in the 1-mW range can be reduced to about 50 μW—a 13-dB loss—by losses in getting the light into the fiber. LEDs are almost never used with single-mode fibers. Loss is lower with an edge-emitting LED.

LASER SOURCES

Semiconductor lasers deliver more power into optical fibers. Their advantages are a smaller emitting area, smaller beam spread, and higher output power. The smaller beams fit better into fiber cores, particularly the tiny cores of single-mode fibers. Where a good LED might couple 50 μW into a 62.5/125-μm graded-index multimode fiber, a good edge-emitting semiconductor laser can transfer a milliwatt or more into a single-mode fiber.

A good laser can transfer a milliwatt into a single-mode fiber.

If this makes lasers seem like better light sources, that's largely because they are. Semiconductor lasers can be modulated faster and deliver more power into a fiber than LEDs. However, cost counts in the real world, and LEDs are cheaper than edge-emitting lasers. They also last longer and generally do not require the cooling and stabilization needed by many edge-emitting laser transmitters.

VCSELs (vertical-cavity surface-emitting lasers) combine many of the advantages of LEDs and lasers. They are long-lived and inexpensive but generate higher power than an LED in a much smaller beam. VCSELs also can be modulated at gigabit rates. This makes them attractive sources for systems transmitting over moderate distances. However, VCSELs emit lower power than edge-emitting lasers and are not as well-developed at 1300 and 1550 nm.

Special high-power lasers, or laser sources followed by optical amplifiers, can deliver powers of around 100 mW into optical fibers. These strong light sources are used when signals must be distributed among many receivers, such as in cable television systems or other networks where couplers split signals among many nodes. Although these lasers or laser-amplifier combinations are considerably more expensive than individual ordinary semiconductor lasers, they are less costly than the multiple lasers that you would otherwise need to serve so many terminals. (If distances are long, high powers can cause nonlinear effects in single-mode fiber.)

FIBER CHOICE

Fibers with larger cores and/or numerical apertures collect more light.

The choice of fiber also affects the light-collection equation. The larger the core diameter and the numerical aperture, the more light a fiber can collect, assuming that the light source emits from a large enough area to fill the larger fiber core. You won't gain anything by switching from a 62.5/125-μm multimode graded-index fiber to a large-core step-index fiber if you have a laser source that emits from a 10-μm stripe.

You can estimate the difference in the efficiency of light collection by a pair of fibers—fiber 1 and fiber 2—using the equation

$$\Delta\text{Loss (dB)} = 20 \log_{10}\left(\frac{D_1}{D_2}\right) + 20 \log_{10}\left(\frac{\text{NA}_1}{\text{NA}_2}\right)$$

where the *D*s are core diameters and the NAs are numerical apertures of the two fibers. You can use the formula as long as the emitting area is larger than the cores of both fibers and no optics are used to change the effective size of the emitting area or the effective NA of the source.

The difference can be significant with a large source. Consider, for example, the difference in how much light a step-index fiber with 100-μm core and 0.3 NA can collect compared to a graded-index fiber with 50-μm core and 0.2 NA. Substituting the numbers gives:

$$\Delta\text{Loss} = 20 \log\left(\frac{100}{50}\right) + 20 \log\left(\frac{0.3}{0.2}\right) = 9.6 \text{ dB}$$

That difference is nearly a factor of 10, well worth considering if you have run out of loss budget.

On the other hand, remember that a larger core reduces transmission bandwidth as well as increasing light-collection efficiency. The sacrifices are largest in moving from single-mode to multimode graded-index fiber and from graded-index multimode to step-index multimode.

SINGLE-MODE FIBERS

Loss in transferring light from an LED into a single-mode fiber is about 19 dB higher than into a 50/125 fiber.

The preceding example was for multimode fibers. Carry it a step further to a single-mode fiber with a nominal core diameter of 10 μm and NA of 0.11, and you will immediately see a big problem. A single-mode fiber collects 19.2 dB less light from an LED than a 50/125-μm fiber.

If your first impulse is to say, "Forget it," you are in good company. However, there are ways to ease that coupling problem. Instead of just butting the fiber end against the LED or aiming LED output in the general direction of the fiber end, developers can focus the light onto the fiber end with tiny optics. Losses remain significant, but they are well below the exceedingly high levels predicted by this simple cookbook formula.

Fibers collect laser output more efficiently because of the smaller emitting area and beam spread.

What about transferring diode laser output into fibers? The huge losses mentioned earlier go away because the light-emitting stripes of edge-emitting lasers are smaller than the cores of single-mode fibers. (Some VCSELs emit from larger areas, but their beams are smaller

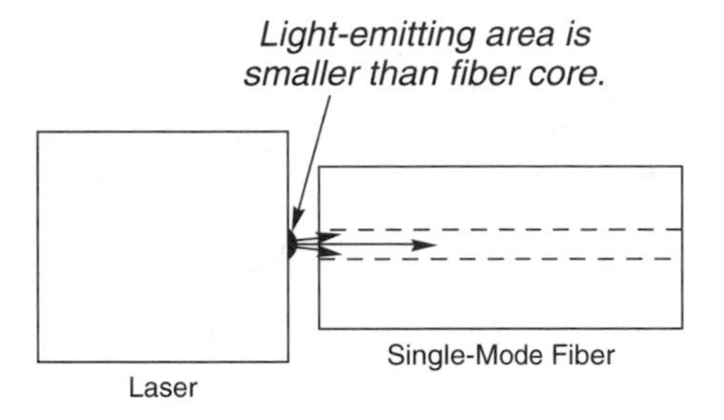

FIGURE 21.3
Laser output couples easily into a single-mode fiber core.

than the cores of standard graded-index fibers.) Laser light also does not spread out as fast as that from an LED. As shown in Figure 21.3, this makes light collection much simpler. In addition, lasers emit higher powers, so an edge-emitting laser can deliver a milliwatt or more into a single-mode fiber. Thus, with very rare exceptions, a laser is the only reasonable choice for use with single-mode fiber.

Fiber Loss

Fiber loss nominally equals the attenuation (in decibels per kilometer) times the transmission distance:

Fiber loss equals attenuation times distance, sometimes plus transient losses.

$$\text{Total loss} = (\text{dB/km}) \times \text{length}$$

However, this is only an approximation for multimode fibers. One problem is that measurements of fiber attenuation in dB/km do not consider transient losses that occur near the start of a multimode fiber. An LED with a large emitting area and high NA excites high-order modes that leak out as they travel along the fiber. Typically this transient loss is 1 to 1.5 dB, concentrated in the first few hundred meters of fiber following the transmitter. This loss becomes less significant after you go a kilometer or two, but graded-index multimode fibers are rarely used over much longer distances. Thus, it's important to remember transient loss and allow for it in your system margin.

An additional problem that can occur with graded-index fiber is uneven and unpredictable coupling of modes between adjacent lengths of fiber. These concatenation effects make loss of long lengths of spliced graded-index fiber difficult to calculate; fortunately, such systems are extremely rare.

Single-mode fibers are much better behaved because they carry only one mode, avoiding differential mode attenuation.

Fiber-to-Receiver Coupling

One of the rare places where the fiber-optic engineer wins is in coupling light from a fiber to a detector or receiver. The light-sensitive areas of most detectors are larger than most fiber cores, and their acceptance angles are larger than those of multimode fibers. Of course, if you're determined to screw things up, you can find a detector with a light-collecting area smaller than the core of large-core multimode fibers. But that isn't likely.

Losses are normally small in transferring light from fibers to receivers.

Receiver Sensitivity

There are trade-offs among received power, speed, and bit error rate or signal-to-noise ratio.

In much of the discussion that follows, receiver sensitivity is taken as a given. That is, I assume that a receiver must have a minimum power input to work properly. Things aren't quite that simple because there are trade-offs between received power, speed, and bit error rate or signal-to-noise ratio. As data rate increases, a receiver needs more input power to operate with a specified bit error rate. If the data rate is held fixed but the input power is decreased, the error rate can increase steeply.

These trade-offs are not always useful. Error rate can increase steeply as input power decreases. At the margin of receiver sensitivity, a 1-dB drop in power can increase error rate by a factor of 1000 or more! You gain more by lowering data rate, particularly near the receiver's maximum speed. However, most system designs do not allow much flexibility in transmission speed. As I describe later in this chapter, some gains in sensitivity are possible by switching bit encoding schemes, but the simplest course may be using a more sensitive receiver or going back to reduce loss or increase power.

Remember, too, that more power is not always a good thing. Too much power can overload the detector, increasing bit-error rate.

Other Losses

Total loss from connectors, couplers, and splices is their characteristic loss multiplied by the number in the system.

Splices, connectors, and couplers can contribute significant losses in a fiber-optic system. Fortunately, those losses are generally easy to measure and calculate. Connectors, couplers, and splices have characteristic losses that you can multiply by the number in a system to estimate total loss. However, there are two potential complications.

One is the variability of loss, particularly for connectors. A given connector may be specified as having maximum loss of 1.0 dB and typical loss of 0.5 dB. The maximum is the specified upper limit for that type of connector; no higher losses should show up in your system (unless the connector was installed improperly or is dirty). The typical value is an average, meaning that average connector loss should be 0.5 dB but that individual connectors may be higher or lower.

You can calculate total loss in two ways for a system with four connectors. The worst-case approach is to multiply the highest possible loss (1 dB) by the number of connectors to get 4 dB. On the other hand, if the average connector loss is 0.5 dB, the most likely total loss is four times that: 2.0 dB. The prudent approach for so few connectors is to take the worst-case value, but it's much more realistic to take the average loss for systems with many connectors or splices. Remember, because detector overload can cause problems, you can run into trouble by seriously overestimating loss as well as by underestimating it.

Transient losses following connectors can further complicate the picture for multimode fibers. Connectors near the transmitter may increase transient losses by effectively stripping away high-order modes, which otherwise would leak out farther along the fiber. However, once light has traveled far enough to reach an equilibrium mode distribution (a kilometer or so), a connector can redistribute some light to higher-order modes, which tend to leak out of the fiber—a milder form of transient losses than experienced with light sources.

Margin

One quantity that always figures in the loss budget is system margin, a safety factor for system designers. This allows for uncertainties in calculating losses, for minor repairs, and for minor degradation of system components. Uncertainties are inevitable because component losses are specified within ranges and because components change as they age and are used. Margin also allows for repairs in case of cable damage, which typically add to cable loss.

System margin is a safety factor to allow for repairs and uncertainties.

Depending on the application, the performance requirements, the cost, and the ease of repair, the loss margin added by designers may be 3 to 10 dB.

Optical Amplifiers

Optical amplifiers can overcome losses by boosting signal strength, but practical concerns may offset this advantage. Optical amplifiers are expensive, are only available readily for the 1550-nm region, and as analog devices inevitably amplify background noise as well as signal. On the other hand, they can amplify gigabit signals and multiple wavelengths in their operating ranges. Thus they are mainly used in high-performance and WDM systems, where their high cost can be spread among many signals being amplified.

Optical amplifiers boost signal strength.

You can use optical amplifiers in several places:

- As postamplifiers after transmitters, to generate high-power signals in fibers
- In the middle of long transmission systems, to boost signal strength for further transmission
- As preamplifiers before receivers, to raise signal strength to the proper level for the receiver
- Before or after couplers, which divide input signals among many outputs

Examples of Loss Budgeting

To see how loss budgeting works, I'll step through three simple examples. Example A, shown in Figure 21.4, is a short system transmitting 100 Mbit/s between two points in a building. Example B, shown in Figure 21.5, is a telephone system carrying 2.5 Gbit/s between two switching offices 300 km apart. Example C, shown in Figure 21.6, is an intrabuilding network linking 10 terminals with each other at a signal speed of 100 Mbit/s. The examples are arbitrary and are intended to show how design works rather than to illustrate actual systems. Note that in considering only the loss budget, you don't directly address whether or not the system can carry the data rate listed. We'll look at that issue later in this chapter.

Example A

Connectors can contribute the dominant losses if several are in a short system.

In Figure 21.4, designers need to transmit signals through 200 m of fiber already installed in a building. That means that they must route the signal through patch panels with connectors. In this example, they have six connector pairs, three on each floor: one linking the terminal device to the cable network for that floor, and one pair on each end of a short cable in the patch panel. (Connectors also attach the fiber to transmitter and receiver, but their losses are included under LED power transfer and receiver sensitivity.) The 50/125 graded-index multimode fiber has attenuation of 2.5 dB/km at the 850-nm wavelength of the LED transmitter. The loss budget is as follows:

LED power into fiber	−16.0 dBm
Connector pairs (6 @ 0.7 dB)	−4.2 dB
Fiber loss (200 m @ 2.5 dB/km)	−0.5 dB
System margin	−10.0 dB
Required receiver sensitivity	−30.7 dBm

The calculation shows that the dominant loss is from the connectors. The fiber loss may underestimate transient loss, but the large system margin leaves plenty of room.

The calculated receiver sensitivity is a reasonable level, and system margin could be improved by picking a more sensitive receiver. This calculation started with a given loss, system margin, and input power, but you could start by specifying receiver sensitivity, system margin, and loss, to calculate the needed input power. Note that the LED transmitter provides a low input power, but that is adequate for this short system.

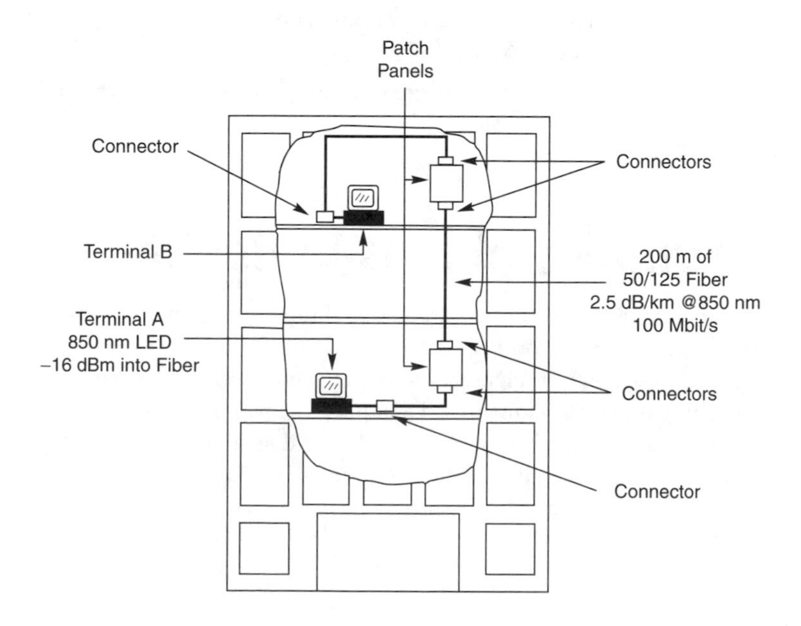

FIGURE 21.4
Example A: Point-to-point link in a building.

Example B

Loss sources in the telephone system shown in Figure 21.5 are quite different. The system spans 300 km, with one splice every 10 km in a single-mode fiber with loss of 0.25 dB/km at 1550 nm. It links two rural areas, carrying a single wavelength at 2.5 Gbit/s. The high speed and long distance demand a laser source and a more sensitive receiver. Signals go through two connector pairs in patch panels at each end. Although the fiber loss is very low at 1550 nm, the long distance makes fiber attenuation the dominant loss. The sample calculation shows the laser transmitter alone does not deliver enough power to span that distance:

Fiber attenuation dominates loss in a 300-km fiber system.

Laser power into single-mode fiber	0.0 dBm
Fiber loss (300 km × 0.25 dB/km)	−75.0 dB
Splice loss (29 × 0.1 dB)	−2.9 dB
Connector pairs (4 × 0.8 dB)	−3.2 dB
Power at receiver	−81.1 dBm
Receiver sensitivity	−32.0 dBm
System power deficit	−49.1 dB

The output power falls far short of system requirements. You need optical amplifiers. Suppose you add a pair of optical amplifiers with 30-dB gain, one at the 100-km point and the second at 200 km. This requires four extra connector pairs (one on each end of each optical amplifier), which replace two splices. The loss budget then becomes:

Laser power into single-mode fiber	0.0 dBm
Fiber loss (300 km × 0.25 dB/km)	−75.0 dB
Optical amplifier gain	60.0 dB
Splice loss (27 × 0.1 dB)	−2.7 dB
Connector pairs (8 × 0.8 dB)	−6.4 dB
Power at receiver	−24.1 dBm
Receiver sensitivity	−32.0 dBm
System power margin	7.9 dB

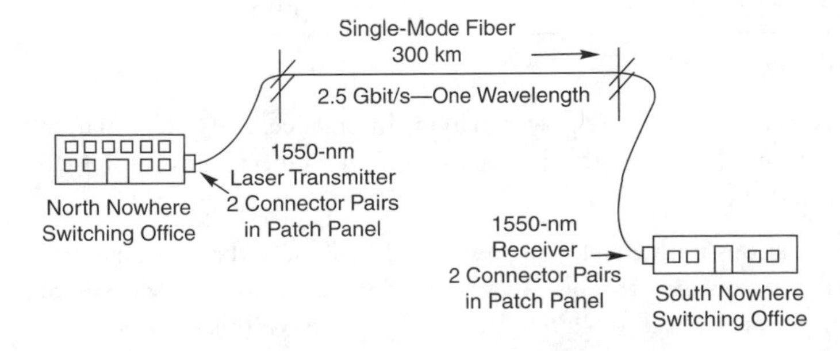

FIGURE 21.5
Example B: 300-km single-mode fiber system.

That looks much better. To verify that the loss budget works for the whole system, you should check the budget for each segment.

Segment 1:

Laser power into single-mode fiber	0.0 dBm
Fiber loss (100 km × 0.25 dB/km)	−25.0 dB
Splice loss (9 × 0.1 dB)	−0.9 dB
Connector pairs (3 × 0.8 dB)	−2.4 dB
Power at optical amplifier	−28.3 dBm
Gain of optical amplifier	30.0 dB
Output of segment 1	1.7 dBm

Segment 2:

Optical amplifier into single-mode fiber	1.7 dBm
Fiber loss (100 km × 0.25 dB/km)	−25.0 dB)
Splice loss (9 × 0.1 dB)	−0.9 dB
Connector pairs (2 × 0.8 dB)	−1.6 dB
Power at optical amplifier 2	−25.8 dBm
Gain of optical amplifier	30.0 dB
Output of segment 2	4.2 dBm

Segment 3:

Laser power into single-mode fiber	4.2 dBm
Fiber loss (100 km × 0.25 dB/m)	−25.0 dB
Splice loss (9 × 0.1 dB)	−0.9 dB
Connector pairs (3 × 0.8 dB)	−2.4 dB
Power at receiver	−24.1 dBm
Receiver sensitivity	−32.0 dBm
System power margin	7.9 dB

For our purposes, 7.9 dB seems an adequate power margin. In practice, the system margin probably will be better, because I have assumed a relatively high connector loss of 0.8 dB and a low laser output of 0 dBm.

Segment-by-segment calculations both check your result and make sure that placement of optical amplifiers doesn't get you into trouble. In practice, optical amplifiers saturate at high powers, so you may get only 25 dB of gain with 20 dBm input. In this example, suppose you put the second optical amplifier at the 170-km point because you happen to have a storage building at that point. Then the calculations for segments 2 and 3 are as follows:

Segment 2:

Optical amplifier into single-mode fiber	1.7 dBm
Fiber loss (70 km × 0.25 dB/km)	−17.5 dB
Splice loss (6 × 0.1 dB)	−0.6 dB
Connector pairs (2 × 0.8 dB)	−1.6 dB
Power at optical amplifier 2	−18.0 dBm
Reduced gain of optical amplifier	25.0 dB
Output of segment 2	7.0 dBm

Segment 3:

Laser power into single-mode fiber	7.0 dBm
Fiber loss (130 km × 0.25 dB/m)	−32.5 dB
Splice loss (12 × 0.1 dB)	−1.2 dB
Connector pairs (3 × 0.8 dB)	−2.4 dB
Power at receiver	−29.1 dBm
Receiver sensitivity	−32.0 dBm
System power margin	2.9 dB

Although the receiver power is above the required level, a system margin of 2.9 dB is inadequate for contingencies. This is a reminder that you can't simply add up the losses and gains of all components without considering the input conditions to components such as optical amplifiers.

Example C

Complications also arise when you have to divide input signals among many outputs, as shown in Figure 21.6. In this case, you need to connect 10 terminals so the output of each one is divided among the receivers of all 10 terminals. You can do this with a 10 × 10 directional star coupler, which divides an input signal from any of the 10 incoming fibers (one from the transmitter end of each terminal) among the 10 output fibers going to the receiver end of each terminal. Assume the coupler divides the signals equally and has excess loss of 3 dB. Because the data rate is a modest 100 Mbit/s, let's calculate the loss budget with an LED source.

Coupling losses are largest in a network distributing signals to many terminals.

LED transmitter (850 nm)	−16.0 dBm
Fiber loss (100 m @ 2.5 dB/km)	−0.25 dB

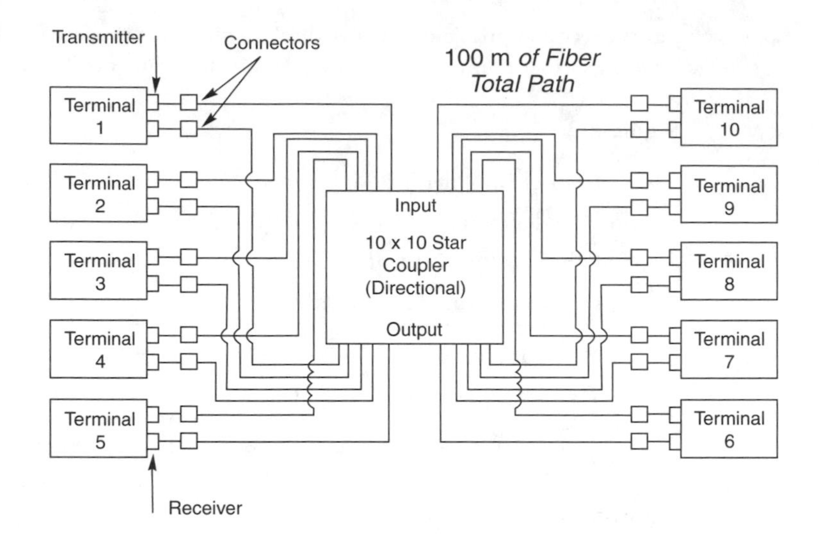

FIGURE 21.6 *Example C: 10-terminal network.*

Connector pair loss (2 @ 0.5 dB)	−1.0 dB
Coupler loss (includes its own connections)	−13.0 dB
Power at receiver	−30.25 dBm
Receiver sensitivity	−30.0 dBm
System power deficit	−0.25 dB

The calculations show that I'm in trouble. The system margin is negative, a power deficit. It's only −0.25 dB, but it's enough that the receiver may not work well under the best of circumstances. Performance will degrade even further if anything reduces receiver power.

A single component, the coupler, dominates the loss budget, but it is not easy to eliminate because it is needed to distribute the signal to all terminals. One way to overcome the power deficit is by replacing the LED with a more powerful edge-emitting laser source. In this case, the power budget becomes

Edge-emitting laser transmitter (850 nm)	0.0 dBm
Fiber loss (100 m @ 2.5 dB/km)	−0.25 dB
Connector pair loss (2 @ 0.5 dB)	−1.0 dB
Coupler loss (includes its own connections)	−13.0 dB
Power at receiver	−14.25 dBm
Receiver sensitivity	−30.0 dBm
System power margin	15.75 dB

This may be too much of a good thing, depending on the power level that overloads the receiver. If the receiver overloads at −15 dBm, you can add a 3-dB attenuator to bring the receiver power down to −17.25 dBm, well within the operating range of −15 to −30 dBm.

Another approach is to use a lower-power VCSEL, with output of −5 dBm. In this, the power budget becomes:

VCSEL transmitter	−5 dBm
Fiber loss (10 m @ 2.5 dB/km)	−0.25 dB
Connector pair loss (2 @ 0.5 dB)	−1.0 dB
Coupler loss (includes its own connectors)	−13.0 dB
Power at receiver	−19.25 dBm
Receiver sensitivity	−30.0 dBm
System power margin	10.75 dB

This should avoid overloading the receiver without the need for an attenuator. In addition, VCSELs are becoming less expensive than edge-emitting lasers, although we aren't considering costs here.

Transmission Capacity Budget

The transmission capacity of a fiber-optic system is the total analog bandwidth or digital data rate it can carry. With wavelength-division multiplexing, total capacity of a fiber is the sum of the capacities of all optical channels the fiber carries. In this section, we will cover only the transmission capacity of a single optical channel; Chapter 22 will cover multichannel systems.

Bandwidth or bit rate depends on fiber, source, and receiver characteristics.

Single-channel capacity depends on how fast all the parts of the system responds to changes in signal intensity. In practice, transmission speed is mainly affected by properties of the transmitter, fiber, and receiver. For simplicity, I will ignore secondary effects such as noisy amplifiers and nonlinear effects that also can restrict data rates.

The simplest way to calculate transmission capacity is from the time response or rise time of the signal in the important components. This corresponds to the rise time of a transmitter or receiver, and dispersion in a fiber. As you will see below, this response is cumulative, and as you saw in Chapter 5, it is the sum of the squares of the responses of the various components.

Both analog bandwidth and digital bit rate are related to the time response or rise time, although the relationships are approximate and depend on details.

Bandwidth and data rate can be calculated from rise time.

For an analog system, the 3-dB bandwidth B in megahertz is inversely proportional to the rise time Δt in nanoseconds

$$B\,(\text{MHz}) = \frac{350}{\Delta t\,(\text{ns})}$$

Thus an analog system with 1-ns rise time has a roughly 350-MHz bandwidth.

For NRZ-coded digital signals, the rise time can be as large as 0.7 times the bit interval. If rise time is in nanoseconds, the maximum data rate in Gbit/s is

$$\text{NRZ data rate (Gbit/s)} = \frac{0.7}{\Delta t\ (\text{ns})}$$

Thus a digital system with 1-ns rise time can support an NRZ data rate of roughly 700 Mbit/s. RZ codes require double the bandwidth or rise times twice as fast:

$$\text{RZ data rate (Gbit/s)} = \frac{0.35}{\Delta t\ (\text{ns})}$$

These relationships are not exact, but they are useful for rough calculations.

The speed limits on digital signals are often easier to understand. You can get an intuitive feel for the impact of response time on a digital signal by thinking about pulse detection. If pulses follow each other at a certain speed, you need to be able to detect a pulse in less time than the interval between pulses.

Overall Time Response

Overall time response is the square root of the sum of the squares of component rise times.

The choice of time response or rise time simplifies calculations. The overall time response of a system is the square root of the sum of the squares of the response times of individual components:

$$\Delta t_{\text{overall}} = \sqrt{\Sigma\ (\Delta t_i^2)}$$

where $\Delta t_{\text{overall}}$ is the overall time response and Δt_i is the time response of each component.

Connectors, splices, couplers, and optical amplifiers do not affect time response significantly in current systems. The important response times are those of the transmitter, receiver, and fiber:

$$\Delta t_{\text{overall}} = \sqrt{\Delta t_{\text{transmitter}}^2 + \Delta t_{\text{receiver}}^2 + \Delta t_{\text{fiber}}^2}$$

That is, the overall response time is the square root of the sum of the squares of transmitter rise time, receiver rise time, and the pulse spreading caused by fiber dispersion.

Transmitter and receiver rise and fall times are listed on data sheets, ready to plug into the formula. Fiber response times must be calculated from the fiber's dispersion properties. You can see how transmitter and receiver properties affect data rate if we assume dispersion is small. Suppose we have a short data link spanning only 20 m of fiber, so we can neglect dispersion. If transmitter and receiver rise times are both 1 ns, the overall response time is:

$$\Delta t_{\text{overall}} = \sqrt{\Delta t_{\text{transmitter}}^2 + \Delta t_{\text{receiver}}^2} = \sqrt{2} = 1.414\ \text{ns}$$

Going back to the earlier formula for maximum data rate, we see it is:

$$\text{Data rate (Gbit/s)} = \frac{0.7}{1.4\ (\text{ns})} = 0.5\ \text{Gbit/s}$$

If one component is much slower than the other, it dominates the response time. For example, if the transmitter had 1-ns rise time but the receiver had a 10-ns response, the overall response time would be 10.05 ns, limiting data rate to 70 Mbit/s.

Fiber dispersion becomes important in longer systems and deserves a closer look.

Fiber Dispersion Effects

Fiber response must be calculated from dispersion.

Fiber response times must be calculated from the fiber length, the characteristic dispersion per unit length, and the source spectral width. As you learned earlier, fibers show modal, material, and polarization-mode dispersion. Which types are most important depends on the type of fiber. In multimode fibers, modal dispersion and chromatic dispersion are important. In single-mode fibers, modal dispersion is zero, but chromatic dispersion and polarization-mode dispersion are significant. (Remember that chromatic dispersion is the sum of material and waveguide dispersion.)

You calculate total pulse spreading caused by dispersion with a sum-of-squares formula similar to that for overall time response:

$$\Delta t = \sqrt{\Delta t^2_{\text{modal}} + \Delta t^2_{\text{chromatic}} + \Delta t^2_{\text{PMD}}}$$

In practice, this can be simplified. For multimode fibers, polarization-mode dispersion is insignificant, so only modal and chromatic dispersions are considered. Single-mode fibers do not suffer modal dispersion, so only chromatic and polarization-mode dispersions are considered. In practice, polarization-mode dispersion is negligible for data rates below 2.5 Gbit/s.

However, dispersion is more complex because other factors enter the equation. Modal dispersion is a characteristic value D_{modal} (specified in ns/km) times fiber length (in km). Chromatic dispersion is a characteristic value $D_{\text{chromatic}}$ (specified in ps/km-nm) times the fiber length (in km) and the spectral width of the transmitter (in nm). Polarization-mode dispersion is a characteristic value D_{PMD} (specified in ps/root-km) multiplied by the square root of the fiber length (in km).

Plugging these quantities into the pulse-spreading equation gives different formulas for multimode and single-mode fibers, based on fiber-dispersion properties, fiber length L, and spectral width of the light source $\Delta\lambda$. For multimode fiber, the pulse spreading is:

$$\Delta t_{\text{multimode}} = \sqrt{(D_{\text{modal}} \times L)^2 + (D_{\text{chromatic}} \times L \times \Delta\lambda)^2}$$

For single-mode fiber, this formula is:

$$\Delta t_{\text{single mode}} = \sqrt{(D_{\text{chromatic}} \times L \times \Delta\lambda)^2 + (\text{D}_{\text{PMD}} \times \sqrt{L})^2}$$

A couple of examples will show how dispersion calculations work.

MULTIMODE DISPERSION EXAMPLE

In Example A, considered earlier in this chapter, an 850-nm LED sends 100 Mbit/s through 200 m of 50/125-μm fiber. Modal bandwidth of a typical commercial 50/125-μm fiber is 400 MHz, which is equivalent to a modal dispersion of 2.5 ns/km. For a 200-m length, that corresponds to modal dispersion of 0.5 ns.

To that, you need to add the chromatic dispersion, calculated from the formula:

$$\Delta t_{\text{chromatic}} = D_{\text{chromatic}} \times L \times \Delta\lambda$$

A typical value of chromatic dispersion is 100 ps/nm·km at 850 nm, which combined with a linewidth of 50 nm for a typical 850-nm LED, gives a chromatic dispersion of 1.0 ns for a 200-m length of fiber. This means that the chromatic dispersion is higher than modal dispersion because of the large LED spectral linewidth.

Adding modal and chromatic dispersions together according to the sum-of-squares formula indicates total dispersion is 1.1 ns. That leaves plenty of room for transmitting 100 Mbit/s, assuming the transmitter and receiver are fast enough.

If you wanted to transmit 1 Gbit/s, you could try using a VCSEL source with a 1-nm linewidth. In that case, total dispersion is

$$\Delta t = \sqrt{(0.5\text{ ns})^2_{\text{modal}} + (100\text{ ps/nm}\cdot\text{km} \times 0.2\text{ km} \times 1\text{ nm})^2} = 0.50\text{ ns}$$

This is essentially equal to the modal dispersion of 0.5 ns and is adequate for gigabit transmission over 200 m.

SINGLE-MODE TRANSMISSION EXAMPLE

In Example B, we considered transmitting a 2.5-Gbit/s signal a total of 300 km through a single-mode fiber at 1550 nm. Assume that chromatic dispersion is specified at 3 ps/nm·km, a relatively low value. Let's consider two cases: a Fabry-Perot laser with linewidth of 1 nm and an externally modulated distributed-feedback laser with linewidth of 0.1 nm. For a first approximation, ignore polarization-mode dispersion.

For the 1-nm laser, chromatic dispersion is

$$\Delta t_{\text{chromatic}} = 3\text{ ps/nm}\cdot\text{km} \times 300\text{ km} \times 1\text{ nm} = 900\text{ ps}$$

This value is much too high for a 2.5-Gbit/s system.

With the distributed-feedback laser, chromatic dispersion is

$$\Delta t_{\text{chromatic}} = 3\text{ ps/nm}\cdot\text{km} \times 300\text{ km} \times 0.1\text{ nm} = 90\text{ ps}$$

Thus using fiber with low chromatic dispersion near 1550 nm leaves plenty of margin. However, you could not get away with using step-index single-mode fiber with dispersion around 20 ps/nm-km at 1550 nm.

$$\Delta t_{\text{chromatic}} = 20\text{ ps/nm-km} \times 300\text{ km} \times 0.1\text{ m} = 600\text{ ps}$$

We also need to consider polarization-mode dispersion. A typical value for new nonzero dispersion-shifted fiber is about 0.5 ps/root-km, so for 300 km of single-mode fiber, the polarization-mode dispersion is

$$\Delta t_{\text{PMD}} = 0.5\text{ ps}/\sqrt{\text{km}} \times \sqrt{300\text{ km}} = 8.7\text{ ps}$$

This is only a tenth as large as chromatic dispersion, so it makes a negligible contribution to total dispersion, and does not affect transmission at 2.5 Gbit/s.

To fully assess performance of the system, you need to consider transmitter and receiver rise time as well. Suppose both have rise times of 100 ps, reasonable for products designed to work at 2.5 Gbit/s. Then the total pulse spreading is

$$\Delta t = \sqrt{100^2 + 100^2 + 90^2} = 168 \text{ ps}$$

Our earlier formula shows that this is adequate for transmitting 4 Gbit/s.

LONG-DISTANCE SINGLE-MODE EXAMPLE

Long-distance systems must span longer distances of several hundred to several thousand kilometers. Dispersion poses more problems in these systems because the total pulse spreading increases with the length of the system. The seriousness of dispersion effects also increases with data rate.

Dispersion poses more problems at long distances.

Suppose you are using nonzero dispersion-shifted fiber with a chromatic dispersion of 3 ps/km-nm and PMD of 0.5 ps/root-km to span a distance of 1000 km. As in the last example, you are using an externally modulated laser with spectral width of 0.1 nm. Plug in the numbers and you find:

$$\Delta t_{\text{chromatic}} = 3 \text{ ps/nm}\cdot\text{km} \times 1000 \text{ km} \times 0.1 \text{ nm} = 300 \text{ ps}$$

$$\Delta t_{\text{PMD}} = 0.5 \text{ ps}/\sqrt{\text{km}} \times \sqrt{1000 \text{ km}} = 15.8 \text{ ps}$$

$$\Delta t_{\text{total}} = \sqrt{300^2 + 15.8^2} = 300.4 \text{ ps}$$

This response time is too long to allow good transmission at 2.5 Gbit/s, according to the criteria described earlier. This means that some form of dispersion compensation is needed to reduce average chromatic dispersion along the length of the fiber so it can carry 2.5 Gbit/s.

Things get even more difficult at 10 Gbit/s. If you flip the formula for maximum NRZ data rate to show the maximum allowable pulse spreading for a given data rate, you get

$$\Delta t_{\text{maximum}} = \frac{0.7}{\text{Data rate}}$$

For a 10 Gbit/s signal, this gives a maximum pulse spreading of 70 ps, which requires considerably more dispersion compensation. Let's assume that the transmitter bandwidth remains 0.1 nm, although in practice 10-Gbit/s modulation can increase this number. Suppose we can reduce the average chromatic dispersion by a factor of 5, to 0.6 ps/nm·km. Using the same assumptions above gives us:

$$\Delta t_{\text{chromatic}} = 0.6 \text{ ps/nm}\cdot\text{km} \times 1000 \text{ km} \times 0.1 \text{ nm} = 60 \text{ ps}$$

$$\Delta t_{\text{PMD}} = 0.5 \text{ ps}/\sqrt{\text{km}} \times \sqrt{1000 \text{ km}} = 15.8 \text{ ps}$$

$$\Delta t_{\text{total}} = \sqrt{60^2 + 15.8^2} = 62 \text{ ps}$$

That figure is adequate if you only consider fiber response, but you also have to consider transmitter and receiver rise times. If they both equal 25 ps—very fast devices—that would bring the total pulse spreading to the 70 ps limit. You might want average chromatic dispersion reduced even further, to about 0.3 ps/nm·km.

So far we have assumed polarization-mode dispersion is fairly benign, by using specifications for new fibers fresh from the factory. However, older fibers or fibers installed under less than ideal conditions may have higher polarization-mode dispersion. Suppose we want to use older fibers with a realistic value of 2 ps/root-km for polarization-mode dispersion. Those calculations give us:

$$\Delta t_{\text{chromatic}} = 0.6 \text{ ps/nm}\cdot\text{km} \times 1000 \text{ km} \times 0.1 \text{ nm} = 60 \text{ ps}$$

$$\Delta t_{\text{PMD}} = 2 \text{ ps}/\sqrt{\text{km}} \times \sqrt{1000 \text{ km}} = 63.2 \text{ ps}$$

$$\Delta t_{\text{total}} = \sqrt{60^2 + 63.2^2} = 87 \text{ ps}$$

This produces polarization-mode dispersion slightly larger than the chromatic dispersion, and yields a total pulse spreading too high for 10 Gbit/s transmission, even ignoring transmitter and receiver rise times.

Dispersion Compensation

Dispersion compensation becomes important at high speeds or over long distances.

As these examples show, dispersion compensation becomes essential as transmission speeds and distances increase.

Compensation is easiest for chromatic dispersion. As you learned in Chapter 5, chromatic dispersion has a characteristic sign that indicates whether the shorter or longer wavelengths have gone farther in the fiber. To compensate for chromatic dispersion, you add a length of fiber (or some other optical component) with dispersion of the opposite sign. If the shorter wavelengths were falling behind, adding a length of fiber that makes the longer wavelengths fall behind serves to compress the length of the output pulse.

One approach to chromatic dispersion compensation is by alternating segments of fiber with a different dispersion sign, as shown in Figure 21.7. In this example, one fiber has a positive dispersion and the other a negative dispersion that is smaller in magnitude. This mixture is possible by using two types of nonzero dispersion-shifted fibers, one with zero dispersion above the 1550-nm band, the other with zero dispersion at longer wavelengths. Optical signals pass through alternating segments of the two fibers, so the cumulative chromatic dispersion is first negative, then shifts positive, and so on. This is called *distributed compensation,* because the fibers that compensate for the pulse spreading are distributed along the fiber segment. In this example, the dispersions of the two fibers do not differ greatly in magnitude, but do differ in sign.

An alternative approach is to use shorter lengths of special dispersion-compensating fibers, which have much larger chromatic dispersion than standard fibers. These fibers do not transmit light as well as standard fibers, but shorter lengths are needed for the offsetting dispersion. Compensating fibers may be added to existing systems as well as designed into

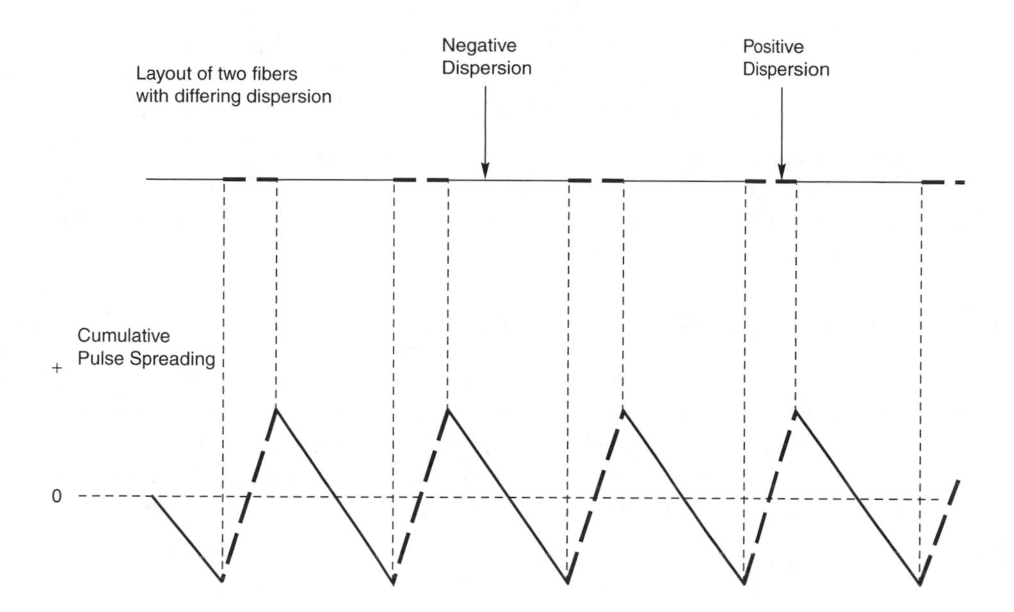

FIGURE 21.7
Dispersion compensation distributed along length of system.

new ones. They also may be packaged as coils of fiber for compensation, rather than being distributed along the transmission line. The two approaches have similar effects on chromatic dispersion, but differ in other effects, depending on the type of fibers used.

In principle, dispersion compensation can cancel the overall chromatic dispersion at one wavelength, to give a net dispersion essentially equal to zero. However, the dispersion of the two fiber types varies with wavelength, so the combination that produces net zero dispersion at one wavelength will leave residual dispersion at other wavelengths.

Techniques for compensating polarization-mode dispersion are in development, but not yet ready for wide use. Thus present dispersion compensation techniques only cancel chromatic dispersion, and cannot effectively reduce polarization mode dispersion.

Transmitter and Receiver Response Times

So far I have concentrated on fiber dispersion, but transmitter and receiver response times also play critical roles in system bandwidth budgets. Just because a transmitter and receiver are rated to operate at 10 Gbit/s in some situations does not mean they can transmit at that speed in *all* systems. For example, a transmitter and receiver, both having 40 ps rise times, can transmit 10 Gbit/s signals through fiber with up to 40 ps of cumulative pulse spreading. However, they couldn't be used in the example I showed earlier, where total pulse spreading in the fiber was 62 ps. That would require a faster transmitter and receiver, with response times around 25 ps.

Transmitters and receivers must be matched to fiber characteristics.

For this reason manufacturers sell different models of transmitters and receivers for transmission at the same data rates through different types of fibers. Transmitter–receiver pairs intended for short-distance use do not have to meet the same stringent speed requirements as those used for long-distance systems. In short, you have to match transmitters and receivers to the fiber system used to assure you get the desired transmission speed.

To show you how time-response calculations work, I have chosen simple examples. In reality there are a few other complications that come from the nature of transmitters and receivers. In particular, the range of wavelengths from transmitters is broadened by a couple of distinct effects.

Transmitter Spectral Broadening

As you learned earlier, directly modulating a semiconductor laser changes its refractive index as the density of current carriers changes. This effect, called *chirp,* causes the resonant wavelength to shift during a pulse, effectively spreading the range of output wavelengths. A laser that emits a bandwidth of 0.001 nm in a continuous beam spans a much larger range when directly modulated. External modulation can avoid chirp.

Modulation broadens the wavelengths in an optical carrier.

External modulation cannot prevent a second type of spectral broadening that arises from any amplitude modulation of a pure carrier signal. The same effect occurs with radio. Modulation generates new frequency components by adding (and subtracting) the frequency of the modulating signal to the carrier frequency, as shown in Figure 21.8. This process generates *side bands* at frequencies both above and below the carrier.

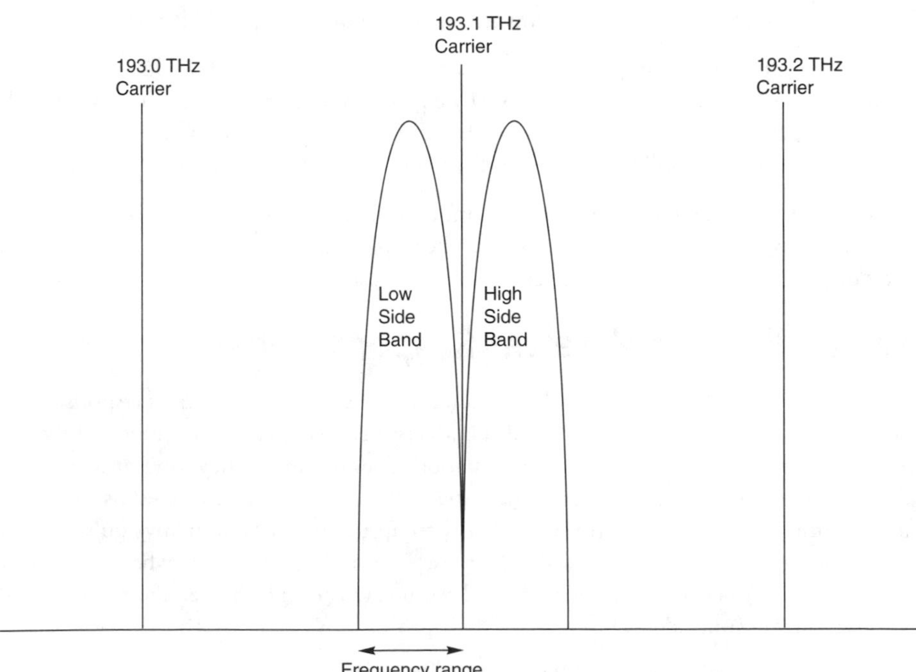

FIGURE 21.8
Modulation broadening is caused by side bands generated by the modulating signal.

The two side bands contain identical information, so radio transmitters generally suppress one side band to conserve scarce frequency space. Side-band suppression is very difficult at optical wavelengths, so both side bands are present. The result is an effective broadening of the transmitter spectrum that increases directly with the bandwidth. If modulation produces a 10-GHz range of frequencies, it generates 10-GHz side bands on each side, for a total spectral range of 20 GHz. Somewhat larger transmission bands are common in practice. This affects both chromatic dispersion in the fiber and the channel spacing possible in DWDM systems. Remember, that the tightest spacings now are 50 GHz between optical channels.

Cost/Performance Trade-offs

So far I have only mentioned in passing one of the most important considerations in real-world system design—cost. Minimizing cost is an implicit goal in all system design. I list some simple guidelines below, but it is impossible to give hard-and-fast rules for the tough job of making trade-offs between cost and performance. Ultimately it is your judgement as a system buyer or designer whether pushing error rate from 10^{-9} to 10^{-12} is worth an extra $1000. The best I can give you are some ideas to apply in working situations.

Users must make cost/performance trade-offs.

Choice of Fiber Type

The choice of fiber type has a tremendous impact on the cost and performance of any system. A fundamental choice is between single- and multimode fiber, but several variations on both types are available. Installation is expensive, so you want to allow room for your system to expand. Bandwidth requirements inevitably increase, like each new generation of software demands more computer memory and hard-disk space.

The choice of fiber type is crucial.

You need single-mode fiber if your system spans more than a couple of kilometers. Premium types, preferred for amplified WDM systems, are nonzero dispersion-shifted fibers, with low dispersion through the 1550-nm erbium amplifier window. Some single-mode fibers are optimized for "metro" applications over distances to a few hundred kilometers; others are optimized for terrestrial long-haul or submarine systems. Step-index single-mode fiber, with zero dispersion near 1310 nm, is still common, but normally is used at shorter distances because of its high dispersion at 1550 nm. Low-water fibers have a broader transmission window where amplification is not needed.

Single-mode fiber sometimes is used at distances shorter than a couple of kilometers, especially for high-speed transmission. For example, 10-Gigabit Ethernet calls for single-mode fiber to transmit more than 100 m.

Graded-index multimode fibers are normally preferred for networks where transmission distances are up to a couple of kilometers, speeds are moderate, and many connections are required. Their big advantages are easier coupling to each other and to low-cost light sources. Bandwidth are significantly higher for 50/125-μm fiber than for 62.5/125-μm fiber. Remember that transmission speeds are sure to increase, so plan for future capacity expansion.

Plastic fibers and large-core step-index multimode fibers have very limited distance ranges. Plastic fibers have high attenuation, and large-core multimode fibers have low bandwidth. Their applications remain limited, but they can be valuable in certain situations, such as short data links inside of equipment.

Other Guidelines

It's impossible to give a comprehensive set of guidelines for fiber system design, but I can give a set of rough-and-ready suggestions, starting with a few common sense rules:

Don't forget to apply common sense in system design. Labor is never free.

- Your time is valuable. If you spend an entire day trying to save $5 on hardware, the result will be a net loss.
- Installation, assembly, operation, and support are not free. For a surprising number of fiber-optic systems, installation and maintenance cost more than the hardware. You may save money in the long term by paying extra for hardware that is easier to install and service.
- It can cost less to pay an expert to do it than to learn how yourself. Unless you need to practice installing connectors, it's much easier to buy connectorized cables or hire a fiber-optic contractor for your first fiber-optic system.
- You can save money by using standard mass-produced components rather than developing special-purpose components optimized for a particular application.

Some basic cost trade-offs are common in designing fiber-optic systems.

- The performance of low-loss fiber, high-sensitivity detectors, and powerful transmitters must be balanced against price advantages of lower-performance devices.
- Low-loss, high-bandwidth fibers generally accept less light than higher-loss, lower-bandwidth fibers. Over short distances, you can save money and overall attenuation by using a higher-loss, more costly cable that collects light more efficiently from lower-cost LEDs. (Because of the economics of production and material requirements, large-core multimode fibers are considerably more expensive than step-index single-mode fibers.)
- The marginal costs of adding extra fibers to a cable are modest and much cheaper than installing a second parallel cable. However, if reliability is important, the extra cost of a second cable on a different route is a worthwhile insurance premium.
- LEDs are much cheaper and require less environmental protection than lasers, but they produce much less power and are harder to couple to small-core fibers. Their broad range of wavelengths and their limited modulation speed limit system bandwidth.

- Fiber attenuation contributes less to losses of short systems than losses in transferring light into and between fibers.
- Topology of multiterminal networks can have a large impact on system requirements and cost because of their differences in component requirements. Coupler losses may severely restrict options in some designs.
- Light sources and detectors for 1300 and 1550 nm cost more than those for the 650- or 800- to 900-nm windows, although fiber and cable for the longer wavelength may be less expensive.
- 1550-nm light sources cost more than 1300-nm sources.
- Fiber and cable become a larger fraction of total cost—and have more impact on performance—the longer the system.
- Balance the advantages of eliminating extra components with the higher costs of the components needed to eliminate them. For example, it's hard to justify the expense of two-way transmission through a single fiber over short distances.
- Optical amplifiers or high-power laser transmitters make sense in systems distributing signals to many terminals.
- Narrow-line distributed-feedback lasers are more expensive than lasers with broader spectral linewidth.
- Compare costs of high-speed TDM on single channels or lower-speed TDM at multiple wavelengths.
- *Dark fibers*—extra fibers installed in the original cable that were never hooked up to light sources—are often available in existing cables.
- Remember that human actions—not defective equipment—cause most fiber-optic failures. Consider ring or mesh topologies that can survive a single break. Take the extra time and spend the extra money to make important systems less vulnerable to damage. This means labeling and documenting the system carefully, as well as not leaving cables where people can trip over them.

Always leave room for future upgrades. Bandwidth requirements are sure to increase, and it costs less overall to install higher-capacity fiber now than to install a cheap one that you have to replace with a more expensive one later.

- Install extra fibers in your cables; they're a lot cheaper than going back later to install more cables.
- Leave margin for repair, such as slack in cables. It costs much less than complete replacement later.
- Leave room for expansion by adding WDM to your system.
- Watch for potential bottlenecks that might prevent future expansion.

As you grow more familiar with fiber optics, you will develop your own guidelines based on your own experience.

What Have You Learned?

1. Design of fiber-optic systems requires balancing sometimes-conflicting performance goals as well as costs.
2. The system loss budget is calculated by subtracting all system losses from the transmitter output power plus the gain of any optical amplifiers. The resulting output power should equal the input power required by the receiver plus system margin.
3. Significant losses can occur in coupling light from sources into fibers. You need multimode fibers to collect light from LED sources and single-mode fibers to collect light from edge-emitting laser sources. Large-core fibers are more efficient for large-area LEDs.
4. Total fiber loss equals attenuation (dB/km) multiplied by transmission distance. Multimode fibers may suffer transient losses in the first 100 to 200 m.
5. Total loss from connectors, couplers, and splices is their characteristic loss multiplied by the number of each in the system. You calculate the most likely loss using average loss and the worst case using maximum specified loss.
6. System margin is a safety factor that allows for repairs and aging of components. Typical values are 5 to 10 dB.
7. Optical amplifiers boost signal strength, but because of their high cost they are best used in long, high-speed systems or systems that distribute signals from one source to many receivers.
8. Transmission capacity budgets calculate bandwidth or bit rate; they depend only on source, fiber, and receiver characteristics. You can estimate capacity by calculating response time.
9. Response time of a system is the square root of the sum of the squares of component response times. Calculations must include transmitter and receiver response times as well as fiber dispersion.
10. Modal dispersion and chromatic dispersion combine to limit capacity of multimode fibers. Because of these capacity limits, multimode fibers are rarely used over more than a couple of kilometers.
11. Chromatic dispersion and polarization-mode dispersion limit capacity of single-mode fibers, which usually transmit over a kilometer or more. Chromatic dispersion depends on source spectral width as well as fiber dispersion.
12. Dispersion compensation becomes important at high speeds or over long distances. The compensating elements can be fibers distributed through the length of the system, or lumped at certain points.
13. Transmitters and receivers must be matched to fiber characteristics.
14. Modulation broadens the range of wavelengths in the optical carrier. The largest effect is chirp for directly modulated lasers, but external modulation also broadens transmitter spectral range.
15. Installation can cost much more than hardware. With demand for transmission capacity rising steadily, you should keep your upgrade paths open.

What's Next?

Chapter 22 covers the fundamentals of optical networking design, including wavelength-division multiplexing.

Further Reading

Gerd Keiser, *Optical Fiber Communications,* 3rd ed. (McGraw Hill, 2000)

Joseph C. Palais, *Fiber Optic Communications,* 4th ed. (Prentice Hall, 1998)

Questions to Think About for Chapter 21

1. Why is the relative light-collection efficiency of fibers in decibels proportion to 20 times, rather than 10 times, the log of the ratio of their diameters?
2. Suppose the amplifiers in a transatlantic cable are limited to 12 dB of gain to limit noise. How far apart can they be spaced if the fiber attenuation averages 0.24 dB? You can neglect splices, and the system contains no connectors.
3. You are installing a fiber-optic data link between a laboratory and a remote data-collection center 5 km away. You want to use a VCSEL transmitter with −5 dBm output and a fiber with loss of 2.5 dB/km at 850 nm. The system includes patch panels on each end, with two connector pairs at each patch panel, and additional connectors on the terminal equipment. If connector loss is 0.5 dB, and you want a 10 dB margin, how sensitive must your receiver be?
4. You have to design a system with 1 Gbit/s data rate using return-to-zero (RZ) digital coding. What is the 3-dB analog bandwidth of this system? What NRZ data rate could this system transmit?
5. You want to transmit 1 Gbit/s through 100 km of single-mode fiber, using a transmitter and receiver that each have response times of 0.4 ns. The transmitter has line width of 0.1 nm. Neglecting polarization mode dispersion, what is the maximum chromatic dispersion allowable in the fiber?
6. Can you get away with using a 1550 nm VCSEL in the system of Problem 5 if the VCSEL has linewidth of 0.5 nm, output of −5 dBm, and the fiber loss is 0.25 dBm? (This assumes you can find such a VCSEL.) Check both the pulse spreading and the power level, assuming receiver sensitivity of −30 dBm.

Quiz for Chapter 21

1. A large-area LED transfers 10 μW (10 dBμ) into an optical fiber with core diameter of 100 μm and numerical aperture of 0.30. What power should it couple into a fiber with 50 μm core and NA of 0.2?

a. 10 dBμ

b. 9.5 dBμ

c. 3 dBμ

d. 1.0 dBμ

e. 0.4 dBμ

2. A connector is specified as having loss of 0.6 dB ± 0.2 dB. What is the maximum connector loss in a system containing five such connector pairs?

a. 0.6 dB

b. 3.0 dB

c. 4.0 dB

d. 5.0 dB

e. None of the above

3. A 100-Mbit/s signal must be sent through a 100-m length of fiber with eight connector pairs to a receiver with sensitivity of −30 dBm. The fiber loss is 4 dB/km, and the average connector loss is 1.0 dB. If the system margin is 5 dB, what is the minimum power that the light source must couple into the fiber?

a. −13.0 dBm

b. −13.4 dBm

c. −16.0 dBm

d. −16.6 dBm

e. −20.0 dBm

4. A system is designed to transmit 622 Mbit/s through 50 km of cable with attenuation of 0.4 dB/km. The system contains two connector pairs with 1.5 dB loss, a laser source that couples 0 dBm into the fiber, and a receiver with sensitivity of −34 dBm. How many splices with average loss of 0.15 dB can the system contain if the system margin must be at least 8 dB?

a. None

b. 10

c. 20

d. 30

e. 40

f. None of the above

5. A 2.5-Gbit/system must span a distance of 2000 km, with optical amplifiers every 80 km. If the fiber loss is 0.3 dB/km at 1550 nm and there is one 0.1 dB splice every 16 km, what must the amplifier gain be if the system is not to gain or lose signal strength across its entire length?

a. 20 dB

b. 24.4 dB

c. 26.4 dB

d. 30 dB

e. 34.4 dB

6. You need to transmit identical 1-Gbit/s signals to 200 homes using a 1310-nm laser source. The homes are 1 to 4 km from your transmitter and use receivers sensitive to 30 dBm. What transmitter power do you need to achieve a 5-dB system margin if your fiber has a 0.4-dB/km loss at 1310 nm, each signal path from transmitter to home includes 6 connectors with a 0.5-dB average loss, and you split the signal in a 1 × 200 tree coupler with no excess loss?

a. 4.6 dBm

b. 9.6 dBm

c. 2.6 dBm

d. 0.0 dBm

e. −0.4 dBm

7. What is the duration of a single-bit interval in a 1.7-Gbit/s signal?

a. 1.7 ns

b. 1 ns

c. 0.588 ns

d. 0.294 ns

e. 0.170 ns

8. What is the response time of a system with transmitter response of 2 ns, receiver response of 1 ns, and 100 m of multimode fiber with dispersion of 20 ns/km (including both modal and chromatic dispersions)?

a. 2 ns

b. 2.236 ns

c. 2.646 ns

d. 2.828 ns

e. 3 ns

9. What is the total dispersion of 10 km of graded-index fiber with modal dispersion of 2.5 ns/km and chromatic dispersion of 100 ps/nm·km when it is used with an 850-nm LED having a 50-nm spectral width?

a. 5 ns

b. 25 ns

c. 50 ns

d. 55.9 ns

e. 75 ns

10. What is the total dispersion of 10 km of single-mode fiber with chromatic dispersion of 17 ps/nm·km and polarization-mode dispersion of 0.5 ps/$\sqrt{\text{km}}$ at 1550 nm when used with a laser source with spectral width of 1 nm?

a. 1.58 ps

b. 10 ps

c. 17 ps

d. 170 ps

e. 172 ps

11. You are designing a 1 Gbit/s system using NRZ-coded signals with a transmitter with 0.3 ns rise time and a receiver that also has 0.3 ns rise time. What is the maximum total dispersion allowable through the entire length of the fiber?

a. 0.1 ns

b. 0.3 ns

c. 0.44 ns
d. 0.56 ns
e. 0.7 ns

12. You generate a 2.5 Gbit/s NRZ signal with rise time of 0.15 ns and spectral width of 0.1 nm. You have to send it through 80 km of nonzero dispersion-shifted fiber with chromatic dispersion of 6 ps/km·nm at the transmitter wavelength. Polarization-mode dispersion is 0.5 ps/root-km. What is the total pulse dispersion in the fiber?

a. 4.5 ps
b. 45 ps
c. 48 ps
d. 52.5 ps
e. 80 ps

CHAPTER

22

Optical Networking System Design

About This Chapter

Optical networking extends the capabilities of fiber optics beyond the point-to-point transmission of the systems described in the last chapter. The central concept behind optical networking is organizing signals in optical form. Wavelength-division multiplexing organizes signals for transmission as optical channels. Optical switching and other techniques direct and process optical channels.

This chapter describes the design of optical networking systems. A major emphasis is on wavelength-division multiplexing of many optical channels through a single fiber. Optimizing WDM systems requires balancing a number of parameters, including data rate, channel spacing, and total power levels at various points in the fiber. I also will cover concepts behind optical switching and other signal manipulation in the optical domain, such as wavelength conversion. These design criteria are not as well-formulated as for WDM, but I will explain the major issues.

WDM versus High-Speed TDM

The trade-offs between wavelength-division multiplexing and higher-speed time-division multiplexing are real, but not as straightforward as you might think.

As you learned in Chapter 21, signal transmission through fiber-optic systems becomes increasingly difficult as the data rate on an optical channel increases. Dispersion effects become more significant at higher speeds, and can limit transmission distances, depending on the type of fiber. The faster the channel rate, the shorter the distance signals can travel. Thus 2.5 Gbit/s signals can go farther than 10 Gbit/s signals, and 40-Gbit/s

Trade-offs between WDM and faster data rates are complex.

signals cannot go as far as 10 Gbit/s. Thus one way to achieve higher overall transmission rates over the same distance is to break the signal into many parallel optical channels and transmit them at different wavelengths through the same fiber.

WDM can simplify signal processing by assigning signals generated by other equipment to their own optical channels. If your long-distance systems receives signals at 2.5 Gbit/s, for example, you could transmit those signals directly in optical form on separate optical channels. This may be simpler and cheaper than time-division multiplexing them to generate single channels at 10 Gbit/s or 40 Gbit/s.

WDM is a fundamentally different way to organize signals.

Another extension of that idea is using WDM to offer many separate optical channels to different customers. Instead of selling transmission capacity at a series of different data rates, a carrier could lease separate optical channels on the same fiber to different customers. This would allow customers to transmit signals in whatever format they wanted, instead of limiting them to formats compatible with the carrier's transmission system.

The important point to remember is that WDM is not merely a way to squeeze more bits per second through the same optical fibers. WDM also is a fundamentally different way to organize signals, which may offer particular advantages.

Density of Optical Channels

WDM transmission capacity depends on channel spacing and data rates.

The transmission capacity of a WDM system depends on the data rates of individual optical channels, the spacing between them, and the total range of usable wavelengths. Channel spacing depends on the optics and the characteristics of each optical channel, particularly its spectral bandwidth. The range of wavelengths available depends on the characteristics of optical amplifiers used in the system.

The total capacity is the product of the number of usable optical channels times the data rate on each channel.

Spectral Bandwidth and Channel Spacing

Faster signals require broader channel spacing.

As you learned earlier, the spectral bandwidth of an optical channel increases with the data rate it carries. This means that high-speed signals cannot be packed together as tightly, as shown in Figure 22.1 for 2.5- and 10-Gbit/s channels. Although technological improvements can reduce bandwidths somewhat, the limitation is fundamental. A higher-speed signal will require more spectral width than a slower signal.

This fundamental tradeoff means that faster signals generally require broader channel spacing than slower signals. The exact spacing depends on the transmitter and the optics. Experiments have demonstrated transmission of 1 Gbit/s on optical channels separated by a mere 1 GHz, and of 40 Gbit/s signals with 100-GHz spacing.

Signals can be packed a little closer together if they are separated in mode or direction. For example, 40-Gbit/s signals have been packed at 50-GHz spacing by transmitting alternat-

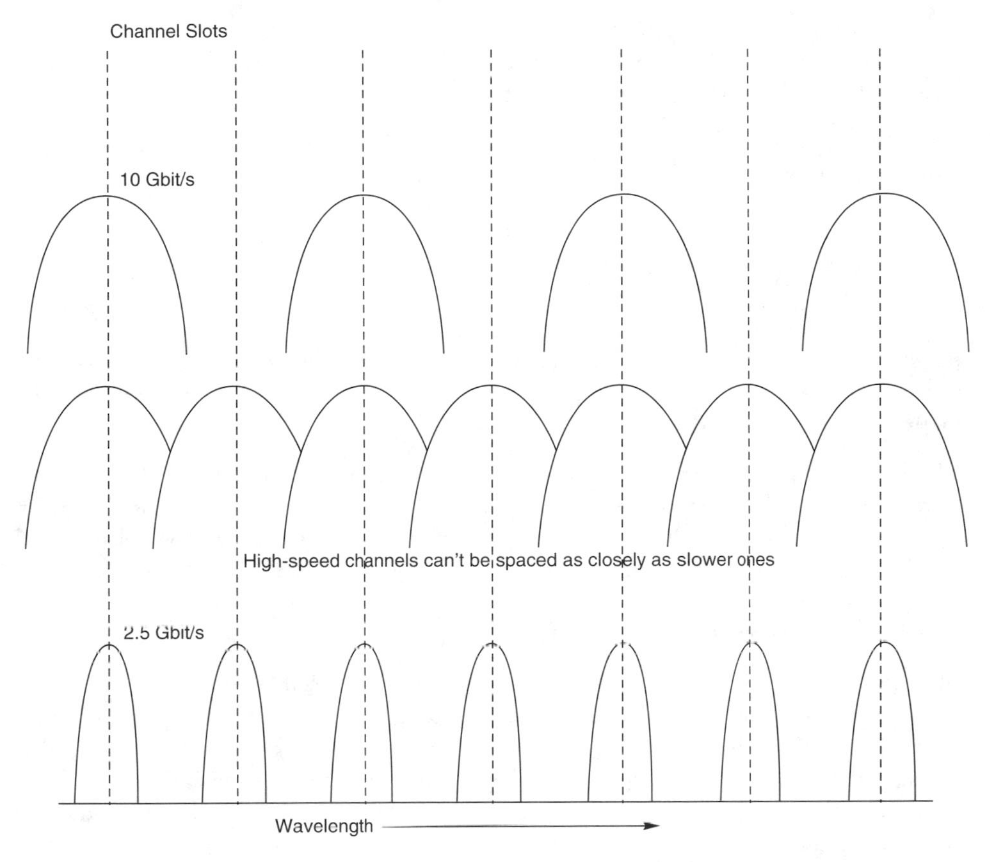

FIGURE 22.1
Higher data rates may require wider channel spacing.

ing channels in opposite directions. That is, the closest spacing of two channels going in the same direction is 100 GHz, but a channel going in the opposite direction is transmitted at a wavelength midway between the two, separated by 50 GHz. This interleaving allows the highest packing density to date, which corresponds to a data rate of about 0.8 bit per second per hertz of transmission bandwidth. That has only been demonstrated in the laboratory.

Spectral Ranges with Optical Amplifiers

The range of usable wavelengths is limited by the optical amplifiers used in the system. As you learned earlier, erbium-doped fiber amplifiers are designed for two different passbands. C-band erbium-fiber amplifiers typically operate from 1530 to 1565 nm, and L-band erbium amplifiers from 1570 to 1610 nm. These transmission ranges can be extended a few nanometers on each end, but are limited by the range over which the amplifiers have reasonably uniform gain.

Optical amplifiers limit the usable wavelength range.

With 100-GHz channel spacing, each erbium-amplifier band can accommodate about 40 channels; with 50-GHz spacing, this doubles to about 80 channels. Altogether, this gives 80 100-GHz or 160 50-GHz channels in the erbium-amplifier range.

Similar spacings should be possible in other wavelength bands, but such high-capacity systems have not been demonstrated.

Unamplified Systems

Short unamplified systems can transmit hundreds of channels.

Systems that do not require optical amplification can transmit signals at a much broader range of wavelengths, offering much higher WDM channel counts. This allows very high transmission capacity over limited distances of tens of kilometers, particularly for use in metro or regional networks.

The usable range for metro WDM systems depends on fiber attenuation and dispersion. Low-water fibers have reasonably low attenuation—0.5 dB/km or less—across a range of about 350 nm, from 1300 to 1650 nm. Other fibers are not usable at wavelengths of roughly 1350 to 1430 nm, centered on the water absorption peak at 1380 nm. No fibers have low dispersion across the entire band from 1300 to 1650 nm, but dispersion has little impact over distances of less than 100 km.

Many scenarios are possible for using this broad band, which spans about 50 THz. Dense WDM could pack 500 to 1000 channels closely together across the entire range, offering the potential of sending tens of terabits per second through a single fiber. An alternative is to use wider channel spacing to spread a few dozen channels across the entire range. This would match the capacity of dense-WDM systems in the erbium-fiber band, but would cut costs by using less-expensive optics and broader-band laser transmitters.

Populating Channels

Not all wavelength slots are populated.

WDM systems typically are designed with *slots* that can accommodate a certain number of optical channels. It is important to distinguish between the channel slots that are available and those that are *populated* with transmitters and receivers.

A telecommunications carrier may initially install a WDM system set up like Figure 22.2. This system potentially has slots for 40 optical channels, but only three of the optical channels are *populated* with transmitters and receivers, with only the receivers shown here. In this example, the carrier has bought boxes, which divide wavelengths into five groups of optical channels, and the WDM optics to divide one of those five groups of optical channels into eight separate channels. So far the carrier has installed transmitters and receivers on only three of the eight channels.

This is typical of the incremental approach carriers often use to provide services. The carrier does not need all the potential capacity immediately, but wants the room for future expansion. Transmitters and receivers are expensive, and their prices are coming down, so the carrier populates only the channel slots needed immediately. As more capacity is needed,

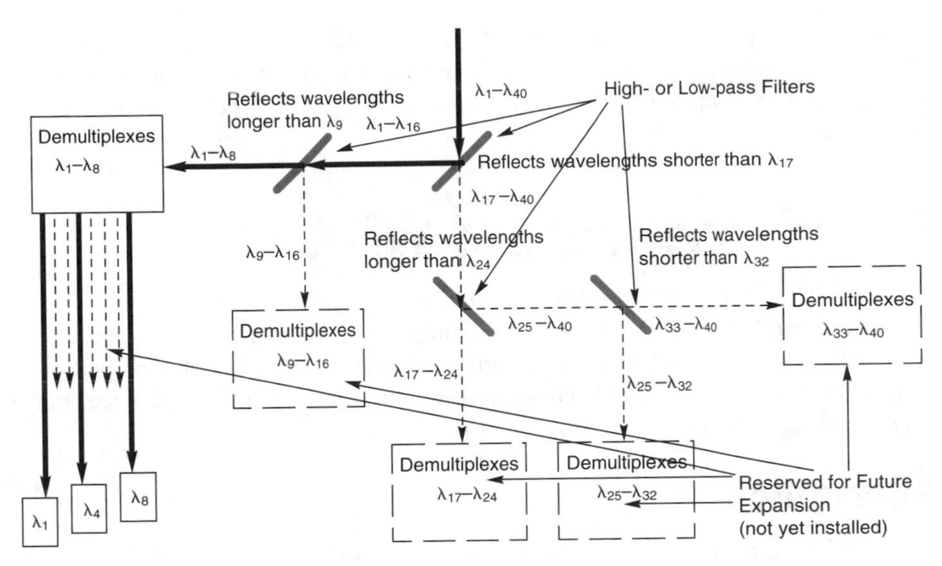

FIGURE 22.2 *Partial provisioning of a 40-channel WDM system.*

more slots can be populated. Traditional learning curves are likely to drive transmitter costs down, so the carrier needs only spend for the capacity it needs today. Later on, the transmitters and receivers should cost less.

Dense and Coarse WDM

Dense wavelength-division multiplexing is important for high-performance systems, but *coarse* or *wide WDM* can be a way of reducing transmission costs or increasing data rates possible over a given distance.

Coarse WDM can reduce transmission costs.

Typically, coarse WDM means systems with optical channels separated by at least 1000 GHz (8 nm at 1550 nm). Often the spacing is 10 to 20 nm near 850, 1310, or 1550 nm for use in glass fibers, but some systems have been developed for use in plastic fiber at visible wavelengths. Wide spacing allows for the use of inexpensive laser sources with minimal temperature stabilization, as well as lower-cost wavelength-separation optics.

Like other types of wavelength-division multiplexing, coarse WDM allows data transmission at higher speeds than possible at a single wavelength. For example, four coarse-WDM channels at 2.5 Gbit/s can carry 10-Gigabit Ethernet through up to 300 meters of 62.5/125 multimode graded-index fiber at 1300 nm, or through up to 10 km of single-mode fiber.

10-Gigabit Ethernet can use four-channel coarse WDM.

The coarse WDM transmitters will be built around uncooled distributed-feedback lasers operating at wavelengths of 1275.7, 1300.2, 1324.7, and 1349.2 nm. The broad range of wavelengths is possible because no amplifiers are needed for the distances envisioned for 10-Gigabit Ethernet. Loose wavelength tolerances and avoiding the need for cooling should increase yields and lower costs. The lasers could operate at temperatures of 0 to 70°C, with wavelength drifting no more than 5 nm, keeping them in the proper coarse-WDM slots. Common transmitters and receivers could be used for either single- or multimode fiber.

Fiber Properties and WDM

You learned in Chapter 21 how fiber attenuation and dispersion affect transmission at a single wavelength. Matters become more complex when the fiber transmits signals at several different wavelengths because attenuation, dispersion, and other fiber properties vary with the wavelength transmitted. Thus signals on one optical channel experience different attenuation and dispersion than those on other channels.

Attenuation

Fiber attenuation varies within transmission windows.

Look back at Figure 5.2, and you can see that fiber attenuation varies significantly across the low-loss window from 1280 to 1650 nm. Loss in the 1310-nm region is about 0.35 dB/km, compared to a minimum near 0.2 dB/km in the erbium-amplifier band. Thus a 100-km system would have 20-dB attenuation at 1550 nm, but 35-dB at 1310, a 15-dB difference at the receiver if transmitter powers were equal. This difference is large enough to impact design of systems operating over the entire range.

The difference is much smaller across the erbium-fiber band, as shown in Figure 22.3. For the fiber shown, attenuation ranges between 0.192 and 0.200 dB/km between 1530 and 1610 nm, rising to 0.205 dB/km at 1620 nm. In this case, the difference in attenuation

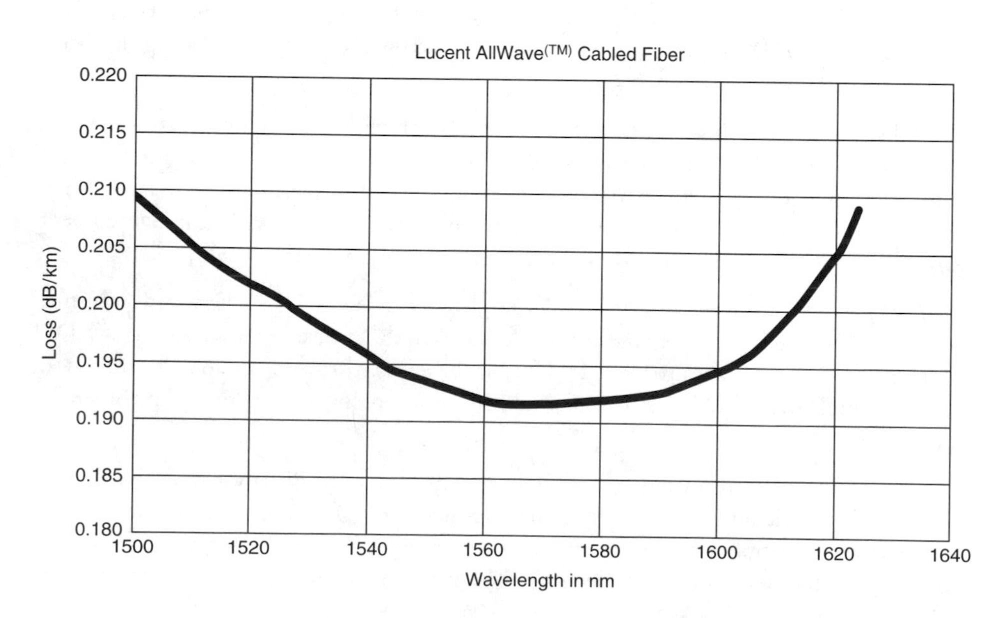

FIGURE 22.3 *Attenuation of low-water AllWave fiber in erbium-amplifier window. (Copyright Lucent Technologies Inc.).*

after 100 km would be small, 19.2 dB at 1570 nm compared to 20.0 dB at 1610 nm, a mere 0.8 dB. The techniques used to equalize amplifier gain over this range can accommodate this small variation in fiber attenuation.

Dispersion Slope and Compensation

In Chapter 21, I described how dispersion compensation can reduce the total chromatic dispersion light experiences in passing through a fiber-optic link. Dispersion compensation is relatively straightforward at a single wavelength. However, as you learned in Chapters 4 and 5, chromatic dispersion also varies with wavelength. This makes it necessary to consider the *dispersion slope*—the rate of dispersion change as a function of wavelength—in designing WDM systems.

Dispersion slope differs among fibers.

With no dispersion compensation, the cumulative dispersion can vary significantly across a range of wavelengths, as shown in Figure 22.4. Dispersion slope is largest in large-effective-area fibers, around 0.086 ps/nm^2-km in the C-band erbium-fiber window. The dispersion slope may be half that value in a reduced-slope fiber, a number that is significantly lower but not negligible.

As long as there is no dispersion compensation, the system must be designed for the wavelength with the worst dispersion. This becomes more of a problem for fibers that transmit both C- and L-bands, because typically total dispersion is higher in the L-band, as you can see in Figure 22.4.

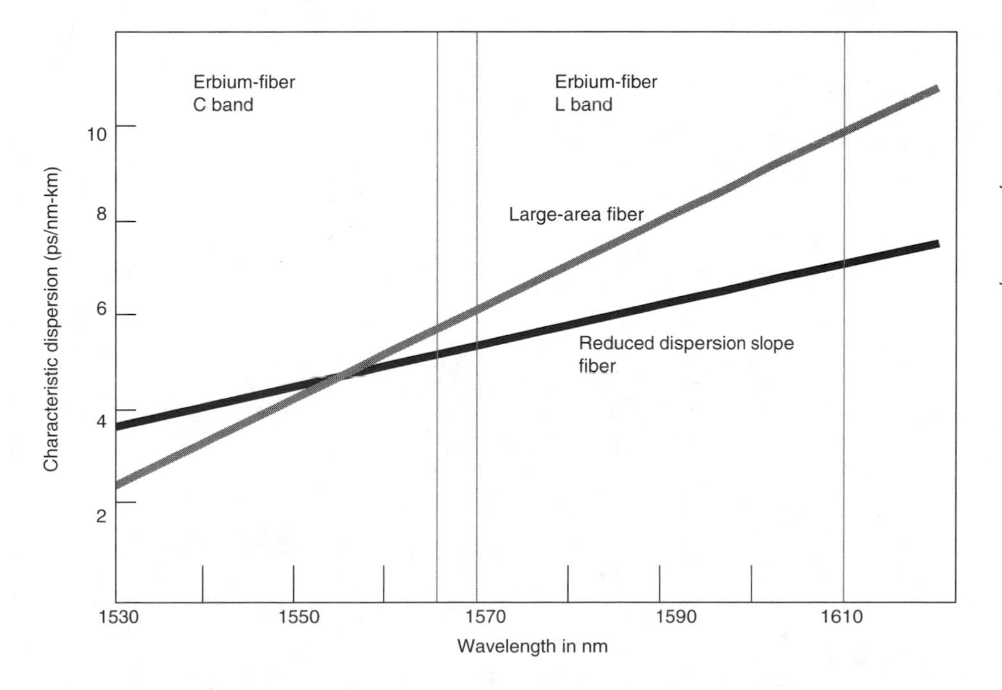

FIGURE 22.4
Dispersion slopes of reduced-slope and large-effective area fiber. Conventional nonzero dispersion-shifted fibers have intermediate slopes.

Dispersion compensation matches fibers to cancel dispersion.

Dispersion compensation depends on using two or more fibers with different chromatic dispersions to offset the pulse spreading that each generates. Figure 22.5 shows one example of how this works. The transmission fiber has a small positive dispersion, which is offset by a compensating fiber with a larger negative dispersion. In this example, the system consists of alternating lengths of transmission and compensating fiber, with 1 km of compensating fiber offsetting the loss for each 5 km of transmission fiber. The result is a large reduction in cumulative dispersion.

Note that the cumulative dispersion is zero at only one wavelength. It is simple to control chromatic dispersion for one wavelength, by balancing the net positive dispersion produced by one fiber with the net negative dispersion from the other. However, as shown in Figure 22.5, the dispersion slopes of the two fibers are not exactly complementary. At shorter wavelengths, the net cumulative dispersion is slightly negative, and at longer wavelengths the cumulative dispersion is slightly positive. Ideally the net dispersion should vary less than the dispersion of one type of fiber, but it will not be uniform with wavelength.

FIGURE 22.5
Dispersion compensation in WDM system.

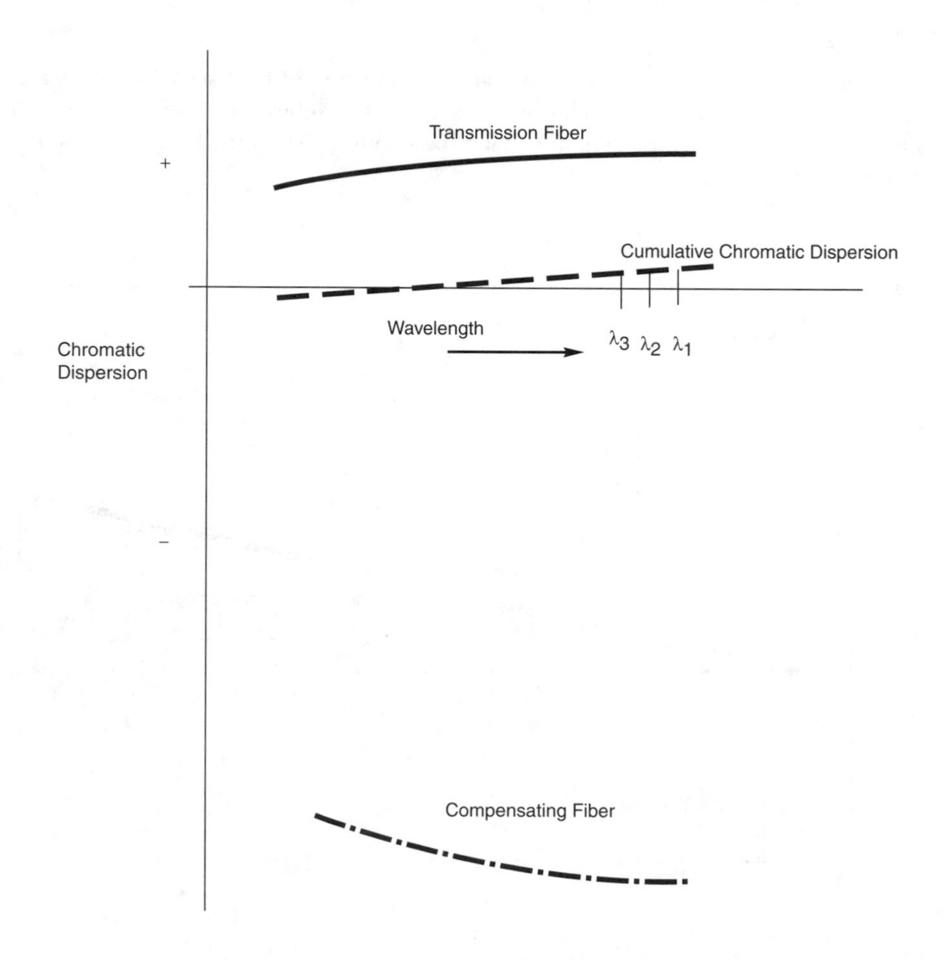

If you view the cumulative dispersion along the length of the fiber, you will see it changes, as shown in Figure 22.6. In this case, you start with a short length of compensating fiber, which produces negative dispersion, then a longer length of regular transmission fiber, followed by a length of compensating fiber, and so on. The most obvious trend is that cumulative dispersion is first negative, then slowly increases to positive values, then drops to negative values in the next segment of dispersion compensating fiber. This produces an overall sawtooth-shaped plot of dispersion along the length of the fiber.

Cumulative dispersion varies along the fiber, and with wavelength.

A more subtle effect is that each wavelength experiences slightly different dispersion. In this case, λ_1 experiences slightly more than the two shorter wavelengths. This shows an imperfection in the compensation scheme, and the net difference in the cumulative chromatic dispersion is a residue that remains at the end of the system. That means that dispersion compensation cannot cancel chromatic dispersion at all wavelengths, only at one wavelength.

It is not necessary to use dispersion-compensating fiber, which has relatively high attenuation and some other drawbacks. All you really need are two types of fiber that have chromatic dispersion of different signs. Submarine cables may use two or more different types of fiber in each span between optical amplifiers. The bulk of the span is nonzero dispersion-shifted fiber with a slightly negative dispersion—about 2 ps/nm-km at 1550 nm—and zero-dispersion wavelength longer than about 1600 nm. That negative dispersion is offset by the positive dispersion of about 17 ps/nm-km of step-index single-mode fiber. (Large-effective-area fiber may also be used near optical amplifiers to reduce nonlinear effects.)

Many refinements are possible in chromatic dispersion compensation for WDM systems. The designer could target nominally zero total chromatic dispersion for a wavelength in the middle of the transmission band, to minimize the total chromatic dispersion at other wavelengths. Compensating fibers could be added at the receiver end, to offset residual chromatic dispersion at some or all wavelengths. The goals are both to reduce total chromatic dispersion, and to minimize chromatic dispersion on the optical channels most strongly

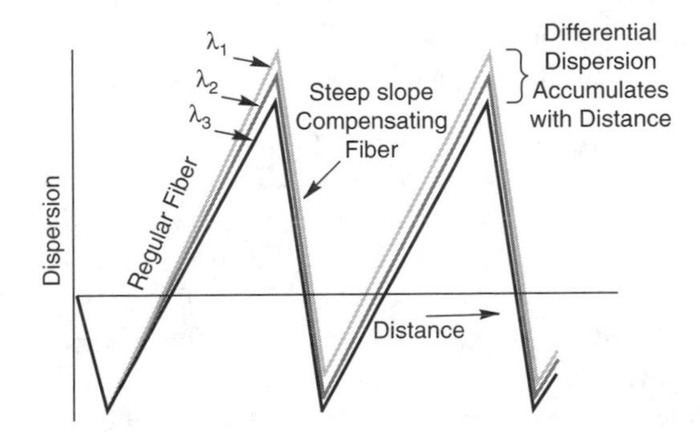

FIGURE 22.6
Dispersion in system with short lengths of dispersion-compensating fiber. Note the dispersion at each wavelength gradually diverges with increasing distance, because compensation differs slightly with wavelength.

affected—that is, farthest from the wavelength of nominally zero cumulative chromatic dispersion.

Polarization-mode dispersion is a different matter, although it can limit data transmission rates. Polarization-mode dispersion has little dependence on wavelength, so it is not a problem peculiar to WDM systems. In addition, techniques to compensate for polarization mode dispersion remain under development.

Nonlinear Effects in WDM Systems

High total power makes WDM systems vulnerable to nonlinear effects.

Nonlinear effects are proportional to the total density of optical power in a fiber, and to the distance the light travels in the fiber. This makes long-haul WDM systems particularly vulnerable because the total optical power transmitted is the average power per channel times the number of channels. The more channels, the higher the power, and the more subject the system becomes to nonlinear effects, which can degrade transmission. A single-channel system normally has no trouble transmitting a power of 3 mW, but if a WDM system tries to transmit 80 channels of 3 mW each, the total power reaches 240 mW, which can produce nonlinear effects.

Four-wave mixing poses particular problems in WDM systems. As you learned in Chapter 5, it occurs when signals at three input frequencies combine to generate a mixed signal at a fourth frequency:

$$v_1 + v_2 - v_3 = v_4$$

(The three input signals need not all be at different frequencies; two of them could be on the same optical channel.) The equal spacing of WDM channels means that the new frequency is likely to fall on another optical channel, producing noise and crosstalk that can interfere with the signal on that channel.

Nonlinear effects are not significant in short metro WDM systems.

The strength of four-wave mixing is proportional to the length of fiber over which the interaction takes place. Nonlinear interactions are relatively weak in glass, but their strength accumulates with distance. In practice, they are not significant in metro systems that run tens of kilometers, but can pose problems in long-haul systems running thousands of kilometers.

The picture is complicated by the fact that fibers have attenuation, which reduces the optical power along the length of the fiber, so the strength of the nonlinear effect declines with distance from the light source. This means that there is a maximum effective length over which nonlinear effects accumulate, which depends on the fiber attenuation. For a typical single-mode fiber with 0.22 dB/km attenuation at 1550 nm, this is about 20 km per span between optical amplifiers or between transmitter and receiver. (The value is smaller for fibers with higher attenuation.) Thus in long-haul systems, nonlinear effects increase with the number of spans between amplifiers.

Chromatic dispersion also plays a role in easing four-wave mixing because it causes signals at different wavelengths to drift out of phase. At the zero dispersion point, signals at different wavelengths remain in phase over long distances, so four-wave dispersion builds up more rapidly. This makes it important for fibers to have a minimum chromatic dispersion in the entire WDM transmission window, typically at least 1 ps/nm-km. (At the same time, too much chromatic dispersion can limit transmission bandwidth, as you saw in Chapter 21.)

Fortunately, dispersion compensation does not enhance four-wave mixing. Compensation schemes include lengths of fiber with positive and negative dispersions that offset each other, so the signal pulses do not remain in phase with each other over long distances. Four-wave mixing occurs where the entire fiber has near-zero dispersion, not where a fiber span contains a mixture of different dispersions, which *average out* near zero, as in Figure 22.6.

The choice of fiber also influences the strength of nonlinear effects, because of their dependence on power density. Power density in the fiber is the total power divided by effective area, so large effective area fibers reduce the power density and thus the severity of four-wave mixing and other nonlinear effects. Typically, the effective area is large in step-index single-mode fibers, somewhat smaller in nonzero dispersion-shifted fibers, and smallest in dispersion-compensating fibers. Other tradeoffs occur between effective area and reduced dispersion slope; the design features that increase effective area tend to increase dispersion slope, and conversely, reducing the dispersion slope also tends to decrease the effective area.

Nonlinear effects decrease with a fiber's effective area.

Different types of fiber can be combined in a particular sequence to minimize nonlinear effects as well as optimize other performance. For example, designers of a submarine cable could use a large-effective-area nonzero dispersion-shifted fiber close to the amplifiers, where optical power is highest. Where power is lower, reduced-slope fiber could be used, and step-index single-mode fiber could be used for dispersion compensation.

Optical Amplification and WDM Design

In Chapter 21, we considered optical amplifiers simply as increasing the power in a single optical channel. However, their role in WDM systems is somewhat more complex. You have already seen how the operating range of a fiber amplifier can limit the range of wavelengths transmitted in an amplified WDM system. Amplifier gain is not uniform across the operating spectrum, so power levels must be adjusted to balance power across all optical channels. In addition, the total power available from a fiber amplifier is limited, and must be shared among optical channels in a WDM system.

Optical amplifiers are analog devices, so noise from amplified spontaneous emission and other sources gradually builds up along a chain of amplifiers. However, with very careful design systems can include over 100 fiber amplification stages. I will concentrate on erbium-doped fiber amplifiers because they are the type now used in WDM systems.

Amplifier Power Levels

Amplifier output is divided among many WDM channels.

Erbium-fiber amplifiers are specified as having a maximum output power, typically 17 to 24 dBm for C-band amplifiers. This is the total output available on all optical channels in the amplifier's range. It is limited by the transfer of energy from the pump laser to the erbium atoms in the fiber, and from the erbium atoms to the amplified beam, which cause the saturation effects described in Chapter 12. This peak output can be concentrated on a single wavelength in a single-channel system, but in a WDM system it must be divided among many channels, limiting the output power available per channel. Thus the gain realized depends on the total power summed over all wavelengths, not on a single optical channel.

One result is that the power per channel decreases with the number of channels in a WDM system. Thus an amplifier operated at its maximum output power of 80 mW could deliver 20 mW on each of 4 optical channels, 10 mW on each of 8 channels, or 1 mW on each of 80 channels.

If the amplifier is operated to maximize output power per channel, saturation effects will reduce the gain per channel. For example, if the input was a single optical channel at 30 μW (−15 dBm), it could be amplified by about 30 dB to 30 mW, or +15 dBm On the other hand, if the input consisted of 30 channels at the same power level, a total of 900 μW, the total amplification would be only about 15 dB, raising total output to roughly the same level, but with only 1 mW per channel.

This decline in amplification per channel can be an important effect. In this example, it means that each amplifier can boost each channel only 15 dB. If the fiber loss is 0.2 dB/km, this limits amplifier spacing to 75 km. Note that this decline in amplification per channel can impact system operation as optical channels are added. Depending on operational details, doubling the number of optical channels could reduce the power level on each channel by 3 dB, equivalent to 15 km of fiber with loss of 0.2 dB/km.

Amplifier power also affects the data rate possible per optical channel. As you learned in Chapter 11, receivers need a minimum number of photons to reliably detect a pulse. This means that the receiver has a higher detection threshold at high speeds, because it must detect more pulses per second, and has briefer intervals to receive the photons from each pulse. In WDM systems, this translates into a direct tradeoff. Dividing power among more optical channels, which can transmit more data in parallel, reduces the power available to transmit signals at higher data rate per channel. In short, when you're operating near system margins, increasing the number of optical channels trades off directly with increasing the data rate per channel.

Gain Flatness and Channel Equalization

Gain varies with wavelength.

Optical fiber amplifiers do not amplify all optical channels equally. Typical specifications are that gain is flat to within 1 to 3 dB across the amplifier's operating range. This sounds good until you start cascading amplifiers. If gain varies by 2 dB across the spectrum, and the effect is consistent between amplifiers, after a chain of five amplifiers the difference can amount to 10 dB, potentially causing the loss of weaker channels.

Gain-equalizing filters can compensate for this by reducing the power on the strongest optical channels, as shown in Figure 22.7. The extra attenuation in the filter offsets the stronger gain at certain wavelengths, making amplifier gain flat across the spectrum.

FIGURE 22.7
Function of optical gain-flattening filter. (Courtesy of Furukawa Ltd.)

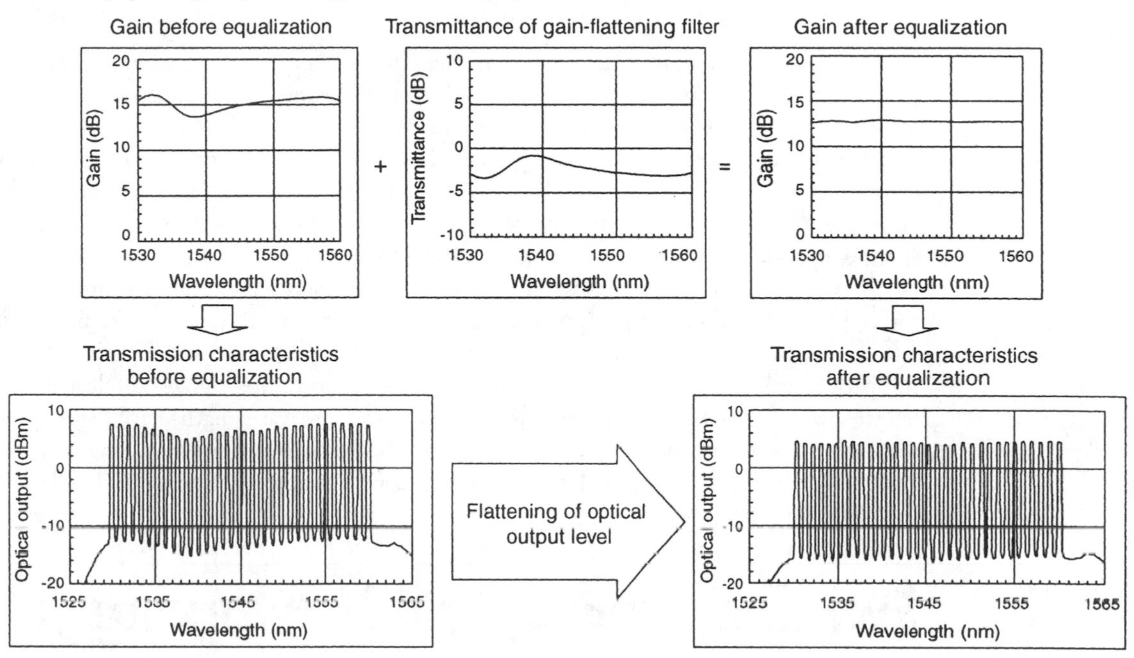

Filters also can compensate for wavelength-dependent variations in fiber attenuation across the amplifier bands, mentioned above. Another alternative is Raman amplification, described below.

Small, random variations in the strength of optical channels arising from imperfect gain equalization and other effects accumulate over long chains of optical amplifiers. However, with careful engineering they can be limited to no more than a few decibels over a chain of more than 100 amplifiers in a transoceanic submarine cable.

Raman Amplification

As you learned in Chapter 12, Raman amplification can supplement erbium-fiber amplifiers by amplifying wavelengths where erbium has lower gain. *Hybrid amplifiers* combine Raman amplification with an erbium-doped fiber amplifier to give more uniform gain across a range of wavelengths.

Raman amplifiers can balance gain of erbium amplifiers.

Raman amplifiers depend on stimulated Raman scattering, which occurs at a range of wavelengths offset from the pump source. The peak Raman gain is about 13 THz from the pump wavelength, a shift of about 100 nm at the erbium-fiber window. This allows Raman amplifiers to be designed for any wavelength where suitable pump lasers are available, including both the C- and L-bands of erbium-fiber amplifiers. To obtain peak gain at 1550 nm, the pump laser would be near 1450 nm; for peak gain at 1600 nm, the pump should be near 1500 nm. Typical pump powers are around one watt in single-mode fiber.

The amount of Raman gain depends on the length of fiber as well as the pump power. It is possible to build discrete Raman amplifiers, but typically they require a kilometer or more of fiber. Raman amplification does not require special fiber, and the current trend is to use standard fiber. Figure 12.10 showed how the final part of a fiber-optic span can serve as a Raman amplifier. A Raman pump laser is installed at the same location as an erbium-fiber amplifier. A remote transmitter generates an optical signal that passes through a long fiber span, eventually reaching a coupler, which directs incoming light to an erbium-fiber amplifier. The same coupler also directs light from the Raman pump laser into the input fiber, in the opposite direction to the input signal. Stimulated Raman scattering in the fiber transfers energy from the strong pump wavelength to the weaker signal wavelengths, offset by the proper amount from the pump. This amplifies the incoming signals at selected wavelengths where the erbium amplifier has lower gain, so those wavelengths have higher power when they reach the erbium amplifier. The pump beam fades as it passes through the fiber, and may eventually be blocked from reaching the transmitter by an optical isolator or wavelength-selective filter.

Switching and Optical Networking

One drawback of transmitting signals in a single stream of high-speed data is managing and processing them. The only way to process part of a high-speed signal is to convert it into electronic form and look "inside" the signal. The details vary with the transmission format. SONET signals can be switched out of the high-speed data stream, but some other formats require that the signal be broken down into slower-speed components for processing.

Optical networking manages signals by optical channel.

WDM and optical networking offer a different way of managing signals—by optical channel or wavelength, sometimes called by *lambda* after the Greek letter λ used to symbolize wavelength. For example, a WDM system can transmit 100 Gbit/s in the form of 40 optical channels at 2.5 Gbit/s each. Such a signal is much more *granular* than a single data stream, because a telecommunications carrier can direct and process each 2.5-Gbit/s optical channel without disturbing any of the other optical channels.

Figure 22.8 is a simple example of how this granularity can operate. A fiber delivers eight optical channels from Omaha to Chicago. An optical switch in Chicago directs these wavelengths to separate outgoing fibers. Two wavelengths go to both Detroit and Indianapolis. One wavelength each goes to Minneapolis, Milwaukee, Chicago suburbs, and St. Louis. In this example, the input signals are organized in Omaha, and the Chicago switch directs them on their way without any further processing. The same fiber could carry other wavelengths going to other destinations, but you wouldn't be able to make sense of the drawing if I tried to include 40 optical channels.

These capabilities are in the early stages of development and implementation, and can take various forms. An add-drop multiplexer is one example of this type of signal management.

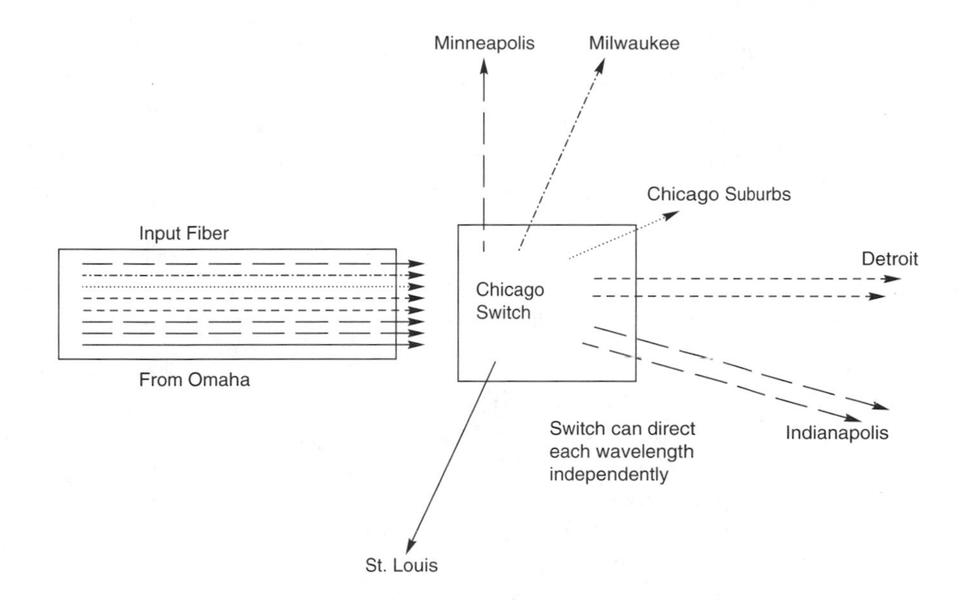

FIGURE 22.8
Granularity of optical channels.

In this case, a fiber carrying 20 optical channels of long-distance traffic from Boston to New York might drop one wavelength in Worcester, another in Springfield, a third in Hartford, and a fourth in Bridgeport, and deliver the remaining 16 channels to New York. This would be a more efficient use of fiber than running separate links from the smaller cities to Boston or New York, then merging their signals in with other traffic.

You learned about the basic concepts of optical switching in Chapter 16. Now let's look at their implications for optical networks and WDM systems.

Transparent and Opaque Systems

One question that has yet to be resolved is the relative merits of *transparent* and *opaque* systems. The question is fairly generic, and may apply to individual switches as well as to entire systems.

Light signals pass unchanged through transparent systems.

In this context, *transparent* means that the equipment transmits the signal unchanged in optical format, so the light could "shine through" the entire system. If you replaced it with a different wavelength, the new wavelength would appear at the other end. In contrast, *opaque* means that the signal is converted from its original optical form into some other form for processing, so the original light does not emerge at the other end. For example, a micromirror switch that reflects the input light signal out another optical port is transparent. An electro-optical switch that converts the signal into electronic form to direct it to its destination is opaque.

Transparency may sound better because it seems simpler to implement than converting signals to different formats or wavelengths. However, it also suffers potential limitations. As the slots for wavelengths fill in optical networks, it will become more difficult to find matching slots to transmit signals. That is, the wavelength used to transmit a signal on one part of its journey may not be available on the next. The signal would have to be converted from one wavelength to another, like you might have to move between lanes on a freeway.

An optical network may have islands of transparency separated by opaque elements.

You can think of an optical network as including "islands" of transparency, where signals are transmitted in purely optical form. A length of fiber between a transmitter and receiver is one such island. Islands can include other transparent components such as optical switches and amplifiers. They might be connected by opto-electronic-optical interfaces, such as a receiver-transmitter pair, or an electronic switch that converts signals into electronic form to reorganize them for another length of optical transmission.

The idea of an *all-optical network* is related to that of the transparent network. Essentially an all-optical network is one that uses only optical components such as moving-mirror switches and optical amplifiers to process signals passing through it. In that sense, it is only processing signals in units of optical channels. The term has become a marketing buzzword for an advanced fiber-optic system that switches light signals as well as transmitting them from point to point, and the meaning is becoming hazy. Some people's "all-optical" networks may include opaque components.

The relative advantages of all-optical and electro-optical switching are sufficiently muddled that at least one company is offering a switch that can work in either mode, depending on which works best for particular connections.

Wavelength Conversion and Tunability

Wavelength conversion is important for managing optical channels.

The ability to shift an optical channel from one wavelength to another will become increasingly important as optical networks mature. Wavelength conversion is a key technology that may tip the scales between transparent and opaque optical networks.

All-optical wavelength conversion is at this point a laboratory technology. One leading approach involves using an optical input signal to modulate the output of a laser emitting at another wavelength. The concept works, but products are not well-developed.

Electro-optical wavelength conversion is conceptually very simple. A receiver converts the input optical signal into electronic form, and that electronic signal drives a transmitter emitting at another wavelength. The idea is the same as an electro-optical repeater, but in this case the transmitter emits at a different wavelength than the input signal.

Another desirable feature is tunable output from the wavelength converter. A fixed-output wavelength converter, like a permanent add-drop multiplexer, always selects the same wavelength. Changing its configuration requires sending a technician to change the equipment. Telecommunications carriers would prefer to avoid that expensive labor, so they want wavelength converters that can be tuned to emit different wavelengths—preferably from a remote location, so they can be operated like a remote switch. This technology is in development along with tunable light sources. It could be implemented by adding a tunable laser to an electro-optic wavelength converter, or by processing optical input signals in suitable ways.

Wavelength Routing

You learned earlier that wavelength routing directs optical signals in certain directions depending on the wavelength. The concept is most valuable when used with tunable light sources or with multiple inputs at different wavelengths.

You can think of a wavelength router as a fixed wavelength-division demultiplexer. If the input signal is at a certain wavelength—say 1542 nm—you know it will be routed to a certain destination, as you saw earlier in Figure 16.5. You could set up an entire system to generate signals of certain wavelengths when directing signals to particular destinations, but it's not clear if this is a practical approach.

A wavelength router directs signals like a wavelength-division demultiplexer.

Another application is to redirect signals among a few possible outputs by tuning a light source. This has the potential for making switches without moving parts, but at this time tunable lasers remain a problem.

Network Layers, Optical Switching, and Routing

As you learned in Chapter 16 there is an important distinction between switching and routing of signals. Packet-switched signals are directed through the Internet by routers, which read headers on the signals and route them to the next node in the direction of their destination. Switches create virtual circuits connecting two points, such as a pair of telephones on which people are having a conversation or sending a fax message. In this sense, optical switches are *switches,* which don't read headers on individual packets, but instead create virtual connections between points. Indeed, true optical routers—which read packet headers in optical form—do not yet exist, and are not necessary with current network architecture. (However, some optical switches are called "routers" for reasons best known to corporate marketing departments.)

Routers and switches look different, but really are complementary devices that can operate on the same signals as they pass through different layers of the network. (Recall the "layering" of standards described in Chapter 20.) As shown in Figure 22.9, routers read packet headers, and "think" they are sending the packets directly to other routers, which are on the same layer of the network.

Routers and switches are complementary.

In reality, switches such as optical crossconnects establish links between routers on lower layers of the network, which the routers don't "see." The switched connections are static while packets are being sent, but can be changed at other times. In that way they are like switched telephone connections, which remain fixed as long as the call continues, but change afterwards. Lower layers in the network are sometimes said to exist in a "cloud" because what's inside doesn't matter to the upper layers. The same is true for your telephone service; you don't see any layers below the voice service layer. If human operators were fast enough, you couldn't tell if your calls went through state-of-the-art digital electronic switches and fiber-optic cables or through an old-fashioned switchboard and copper wires.

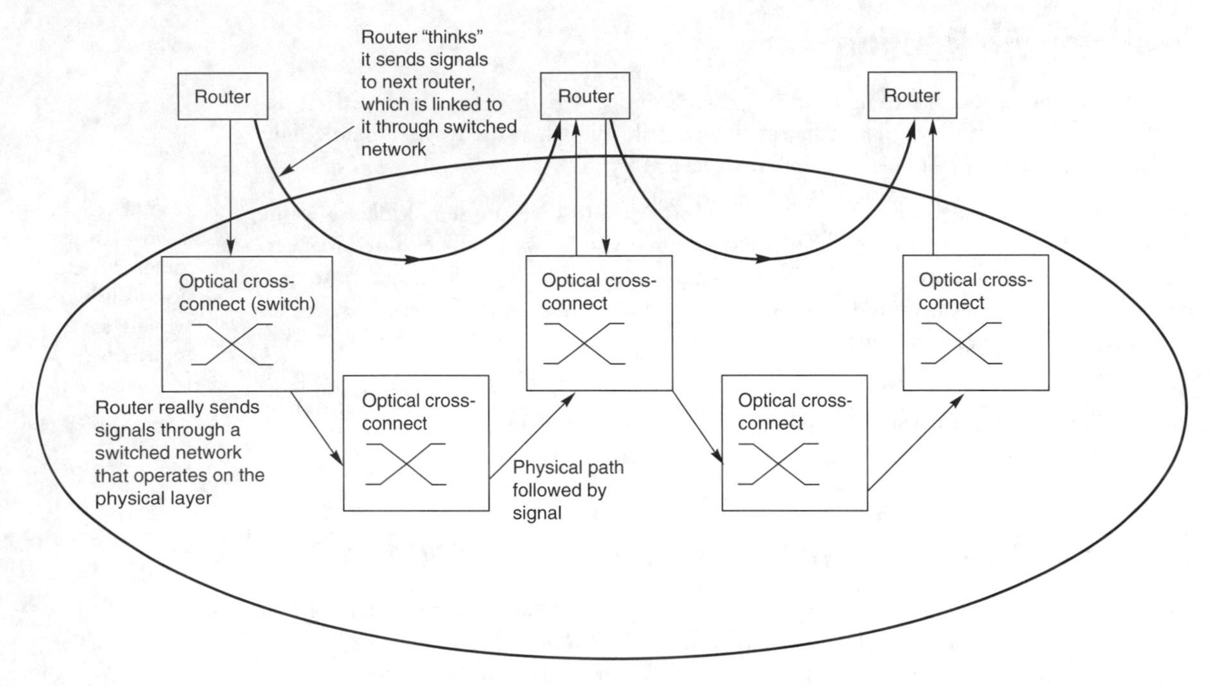

FIGURE 22.9
Routers and switches work together on different layers of the network.

Thus routers operate on a higher layer of the network than switches. Routers are electronic systems that work with signals in electronic form, as they are being processed for optical transmission on the physical layer. Electronic switches work on the physical layer, where they provide interfaces between electronic and optical signals. Optical switches operate on the optical layers that lie below the physical layer. They make physical paths for the signals to follow on their way between routers. They may direct individual wavelengths or the transmission load of entire fibers. (I should note that some electronic *telephone* switches do operate on the same layer of the network as routers, but they operate on circuit-switched telephone signals, not on packets.)

Layering simplifies individual network design tasks.

This layered structure is complicated, and often obscured by industry jargon, but as you learned in Chapter 20, it can make individual design tasks easier. Engineers providing various types of services need only worry about what happens on their layer of the network. Your home telephone has to deliver signals in a certain format, but it doesn't care what happens to the signals between its output and the input of another telephone. Likewise, the engineers who design circuit switches don't care what your phone looks like—only about the form of the signals it generates, and the format of the signals they convert it to. Engineers designing fiber-optic transmission lines likewise don't care what the bits they transmit mean, only that they enter and leave their system in a standard format. This approach simplifies individual design tasks by compartmentalizing them.

Types of Optical Switches

Optical switches can serve two distinct functions in WDM networks. They may switch all the signals in a fiber simultaneously to another fiber, or they may separate the optical channels and switch them individually. The switch in Figure 22.8 is an example of one that switches individual channels to different points. You can think of it as a fiber followed by a wavelength-division demultiplexer, which separates the optical channels for separate processing.

The same type of hardware can serve both functions. A single movable MEMS micromirror can switch all wavelengths carried by a fiber if they are not separated, or a single wavelength after optical demultiplexing. On the other hand, switches that are wavelength-sensitive generally cannot switch many wavelengths simultaneously because the signals might not go to the right output ports.

System Interfaces and Regeneration

The optical network has interfaces with electronic signals at input and output ends. The input electronics typically organize the input electronic signals in some way, such as time-division multiplexing to combine signals for high-speed transmission, or regrouping signals from other inputs.

The optical network terminates at electronic transmission centers.

Interfaces generally are at transmission centers, nodes, or hubs where many transmission lines come together. Typically, local signals feed into these nodes, where they are combined with signals arriving from other points and transmitted on outgoing lines. Normally, optical signals are only regenerated electronically at these points, not at intermediate points along long-distance transmission lines.

Optical regeneration has been demonstrated in the laboratory, but has yet to find practical applications. System requirements for optical regeneration are not yet clear.

What Have You Learned?

1. WDM is a fundamentally different way to organize signals than time-division multiplexing.
2. WDM transmission capacity depends on channel spacing and data rates. Faster data rates require broader channel spacing.
3. Fiber amplifiers limit wavelength ranges of long-distance systems. The erbium-fiber C-band spans 1530 to 1565 nm, and the L-band spans about 1570 to 1610 nm.
4. Short unamplified systems can transmit hundreds of channels because their bandwidth is not limited by optical amplifiers.
5. Not all wavelength slots in WDM systems are populated with signals.
6. WDM systems are dense or coarse depending on channel spacing. Coarse WDM may be less expensive than transmitting one higher-speed channel.

7. Combining two or more fibers with different dispersion properties can reduce total dispersion.
8. Cumulative dispersion varies along the length of a compensated fiber system.
9. WDM systems are particularly vulnerable to nonlinear effects because they carry many optical channels and have high total power.
10. Fibers with high effective area are less vulnerable to nonlinear effects.
11. Optical amplifiers have limited total output power, which must be divided among many optical channels in WDM systems. This means WDM systems have less power per channel than single-channel systems.
12. Gain of erbium-fiber amplifiers depends on wavelength, so all channels are not amplified equally.
13. Raman amplification can be combined with erbium-fiber amplification to give more uniform gain across a range of wavelengths.
14. Transparent optical systems transmit signals unchanged in optical format, so light shines through. Opaque systems convert optical signals to other forms for processing. Wavelength conversion typically is opaque.
15. Routers and switches are complementary devices that can operate on the same signals as they pass through different network layers. Optical switches are circuit switches, which make connections.
16. Optical switches may switch all optical channels in a fiber individually or collectively.

What's Next?

In Chapter 23 you will learn about the structure of the global telecommunications network.

Questions to Think About

1. Your design allows you to pack 2.5 Gbit/s optical channels only 50 GHz apart, but 10 Gbit/s signals require 100 GHz spacing to give adequate performance margins. Which allows you to transmit more data if you populate every available slot in the erbium-amplifier C-band? By how much?
2. You have wavelengths of 1450 to 1625 nm available in an unamplified metro WDM system. How many optical channels can you transmit with 10-nm spacing?
3. You have two types of fiber available for dispersion compensation, step-index single-mode fiber with dispersion of +17 ps/nm-km at 1550 nm, and a reduced-slope nonzero dispersion-shifted fiber with -2 ps/nm-km at the same wavelength. How much of each do you need for a 95-km span with zero cumulative dispersion at 1550 nm? Which should you use closer to the transmitter?

4. An erbium-fiber amplifier has a peak output of 20 dBm. You use it to transmit 40 optical channels. What is the output power on each? What happens to the output power per channel if you double the number of channels?
5. You need to send 20 optical channels through a series of 50 erbium-fiber amplifiers, with the output signals differing by no more than 6 dB at the end of the amplifier chain. How precisely do you need to equalize gain if all that difference is due to unequal gain across the range of optical channels?
6. The input to an all-optical switch is a cable containing 48 fibers. Each fiber can transmit up to 40 optical channels. How many optical channels can the switch direct if it has 128 input and 128 output ports?

Quiz for Chapter 22

1. How many 2.5 Gbit/s optical channels are equivalent to one 40 Gbit/s TDM channel?
   a. 1
   b. 4
   c. 16
   d. 40
   e. 100
2. Your WDM system is limited to 40 optical channels spaced 100 GHz apart in the 1550-nm region. What is the approximate total spectrum used by the system?
   a. 40 GHz
   b. 100 GHz
   c. 400 GHz
   d. 4 THz
   e. 193.1 THz
3. How is the spacing required between optical channels related to the data rate transmitted on the channel?
   a. The spacing increases with the data rate; the required spacing in gigahertz is larger than the data rate in gigabits.
   b. The spacing equals the signal speed in gigahertz.
   c. The spacing in nanometers is equal to the data rate in gigabits per second.
   d. The spacing decreases with the data rate; it is proportional to the length of data pulses.
   e. The relationship is impossible to state because data rate is digital and frequency spacing is analog.
4. Which of the following systems can accommodate the most optical channels?
   a. Coarse WDM in a system that includes several erbium-doped fiber amplifiers
   b. Dense WDM in a system that includes several erbium-doped fiber amplifiers
   c. Coarse WDM in a system without optical amplifiers
   d. Dense WDM in a system without optical amplifiers
   e. The question does not give enough information to tell

5. You need to provide dispersion compensation for a 60-kilometer length of fiber with chromatic dispersion of −3 ps/nm-km. How much fiber with chromatic dispersion of +15 ps/nm-km at the same wavelength (1550 nm) do you need?
   a. 5 km
   b. 12 km
   c. 15 km
   d. 25 km
   e. 60 km
6. Your fiber has a dispersion slope of 0.08 ps/nm^2-km in the C-band of erbium-fiber amplifiers. Assuming that slope is a straight line, how much does the dispersion change over the width of the C-band?
   a. 0.08 ps/nm-km
   b. 0.8 ps/nm-km
   c. 2.8 ps/nm-km
   d. 8 ps/nm-km
   e. 28 ps/nm-km
7. You have a mixture of three types of fiber in a span running between two fiber amplifiers. Which type should you put closest to the output of the first amplifier to reduce four-wave mixing?
   a. Step-index single-mode fiber with effective area 80 μm^2 and dispersion +15 ps/nm-km
   b. Reduced-slope single-mode fiber with effective area 50 μm^2 and dispersion −2.5 ps/nm-km
   c. Large effective area single-mode fiber with effective area 68 μm^2 and dispersion −20ps/nm-km
   d. Either of the low-dispersion fibers
   e. It doesn't matter
8. In what fiber is four-wave mixing the largest?
   a. Step-index single-mode fiber with effective area 80 μm^2 and dispersion +15 ps/nm-km
   b. Reduced-slope single-mode fiber with effective area 50 μm^2 and dispersion −2.5 ps/nm-km
   c. Large effective area single-mode fiber with effective area 68 μm^2 and dispersion −2.0 ps/nm-km
   d. Zero dispersion-shifted single-mode fiber with effective area 75 μm^2 and dispersion 0 ps/nm-km
   e. 50/125 graded-index multimode fiber
9. Your optical amplifier has been generating 2 mW on each of 40 optical channels. An optical switch downstream diverts half of the optical channels to another cable. If everything else remains constant, what is the power on the remaining channels?
   a. 2 mW
   b. 3 mW
   c. 4 mW
   d. 5 mW
   e. The amplifier will not work
10. What is the difference between a transparent and an opaque optical switch?

a. A transparent switch allows light signals to pass through it in optical form; an opaque switch converts it to some other form.

b. A transparent switch is made of clear glass; an opaque switch is made of metal.

c. A transparent switch does not change the data rate of the signal; an opaque switch changes the data rate.

d. A transparent switch transmits a signal; an opaque switch blocks the signal, turning it off.

11. What type of device reads header bits in a signal and then decides where to transmit them?

a. Wavelength router

b. All-optical switch

c. Router

d. Electro-optical switch

e. Raman amplifier

12. How many optical channels can an all-optical switch process by reflecting light from an input fiber?

a. 1

b. 10

c. 40

d. 80

e. As many as the fiber can carry

CHAPTER

23

Global Telecommunications Applications

About This Chapter

Now that you have learned about fiber-optic hardware, standards, and system design, it's time to look at the major applications of fiber-optic systems. Changing technology and regulations are eroding traditional divisions, but it is still useful to separate telecommunications into a few sectors. The largest in scale is the global telecommunications network, the backbone of international telecommunications, including long-distance transmission under the oceans and on land. Other sectors are regional or metro networks, and distribution networks for voice, video, and computer data services. You learned a little about these ideas in Chapter 3; now it's time to take a closer look.

In this chapter, I will describe the long-distance fiber-optic transmission systems that carry data, voice, video, and other signals around the world. They include intercontinental submarine cables, as well as national and international systems on land. They are the world's biggest telecommunications "pipelines," and they are designed to maximize both transmission speed and distance. Fiber-optic technology has dominated these systems for over a decade, first with single-channel single-mode transmission, and now with high-speed DWDM systems.

These long-distance networks feed into regional or metro networks, which in turn connect switching offices and distribution networks, which deliver services to individual homes and offices. Later chapters will look at those networks.

Defining Telecommunications

The term *telecommunications* is deliberately broad. It dates back to the era when communication specialists were trying to group telephones and telegraphs under one heading. As the telegraph industry faded away, telephony become dominant, but the new word had caught on. It was useful because new communication services were emerging. Radio and television networks spread, broadcasting voice, music, and pictures. Telex relayed printed messages around the world. Facsimile systems transmitted images of documents. Computer data communications grew rapidly. Wireless telephones and pagers spread. They all fell under the broad heading of telecommunications.

Telecommunications encompasses voice, data, facsimile, video, and other forms of communications.

Different types of telecommunications had different origins, but most are *converging* toward a common network that delivers many different services. The main reason is that it costs less to build one versatile network than many specialized ones. Convergence will never be complete because some services inherently differ from others. A mobile phone small enough to fit into your pocket can display brief text messages, but not complex Web pages or large-screen broadcasts of a football game.

Convergence is strongest in the global backbone network designed to carry digitized signals around the world. This is dominated by fiber-optic information pipelines that form two somewhat distinct networks: the traditional circuit-switched system that has grown from the telephone network, and the Internet. Note that I say *somewhat* distinct, because the two networks carry similar information and may overlap in some areas.

The main distinctions between the Internet and the traditional telecommunication backbone are the transmission format and the labels the systems carry. As you learned earlier, the Internet carries packet-switched data, and the traditional telecommunications network is circuit-switched. In fact, the distinction is not rigid, because Internet Protocol data can be encoded into SONET frames and transmitted on circuit-switched fiber. That is, a signal may be encoded as IP packets in one layer of the transmission system, which are then transmitted in circuit-switched form in the physical or optical layers. Indeed, look closely at some Internet service maps, and you can find places where Internet links go over T1 or T3 telephone lines.

Satellites transmit video signals to cable networks and home subscribers; they also transmit to pagers.

There are also other types of long-distance transmission. Communication satellites relay video signals to television broadcasters, cable-television companies, and subscribers to direct-broadcast services. Satellites also can provide paging services and mobile data transmission. You should know they exist, but I won't cover them because they aren't part of the fiber-optic network.

The telecommunications industry is changing at the same time as the technology. The once-monolithic telephone industry has become a collection of many companies that form shifting alliances and tend to merge or divorce at the whims of Wall Street. Competition is real in the long-distance market, where many carriers have separate networks spanning part of—or in a few cases, most of—the world. Competition is more dubious on the local level. Mergers seem to change the name of your local telephone or cable-television company more often than most people change local carriers, but cable and telephone companies do compete with each other to offer data and—sometimes—telephone services Everything has to interconnect for the system to function, and most of the time most of the network does.

Let's look briefly at the elements of the global telecommunications network before focusing on long-distance transmission.

The Telephone Network

The telephone network spread around the globe in the twentieth century, supplanting electrical telegraphs to become the backbone of the international telecommunication system. Its original purpose was to carry conversations between any two phones connected to the network. To do this, it has a network of connections extending to individual homes and offices around the world. Signals from individual conversations—and from other services that share phone lines—are combined and directed to their destinations at telephone switching offices. Once mechanical switches connected electrical wires; now electronic and optical switches route signals through optical fibers and other media.

The telephone network evolved into the backbone of the traditional global telecommunications system.

Local and regional telephone systems interconnect with each other and with long-distance and international carries to offer service around the block and around the planet. Telephone numbers provide the information needed for switching signals. In much of North America, you can direct calls within your area code by dialing seven digits. Long-distance calls within the United States, Canada, and parts of the Caribbean require dialing a long-distance code (1), a three-digit area code (XXX), and a seven-digit local number. (You must always dial the area code in places where two or more area codes are overlaid in the same area.) To make overseas calls, you dial an international code (011), a 1- to 3-digit country code (e.g., 44 for Britain or 81 for Japan), usually a city code or other regional code (e.g., 207 for inner London or 3 for Tokyo), then a local number (usually 6 to 8 digits). Thanks to this system, you can call most of the phones in the world from your home or office, although you may regret it when you get the bill.

Each traditional twisted-wire-pair telephone line carries only a modest amount of information. A standard analog phone line carries sound frequencies of 300 to more than 3000 Hz, which the industry calls POTS, for Plain Old Telephone Service. (Phone lines can carry frequencies to 4000 Hz, but the upper frequencies are used for control signals.) That is enough for intelligible conversations, but it is far short of the 20- to 20,000-Hz range of the human ear. Pulse-code modulation converts the analog signal to digital format, with one voice channel equal to 64,000 bit/s in North America.

The telephone system originally included only fixed telephones that provided voice service. Because the phone network extended around the world, it was adapted for other services as well. Fax machines, mobile phones, pagers, and dial-up computer modems all work within the confines of the telephone network. They dial numbers in the same way, and must transmit signals within the same frequency range the telephone network uses for voice signals. Mobile phones and pagers generally are served by separate switching centers, but faxes and dial-up modems are hooked up to ordinary telephone switches.

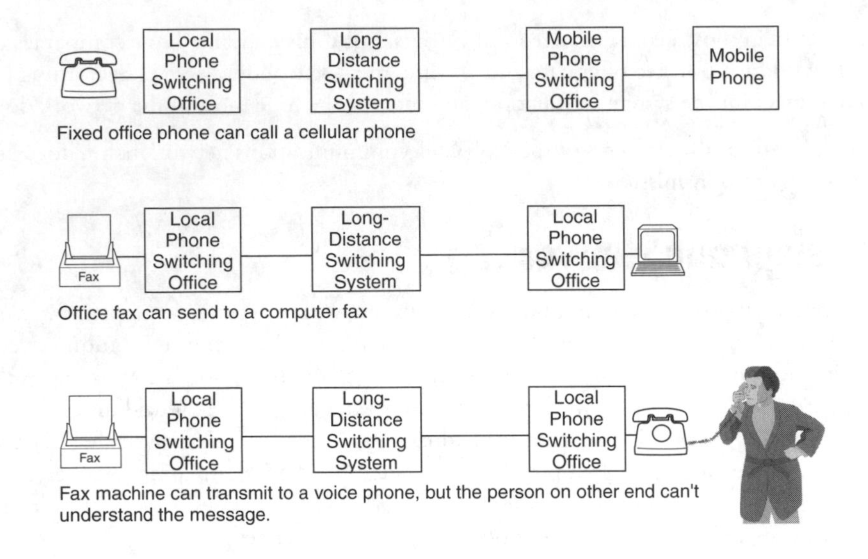

FIGURE 23.1
Voice, fax, and other signals share the telephone network.

Voice phones, faxes, and dial-up modems share a common dialing system.

All this equipment shares the common dialing system, as shown in Figure 23.1. Ideally, people always call voice lines or pagers, and fax machines always call other faxes. However, mistakes can happen, and nothing in the telephone numbering system warns you that a number is assigned to any particular type of equipment. If somebody on the other side of the world tries to send a fax to your home (voice) phone in the middle of the night, their fax machine will keep on trying for the programmed number of redials. It's like calling a wrong number and getting someone who doesn't speak English—the parties on the two ends of the line don't speak the same language.

The Internet and Computer Networks

The Internet carries digital data more efficiently than phone lines.

The Internet began as a network linking the computer networks at major research universities and laboratories but has since expanded to link many of the world's computers. It generally carries digital data more efficiently than phone lines. As with the telephone network, you can divide the Internet into a long-distance backbone system and regional and local transmission systems.

In practice, there is considerable overlap between computer networks and telecommunications networks. You can connect a personal computer to the following:

- A dial-up modem linked to an analog telephone line, which sends data to an Internet service provider over phone wires.
- A wireless modem linked to the cellular telephone network. A new generation of mobile phones will include facilities for data transmission.
- A cable modem connected to the cable-television network, which the cable company links to the Internet.

- A Digital Subscriber Line (DSL) connection over copper phone lines, which transmits digital data at frequencies higher than the voice band. The same wires may simultaneously transmit analog voice signals, as described in Chapter 25.
- A corporate or university network that links to the Internet.
- Direct to the Internet, usually via high-speed digital lines linked to a regional Internet node, if you have a very powerful computer.

All these services connect you to the Internet, and direct your signals through the switches and routers that process Internet traffic. You will learn more about computer networking in Chapter 26, but the details of Internet transmission are beyond the scope of this book.

Video Communications

Video signals are broadcast through the air by ground or satellite transmitters, or are transmitted through cables. There are three main types of video distribution—long-distance distribution from central sources to individual broadcasters and cable companies, direct broadcast from satellites to home receivers, and local distribution from cable companies and broadcast stations to individual homes. You will learn more about these in Chapter 27.

Cable-television systems receive signals from distant sources for local distribution.

Traditionally, video distribution has been separate from the global telecommunications network, but this is changing. Much distribution of programs to broadcast stations or cable head-ends is via satellite, but some video signals are distributed over fiber. Cable companies link to the Internet to provide cable modem service and to the telecommunications network if they compete with the local telephone company to offer telephone service.

Television is starting to make a transition from analog to digital technology that will strongly impact video transmission. Unprocessed digital high-definition video signals require much more transmission capacity than lower resolution analog signals—but digital signals can be compressed very efficiently, to occupy about the same bandwidth as analog video signals. What this will mean in practice for video transmission remains to be established.

Most video-signal distribution is one way, but video conferencing requires two-way transmission. Although video conferencing remains a limited application, streaming and two-way video signals can be routed through the Internet as well as through phone lines.

Other Communications

Some communication systems don't fit neatly into the categories described so far. For example, military organizations have their own high-speed networks designed to survive hostile attack. Utilities and railroads often have their own dedicated systems along their rights of way to meet special needs, such as monitoring their own operations. Often these networks interconnect with the global telephone network and/or the Internet. Many utilities and railroads lease telecommunications capacity along their rights of way to other companies, such as long-distance phone and Internet carriers.

The Global Telecommunications Network

Diverse communication systems are interconnected to form a global network.

The pieces I have described so far are linked together, as shown in Figure 23.2, to form the global telecommunications network. Although many components were built separately, they have been interconnected, because connectivity is vital to communications.

The global telecommunications network operates on many levels and through many media. At the highest level, it connects national telecommunication networks. Submarine cables and satellite links cross oceans, linking continents at tens or hundreds of gigabits per second. Long-distance land cables connect adjacent countries. The nature of international links varies greatly, depending on geography, history, politics, economics, and a host of other factors. Closely allied countries, like the United States and Canada, are closely tied

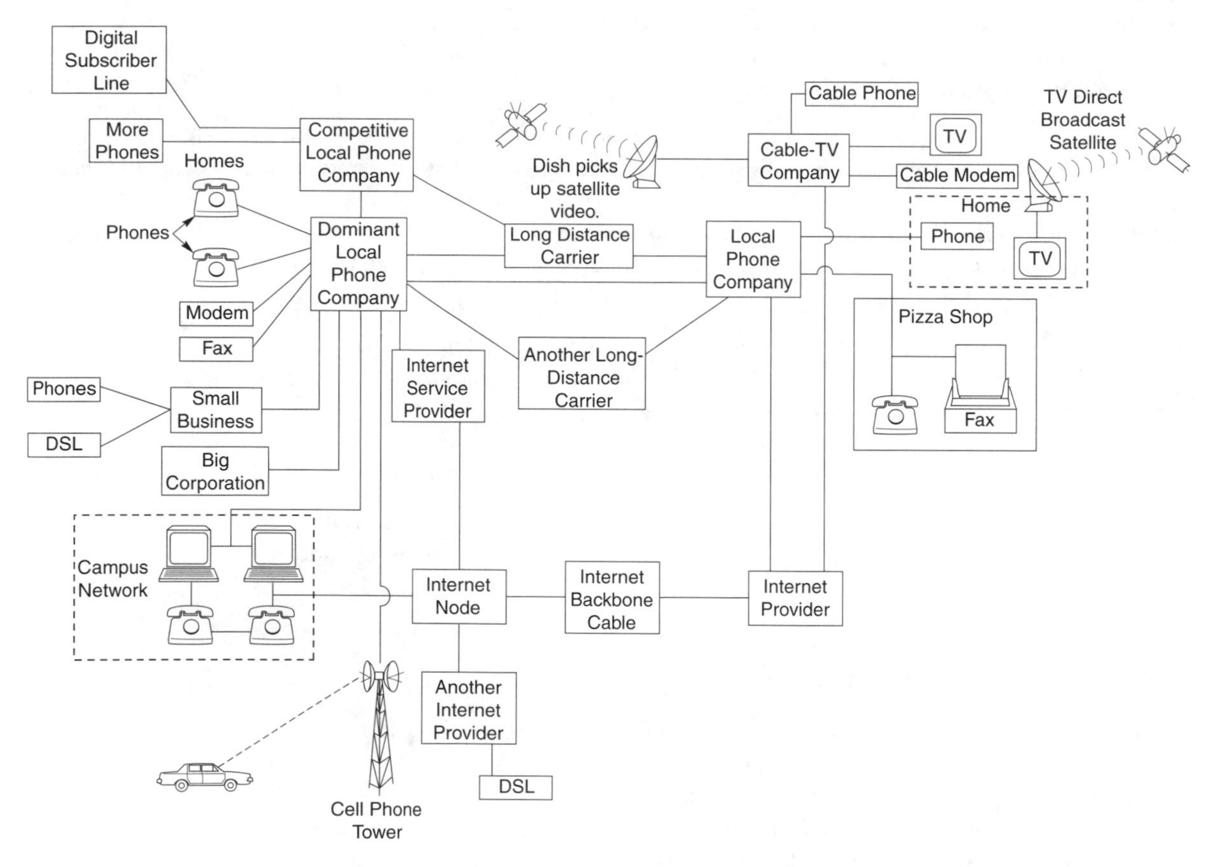

FIGURE 23.2
Interconnection of telecommunication networks.

(to the extent that they share the same system of area codes). The countries of western Europe likewise are closely tied. On the other hand, only one submarine cable links the United States and Cuba, reflecting decades of political hostility. High-capacity land and submarine cables span heavily trafficked routes, such as between London and New York or London and Moscow. Networks branch out from major nodes to serve nations, states, or regions. A call from London to Rio de Janeiro may pass through a transatlantic cable to New York and then through another submarine cable crossing the Atlantic to South America. Alternatively, it might be routed from London to southern Europe and from there across the Atlantic to the Caribbean and south to Brazil. State-of-the-art submarine cables carry 40 Gbit/s on each fiber pair; systems now in the planning stages will carry up to 640 Gbit/s on each fiber pair and up to 7.68 Tbit/s in a single undersea cable.

Satellite links distribute video signals and provide telephone service to remote sites impractical to reach by cables. Calls to the Falkland Islands off the southern tip of South America or to Tuvali in the mid-Pacific will go via satellite. CNN reporters reporting live from remote battlefields send signals via satellites. Cable companies typically receive their video feeds by satellite. Satellite systems also can provide mobile phone and data services around the world.

National telecommunication networks differ in scale with the sizes of countries, but also usually operate at hundreds of megabits to hundreds of gigabits per second. The United States network rivals that of western Europe in size, for example. Submarine cables play only minor roles in national networks (except for countries made up of many islands). The backbones of the telecommunication networks that carry telephone and data traffic are land-based fiber-optic cables. Satellites broadcast signals to multiple points (particularly video feeds to cable television operators), and carry limited other signals, particularly to remote sites. Land-based microwave radio and coaxial cable systems are becoming rare, made obsolete by high-speed fiber-optic systems.

Fiber optics are the backbones of national telecommunication networks.

Regional and metro networks are the next step down. In the United States, high-speed fiber-optic cables link major cities and then spread out to serve the surrounding area. Typically they branch out from points on the national backbone system, as you will see in Chapter 24.

Local telecommunication networks branch out from regional systems and in practice can be complex. As you saw earlier, there are actually multiple local networks that are to some extent linked together. I won't talk much about local broadcasting of television and radio signals or about cellular telephones, but Chapters 25 through 27 will cover fiber-based services: telephone, digital data, and cable television. The technology behind these services is converging—and the telephone and cable television networks are coming to compete with each other in some ways.

Local telecommunication networks branch out from regional systems.

Putting Networks Together

Putting the pieces together to make a single functional network requires combining many signals from diverse sources. There are two crucial concepts involved: multiplexing or combining

Signals must be combined and put into a common format for transmission through the telecommunication network.

many signals for transmission through a single carrier and conversion of signals into a common format. These can be addressed in various ways, but the basic concepts and reasoning are the same. It's cheaper and easier to combine signals traveling the same path than to send them separately. Likewise, all the signals traveling through the system have to be converted to the same format.

You can compare the telecommunication network with the circulation system of your body. Blood flows from tiny capillaries to larger veins, which in turn feed larger veins that can carry more blood. After the blood passes through your heart and lungs, it is divided up into smaller and smaller arteries and ultimately reaches the tiny capillaries. Individual phone lines are the capillaries of the old telephone network, upon which the modern telecommunication system is based. Low-speed lines feed into higher-speed systems, which go longer distances. There is a standardized hierarchy of transmission rates, described earlier in Chapter 20.

All the information handled by any transmission system has to be translated into the same format. As you will see later, the format is changing as the network evolves, but it has remained compatible with older hardware. With a pair of wire cutters and a screwdriver, you can attach a massive 1950-vintage dial phone in basic black to the same standard analog telephone line as the latest 56-kbit/s fax-modem.

The Digital Telephone Hierarchy

Telephone-based systems are organized in a hierarchy of transmission rates.

The digital transmission hierarchy is itself evolving, with a new family of standards emerging. One old standard, still in wide use, is the North American digital telephone hierarchy of Figure 23.3, which was largely established in the days when AT&T was *the* telephone company. The figure shows how low-speed signals (in units of voice channels) are time-division multiplexed to higher speeds at each step of the hierarchy. In practice, all those slots in the higher-speed signals may not be filled. For example, rural areas may have too few telephone subscribers to use all the available slots. In addition, most phone companies now skip directly from the 1.5-Mbit/s T1 rate to the 45-Mbit/s T3 rate, not bothering with the intermediate T2 rate.

Figure 23.3 shows only the first three levels of the North American digital telephone hierarchy. Standards have evolved as transmission speeds have increased. In practice, there are three major families of standards, listed in Appendix C: the North American Digital Hierarchy, the international telephone standard developed by the International Telecommunications Union to cover similar transmission rates, and the SONET/SDH hierarchy, which is essentially uniform around the world.

NORTH AMERICAN HIERARCHY

The North American Digital Hierarchy starts with 64,000 bits/s as a single digitized voice channel. The most important rates in use are the 1.5-Mbit/s T1 rate and the 45-Mbit/s T3 rate. The 6.3-Mbit/s T2 rate is like second gear in a sports car, rarely used or quickly shifted through.

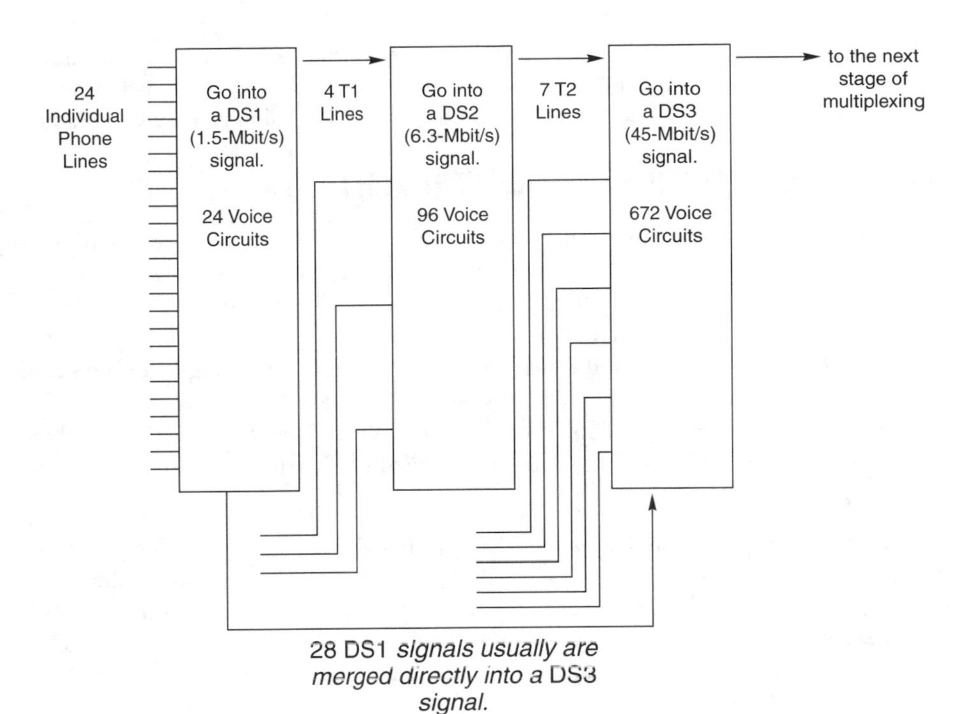

FIGURE 23.3
Time-division multiplexing in the North American digital telephone hierarchy.

The original specifications called for a 274-Mbit/s T4 rate, but it never found wide application and is now ignored. During the 1980s, a variety of other time-division multiplexing rates were introduced in North America. The major family was based on a nominal data rate of 405 Mbit/s (equivalent to 6048 voice channels, or nine T3 carriers) with additional rates of 810 Mbit/s, and 1.7 and 2.4 Gbit/s that were multiples of the 405 Mbit/s rate. These systems were not formally accepted as standards, and have since been supplanted by SONET and related systems.

INTERNATIONAL/ITU TELEPHONE HIERARCHY

Europe uses the ITU digital telephone hierarchy.

The International Telecommunications Union adapted its own digital telephone hierarchy, also based on 64,000 bit/s digitization of phone lines. However, its basic time-division multiplexing step was combining 30 digitized voice lines to give a *Level 1* channel at 2.048 Mbit/s. The next step combined four Level 1 signals to make an 8.448 Mbit/s Level 2 signal (120 voice channels). Further multiplexing also combined four lower-level channels, giving Level 3 (480 voice channels) at 34.304 Mbit/s, Level 4 (1920 voice channels) at 139.264 Mbit/s, and Level 5 (7680 voice channels) at 565.148 Mbit/s. All are listed in Appendix C. They are sometimes called the CCITT rates, from the French translation of the International Consultative Commission of Telephone and Telegraph, the ITU body that developed them.

These ITU standards are mainly used in Europe, so they sometimes are considered European rates. As with the North American digital hierarchy, the highest rate, 565 Mbit/s, has fallen out of use today. (A few 565 Mbit/s systems were installed in North America during the 1980s, when the demand for transmission capacity had leaped ahead of North American standards.) Level 4 systems now feed into the Synchronous Digital Hierarchy set of rates.

SONET/SYNCHRONOUS DIGITAL HIERARCHY

SONET packages input signals into frames, to guarantee capacity.

In the 1980s, the telecommunications industry wrote a new set of time-division multiplexing standards for higher-speed transmission of mixed voice and data. These became the Synchronous Optical Network (SONET) and Synchronous Digital Hierarchy (SDH) standards covered in Chapter 20. The two standards are quite similar, but differ in details important for compatibility with the older North American and ITU digital telephone hierarchies. As you learned in Chapter 20, they package data into frames, which allocate specific slots for input channels, making them function as circuit-switched systems. SONET and SDH can also transmit packets, but not as efficiently as a true packet-switched system such as the Internet Protocol.

The base of the SONET standard is the 51.84 Mbit/s OC-1 (Optical Carrier-1); an electronic signal at that rate is called STS-1. The OC-1 rate was designed to handle the T3 input signals common in North America, but it can also handle input signals in ATM or IP formats. The next step up the SONET/SDH ladder is OC-3, 155.52 Mbit/s, which in SDH is designated as STM-1. This format was designed to match Level 4 of the ITU digital hierarchy, as shown in Figure 23.4.

Overhead consumes about 10% of SONET/SDH capacity.

If you compare data rates, you will notice that the SONET/SDH rates are about 10% higher than the rates quoted for the input signals from the digital telephone hierarchies. This reflects additional "overhead" data needed to manage and monitor signal transmission. The SONET standard includes capabilities to spot failures and redirect signals to

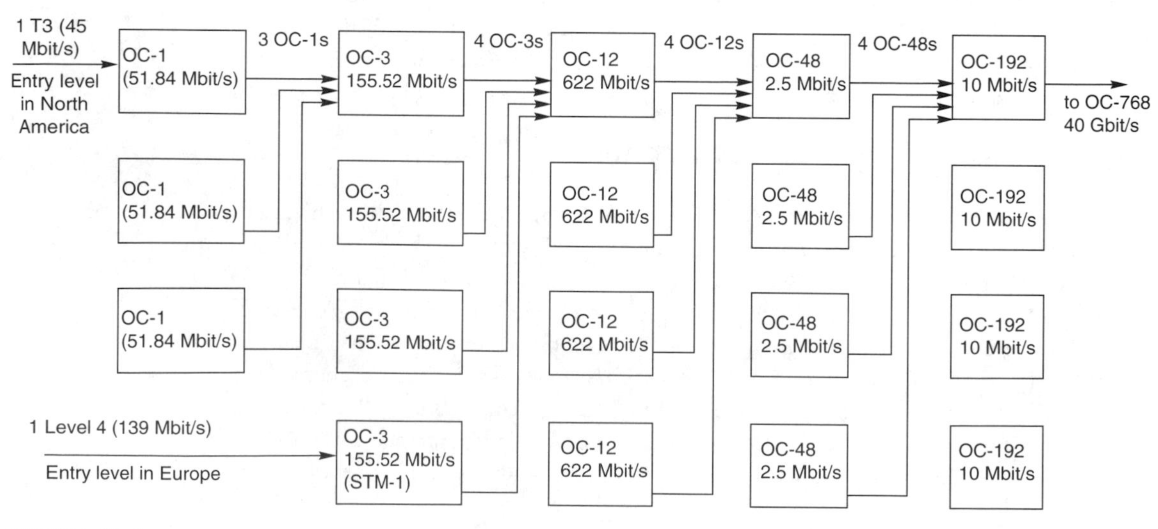

FIGURE 23.4
SONET/SDH time-division multiplexing hierarchy.

maintain service. The SONET and SDH standards also specify network structures, that make recovery from failure easier.

From 155.52 Mbit/s, both SONET and SDH step up by factors of 4, to 622 Mbit/s, then to roughly 2.5, 10 and 40 Gbit/s. So far the 10 Gbit/s is the upper limit in most practical systems, although developers have begun to offer 40 Gbit/s equipment.

SONET and SDH are time-division multiplexing formats, which cover the speed and sequence of bits transmitted in a single data stream. This means they are designed for use on a single optical channel. You can stack multiple SONET or SDH signals (or mixtures of the two) onto a fiber using wavelength-division multiplexing.

Internet Transmission

In practice, Internet signals typically are transmitted at the same data rates and in the same formats used for telephone traffic. Look at a map of an Internet backbone system, and you probably will find capacity specified in T1, T3, and SONET OC-3 through OC-192 data rates. Some of these signals go through lines leased from telephone carriers; others are dedicated for Internet transmission.

Internet data typically is transmitted at the same speeds as telephone traffic.

No law states that you must transmit high-speed signals through fiber in SONET or SDH formats. It's done because it's easy with present equipment. SONET transmitters and receivers are readily available, and existing long-distance networks typically lease capacity at SONET rates. Except at the highest speeds, Internet signals must be in a format compatible with other traffic. A 155-Mbit/s stream of Internet data is unlikely to have a fiber all to itself, so it must be in a format that can be time-division multiplexed where necessary.

There is growing interest in other transmission formats. For example, some companies that supply fiber-optic systems talk about "IP direct over fiber," without using SONET format. Their goal is to save the high cost and overhead of SONET and ATM equipment, and use a high-speed Internet Protocol input to drive a fiber-optic transmitter. When this is implemented, you might have a box that takes IP input and generates a stream of data at 2.5 or 10 Gbit/s to drive a laser transmitter.

Optical Channels

As you learned earlier, wavelength-division multiplexing allows a single fiber to carry many optical channels, multiplying total transmission capacity. Standards specify distinct center frequencies for each channel in a WDM system.

The erbium-fiber amplifier C- and L-bands can each transmit about 80 optical channels with 50 GHz spacing, or 40 channels each with 100-GHz spacing. This packing corresponds to a total capacity of 1.6 Tbit/s per fiber if each optical channel transmits 10 Gbit/s. Even higher data rates, to 7.7 Tbit/s, have been achieved in laboratory experiments, and companies have claimed they can deliver systems transmitting 1.6 Tbit/s or more on a single fiber. So far, however, none have been installed and operated at full capacity. Most have only a few operating channels.

WDM also makes it possible for one fiber to carry many different types of traffic. For example, a single fiber may carry separate optical channels that transmit long-distance telephone, video, and Internet traffic, each at its own wavelength.

Shared Transmission

Fiber transmission capacity usually is shared among users.

One important commercial reality is that transmission capacity on a cable or even on an individual fiber is often shared. There are many possible arrangements. Some companies may install a cable containing up to several hundred fibers and lease individual fibers to telecommunications carriers. Companies may trade capacity on different routes, for example, swapping two fibers in a cable from Chicago to Cleveland for two fibers in a cable from Chicago to St. Louis. Some companies may lease individual wavelengths on a fiber. Some carriers subdivide their capacity, signing contracts to supply OC-3 services to four separate companies, and combining the signals to make a single OC-12 data stream.

This makes the business of owning and operating cables quite complex. For example, company A could lease a right of way from a railroad company and install a dozen underground ducts along the route. One cable could contain fibers leased to companies B, C, D, and E, plus the regional telephone company. Company E, in turn, could lease four wavelengths on one fiber pair to a competitive local phone company, and four other wavelengths to a company competing with company C. Thus competitors could wind up sharing the same cable, and you could find route maps that show different companies using the same route. Likewise, if you look carefully, you could find that several Internet backbone companies and long-distance telephone carriers happen to share routes through the same cable.

Submarine Cables

Submarine fiber cables are the backbones of intercontinental telecommunications.

The largest links in the global telecommunication network are submarine fiber-optic cables, which cross oceans to link continents. They have been vital links since the age of the telegraph, shrinking the world. In the rest of this chapter, we will look at them and their long-haul terrestrial cousins that form the backbone of telecommunications on land. Later chapters will look at regional and local systems.

Submarine cables come in many types. Some cross the few kilometers of seawater separating an island from the mainland; one of the first to use fiber was an 8-km cable from Portsmouth, England, to the Isle of Wight off the English coast. Many cross tens or hundreds of kilometers of sea; the Mediterranean and Caribbean seas are crisscrossed with submarine cables. Some span thousands of kilometers of ocean; the longest run across the bottom of the Pacific and from Europe to Japan.

Submarine cables must meet extremely tough requirements. Their transmission capacity should be as high as possible, because the cables are costly to make, lay, and operate. The cable, and any optical amplifiers or repeaters, must withstand harsh conditions on the bottom of the ocean for a design life of 25 years. Components must be extremely reliable, because it is very expensive to recover the cable from the sea floor and haul it to the surface for repairs. The cable should transmit digital signals cleanly to be compatible with modern equipment. These specifications veritably call out "fiber optics," and since the 1980s fibers have been standard for submarine cables. Figure 23.5 shows how they have spread around the world.

The Impact of Submarine Cables

Submarine cables date back to the days of the electrical telegraph, and for well over a century they have played a vital role in binding the world together. Undersea telephone cables came long after telegraph cables, and the first transatlantic fiber-optic cable was not laid until 1988. However, since then the technology has grown at amazing speed.

The first submarine telegraph cable was laid in 1850.

The first submarine telegraph cable was laid in the English Channel in 1850, as Europe expanded its electrical telegraph system. It carried only a few messages between England and France before a fisherman snared it and hauled a piece to the surface. He thought it was a strange type of seaweed. That experience taught submarine cable engineers an important lesson—waterproof isn't enough. Fishing trawlers and ship anchors remain the biggest threat to cables in shallow water, so modern cables are buried a meter below the sea floor except in ocean depths below a few hundred meters.

The first attempt to lay a transatlantic telegraph cable failed in 1857, and it was not until 1866 that a reliable cable began operating under the Atlantic. Very long telegraph cables were possible in the nineteenth century because mechanical relays could amplify their dots and dashes. Long-distance telephone transmission proved much harder, because it required electronic amplifiers. Vacuum tubes relayed signals on land, but transatlantic telephone calls had to rely on short-wave radio links until 1956, when the first transatlantic telephone cable, TransAtlantic-1 (TAT-1), began operation between Britain and Canada. It included vacuum-tube amplifiers sealed into special cylinders built to withstand the tremendous pressure at the ocean bottom.

TAT-1 was made of copper coaxial cable, which offers the highest bandwidth of any standard metal cable. However, coax attenuation increases with the square root of transmission frequency ν, and can only be reduced by increasing the inside diameter D of its outer conductor.

$$\text{Attenuation} = \frac{\text{constant} \times \sqrt{\nu}}{D}$$

This means that to transmit higher frequencies, coaxial cables must be made fatter, or have repeaters spaced closely together. Later engineers were able to squeeze up to 4200 telephone circuits onto a single coaxial cable, but to span the Atlantic it had to be an unwieldy 5.3 cm in diameter and required one repeater every 9.5 km—a total of 664.

Switching to fiber increased cable capacity by a factor of 8 over coax.

By the mid 1970s, it was clear that coax had reached the end of the line, and satellites looked like they would eventually dominate intercontinental communications. However, Bell Labs turned to fiber, and in 1980 announced a design for the first transatlantic fiber-optic cable, TAT-8. By using single-mode fiber transmitting at 1300 nm, they calculated

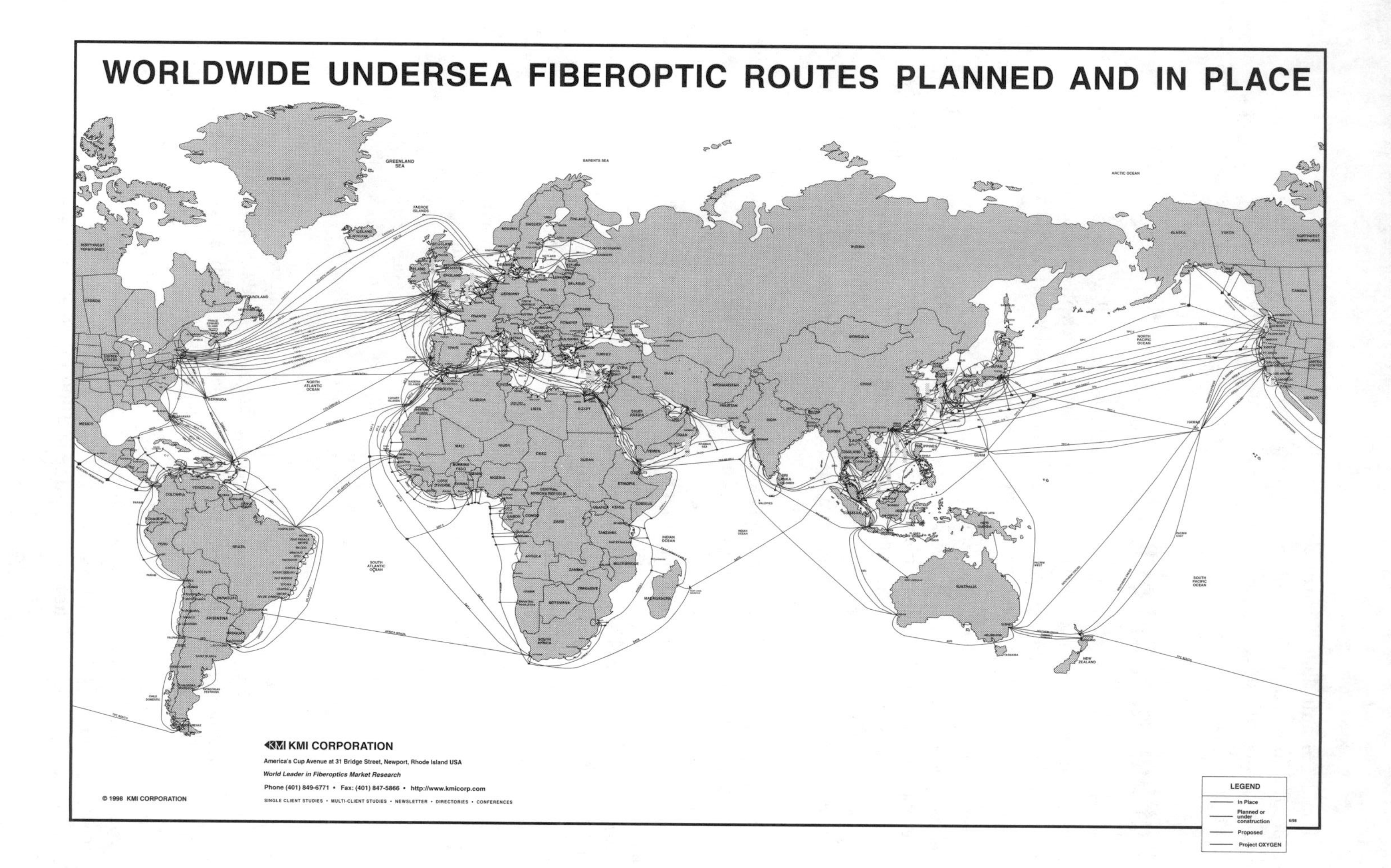

FIGURE 23.5

Worldwide undersea fiber-optic cable routes. (Courtesy KMI Corp., Newport, RI)

they could transmit 280 Mbit/s on each of two fiber pairs, with a third fiber pair kept in reserve. With digital data compression, that was equivalent to a total of 35,000 voice channels, over eight times more than TAT-7. After eight years of hard work, TAT-8 began service at the end of 1988 with repeaters averaging more than 50 km apart.

That began an era of explosive submarine fiber-optic cable growth that continues today. Repeaters have given way to optical amplifiers, and wavelength-division multiplexing allows each fiber to carry many optical channels. Looking back, you can see that TAT-8 carried a thousand times more telephone circuits than TAT-1 did 30 years earlier. Looking only a little bit forward, you can see that newly installed cables have the capacity to carry 1700 times more data than TAT-8—although it may be several years before all the transmitters needed to realize that capacity are installed. Figure 23.5 shows major submarine cable routes.

Submarine Cable Basics

The design of submarine cables is shaped by the submarine environment, which is very different than the environment for terrestrial cables. The underwater environment is very stable, very extreme, and very hard to reach. Repairs of undersea cables are difficult and expensive, so systems are designed to operate without service for long periods. Specifications usually call for no more than two underwater repairs in a cable's nominal 25-year lifetime, and the target is no repairs.

Submarine cables are designed for long life, high pressure, and high speed.

Electrical power must be transmitted from the cable termination points on land, so power is at a premium. Early systems used repeaters, but since the mid-1990s all submarine cables have used only optical amplifiers underwater. Full "three-R" regeneration—reshaping and retiming pulses as well as reamplifying them—is done only at the cable termination points on shore. For intercontinental cable systems spanning thousands of kilometers, this imposes very stringent requirements on the levels of noise, dispersion, and nonlinear effects in the transmitting fiber and the optical amplifiers.

As you learned in Chapter 8, the fiber-optic cables used in submarine systems are highly specialized, with fibers embedded deep in the core of a pressure-resistant structure. The outer layers of deep-sea cables are medium-density polyethylene. Heavy metallic armor covers the polyethylene in shallow-water cables, which are buried to protect them from fishing trawlers and ships' anchors—the undersea counterparts of backhoes. If undisturbed, the cable structure should withstand intense pressures and exclude salt water for decades.

Optical amplifiers are mounted inside pressure-resistant cases originally developed to house repeaters. They are built into the cable but are larger in diameter, so at first glance they resemble a rabbit swallowed by a python. Submarine cable developers still call these cases "repeaters," but don't be fooled—repeaters have not been used on submarine cables for several years (although the repeaters on old cables have not been replaced).

Submerged amplifiers are housed in pressure-resistant cases.

Submarine cables fall into two broad classes, *unrepeatered* and *repeatered*. In the world of submarine cables, these terms define whether or not the system includes optical amplifiers with their pump lasers in the same underwater housing as the optical amplifier. Underwater pump lasers mark a key dividing point because they are electronic components subject to failure, and because they require electrical power to be transmitted through the cable. These two types can be further subdivided according to the distance they span and their configuration, but we will concentrate on the basic categories. As you will see later in this chapter and in Chapter 24, these systems have counterparts on dry land.

UNREPEATERED UNDERSEA CABLES

Unrepeatered cables can span up to 400 km.

The length of unrepeatered cables is limited by the need to sustain sufficient optical power to avoid repeaters. In simplest form, this generally means distances less than a couple hundred kilometers from end to end, so the transmitter output is not attenuated below the receiver threshold. This distance can be stretched to as much as 350 or even 400 km by adding a postamplifier after the transmitter, a preamplifier before the receiver, and remotely pumped amplifiers, as shown in Figure 23.6.

On-shore or remotely powered amplifiers can stretch unrepeatered transmission distances.

Transmission distance in a repeaterless system normally is limited by the system power budget, which depends on the transmitter output power and the receiver sensitivity. For example, a system in which the transmitter output is 28 dB above the receiver sensitivity normally can span about 130 km without amplification. Longer distances are possible by increasing transmitter power and/or receiver sensitivity.

As you learned earlier, a postamplifier can increase the output power from a transmitter. Likewise, a preamplifier can increase receiver sensitivity. It's possible to stretch "repeater-

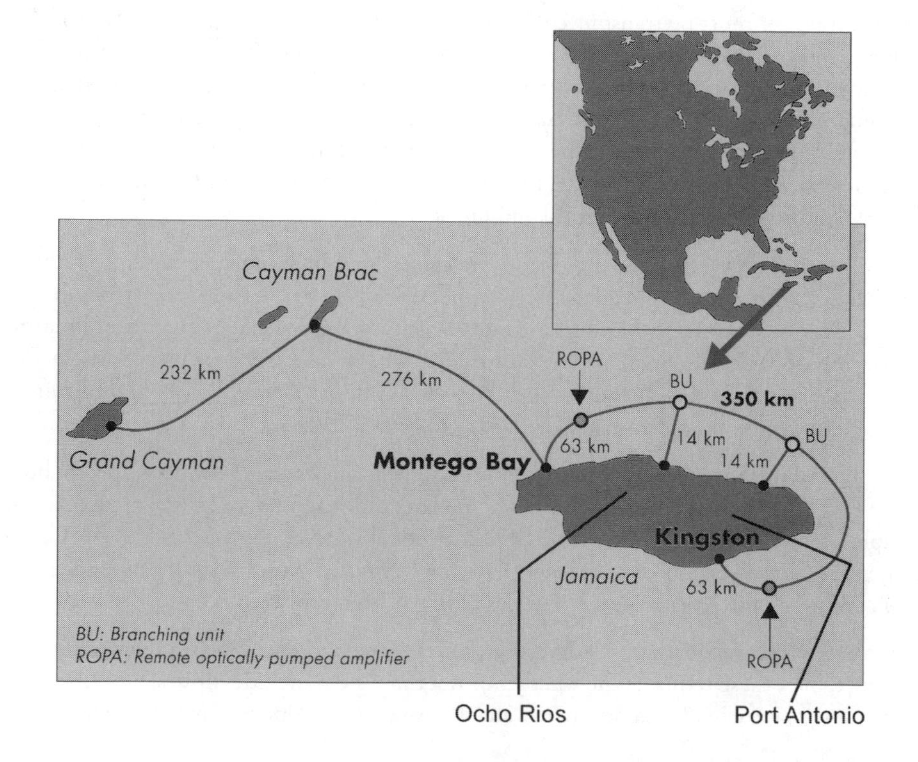

FIGURE 23.6
Unrepeatered systems can hop a series of islands to span long distances. Use of optical postamplifiers, optical preamplifiers, and remotely pumped optical amplifiers can stretch single spans. (Copyright Alcatel)

less" transmission further by siting *remote optically pumped amplifiers* offshore. In these systems, the cable from shore to the offshore amplifier includes a separate fiber that carries light from an on-shore pump laser, which then does not require submerged pump electronics. Fiber attenuation makes it impossible for a remote pump laser to deliver as much pump power as a pump laser could if it was in the submerged housing, so the amount of amplification is less than with a conventional optical amplifier. The design also requires dedicating one fiber slot in the cable to a pump fiber for each remote optical amplifier. However, post- or preamplification with optically pumped amplifiers can stretch the power budget to as much as 88 dB for single-channel transmission at 2.5 Gbit/s. In Figure 23.6, the use of two remote optically pumped amplifiers allows the cable around Jamaica to stretch a total of 350 km. (Transmitter powers can be higher than on longer cables because nonlinear effects do not build up over the short distance.)

Using remote amplifiers limits the number of fiber pairs that can carry traffic, because repeater housings can contain only a limited number of amplifiers. Thus a 100-km system with no amplifiers can carry signals on as many fibers as can fit in the cable (a few dozen in current designs), but a 300-km system would be limited to fewer fibers.

Unrepeatered submarine cables are widely used to connect the mainland with offshore islands, link islands with each other, or loop along the coast of a continent or large island. Most run only a few kilometers to tens of kilometers between islands or from the mainland to an offshore island. Examples are across the English Channel, or between islands in Hawaii, Denmark, Japan, or Indonesia.

Repeaterless systems can span longer distances by island-hopping, as shown in Figure 23.6. Another approach is to run a series of unrepeatered cables between coastal cities in a *festoon system,* such as the one around Italy shown in Figure 23.7. Laying offshore festoon cables to link coastal cities may cost less than burying cables on land—particularly where cities are along the coast and the terrain is mountainous, as in Italy.

REPEATERED SUBMARINE CABLES

Repeatered cables can fit fewer active fiber pairs.

Transmitting signals over spans longer than about 300 to 400 km requires submerged repeaters, which put important constraints on repeatered systems. The cable itself must carry electrical power from the termination points on shore. This electrical power transmission capability is limited, restricting the total number of amplifiers in the chain, and thus both the total distance spanned and the number of usable fiber pairs. Typical repeater housings can hold only 8 to 16 fiber amplifiers, also limiting the number of fiber pairs.

All repeatered submarine systems are effectively long-distance systems, although the total distances range from hundreds of kilometers to a total of more than 20,000 km in the longest systems. They are designed in various configurations. Some run thousands of kilometers between two points on opposite sides of an ocean, such as the east coast of the United States and the west coast of Europe. Others are loops or rings. Many large systems run along coasts, landing at a number of points, such as the SEA-ME-WE-3 (Southeast Asia-Middle East-Western Europe) system shown in Figure 23.8. They may include underwater branching points, where optical channels or whole fibers worth of signals are added

FIGURE 23.7
Submarine fiber cables link coastal cities in Italy. They are part of a network that includes land lines. (Courtesy of Corning Inc.)

and dropped. Many long systems like SEA-ME-WE-3 have intermediate landing points where signals can be regenerated if needed.

Submarine cables are designed from end to end as a unit.

The terminal points of submarine cables link with long-distance terrestrial cables and often with other submarine cables as well. One important difference between submarine and terrestrial long-distance systems is that the submarine systems are designed from end to end to function as units, while terrestrial links are part of larger networks built up of many cable spans. This mean that submarine designers can count on complete control of their entire cable system, and deploy different fibers along its length to optimize performance. As you learned in Chapter 22, this makes it possible to install fiber with large effective area near the

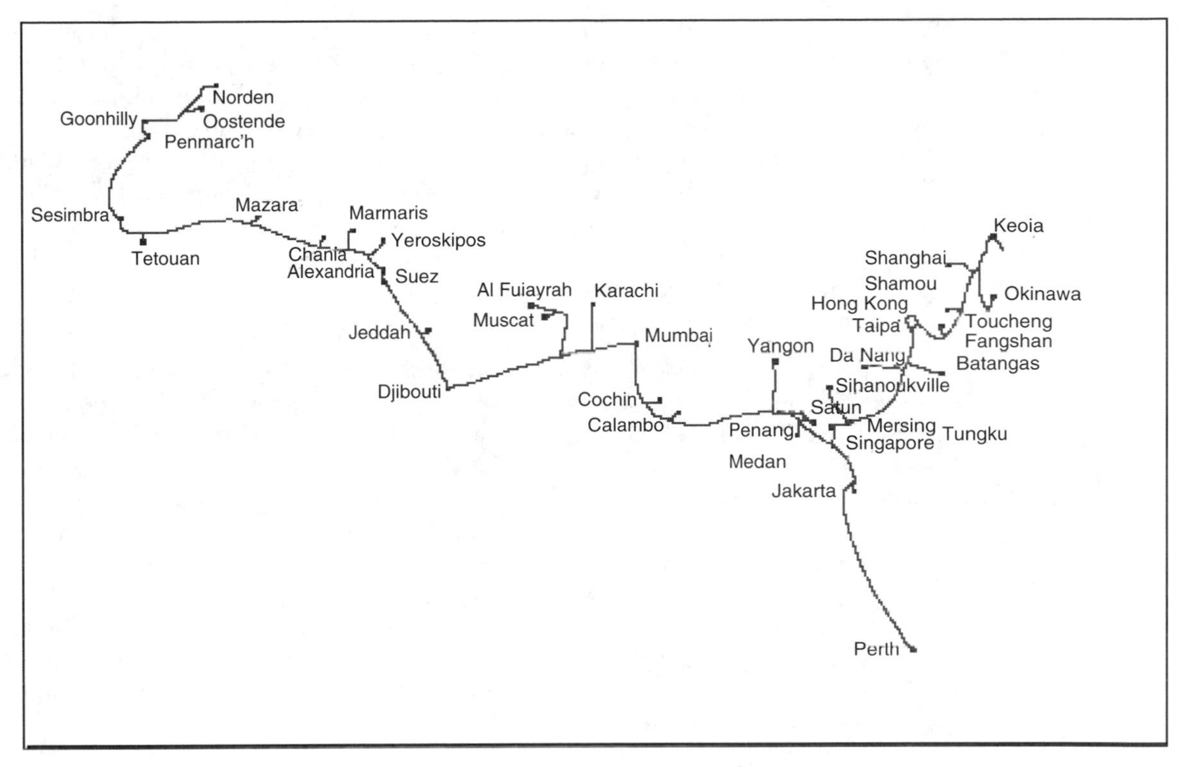

FIGURE 23.8

SEA-ME-WE-3 submarine cable system lands at many points from western Europe to Southeast Asia, spanning a total length of 27,000 km. (Copyright of Alcatel)

transmitter, to control nonlinear effects, while using other fibers elsewhere along a span. Designers of most terrestrial cables cannot count on this flexibility.

Another important difference is that submarine cables link points on islands and continents, but not in the ocean. All drops go to land. Terrestrial cables go through many sparsely populated regions, but farmers, ranchers, and rural villages need much more bandwidth than fish. Networks of terrestrial cables link many points throughout the regions they serve; submarine cables link points on the edges of the oceans.

Design constraints are tight for high-speed repeatered submarine cables.

Repeatered submarine cables spanning a few thousand kilometers face stringent design constraints to maintain signal quality. Gains of optical amplifiers in submarine cables are kept low—typically around 10 dB—to control noise from amplified spontaneous emission and nonlinear effects, which accumulate along the length of the cable. This limits optical amplifier spacing to about 50 km, roughly half the distance in terrestrial long-haul systems. Using higher input power and lower gain also helps to equalize power across the spectral range of the optical amplifier, which is critical in systems that may contain 100 or more amplifiers in series.

Precise dispersion management is crucial. In state-of-the-art submarine systems, a cable run between amplifiers includes three types of fiber. Large-effective-area fiber is used for the first part of the run after the optical amplifier to minimize nonlinear effects arising from the high-power levels. Typically a length of nonzero dispersion-shifted fiber designed for submarine use follows, with zero dispersion shifted to a wavelength longer than the erbium-fiber amplifier band. Then comes a length of standard step-index single-mode fiber. The dispersion in each fiber segment is large enough to limit four-wave mixing, but the overall dispersion is low enough to allow high-speed transmission across the erbium-fiber amplifier band. Raman amplification in the final fiber segment both preamplifies the signal for the optical amplifier and equalizes gain across the WDM spectrum.

The capacity of repeatered submarine cables has increased steadily, as shown in Table 23.1. The first systems transmitted at 1300 nm with 1550-nm single-channel transmission, starting in the early 1990s. WDM systems followed, with the first carrying 4 or 8 channels per fiber at 2.5 Gbit/s, and total capacity of 40 Gbit/s for the entire cable. The recently installed Pan American Crossing system has the capacity to carry 32 channels per fiber at 10 Gbit/s, but as in terrestrial systems not all the channels are populated with transmitters and receivers. So far, all systems work in the erbium-fiber C-band, where gain is flattest. The latest systems in development can carry up to 64 optical channels per fiber at 10 Gbit/s in the C-band amplifier window, for total data rates to 640 Gbit/s per fiber. For 8-fiber cables, that means maximum capacity of 7.68 Tbit/s. Upgradability is a big advantage in the WDM systems; operators can add transmitters and receivers at the terminal points to increase the number of optical channels as demand increases.

Table 23.1 concentrates on the largest and highest-capacity repeatered submarine cables. Many shorter repeatered systems have been installed in regions such as the Mediterranean and Caribbean. For example, the 1300-km Black Sea Fiber Optic Cable System links Bulgaria, Ukraine, and Russia with two fiber pairs operating at data rates of 2.5 Gbit/s with provision for WDM transmission. Their shorter lengths relax design criteria for these systems, because they have less opportunity to accumulate large cumulative dispersion and nonlinear effects. Generally these cables are international.

Undersea Optical Networking

Undersea branch points divide cables between multiple landing points.

Undersea branch points have long been used to divide submarine cable capacity between multiple landing points. TAT-8 divided off the French coast, splitting signals between Britain and France. Offshore branching points are used in cables such as SEA-ME-WE 3. The use of wavelength-division multiplexing makes it possible to add and drop individual optical channels, as shown in Figure 23.9, when an add-drop multiplexer is offshore.

This design is being used in the Africa One cable, which will loop around the continent as shown in Figure 23.10. When operations begin in 2002, the cable will initially carry two 10-Gbit/s optical channels on each of four fiber pairs, but can be upgraded to carry more channels as African requirements grow. The entire cable will come ashore at seven points—five in Africa, one in Portugal, and one in Italy—which are called "central offices." These are

Table 23.1 Initial capacities of some major undersea fiber cables.

System	*TAT-8*	*TAT-10*	*TAT-12/13*	*SEA-ME-WE-3*	*Atlantic Crossing 1*	*Pan American Crossing*
Operational	Dec. 1988	1992	1996	1998	1998	2001
Location	US–UK and France	US–Germany	US–UK–France–US loop	Germany to Singapore	US–UK–Netherlands–Germany loop	California–Mexico–Panama, Venezuela, St. Croix
Initial Data Rate per Fiber Pair	278 Mbit/s	565 Mbit/s	5 Gbit/s	2.5 Gbit/s per optical channel	2.5 Gbit/s per optical channel	10 Gbit/s per optical channel
Working Pairs	2	2	2 each half of loop	2	4 each half of loop	3
Fiber	Single-mode	Single-mode	Dispersion-shifted to 1550 nm	Zero dispersion at 1580 nm	Zero dispersion at 1580 nm	Nonzero dispersion-shifted
Repeater spacing	Over 50 km	Over 100 km	None	None	None	None
Wavelength	1300 nm	1550 nm	1550 nm	Up to 8 near 1550	4 near 1550 nm	Up to 32 near 1550 nm
Optical amplifiers	None	None	Yes	Yes	Yes	Yes
Total cable capacity	560 Mbit/s	1130 Mbit/s	10 Gbit/s	Up to 40 Gbit/s	40 Gbit/s	960 Gbit/s
Notes				Optical add-drop capability	Loop, upgradable	Not all channels installed initially

essentially terminations of segments of the cable. All other landings will be branches, with signals diverted from two fibers by a remotely controlled optical add-drop switch at an offshore branching node. A cable will run from that point to a cable station on shore.

Long-Haul Terrestrial Systems

Terrestrial cables are not as long as the longest submarine cables.

Long-haul terrestrial telecommunication systems, like long-distance submarine cables, carry high-speed signals and serve as backbones of the global telecommunications

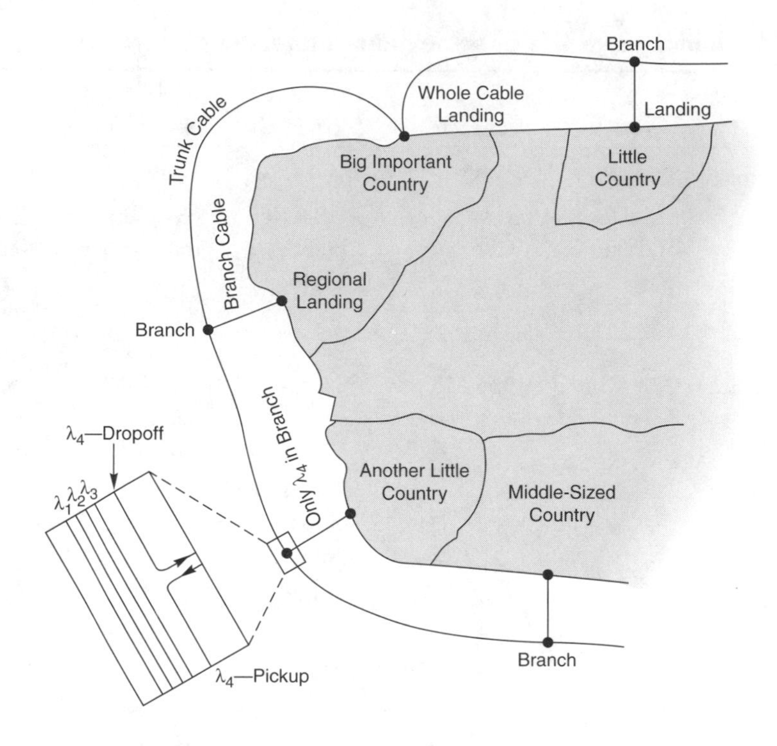

FIGURE 23.9
Undersea cable branch connection is an offshore add-drop multiplexer.

network. The same principles underlie the operation of submarine and terrestrial systems. The main differences are in the details. Terrestrial systems generally are part of a network mesh connecting major urban centers or telecommunication transport nodes, which are scattered across continents. This means that most long-haul terrestrial systems do not have to span the intercontinental distances of long-haul submarine cables. It also means that terrestrial networks often are installed piece-by-piece, rather than as entire systems. Terrestrial environments are much more accessible, making repairs and powering amplifiers much easier.

This gives long-haul terrestrial telecommunications systems a distinct topology, as shown in Figure 23.11 for a representative network operated by Qwest. Like a national railroad or interstate highway map, the main routes connect major population centers. In fact, many long-distance telecommunication lines run along railroad or highway rights of way. The nodes indicated are major interfaces with regional and local telecommunications networks, some of which are shown in the western United States where Qwest provides local telephone service.

It's worth comparing the long-distance backbone network with Qwest's Internet backbone, shown in Figure 23.12. The Internet backbone looks different because it represents Internet Protocol connections on a higher layer (layer 3) than the physical connections made through fiber in the long-haul network. The long loops between points such as Los Angeles and Dallas show that a connection is made between the two points, but don't try to show the physical route the signal takes. You can also see that IP signals may need to make two or

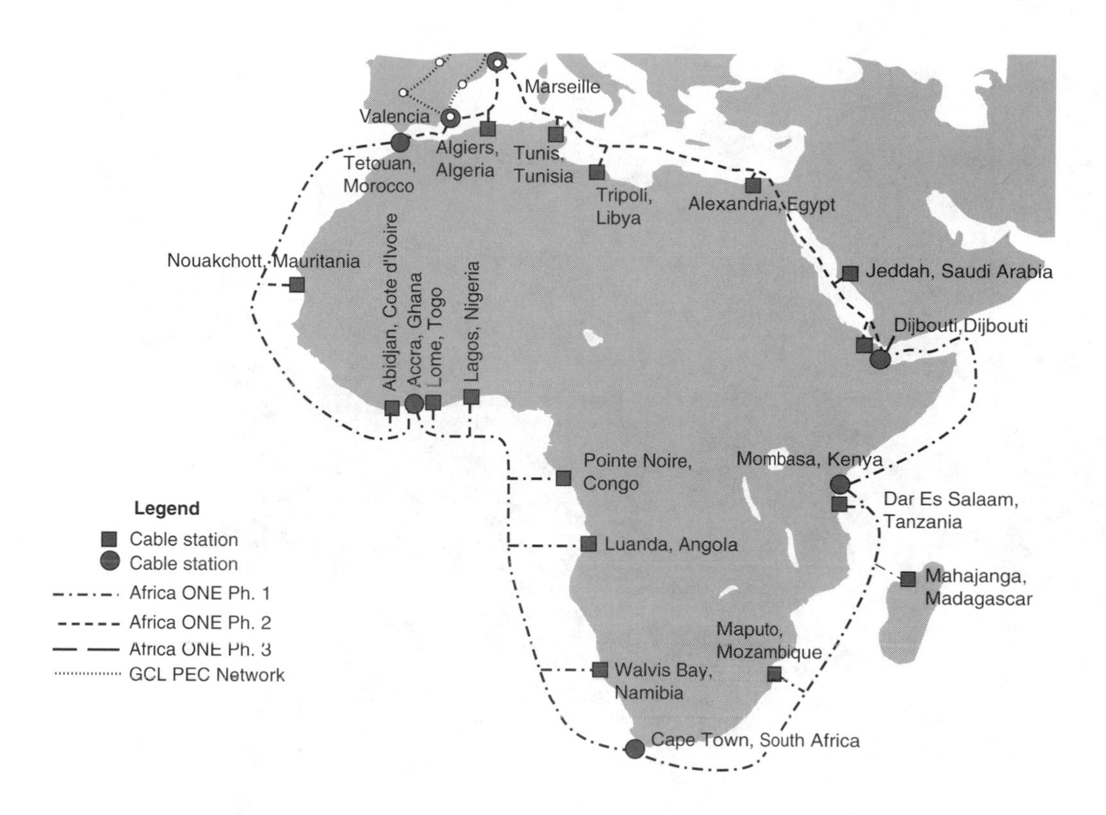

FIGURE 23.10

Africa One network circles Africa. (Courtesy of Africa One)

more hops even between major hub cities. For example, messages routed from New York to Seattle would have to go through Chicago or San Francisco. The routes shown on the map transmit at 155 and 622 Mbit/s, and 2.5 and 10 Gbit/s.

Transmission Rates and Distances

As you can see from Figures 23.11 and 23.12, long-haul terrestrial systems generally span shorter distances than submarine cables. Typical lengths are a few hundred to a thousand kilometers between hubs or nodes where the signals are regenerated and redistributed. All the points in Figure 23.11 are not terminal modes; most are nodes where there are add-drop multiplexers or other connections that allow the system to pick up and drop signals. Signals going across the country normally go straight through these points, without being converted to electronic form except at the largest nodes.

The shorter spans of land cables relax requirements on optical amplifiers.

These shorter cable runs relax many requirements on optical amplifiers, because noise, pulse dispersion, and differences in gain as a function of wavelength do not accumulate as much over such distances. Terrestrial cables may be able to use 25 to 30 dB of optical

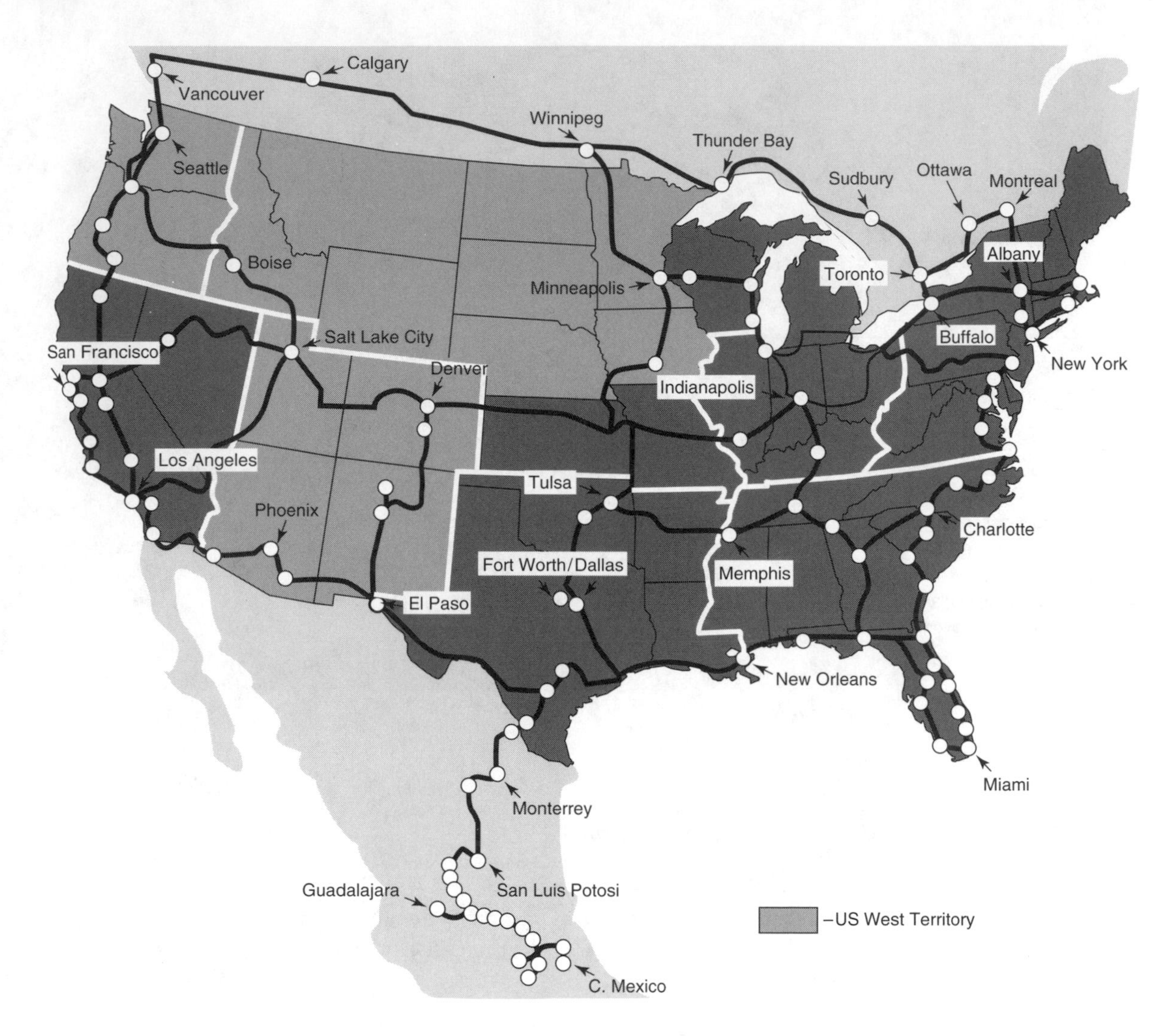

FIGURE 23.11

Qwest's North American fiber-optic backbone system, showing connections with the company's local service area in the western United States. (Courtesy of Qwest)

amplifier gain, instead of the 10 dB limit in transoceanic submarine cables. Thus instead of one amplifier every 50 km, you could have one every 100 km or more. The shorter distances reduce the accumulation of nonlinear effects. The shorter distances also ease the requirements on gain uniformity, so terrestrial systems can use a wider range of wavelengths in WDM systems, and thus transmit more optical channels.

Terrestrial systems also can accommodate more fibers because they do not suffer the electrical power and amplifier number limitations of submarine cables. The only limitations on the number of amplifiers are space for equipment racks and the ability to pump more elec-

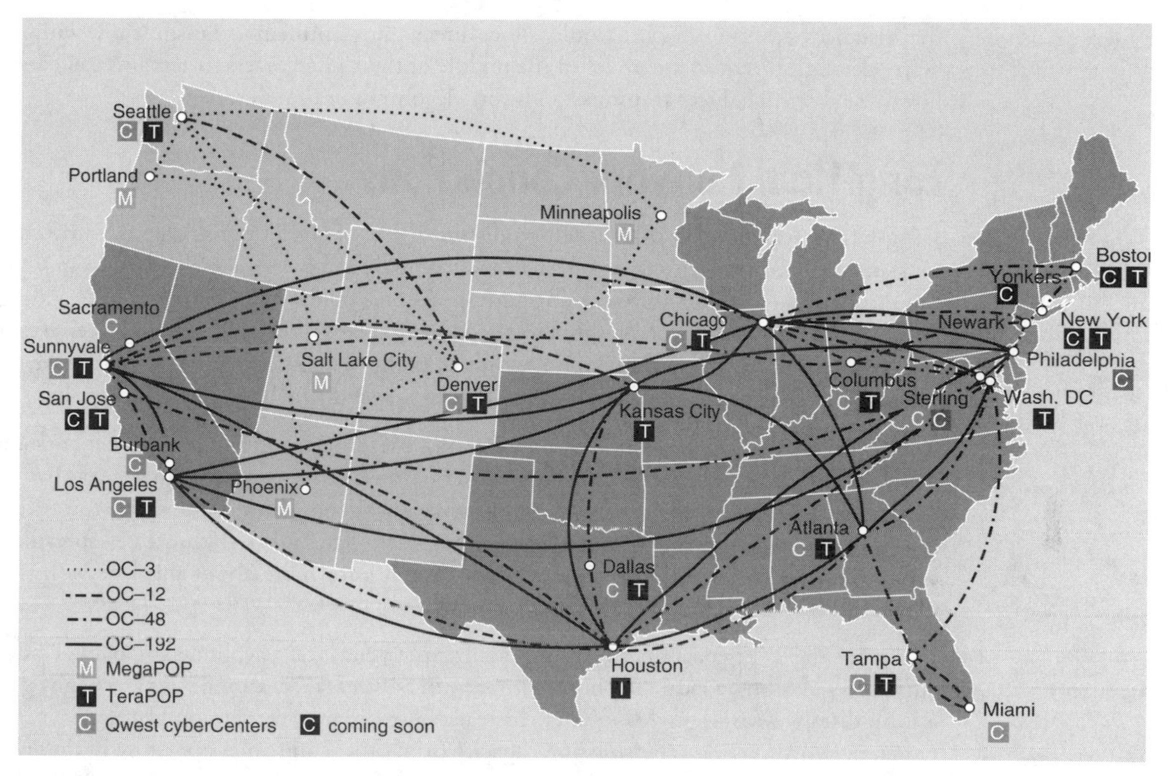

FIGURE 23.12

Qwest IP network in the United States. (Courtesy of Qwest)

trical power into the building housing them. This further raises maximum possible transmission capacity of a cable.

As in submarine cables, the optical channels in terrestrial systems operate at high speeds. These speeds have increased steadily since the first single-channel long-distance fiber networks were installed in the early 1980s. Those systems transmitted 400 Mbit/s at 1300 nm. By the early 1990s, the state of the art in commercial systems was 2.5 Gbit/s at a single wavelength. Today, long-distance terrestrial systems transmit 2.5 or 10 Gbit/s on each of several optical channels, the same speeds used in long-distance submarine cables. There is much talk of 40-Gbit/s channels, but they are likely to come first in the metro networks covered in Chapter 24, because fiber dispersion can limit transmission to relatively short distances. New high-speed systems typically use nonzero dispersion-shifted fiber to allow high-speed transmission on many optical channels.

If you count the total data rate through the cable, terrestrial long-distance cables carry more traffic than their submarine counterparts. However, the data rates on individual optical channels are the same.

The manufacturers of long-distance submarine cables are adapting their technology for transcontinental transmission on land. In principle, you can use the same system technology to transmit signals from Boston to San Francisco as from Boston to London because the distances are similar. This would allow direct transcontinental transmission, without going through intermediate nodes in the middle of the United States. It remains to be seen how desirable such direct transmission is for telecommunications carriers.

Long-Haul Network Connections

Long-haul land cables link to international and regional networks.

Long-haul terrestrial networks have two distinct types of connections. One is with other long-distance networks, such as international submarine cables. To make calls from Chicago to Berlin, for example, you need a terrestrial connection from Chicago to the landing point of a transatlantic cable, a submarine connection across the Atlantic, and a terrestrial connection from the European landing point to Berlin. These connections typically are on the east and west coasts, and on southern borders, depending where the traffic is going.

The other type of connection is to regional and metro telecommunications networks, a few of which are shown in parts of Figure 23.11. You'll be well aware of this if you live in South Dakota—the map shows no long-distance backbone nodes in the state. You access the long-distance network through the regional network shown as thin lines spreading across the state and linking to the backbone. You'll learn more about these networks in Chapter 24.

Note that these "regional" networks are not simply your local telephone company. Under current regulations, a local telephone company in South Dakota cannot transmit signals to a long-distance carrier in Minnesota. Instead, long-distance carriers build their own regional networks, or lease transmission capacity from other companies operating in the area. These fine tendrils of the network don't show up on the scale of these maps.

If you look closely at Figure 23.11, you will note that every major node in the United States is on a ring of cable. These are SONET-type rings, which provide insurance in case of equipment failures or cable breaks. For example, if a flash flood east of San Diego washed out the main cable from Los Angeles to Phoenix, traffic between the cities could be redirected through Salt Lake City, Denver, and El Paso.

Add-Drop Multiplexing and Wavelength Conversion

Long-distance terrestrial cables may add and drop signals at intermediate points.

Unlike long-haul submarine cables, long-haul terrestrial cables may need to make connections at intermediate points along their routes. Typically these are cities large enough to generate significant traffic, but not large enough to be hubs. This is done with add-drop multiplexers, which you learned about earlier.

Add-drop multiplexers can take various forms. They may be static, always directing signals in the same ways, or dynamic, able to switch signals in different directions. They also may split off the contents of an entire fiber, or individual optical channels in a fiber carrying WDM traffic. The choice depends on the type of system and the amount of traffic.

Normally signals are dropped at the intermediate location, and others added in their place. In WDM systems, this may require converting the wavelengths of some signals to wavelengths that are available in the through cable.

Wavelength conversion also may be necessary at hubs, where signals are switched in different directions and reorganized. For example, signals from both San Francisco and Portland may reach Salt Lake City at 1540 nm, but only one 1540-nm channel may be available to Denver. One of those signals must be converted to a different wavelength.

Types of Long-Distance Services

So far I have concentrated on the technology of piping high-speed data over long distances. You may be wondering about the structure of the industry that handles the job. That structure has been changing thanks to new regulations that have broken up old monopolies, and the growth of many new companies. A few years back, you could separate carriers into local, long-distance, and international. Now this is no longer possible. Some international carriers also own regional telephone companies and provide long-distance service in the United States. Companies have been sold and merged at a dizzying rate.

This isn't a book about the telecommunications business, so I won't go into detail, but you should recognize a few distinct services:

- The *public switched telecommunications network,* which has grown from the telephone network to provide service on a call-by-call basis. You use it to make long-distance phone calls and to send faxes over phone lines. Generally long-distance calls pass through two or more companies, and you shouldn't notice the difference.
- The Internet, which transfers data packets among users around the world. Most of its long-distance traffic goes over a set of fibers dedicated to Internet transmission.
- Private leased lines, which are transmission capacity that businesses lease on fibers from carriers whose business is providing that capacity. This can get complicated, because some carriers actually lease lines to provide part or all of their capacity. For example, long-distance calls from South Dakota to Minneapolis might go through a line that a long-distance carrier had leased from another company, which laid cable along the right of way of a gas pipeline. Sometimes carriers will even lease lines from each other to avoid the costs of building a pair of separate parallel transmission lines.

All these types of services also exist in regional telecommunications networks, covered in the next chapter.

What Have You Learned?

1. Telecommunications encompasses voice, data, facsimile, video, and other forms of communications, which are carried by a global network that includes fiber-optic systems, satellites, and other media.

2. The telephone network evolved into the backbone of the traditional global telecommunications network. It connects local, regional, and long-distance networks so you can dial phones around the world. Voice phones, faxes, dial-up modems, and pagers share a common dialing system.
3. The Internet was developed to connect computer networks. It transmits bursty digital data more efficiently than circuit-switched telephone lines.
4. Cable-television systems receive signals from distant sources for local distribution; some come over fiber, but most come from satellites.
5. Optical fibers form the backbone of the global telecommunications network.
6. Telephone systems transmit digital signals at a hierarchy of increasing data rates established by standards.
7. SONET packages input signals into digital frames, which are arranged to guarantee circuit capacity for telephone signals, and to accommodate computer data as well.
8. Internet data typically is transmitted at the same speeds as telephone traffic because the equipment is readily available.
9. Fiber transmission capacity usually is shared, with different users sharing the same fiber or even the same optical channel.
10. Long-distance submarine fiber-optic cables are the backbones of intercontinental telecommunications. The first was laid in 1988.
11. The undersea environment shapes the design of submarine fiber-optic cable. The cables must withstand high pressure, be extremely reliable, and have very high transmission capacity.
12. Unrepeatered submarine cables are the simplest types because they avoid submerged amplifiers. On-shore or remotely powered amplifiers can stretch transmission to reach 350 to 400 km. They typically link islands to each other or the mainland.
13. Repeatered submarine cables can span many thousands of kilometers. Limited electrical power and space in repeater housing restrict the number of active fibers. They often link continents.
14. Constraints on design of repeatered submarine cables are tight, limiting repeater spacing to about 50 km, and requiring precise dispersion compensation. Their capacity has increased steadily.
15. Submarine cables can include undersea branching points, which divide cable capacity among two or more landing points.
16. Long-haul terrestrial cables resemble long-haul undersea cables, but differ in details because of their environments. Generally, long-haul land cables span shorter distances than submarine cables, and their network topology is different.
17. Terrestrial cables can accommodate more parallel fibers than submarine cables.
18. Long-haul terrestrial networks link to regional networks and to international submarine cable networks.

What's Next?

In Chapter 24, you will learn about regional and metro telecommunication networks.

Further Reading

If you want to look at submarine and terrestrial cable networks in more detail, check out *http://www.cybergeography.org/* and its links.

For more on the telecommunications network as a whole, read

Roger L. Freeman, *Fundamentals of Telecommunications* (Wiley-Interscience, 1999)

To learn about design of the first transatlantic fiber-optic cable, see

Peter K. Runge and Patrick R. Trischitta, eds., *Undersea Lightwave Communications* (IEEE Press, 1986)

For more on modern submarine fiber-optics, see

Patrick R. Trischitta and William C. Marra, "Applying WDM Technology to Undersea Cable Networks," *IEEE Communications Magazine,* Vol. 36, February 1998, pp. 62–66

Questions to Think About for Chapter 23

1. The data rate of a SONET OC-12 carrier is 622.08 Mbit/s. If that corresponds to 8064 voice channels, and each voice channel is 64,000 bits per second, how much of that signal is overhead?
2. SONET frames are a fixed length. Internet packets are a variable length, with the header indicating the packet size. How does this difference relate to the difference between circuit- and packet-switching?
3. TAT-8 transmits 560 Mbit/s and began operation at the end of 1988. Atlantic Crossing 1 began operation in 1998 transmitting 40 Gbit/s on a parallel route. How much did data rate increase over that decade?
4. Remember that "Moore's Law" says that the capacity of integrated circuits doubles every 18 months. Judging from the increase in data rates on transatlantic cables in Question 3, what is the doubling time of fiber-optic capacity?
5. Using the results of Questions 3 and 4, if fiber-optic capacity continues to expand at its present rate, what will be the capacity of a transatlantic fiber-optic cable in 2005? How does that compare with the 960-Gbit/s capacity possible if all optical channels are activated on the PanAmerican Crossing cable?
6. In 1983, the peak data transmission rate in a terrestrial fiber-optic cable was 400 Mbit/s on a single fiber. In 2000, you could buy a system capable of transmitting

160 optical channels at 10 Gbit/s through a single fiber. How much an increase does that represent, and what's the doubling time?

7. Nobody actually installed 160 optical channels on a single fiber in 2000, but some carriers did install 32. What's the increase and doubling time for the installed systems?

Quiz for Chapter 23

1. What types of signals travel on the global telecommunications network?
   a. Voice telephone
   b. Digital data
   c. Facsimile
   d. Video
   e. All of the above
2. How are signals carried on the global telecommunications network?
   a. They are digitized and multiplexed to generate high-speed signals that can be routed long distances
   b. Analog and digital signals are carried on separate networks
   c. Local networks feed signals to regional networks, which route them to national backbone systems and international lines
   d. a and c
   e. a, b, and c
3. How does the Internet relate to the global telecommunications network that evolved from the telephone system?
   a. The two interconnect, and both carry digital data along separate paths.
   b. The telephone network carries only analog signals; the Internet transmits only digital data.
   c. The two are identical.
   d. The Internet has replaced the global telecommunications network.
   e. Only the Internet can carry packet-switched signals.
4. A single voice channel in the North American Digital Hierarchy corresponds to a speed of
   a. 4000 Hz.
   b. 4000 bit/s.
   c. 14,400 bit/s.
   d. 64,000 bit/s
   e. 1.5 Mbit/s.
5. How many T1 signals go into a SONET OC-3 signal?
   a. 84
   b. 96
   c. 672
   d. 2016
   e. 155 million
6. Which of the following signals can feed a SONET OC-3 system?
   a. ATM format
   b. Packet-switched Internet Protocol
   c. T3 from the Digital Telephone Hierarchy
   d. Multiple T1 circuits
   e. All of the above

7. How do unrepeatered submarine cables differ from repeatered cables?

a. Only repeatered cables contain electro-optic repeaters

b. Unrepeatered cables do not include any optical amplifiers

c. Unrepeatered cables do not include any optical amplifiers powered by pump lasers under water

d. Unrepeatered cables can transmit signals farther

e. Unrepeatered cables include copper wires, which transmit electrical power

8. Unrepeatered cables *cannot* be used for which of the following?

a. Festoon systems along the coast of a country

b. Transatlantic cables

c. Links across the English Channel

d. Island-hopping systems

e. A cable crossing 200 km of ocean

9. Which of the following techniques can stretch transmission distance of an unrepeatered submarine cable?

a. Optical postamplifier on shore to boost output power of the transmitter.

b. Remote optical amplifier powered by light from an on-shore pump laser.

c. Preamplifier onshore to boost input power before the receiver.

d. a and c

e. All of the above

10. Which of the following factors limits the number of fiber pairs usable in a repeatered submarine cable?

a. Fiber attenuation and dispersion

b. The need for dispersion management on all fibers

c. Limited space in repeater housings and limited electrical power for optical amplifiers

d. The core of the cable cannot be larger than a certain size

e. The number of optical channels used for wavelength-division multiplexing

11. A submarine cable has four fiber pairs, each carrying 2.5 Gbit/s at each of four wavelengths. What is the cable's total data rate?

a. 2.5 Gbit/s

b. 10 Gbit/s

c. 40 Gbit/s

d. 80 Gbit/s

e. 160 Gbit/s

12. Why is the amplifier spacing in a transatlantic fiber-optic cable limited to 50 km?

a. To limit the noise and unequal amplification of optical channels that accumulate over transatlantic distances

b. Electrical power for pump lasers is not available under water

c. Water pressure reduces the gain that can be achieved

d. Water pressure increases fiber attenuation

e. All of the above

13. Why would different types of fiber be used in the same submarine fiber-optic cable?

a. To reduce attenuation

b. To compensate for chromatic dispersion

c. To increase power levels that can be transmitted

d. To balance attenuation at different wavelengths

e. Because the factory ran out of the first type part-way through the cable

14. All WDM channels in a 5000-km fiber system must have power within 6 dB of each other in order for the system to operate properly. If amplifiers are spaced every 50 km, how uniform must their gain be across the WDM spectrum, assuming differences accumulate uniformly over the cable length?

a. 0.05 dB

b. 0.06 dB

c. 0.12 dB

d. 0.5 dB

e. 0.6 dB

15. How does the terrestrial long-distance network differ from repeatered submarine cables?

a. Terrestrial systems do not require optical amplifiers.

b. Terrestrial systems form a grid interconnecting many points on land.

c. Terrestrial systems connect termination points on the coast.

d. Terrestrial systems do not require electrical power.

e. Only terrestrial systems have branching points.

CHAPTER 24

Regional and Metro Telecommunications

About This Chapter

Regional and metro telecommunications bridge the gap between local networks that serve your community and long-distance telecommunication systems that form the backbone of the global network. Like unrepeatered submarine cables, they make relatively short connections on distance scales from a few kilometers to hundreds of kilometers. In this chapter, you will learn about their design, and how they link distinct local services such as fixed and mobile telephones with each other and the long-distance network.

Regional Network Structure

A regional or metro network is a distribution system, not a simple information pipeline between a pair of points. You can think of the regional telecommunication network as similar to the grid of main streets, which link neighborhood roads to limited access highways. The analogy works best if you think of rural areas, where small towns have a few local streets, numbered surface highways between towns, and the nearest interstate highway is an hour's drive away. The regional telecommunication network is the equivalent of the numbered country roads, which carry traffic between towns or to the nearest city.

Figure 24.1 shows how a regional telephone network might look in a rural area. The network is a grid of cables connecting the telephone switching office in each town. Low-capacity cables run between small towns, and from small towns to their larger neighbors. Larger-capacity cables run from the larger towns to the regional hub, a small city, where they connect with long-distance lines. The network resembles a hub-and-spoke

FIGURE 24.1
Regional telecommunication network in rural area.

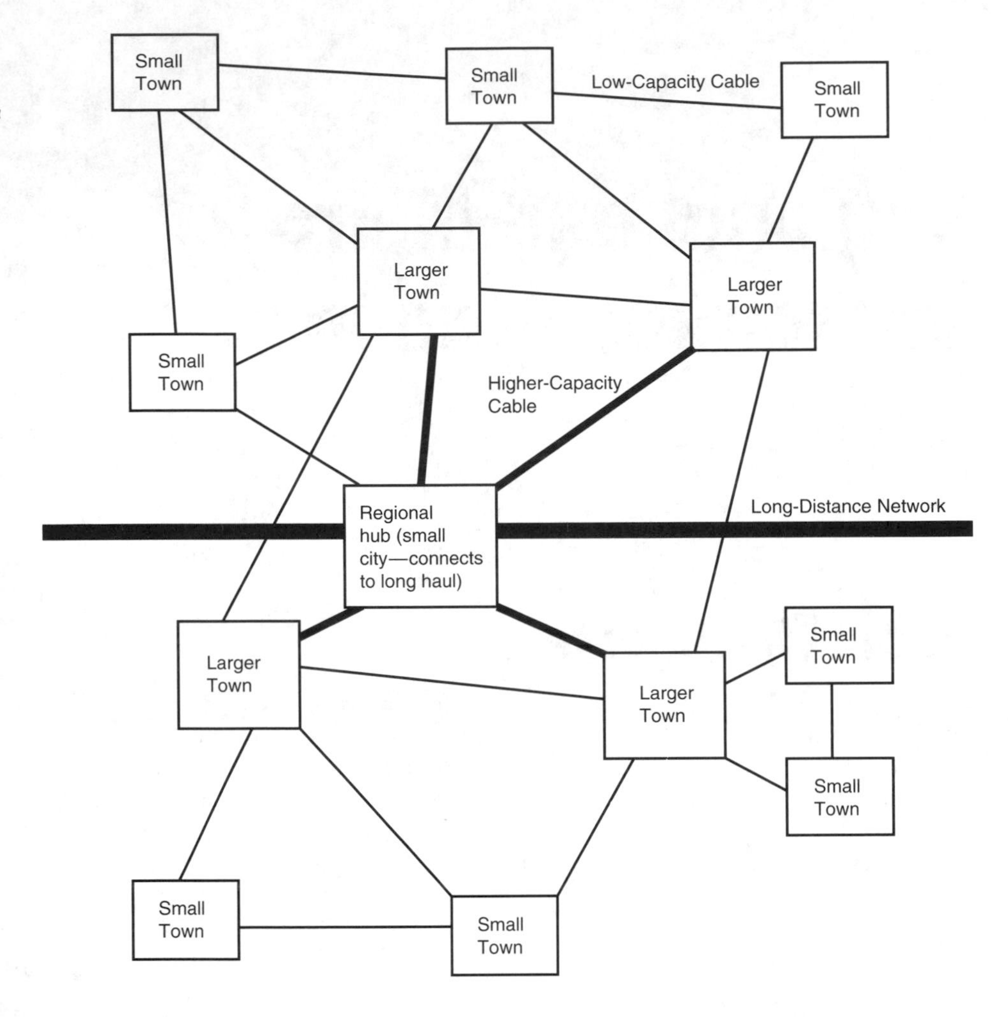

Regional phone networks connect local phone switches with each other and the long-distance network.

system, feeding traffic to larger and larger switches, but also includes cross-links and rings, which prevent single-point failures from knocking out service.

A regional network may have different levels, with a regional backbone spreading out to small cities from which services are distributed to towns. Figure 24.2 shows an example, the planned State of Oregon Enterprise Network, intended to serve both rural and urban areas. Smaller-scale networks like the one in Figure 24.1 spread out from county seats in rural areas. These branching networks have lower transmission capacity and operate at lower data rates.

Regional networks connect to local phone lines at telephone switching offices, which are the hubs for local phone service. As you will see in Chapter 25, you can trace all local phone lines to a switching office or central office, where they make connections with other local phone lines and with regional and long-distance networks. Regional networks also

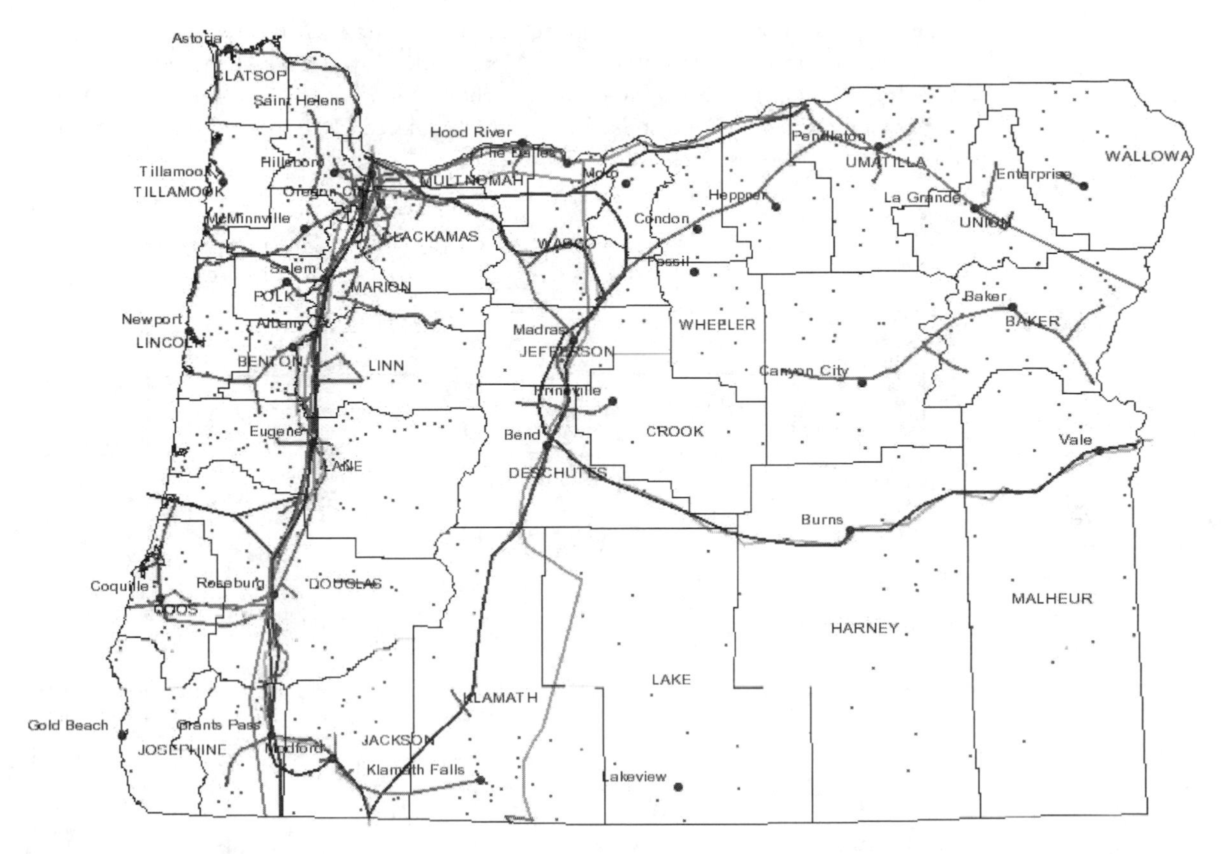

FIGURE 24.2
State of Oregon Enterprise Network is planned to provide links throughout the state. (Courtesy of Oregon Economic & Community Development)

connect with other local services, including cable television companies, mobile phone systems, and Internet Service Providers that rely on telephone lines for connections with the Internet backbone.

Regional network concepts also apply to urban and suburban areas, but in that case the transmission lines link adjacent suburbs, and feed into the urban center, where they connect with long-distance lines. The result is a network that looks like a more elaborate version of Figure 24.1, but with the nodes labeled suburbs and city center. The city center is the hub, with individual suburbs on spokes, and cross-links between adjacent suburbs.

A metro network interconnects many points in a metropolitan area.

You can think of a *metro network* as an updated vision of a regional network, nominally designed for use in large metropolitan areas. In a sense, it's just a new name for a regional network devised by trendy corporate marketing departments, but there are some distinctions. Unlike a classic regional telephone network, a metro network is a high-capacity system that interconnects with many points beside telephone switches, such as corporate and university campuses and Internet Service Providers. Metro networks often are structured as loops with add-drops at nodes of similar importance, as shown in Figure 24.3. Regional networks tend to have a modified hub-and-spoke network with links added to make them function like a mesh or array of rings, as shown in Figure 24.1.

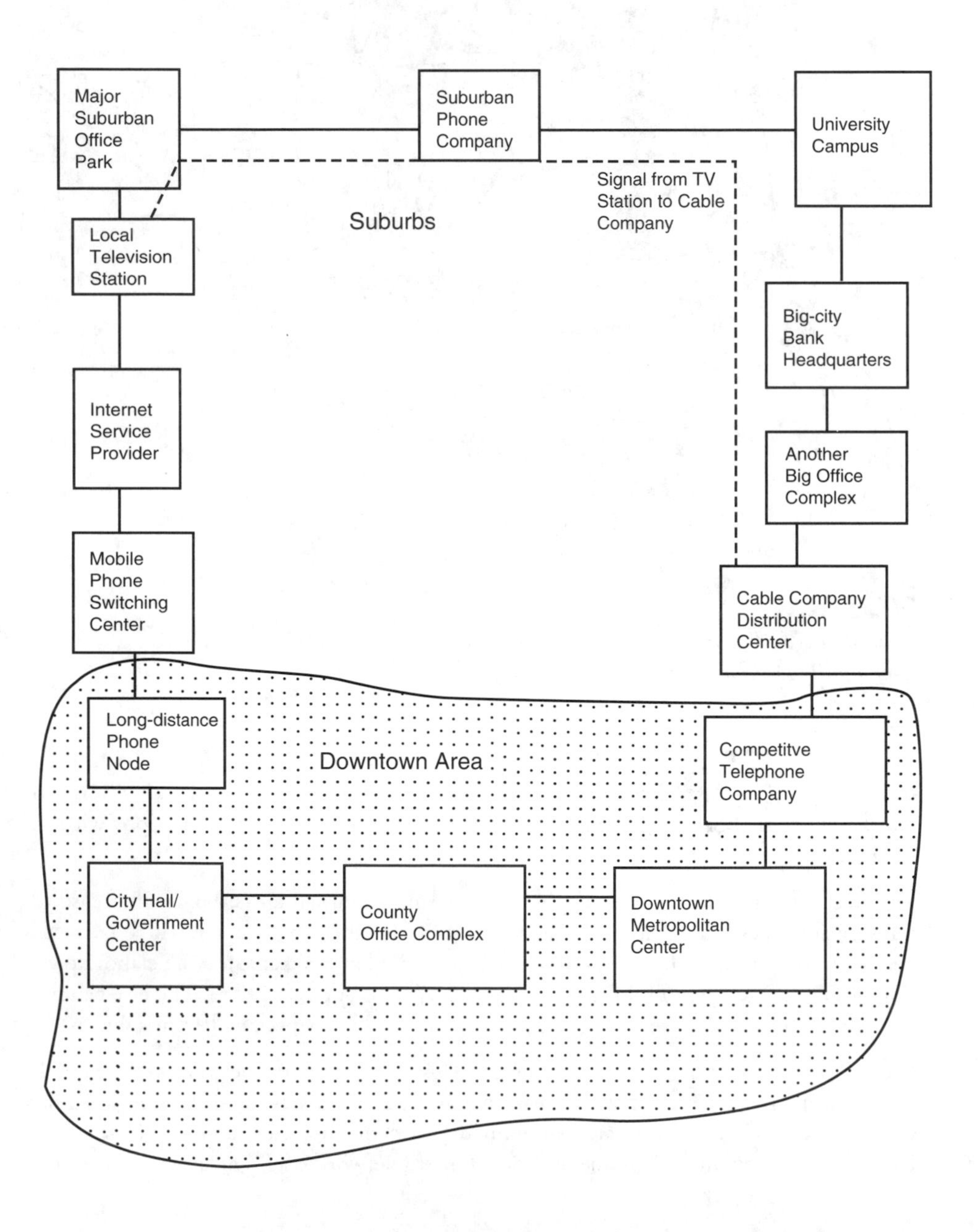

FIGURE 24.3
A metropolitan network links many sites.

Typical regional networks were installed piece by piece over many years by an established telephone company. They are part of the public switched telephone system, and their main business is carrying ordinary voice phone calls. In urban areas, they are becoming metro networks, with connections that go beyond the traditional telephone system. Meanwhile, other companies are building their own metro networks that provide generalized telecommunications services—that is, transporting any kind of signal from point to point in the area. The same cable may carry data from a big bank to its Internet Service Provider, phone calls from urban and suburban phone companies to a long-distance node, and video signals transmitted from a television station to a cable company that distributes video signals over its local network, as shown in Figure 24.3. The signals may go over the same fiber at different wavelengths or over different fibers in the same cable. Some may be combined into a TDM signal transmitted on one optical channel.

Metro networks are new systems; regional networks grow piece by piece.

The borderlines between "metro" and "regional" networks are hazy because both serve the same purpose of making connections within a region. The same building blocks can be used for both types, but systems in metropolitan areas typically require higher capacity to serve their customers. The rest of this chapter will look more closely at established and new regional and metro networks, then turn to the fiber-optic hardware used in short terrestrial systems.

Established Regional Telecommunication Networks

A regional telecommunications network both collects and distributes signals. On the collection side, it picks up signals from many points, packages them together where necessary, and delivers them to their destinations, which may be inside the regional network or elsewhere. On the distribution side, it receives incoming signals, sorts them out, and delivers them to individual destinations. Thus the regional network is more than an information pipeline; it's also a distribution system.

Regional networks are distribution systems.

A classic regional network began as part of the telephone system, which distributed phone calls in the area and still handles a large fraction of telephone traffic. It may also carry other signals, typically on different lines. These networks originally consisted of copper cables, but now are largely fiber-optic.

Regional Telephone Connections

The traditional telephone system was strongly hierarchical, and that legacy remains strong in regional network organization. In a conventional public switched telephone system, all phone lines in a community feed into a *central office* or *switching office* that switches signals among subscribers in that area, and directs signals to other areas to the central offices in those regions. The lines connecting central offices were traditionally called *trunk lines,* a term that is still in use. (Interestingly, the first major use of fiber optics in the phone system was as trunk lines transmitting 45 Mbit/s T3 signals, a speed for which other media were

Regional networks link switching offices; they tend to focus more traffic on the largest switches.

not well suited. The first systems, installed in the late 1970s, used graded-index fiber and transmitted near 850 nm, operating over 10 km or less.)

Originally, central offices were arranged in a hierarchy that depended on their size. Small central offices served small communities linked to large ones, which in turn linked to larger ones. Only the largest central offices connected to the long-distance network. This hierarchy is not as rigid today, but it remains evident in the rural network of Figure 24.1.

Changes in the telephone industry have complicated this picture in the United States. The regulations that broke up AT&T for the first time at the start of 1984 separated long-distance calls from "local" calls according to the area codes in effect at the time; most of those area codes have now been split into two or more area codes. Phone companies also are offering services beyond the traditional voice telephone lines, and these services are being merged into the digitized data streams generated by ordinary voice telephone traffic. New companies are offering local telephone service, and those services have to be transmitted through new or existing regional networks. Many local phone companies cannot offer long-distance service, so calls have to be transferred to long-distance companies to go from state to state and often shorter distances as well. Nonetheless, regional networks continue to perform the fundamental task of switching and distributing phone calls.

Legacy and New Networks

Regional networks are upgraded gradually to higher capacity.

Regional networks inevitably contain a mixture of old ("legacy") and new equipment. The old stuff generally isn't as old as it is in local phone systems, but regional carriers generally do not replace systems in good working condition as long as their transmission capacity is adequate. Normally they replace one segment of the network at a time, both to limit expenses and to avoid disrupting service.

Replacements generally occur where old equipment fails or more capacity is needed. Carriers try to avoid new construction wherever possible, so they may prefer to lay new cables in existing underground ducts, or add wavelength-division multiplexing and new optical channels to existing systems transmitting only one wavelength.

Generally, regional networks require less transmission capacity than long-distance systems. Thus regional networks make much less use of WDM technology than long-haul systems. Upgrades also are not evenly distributed. Although users are steadily increasing the demand for transmission capacity, the growth is not even. New capacity is most likely to be needed in areas where new development has outstripped existing systems.

Other Connections

Regional networks interconnect with other systems.

The regional networks built for fixed ("wire-line") telephone signals now carry a variety of other signals and must connect with other equipment. Mobile telephones have become common, so regional networks must connect with cell-phone towers and switching centers; this means links in each region ("cell") for each mobile phone carrier serving the region.

Competitive telephone companies also require connections. Many of them use the same local wiring as the dominant local carrier, but have separate switches, either at the dominant phone company's central office or another site. Regional networks also have to connect to telephone services provided by local cable-television companies.

Data services also require special connections. Internet Service Providers need to lease lines to make connections to the Internet. So do data-processing and computer centers at large companies and universities. Large companies also may lease lines to connect with other facilities in the area, which don't have to go directly through the switched telephone network.

Transmission Requirements

Transmission requirements in regional networks cover a much broader range of data rates and distances than in long-distance systems.

Regional networks transmit at a wide range of data rates.

The economics of long-haul transmission dictate that signals be packed as tightly as possible. You trade off the cost of the equipment to package signals against the cost of transmission. That means you have to amortize the extra money you spend at the transmitter and receiver against the money you save on fiber, cable, and optical amplifiers. If you have to send signals 5000 km across the ocean, you can afford to spend much more at the transmitter and receiver end than if the signal is only going 50 km.

Individual telephone circuits are grouped together for efficiency, but the degree of grouping depends on the traffic level. Small towns may generate only enough traffic to fill a few 1.5 Mbit/s T1 lines on the busiest days. Larger towns will generate signals at the 45 Mbit/s T3 rate or higher. Regional networks carry signals at data rates from the 1.5 Mbit/s T1 speed up to 2.5 or 10 Gbit/s, depending on customer requirements. For long-haul systems, it usually makes sense to time-division multiplex slow signals, combining multiple inputs to make a single higher-speed data stream. Regional networks usually multiplex to lower rates.

Metro Networks

Metro networks serve large metropolitan areas. Many have grown from traditional telephone networks. Others are new builds by nontraditional telecommunications companies with somewhat different business plans.

Metro telephone networks largely link local switching offices, from which telephone lines branch out to individual customers as you will learn in Chapter 25. Blocks of telephone lines leased by large business customers usually run through a switching office to the telephone network. The part of the network from the switching office to the customer is called the *access* network, while that from the switching office inward is called the metro.

Telephone networks in larger cities may have two layers, as shown in Figure 24.4. In this case *metro access* loops connect groups of switching offices, and these loops, in turn, link to a *metro core* network that serves the whole metropolitan area.

Other companies are building their own versions of metro networks. Many lease bulk capacity on their fibers to other companies in the area. As in the example of Figure 24.3, this

FIGURE 24.4
Metro core and access network in big city.

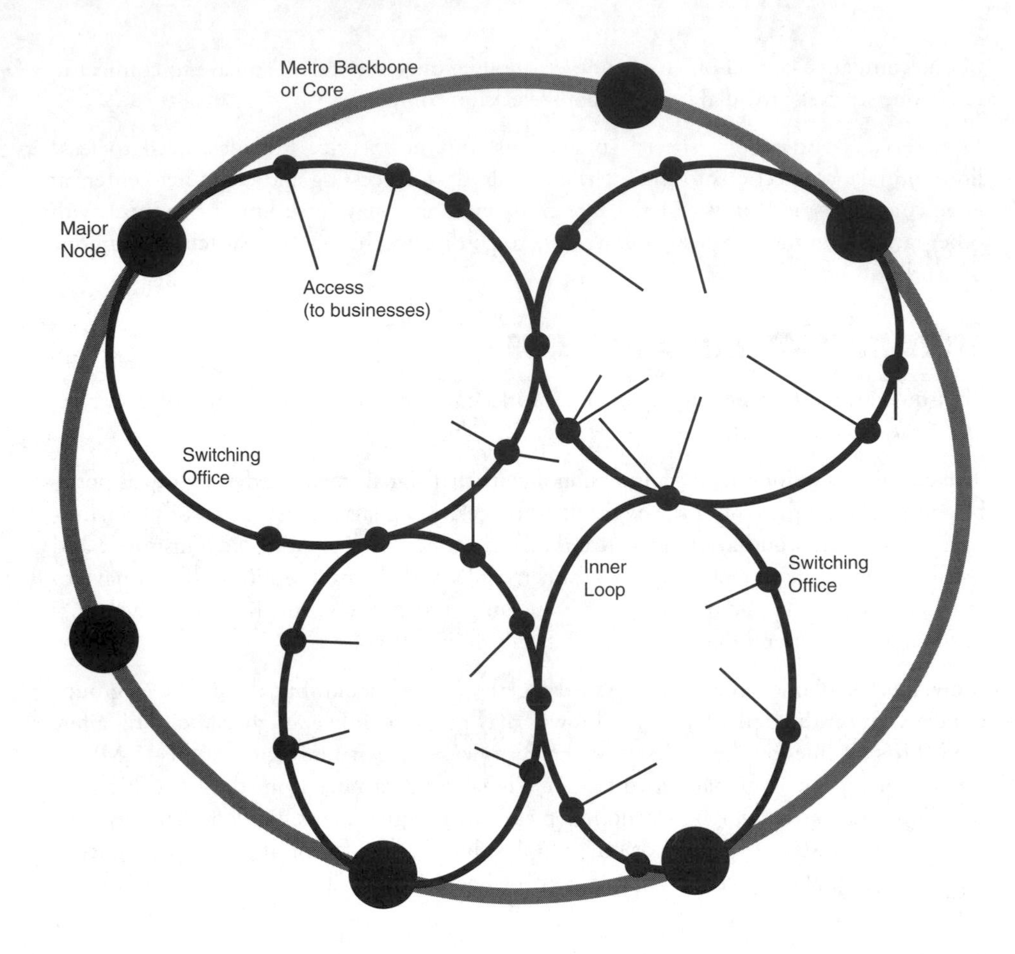

can include phone companies making connections among their own offices or to long-distance carriers, cellular phone companies, companies leasing private lines, or competitive phone companies that don't want to build their own networks.

Metro networks may transmit between pairs of points on the system.

Transmission often runs between two points on the metro network. In Figure 24.3, the cable company has leased capacity between its operational headquarters and a local television station. The television station sends its signal directly through the fiber, giving the cable company a clean signal for distribution through its cable network. The cable company may have leased lines between its headquarters and other local television stations; it may also have leased capacity to an antenna farm in the suburbs, which picks up satellite signals. Those signals always travel the same routes; they never need to go to other destinations.

Metro networks also can lease temporary capacity when customers want it. If the cable company wants to cover an event in City Hall, it can lease capacity for the day. If the local television station is covering a big trial at the county courthouse, it can lease capacity week by week.

Established phone companies build their metro networks incrementally, but other carriers may install their metro networks in one major construction project. New cables may contain hundreds of fibers, to provide the maximum capacity per construction dollar.

Metro networks typically lease capacity in larger blocks than regional networks. Some new carriers only offer high-speed services, from 155 Mbit/s (OC-3) to 10 Gbit/s (OC-192). Some metro networks have chosen to lease capacity as optical channels in WDM systems or as whole fibers, which customers can use to transmit data at whatever rate they choose.

Metro networks lease capacity in large chunks.

Regional/Metro System Design

The design of metro and regional systems differs in important ways from that of long-distance systems. Some obvious differences arise from the network structure. Metro and regional systems span much shorter distances, so they rarely need optical amplifiers except to compensate for lossy component such as add-drop multiplexers and couplers. This also allows the use of wavelengths where optical amplifiers are not available.

Metro/regional system design differs from long-distance systems.

Other differences are more subtle, such as the nature of connections and the use of WDM technology. The dichotomy between metro and regional systems described earlier also leads to some design choices.

Network Access

There are two distinct ways of accessing a metro or regional network, through a hub and through an add-drop multiplexer. We have glossed over these differences so far, but they impact system design.

Hubs or add-drops can provide network access.

A *hub* is a point where all (or most) of the signals in a system are switched and organized. Hubs include local switching offices, and correspond to the terminal points on submarine cables or long-distance systems. In the metro network of Figure 24.5, there are two hubs: the network operation center at the top and the urban telephone switching office at the lower right. Other metro networks may include several switching offices.

Add-drops are points where some signals are picked up and others dropped. An add-drop multiplexer along the network diverts only part of the signals to the node. In Figure 24.5, the only signal picked up from the television station is one video signal directed to the cable-television company, which will distribute the signal to its subscribers. The figure also shows a two-way link between City Hall and an Internet Service Provider, which operates the city's Web site and provides officials with Internet access.

The metro network in Figure 24.5 is a ring that contains many add drops and a few hubs. This means it functions largely as a network, interconnecting many nodes and directing signals between pairs of nodes. In a metro network, the signals often are transmitted separately, at different wavelengths or on different fibers. They may pass through add-drops, but not through the nodes attached to the network. In this they differ from other networks, which route a combined signal through all nodes, but each node only receives the signals

Metro networks typically are rings with many add-drops.

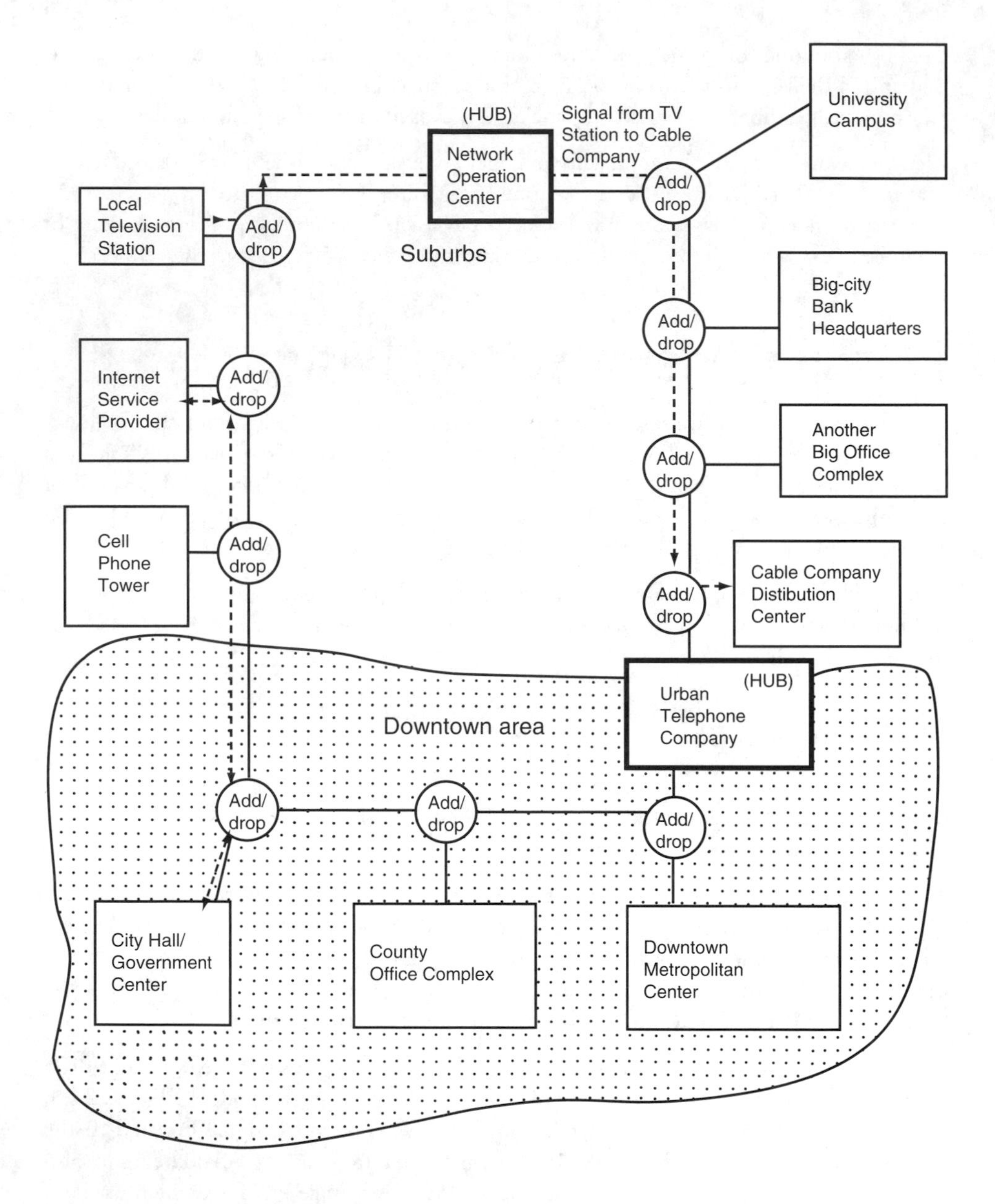

FIGURE 24.5
Hubs and add/drop nodes in a metro network.

directed to it. This subtle distinction can be important to some users, such as banks, that are particularly concerned with security.

Other metro and regional networks may contain mostly hubs. In Figure 24.1, you could consider all the towns to be hubs, some smaller than the others. You can also think of a regional telephone network as an array of point-to-point links, with each terminating at a telephone switch in a different location. Telephone companies often treat their regional or metro networks as a grid of point-to-point links, so they upgrade parts at different times.

Speed and Capacity Requirements

Transmission requirements and speed of regional and metro networks generally are smaller than those of long-distance systems, although data rates are steadily increasing, and the difference is shrinking.

A traditional regional telephone network is built on the assumption that signals come in small chunks—voice channels—which are combined by time-division multiplexing to make higher-speed signals. Small towns have a few T1 connections, larger towns have a T3 line, cities have OC-3 or OC-12 links, and so on up the hierarchy of data rates. The highest rates in regional networks are likely to be 2.5 Gbit/s OC-48 lines.

Metro networks are less granular in the sense that their input signals come in larger blocks. This reflects the scale of urban commerce. If you need only a single-voice phone line, you deal with the phone company. If you need a big data pipeline to carry 1 Gbit/s between your company headquarters and a remote order-processing facility, you lease that capacity in bulk.

Metro networks can divide their capacity in various ways. Some may combine signals at low data rates using time-division multiplexing, like a telephone network. Alternatively, they may lease customers capacity in units of one or more optical channels, or as "dark fibers," with the customers supplying transmitters and receivers.

WDM and Transmission Formats

The trend toward leasing optical channels or separate fibers in metro systems opens up the possibility of transmitting signals in different formats.

Traditional regional networks, like the telephone network as a whole, are based on transmitting all signals in one standard set of compatible formats. Each is designed to feed into the next in a hierarchy of transmission rates. This approach was designed for networks where transmission capacity was at a premium, so signals had to be packed as tightly as possible.

Wavelength-division multiplexing and installation of high-fiber-count cables shift the balance in metro networks. With transmission distance short and capacity relatively inexpensive, the interface electronics needed to convert signals to different formats may cost more than their improved efficiency would save. This has led to what are sometimes called *protocol-agnostic WDM systems,* which assign each customer an optical channel that they can use in whatever format they choose. In Figure 24.6, four customers each pick different formats: 2.5 Gbit/s OC-48, digital video, Gigabit Ethernet, and Fibre Channel. Each uses a fiber transmitter to generate a signal on the assigned wavelength, but the transmission formats differ.

Metro WDM systems may transmit different formats on different channels.

This can save money by avoiding the need for conversion on both ends. Suppose, for example, the television station in Figure 24.5 wants to relay digital video signal to the cable network. By leasing a wavelength of their own, they can avoid the cost of converting to a standard telephone industry format for fiber transmission. How much of an advantage this is depends on the situation.

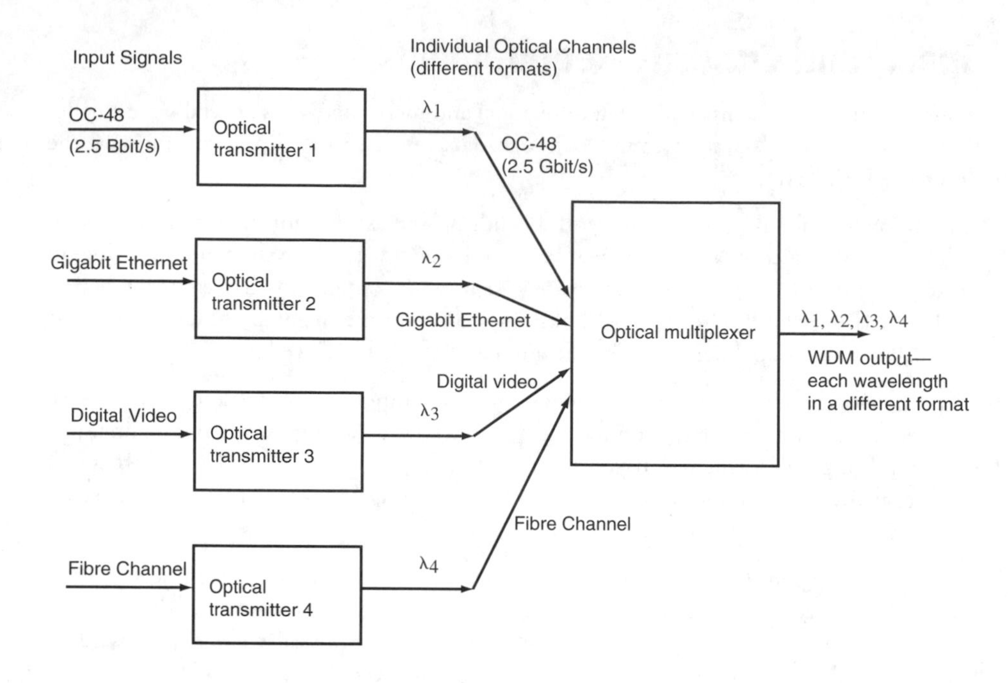

FIGURE 24.6 *Metro WDM system transmits optical channels in different formats.*

METRO AND REGIONAL TRANSMISSION DISTANCES

Metro and regional transmission distances are under 200 km.

Transmission distances in metro networks range from a few kilometers to about 200 km. Regional networks may span greater distances, especially in sparsely populated rural areas, but few individual links are much longer than 200 km, and most are much shorter. Most spans in both types of systems are under 100 km, and thus do not require optical amplifiers unless coupling losses are unusually high.

Higher than normal losses are possible if systems include many add-drop multiplexers on a single fiber. This can happen if WDM networks drop many wavelengths from the same fiber at different locations. This could require amplification, but such systems are not yet common enough for this to be a widespread problem.

Avoiding the need for amplification allows operation outside the erbium-fiber band, at wavelengths from about 1300 to 1530 nm where amplifiers are not now readily available. This gives designers the option of either packing the fiber with more wavelengths using dense-WDM, or spreading the same number of wavelengths across a broader range with less-expensive optics and transmitters. Most single-channel metro and regional systems operate at 1310 nm.

Dispersion effects are low because metro and regional systems are short.

Reducing overall transmission distance also reduces both cumulative nonlinear effects and cumulative dispersion. Decreasing the vulnerability to nonlinear effects allows higher transmission powers per optical channel and more optical channels. (Recall that nonlinear effects are proportional to both total power level and the distance the light travels at that power.) Reducing cumulative chromatic dispersion allows higher data rates. Chromatic dispersion increases linearly with distance, so a 100-km metro system experiences only one-tenth as much dispersion as a 1000-km long-distance system. Polarization-mode dispersion is proportional to the square-root of distance, so the decrease is only one-third for the metro system, but that can be an important improvement.

The lower dispersion and nonlinear effects combine to allow some metro networks to transmit at data rates to 40 Gbit/s. At this writing, that speed has yet to come into practical use in metro systems, but it appears to be technically feasible. The main drawbacks are the costs of the high-speed transmitters and receivers. Cumulative dispersion makes 40-Gbit/s transmission over long distances much more difficult, although impressive results have been demonstrated in the laboratory. On the other hand, the demand for such high capacity is limited, with many metro and regional systems transmitting at 622 Mbit/s (OC-4) or below.

SPECIAL-PURPOSE METRO EQUIPMENT

Special fibers are made for metro applications.

Some components are being fine-tuned for metro and regional applications. Most current systems use standard step-index single-mode fiber, but new fibers have been developed specifically for metro applications. One type has dispersion slope reduced to as little as 0.05 ps/nm^2-km, limiting the variation in dispersion across the operating range. The zero-dispersion point may be kept at 1310 nm, or moved to wavelengths longer than 1625 nm. Another type removes virtually all traces of hydrogen, to nearly eliminate the OH or water absorption peak that blocks transmission around 1380 nm in conventional fibers.

Paying a premium for fiber may help reduce other costs. Since metro DWDM systems operate at slower data rates than long-distance systems, controlling fiber dispersion should allow the use of lower-cost directly modulated laser transmitters.

What Have You Learned?

1. Metro and regional networks are distribution systems, not information pipelines. They serve similar purposes, but differ somewhat in detail. "Metro" has become a hot buzzword.
2. Regional phone networks connect local phone switches with each other and long-distance networks. They operate on different levels, ranging from low-capacity connections to small towns to high-capacity connections between suburbs and big cities.
3. A metro network interconnects many points in a metropolitan area. These include telephone switching centers and other facilities that require high-speed connections, such as large businesses, cable television companies, universities, and government offices.

4. Some metro networks are whole new systems; others have been built up piece by piece.
5. Regional networks link switching offices; they tend to concentrate traffic to the largest switches. They transmit signals at a wide range of data rates.
6. Originally built for fixed telephone service, regional and metro networks now carry a variety of other signals and must connect with other equipment, including pagers, cell phones, Internet Service Providers, and competitive phone companies.
7. Metro networks transmit high-speed traffic between pairs of points on the system.
8. Network access can be through hubs or add-drop multiplexers. Many metro networks are rings with many add-drops and few hubs. Regional networks typically have many hubs.
9. Regional networks generally transmit signals at a series of successively higher time-division multiplexed data rates. Much of a regional network operates at slower speeds than long-distance networks, but some links may operate at comparable rates.
10. Metro networks typically operate at higher data rates than regional networks.
11. Because metro and regional links are shorter than long-distance systems, you don't want to spend as much money on transmitters and receivers that pack the signals into the most compact format to save transmission capacity.
12. Metro WDM systems may transmit signals in different formats at different wavelengths.
13. Transmission distances in metro and regional networks typically are a few kilometers to about 200 km; most links are shorter than 100 km. Amplifiers are rare, but may be needed in sparsely populated rural areas or if coupling losses are high.
14. Regional and metro networks do not require optical amplifiers, so they can operate at wavelengths outside the erbium-fiber range of 1530 to 1610 nm.
15. Dispersion effects are low because metro and regional systems are short.

What's Next?

In Chapter 25 you will learn about local-telephone networks. Chapters 26 and 27 cover video and data networks.

Further Reading

John C. Ballamy, *Digital Telephony,* 3rd ed. (Wiley-Interscience, 2000)

Yi Chen et al., "Metro Optical Networking," *Bell Labs Technical Journal* (January–March 1999) pp. 163–186.

Roger L. Freeman, *Fundamentals of Telecommunications* (Wiley-Interscience, 1999)

Questions to Think About for Chapter 24

1. Telephone calls that go outside of your community pass between your local-telephone switching office and other switching offices on the regional telephone network. What makes some of these long-distance calls?
2. A fire destroys the local switching office in one of the "larger towns" of Figure 24.1. How can the smaller towns connected to it still receive phone service?
3. Your company needs more transmission capacity between two offices 20 km apart. You can lease wavelengths at $20,000 per wavelength per year on a metro network with a five-year lease. You need to transmit three channels, two containing Gigabit Ethernet, and one containing TDM phone signals at 622 Mbit/s. They could fit on a single 2.5 Gbit/s line, but you would need to buy a time-division multiplexer and demultiplexer for the job. What's the most you could pay if you have to amortize costs over 5 years?
4. You're dealing with the same metro network as in Question 3. Their base rate is $1000 per kilometer per wavelength per year. What's the most you would pay for a time-division multiplexer and demultiplexer if you were sending signals only 5 km?
5. A metro network uses low-water single-mode fiber with zero dispersion at 1310 nm, and dispersion of 17 ps/nm-km at 1550 nm. If it transmits signals 100 km, what is the maximum data rate it can handle in the 1550-nm window with a transmitter having linewidth of 0.1 nm? Assume polarization-mode dispersion is

   insignificant, and that the maximum data rate is

$$\text{Data rate} = \frac{0.7}{\Delta t_{\text{maximum}}}$$

6. You're using nonzero dispersion-shifted fiber in a metro network. If the chromatic dispersion is 1 ps/nm-km and the source bandwidth is 0.1 nm, how far can you transmit signals at 40 Gbit/s? Assume you can neglect polarization-mode dispersion.

Quiz for Chapter 24

1. The best analogy for a regional telephone network is
   a. a pipeline pumping information from coast to coast with no detours.
   b. a superhighway crossing a large unpopulated area with few off-ramps.
   c. surface main streets connecting rural towns.

d. residential streets in suburbia.

e. a radio station broadcasting signals in all directions.

2. Which of the following is *not* connected directly to regional telephone networks?

a. Home and office telephones

b. Telephone switching offices

c. Cellular telephone networks

d. Long-distance phone lines

e. Internet service providers

3. A metro network is a(n)

a. high-speed subway system in Washington, DC.

b. mesh connecting many telephone switching offices.

c. mesh connecting major cities across the country.

d. loop interconnecting many points in an urban area.

e. antiquated idea that preceded regional telephone networks.

4. The hubs in a regional telecommunications network are

a. connections to international phone lines.

b. points from which cable television signals are distributed.

c. switching offices in towns and cities served by the network.

d. customers connected by add-drop multiplexers.

e. individual phone lines.

5. A regional telecommunications network serves

a. small towns in rural areas.

b. small cities and county seats in rural areas.

c. suburban communities.

d. large cities.

e. all of the above

6. Subscribers to metro networks may include

a. cable television operators.

b. regional telephone companies.

c. large businesses.

d. state and local governments.

e. Internet service providers.

f. all the above, if they have enough money.

7. A regional telecommunication network is a(n)

a. mesh of point-to-point links between hubs.

b. loop circling a large city.

c. hub-and-spoke system with individual towns connected only to the central point.

d. single loop circling a state.

e. obsolete concept.

8. A typical distance spanned by a metro network is

a. under 5 km.

b. a few to 200 km.

c. 100 to 200 km.

d. at least 200 km.

e. 200 to 1000 km.

9. Which of the following is true for the structure of signals on a regional network?

a. All signals must be at the same speed

b. Signals can be transmitted at a hierarchy of speeds from T1 up

c. Signals cannot be transmitted in different formats on the same fiber
d. Signals always are transmitted at 2.5 Gbit/s
e. All signals are time-division multiplexed together for transmission in a single data stream

10. What does "protocol agnostic" refer to?

a. A metro WDM system that transmits signals in different formats on different wavelengths through the same fiber, depending on customer requirements
b. A metro WDM system in which all signals in each fiber are transmitted in the same format on different wavelengths
c. Simultaneous transmission of the same signal on many different wavelengths for redundancy
d. A regional network that accepts many formats and converts them all to a common form for transmission
e. An engineer who believes all transmission formats are equivalent and can't decide which one to use

11. A metro DWDM network operates over a cable containing 864 individual fibers. Each fiber can transmit 100 optical channels at 2.5 Gbit/s. What is the maximum capacity of the entire cable for two-way transmission?

a. 250 Gbit/s
b. 2.16 Tbit/s
c. 86.4 Tbit/s
d. 108 Tbit/s
e. 216 Tbit/s

12. What makes it possible to transmit 40 Gbit/s on one optical channel in a metro network?

a. Dispersion compensation
b. Short transmission distances
c. Low nonlinear effects
d. Transmitting only a single wavelength in the fiber
e. Nothing; it's impossible

CHAPTER 25

Local Telephone or "Access" Networks

About This Chapter

The most visible part of the telephone system is the local network, connecting individual phones to the local switching office. Technological and industrial changes are reshaping the local phone network, as well as the rest of the telephone system. This chapter introduces you to the local phone system, sometimes called the access network, and shows how it serves home, business, and other users. It will explain the changes gradually spreading through the system, and the increasing role of fiber optics in delivering services first to large customers and eventually to homes.

Structure of the Local Phone Network

The local telephone network distributes signals to and from individual users. Traditionally, it was called the *subscriber loop,* industry jargon for the wires that form a circuit or loop from the local switching office to your home phone. At the switching office, local lines connected to a switch that could route signals to other local lines, the regional phone network, or the long-distance network.

That picture has become more complicated as the network has grown. The modern version is sometimes called the *access network,* because it gives individual subscribers access to the global telecommunications network. It does not always route all the signals through a traditional switching office. Much of the traffic may not even be traditional voice-switched telephone signals. In this modern topology, the access part of the network runs between the individual subscriber and the *network edge*—the point where signals enter the regional or global network.

The subscriber loop distributes telephone signals to individual users.

The main difference between the two views is that the access network includes more options and services. In fact, the access network is no longer purely a telephone network because it carries other services besides traditional voice telephony. However, it is largely an outgrowth of the old local telephone network, and retains a common structure that goes deeper than the old black dial telephone that has been sitting in your grandmother's parlor since 1952. To understand that structure, let's take a look at it step by step, moving out toward the user.

The Network Edge

The network edge is the interface where calls are directed into the global network.

The "edge" of the network is the point where signals enter the regional or global telecommunications network. As Figure 25.1 shows, the edge includes traditional local telephone switching offices as well as mobile telephone switches, Internet Service Providers, and large organizations that make heavy use of telecommunication services. The figure greatly oversimplifies, by hiding the core of the network and covering the edge in only one region. The central point is that equipment on the edge serves as a link between individual users and the telecommunications network as a whole, whether the user has a mobile phone, is a student in a dormitory, or is an employee of a big company.

The edge is an interface from which calls are directed. Switches in a central office direct individual telephone calls elsewhere in the telecommunications network. Switches operated by mobile phone carriers and competitive phone companies serve the same function. Like-

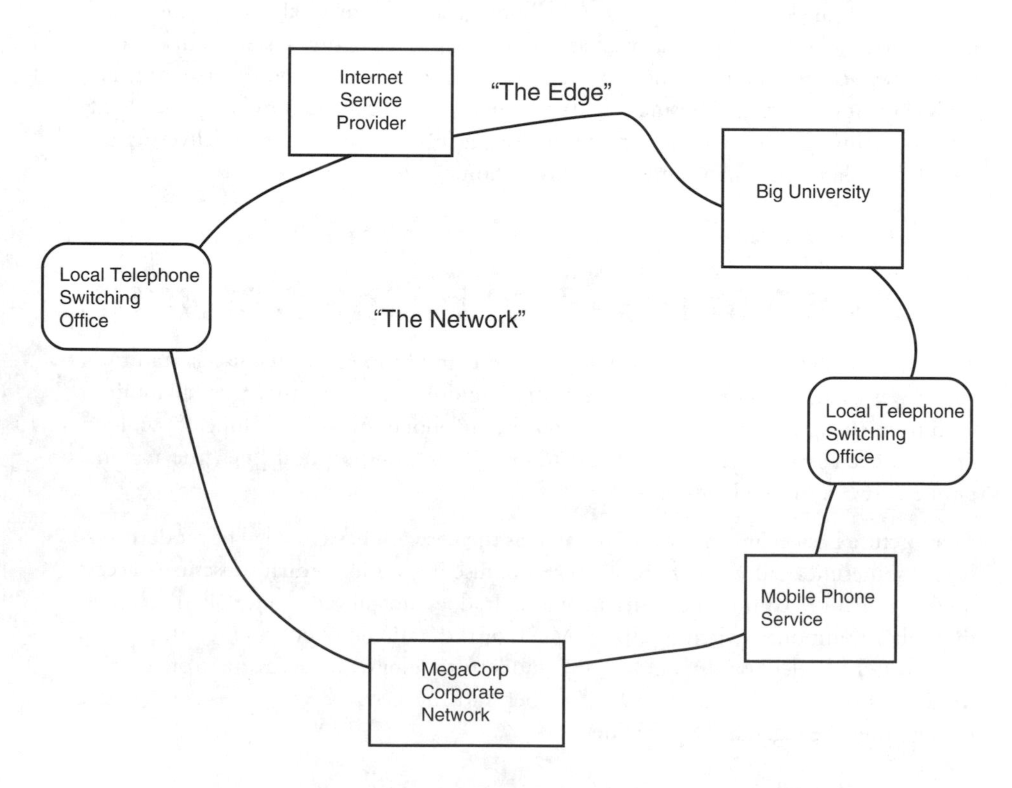

FIGURE 25.1
The network edge.

wise, an Internet Service Provider directs data packets, typically to routers somewhere inside the network. Corporate and university networks have come to serve similar functions internally rather than directing signals through a telephone switching office or Internet service provider.

The Telephone Switching Office

The telephone *switching office* or *central office* is the telephone company facility that provides connections to and from individual telephone customers. You can think of it as a central point in a community from which telephone services radiate outward to individual users. Figure 25.2 shows the concept, which dates back to the early days of the telephone industry.

A switching office makes connections to individual customers.

Cables from homes run directly to the switching office or to an intermediate node. At nodes, individual phone lines may be multiplexed together to generate a single digital signal for transmission from the node to the switching office over fiber or over copper. Typically homes near the switching office are served directly from there, but homes farther away—particularly clustered in new developments—may be served from a local node or concentrator that distributes signals from a remote point.

The small businesses in the business district receive their service through a local distribution node. The insurance office needs the capacity of a T1 line but most other businesses, like the shoe-repair shop, need only one or two ordinary phone lines. The Town Hall has its

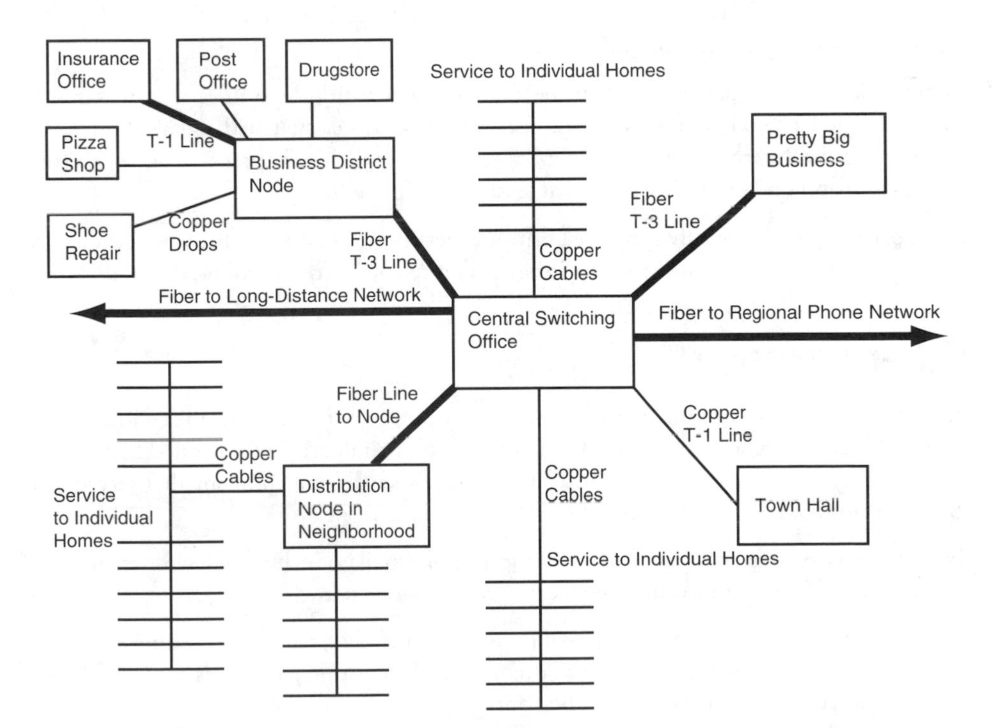

FIGURE 25.2
A telephone switching office has a central role in distributing signals.

own T1 line direct from the central office. The biggest business in our little town has its own T3 fiber-optic connection to the switching office.

All these services are processed through the central office. If the town manager calls the shoe repair shop to see if her shoes are finished, the call goes through the central office, where a switch connects one of the town phone lines to another phone line going to the shoe shop. If someone wants to call overseas, the switching office directs the call to the long-distance network, which makes the overseas connection. Incoming calls from the regional phone network go through the switching office, which makes the connections needed to direct them to the proper destination.

Telephone switching offices do not have one outgoing line for every customer because normally all phone lines are not in use simultaneously. Telephone companies decide how many output lines to allocate to a switch, based on statistical averages of usage.

"Access" Customers

All local traffic does not have to pass through a traditional switching office.

Changes in the telecommunications network mean that local traffic need not pass through a traditional telephone switching office. Figure 25.1 showed a number of other organizations on the edge of the network, such as large corporations and universities. These organizations could have their own switches and be hooked directly to a metro network, as seen in Figure 24.3. Their telephone traffic could go directly to long-distance carriers, to regional carriers, or over lines that they leased from regional carriers for corporate use. For example, a state university might have leased lines from administration offices in the state capitol to campuses around the state.

Internet Service Providers are also on the edge of the network. Instead of sending circuit-switched signals to telephone switches, they generate packet-switched signals, which are routed over the Internet Protocol network. As you learned in Chapter 23, some Internet traffic winds up on leased phone lines, at least for the present.

Large organizations typically generate both Internet and telephone traffic, which they may transmit over metro networks to their Internet and telephone connections.

Other Carriers

Cell phone and competitive phone companies have their own switching offices.

Back in the days of telephone monopolies, every community had exactly one central office, operated by *the* telephone company. In urban and suburban parts of the United States, that was generally AT&T, but in small towns it was often a small local company that served just that community, and maintained links to AT&T's regional network.

Two trends have changed that pattern: the spread of mobile or cellular telephones, and the introduction of competition into the local fixed telephone network.

Mobile telephone carriers have their own networks. Instead of cables distributing signals to each phone, they have many towers or antennas, each transmitting signals through the air in its own "cell." Cables, generally fiber-optic, connect each antenna in the region to a switching center, as shown in Figure 25.3. The antennas function like distribution nodes in

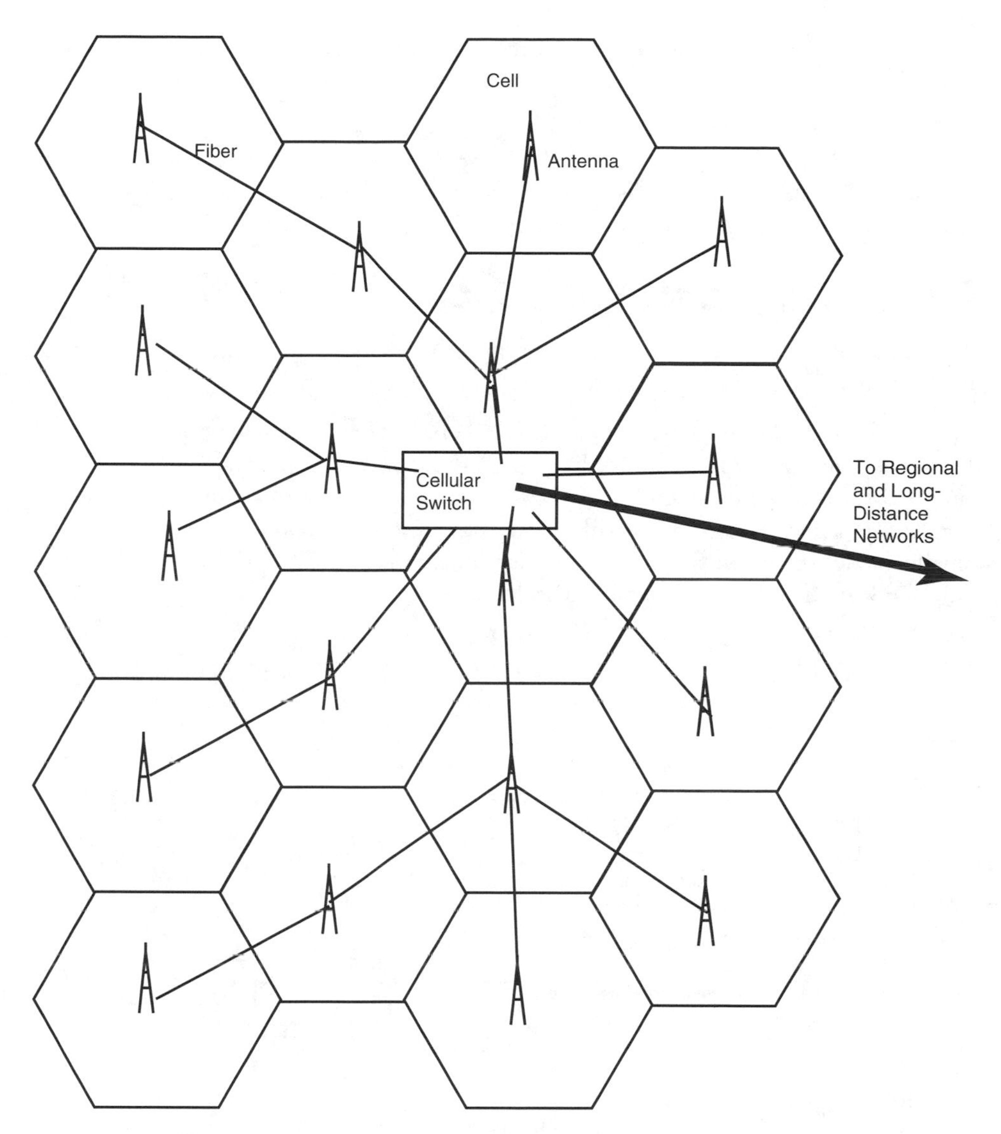

FIGURE 25.3 *Cell-phone network.*

the fixed subscriber loop, receiving signals from the switch and relaying them to individual destinations. The cellular network gets more complicated if you're moving, because the system must coordinate signals in adjacent cells to transfer your conversation smoothly, but otherwise the operations are similar. Because this is a book about fiber optics, we won't look closely at the cellular network.

Competitive telephone carriers may or may not have their own networks. Most of them actually lease lines to homes and businesses that were installed by the dominant local carrier. (The Federal Communications Commission has rules requiring and regulating the practice.) These lines feed into the dominant telephone company's central office. Other lines also feed into the central office; however, they connect to different switches, either in that building or in a separate building. The switches may not be as big and complex as those in a standard central office, but they serve the same function.

A few competitive carriers build their own separate networks running along the same poles or underground ducts as the established telephone company's lines. As you will learn in Chapter 27, cable television companies can adapt their cables to carry voice telephone signals and data services, creating a separate physical telephone network.

Distribution, Concentrators, and the Subscriber Loop

Cables distribute signals from the central office to subscribers.

Twisted pairs of copper wires connect to telephones.

A network of cables distributes signals from the central office to subscribers. The structure of this network varies from place to place.

Figure 25.4 shows a typical network structure. Homes near the switching office are served by copper wire pairs running directly to the switch, as shown at top. Thick cables with hundreds of wire pairs run along main streets to service boxes or manholes, where the cables are split into smaller cables with dozens of wire pairs, as shown at bottom left. The smaller cables run down side streets and distribute service to individual homes. The wires running from these cables to homes are called *drops.*

The copper wiring used in these cables, called *twisted pairs,* traditionally are a pair of wires twisted together to reduce the noise they pick up from other sources. Flat ribbon cables

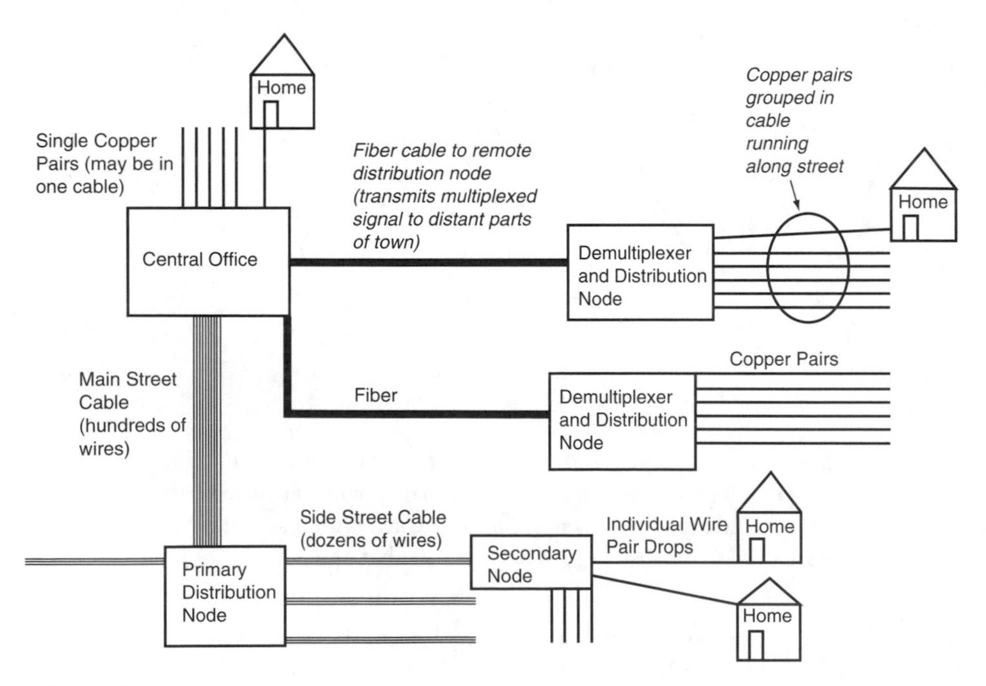

FIGURE 25.4
Subscriber loop distribution.

with two or four lines are used over short distances, such as in household telephone wiring, but long runs of untwisted wiring can act as antennas, and pick up noise like radio signals. Each wire pair carries traffic for a single telephone circuit or phone line.

Distant homes are served by multiplexing signals at the central office and sending the multiplexed signals through fiber-optic cables to remote distribution nodes, as shown at the right in Figure 25.4. Although the scale of the figure doesn't show it, typically, these fiber cables run a few miles. Copper cables run from the remote distribution node to homes, with fat cables subdividing into smaller cables with lower wire counts on side streets, as shown near the central office in the figure.

Copper wires can carry voice telephone signals several miles without serious degradation. However, the maximum transmission distance is shorter for other services, particularly Digital Subscriber Line (DSL), which I will cover later in this chapter. That has led to some refinements in design of subscriber loops.

Copper wires can carry voice, but not DSL, for several miles.

In the 1980s and early 1990s, telephone companies installed systems called *digital loop carriers* that multiplexed together many voice channels and distributed them at remote locations, as shown at the right of Figure 25.4. This seemed like a good idea at the time, because a single fiber transmitting at the modest 6.3 Mbit/s T2 rate could carry 96 telephone circuits. The fiber cables were much smaller, cheaper, and more sturdy than copper cables with equivalent capacity. A single, 12-fiber cable could easily carry 500 phone lines at 6.3 Mbit/s per fiber, so these systems were installed to new developments or during upgrading of existing phone systems.

However, old digital loop carriers posed a problem when telephone companies started thinking about DSL. Digital loop carriers transmit only the voice frequencies carried by phone lines, but DSL relies on transmitting higher frequencies that are lost in multiplexing. Thus the old digital loop carriers can't deliver DSL to phone lines they serve. It's hardly the only case where yesterday's bright idea becomes tomorrow's bottleneck.

To get around this bottleneck, telephone carriers must install modern fiber systems that transmit signals differently. Data to be delivered over DSL can be separated from digitized voice signals at the central office, then multiplexed in separate data streams. At the remote demultiplexing node, the two signals can be combined into an analog voice/DSL line, as they would be at the central office if the DSL subscriber were closer. The important difference is that the modern systems can move more functions of the switching office to the remote node; the older systems can only multiplex voice signals.

Business subscribers are served the same way as home telephone subscribers. The larger the business, the bigger the information pipelines needed to serve them. You can see this if you look back to Figure 25.2. The smaller businesses like the shoe-repair shop need just a phone line or two. Businesses that need more phone lines or high-speed Internet access may lease blocks of lines from the phone company. The insurance office and town hall both have T1 lines. The "pretty big business" has a 45-Mbit/s T3 line. The higher-capacity services may be delivered partly or entirely over fiber. A T3 fiber-optic cable runs from the switching office to the node serving the business district, and a short copper cable delivers T1 service from that node to the insurance company.

Subscriber and Access Services

Traditional phone lines were designed to carry only analog voice signals.

The traditional subscriber loop was designed to carry voice telephone signals, called *Plain Old Telephone Service* or *POTS* in the industry. The local telephone network now provides many other services because it can transmit signals two ways, and make connections between any pair of phones attached to the system. When facsimile machines came into use, it was much easier to add them to the existing telephone network than to build a separate fax network. All you did was install a new phone line, plug in the fax machine, and tell people to send faxes to the new phone number. Likewise, dial-up modems take advantage of existing phone networks between your home or office and your Internet service provider instead of requiring a complete new network. Digital Subscriber Line technology expands on this trend by enhancing the transmission capacity of voice phone lines.

A broader range of services is offered to business and access customers, whose demand has steadily grown from multiple voice phone lines to large numbers of phone lines and high-speed data links. We'll look briefly at the most important services.

Leased Lines

Businesses lease phone lines for high-volume service.

Telecommunication users who need a large volume of service often *lease lines,* renting transmission capacity in bulk from a carrier. When dealing with telephone companies, they generally lease bulk capacity in a standard telephone-industry format, such as a T1 or T3 line, or an OC-3 carrier. The lines may run between user facilities, such as between buildings used by a large company or between a city's data-processing center and city hall. They also may run from the user's facility to another point, such as between a university and a regional Internet node or a long-distance carrier.

Although the service is called a *leased line,* the signals may not be carried over a physically separate wire, fiber, or cable along the entire route. If a company leases a T3 line between a downtown office building and a suburban factory, it is buying a guaranteed capacity of 45 Mbit/s on that route. The telephone company may time-division multiplex that 45-Mbit/s signal into a 2.5-Gbit/s OC-48 signal that a fiber transmits from downtown to the suburban central office, then run a separate fiber pair to the plant. The user sees no difference between that service and a separate pair of fibers—but the phone company can lease the capacity more cheaply.

Access lines generally serve the same purpose, but may be arranged differently. For example, the company may rent one optical channel on a fiber in a metro network that runs from downtown to near its suburban plant. The carrier may transport the signal in OC-1 or OC-3 form through the fiber, using its own equipment. Alternatively, the company could supply its own transmission equipment and send the signal in whatever form it wanted. If the line is intended to link the local-area networks in the factory and company headquarters, the signal might be in Gigabit Ethernet form.

Access lines also can go beyond the region. If a large magazine publisher has offices in Boston and Chicago, it might lease an OC-12 circuit between the two cities, so it can transmit signals at up to 622 Mbit/s. The publisher might also lease OC-3 lines between both magazine offices and its printer in Mississippi. The company's signals can be combined with other signals and multiplexed to higher speeds for part of the route, but the user would not see any difference. Alternatively, users could lease wavelengths or dark fibers and have all the capacity they could use for themselves between two points.

Telephone Lines

Standard phone lines were designed to carry voice frequencies of 300 to 3000 Hz. This isn't high fidelity, but it's adequate for intelligible speech. The bandwidth is limited by the attenuation of copper wire pairs, which increases with frequency. As with fiber, the total attenuation of a copper-wire subscriber loop depends on the length as well as the wire properties. Traditional phone companies long ignored higher frequencies.

Fax machines and dial-up modems transmit digital data, which they encode in the form of tones at different frequencies in the 300 to 3000 Hz voice telephone range. This allows them to transmit signals over ordinary analog phone lines. The signals must be robust enough to survive conversion to digital form for regional and long-distance transmission. Standard level 3 fax signals can transmit up to 14,400 bits per second. Dial-up modems can transmit up to a nominal 56,000 bits per second, although current regulations limit the speed to about 53,000 bits per second to avoid overloading phone lines; actual data rates tend to be lower.

Fax machines and dial-up modems code digital data as tones at voice frequencies.

INTEGRATED SERVICES DIGITAL NETWORK (ISDN)

Many years ago the telephone industry realized the potential for digital services and developed a standard for digital voice and data transmission called the *Integrated Services Digital Network (ISDN)*. ISDN never went very far, and for many years the telephone industry kept it on the shelf.

ISDN offers two 64-kbit/s digital voice channels and a 16-kbit/s data channel.

Instead of digitizing voice signals at the switching office, ISDN converts voice signals to digital format right at the phone. An ISDN line can transmit a total of 144 kbit/s over twisted-wire pairs. The ISDN standard divides that signal into two voice lines at 64 kbit/s each and one data channel at 16 kbit/s. (The leisurely data channel speed testifies that the standard is old.) Twisted-wire pairs used for present voice lines should be able to carry ISDN signals at least 5.5 km (3.4 mi) from the central office, but they may require "conditioning" to improve their transmission quality.

ISDN was intended to digitize voice signals right at the entry to the local telecommunication network to make switching more efficient. However, that required special equipment both in the subscriber's home or office and in the switching office. After many years of delay, phone companies finally installed the required equipment in their switching offices, but the special digital phones needed by customers remain quite expensive. As a result, ISDN has found few telephone customers, particular in the United States.

ISDN also can be used as a 128- or 144-kbit/s digital pipeline from home or business to the Internet. In practice, this ISDN data service is offered as another "flavor" of the digital subscriber line service described next. ISDN rates are lower, and the transmission format different, so ISDN signals can reach homes outside the distance limit of DSL. (The nominal design distances are the same as for some types of DSL, but ISDN signals are more robust.) When ISDN is sold as a form of DSL to people too far from the central office, it is often called *IDSL* so it sounds like a member of the same family of services, although it transmits signals differently and can stretch to about 25,000 feet.

DIGITAL SUBSCRIBER LINE (DSL)

DSL transmits data on phone lines at frequencies above the normal voice range.

Digital Subscriber Line (DSL) transmits digital data over copper wires at frequencies higher than those used for analog voice transmission. Ideally, DSL signals can be transmitted over the same twisted-pair lines used for voice conversations, with a standard phone responding only to the voice signals and the DSL modem responding only to the data. In practice, it often isn't that simple, and may require a device, called a *splitter,* that separates the low analog voice frequencies from the higher frequencies carrying the digital data.

There are several versions of DSL. The data rates they can transmit depend on the quality of the phone lines and the length of wire separating the subscriber from the switching office, as well as on the design. Load coils used to improve the quality of analog voice transmission also attenuate the high frequencies that carry DSL signals, so they can't be used with DSL. Some phone lines can't carry high frequencies well enough. Even in the best phone lines, attenuation increases with frequency, so the data rate possible decreases with transmission distance, as shown in Figure 25.5 for two types of DSL.

Table 25.1 lists several DSL-related technologies and their nominal data rates and maximum transmission distances. The actual data rate achieved at the maximum distance typically is well below the maximum rate, unless noted. The limiting distance is measured along the cable route, which is not a straight line, so many homes fall outside the limit even in cities or inner suburbs. Because data rate decreases with distance, it generally is impossi-

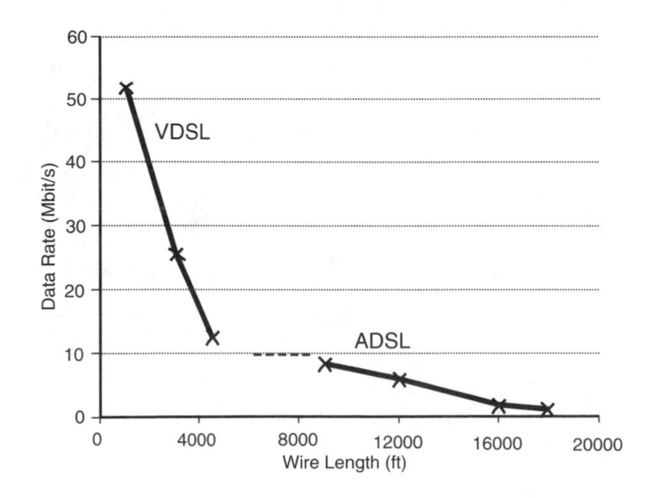

FIGURE 25.5
DSL data rates vs. wire length.

ble to achieve the maximum speed over the maximum possible distance. For example, the Asymmetric Digital Subscriber Line (ADSL) format is rated to send 8.448 Mbit/s to a terminal 9000 ft (2.7 km) away, but only 1.5 Mbit/s to a terminal at 18,000 ft (5.5 km).

Values quoted for the maximum data rate over a given distance differ significantly among sources. This reflects wide differences in the quality of the phone lines, immaturity of the

Table 25.1 Types of digital subscriber line services

Technology	*Standards*	*Nominal Date Rate*	*Specified Maximum Distance*
ISDN	ANSI/ITU	128 or 144 kbit/s both ways	18,000 ft (5.5 km) (longer distances possible)
G.Lite ("Splitterless" DSL)	ITU	1.5 Mbit/s downstream, 384 kbit/s upstream	18,000 ft (5.5 km)
ADSL (Asymmetric Digital Subscriber Line)	ANSI	8 Mbit/s 640 kbit/s downstream, upstream	9,000 ft (2.7 km) (at 8 Mbit/s)
ADSL	ANSI	1.5 Mbit/s downstream	18,000 ft (5.5 km) (at 1.5 Mbit/s)
G.SHDSL (Symmetric high-rate DSL)	ITU	2.3 Mbit/s both ways	6,500 ft (2 km) (at 2.3. Mbit/s)
G.SHDSL	ITU	394 kbit/s both ways	15,500 ft (4.7 km) (at 394 kbit/s)
T1	Digital Telephone Hierarchy	1.5 Mbit/s both ways	3,000 ft (900 m)
RADSL (rate adaptive DSL)	—	Adaptive—to 9 Mbit/s downstream, 1 Mbit/s upstream	12,000 ft (3.6 km)
VDSL (very high DSL)	ANSI	13 to 52 Mbit/s down, 1.5 to 2.3 Mbit/s upstream	4,500 ft (1.4 km) at 13 Mbit/s; 1000 ft (300 m) at 52 Mbit/s

technology, and considerable marketing hype. Published reports indicate that DSL cannot achieve the specified speeds over a significant fraction of phone lines, and may not be usable over some existing phone lines within the distance limitations. Problems arise from both condition of existing outdoor phone wires and from the wiring and equipment inside homes. The stated maximum distances are indeed the upper limits; many companies limit their offering of DSL to subscribers closer than the listed maximums.

Most estimates indicate at least 20% of telephone users can't receive DSL because of phone-line length or quality; some estimates range as high as 50%. Other potential customers are limited to the lowest speed, 128 kbit/s, which may not justify the expense. This is an unusual failing for any industry that at least nominally offers its services to everyone. (As you may suspect, my house is just outside the DSL range—18,050 feet from the switching office.)

As Table 25.1 shows, some DSL formats, notably VDSL, can't go far. Instead of delivering service over wires running from the central office, VDSL is intended to deliver service over a short wire running from a nearby fiber-optic terminal, as described later in this chapter. At the highest speeds, the terminal would have to be on your block.

This tradeoff between transmission speed and distance is fundamental, and inevitable when using copper. Higher speeds require different transmission media.

Emerging Services and Competing Technologies

Follow telecommunications technology for a while and you learn that new services and technologies are always "just around the corner," and that some of them stay "just around the corner" for many years before quietly evaporating. One of them was the video-telephone, which started as the stuff of pulp science fiction in the 1920s (see Figure 25.6). AT&T introduced its version, called Picturephone, in 1970, but it quietly faded away. Software and hardware for videoconferencing over the Internet exists today, but few people bother.

Nonetheless, it's important to look a little ways into the future. Telecommunications technology, particularly fiber optics, is moving very fast. The present local phone network contains relatively little fiber, but fiber is likely to spread as the demand for bandwidth continues to increase. Let's look at emerging services and competing technologies before turning to ways that fiber is likely to spread in the local telephone/telecommunications network.

Voice on Internet Protocol and DSL

Digitized voice can be transmitted via Internet Protocol or DSL.

Current industry magazines contain many articles about delivering telephone-grade voice services using Internet Protocol or DSL lines, instead of the more conventional switched circuits. The idea seems attractive because IP transmission is relatively cheap and DSL allows a single phone line to carry the equivalent of many digitized voice lines. However, it remains to be seen how widely either of these ideas will be accepted.

FIGURE 25.6
Videophones were part of the background that Hugo Gernsback, publisher of the first science-fiction magazines in the 1920s, used for his first science fiction novel. However, the cover artist's vision in this paperback still included a dial. (Courtesy of Fantasy Books)

Software is available that allows you to place phone calls over the Internet using a personal computer. The price is attractive because the calls are essentially free if you have an unlimited-time Internet connection. Unfortunately, voice on IP (VoIP) quality tends to be poor because voice is broken down into packets that tend to take slightly different times to travel through the network and some packets can get lost. The overall result now is often a poor quality phone line. It might be acceptable when calling a friend overseas to avoid huge bills, but otherwise you would probably complain.

Voice on DSL (VoDSL) is not yet readily available so it's hard to tell what shape it will take. If the DSL line is used to transmit time-division multiplexed voice, it should give better quality than if transmission is in IP format. One potential drawback is the need for analog-to-digital conversion in the telephone or at the input of the DSL line.

Video Services

Today most video programs reach homes by broadcast from local transmitters or through the cable-television networks described in Chapter 27. As you will learn in Chapter 27, cable networks parallel telephone networks in cities, suburbs, and most towns; modern cable systems—like phone systems—can deliver voice and data services.

Two alternative approaches to video distribution are emerging. *Direct broadcast satellites* are television transmitters that orbit the earth once a day; thus, they appear to stay over the same position on the equator. These satellites transmit hundreds of digital television channels at microwave frequencies to small antennas used by people who subscribe to the services. DirecTV is an example. *Fixed terrestrial wireless* systems are low-power microwave transmitters on the ground that broadcast television signals to serve a small region. They are designed to provide the equivalent of standard cable-television service without cables. They also can transmit digital data and could be used to transmit phone signals. This technology is not widespread.

DSL lines can transmit switched digital video signals.

Switched digital video services deliver user-selected video channels, sometimes called *video on demand,* to customers. Copper wire pairs can deliver one video channel at a time, using the same technology as DSL. In fact, the digital transmission techniques used for DSL were originally developed for switched digital video, when phone companies were trying to offer them in competition with cable companies. Finding little demand for switched video, the phone companies instead adapted the technology for high-speed data transmission.

Wireless Networks

Mobile phones are an alternative to wired phone lines.

As the price of cellular telephone service drops, some observers believe that wireless phones may come to replace wired phones (sometimes called "wire lines" or "land lines"). Instead of having separate home and mobile phones, you would have a single mobile phone that you would use at home as well as on the road. New *third-generation* or *3G* mobile phones will serve as data terminals as well as voice phones. Another part of this vision involves replacing wired or cabled Internet links with wireless data services. Data transmission could be obtained from local wireless terminals or from satellites.

A few people have made this change, largely outside the United States. It is not clear how far this trend can go, but there are limits. Radio spectrum is scarce, and the quality of cellphone reception can be spotty. Even if you don't use a cell phone, you've seen callers pacing restlessly as they try to find a sweet spot for reception.

Satellite telephones and data links were a much-heralded idea just a few years ago. However, the commercial failure of the expensive Iridium satellite phone system has dampened interest.

Fiber to the Curb

Fiber is spreading gradually through the local telephone system as bandwidth requirements increase. In Figure 25.4, you saw how fiber may serve remote neighborhoods in current distribution systems. Continuation of this trend will bring fiber ever close to homes.

The next logical step is *fiber to the neighborhood* (FTTN), shown in Figure 25.7. High-speed fiber lines distribute signals to neighborhood nodes from which copper wires spread up and down local streets to distribute signals to individual homes. Such a fiber node might serve several hundred telephone subscribers, including small local businesses. Similar fiber nodes serve business and government, like the City Hall at the lower right in Figure 25.7.

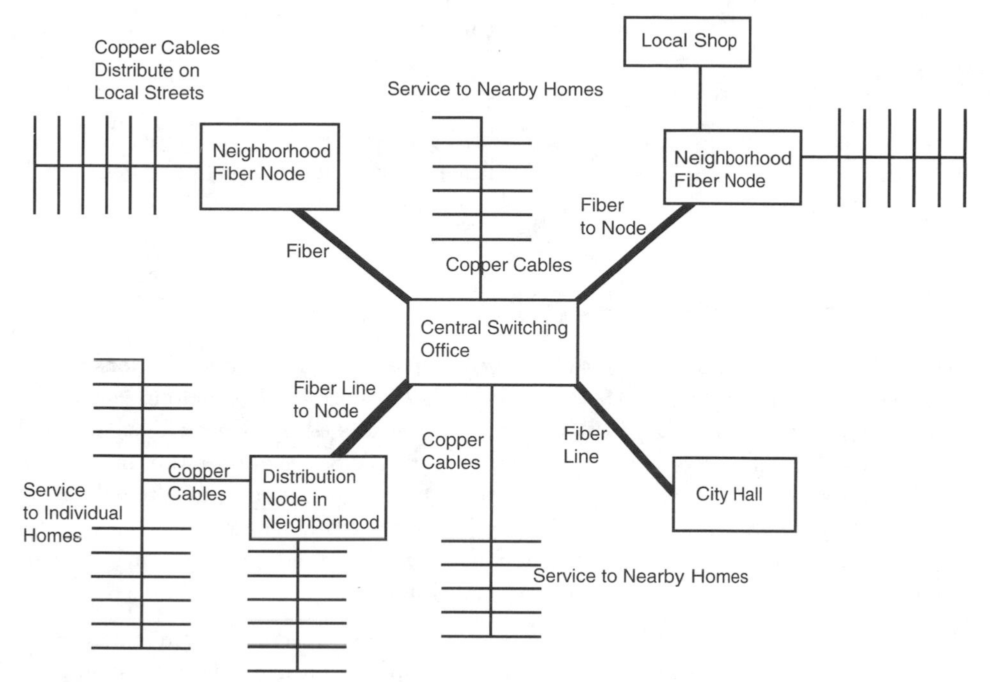

FIGURE 25.7
Fiber to the neighborhood.

Copper cables running from the central office distribute signals in its neighborhood. Fiber to the neighborhood (or node) is not standard, but it is used in some areas.

A few regional telephone companies have started installing *fiber to the curb (FTTC),* which runs fiber cable down every street. Fibers end at a distribution node on each block, in front of homes, and drop cables run from this node, as shown in Figure 25.8. In this example, an add-drop multiplexer drops a signal from a buried fiber cable at a curbside service node, typically a box a little bigger than a standard television. Buried copper cables run from the box to nearby homes, typically serving the homes on that block. Short runs of twisted-pair

Fiber to the curb runs fiber down every street and drops signals on copper wires at homes.

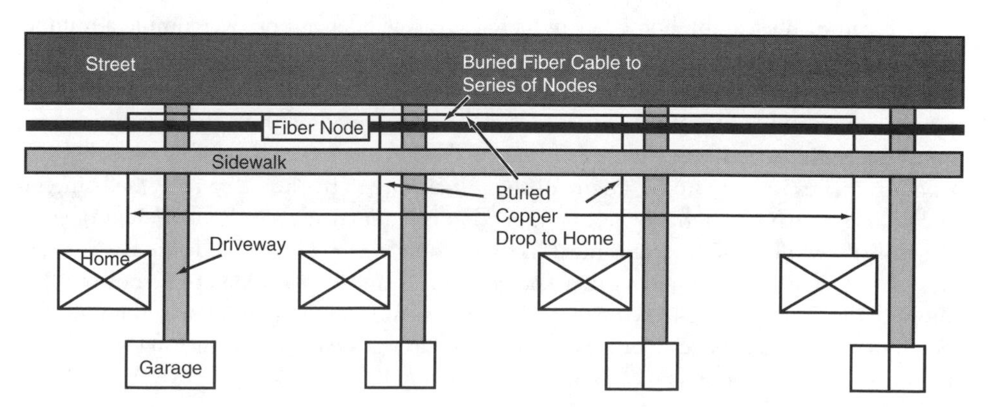

FIGURE 25.8
Fiber to the curb.

wiring can handle VDSL signals, delivering much higher bandwidth to the home than longer twisted-pair links. The same approach can be used in areas with overhead utilities, by mounting the service node and drop cables on poles.

Fiber to the curb systems also can deliver cable television services, if the carrier decides to offer them. In this case, the fiber cable also delivers cable television service to nodes, with coaxial cable drops to each home connected to the cable.

Fiber to the Home (FTTH)

The ultimate leap for fiber optics is all the way to the home. Telecommunications visionaries have been thinking about *fiber to the home (FTTH)* since 1972 when John Fulenwider first suggested the idea for a wired-city project being studied by General Telephone and Electronics. The first experimental system, called Hi-OVIS, began operation in Higashi-Ikoma, Japan, in 1978. Canada, France, and a few U.S. telephone companies tested fiber to the home in the years that followed. In most cases, the technology worked, but the economics didn't. No one found a combination of services that could generate enough revenue to pay for the high cost of installing fibers to every home.

Fiber to the home could meet the demand for broadband Internet links.

That is changing, however, with the growing demand for broadband Internet access and the declining cost of fiber-optic hardware. Bandwidth is a bit like a drug—many people can't get enough of it. Over the last 15 years, Internet users have moved from 1200-bps dial-up modems to 56,000-bps modems, and now to cable modems and DSL at user speeds of up to a megabit per second. The tremendous increase in raw speed has been offset by the bloating of computer files and the increasing sophistication of graphics and software on the World Wide Web, thus requiring ever more transmission capacity. Streaming video and downloads of MP3 music files gobble bandwidth. Like the household clutter that fills your closets, your information needs expand to fill the capacity available.

A handful of companies have begun offering fiber to the home services. They generally look like the fiber to the curve services shown in Figure 25.8, but with fiber replacing the copper cable drops. Some installations are in new upscale developments; others in rural areas where fibers can span long distances more effectively than copper cables.

The spread of fiber to the home beyond a few areas will depend on overcoming a number of barriers. The biggest is financial. Fiber-optic equipment is more expensive than copper; the major cost is in the labor required to install the new equipment. Construction is expensive, particularly where utilities are buried. Neighborhoods with the eyesore of overhead wires can console themselves with the thought that it costs less to hang fiber on poles than to dig trenches to every home. Many of the initial fiber to the home systems are being installed in new developments, where construction is required anyway. Replacement of existing utilities will be slower, and is most likely to come when existing hardware has come to the end of its lifetime or when other construction is underway. (One episode of the PBS show "This Old House" showed new fiber being installed in the walls of an old home. The neighborhood didn't have fiber service yet, but the walls were open, making installation cheap, and the renovators expected fiber would come.)

Supplying power to telephones attached to fiber to the home systems is another mundane but important issue. Copper phone lines traditionally deliver power for the phone as well as signals (except for cordless sets and many office phone systems). The most logical approach would be to get power from local electric lines, but that would make phone lines go down when the power failed, potentially leaving people stranded without a way to report a power outage. The spread of cellular phones may ease this concern, because unlike cordless phones they don't require a powered base station, and can be used when the power is out. Another possibility is leaving wires to serve one or more standard voice phones as a "lifeline" service.

Fibers can't supply electric power to phones.

Two families of technologies are in the running for future fiber to the home systems, passive optical networks and Gigabit Ethernet. We'll look at them in turn.

Passive Optical Networks

One way to design low-cost, low-maintenance fiber-optic systems that serve individual subscribers is to eliminate active components between the transmitting terminal and the subscriber. This approach is called a *passive optical network,* or *PON.* In its simplest form, a single transmitter generates signals that are transmitted through a branching network of fibers that divide the signals among a number of terminal points. As in local-area networks, all signals reach all terminals, but each terminal only pays attention to the signals addressed to it. Each receiving terminal also has its own lower-speed transmitter that can direct signals back to the head end through the same branching fiber network, or through a different set of fibers.

The part of the network between the head end and the subscriber includes only passive components: couplers that divide signals, plus fibers, connectors, and splices. These components are robust, so they can withstand the environment in an outdoor case or housing. They also require no electrical power, so housings that contain couplers do not require power. The expensive high-speed transmitter is shared among many individual subscribers.

Passive optical networks have no active components between switch and subscriber.

A consortium of telecommunications carriers and equipment manufacturers are standardizing a passive optical network design for subscriber service. The proposed standard being considered by the International Telecommunications Union G.983 panel is called the *Full Service Access Network (FSAN).*

The overall approach is a common design that could serve a range of applications including fiber to the neighborhood, fiber to the curb, fiber to the business (or building), and fiber to the home. Figure 25.9 shows the basic approach. Fiber service originates from an optical line terminal (at the left). This equipment may be installed in the local central office or at some intermediate point. It collects input signals from all services distributed over the fiber network, including data, voice, and video, and codes them into a signal for distribution over the fiber.

The Full Service Access Network is a passive optical network standard.

The FSAN standard calls for transmission downstream to the subscriber at either the 155-Mbit/s OC-3 rate or the 622 Mbit/s OC-12 rate in the 1.5-μm window, using

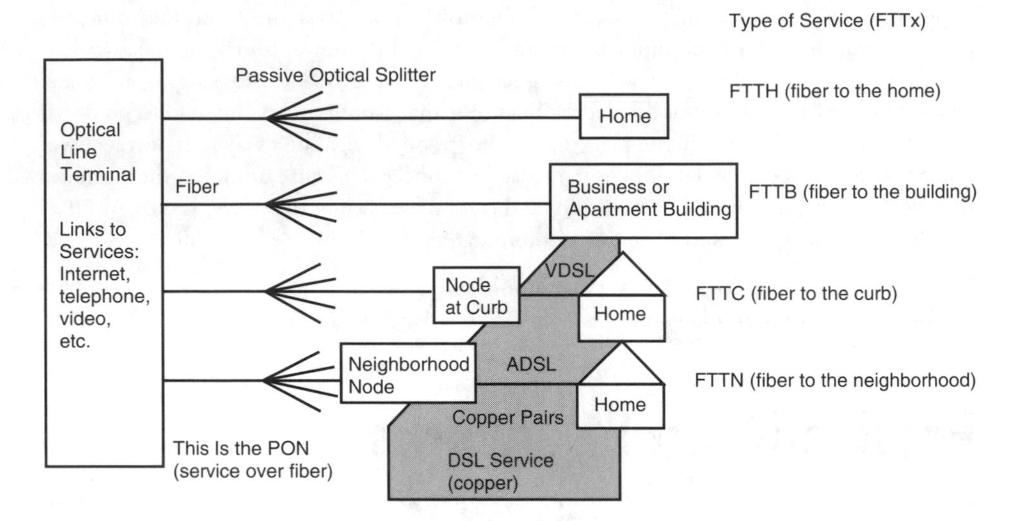

FIGURE 25.9 *Passive optical network for FTTx services.*

Asynchronous Transfer Mode format. The maximum distance is 20 km of single-mode fiber.

Transmission starts along a single fiber, and at some intermediate point a passive coupler divides the optical signal. The FSAN standard specifies splitting the signal among up to 32 fibers. Figure 25.9 shows only one fiber beyond this point, for simplicity.

How far the fiber carries the signal it is distributing depends on the design. In the fiber to the neighborhood and fiber to the curb systems (at bottom of figure), the optical signal stops at a local node that converts the optical signal to electronic form and distributes it over copper. If the node serves a neighborhood, data would be distributed over DSL links. For a curbside node serving an individual block, the wires could carry high-speed VDSL.

Fiber to the building brings fibers to a business or apartment building.

Alternatively, the fiber can go all the way to the destination. A *fiber to the business* or *fiber to the building (FTTB)* system delivers signals directly to a business or apartment building. A node inside the building converts the optical signals to electronic form and distributes them to individual terminals or subscribers. In fiber to the home systems, the fiber goes to a node in the house, where the signal is converted to electronic form and distributed.

The optical nodes (at the right in Figure 25.9) also can transmit signals in the other direction. The FSAN standard calls for return transmission at 1.3 μm through the same fiber at total rates to 155 Mbit/s. The system allocates separate time slots to individual return transmitters, so that their outputs interleave where the fibers combine on the return path. The receiver separates the inputs by time-division demultiplexing. Normally each return transmitter operates at only a fraction of the total rate, but the system can allocate extra capacity when it's needed.

One important feature of this standard is that a common architecture serves all levels of fiber distribution, from fiber to the neighborhood to fiber to the home. Thus the network can evolve as demand changes.

Gigabit Ethernet and Internet Protocol

Fibers to the home can carry Gigabit Ethernet.

An alternative approach to fiber to the home is using a Gigabit Ethernet network to transmit data in Internet Protocol format. The idea is to adapt local-area network technology to distributing signals to homes over fiber. Instead of using new technology, this approach uses hardware already developed for Gigabit Ethernet. The signals are transmitted over fiber to individual subscribers.

As you will learn in Chapter 27, cable-television networks use Ethernet to transmit data to homes through cable modems. The advantage of using Ethernet is that it is a widely accepted standard. Some personal computers come with built-in Ethernet ports; add on Ethernet cards are inexpensive are readily available for other computers. All you need is an interface box such as a cable modem.

Gigabit Ethernet subscriber links are being developed on the same premise of using a widely accepted standard to take advantage of inexpensive hardware. As Figure 25.10 shows, a network interface unit is located at a remote site, with fiber linking it to the Internet, and possibly to other services. The interface unit receives signals from the network at 1 Gbit/s, then distributes them to the local-area network—that is, to the fibers attached to the unit. Individual subscriber terminals can receive signals at peak rates to 1 Gbit/s, but the total data rate over the entire network cannot exceed that rate. As in the computer local-area networks described in Chapter 26, data flow to individual terminals varies widely.

In present designs, the network interface unit is a rack of equipment that fills a closet. It can be installed in any building with electrical power. In urban areas it could be a school or public building; in rural areas it might be a general store. The long reach of fibers makes this design suitable for use in rural areas where population density is low and long runs of cable are needed for connections. The individual subscriber units are small boxes the size of set-top cable boxes.

The technology is emerging, and has yet to take final form. The main initial interest is in high-speed Internet connections, but voice telephone signals also can be encoded

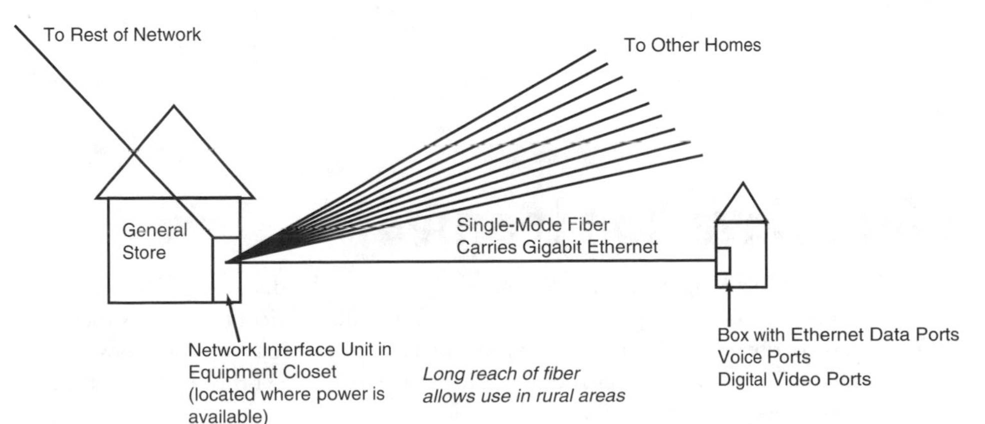

FIGURE 25.10
Gigabit Ethernet to the subscriber.

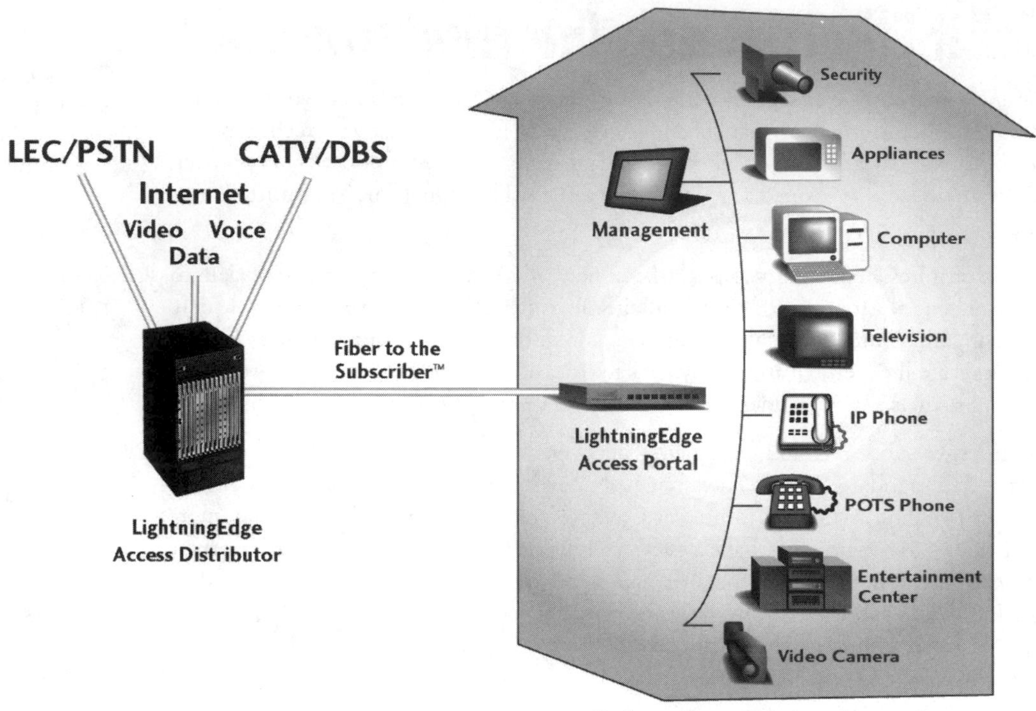

FIGURE 25.11
Multiple services delivered over Gigabit Ethernet link. (Courtesy of World Wide Packets)

for Internet transmission. Potentially the technology could also handle video transmission, security monitors, video cameras, and Internet links to appliances, as shown in Figure 25.11.

Ethernet to the subscriber will compete with passive optical networks for other types of signal distribution as well as fiber to the home. Some of the earliest Gigabit Ethernet networks were installed by Canadian school systems as part of a program to enhance Internet connectivity to schools.

What Have You Learned?

1. The subscriber loop distributes telephone signals to individual users. The access network is a new term for the network that distributes services to business users.
2. The subscriber loop distributes signals from a switching office or central office to individual users over copper and fiber cables. The switching office is the interface between individual phone lines and the global network.
3. Cell phone and competitive phone companies have their own switching offices.

4. Twisted pairs of copper wires are widely used in the subscriber loop. They can carry voice signals several miles, but DSL is limited to shorter distances.
5. Traditional phone lines were designed to carry only analog voice signals at 300 to 3000 Hz called Plain Old Telephone Service (POTS).
6. Fax machines and dial-up modems transmit digital data by converting the data to analog tones and transmitting the tones over voice phone lines.
7. Businesses lease phone lines to buy capacity in bulk at standard transmission rates. An alternative is to lease dark fiber or optical channels.
8. ISDN was designed to carry two 64-kbit/s digital voice channels and a 16-kbit/s data channel. When used for data only it carries 128 kbit/s.
9. Digital Subscriber Line (DSL) transmits data on twisted-pair phone lines at frequencies above the normal voice range.
10. Several types of DSL transmission have been developed; they differ in nominal data rate and maximum transmission distance. The maximum speed drops with total transmission distance, measured along the cable route.
11. DSL lines can transmit switched digital video.
12. FTTx is a family of services including fiber to the neighborhood, fiber to the curb, and fiber to the home.
13. Fiber to the curb runs fiber down every street to nodes at the curb. Copper wires run from the nodes to individual homes.
14. Fiber to the business or building serves business customers. It also describes fiber connections to apartment buildings.
15. The growing demand for high-speed Internet access is increasing interest in fiber to the home.
16. Passive optical networks can be used to provide fiber to the home service. They include no active components between the transmitter and the subscriber. The leading standard for passive optical networks is the Full Service Access Network.
17. Gigabit Ethernet is an alternative way to offer fiber to the home. It makes homes part of a high-speed local-area network.

What's Next

In chapter 26 you will learn about computers and local-area networks.

Further Reading

DSLReports: *http://www.dslreports.com/*

Roger L. Freeman, *Fundamentals of Telecommunications* (Wiley-Interscience, 1999)

Gigabit Ethernet to the home: see *http://www.canarie.ca*

Gary M. Miller, *Modern Electronic Communication,* 6th ed. (Prentice Hall, 1999)

Passive Optical Networks/Full Service Access Network: *http://www.fsanet.net/fsan/*

Virata Corp., "Personal Broadband Services: DSL and ATM," available at *http://www.virata.com/pdf/virata_dsl2.pdf*

Questions to Think About for Chapter 25

1. A switching office serves 5000 voice telephone lines. It is designed so that at peak usage 20% of the lines can be connected—a total of 1000 phone lines. Residents discover the Internet and 500 of them buy dial-up modems and install new phone lines so they can leave their modems on all of the time. How much more switching capacity does the phone company have to install, both in number of lines and percent?
2. Suppose that instead of installing new phone lines, the residents of the town in Question 1 hooked their modems up to existing phone lines. If 500 people buy modems and half of them go on the Internet in the evening at a time of peak residential calling, how much does the phone company have to increase its switching capacity? How much must switching capacity be increased to accommodate half the households buying modems and half of the modem users going on the Internet in the evening? Assume the rate of voice calling does not change.
3. Check to find how far your residential phone line is from your local phone company's switching office at *http://www.dslreports.com.* What DSL rates are available? If you're in a class, compare the rates and distance with those of other students.
4. A passive optical network splits signals among 32 subscribers. If the data rate over the whole fiber line is at the OC-3 rate, what is the difference between the data rate a user will see if everyone is on the line and if only that user is on the line?
5. The transmitter for a passive optical network generates a 1-mW signal that is divided equally among 32 users. If the cable loss is 10 dB and the couplers have no excess loss, what is the signal that reaches each user?
6. A Gigabit Ethernet signal is split among 32 subscribers. Neglecting losses arising from congestion, what is the maximum data rate if all are receiving signals at equal capacity?

Quiz for Chapter 25

1. Which of the following is at the network edge?
   a. Individual telephone subscribers
   b. Digital Subscriber Line
   c. Individual telephones
   d. Telephone switching offices
   e. International connections from national telecommunication networks
2. The network edge is the
   a. interface at which calls are directed.
   b. point where digital signals stop.
   c. point where telephone signals stop at a telephone.
   d. point where signals are transferred between regional and long-distance networks.

e. point where signals are transferred between long-distance and international networks.

3. What connects to standard voice telephones?

a. Optical fibers

b. Twisted-wire pairs

c. Single copper wires

d. Coaxial copper cable

e. Special hybrid cables with one fiber and one copper wire

4. What transmits digital subscriber line (DSL)?

a. Optical fibers

b. Twisted-wire pairs

c. Single copper wires

d. Coaxial copper cable

e. Special hybrid cables with one fiber and one copper wire

5. How does the use of digital loop carriers limit DSL over voice phones lines?

a. Not at all

b. The DSL subscriber can be no more than 18,000 feet from the switching office

c. The digital loop carrier does not transmit the high frequencies that carry the DSL signal

d. The digital loop carrier cannot carry analog signals, so subscribers can receive only the digital DSL signals

e. Digital loop carriers require the use of load coils to separate DSL from voice signals

6. What digital service do phone companies usually offer to residential subscribers beyond the standard limit for DSL?

a. G.lite DSL

b. IDSL, the renamed version of ISDN

c. T1 lines

d. Fiber to the home

e. Enhanced dial-up modems

7. Which of the following can limit the availability of DSL services?

a. Load coils

b. Distance from the central switching office

c. Quality of phone lines

d. Installation of digital loop carriers

e. All of the above

8. You are in charge of telephone operations for a resort town. The CEO of your company has bought a vacation home in a new development outside of DSL reach, but wants at least 1.5-Mbit/s ADSL service to keep in touch with corporate headquarters. What's the best way to upgrade service to the whole new development and please the big boss?

a. Run fiber-optic cable to the CEO's door

b. Run fiber to the center of the new development and build a new node to distribute DSL and other services there

c. Install IDSL because it can reach farther than ADSL

   d. Run VDSL from your switching office
   e. Start looking for a new job because you're not going to be able to do it

9. What services besides POTS are transmitted in the low-frequency analog band of copper twisted pair?
   a. Digital switched video
   b. DSL
   c. Fax and dial-up modem signals
   d. Gigabit Ethernet
   e. Passive optical networks

10. Each node in a fiber to the curb system would serve about how many homes?
    a. 1
    b. 10
    c. 100
    d. 500
    e. Over 1000

11. Each node in a fiber to the neighborhood system would serve about how many homes?
    a. 1
    b. 10
    c. 100
    d. 500
    e. Several thousand

12. A passive optical network can be
    a. used only for fiber to the home.
    b. used only for fiber to the curb.
    c. adapted for a range of fiber services as demand grows.
    d. upgraded for higher-speed transmission by installing active components between the transmission node and the home.
    e. installed inside homes to network appliances.

13. The Full Service Access Network is a(n)
    a. standard for passive optical networks.
    b. standard for adapting Gigabit Ethernet for fiber to the home.
    c. standard for the global fiber-optic network.
    d. abandoned form of DSL.
    e. standard for transmitting video and data services over DSL.

14. Where could Gigabit Ethernet be used in the local telephone network?
    a. For transmission inside the telephone switching office
    c. For local-area networks inside businesses
    c. It is required for any transmission on optical fiber
    d. For transmission to homes and schools
    e. For connection to the long-distance telephone network

15. What connects to the access network?
    a. Mobile telephone companies
    b. Large businesses
    c. Telephone switching offices
    d. Internet service providers
    e. All of the above

CHAPTER

26

Computers and Local-Area Networks

About This Chapter

The Internet links computers and local-area networks like the global telecommunications network links telephones. As you have learned, the two networks share some connections, but the data and phone networks differ in important ways. This chapter covers computer networking, with particular emphasis on local-area networks, where the differences between voice and data networks are the largest. I will start with the Internet and work down to smaller computer networks, then cover important networking equipment and key standards important for fiber-optic data communication.

Computer and Phone Networks

The Internet and the global telecommunications network have many similarities. The Internet originated from the idea of "internetworking," or linking separate computer networks so they could communicate with each other. Similarly, the global telephone network had its roots as separate telephone systems serving many individual cities.

The telephone network came together slowly. The first urban phone systems were installed in 1877, but it took time to develop technology to span long distances. The first long-distance lines were between nearby cities, and it wasn't until 1915 that the first transcontinental telephone line spanned the United States.

Likewise, the first connections between computers were strictly local. Universities and government agencies installed remote terminals so users could log in to remote mainframe computers. At first the terminals were merely a convenient way to avoid delivering decks of punched cards to the computer operating room, but soon they became

The Internet grew from links between local-area networks.

more interactive. Early networks were little more than keyboards and remote electric typewriters all linked to a single mainframe. As time passed, universities and government agencies bought more computers and linked them to each other as well as to the terminals. Later, scientists wanted to send messages directly between computers at different places, so they began building long-distance connections. In time, these connections spread to allow "internetworking"—communication between different networks—and the overall network became the "Internet."

Packet and Circuit Switching

Although the ideas behind the Internet and the telephone network seem similar, the details are very different. Telephone connections require a constantly open line so people can hear each other talking without delay. Voice transmission does not require a tremendous capacity, but it does require a steady capacity. That is, you always have the same bandwidth because the phone system can't predict when you or the other person will start talking. This is the reason for the circuit switching that you learned about earlier.

Computers do not transmit data at uniform rates.

In contrast, computers do not transmit data at uniform rates. Back in the days of mainframe computers, the computer could spend a couple of hours analyzing your data, sending no information at all back to the terminal. Then it could transmit the results of its calculations as fast as the terminal could handle it before lapsing into silence. The same uneven flows occur today. When you request information from a Web site, the host computer sends you a series of chunks of information. Then you spend a minute reading it and send a request to collect information from another Web page, as shown in Figure 26.1. Engineers who analyzed these patterns decided the most efficient way to handle computer data was packet switching, which you also learned about earlier.

For computer networks to talk to each other, they must agree on standard formats to transmit signals. The Internet Protocol is the best-known of these computer protocols. It breaks data into packets, and assigns *headers* to each packet that specify its destination and the size of the packet.

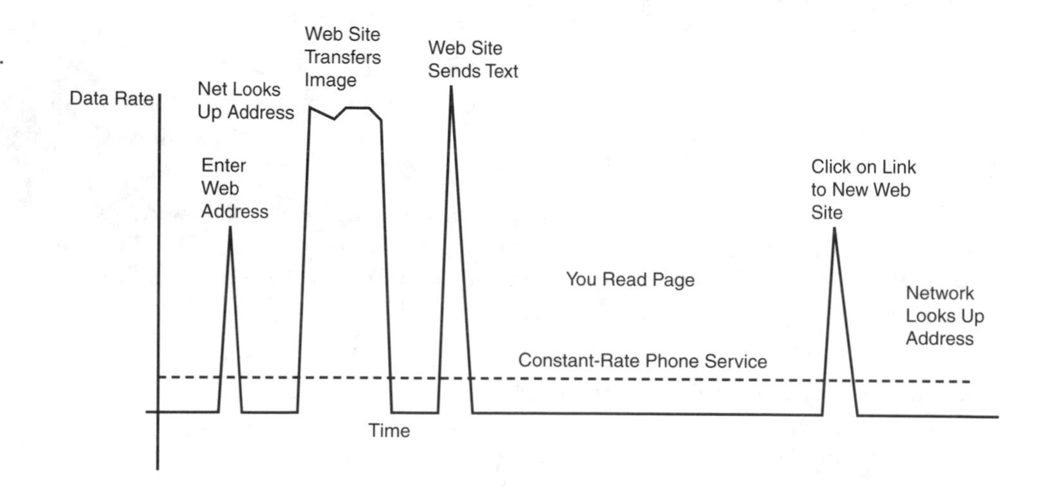

FIGURE 26.1
Bursty computer data on a Web link.

Layering of Computer Networks

In Chapter 20 you learned how standards are arranged in layers that define interfaces between various operations. In that chapter, I concentrated on the general concept of layering. Here we will see how layering works in computer networks.

Figure 26.2 shows the stack of layers in a computer network. When you send electronic mail, an application on your computer prepares the message and monitors the flow of messages to and from your computer. If you look behind the scenes, you may find settings in the program that specify a server that handles SMTP (simple mail transfer protocol). Your computer operating system also specifies TCP (transmission control protocol) settings that cover how signals are transmitted between your computer and what you see as your Internet connection.

Computer networking standards are layered.

That's normally all you see. The network then uses the Internet Protocol to package the data into packets, the length of which varies. Those packets then may be repackaged into blocks of constant length called *frames,* also described in Chapter 20. Ethernet packages data for network transmission; SONET packages bits for transmission on the physical layer.

Layer Designation	Protocols
Application	Electronic mail, File transfer, etc.
Presentation	Various Protocols: SMTP (simple mail transfer protocol), FTP(file transfer protocol), etc., span both layers
Session	
Transport	TCP (transmission control protocol)
Network	Internet Protolcol (and others)
Data Link	Network specified protocols (Ethernet, etc.)
Physical Layer	Stream of bits (SONET, etc.)
WDM down here	

FIGURE 26.2
Layering and computer network protocols.

Data from one layer may be repackaged for transmission on another. For example, the TCP protocol adds a header to data it receives, then passes it to the IP layer, which treats the whole TCP packet (header and all) as data, and adds its own header. Then the IP packets may be repackaged again into a SONET bit stream.

Internet Structure

The Internet has a complex structure that resembles that of the global and national telecommunication networks you saw in Chapter 23. The details are beyond the scope of this book, but you should get a general idea of how it works.

The Internet Backbone and POPs

The Internet backbone links POPs.

The Internet backbone is a network of high-speed transmission lines between major nodes called *Points of Presence,* or *POPs.* Figure 23.12 showed one backbone system. Figure 26.3 shows another.

Routers at the POPs direct packets from their POP where they enter the backbone to another POP nearer their destination. The routers work at the network layer on IP packets, reading the headers and passing them on to the physical layer, where they are converted to a bit stream and transmitted to the destination. Then they must be converted back to IP packets so the next router can read their headers and direct them to their destination. Because the system is packet switched, this is done packet by packet, which takes more effort on the part of the routers, but uses transmission lines more effectively than circuit switching.

Comparing the Internet backbone map of Figure 23.12 with the same carrier's telecommunications backbone map in Figure 23.11, you can see that Internet POPs are analogous to regional links to the long-haul network. The telecommunications network is on the physical layer, and actually has more nodes through which Internet signals are routed on their way to their destinations. There are many Internet carriers, and their networks interconnect at many points. Users never notice them unless something goes wrong and data packets can't flow from one to the other properly.

Internet Service Providers and Internet Links

The local operating centers of Internet service providers are analogous to the local switching offices of telephone companies. They collect signals from individual telephone subscribers and package them for transmission on the Internet.

University and corporate data centers serve the same function. These private servers are much more common than in the telephone network.

All these services and centers connect to each other and to backbone networks through a complex mesh of links. Many of these services have *peering agreements* with each other, which essentially specify that "if you carry my data, I'll carry yours." The arrangements are surprisingly fluid.

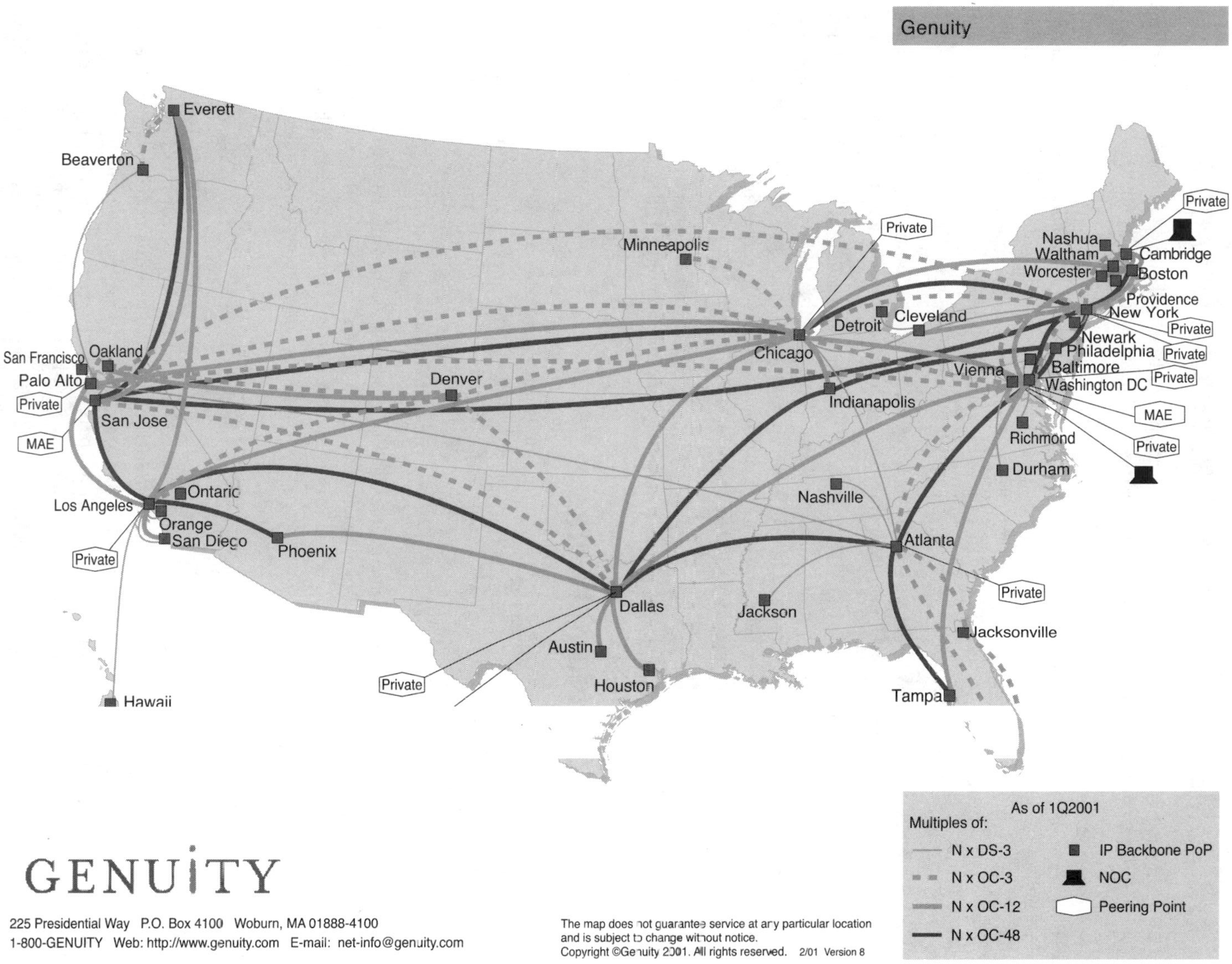

FIGURE 26.3
UUNet's Internet backbone in the United States. (Courtesy of UUNet)

Metro Networks and Wide-Area Networks

In Chapter 25, you learned about metro networks. People involved with computer networking often use the term *metropolitan-area networks,* or *MANs,* for networks that serve the same purpose.

A metropolitan-area network interconnects networks in the same area.

A metro network or MAN connects networks in the same geographic area. A variety of networks can serve this function. One class of metro networks is built by independent carriers that lease space to many different organizations that use the carrier's fiber to transmit their own signals. For example, a city or county government may lease lines to link government office buildings and schools, or a large corporation may lease lines to connect several plants.

The term *MAN* is sometimes used for networks that span a corporate or university campus. Such a network might link the separate networks operated by each department in a university.

The *wide-area network,* or *WAN,* also can have varied meanings. Originally it applied to networks that linked local-area networks in a large organization, typically spanning a number of buildings. However, now it is often used for networks that span areas larger than a metropolitan area, such as a state or national network. By this definition, a nationwide network like the one shown in Figure 26.3 is a WAN.

Although the geographic definitions of MAN and WAN may be a bit hazy, their functions are not. Both MANs and WANs make interconnections among smaller networks.

Local-Area Networks

A local-area network interconnects computers, terminals, and devices.

A *local-area network,* or *LAN,* connects individual computers, terminals, and devices in a certain area. What's "local" is a matter of definition, which can vary considerably. A local-area network may serve a work group, department, a small company, or occupants of a whole building. The definition is not rigid, and a LAN could encompass a number of buildings. Departmental LANs also may interface with each other, creating corporate networks.

Figure 26.4 shows the elements of a small LAN serving a departmental work group. In this case, several computers link to each other, a file server, a more powerful workstation, and shared peripherals such as a laser printer and fax modem. The fax modem is linked to the telephone network, and the whole LAN links to the corporate network and—through the corporate network—to the Internet.

Details vary widely among local-area networks, but the key idea is that all devices can interchange data with each other. A single medium—wires, optical fibers, or radio waves—carries signals to all devices on the system. Data packets carry header information to route them to particular devices. Many terminals can use the network at the same time. For example, the users of computers 1 and 2 could each retrieve data from the file server at the same time, computer 3 could print a report on the laser printer, and computer 4 could access the corporate network to communicate with another department.

Common LAN types are the ring, the star, and the bus.

Local-area networks can take a variety of forms. Three basic approaches to LANs are shown in Figure 26.5. In the star topology all signals pass through a central node that may be

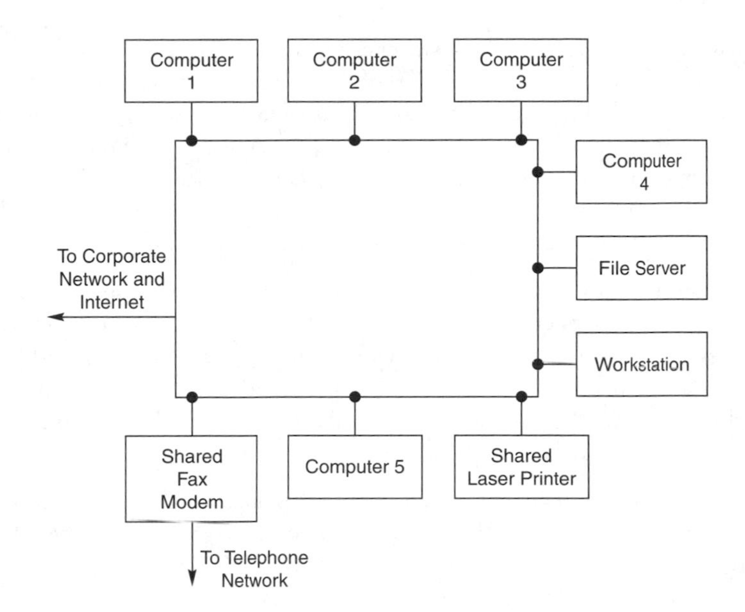

FIGURE 26.4
A LAN interconnects many devices that can send messages to each other and to external devices.

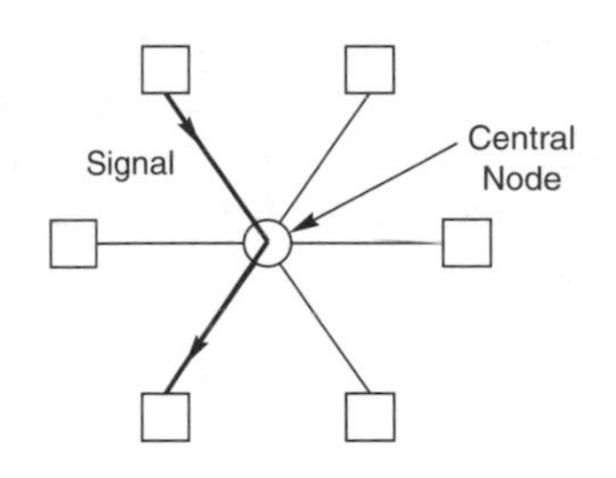

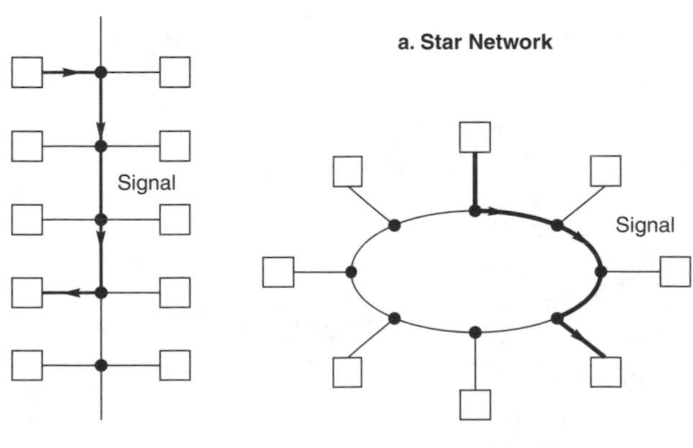

FIGURE 26.5
Star, bus, and ring LAN architectures.

active or passive. (An active star, which switches signals to particular nodes, functions like a telephone switch.) In a ring network, the transmission medium passes through all nodes, and signals can be passed in one or both directions (sometimes over two parallel paths for redundancy). In the data-bus topology, a common transmission medium connects all the nodes but is not closed to form a loop (i.e., the signal does not pass through all nodes in a series). Nodes ignore signals not addressed to them. Variations on these approaches make classification more complex than it might sound, but I will ignore those here.

The major media used for LANs today are copper wire pairs, coaxial cable, radio waves, infrared light, and optical fiber. Copper wire pairs include both shielded and unshielded types, and new high-bandwidth versions such as Category 5 cable. Coaxial cable has been losing favor, but remains in wide use. Both radio waves and infrared light are used for wireless networks such as Apple Computer's AirPort. Avoiding the need for cabling saves installation time and money, but transmission speed and distance are limited.

Optical fibers offer much higher speed and longer distance transmission than wireless or copper links, but they cost more. Fiber is mainly used in the growing number of networks that require high speed or longer distance than that available from copper. A new alternative is "fiberless laser" links, a technology that transmits optical signals through free space instead of through a fiber, described later in this chapter.

Point-to-Point Transmission

Links to peripheral devices are point-to-point.

Much data transmission simply moves information between pairs of devices (e.g., from a personal computer to a printer or external hard disk), as shown in Figure 26.6. A computer surrounded by peripheral devices may look like a miniature local-area network, but it doesn't work like one. Likewise a computer connected to the Internet over a cable or DSL modem is not really "networked" unless it allows file sharing with other computers on the network, which isn't advisable because the "other" users may include hackers.

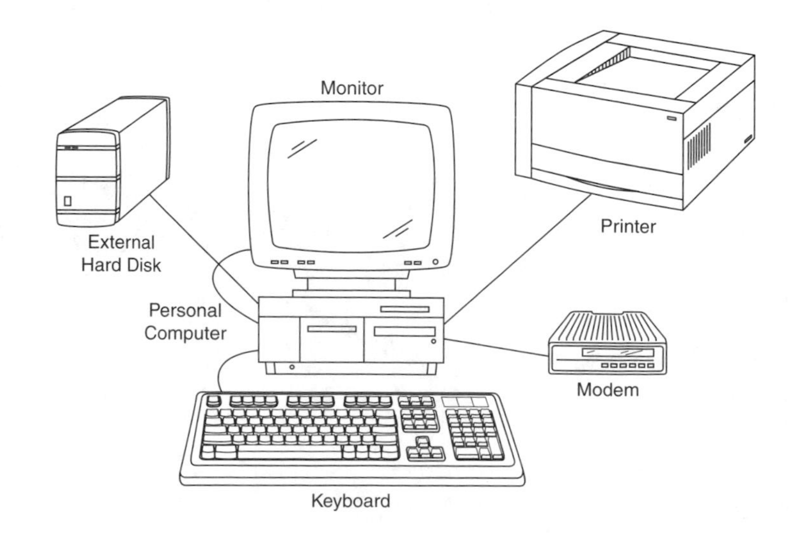

FIGURE 26.6
Point-to-point connections between a personal computer and individual external devices.

Point-to-point transmission is simple for fiber optics because it requires only a transmitter, a receiver, and some fiber. However, in most cases wires are even simpler because they can carry electrical signals between devices without special transmitters or receivers. Wires can carry very high-speed signals for a few feet; some devices have infrared ports for slower signals. Optical fibers may be used when data rates become too high, distances are too long, the environment is too noisy, or other factors make wires undesirable.

Optical Interconnection

All data transfer is not between separate boxes. Data must be moved between circuit boards in computers, between chips on circuit boards, and even within chips. In some cases, such as supercomputers, data must be transferred at very high speeds, and wires and other circuit components must be very tightly packed. Fibers can help both by allowing faster data transmission and by avoiding interference between adjacent wires. Gigabit data rates may require fiber optics over all but very short distances.

Optical signals can make interconnections within computers and even between chips.

Optical links may also carry information directly to and from integrated-circuit chips. Electronic connections grow harder as the scale of integration increases, because the number of circuit elements on the chip surface can increase much faster than the space available to make electrical connections along the edges of the chip. Optical interconnection offers a way around this problem.

One possibility is to multiplex several digital signals together, forming a single high-speed signal that could be transmitted through one fiber rather than several wires. Other alternatives involve optical devices that emit and receive light from the chip surface, with or without fibers. Vertical-cavity semiconductor lasers (VCSELs) are attractive because they emit light from their surfaces. External optics or fibers could collect this light and focus it onto a detector on another chip that would convert the signal into electrical form.

Some high-performance computers already use high-speed fiber-optic connections between internal components. So do electronic telephone switches that are really special-purpose computers. These applications are likely to increase as computer speeds continue to rise.

Atmospheric Optical Links

All optical data links do not transmit signals through fibers. Another option is to transmit light through air. The most familiar examples are infrared ports on electronic equipment, particularly remote controls for televisions and other home entertainment equipment. They use low-power infrared LEDs, transmit low-speed signals, and don't go very far.

A new type of optical data network transmits high-speed laser signals through thin air. The transmitters are similar to those used in fiber systems, but the output optics focus light into a wide beam to pass through the air instead of coupling it into a fiber.

Laser beams can transmit data through air in fiberless networks.

Although air looks clear, it is not a reliable transmission medium because occasional clouds, fog, or precipitation can block light, particularly over long distances. However, sending the beam through air offers important savings by avoiding the need for laying fiber. Laser transmission must meet eye safety regulations, but unlike radio links are not regulated by the Federal Communications Commission.

Atmospheric laser links are most attractive for sending signals between buildings in downtown areas, as shown in Figure 26.7. The buildings are close together and multiple trans-

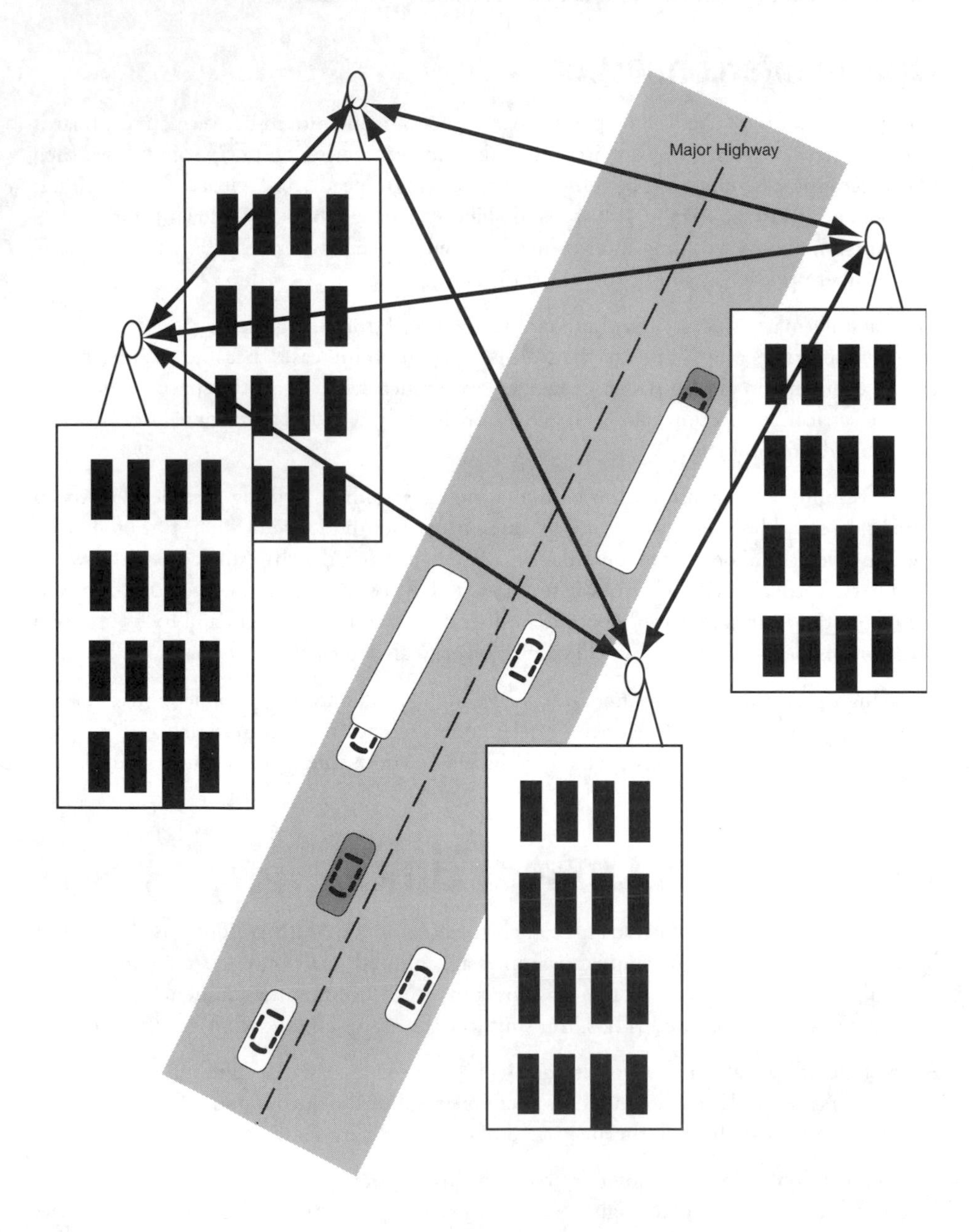

FIGURE 26.7
Laser links through the air in downtown area.

mitters and receivers can make an interconnecting mesh, so buildings won't lose their connections if one beam is temporarily blocked. Fibers also can make such connections, but duct space in urban areas is scarce and expensive, and installing the links can take considerable time. Putting laser transmitters and receivers on the roof is faster and cheaper, as long as there is a clear line of sight between them. Typically individual links in the network mesh run a few hundred meters.

Fiber versus Copper Links

Copper wires are adequate for most computer data transmission because they can transmit high-speed signals a short distance and slower signals much farther.

Wires have several advantages over fiber links. They can directly carry electrical signals from computers and peripheral devices without the need to convert them to optical form. This makes them compatible with existing equipment that transmits electrical signals. Electronic couplers are much simpler and less expensive than optical couplers. Electronic wiring is easy to install, and more technicians know how to install it.

Wire links are cheaper than fiber.

On the other hand, fibers offer their own advantages, starting with their ability to transmit signals over longer distances and at higher speeds. Let's look briefly at these advantages, which can offset the higher cost of fiber equipment.

High-Speed and/or Long-Distance Transmission

Where fiber shines is at high speeds and over long distances. The attenuation of copper wires increases with frequency, as shown in Figure 26.8. The figure plots typical loss per 100 m (*not* per kilometer, as is usual for fiber) as a function of signal frequency for four types of unshielded twisted-pair wires designed for data transmission. Categories 3 and 5 are unshielded twisted pairs meeting standards established by the TIA and EIA; the shielded twisted pair is a cable from the same manufacturer in which the wire pairs are shielded by a metal foil layer. As you learned earlier, fiber attenuation does not increase in the same way.

Category 5 cable was designed specifically for short-distance, high-speed data transmission, and it can indeed carry signals up to 1 Gbit/s over short distances. One qualification is that installation must confine to rigid rules that limit such factors as how much wire can be exposed at the end of the cable when stripping insulation. These factors are not important at low frequencies but become significant at high frequencies and, in practice, can affect transmission at high speeds. In short, Category 5 can work, but you have to do it just right.

Fibers get around these distance limits, whether for individual remote terminals requiring high-speed transmission or for backbone links between local-area networks at different locations. You don't have to install an entire fiber-optic network; as with residential telephone systems, you can install fiber only for the longer, higher-speed parts of the network. For example, in a large office building high-capacity fiber cables may run to each floor and then branch out on each floor to serve several local nodes—like fiber lines to telephone

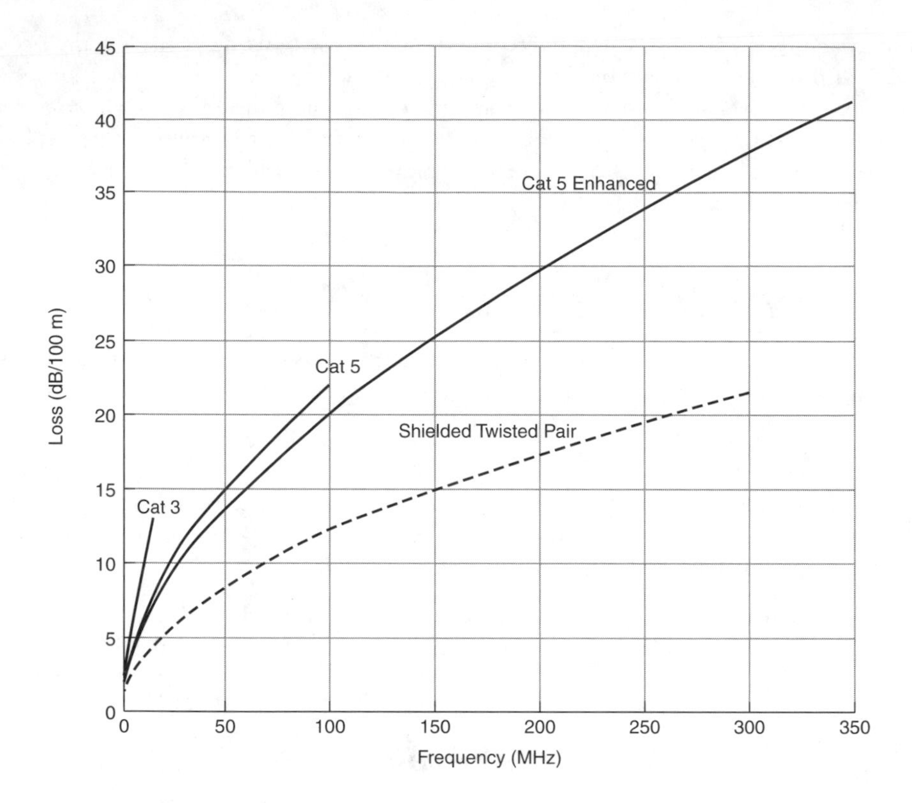

FIGURE 26.8
Typical loss of twisted-pair wiring designed for high-frequency use.

distribution nodes. Copper wires can branch out from local nodes to distribute signals the short distance to desktops. Alternatively, easy-to-install plastic fibers with limited bandwidth can be used for the desktop connections.

Upgrade Capability

Fiber-optic cables make future upgrades easier.

Even if the high capacity of fiber is not absolutely necessary now, you may want it in the future. You aren't going to rip out a functioning copper system to make room for fiber that will be needed at some indefinite date in the future. However, fiber cable may be built into new facilities to allow for future expansion. Labor accounts for a large part of the cost of cable installation, and it takes much less labor to install cable in a new building than to string new cable through existing walls, where access is limited.

Suppose, for example, you are moving into a new building that you expected to occupy for 25 years. You can get away with copper cables now, but you expect to need the higher capacity of fiber within 5 to 10 years. Assume that the fiber hardware costs 50% more than copper cabling and that fiber installation costs an added 20%. You know that fiber costs are coming down, but you also know that installation in a finished building is going to be twice as expensive as installation in a new plant. Table 26.1 shows that copper is much cheaper today, but you'll pay a lot more altogether if you have to go back and retrofit fiber cables later. (These figures are purely illustrative, not from actual installations.)

Table 26.1 Hypothetical costs of installing copper and fiber in new building

Costs	*New Copper*	*New Fiber*	*Retrofit Fiber Later*
Hardware	$5,000	$7,500	$5,000
Labor	$12,000	$14,400	$28,800
Total costs	$17,000	$21,900	$33,800

Immunity to Electromagnetic Interference

Optical transmission is immune to electromagnetic interference (EMI), a common source of noise. EMI arises because changing currents generate magnetic fields, and magnetic field lines induce an electrical current when they cut across a conductor. This can generate noise, as shown in Figure 26.9, when two wires come close together. The strong current in the upper wire induces a current in the lower wire, which adds to the input signal to become noise. If the noise power is high enough, it can overwhelm the input signal. A noise spike, caused by a sudden surge of current, can add a spurious data bit to a digital signal.

Fibers are immune to electromagnetic interference, unlike wires.

You don't need two wires next to each other to get EMI. All you need is passage of magnetic flux through a conductor. Shielding, as in coaxial cables, can reduce electromagnetic interference by weakening the magnetic field that reaches the inner conductor, but even coax

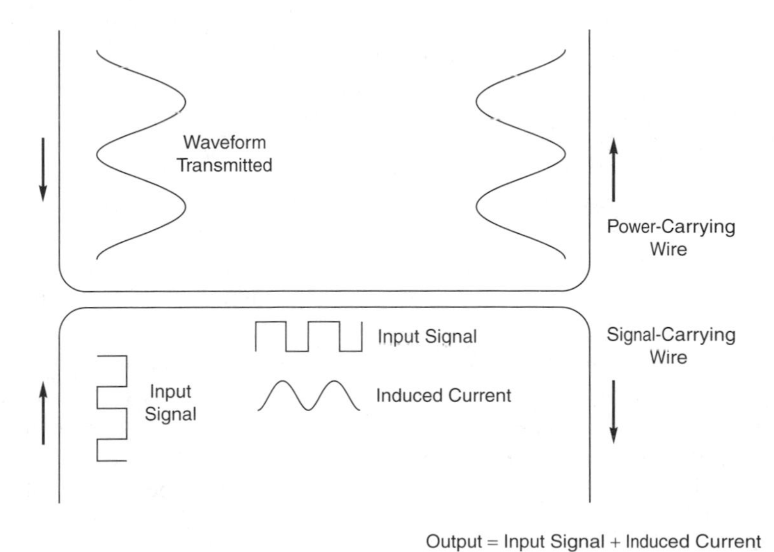

FIGURE 26.9
Induction of noise currents in a wire carrying an electrical signal.

isn't immune to EMI. Optical fibers are immune because they transmit signals as light rather than current. Thus, they can carry signals through places where EMI would otherwise block data transmission—from power substations to hospitals with radio paging systems. (Note, however, that fiber-optic transmitters and receivers can pick up EMI.)

Data Security

Fibers do not emit electromagnetic fields that can be tapped by eavesdroppers.

Magnetic fields and current induction also can let information leak out of a signal-carrying conductor. Changes in signal current cause fluctuations in the magnetic field outside a conductor that carries the same information as the current passing through the conductor. That's good news for potential spies because they can eavesdrop on these magnetic fields without cutting the cable. Shielding the wire, as in coaxial cable, can alleviate the problem, but signals can leak out at bends and imperfections.

Optical fibers confine electromagnetic fields, so there are no radiated magnetic fields around a fiber. That makes it impossible to tap the signal being transmitted through a fiber without bending the fiber (to induce bend losses) or cutting into the fiber. Either of these effects would increase fiber loss sharply, in a way easy for users of the communication channel to detect, making fibers a much more secure transmission medium. This has led secure government data networks to use many fiber-optic links.

Nonconductive Cables

Nonconductive fiber-optic cables are immune to ground loops and resistant to lightning surges.

Subtle variations in electrical potential between buildings can cause problems in transmitting electrical signals. Electronic designers assume that ground is a uniform potential. That is reasonable if ground is a single metal chassis, and it's not too bad if ground is a good conductor that extends through a small building (e.g., copper plumbing or a ground wire in electrical wiring). However, the nominal ground potential can differ by several volts between different buildings or sometimes even different parts of the same building.

That doesn't sound like much—and in the days of vacuum tube electronics, it wasn't worth worrying about. However, signal levels in semiconductor circuits are just a few volts, creating a problem called a ground loop. When the difference in ground potential at two ends of a wire is comparable to the signal level, stray currents start to cause noise. If the differences grow large enough, they can damage components. Electric utilities have the biggest problems because their switching stations and power plants may have large potential differences. Fiber optics can avoid these problems, as long as the nonconductive glass fiber is enclosed in a nonconductive cable with no metal elements. Many sellers of computer data links recommend that fibers be used for interbuilding connections to avoid noise and ground-loop problems. Nonconductive cables can also alleviate a serious concern with outdoor cables—lightning strikes that can cause power and voltage surges large enough to fry electronics on either end.

Sparks from electrical wires can be dangerous in explosive atmospheres.

In some cases, transmitting signals electrically can be downright dangerous. Even modest electric potentials can generate small sparks, especially at switches. Those sparks ordinarily pose no hazard, but they can be extremely dangerous in an oil refinery or chemical plant where the air contains potentially explosive vapors. One tiny spark could make one very large and deadly boom. Again, nonconductive fiber-optic cables can avoid this hazard.

Ease of Installation

Increasing transmission capacity of wire cables generally makes them thicker and more rigid. Shielding against EMI or eavesdropping has similar effects. Such thick cables can be difficult to install in existing buildings where they must go through walls and cable ducts. Fiber cables are easier to install because they're smaller and more flexible. This helps reduce installation costs, especially in cases where only fiber cables are small and flexible enough to fit through existing ductwork in buildings or underground.

Fiber cables can be much easier to install than metal cables.

The small size, light weight, and flexibility of fiber-optic cables also make them easier to use in temporary or portable installations.

Fiber-Optic Data Link Design

The design of fiber-optic data links differs from that of telecommunication systems because most data links are much shorter, running between terminals in a building rather than between buildings. The shorter cable runs make dispersion and attenuation less important than coupling losses. The short transmission distances also make transmitters and receivers a relatively large share of the total system cost, so trade-offs favor reducing transmitter and receiver costs. As a result, data links typically use graded-index fiber, and step-index multimode fiber is not out of the question.

Several types of systems may be used in LANs and data links spanning up to a couple of kilometers, including:

- Plastic fiber transmitting at 650 nm (inside building only)
- Graded-index plastic fiber transmitting in the visible (inside building only)
- Coarse WDM over plastic fiber in the visible (inside building only)
- 850-nm LEDs over graded-index glass fiber
- 850-nm VCSELs over graded-index glass fiber
- 1300-nm LEDs over graded-index fiber
- 1300-nm lasers over graded-index fiber
- Wide WDM with lasers near 1300 nm over graded-index fiber

At this writing, the most common fiber LANs and data links transmit at 850 or 1300 nm through graded-index fiber. At 1 Gbit/s, a single optical channel can be transmitted only about half a kilometer through graded-index fiber; at lower speeds, laser transmitters can send signals a couple of kilometers.

Graded-index fiber is common in LANs.

Single-mode transmission is common at higher speeds or over longer distances. Single-mode fibers are standard in MANs, WANs, and metro networks, where they can transmit to 40 km or more.

Wavelength-division multiplexing is rare in LANs. Some new systems use coarse WDM at high speeds, notably 10-Gigabit Ethernet. WDM is likely to be used in metro networks and WANs, which span long distances. The spread of DWDM in metro, regional, and long-distance systems has led to the leasing of optical channels that may serve as part of MANs or WANs. Thus some networked signals may be transmitted via DWDM "pipes," but the networking layers aren't aware of WDM.

Plastic fibers and LED sources are limited to lower speeds and shorter distances. They compete against copper cable systems, not other fiber technology, and so far have found only limited applications in the data communications market.

Fiber in Standard Data Networks

Many network standards include options for fiber transmission.

Almost all local-area networks are based on industry standard designs and interfaces, such as the highly successful Ethernet family. Links between computers and peripherals also are based on standards. These standard interfaces specify data formats, connector types, and transmission media, so many different devices can be interconnected with minimal trouble. As transmission rates have moved to higher speeds, many of these standards have come to include options for fiber-optic transmission.

The growth in data rates has been steady. The first generation of local-area networks, such as the original Ethernet (10 BaseT) and the IBM Token Ring network, transmitted at about 10 Mbit/s. Higher-speed networks, originally developed for interconnecting local-area networks within an organization, have been adopted for LANs as well. A 100-Mbit/s Fast Ethernet (100 BaseT) followed the original Ethernet, followed by Gigabit Ethernet (1000 BaseX) in 1998 and the 10-Gigabit Ethernet now being finalized. Fibers are important in the higher-speed Ethernet standards, and in the Fibre Channel and Fiber Distributed Data Interface (FDDI) standards.

These data-transmission standards are finding many applications outside of traditional local-area networks, as the growth of the Internet makes computer communications more important. Standard 10-Mbit/s Ethernet is a common interface between personal computers and cable modems. Gigabit Ethernet is being used in fiber to the home systems, as well as to transport signals in metro networks. Let's look at the most important of these standards.

10-Mbit/s Ethernet

The 10-Mbit/s Ethernet LAN was designed for coax, but fiber can be added.

The first LAN to gain much acceptance was the original Ethernet standard codified as IEEE (Institute of Electrical and Electronics Engineers) standard 802.3. It distributes digital data packets of variable length at 10 Mbit/s to transceivers dispersed along a coaxial cable bus, as shown in Figure 26.10. Separate cables up to 50 m long, containing four twisted-wire pairs, run from the transceivers to individual devices (e.g., personal computers, file servers, or printers). The network can serve up to 1024 terminals.

An Ethernet network has no overall controller; control functions are handled by individual transceivers. If a terminal is ready to send a signal, its transceiver checks if another signal is going along the coaxial cable. Transmission is delayed if another signal is present. If not, the

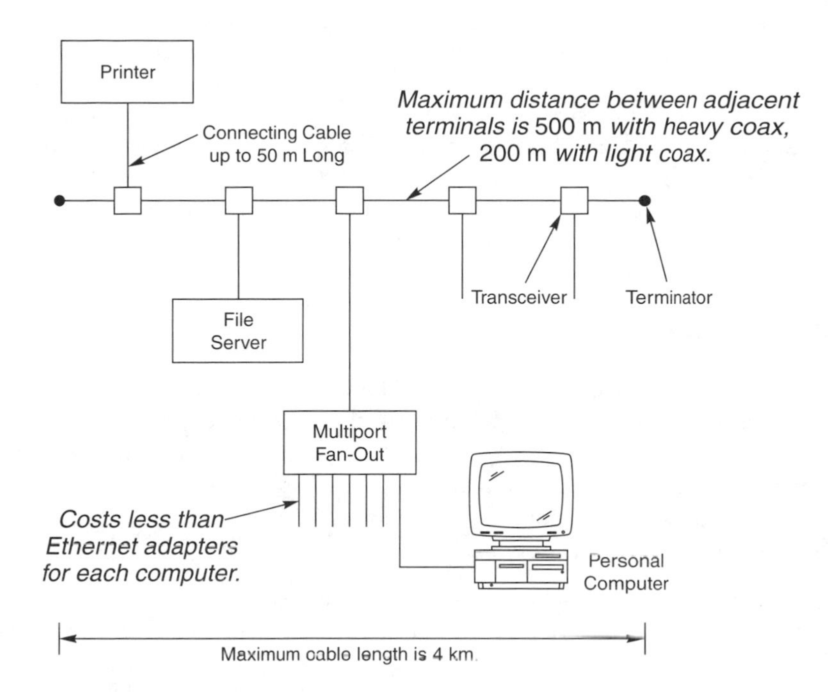

FIGURE 26.10 *Basic elements of 10-Mbit/s Ethernet.*

terminal begins transmitting and continues until it finishes or detects a collision—the transmission of data at the same time by a second terminal. Such collisions happen because it takes time—several nanoseconds a meter—for signals to travel along the coax. If the delay is 6 ns/m, a collision would occur if two terminals 300 m apart on the coax started sending within 1.8 μs of each other. The terminal stops transmitting if it detects a collision and waits a random interval before trying again.

An address header specifies the destination for every data signal. All the transceivers on the network see every data signal, but they ignore the signals not directed to them. A transceiver relays to the attached terminal only those signals addressed to it.

There are some important variations on the basic 10-Mbit/s Ethernet design. The original standard heavy coaxial cable allows transceivers to be up to 500 m apart, but it is expensive. Substituting a lighter grade of coax limits transceiver spacing to 200 m, but this "thin" Ethernet is adequate for most purposes. Another alternative is using twisted-wire pairs that can carry signals up to about 100 m. In addition to the data bus configuration shown in Figure 26.10, Ethernet often is arranged in a star configuration, with cables radiating outward from a hub, which relays signals to other terminals.

Optical fibers can stretch transmission distances beyond the limit imposed by the loss of coaxial cable to distances limited by other factors, such as the time signals take to travel through the network. A point-to-point fiber link may connect two coaxial segments of an Ethernet or a remote terminal with a central Ethernet. This allows a single Ethernet to link terminals in different buildings, which is difficult with the 500-m limit of coax.

The maximum transmission distance depends on whether the network is operating in half-duplex mode where terminals either transmit or receive at any one time, or in full-duplex mode where they simultaneously send and receive data. In half-duplex mode, the maximum distance is 2 km for either multimode or single-mode fiber. In full-duplex mode, multimode fiber allows cable runs to 2.5 km and single-mode allows spacing to 15 km. (For reference, light takes roughly 75 μs to travel a 15-km fiber, meaning the terminal on the end of a 15-km fiber lags 75 μs behind the rest of the network.)

Fast Ethernet (100 Mbit/s)

Fast Ethernet operates at 100 Mbit/s.

As the name implies, Fast Ethernet (100 BaseT) is a faster version of Ethernet, using interface cards that operate at 100 Mbit/s but retain the same frame format and transmission protocols as the original 10-Mbit/s Ethernet. The Fast Ethernet standard was approved in 1995. It uses the same network configurations and cabling as 10-Mbit/s Ethernet; the major change is replacing 10-Mbit/s Ethernet cards with Fast Ethernet cards. However, the faster speed limits coax runs to 100 m.

The Fast Ethernet specification limits half-duplex transmission to 412 m over either single- or multimode fiber, a travel time of 2 μs. Full-duplex transmission stretches the maximum distance to 2 km for multimode fiber and 10 km for single-mode. The differences arise because of differences in the nature of half- and full-duplex transmissions.

Gigabit Ethernet (1 Gbit/s)

Gigabit Ethernet needs fiber to transmit beyond 100 m.

The next step is Gigabit Ethernet, which, as you might expect, operates at 1 Gbit/s. This standard uses the same protocols and frame format as slower Ethernets, but with only full-duplex transmission. Because of the high speed, Gigabit Ethernet node spacing is shorter. A special *twinax* cable—a coaxlike cable with a twisted pair at the center instead of a single metal wire—is used for jumper cables to 25 m. Splitting the 1 Gbit/s among the four twisted-wire pairs in a Category 5 cable allows it to span distances to 100 m.

Fiber is the backbone for Gigabit Ethernet, with copper used only for short connections. Table 26.2 shows the transmission distances specified for various types of fiber at different wavelengths. Note that the bandwidth specification for graded-index fiber varies with wavelength.

The wide variation in fiber transmission distances is due primarily to differences in dispersion among fibers. New broad-bandwidth graded-index fibers have extended the range of multimode fibers. The two long-haul laser specifications are not formally part of the IEEE standard, but have been approved by a group of companies. They are intended largely for use in metro networks, where distances exceed the nominal 5-km limit of 1300-nm single-mode systems. Care must be taken with those systems not to overload the receiver with excess laser power.

The signaling standard and much specific hardware for Gigabit Ethernet come from the Fibre Channel specification for 1-Gbit/s transmission.

Table 26.2 Fiber transmission formats for Gigabit Ethernet

Transceiver	*Fiber Types*	*Bandwidth (MHz-km)*	*Range (m)*
850 nm laser (1000Base-SX)	62.5/125	160	2–220
850 nm laser	62.5/125	200	2–275
850 nm laser	50/125	400	2–500
850 nm laser	50/125	500	2–550
1300 nm laser (1000Base-LX)	62.5/125	500	2–550
1300 nm laser	50/125	400	2–550
1300 nm laser	50/125	500	2–550
1300 nm laser	9/125 (single-mode)	—	2–5000
1300-nm long-haul laser (1000Base-LH)	9/125	—	1000–50,000 (1–49 km)
1550 nm long-haul laser (1000Base-LH)	9/125	—	50,000–100,000 (50–100 km)

10-Gigabit Ethernet

10-Gbit/s Ethernet is mainly for MANs and WANs.

The next logical step is to 10-Gigabit Ethernet, a standard that at this writing has been drafted but has yet to receive its final approval. 10-Gigabit Ethernet is intended largely for metro and wide-area networks, where it will provide backbone links between other networks. It also will provide relatively short connections between high-speed equipment in the same building. The transmission format is compatible with the 10-Gbit/s OC-192 SONET rate, so a 10-Gigabit Ethernet output could drive one optical channel in a DWDM system at 10 Gbit/s for long-haul transmission.

As a high-speed network intended for interconnection of other networks, the initial version of 10-Gigabit Ethernet has no provisions for copper connections, which would be problematic at that speed. Instead, the proposed standard covers several distinct fiber transmission formats, including 10-Gbit/s transmission on a single optical channel, and 2.5 Gbit/s transmission on each of four widely spaced WDM channels. Table 26.3 shows the recommended transmission formats.

The difference between transmission distances through graded-index fiber at 850 and 1310 nm may seem surprising given the low bandwidth of the 62.5/125 fiber; however, this is because the two alternatives are intended for different applications. The 850-nm

Table 26.3 Fiber transmission formats for 10-Gigabit Ethernet

Transmitter	*Fiber Types*	*Fiber Bandwidth*	*Distance Limit*
850-nm single-channel	50/125 graded-index	500 MHz-km	65 m
1310 nm Wide WDM	62.5/125 multimode	160 MHz-km	300 m
1310 nm Wide WDM	9/125 single-mode	NA	10 km
1310 nm single-channel	9/125 single-mode	NA	10 km
1550 nm single-channel	9/125 single-mode	NA	40 km

standard calls for modulation of a single transmitter at 850 nm to transmit through newly installed graded-index fiber with a relatively high bandwidth. The 1310 nm standard is intended for use with low-bandwidth, graded-index fiber that had been installed previously; it transmits 2.5 Gbit/s at four separate wavelengths so the old fiber can be used.

Fibre Channel

The Fibre Channel standard covers a range of signal transmission, as you saw in Chapter 20. In addition to its use as a storage-area network and point-to-point transmission standard for computer systems, it can be used for backbone networks and switched transmission. Hubs connect nodes to form loops; switches are interconnected to make a fabric that functions somewhat like the phone system in directing signals between devices.

Fibre Channel uses 10-bit coding for each byte; top speeds are 4.25 Gbit/s.

Fibre Channel uses a 10-bit coding for each 8-bit byte. The bits can enter the system in parallel, but Fibre Channel transmits them in series. Data rates can be specified either as megabits per second (Mbit/s, sometimes abbreviated Mb/s) or megabytes (Mbytes, sometimes abbreviated MB), which can be confusing. Overhead bits increase the bit rate by 6.25% above the 10-bit coding. Table 26.4 lists the speeds in both formats.

Table 26.4 Fibre Channel speeds

Assigned Number	*Megabytes/Second*	*Megabits per Second (with Overhead)*
12.5	12.5	133
25	25	266
50	50	531
100	100	1062
200	200	2125
400	400	4250

Like Gigabit Ethernet, Fibre Channel allows transmission over copper as well as fiber. Twisted pair, coax, and twinax are the major alternatives for speeds of 1 Gbit/s and below; at the higher rates only fiber is specified. Transmission can be in the short- or long-wavelength bands.

Like Ethernet, the Fibre Channel protocol limits the numbers of terminals in a loop and the arrangement of hubs and switches. The details are beyond the scope of this book.

Fiber Distributed Data Interface (FDDI)

The Fiber Distributed Data Interface (FDDI) network transmits user data at 100 Mbit/s. The FDDI standard calls for the ring topology shown in Figure 26.11, with two rings that can transmit signals in opposite directions to a series of nodes. It also specifies concentrator-type terminals that allow stars and/or branching trees to be added to the main FDDI backbone ring. Normally one ring carries signals while the other is kept in reserve in case of component or cable failure. The maximum distance between nodes is 2 km over multimode fiber at 1300 nm.

The 100-Mbit/s FDDI standard LAN is based on fiber optics.

FDDI transmission is controlled by a scheme of "token passing" used in slower-speed token ring networks, covered by the IEEE 802.5 standard. Terminals do not contend for space to send signals, as does Ethernet, but instead pass around the loop an authorization code called a *token.* When a node with a message to send receives the token, it holds the token and sends the message, which includes a code identifying its destination. All other nodes ignore the message, which is canceled when it completes its path around the ring. Then the terminal that sent the message begins passing the token around the ring again. You can think of this as a somewhat more orderly scheme than Ethernet transmission, although it has its own limitations.

FDDI uses a 4 of 5 transmission code that adds one extra bit for every four data bits, so the line rate is 125 Mbit/s. The standard was developed around fiber-optic transmission, but copper wires can be used over short distances. Wired versions of FDDI are sometimes called CDDI, with the C from copper substituted for the F from fiber.

Copper wires can carry FDDI signals short distances.

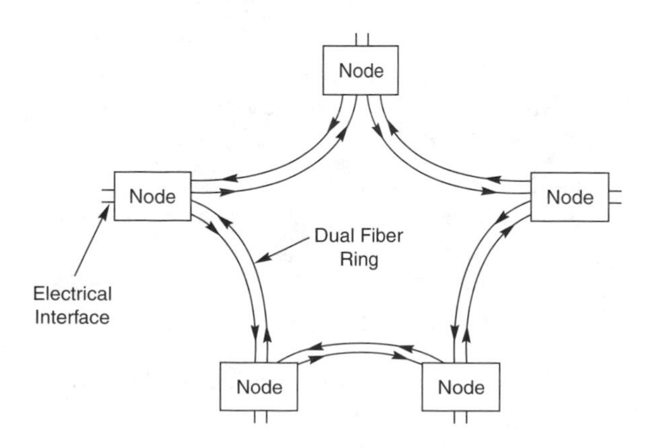

FIGURE 26.11
FDDI dual-fiber ring.

What Have You Learned?

1. The Internet grew from links between local-area networks.
2. Computers do not generate output at uniform rates, so computer data lends itself to packet switching, with routers directing packets to their destinations.
3. Computer networking standards are structured in layers, like other telecommunication standards.
4. The Internet backbone system links Points of Presence (POPs), which are comparable to the major nodes on the long-distance telecommunications network.
5. A metropolitan-area network (MAN) or metro network interconnects networks in the same area. A wide-area network (WAN) typically interconnects networks over a wider area.
6. Computers exchange data over point-to-point links between pairs of devices or over local-area networks (LANs) that interconnect computers, terminals, and devices. A LAN serves devices on one place, typically a building or organization.
7. Common LAN configurations are stars, rings, and buses.
8. Laser beams transmit data through air in wireless data networks, which are used to avoid the cost of laying cables between buildings in downtown areas.
9. Copper wires are cheap for data transmission, but they cannot carry signals over as long a distance or as at high a speed as can fiber. Fibers offer potential room for expansion of services.
10. Fibers are immune to electromagnetic interference. Fibers do not radiate electromagnetic fields, making data transmission on fiber more secure than on copper.
11. Graded-index fiber is common for LAN transmission spanning a short distance. Single-mode fiber is needed for distances beyond 0.5 km at 1 Gbit/s, and beyond about 2 km at lower speeds.
12. Some networks use coarse WDM, but DWDM is rare inside networks. However, single optical channels on a DWDM fiber may serve as links in MANs or WANs.
13. Many network standards allow for fiber transmission, particularly at high data rates.
14. The original Ethernet transmitted 10 Mbit/s. Fast Ethernet transmits 100 Mbit/s, Gigabit Ethernet 1 at Gbit/s, and 10-Gigabit Ethernet at 10 Gbit/s. The faster standards rely heavily on fiber transmission. Gigabit Ethernet uses components developed for Fibre Channel; the 10-Gigabit Ethernet standard is new.
15. Fibre Channel includes point-to-point, loop, and switched transmissions. Fiber transmission is needed at 1 Gbit/s and above, but copper can be used at lower speeds.

What's Next?

In Chapter 27, I will cover fiber-optic video transmission, with particular emphasis on cable-television networks.

Further Reading

Fibre Channel Association: *http://www.fibrechannel.com/*

Gigabit Ethernet Alliance: *http://www.gigabit-ethernet.org/*

Joseph H. Levy and Glenn Hartwig, *Networking Fundamentals,* 2nd ed. (IDG Books, Foster City, CA, 1998)

Peter Rybaczyk, *Novell's Internet Plumbing Handbook* (Novell Press/IDG Books, San Jose, 1998)

10-Gigabit Ethernet Alliance: *http://www.10gea.org/*

Bruce Tolley, "Gigabit Ethernet Comes of Age," *http://www.3com.com/technology/tech_net/white_papers/503003.html*

Questions to Think About for Chapter 26

1. If the Internet is the computer equivalent of the long-distance backbone system for telephony, what is a good analogy for the regional telecommunications network?
2. A cable modem shares a loop of cable with other subscribers in your neighborhood, in effect making all of you share a local-area network run by the cable company. Suppose that the modem is able to deliver 10 Mbit/s to your computer's Ethernet port if nobody else is online. Your next-door neighbor signs up after you say how great it is, then gets eight more neighbors to sign up. If the capacity available to you depends only on the number of users, how much of the original capacity is left?
3. A fiberless optical network can transmit signals between any two office buildings 95% of the time on foggy days. If you have a single laser link to your building, how much time would you expect to be down during a 24-hour interval of foggy weather? Suppose you could add two more laser links that are completely independent of the first and of each other, and would connect you to the same network. How much would your service improve?
4. Your company has just signed a 10-year lease on a building being renovated. The local-area network in your old building was a standard 10 Mbit/s Ethernet, but traffic was doubling every year and the network is at capacity. You want to stay with Ethernet standards. How long can you go before you run out of capacity on Fast Ethernet? How long before even Gigabit Ethernet won't do?
5. You need to transmit Gigabit Ethernet between two buildings separated by 400 m. What are your options? What are your options if you expect to have to upgrade to 10-Gigabit Ethernet?
6. Why would you *not* want to use DWDM to increase transmission capacity of a local-area network?
7. Why is the Internet sometimes called a wide-area network?

Quiz for Chapter 26

1. Points of Presence (POPs) on the Internet are
   a. phone numbers called by dial-up modems.
   b. addresses of Web sites.
   c. major nodes where Internet backbone systems transfer signals.
   d. long-distance data transmission lines.
   e. computers with Internet access.

2. Data flow to and from computers
   a. varies over time in a regular and predictable way.
   b. occurs in irregular bursts.
   c. is at a constant speed determined by your modem.
   d. is at a constant speed determined by your computer.
   e. is at a constant speed from your computer but incoming data rates can vary.

3. Which of the following is a local-area network (LAN)?
   a. A system that interconnects many nodes by making all signals pass through a central node
   b. A ring network with a transmission medium that passes through all nodes
   c. A common transmission medium or data bus to which all nodes are connected but which does not form a complete ring
   d. All of the above
   e. None of the above

4. What makes optical fibers immune to EMI?
   a. They transmit signals as light rather than electric current
   b. They are too small for magnetic fields to induce currents in them
   c. Magnetic fields cannot penetrate the glass of the fiber
   d. They are readily shielded by outer conductors in cable

5. The most important drawback of optical fibers for point-to-point data transmission is
   a. that they require switches.
   b. the higher costs of fiber equipment.
   c. that fiber cannot provide electrical grounding.
   d. that they do not operate properly at low data rates.
   e. that it is difficult to upgrade.

6. Why would you install a fiber-optic network in a new building rather than a less expensive wire system if both could meet current requirements?
   a. To avoid ground loops
   b. Fiber is simpler to install in a new building
   c. To provide future upgrade capability at a cost much lower than retrofitting later

d. To get a big kickback from the fiber supplier

e. Fiber costs are always lower

7. Why are fiber-optic cables recommended for connections between buildings?

a. To avoid ground loops

b. Fiber is simpler to install between new buildings

c. To provide future upgrade capability

d. To avoid eavesdropping by industrial spies

e. Fiber costs are always lower

8. When would optical fibers be used in a standard (10-Mbit/s) Ethernet?

a. Never, the standard requires coaxial cable

b. To extend transmission distance to reach remote terminals

c. Routinely, all Ethernet standards require fiber for distances beyond 10 m

d. When the network includes a hub

e. When the stockroom is out of coaxial cable

9. When would optical fibers be used in a Gigabit Ethernet?

a. Never, the standard requires coaxial cable

b. Always, the standard requires fiber-optic cable at all points

c. In most cases beyond short distances

d. Only if the Gigabit Ethernet network had to be connected to an FDDI network

e. Only when making connections to the Internet

10. Gigabit Ethernet components are adapted from those developed for which other standard?

a. FDDI

b. 10-Mbit/s Ethernet

c. ATM

d. SONET

e. Fibre Channel

11. Fibre Channel can be used for which type of transmission?

a. Point-to-point

b. Ring or loop

c. Switched

d. None of the above

e. All of the above

12. How can you transmit Gigabit Ethernet through Category 5 cable?

a. Under all conditions; it is required by the standard

b. By adding fiber pairs to the Category 5 cable

c. By electrically insulating the Category 5 cable

d. By dividing the signal among four twisted pairs in the Category 5 cable, up to 100 m

e. Only through a miracle

13. One data rate is specified as a standard in Fibre Channel, FDDI, and the various Ethernet standards. It is

a. 10 Mbit/s.

b. 100 Mbit/s.

c. 250 Mbit/s.

d. 500 Mbit/s.

e. 1 Gbit/s.

14. Where could graded-index fibers be used in 10-Gigabit Ethernet?

a. Only for distances less than 50 m

b. For distances of 65 to 300 m

c. For distances up to 300 m

d. For distances up to 2 km

e. For distances beyond 300 m

15. What is the highest-speed type of Ethernet you can transmit over a 500-m run of graded-index fiber?

a. 10-Mbit/s Ethernet

b. Fast Ethernet

c. Gigabit Ethernet

d. 10-Gigabit Ethernet

CHAPTER

27

Video Transmission

About This Chapter

Fiber optics play an important role in cable-television transmission and in behind-the-scenes roles in some other video transmission. As cable-television networks have spread, they have converged with other communications technologies. Modern cable systems now can transmit computer data and telephone conversations, as well as the traditional video programs. This *convergence* is creating competition between cable and telephone networks.

This chapter will explain the basics of video transmission, concentrating on the cable systems that are the main users of fiber optics. It will describe the role of fiber in cable distribution, the evolution of cable systems, and the large uncertainties created by plans to change television transmission from analog format to digital format.

Video Basics

Video signals must encode changing images, usually in color, as well as sound. This makes them more complex than voice signals, and different from data signals. Analog voice telephones convert the continuous variations of the sound waves in your speech into continuous variations of an electrical signal transmitted down the telephone wires. Digital systems translate this time-varying signal into a series of 1s and 0s, which digitally represent the voice signal.

Video has to encode pictures as well as sounds. It does this by breaking the image into a series of parallel horizontal lines, a process called *raster scanning.* You can see the lines if you look closely at a television screen. The color and brightness vary along each line; your eye blends the lines into a composite image. Figure 27.1 shows how raster scanning represents a simple black-and-white pattern, with the background lines left to show how they build the image. Video signals must continuously transmit new images of changing

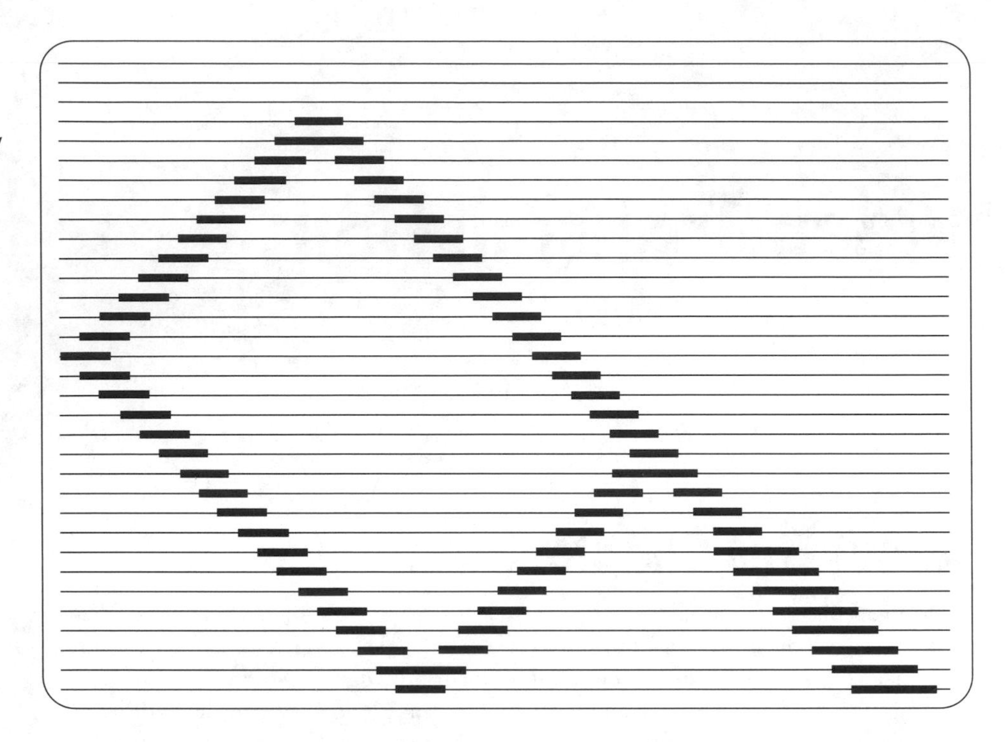

FIGURE 27.1
Raster scanned image is made from many parallel lines.

Video signals encode changing pictures and sound.

scenes, generating new images fast enough that the eye sees no sudden shifts between successive frames. Color television is more complicated because it requires writing images in all three primary colors—red, green, and blue.

Standard formats have been developed to transmit video signals by encoding the information on color, brightness, and sound in different ways. As in other communication systems, both transmitter and receiver must use the same format. The most widely used current standards are based on analog transmission, but digital standards have been developed and efforts are being made to phase them in.

Video-Transmission Requirements

Video requires much more transmission capacity than sound or equivalent digital data.

Video requires much more transmission capacity than sound or equivalent digital data. It's often said that one picture equals a thousand words, but Table 27.1 shows that the picture requires considerably more transmission capacity than 1,000 words of written text. Sophisticated digital compression can reduce data rates by factors of 10 to 60 for video transmission, but they depend on transmitting a series of pictures, so single images cannot be compressed as much. Note that the spoken word is much less efficient when converted into digitized sound, and a thousand digitized spoken words require about as much digital storage space as one HDTV frame.

You can appreciate what these differences mean if you use a 56-kbit/s modem to access the World Wide Web with your personal computer. A page of pure text loads quickly. A page

Table 27.1 Comparison of approximate video, voice, and text transmission requirements

Transmission	*Analog Equivalent*	*Digital Equivalent (No Compression)*	*Digital Equivalent (Compressed)*
Standard U.S. analog television (NTSC)	6 MHz	270 Mbit/s	2–10 Mbit/s
HDTV (U.S. format)	~100 MHz	1.5 Gbit/s	19.3 Mbit/s
Voice telephone	3 kHz	64 kbit/s	~10 kbit/s
One standard TV video frame		9 Mbits	
One HDTV video frame (1/60 s)		25 Mbits	
1000 spoken words (5 min on phone)		20 Mbits	
1000-word text file		60 kbits	

with a fair-sized static image may take half a minute. And you know you've hit the World Wide Wait when you find a page with a video clip that takes many minutes to download, even though the image is small and the sequence lasts less than a minute. You don't find speech much on the Web, because digitized audio files also are large.

Transmission Standards

Video signals are transmitted in standardized formats so transmitters can talk to receivers. These formats have evolved for historical reasons and are often not ideal. One difference in analog video formats is the number of frames per second, originally chosen to be half the frequency for alternating current (60 Hz in North America and 50 Hz in Europe). The North American standard for analog broadcast color television was chosen in the 1950s to be compatible with older black-and-white receivers, because broadcasters did not want to lose that audience.

New standards have been developed for digital television. The original goal was higher screen resolution, or *high-definition television* (HDTV), for large screens, but the standards were broadened to cover *advanced television* (ATV), for sets with smaller screens. The technology comes in part from high-resolution computer monitors. Importantly, the digital standard signals are *not* compatible with existing analog television sets, which will require adapters if and when television broadcasts shift completely to digital television.

Other video standards are also in use, such as those for computer monitors, but I will concentrate mostly on television standards.

Standard Broadcast Analog Video

NTSC video displays 30 analog 525-line frames a second.

The present analog standard for broadcast television programs in North America is the NTSC format, from the National Television System Committee. The analog signals carry information representing the lines that compose the screen images. Pictures are displayed as 525-line frames (although a few lines do not actually show up on the screen). Nominally, NTSC shows 30 frames a second, but to keep the image from flickering to the eye, NTSC uses an interlaced scanning technique. First it scans odd lines on the display, then the even lines, then the odd lines again, as shown in Figure 27.2. Technically, this interlaced scan displays 60 half-images (called fields) a second, with only 267 lines of resolution each, but this fools the eye, giving the appearance of high resolution while avoiding the flicker of slower scanning speeds.

Nominally, the NTSC video bandwidth is 4.2 MHz, with the sound carrier at a higher frequency. However, for broadcast the video signal is used to modulate a radio-frequency signal, a process that increases overall bandwidth to 6 MHz, which is the amount of radio-frequency spectrum allocated to each broadcast television channel in the United States. Figure 27.3 shows the structure of this signal, which extends from 1.25 MHz below the carrier frequency to 4.75 MHz above the carrier. The NTSC format is used in North America, Japan, Korea, the Philippines, and much of South America.

PAL and SECAM are interlaced scanning systems showing 25 frames of 625 lines each per second.

Two other broadcast television standards are in wide use: PAL and SECAM. Both are interlaced scanning systems that show 25 frames (50 fields) per second, each frame with 625 lines. These have nominal video bandwidth of 5 to 6 MHz and broadcast channel bandwidth of about 8 MHz. PAL and SECAM systems are used in Europe, mainland Asia, Africa, and parts of South America.

These standards were set for television broadcasting from ground-based transmitters. National and international standards set aside specific frequencies for television broadcasting, with each channel allocated the required bandwidth (6 MHz for NTSC channels). These

FIGURE 27.2
Interlaced and progressive scanning.

Interlaced Scanning
1 3 5 7 9 11
2 4 6 8 10 12

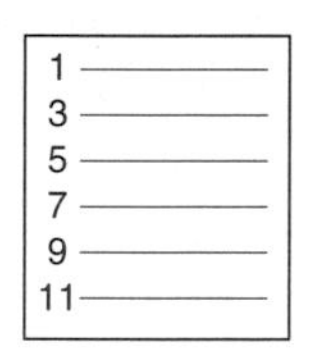

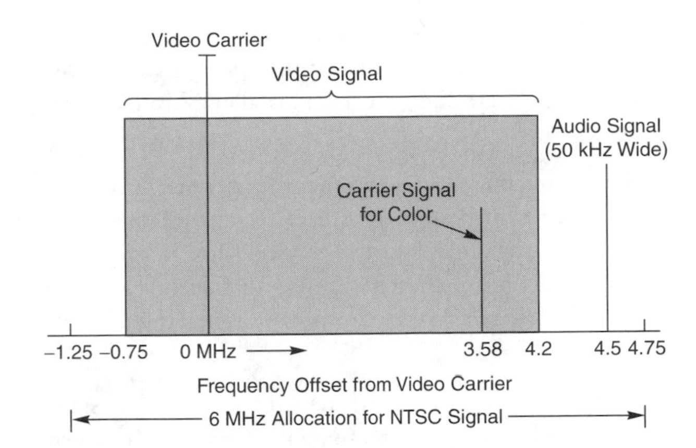

FIGURE 27.3
Structure of NTSC broadcast video signal.

standards have come into wide use for other types of video because NTSC, PAL, and SECAM equipment is readily available.

Remember that broadcast video standards were established decades ago, when color television came on the market. (The NTSC standard was a modified version of the original North American standard for black-and-white television, which goes back to 1948.) This means that these standards were developed for the vacuum-tube technology available in the electronic stone age.

Standard video formats are decades old.

Computer Displays and Video Formats

It might seem logical to use standard television displays for computers, but the two technologies are not readily compatible. Television sets are adequate displays for many computer-based video games, and sufficed for some early personal computers. However, text displayed on a screen with interlaced scanning does not show up well, because the interlacing effectively mixes information from successive frames. The best displays for computers use progressive scanning in which all lines are scanned one at a time, then the entire screen is rescanned, as shown in Figure 27.2.

Computer displays require progressive scanning to show text clearly, not NTSC format.

Progressive scanning demands more bandwidth and faster electronics than interlaced scanning—because it transmits 60 (or sometimes more) complete frames a second to avoid flicker. NTSC video transmits 60 interlaced half-frames or fields, so its resolution is lower.

Multimedia or interactive video displays are hybrids, based on a combination of computer and television technologies. At this writing, they don't have their own special display standards; those that play on computers use computer formats, those played on television sets use television formats. The American HDTV standard allows both progressive and interlaced scanning.

Digitized Video and Compression

Advanced television systems will use compressed digital video.

Like voice signals, video can be digitized. The raw data rate is considerably higher than the corresponding analog bandwidth, as shown in Table 27.1. However, raw image data contains much redundant information, so it can be compressed to occupy less space.

All files can be compressed somewhat by suitable software. One simple approach is to replace long strings of identical bits or bytes with shorter codes containing the same information, such as a code indicating that the next 20 digits are 1s. This allows limited compression of a single image or data file.

Video signals can be compressed much more efficiently than other files because successive frames contain similar information. Many compression techniques compare successive frames and transmit only the information that changes between frames. Most images don't change much, so this takes only a small fraction of the bits needed to transmit a whole frame. Full frames are transmitted at regular intervals so the compressed signal doesn't drift too far from the original.

Impressive progress has been made in video compression, and the American HDTV standard is based on compressing signals from the raw 1.5 Gbit/s to about 19.3 Mbit/s, which can fit into the 6-MHz bandwidth allocated for broadcast video. Compression inevitably degrades the signal slightly, with the damage depending on the extent of compression and the type of signal. Compression works best for images that change little between frames, such as videoconferences or news broadcasts showing talking heads. Rapid motion and changing scenes, such as in broadcasts of sports, are the most degraded.

Studios use various levels of compression at different stages of production, as shown in Table 27.2. Video producers try to maintain the highest possible quality during production but accept lower quality for transmission. In addition, they must avoid compression techniques that depend on a sequence of frames, because editing could change the sequence.

High-Definition Television (HDTV)

HDTV will greatly increase resolution for large-screen sets.

Since the 1980s, the electronics industry has been pressing for a new generation of television technology. Their goal is to offer larger, wider images of much better quality and—not incidentally—to make more money by selling a new and more expensive generation of large-screen television sets. After initially proposing an analog standard, the industry switched to a digital standard, which the Federal Communications Commission accepted in the United States. European countries have developed their own standard, based on similar ideas. (Japan developed an analog HDTV system called MUSE, which found limited use in video production, but plans to switch to a digital system.)

Officially, the U.S. standard is known as the *Digital Television* (DTV) standard of the Advanced Television Systems Committee. It includes 18 distinct digital video formats with

Table 27.2 Compression levels for HDTV

Task	*Compression Ratio*
Video production	4:1
Archival storage	25:1
Transmission	75:1

different numbers of scan lines and screens per second. Six are classed as high definition and the other 12 are standard definition television (SDTV), which offers somewhat better image quality than NTSC analog television. The HDTV formats are intended for large-screen sets; the SDTV formats are for smaller screens. Table 27.3 lists these formats.

A single digital signal will support all these formats. Digital electronics in new sets will decode the signal to produce a display that matches the screen. This compromise allows both the interlaced scanning preferred by the television industry and the progressive scanning preferred for computer displays. The different picture sizes accommodate the variety of television set sizes.

Digital television supports several levels of screen resolution.

Some heavy politics went into persuading broadcasters to make the switch. Congress agreed to give every station a free digital channel, providing they returned their analog channel after switching to digital broadcasts. The nominal deadline is 2006—but only if 85% of viewers have switched to digital television, which is not likely to happen by then.

Several problems have slowed the switch to HDTV. Engineers have found problems with the U.S. broadcast standard. Large-screen digital sets are very expensive, and sales have been very slow. Few programs are broadcast in digital format, and many broadcasters would rather use their digital channels to transmit multiple standard-definition digital video signals than a single HDTV signal. The cable television industry has ignored digital television, which poses a serious problem because about two-thirds of American households get their television signals over cable. At this writing, it's not clear what is going to happen—stay tuned.

Table 27.3 Advanced television digital formats

Picture Size (Lines High by Pixels Long)	*Frames per Second*	*Aspect Ratio*
1080 × 1920	60 interlaced	16:9 (wide-screen)
(HDTV)	30 progressive	
	24 progressive	
720 × 1280	60 progressive	16:9 (wide-screen)
(HDTV)	30 progressive	
	24 progressive	
480 × 704	60 interlaced	16:9 (wide-screen)
(SDTV)	60 progressive	4:3 (conventional)
	30 progressive	
	24 progressive	
480 × 640	60 interlaced	4:3 (conventional)
(SDTV)	60 progressive	
	30 progressive	
	24 progressive	

Transmission Media

So far I haven't said much about how video signals are distributed. Because this is a book on fiber optics, our main interest is in the use of fiber in cable television systems. However, you also should know a bit about other important transmission technology: ground-level broadcast at radio or microwave frequencies, satellite broadcast at microwave frequencies, and coaxial cables.

Terrestrial Broadcast

Local stations broadcast at radio frequencies on assigned channels.

Television began as a broadcast medium, with local stations transmitting radio-frequency signals from tall antennas. NTSC, PAL, and SECAM standards were all based on the assumption that signals would be broadcast from terrestrial towers to antennas connected to home receivers. The radio waves induce small electrical currents in the antennas and receivers amplify and process the signal to extract the video. Government agencies assign specific frequencies for stations to use at their locations. Other stations may use the same frequencies if they are far enough away that the signals will not interfere.

The U.S. HDTV signal was also designed for broadcast, in the same 6-MHz frequency bandwidth assigned to an analog broadcast. The Federal Communications Commission assigned each station a second channel for digital broadcasts, at a different frequency than any analog channel in the area, but in the same band of channels assigned to television broadcast. Digital signals are not supposed to interfere with analog broadcasts.

Microwave Rebroadcast Services

Your television antenna can only receive broadcast signals from stations near you. Some services use microwave frequencies to distribute signals from distant stations and premium services, such as Home Box Office, to paying subscribers. These services include direct broadcast satellites and local multipoint distribution services.

Satellites broadcast microwave signals that carry many channels.

A *direct broadcast satellite* transmits many video signals simultaneously in the microwave band to subscribers with small dish antennas and decoders. The transmitting satellite is placed in an orbit above the equator so it circles the planet exactly once every 24 hours. This makes it appear to stay in place above the same point. Its transmitting antenna beams microwaves to the area it serves. This design allows a single transmitter to cover most of the United States, including rural areas where cable service is usually unavailable.

A *local multipoint distribution service* (LMDS) is a fixed wireless system that transmits microwave signals from many small terrestrial antennas distributed in its service area. Each antenna broadcasts signals over a small area, like a cell phone, but it distributes signals to fixed receivers. This distribution of antennas allows the system to offer two-way services such as high-speed data, teleconferencing, interactive video, and voice telephone service. The service also can distribute video signals to subscribers in the same way as direct broadcast satellites.

So far, direct broadcast satellites are much better developed and are posing a serious challenge to cable television. Few fixed terrestrial systems have been installed. Direct broadcast satellites transmit analog or digital video, encoded for microwave transmission.

Cable Television

Cable TV systems carry dozens of analog video signals on fiber and coaxial cables.

Cable television began as community antenna television (CATV, an acronym still used by the industry) to serve areas not normally reached by broadcast television signals. The idea was to build one big antenna to pick up remote broadcasts and then distribute the signals via coaxial cables to local homes. Eventually the concept spread to urban and suburban areas where broadcast quality was better, but the choices were limited. Economics and interference limit the number of broadcast channels in a metropolitan area, but cable systems can pick up many more channels (from satellites and distant stations) and distribute them to homes along with signals from local stations. Cable systems can also offer extra-cost "premium" services to customers who rent special decoders. (The signals are scrambled and sent to all subscribers, but only those who pay the premium for the decoders can unscramble them.) Existing cable television systems serve essentially the same function as broadcast television—they distribute the same signals to everyone.

Present cable systems carry analog video signals in NTSC format. Each channel is assigned a frequency slot and signals are multiplexed together for transmission over the cable. Typical cable systems carry dozens of 6-MHz channels, but some new ones can carry up to 100. The signals are carried through fiber and coaxial cables at a broad range of radio frequencies to about 750 MHz. Set-top cable boxes demultiplex the signals, picking out the one selected by the viewer.

Cable systems have evolved considerably in the past several years. Many now have some capability to transmit signals from hences as well as to homes; some offer digital data transmission, as I will explain later in this chapter. However, it remains to be seen how cable systems will handle HDTV signals.

Fiber optics has come to play an important role in cable television, and that role is likely to expand as systems continue to evolve.

Cable Television Networks

Hybrid fiber/coax combines fiber and coax.

The dominant type of cable-television distribution network today is called *hybrid fiber/coax (HFC)*. It is based on a combination of fiber and coaxial cables, designed to take advantage of the best features of each. Like other aspects of cable-television, the distribution network has evolved over the years, so the network is in many ways a hybrid.

Originally, cable television was a purely one-way tree system that collected television signals and distributed the same signals to all subscribers. Premium channels were scrambled electronically so subscribers had to rent special equipment to descramble and view them. Over the years, cable operators have added more two-way and interactive services to their networks, so they can deliver pay-per-view programs, and provide telephone and data services.

The standard design of cable networks offers them much greater bandwidth than telephone networks. However, telephone networks retain much greater switching capacity.

Basic Cable Architecture

The head end is the control center of a cable system.

The traditional architecture for a cable television system is a tree network that distributes signals from a control center called the *head end,* as shown in Figure 27.4. The head end receives signals from satellite feeds, local television stations, and community access centers. Head ends operated by the same cable company in nearby communities may transmit signals between each other. If the same company has cable franchises in several adjacent suburbs, it may put all its satellite down-links at one location and transmit signals over fiber from there to other head ends.

The head end multiplexes all the video inputs to generate an analog signal spanning a range of frequencies from a few tens of megahertz to several hundred megahertz. This signal modulates an analog laser transmitter, and single-mode fiber *trunk lines* distribute the optical signal to distribution nodes in the community. This effectively breaks the cable system into smaller systems, each serving a small area from its own distribution node, as shown in Figure 27.4. Some systems use an intermediate level *hub* between the head end and distribution node, with each hub transmitting signals over fiber to a number of distribution nodes. This approach is more likely in larger communities.

Early cable television systems were built entirely from coaxial cable, but the high attenuation of coax required amplifiers or repeaters about every 600 m (2000 ft). Long chains of amplifiers could introduce noise or distortion. They also created reliability problems because a single amplifier failure knocked out service to everyone whose signals had to pass through that amplifier.

Linear laser transmitters are needed for fiber trunk lines.

Cable television companies were quick to test fiber-optic trunk lines that could eliminate the need for amplifiers between the head end and the hub or distribution node. However, early laser transmitters did not have the highly linear output characteristics needed for analog transmission, so cable companies were slow to adopt fiber. Fiber started spreading in cable systems after the development of highly linear distributed-feedback laser transmitters, which are now standard for transmission between head end and distribution nodes.

Each node converts the optical signal into electronic form, and distributes it through a network of coaxial cables to households. Typically each node serves 500 to 2000 households. Thick coaxial cables carry signals most of the way to subscribers; lighter, flexible cables serve as drops to individual homes. The distribution cables may require a few amplifiers, but not the large numbers that caused reliability problems and noise in the older systems.

Hybrid Fiber/Coax Transmission

Hybrid fiber/coax allows subscribers to send signals as well as to receive them. They divide the available spectrum into four segments, shown in Figure 27.5. These are not rigid industry standards, so some systems may differ.

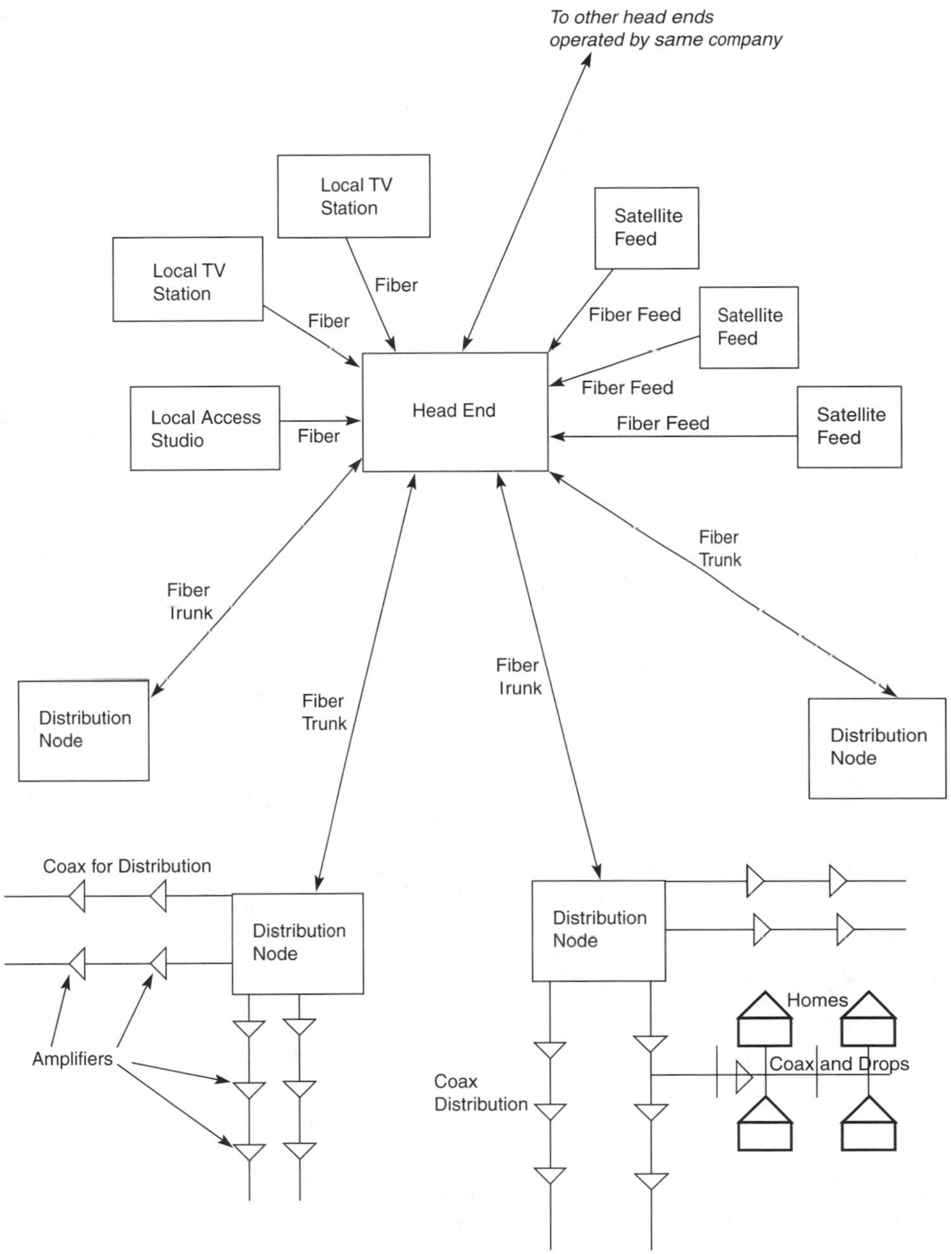

FIGURE 27.4
Hybrid fiber/coax.

Frequencies of 5 to 40 MHz are allocated for upstream signals from subscribers, including data, video, and telephone traffic. These frequencies are lower than the frequencies used for downstream video transmission on cable. Standard NTSC video is transmitted at 50 to 550 MHz, a range that includes the broadcast frequencies of many analog channels. Cable systems often transmit many analog channels at the same frequency that they broadcast, if

Upstream signals are transmitted to 5 to 40 MHz.

FIGURE 27.5
Spectrum of hybrid fiber/coax.

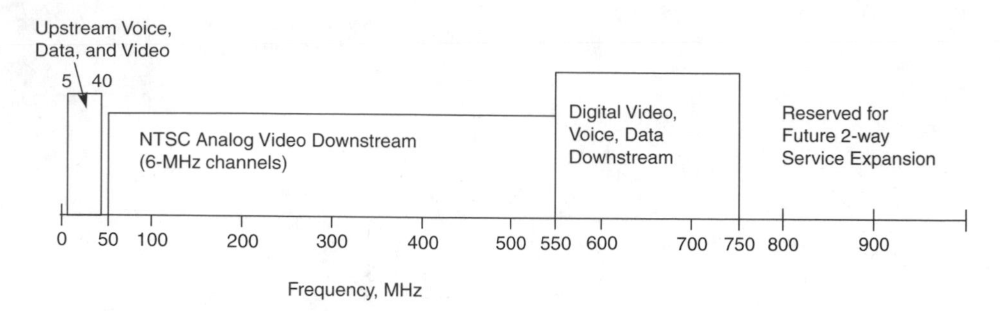

the frequencies are available on cable. Frequencies of 550 to 750 MHz are allocated for downstream digital data, including telephone and data as well as for any video channels that might be transmitted in digital form. At this point they are mainly used for cable modems and cable-based telephone service. The space above 750 MHz is reserved for future two-way service, as yet unidentified. Potential system bandwidth should extend beyond 1 GHz.

Note that upstream and downstream portions of cable modem and telephone service are transmitted at widely separated frequencies. More bandwidth is available for cable modem transmission to homes at high frequencies than from homes at low frequencies, which is why cable modems normally download at higher speeds than they upload.

The upstream and downstream signals are transmitted simultaneously in different directions in the coax part of the network. The signals are separated in the fiber portion of the network, either by transmitting them through different fibers or by transmitting at different wavelengths through the same fiber. Systems may transmit signals between the head end and nodes at 1310 and/or 1550 nm. Two optical channels are the most wavelength-division multiplexing used in current hybrid-fiber coax systems.

Optical signals from subscribers are collected at the node and transmitted upstream on the fiber to the head end. Each node receives different signals from subscribers, so it delivers different signals to the head end.

Optical signals from the head end are interpreted at each node. The signals can be distributed either by broadcasting with a high-power transmitter or *narrowcasting* with a lower-power transmitter. The two approaches can be combined, as shown in Figure 27.6.

In broadcasting, the same signals are distributed to all nodes, often by splitting signals from one transmitter. This saves money by allowing one powerful transmitter to serve many nodes. The analog video signals must be distributed to all nodes. In broadcast mode, all digital downstream signals are transmitted to all nodes, but each node ignores all signals that are not directed to it. The node takes signals directed to it and distributes them to subscribers in the same way. As in a local-area network, all signals reach each subscriber terminal, but each terminal pays attention only to signals directed to it. This requires a single laser transmitter, which generates a high power that is split among many outputs. One high-power transmitter is less expensive than many lower-power transmitters.

Narrowcasting generates separate signals for each distribution node.

In narrowcasting, the head end generates a separate signal for each node, which drives a laser transmitter that sends the signal to that node. In this case, signals going to a particular

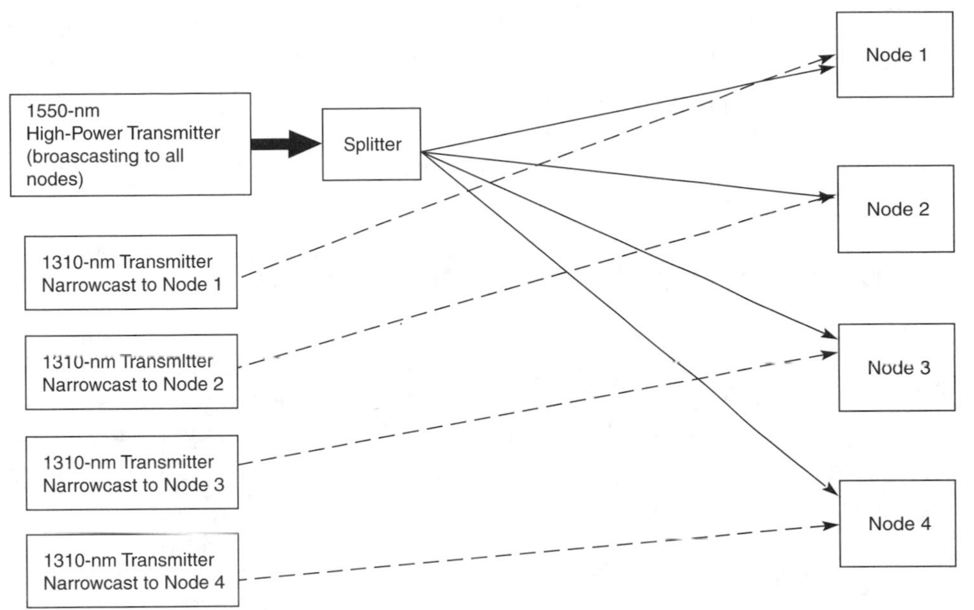

FIGURE 27.6
Broadcasting and narrowcasting with hybrid fiber/coax.

node are transmitted only to that node. This approach requires many low-power transmitters, which raises the cost. However, it also allows cable companies to target specific content to certain parts of their neighborhood. For example, they might direct advertisements for the local BMW dealer in signals distributed to the affluent side of town, and advertisements for a local used-car dealer in signals to a working-class area.

The two approaches also can be overlaid using wavelength-division multiplexing. For example, a high-power 1550-nm signal from an optical amplifier can carry the analog video channels distributed to all customers, and lower-power 1310-nm transmitters can distribute different signals to each node.

Cable Modems and Telephone Service

Data channels delivered from the head end provide downstream data for cable modems. Standard cable modems are really nodes on an Ethernet local-area network. You share that network capacity with your neighbors. The number of households connected depends on the system design and the number of people who subscribe to the service.

A cable modem is a node on Ethernet distributed from a cable node.

The distribution nodes receive downstream data and distribute those among one or more Ethernet networks on the cable. If the node serves 500 homes but only 2% have cable modems, it may deliver data to all 10 of you through one local-area network. If 20% of subscribers sign up for cable modems, the cable company typically would arrange its cables to split its 100 customers among multiple LANs.

Like other LANs, cable modems can carry data all the time, without tying up a phone line. Although the network is shared, downloads can reach peak speeds in the megabit range as long as too many other users don't try to download at the same time. Access speeds also can be limited by traffic jams at other points on the Internet, not just by the cable modem's capacity.

Cable networks also can transmit telephone signals in digital form, with special hardware converting the digital signals to analog form so you can use standard telephones over the cable network. The signals are transmitted in the same way as cable modem signals, but typically at other frequencies.

Evolving Cable Networks and Bandwidth

Interactive services such as telephone, cable modems, and two-way video increase the demand on distribution nodes. Present nodes are very efficient at distributing identical signals to several hundred subscribers. They are less efficient in distributing different services to each of those subscribers. The more bandwidth each subscriber needs, the more serious the problem.

One way to enhance the bandwidth per subscriber is to split nodes so each one serves fewer households. Splitting nodes also can extend fiber farther into the community and closer to the home. Such extensions will be needed if cable networks are to compete with fiber to the curb and fiber to the home telephone networks for home users who want broadband net access.

If both cable and telephone networks continue the trend of stretching broadband services farther toward homes, the two will wind up looking increasingly similar. The key difference may be in the type of access they offer, because cable systems do not provide the same true circuit-switched service as phone networks.

Digital Television and Cable Systems

Digital television poses problems for cable systems.

The introduction of digital television created some big problems for cable television, which have yet to be resolved. Cable systems have only a limited number of channel slots available, and most of the slots are filled with existing analog channels. Promoters of digital television want cable companies to carry digital channels. It makes sense because the people most interested in television typically subscribe to premium cable or satellite services to get the best available video quality. However, cable companies respond that very few of their customers have sets able to receive digital television. They can't afford to snub the vast majority of their customers with analog sets, and the nature of their networks dictates that they distribute the same video signals to everyone.

Digital television can be encoded for cable transmission but HDTV signals consume more bandwidth than lower-definition digital video. Cable companies could squeeze more lower-definition digital channels through their existing networks, but that would defeat the original purpose of introducing HDTV. Digital transmission also would require new set-top boxes for cable systems. The new boxes would cost money, and probably would need both analog and digital outputs, adding to the expense.

The FCC has decided it cannot force cable systems to carry both analog and digital channels simultaneously, but is considering requiring that some television sets be equipped for digital reception. It is far from clear how cable systems are going to convert from analog to digital transmission. Nor is it clear that the whole scheme will work in the near future without strong regulations and large expense. New technology will be needed down the line, but the immediate future is not clear.

Other Video Applications

Cable television is the largest-volume video application for fiber optics, but there are many other cases where fiber is used for video transmission. Table 27.4 lists a sampling of important applications, with brief descriptions. Most involve point-to-point transmission.

Small, light, and durable fiber cables are valuable for portable systems.

Transmission requirements vary widely for these systems. Although many require high transmission quality, security video systems must be low in cost. Although metal cables can do some of these jobs, fibers offer benefits of lighter weight, smaller size, higher signal quality, longer transmission distances, immunity to electromagnetic interference, better durability, and avoidance of ground loops and potential differences.

Small, light, and flexible fiber cables offer important benefits where portability is important, such as in remote news gathering and when covering special events. Many systems use rugged cables and connectors developed to meet rigid military specifications for durability. Whenever cables are strung anywhere, they are vulnerable to damage.

Fiber transmission also offers more subtle advantages, notably avoiding the need to adjust transmission equipment to account for differences in cable length. Television studio amplifiers are designed to drive coaxial cables with nominal impedance of 75 Ω. However, actual impedance of coaxial cables is a function of length. As cable length increases, so does its capacitance, degrading high-frequency response if the cable is longer than 15 to 30 m (50 to 100 ft). Boosting the high-frequency signal, a process called equalization, can compensate for this degradation, but proper equalization requires knowing the cable's length

Table 27.4 Other video-transmission applications for fiber optics

Application	*Requirements*	*Special notes*
Electronic news gathering, special-event coverage	Light, durable cable to link mobile camera to fixed equipment	Camcorder an alternative
Security video	Vary; low cost important	Often low resolution
Studio and production transmission	High-quality link inside studio	1.5 Gbit/s for HDTV
Feeds to and from remote equipment (e.g., antennas)	High transmission quality	

and attenuation characteristics. Compensation also becomes harder with cable lengths over 300 m (1000 ft) and is impractical for cables longer than about 900 m (3000 ft). There is no analogous effect in optical fibers, so operators need not worry about cable length.

What Have You Learned?

1. Video signals encode continuous changing pictures and sound. They are transmitted in standard formats and require considerably more capacity than voice or digital data.
2. Analog NTSC video displays 30 analog 525-line frames a second with interlaced scanning. Each NTSC channel requires 6 MHz of broadcast spectrum. PAL and SECAM are interlaced scanning systems that each second show 25 analog frames of 625 lines each. These formats are decades old.
3. Computer displays need progressive scanning to show text clearly, not the interlaced scanning of NTSC, PAL, or SECAM. Progressive scanning demands more bandwidth and faster electronics.
4. Digitized video signals can be compressed by a factor of 60 without seriously degrading quality.
5. Digital television standards cover both high-definition (HDTV) and standard-definition (SDTV) video in 18 distinct formats. The HDTV formats have 720 or 1080 lines and a wide-screen format.
6. Digital television broadcasting is being phased in to replace analog broadcasts in the United States, but the change probably will take much longer than had been planned.
7. Video signals can be broadcast from a ground station to serve a local area. Microwave transmission from direct broadcast satellites can serve a much larger area; customers need satellite dishes and converters.
8. Modern cable television systems now carry dozens of analog NTSC video channels over fiber-optic and coaxial cables; the fiber runs from the head end to distribution points or optical nodes. Coaxial cables run from those points to homes. Customers need set-top converters to access premium channels.
9. Hybrid fiber/coax systems transmit NTSC video to subscribers at 50 to 550 MHz. Digital services are transmitted to optical nodes at 550 to 750 MHz, and signals from subscribers return at 5 to 40 MHz. Each optical node serves about 500 homes.
10. Hybrid fiber/coax can deliver services including Internet connections, telephony, and subscription video services. Internet connections via cable modem work like local-area networks.
11. Hybrid fiber/coax can be upgraded by splitting optical nodes to serve fewer subscribers.
12. Video transmission generally is over single-mode fiber at 1300 or 1550 nm.
13. Small, lightweight fiber cables are valuable for portable news gathering and sports event coverage.

What's Next?

In Chapter 28 you will learn about the role of fiber optics in vehicles and other mobile communications for civilian and military applications.

Further Reading

Analog television: *http://www.ntsc-tv.com/*

Walter Ciciora, James Farmer, and David Large, *Modern Cable Television Technology: Video, Voice and Data Communications* (Morgan Kaufmann, San Francisco, 1999)

Digitaltelevision.com: *http://www.digitaltelevision.com/*

Gary M. Miller, *Modern Electronic Communication,* 6th ed. (Prentice Hall, 1999); see Chapter 7 on Television

Questions to Think About for Chapter 27

1. Analog-to-digital conversion generates lots of extra bits, so it isn't fair to say that a 25-megabit HDTV frame contains only 2.8 times more information than an NTSC frame digitized to give 9 megabits. It's fairer to compare the number of lines of resolution and the width of the screen. Using those guidelines, how much more information does a 1080-line HDTV image contain than a 525-line NTSC image? Remember that the HDTV image has a 16:9 aspect ratio, while the NTSC image is only 4:3. (*Hint:* calculate the number of picture elements or pixels.)
2. The highest resolution possible for digital television is 1080 lines by 1920 pixels, in 60 interlaced frames per second. The lowest is 480 lines by 640 pixels in 24 progressive scans per second (corresponding to a digitized movie). How do the numbers of pixels per second compare? (Note that multiple bits encode each pixel, so this is not the data rate.)
3. Digitizing voice and video both produce data streams with much higher numbers of bits per seconds than the bandwidth in hertz. Compare the ratios of bits per second per hertz for voice and video. What might cause the difference?
4. A broadcast transmitter in a hybrid fiber-coax system generates output power of 10 dBm (10 mW). Analog receivers require an input power of 5 μW (−23 dBm) for adequate signal-to-noise ratio. If the transmission loss between head end and distribution node is 10 dB (not counting the splitter), and system margin is 10 dB, how many nodes can this transmitter support? How many could you serve by reducing the system margin by 3 dB?
5. You need narrowcast transmitters for the same system. What power level do they require if system margin, receiver sensitivity, and cable loss are the same?

6. You need to lease capacity on a metro network to transmit one channel of studio-quality HDTV from a studio to a television transmitter. The network operator has four types of transmitters available, which operate at rates to OC-3, OC-12, OC-48, and OC-192. Which one offers the capacity you need without too much excess?

Quiz for Chapter 27

1. What is the analog bandwidth of one standard NTSC television channel?
   a. 56 kHz
   b. 1 MHz
   c. 6 MHz
   d. 25 MHz
2. How many lines per frame do standard analog European television stations show, and how many full frames are shown per second?
   a. 525 lines, 25 frames per second
   b. 625 lines, 25 frames per second
   c. 625 lines, 30 frames per second
   d. 1125 lines, 25 frames per second
3. The HDTV standard in the United States transmits about 20 Mbit/s after digital compression. How much compression is used, and what would the data rate be without it?
   a. 4-to-1 compression, 80 MHz
   b. 10-to-1 compression, 200 Mbit/s
   c. 13-to-1 compression, 270 Mbit/s
   d. 75-to-1 compression, 1500 Mbit/s
   e. None of the above
4. What key development made the quality of analog fiber-optic transmission adequate for cable television trunks?
   a. Highly linear distributed-feedback lasers
   b. Inexpensive single-mode fiber
   c. Dispersion-shifted fiber
   d. Digital video compression
   e. Optical amplifiers for 1550 nm systems
5. What is the most important advantage of fiber optics over coax for distributing cable television signals from head ends to optical nodes?
   a. Fiber optics are hard to tap, so they reduce signal piracy.
   b. Fiber repeater spacing is much longer, avoiding noise and reliability problems with coax amplifiers.
   c. Fiber can be extended all the way to subscribers.
   d. Fiber cables are less likely to break.
6. How are analog video signals distributed to subscribers on present cable television systems?
   a. All subscribers receive the same signals, which require

set-top decoders to show premium services.

b. Signals from set-top controls are used to switch designed signals to the home.

c. Equipment at the head end switches selected services to each subscriber.

d. One pair of optical fibers runs directly from head end to home.

7. What signal format is used by present cable television systems?

a. Each system has a proprietary format

b. Analog NTSC signals, with 6 MHz bandwidth, assigned to radio frequencies between 50 and 550 MHz

c. Digitized compressed video at 20 Mbit/s

d. Analog PAL format in North America

8. What frequencies are used for signals from the subscriber to the head end in hybrid fiber-coax?

a. 50 to 550 MHz

b. 0 to 1 GHz

c. 550 to 750 MHz

d. 5 to 40 MHz

e. None of the above

9. How do cable modems work on hybrid fiber-coax networks?

a. They switch signals directly from the head end to individual subscribers

b. They transmit signals in one direction only

c. They function like a local-area network, addressing high-speed signals to one of many subscriber terminals served by the same network

d. They digitize video images for videoconferencing but cannot be used for other purposes

e. They are incompatible with hybrid fiber-coax

10. Which format is used for digital television displays?

a. 1080 lines, 1920 pixels, 60 interlaced frames per second

b. 1080 lines, 1920 pixels, 30 progressive scan frames per second

c. 720 lines, 1280 pixels, 60 progressive scan frames per second

d. 480 lines, 640 pixels, 60 interlaced frames per second

e. All of the above

11. How many analog video channels are required to transmit a full HDTV digital signal?

a. 1

b. 2

c. 4

d. 5

e. 6

12. A fundamental difference between cable-television and telephone networks is that

a. cable networks can't carry two-way telephone traffic.

b. only cable networks can carry high-speed data.

c. cable networks do not use circuit switching.

d. only cable networks use single-mode fiber.

e. there are no differences left.

CHAPTER

28

Mobile Fiber-Optic Communications

About This Chapter

The past few chapters have described the many applications of fiber optics in fixed telecommunication systems. Fibers also are used in a variety of mobile systems for civilian and military systems. Fiber cables can be used for remote control of robotic vehicles and guidance of tactical missiles. Fiber cables also are used inside vehicles ranging from battleships to private automobiles. This chapter briefly surveys these diverse applications, explaining how and why fibers are used.

Mobile Systems

Mobile systems differ in important ways from fixed systems, with the details depending on the application. This leads to some constraints on system design.

Some fiber-optic cables are used as tethers or connections for transmitting signals to and from moving objects. An example is a cable connected to a remotely operated vehicle that may venture into extreme environments where humans can't go. These cables have to survive whatever environment they pass through, so they are specially designed for that environment. The requirements can vary widely. Cables that tether a robotic mini-submarine to its human operators in a ship have to be strong and rugged. In contrast, the single fiber that connects to a fiber-guided missile during its brief flight must be light and flexible as well as strong, and is used only once.

Connections inside vehicles generally are short, with the exception of those in ships. Links inside a plane or car run no more than tens of meters, and often only meters, so graded-index fiber is often used, although single-mode fibers are used in some places.

Connections in vehicles generally are short.

Often these are miniature dedicated local-area networks that interconnect the growing variety of electronic systems in the vehicle. Copper cables often can carry data the required distances, but fiber cables are lighter and smaller. The immunity of fiber to electromagnetic interference is a big plus in vehicles where electrical, electronic, and mechanical systems are packed tightly together.

Vehicles are more hostile environments than offices.

Environmental requirements generally are much more stringent for equipment installed in a vehicle than in an office. Connections to your desktop computer don't have to withstand the constant vibration of a moving car or flying airplane. Some common fiber connectors that work perfectly well in an office building or telecommunications switching center might work loose in moving vehicles that are exposed to outdoor temperature extremes.

Military equipment must be rugged and repairable in the field.

Military equipment must meet special requirements for ruggedness and field repairability, and some must meet radiation-hardening specifications. Recent changes allow the use of some off-the-shelf commercial components that meet most military requirements.

Military systems share some other common features with civilian aircraft and automotive systems. They tend not to adapt cutting-edge optical and electronic technologies. Design cycles and production cycles are usually much longer than for telecommunications or computer equipment. Many systems are critical for safety and have to pass stringent testing requirements. You can survive a system error on your personal computer, but you might not survive if your car's computerized braking system froze when you stomped on the brakes in an emergency. Military and civilian aircraft are designed to operate for a dozen years or more, so their control systems have to meet the same requirements. Automotive systems are supposed to last for many years, but also must be mass producible at low cost, leading to long design lead times.

This chapter will give you a brief overview of these special fiber systems, emphasizing how they resemble and differ from other fiber communications equipment.

Remotely Controlled Robotic Vehicles

When we think of remotely controlled vehicles, most of us think first of radio-controlled toys that zip across the floor until the batteries run down. Radio controls are cheap and simple, but limited. You can command your radio-controlled car to go faster, slower, forward, backward, or turn right or left—but not much more.

Fiber-optic cables carry signals to control robotic vehicles.

Control of advanced robotic vehicles is a far more demanding job. The operator needs video transmission from a camera in the robot to see the local environment. Other environmental sensing information may also be needed, such as temperature and pressure readings. Signals must flow in the opposite direction so the operator can control the vehicle. Fiber-optic cables carry signals in both directions in a variety of remotely controlled vehicles, often sending two signals in opposite directions through a single fiber at different wavelengths. Although care must be taken to protect them, fiber-optic cables can work in places where radio signals cannot, including underwater and in electromagnetically noisy environments. Fiber cables can also be made quite rugged and special ruggedized connectors are made for military systems. Other fiber advantages include their ability to carry high-bandwidth signals over greater distances and their light weight.

Remotely controlled robots can go into places unsafe for humans. Robots can probe the radioactive parts of nuclear reactors to take measurements or make repairs, or to dissemble old reactors at the ends of their operating lifetimes. Robots can descend deep into the ocean or explore the surface of the moon or Mars. Robots can be scouts for armies, and they can even deliver weapons to their target (we call them guided missiles).

Fiber-Optic Guided Missiles

Guided missiles are, in a sense, simple robots with rather deadly missions—to deliver bombs to their targets. The Pentagon has developed a system that uses a ruggedized optical fiber to carry control signals to and from a short-range missile on the battlefield, letting a soldier guide it to its target from a safe hiding place. Called FOG-M, for fiber-optic guided missile, the program gives a good idea of how remote control through optical fibers works.

A fiber can send images from a television camera in a missile back to a soldier guiding the missile to its target.

The idea of FOG-M, shown in Figure 28.1, is to send images from a video camera in the missile to a soldier on the ground, who guides the missile to its target. The missile is launched toward the target with a single ruggedized bare optical fiber trailing from it to the launcher. Looking at the video image, the soldier operates controls that direct the missile to its target. The soldier can follow the missile right to impact.

Military agencies like FOG-M because the soldier operating it is kept safe under cover. This is an advantage over laser-guided bombs, where the laser operator must have a line of sight to the target to project a laser spot onto it. Missiles can be guided by wire, but the wires can't carry video images, which allow more accurate guidance. Only an optical fiber has the combination of small size, light weight, strength, low attenuation, and bandwidth needed to transmit video signals over the missile's 15-km range.

Fiber unwinds from a special reel on the missile.

The basic hardware for FOG-M is shown in Figure 28.2. The missile contains a video camera, a fiber-optic video transmitter, a low-bandwidth receiver, and a special reel of fiber.

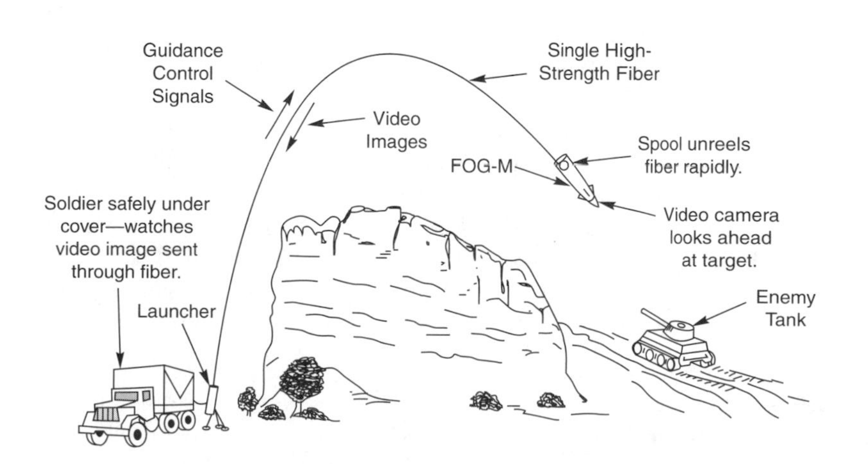

FIGURE 28.1
A concealed soldier guides a FOG-M missile to its target.

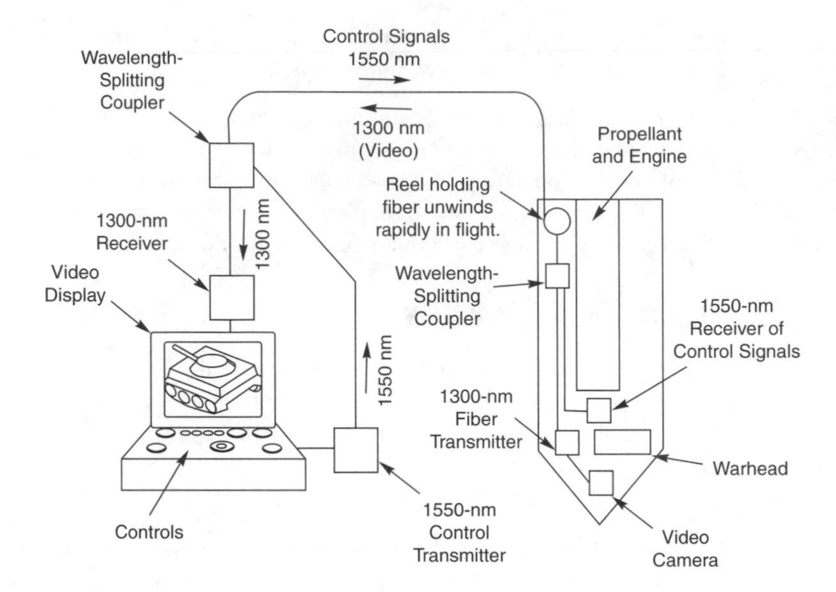

FIGURE 28.2
FOG-M components.

One end of the fiber is mounted on the launcher, so it remains behind when the missile is fired. As the missile speeds toward its target, the fiber rapidly unwinds from the reel, forming a long arc over the battlefield. The reel is a critical component, because the fiber will tangle or break if not unwound at the right rate. The camera transmits a video image to a soldier at the launcher, who sends control signals back through the fiber to pinpoint the missile onto its target in the proper place in his field of view.

FOG-M uses wavelength-division multiplexing to send signals in opposite directions. In this drawing, video signals from the missile are sent at 1300 nm through fibers with low dispersion at that wavelength. Low-bandwidth control signals are sent at 1550 nm in the opposite direction.

Robotic Vehicles on Land

Fiber cables can control robotic land vehicles.

Fiber-optic cables also can guide robotic vehicles on land. Military agencies have done extensive testing of unmanned ground vehicles that are remotely controlled in various ways, including by fiber. Radio signals have the obvious advantage of not requiring a physical connection to the vehicle. Cables have to be ruggedized, because the robots are likely to run over them. However, fiber cables are immune to electromagnetic interference, electromagnetic pulse effects, and jamming that could block radio communications. Fibers also offer high bandwidth, so operators can obtain detailed information from the vehicle.

The robots are intended for hazardous duty. For example, the Naval Sea Systems Command has developed the Remote Ordnance Neutralization System, which runs on pivoting tracks and has a robotic arm that can defuse explosives. A soldier can control the robot over a radio link or fiber cable, staying safely out of harm's reach in case anything goes wrong. Similar robots could be used to detect and remove land mines, which continue to take a heavy toll of civilians long after the wars are over. Other robots might fire weapons at dangerous targets.

The high bandwidth of fiber cables makes it possible to consider using virtual reality techniques to remotely operate robotic vehicles. Sensors on the vehicle would scan the area, serving as the operator's "eyes," while other sensors would listen for sounds and "feel" the terrain. The sights, sounds, and feel would be conveyed to the operator—far from the vehicle—using screens, speakers, and perhaps virtual-reality gloves. The goal would be to keep the operator safe while feeling as if he or she were driving the vehicle.

Remote-controlled robots could also serve many nonmilitary purposes in dangerous environments. Robots could inspect the "hot" interiors of nuclear power plants or perform needed repairs inside the reactor. The robots could be left inside the reactor permanently if they became contaminated. Specialized robots could be used to dismantle old reactors, without exposing people to the highly radioactive materials inside. Likewise, remotely controlled robots could clean up hazardous wastes. Scientists used a fiber-optic cable to control a multilegged robot as it climbed into a hazardous Antarctic volcano to collect data.

Remotely controlled robots could clean up hazardous wastes and dismantle old nuclear reactors.

Submersible Robots

Radio links can substitute for cables in many land applications, but most radio waves don't penetrate far into water. Cables or acoustic signaling are required to maintain contact with submerged vessels. Although crewed submersibles can operate without a continuous link to the surface, only cables can provide the transmission capacity needed to remotely operate a sophisticated submerged vessel. Fiber cables are preferred because of their high bandwidth and durability.

Hybrid fiber-electrical cables carry signals and power to remotely operated submersibles.

Hybrid fiber and electrical cables carry the signals and power needed to steer and accelerate submerged vessels, as well as bring video signals and telemetry to the surface, where shipboard operators can monitor them.

Fiber cables also allow operators aboard a submersible to control robotic vehicles that can be sent into small spaces or dangerous areas. The most famous example came when Robert Ballard's team from the Woods Hole Oceanographic Institution discovered the sunken wreck of the *Titanic* in 1985. The scientists discovered the wreck with *Alvin,* a submersible that carried three people. However, they did not dare to explore the inside of the deteriorating wreck. Instead, they used a 250-lb robot, tethered to *Alvin* with a fiber-optic cable that carried control signals. It was this fiber-controlled robot that photographed details of the dark interior of the wreck. Ballard remains enthusiastic about the use of fiber-controlled robot submarines.

Fibers in Aircraft

Fibers can serve as the control networks for aircraft.

It was not too long ago that most aircraft were controlled by hydraulic systems. When the pilot moved a lever, it would cause hydraulic fluid to move a control surface (e.g., a wing flap), much as hydraulic brakes work in an automobile. Newer planes have fly-by-wire electronic controls that send electronic signals to motors that move control surfaces. Modern aircraft—particularly military planes—also use many electronic systems and sensors, adding to signal transmission requirements. These include radars, navigation and guidance systems, and—in military planes—weapons systems with automatic targeting capabilities and electronic countermeasure equipment.

Like many other users faced with increasing communications requirements, the aerospace industry and the Air Force began investigating fiber optics. Pentagon research programs date back to the 1970s. Aircraft performance depends on weight, and military engineers wanted to reduce the load of heavy, metal cables. They also wanted to protect their avionic systems from electromagnetic interference, electromagnetic pulse effects, and potential enemy countermeasures.

New tactical aircraft use fiber-optic links.

Fiber is used extensively in the latest generation of tactical aircraft, such as the U.S. Air Force F-22 Raptor, the Eurofighter, and the U.S. Army's RAH-66 helicopter. Multimode fibers transmit signals for radars, weapons control systems, and electronic warfare systems. The use of fiber greatly reduces the vulnerability to enemy countermeasures.

Some older military aircraft have been retrofit with fiber for high-bandwidth applications. An example is the modernization of computer systems in the Air Force's AWACS (Airborne Warning and Communications System) fleet. Fiber systems transmitting 1 Gbit/s using the Fibre Channel standard are replacing communications equipment developed in the 1970s.

Another addition to existing airborne systems is the Fiber Optic Towed Decoy, which helps protect aircraft from enemy missiles. When instruments detect a potential threat, the aircraft releases the radar-emitting decoy and tows it behind the plane on a fiber-optic cable. The system identifies the type of threat, and commands the decoy to emit a radio signal designed to fool that particular enemy weapon system. The commands go through the fiber cable, which emits no radio waves, and the aircraft's radio emissions are limited, so the enemy weapon is fooled by the decoy's emissions and homes in on it. The decoy is part of a new countermeasure system being integrated into F-15 and F-18 fighter planes.

Fiber-optic aircraft links use multimode fiber.

Military systems have long development times, and most military aircraft use multimode fibers to transmit no more than a few hundred megabits per second over the short distances inside a plane. However, some new systems use advanced fiber technology. The towed decoy borrows techniques developed to transmit analog microwave signals in cable television and microwave antenna systems. High-speed detectors convert the analog signal from an externally modulated laser back to microwaves, which the decoy then emits. The special towing cable is built around single-mode fiber designed to be particularly insensitive to bending, which is hard to control in a cable hanging from a fast-moving jet.

Civilian airliners also use fibers in control systems. After testing "fly-by-light" controls in eight 757 jets, Boeing included a 100-Mbit/s fiber-optic local-area network in the control system of its 777 airliners. Boeing engineers calculated that they could have saved over 1300 kg if they used only fiber instead of copper wiring for the communication system in the plane, but they did not go that far. Airlines are likely to turn to fiber for higher bandwidth as they upgrade in-flight entertainment systems.

Fiber-optic systems must meet some special requirements for aircraft use. Temperatures can vary widely, and planes suffer continual vibrations in flight. Many connectors that work

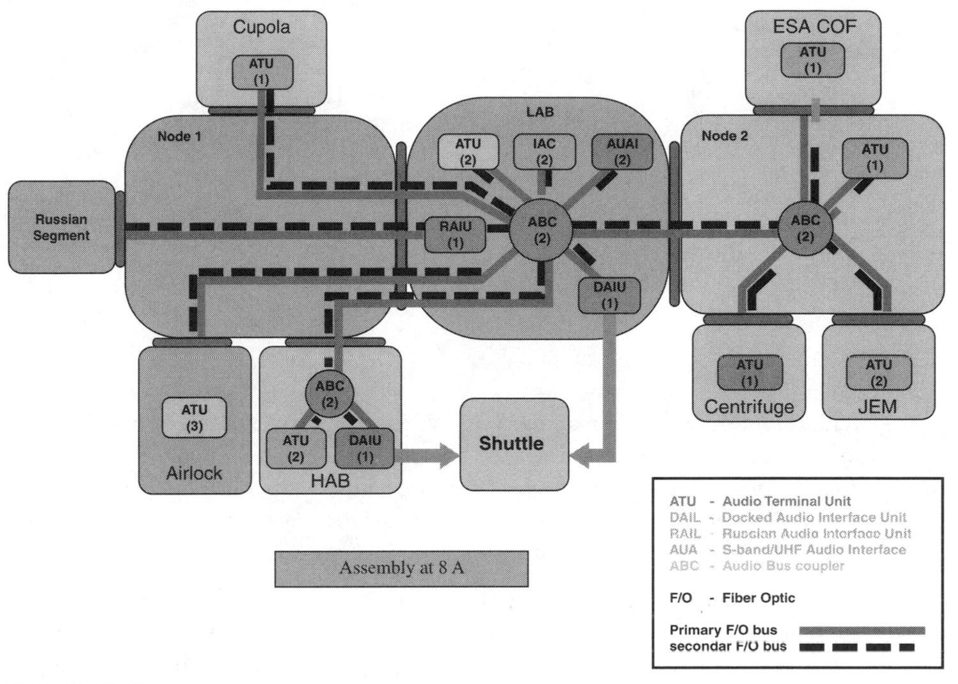

FIGURE 28.3

Fiber-optic audio network in the International Space Station. (From NASA "International Space Station Familiarization" government work not subject to copyright)

fine in ground-based equipment, such as the FC, ST, and SC, are not recommended for aircraft use. However, suitable connectors are available, including some designed to meet military environmental requirements.

Light weight also tips the balance toward fiber in spacecraft, and the International Space Station includes fiber-optic networks for audio and video transmission. Figure 28.3 shows the layout of the station's fiber-optic audio network.

Shipboard Fiber-Optic Networks

The communication requirements of a big ship rival those of an office building. Ships have their own telephone networks to keep officers and crew in contact, as well as communication systems that link them to the outside world. Military ships carry a variety of weapon systems, as well as radars, sonar systems, and other sensors. Modern military ships have computer rooms, both to control on-board weapon systems, and to analyze incoming information. Computers have become so important to naval operations that many ships in the U.S. Navy are now equipped with fiber-optic local-area networks. Figure 28.4 shows one example of how fiber-optics can interconnect shipboard systems.

Big ships have massive communication requirements.

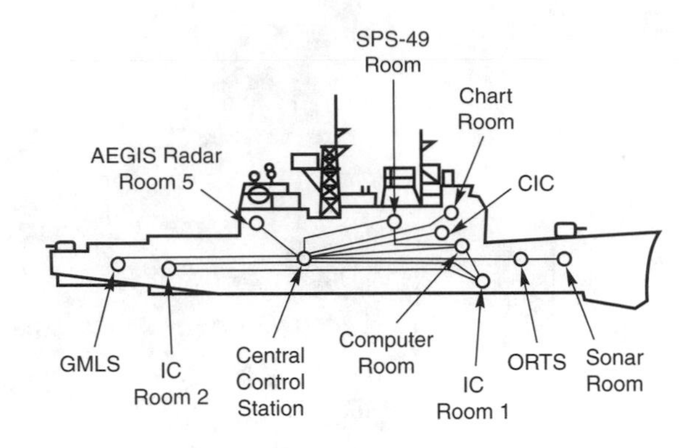

FIGURE 28.4
Cabling in a large ship. (Courtesy of AT&T)

Radars and other imaging and sensing systems often collect large volumes of data, requiring the high transmission capacity of fiber links. Weight is not as critical on ships as it is on aircraft, but scrapping metal cables can save valuable space. Fibers are immune to electromagnetic interference, which can be a problem when mechanical and electronic systems are packed tightly in the close quarters of ships.

Early naval systems used multimode fiber, but newer systems include some single-mode fibers for high-capacity systems. Recent developments include a standardized ship-to-shore cable for in-port connections, which includes eight multimode fibers and four single-mode fibers.

Automotive Fiber Optics

Engineering in the automobile industry, as in the military, involves an odd mixture of conservative and advanced design. The latest luxury cars include some surprisingly state-of-the-art innovations, such as navigation systems based on the Global Positioning System. At the same time, internal communications are through wiring harnesses full of copper cables that have changed surprisingly little for decades. Fiber optics keeps threatening to spread in automotive systems, but as of the 2001 model year, most cars still make very little use of fiber.

Cars can be extreme environments for communication systems.

The problem is not that automotive engineers are hopelessly conservative about fiber. The automobile is a tough environment, automotive equipment must be easily repairable, and safety concerns must be met. Modern cars have to meet hostile conditions ranging from withstanding North Dakota winters to baking in south Texas parking lots in mid-summer. Cars have to be repairable by technicians with a wide range of skills, which are not likely to include fiber optics. Safety-critical systems, such as steering, brakes, and wipers, must withstand minor failures. Planning cycles for major new equipment are long, with planning typically starting about five years before introduction in a new model. And unit costs have to be low. Engineers have talked for years about fiber optics, but have insisted that fiber must meet these rigorous requirements.

Now improved fiber technology and the proliferation of automotive electronics is pushing the auto industry toward the fiber-optic threshold. Auto engineers have developed a standard fiber-optic data bus for cars called Media Oriented Systems Transfer (MOST), which transmits data at rates to 25 Mbit/s. That doesn't sound like much by telecommunication standards, but it's "high-speed" for cars.

MOST runs on plastic fiber, which meets industry desires for low cost and easy repair. The biggest concern has been how well plastic fiber can withstand high temperatures, but the highest temperatures are in the engine compartment, and most of the new electronic equipment goes elsewhere.

The MOST automotive fiber-optic data bus transmits 25 Mbit/s on plastic fiber.

Three companies in the group that developed MOST—BMW, Audi, and Daimler-Chrysler—have said they plan to use the technology in future production cars. Other automotive data busses that don't use fiber are in the works, so there will be competition. If there is a transition, it's likely to take time. The spread of fiber also may be affected by plans to change automotive electronics by shifting from 12-volt to 42-volt batteries in future models. Over the years a number of plans to add fibers to automobiles have failed to come to fruition; it seems likely to happen eventually, but don't hold your breath.

What Have You Learned?

1. Mobile systems must operate in different environments than fixed systems. Vehicle systems face more difficult conditions than those used in offices.
2. Vehicle communication systems generally span short distances.
3. Military equipment must meet special requirements for ruggedness and field repairability. New military systems can use some commercial equipment that meets their requirements.
4. Military systems have long design cycles. Military and aerospace systems typically have lifetimes of over a dozen years.
5. Optical fibers can carry signals to control robotic vehicles on land, in air, or in the water. Fibers' advantages are their small size, light weight, immunity to EMI, and high bandwidth. Ruggedization of cables is critical for vehicle applications.
6. A fiber can transmit images from a television camera in a missile back to a soldier guiding the missile to its target. Fiber unwinds from a special reel on the missile.
7. Remotely controlled robots could defuse bombs. In civilian applications, remotely controlled robots could clean up hazardous wastes and dismantle old nuclear reactors.
8. Fibers can be used for signal transmission in aircraft because of EMI immunity, small size, and light weight.
9. Fiber-optic communication networks are used on military ships, which have large communication needs.
10. The automotive industry has been slow to adopt fiber optics because cars are difficult environments. The increasing use of automotive electronics has pushed new interest in fiber.
11. MOST is an automotive fiber-optic data bus that transmits 25 Mbit/s on plastic fiber. It is planned for future models.

What's Next?

Chapter 29 covers applications of fiber optics in sensing.

Further Reading

Automotive Fiber Optics: *http://mostnet.de/main/index.html*

Enhanced Fiber-Optic Guided Missile: *http://efogm.redstone.army.mil/*

Questions to Think About for Chapter 28

1. Military research agencies in Britain and the United States were among the first sponsors of fiber-optic development. Yet military equipment has lagged far behind civilian telecommunications in deploying fiber-optic systems. Why?
2. One of the first fiber-optic systems deployed by the military was a portable battlefield communications network. The lightweight fiber cables replaced thick copper cables, which had proved very vulnerable to damage in handling. Recently, wireless systems have replaced the fiber network. Why do you think fiber optics were deployed earlier in battlefield networks than in other systems?
3. Why can't radio-controlled vehicles be used underwater?
4. A Nimitz-class nuclear aircraft carrier is 1092 ft (333 m) long and requires a crew of 3300 people to sail. Could you run a gigabit Ethernet link the length of the ship with 62.5/125-μm graded-index fiber at 850 nm? If not, could you use other types of multimode fiber?
5. A new car is 20 ft (6 m) long. A step-index plastic fiber has bandwidth of 10 MHz/km. Can you use this fiber to run a 25-Mbit/s MOST link the length of the car?
6. How far can you transmit a 25-Mbit/s NRZ signal through the step-index fiber of Question 5, considering only bandwidth? If attenuation is 200 dB/km, and the link margin is 20 dB, what is the limit imposed by distance, neglecting connector and coupling losses?

Quiz for Chapter 28

1. What kinds of remotely operated vehicles cannot be controlled by operators through fiber optics?
   a. Guided missiles
   b. Submersibles
   c. Munition-defusing robots
   d. Robots for nuclear waste cleanup
   e. Satellites

2. What signals are transmitted from fiber-guided missiles to the operator?
   a. Video images of the target
   b. Control commands
   c. Data on temperature and pressure
   d. Data on fiber attenuation

3. How are signals transmitted to and from a fiber-guided missile?
   a. Separately through two fibers in a single cable
   b. Bidirectionally through one fiber by time-division multiplexing
   c. Bidirectionally through one fiber by wavelength-division multiplexing
   d. Signals are transmitted only one way
   e. From the missile through the fiber; to the missile via radio

4. Which of the following attributes of fiber optics are important for remote control of land vehicles?
   a. Secure data transmission
   b. Lightweight, durable cable
   c. EMI immunity
   d. b and c
   e. a, b, and c

5. Which of the following reasons do not influence the use of fibers for signal transmission in aircraft?
   a. Optical fibers are immune to EMI.
   b. Optical fibers are lighter than wires.
   c. Aircraft lack adequate power supplies for wire-based communications.
   d. Military aircraft must be hardened against enemy electronic countermeasures.
   e. Fiber optics can help reduce aircraft visibility to radar.

6. Why do many military in-vehicle systems use graded-index fibers?
   a. Graded-index fiber is cheaper than single-mode.
   b. They don't span long distances.
   c. Graded-index fiber is more resistant to vibration.
   d. Military specifications do not allow use of single-mode fibers.
   e. None of the above

7. Aircraft fly-by-light systems use optical fibers
   a. to illuminate cockpit instruments.
   b. in high-strength cables to pull mechanical actuators.
   c. to carry control signals to motors that move mechanical parts.
   d. only to transmit data from sensors to cockpit instruments.
   e. to deliver bright flashes of light that ignite fuel in the engines.

8. Single-mode fibers are used
   a. to transmit radio-frequency signals to radar decoys towed by aircraft.
   b. in local-area networks on submarines.
   c. in cables controlling mobile ground robots.

d. in data links on board aircraft.
e. in automotive systems.

9. What type of system uses plastic fibers for data transmission?
a. Fiber-guided missiles
b. Radar decoys towed by aircraft
c. Local-area networks on board ships
d. Data links on board aircraft
e. MOST-format data links for automobiles

10. The MOST standard for automotive data networks specifies what data rate?
a. 25 Mbit/s
b. 100 Mbit/s
c. 155 Mbit/s
d. 622 Mbit/s
e. 1 Gbit/s

CHAPTER

29

Fiber-Optic Sensors

About This Chapter

So far, I have concentrated on how optical fibers are used for communications. However, fiber optics also have other important uses. This chapter will show how fibers are used as sensors. Fiber sensors work in a variety of ways, sometimes just using fibers to deliver light, other times monitoring changes induced in light transmission caused by external effects. Fiber sensors can measure pressure or temperature, serve as gyroscopes to measure direction and rotation, sense acoustic waves at the bottom of the sea, and do many other tasks.

Fiber-Sensing Concepts

The label *fiber sensors* covers a broad range of devices that work in many different ways. The simplest use optical fibers merely as a probe, to detect changes in light outside the fiber. The fiber may collect light from a given point, to see if an object (such as a part on an assembly line) is present or not. The fiber also may collect light from another type of optical sensor that responds to its environment in a way that changes the light reaching the fiber. For example, a prism in a tank of liquid may start reflecting light back into a fiber probe if the liquid level drops below the prism's reflective surface, exposing it to air so total internal reflection occurs.

Other fiber sensors detect changes in light passing through a fiber that is affected by changes in the outside world, such as the temperature or pressure. That may seem strange if you're used to communications fibers, which generally do not respond significantly to outside effects. However, you can design special fibers or special structures within fibers to respond more strongly to outside effects. You also can use optical effects such as interference to detect small effects that accumulate over long lengths of fiber. In these ways, fiber sensors can detect changes in quantities such as temperature, pressure, and rotation.

Fibers can serve as probes or as sensors themselves.

There is an amazing multitude of fiber sensors, most used only for a few special purposes. This book can't cover them all. Instead, I will concentrate on simple examples and important types of sensors. I will first survey simple fiber sensors, where the fiber merely probes the environment. Then I will describe sensing mechanisms and some important types of fiber sensors and their applications.

Fiber-Optic Probes

Fiber-optic probes collect light from remote points, often sampling light that was delivered there by fibers. They come in two broad families that perform different sensing functions. The simpler ones look to see if light is present or absent at the point they observe. Others collect light from remote optical sensors, bringing it back to a place where it can be analyzed.

Simple Probes

Fiber probes can detect objects when they block or reflect light.

Figure 29.1 shows a simple fiber-optic probe checking for parts on an assembly line. One optical fiber delivers light from an external source, and a second fiber collects light emerging from the first, as long as nothing gets between the two. When a part passes between them on the assembly line, it blocks the light. Thus light off indicates that a part is on the assembly line, and light on indicates that no part is passing by.

This concept can be used in many ways and is not new. One early example was reading holes in the punched cards used to input data to early mainframe computers, although in this case a detector behind the card directly sensed the transmitted light without a light-collecting fiber. The card passed an array of fibers at a fixed speed, and detectors monitored light transmission as a function of time. When a hole passed the end of the fiber, light reached the detector. When there was no hole, the card blocked the light. The technique was simple and effective, but punched cards are now museum pieces.

More refined variations are possible, such as measuring the size of parts to make sure they meet tolerances. An array of fibers can be mounted beside the production line, so passing parts block the light to some of them. The parts pass inspection if all the fibers above the

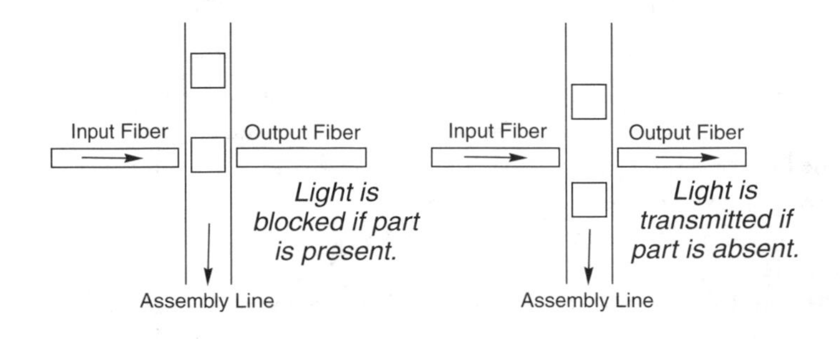

FIGURE 29.1 *Fiber-optic probe checks for parts on an assembly line.*

maximum height receive light and all those below the minimum height do not. Parts that are too small or too tall are rejected when light reaches fibers that are supposed to be dark or does not reach fibers that are supposed to be illuminated.

Optical Remote Sensing

Fibers can collect light from other optical sensors.

Fiber probes can also collect light from other types of optical sensors. In this case, the fibers function like wires attached to an electronic sensor. The optical sensor (which is not a fiber) responds in some way to the environment, changing the light that reaches the fiber probe. The fiber carries that light to a detector, which senses the change.

One example is the liquid-level sensor shown in Figure 29.2, which senses when the gasoline in tank trucks reaches a certain level. Many tank trucks are filled from the bottom so vapor left in the tank can be collected to control pollution, and the liquid level must be sensed to prevent overfilling. One fiber delivers light to a prism mounted at the proper level. If there is no liquid in the tank, the light from the fiber experiences total internal reflection at the base of the prism and is directed back into the collecting fiber. If the bottom of the prism is covered with gasoline, total internal reflection cannot occur at the angle that light strikes the prism's bottom face, and no more light is reflected back into the fiber. When the light signal stops, the control system shuts off the gas pump.

Another example senses temperature changes by observing the response of a phosphor in a glass blob at the end of a fiber. Ultraviolet light transmitted by the fiber stimulates fluorescence from the phosphor at several wavelengths. The ratio of fluorescence at the different wavelengths changes with temperature. The fiber collects the fluorescent light and delivers it to an optical analyzer that compares intensities at different wavelengths and thus measures the temperature.

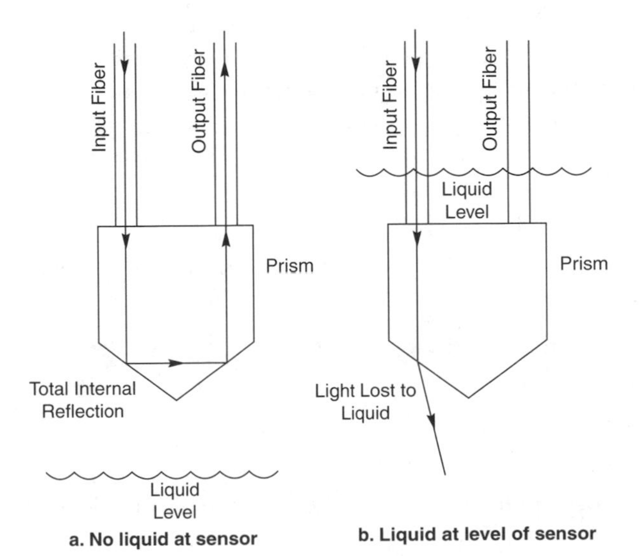

FIGURE 29.2
A liquid-level sensor.

Fiber-Sensing Mechanisms

Some effects can change how fibers transmit light.

Outside influences can directly affect fiber transmission in a variety of ways, depending on the type of fiber and how the fiber is mounted. Communication fibers and cables are designed to be isolated from the environment. For sensing, you design fibers to respond as strongly as possible. For example, you may dope fibers with materials that change their refractive index as temperature or pressure change. Or you may mount fibers between grooved plates, so increasing pressure on the plates causes microbending.

Countless fiber sensors have been demonstrated in laboratories around the world. Some are used for practical measurements, although none are in true mass production. It's impossible to cover all the diverse types of fiber sensors here, but I will give you an overview of the basic principles of intrinsic fiber sensing, which depends on properties of the fiber.

The Idea of Sensing

Sensors convert something hard to measure into units easier to observe.

A sensor converts a physical effect you want to observe into a form you can measure. Let's start with a familiar sensor, a thermometer filled with mercury or some other liquid. As temperature changes, the liquid expands. Most liquid in the thermometer sits in the hollow bulb at the bottom; the hollow tube calibrated with temperatures has a much smaller volume. (It looks big because the glass or plastic cylinder magnifies the apparent width of the tube.) The engineers who design thermometers know how much the liquid expands per degree, so they can calculate how much liquid they need to expand to fill the extra tube.

Suppose, for example, mercury expands 0.01% per degree Celsius. Then, if you start with a volume of 1 cm^3 of mercury, it grows 0.0001 cm^3 larger for each degree it is warmed. To make a mercury thermometer, you can attach a bulb containing 1 cm^3 of mercury to a thin tube marked with lines that indicate 0.0001-cm^3 units of volume inside the tube. If the tube's cross-sectional area is 0.001 cm^2, each 0.1 cm—or 1 mm—represents a 1° temperature change. It isn't quite that simple, because a careful designer must consider thermal expansion of the tube itself, but that's the basic idea. The thermometer converts a hard-to-measure unit, temperature, into one that is easier to measure, the length of a column of mercury.

Fiber sensors work in the same way, but they measure properties like temperature by observing the light transmitted through the sensor. They make the property they are trying to measure modulate the light in some way.

Most fiber sensors work by modulating the light passing through them in one of three ways:

- Directly altering the intensity
- Affecting the polarization of the light
- Shifting the phase of the transmitted light

To actually measure that modulation, you have to convert those changes to variations in intensity.

Direct Intensity Modulation

Some fiber sensors directly modulate transmitted light intensity.

Sensors that directly change light intensity are conceptually simple. The simplest of all is a crack sensor based on a fiber embedded in a material. As long as the material is intact, the fiber transmits light without impediment. A crack breaks the fiber, reducing light intensity or cutting the light off altogether, depending on how large the crack is. You can think of it as a simple on-off sensor. If the light is on, you can drive a heavy truck across the bridge, but if the light stops coming through the fiber, you need to check the structure.

A more subtle type of intensity sensor depends on microbending. Figure 29.3 shows a pressure sensor based on a fiber passing between a pair of grooved plates. If there's no pressure on the plates, the fiber remains straight, and light passes through it. Pressure on the plates causes microbending—the more pressure, the more microbending—and microbending makes light leak from the fiber core. The more pressure, the less light is transmitted through the fiber.

Polarization Sensing

Some fiber sensors affect light polarization.

Other sensors affect the polarization of light in the fiber. There are a number of possible variations. One example is sensing of magnetic fields, using a process called Faraday rotation, which rotates the plane of polarized light by an angle proportional to the strength of the magnetic field. If you send vertically polarized light through a sensitive fiber, you can measure the magnetic field by measuring the angle the polarization is rotated.

In practice, you don't directly measure the angle of polarization, however. You actually measure the changes in the intensity of light transmitted by another polarizer. If the

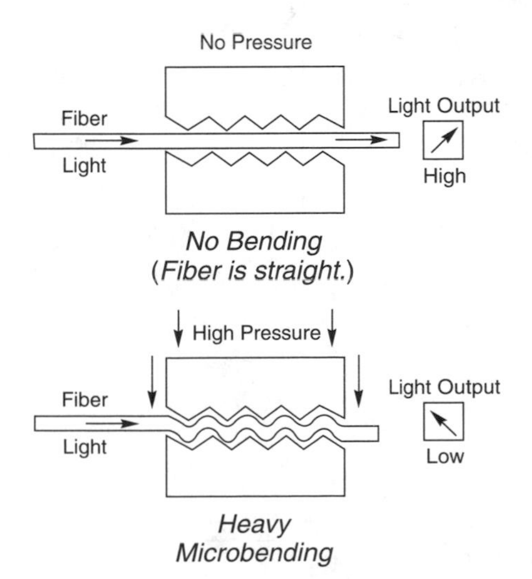

FIGURE 29.3
Increasing pressure on the plates causes microbending, reducing light transmission through the fiber.

second polarizer is also vertical, the decrease in transmitted light intensity measures the degree of rotation. This converts a change in polarization to a change in intensity, which is easier to measure.

Other fiber sensors produce effects that affect light of different polarizations differently. For example, pressure may change the refractive index for vertically polarized light differently than that for horizontally polarized light. This leads to a phase change in the intensities of the light in different polarizations, which requires another kind of measurement, as I describe next.

Phase or Interferometric Sensing

Interferometric sensors can detect very small changes.

Sensors also can modulate the phase of light to cause interference effects that modulate light intensity. To understand how this works, let's continue with the example of the pressure sensor that changes the phase of polarized light. By changing the refractive indexes of different polarizations by different amounts, the sensor effectively delays one polarization relative to the other. To measure this, you can separate the two polarizations at the output end, rotate one by 90°, equalize the path lengths, and then mix them together, as shown in Figure 29.4. If the two polarizations are in phase—that is, there is no delay between the two of them—the output is high; if one is 180° behind the other, the output is low.

Interferometric sensors are very sensitive to small changes, but they have some limitations. One is that the light has to be coherent enough that interference occurs; thus, you need either laser sources or very carefully equalized path lengths. In addition, there is an inherent

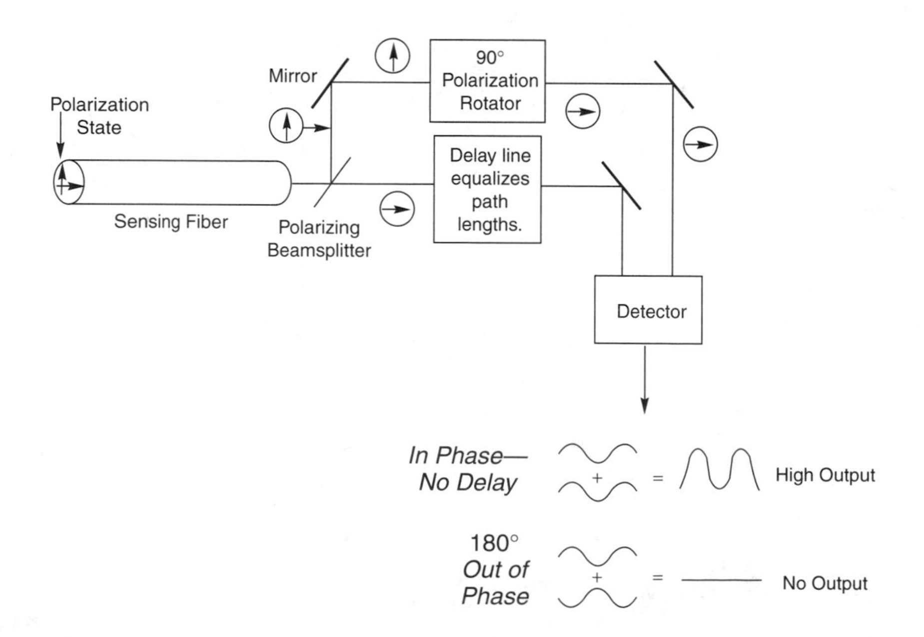

FIGURE 29.4
Polarization-delay fiber sensor.

ambiguity because a 360° delay produces the same effect as no delay or a 720° delay. You have to keep track of how many cycles of shifting occur or just measure a small shift.

Note also that to convert the phase shift to a change in intensity for this sensor, you need to compare two signals. In the case of the polarization sensor, these signals are two polarizations of light affected differently by pressure-induced changes in refractive index.

Another approach is to compare the phases of light passing through two fibers, one isolated from the environment and the other exposed to it. If the effective length of the fiber exposed to the environment changes, the phase changes, which can be measured by mixing light from the two fibers in an interferometric detector.

A third approach is to make a sensor that is itself an interferometer, which changes its resonance wavelengths as pressure, temperature, or other conditions change. I'll describe an example of that type of sensor later.

Constant versus Changing Measurements

One of the many subtleties of sensing is the difference between measuring long-term values and changing quantities. From a physical viewpoint, sound waves are really short-term variations in atmospheric pressure. However, you can't use the same instruments to measure the two. A microphone picks up sound waves but not atmospheric pressure. On the other hand, a barometer measures pressure but not sound waves.

The same is true for fiber sensors. Acoustic sensors work on different principles than pressure sensors. An interferometric fiber sensor on the seabed could pick up undersea sounds, but you'd need a different sensor to measure the pressure there.

Some Fiber Sensor Examples

Now that you've learned the basic principles of fiber sensing, let's look at a few examples. I will first cover a few general examples, then look at some promising specific cases.

Microbending Sensors

One attraction of microbending sensors is their simplicity. Microbending directly affects loss of a fiber; the more microbending, the higher the loss and the less light transmitted. Therefore, microbending sensors require only a simple measurement of light intensity, not a sophisticated interferometric setup to measure phase.

Pressure is the most straightforward quantity to measure with a microbending sensor, as shown in Figure 29.3. You can adapt microbending sensors to measure both static pressure and acoustic waves by designing and calibrating them differently. For total pressure—such as detecting whether a seat is occupied—you could use a fairly small sensor that would not respond to a 10-lb briefcase but would respond to a small 80-lb person. On the other hand, you would use a longer length of more sensitive fiber to detect acoustic waves, monitoring output continuously to detect their variation in time.

Length and Refractive Index Changes

Changing length or refractive index causes a phase shift.

A large family of sensors depend on changes in the effective length of the sensor, which depends on both the refractive index and the physical length. Recall that the time, t, it takes light to travel through a length, L, of material with refractive index n is

$$t = \frac{nL}{c}$$

where c is the speed of light in a vacuum. You can think of nL as the "effective length" of the material. A change in temperature can affect both refractive index and physical length, giving

$$t = \frac{(n + \Delta n)(L + \Delta L)}{c} \approx \frac{nL + n\Delta L + L\Delta n}{c}$$

as long as the changes are small. The result is a change in transit time,

$$\Delta t = \frac{n\Delta L + L\Delta n}{c}$$

which is equivalent to a phase shift in the light emerging from the sensor.

Interferometric detection can sense this phase change. Note that the principles of operation are the same whether the sensor is detecting a temperature change that affects only physical length of the fiber, a pressure change that affects only its refractive index, or something that affects both. (In practice, temperature change may affect refractive index as well as physical length.)

Changes in Light Guiding

Refractive index change also can be measured if it affects light guiding in the fiber. Suppose, for example, the refractive indexes of core and cladding vary with temperature in different ways. At 0°C, the core index is 1.50 and the cladding index is 1.49. As temperature increases, the core index decreases by 0.0005 per degree, but the cladding index decreases by 0.0004 per degree. At 100°C, the two refractive indexes would both equal 1.45. At that point, the fiber would stop confining light to the core, so output light intensity would drop to near zero.

In practice, light intensity might decrease as the core and cladding indexes approach each other, because at smaller index differences total internal reflection would trap an increasingly narrow range of light rays in a large-core fiber. However, the principle has been demonstrated in a sensor that can measure temperature within a few degrees.

A fiber Fabry-Perot interferometer detects a phase shift in a resonant cavity in the fiber.

Fiber Fabry-Perot Interferometer Sensors

The fiber Fabry-Perot interferometer is a sensor that detects a phase shift within a resonant cavity rather than by comparing the phase shifts of light taking two different paths. The sensing element is a section of fiber that has reflective layers on each end, as shown in Figure 29.5.

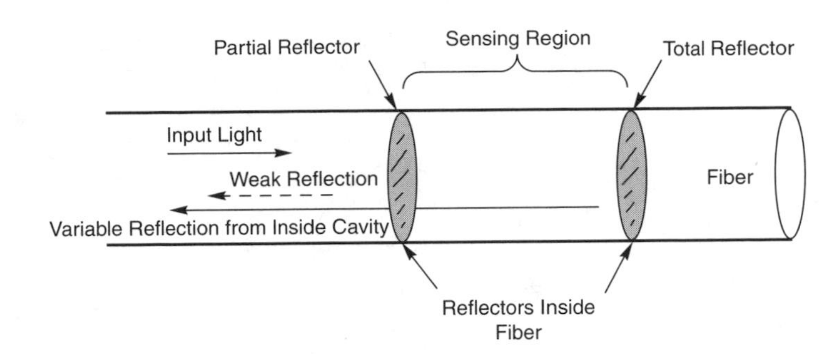

FIGURE 29.5
Fiber Fabry-Perot interferometer sensor.

Light passes through a partly reflecting layer and is reflected by a totally reflecting mirror some distance behind it. The layers can be made by splicing fiber segments together at those points.

These two mirrors form a Fabry-Perot interferometer, which has a series of resonances at characteristic wavelengths defined by the cavity length and refractive index. Recall that at a resonance the round-trip distance must equal an integral number of wavelengths in the material, with refractive index included to account for the difference between the vacuum wavelength, λ, and the wavelength in the material, λ/n.

$$N\lambda = 2Ln$$

If the wavelength stays fixed and the cavity is long compared to the wavelength, the intensity of the reflected light changes with variations in length or refractive index. In temperature sensors, the change in refractive index is about 20 times larger than the change in length, so it dominates the phase shift. The same approach can be used to sense pressure and strain.

Fiber Grating Sensors

The fiber gratings you learned about in Chapter 7 also can be used in sensors. As in a fiber Fabry-Perot interferometer, changes in the refractive index change reflective properties, such as the wavelength of peak reflectivity.

Fiber-Optic Gyroscopes

The fiber-optic gyroscope is probably the most successful fiber sensor so far. It relies on optical processes to sense rotation around the axis of a ring of fiber. Rotation sensing is vital for aircraft and missiles, which have traditionally used gimbaled mechanical gyroscopes as references. Fiber gyroscopes (and laser gyroscopes that serve a similar purpose but operate on different principles) offer a number of advantages, including no moving parts, greater reliability, and no need for a warm-up period to start the gyro.

A fiber-optic gyroscope measures rotation interferometrically.

Figure 29.6 shows the effect that is the basis of a fiber gyro. Light from a single source is split into two beams directed into opposite ends of a single-mode fiber. In actual sensors, the fiber is wound many times around a cylinder, but the drawing shows only one turn for

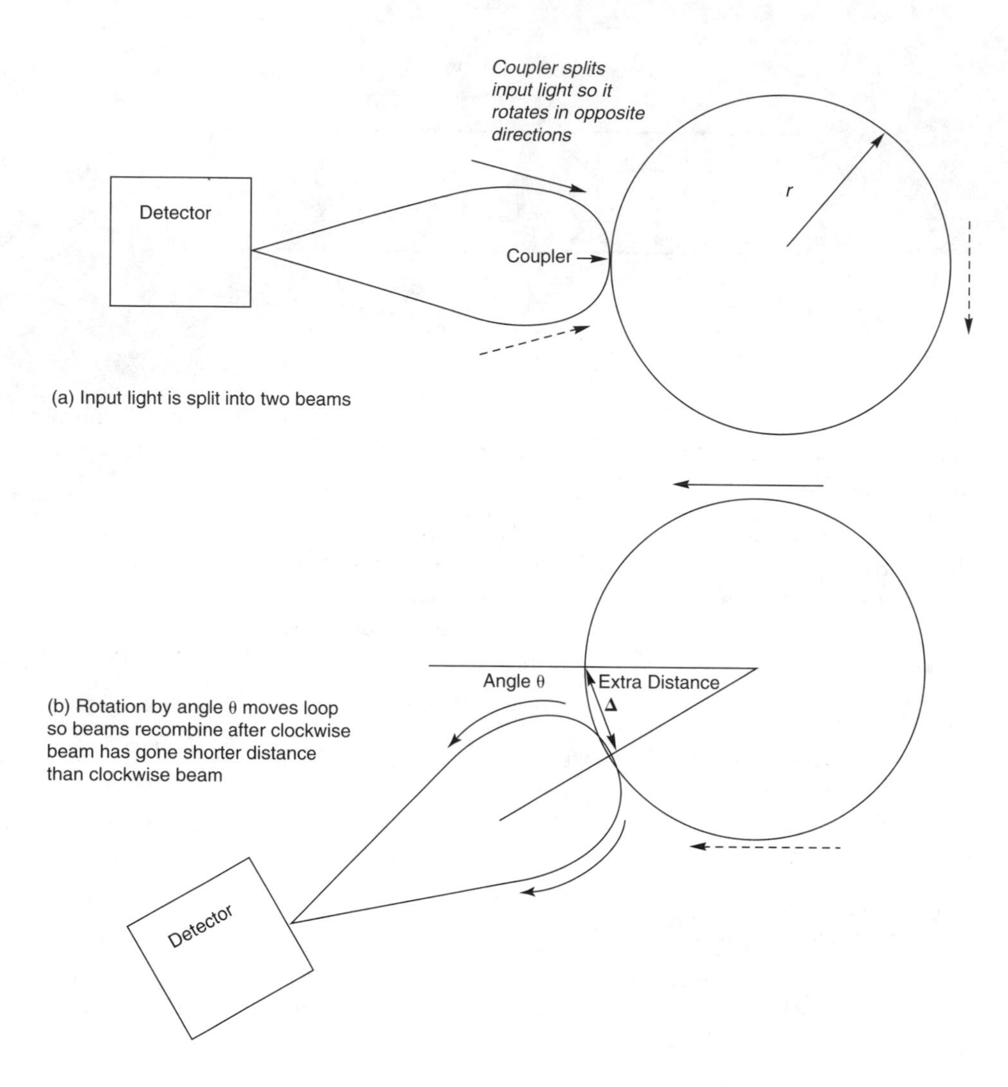

FIGURE 29.6
Fiber-optic gyroscope.

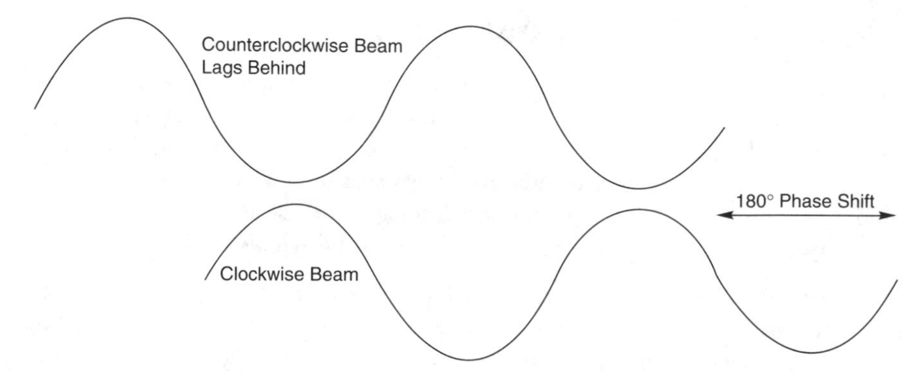

FIGURE 29.6
Continued

clarity. Light takes a finite time to travel around a fiber loop, and in that time the loop can rotate a small amount. When the beams return to their starting point, that point has moved, and the two have traveled slightly different distance.

For simplicity, assume the loop is a circle with radius r and circumference $2\pi r$. During the interval that light travels around this loop, it rotates an angle θ, which is exaggerated in Figure 29.6b. To come back to the coupler where the beams combine, the counterclockwise beam must go an extra distance, Δ, from the point where it started, a total of $2\pi r + \Delta$. However, the clockwise beam goes a shorter distance, $2\pi r - \Delta$. The result is a phase shift of the two beams when they come to the coupler by an amount 2Δ. Figure 29.6c shows this as a total phase shift of 180°. This phase shift, called the Sagnac effect, can be detected with an interferometer for two beans passing in opposite directions through a suitable single-mode fiber.

The distance Δ is a function of the rotation θ, given by

$$\Delta = \left(\frac{\theta}{360°}\right) \times 2\pi r$$

The actual phase shift is 2Δ or twice that value, which you have to remember when you calculate the rotation. In practice, the angle of rotation is calculated from the cumulative phase shift from the time the fiber gyro starts operating. If you start out heading north, and have a cumulative phase shift of 90° to the right, you wind up heading east.

Fiber gyros can be used as part of an inertial navigation system, which keeps track of a vehicle's position. Three separate fiber gyros keep track of the vehicle's angular direction on three perpendicular axes. To know its position, you also have to keep track of the time, so you know when the vehicle made a particular turn. You also need a separate accelerometer to measure acceleration and thus deduce how fast the vehicle is moving and—by keeping track of time—how far it has traveled. Current fiber gyros maintain direction accurately to around one degree per hour.

Fiber gyros are not as accurate as laser gyros, but they are less expensive and are entirely solid-state, unlike laser gyros that use gas lasers. Their low cost and durability make fiber gyros attractive for applications such as guiding missiles (which you don't want to load with lots of expensive equipment) and short-range aircraft. Without any moving parts, they are inherently more reliable than mechanical gyros. Fiber gyros were proposed for use in cars, but global positioning system (GPS) receivers are less expensive and have gained the lead in automobile navigation systems.

Smart Skins and Structures

Fiber sensors can be embedded in composites and other materials such as concrete to create "smart structures" or "smart skins." The goal is to create a structural element (including the skins of aircraft) equipped to monitor internal conditions. The initial emphasis is on verifying that components meet initial structural requirements, but the fiber sensors could be used throughout the life of the component. Figure 29.7 shows how fibers can be embedded between layers of a composite material; in this case, they are encased in an epoxy layer.

Fiber sensors are embedded in composites to make smart structures and skins.

One use of embedded fiber sensors is to monitor fabrication and curing of the composite. The fiber sensors can monitor temperature to be sure curing conditions meet requirements. They can also monitor strain, to verify that the component is not stressed excessively. Detection of cracked fibers indicates serious stress problems. Later, the fiber sensors can provide data on stresses and strains that occur after the composite component is mounted in its final position. Eventually this information may be used by operating engineers, but currently its main use is in studying properties of structures and aircraft.

Once a smart-skin or smart-structure system is in operation, engineers could use the fiber sensors for periodic checks of performance and structural integrity. For example, fibers embedded in aircraft wings could be plugged into monitoring equipment in the service bay, to make sure they suffered no invisible internal cracks that could cause catastrophic failure. Dams and bridges likewise could be monitored by embedded fibers.

Some military planners think the ultimate step would be to plug the fiber sensors into a real-time control system designed to optimize performance. The performance limits of aircraft materials and structures are not known precisely, so engineers err on the side of safety and avoid pushing too far. Real-time fiber monitors could tell computers how well components were withstanding stresses in operation. Ultimately, perhaps, the computers could use the sensor data to apply corrections in real-time that would push the performance envelope further, without endangering pilots or aircraft.

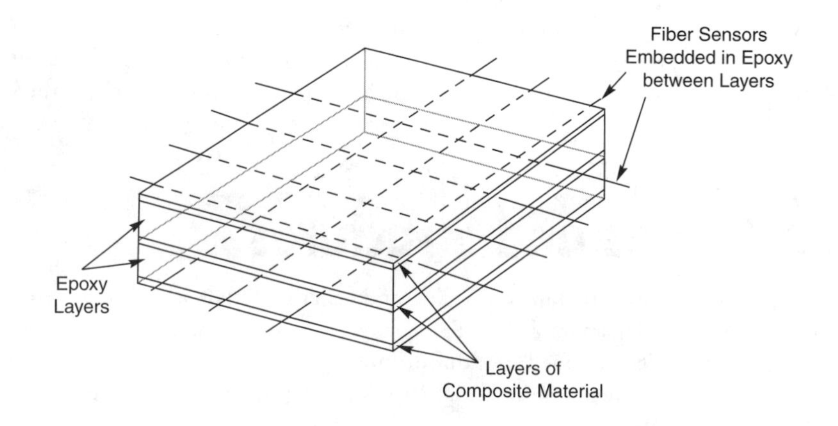

FIGURE 29.7
Sensing fibers in a smart skin are embedded in an epoxy matrix between layers of a composite material.

What Have You Learned?

1. Fibers can serve as probes that collect light for sensing. Fibers can also function as sensors themselves.
2. Fiber probes can detect objects that block or reflect light. This lets them measure shapes, count parts, and do other simple tasks.
3. Fiber optics can collect light from remote optical sensors so it can be measured.
4. Sensors convert something hard to measure into units easier to observe.
5. Intrinsic fiber sensors detect changes in the way fibers transmit light. Unlike communication fibers, these fibers are designed to respond to changes in the environment.
6. Fiber sensors work by modulating light they transmit, either by changing light intensity, by affecting polarization, or by shifting the phase of the light.
7. Interferometric sensors measure phase changes; they can detect very small shifts.
8. Placing a fiber in a place where pressure can cause microbending allows the fiber to sense pressure; the more pressure, the more light lost from the fiber.
9. Temperature and pressure can change the refractive index of glass in a fiber, causing phase shifts and other effects.
10. A fiber Fabry-Perot interferometer detects a phase shift within a resonant cavity in a fiber.
11. Loops of fiber can measure rotation by sensing differences in the time light takes to travel in opposite directions around the loop. Such fiber gyroscopes can be used in guidance systems.
12. Fiber sensors can be embedded in composite materials to make smart structures and smart skins.

What's Next?

In Chapter 30, I will look at other noncommunication applications of fiber optics.

Further Reading

David A. Krohn, *Fiber Optic Sensors: Fundamentals and Applications* (Instrumentation, Systems, and Automation Society Press, Research Triangle Park, NC, 2000)

Herve C. Lefevre, *Fiber-Optic Gyroscope* (Artech House, 1993)

A. Selvarajan, "Fiber Optic Sensors and Their Applications" *http://www.ntu.edu.sg/mpe/research/programmes/sensors/sensors/fos/foselva.html*

Eric Udd, ed., *Fiber Optic Sensors: An Introduction for Engineers and Scientists* (Wiley, 1991)

Questions to Think About for Chapter 29

1. A crack sensor uses a step-index multimode fiber with 100-μm core to detect structural failure. It sets off an alarm when light intensity drops 10 dB. If a crack splits the block of material containing the fiber and causes one side to drop, estimate how far it must drop to set off the alarm. Ignore end reflection effects.
2. Temperature causes the refractive index of a fiber to increase by 0.001% per degree. Two arms of an interferometric fiber sensor each contain 1 cm of fiber. One arm is exposed to the environment, the other is kept at a constant temperature. If you use 1-μm light, how much temperature change is needed to produce a 180° phase shift?
3. There are two different ways you could increase the sensitivity of the interferometric sensor in Question 2 so the sensor measures a 1° temperature change with a 180° phase shift, without changing the glass used in the fiber. What are they?
4. A fiber-optic gyroscope includes a 1-m loop of fiber and a laser light source emitting at 1 μm. How much rotation does it take to cause a 180° phase shift between the two counter-propagating beams?
5. Recall that light travels roughly 3×10^8 m/s. What rotation rate does the 180° phase shift in Question 4 correspond to if it's detected over the time circles through the fiber once?

Quiz for Chapter 29

1. How do fiber-optic probes work?
   a. They detect the presence or absence of light at a point.
   b. They detect the pressure of objects placed on top of them.
   c. Changes in temperature make them expand or contract.
   d. None of the above
2. Which of the following is an example of a fiber collecting light from a remote optical sensor?
   a. Fiber-optic gyroscope
   b. Liquid-level sensor based on total internal reflection from a prism
   c. Acoustic sensor based on microbending
   d. Fiber grating used as a pressure sensor
   e. Smart skins
3. How can microbending effects be sensed?
   a. By observing tension along the length of the fiber
   b. By monitoring changes in light transmitted by the fiber

c. By looking for changes in data rate of a signal transmitted through the fiber

d. By measuring light emitted by the fiber

4. Which of the following can change the refractive index of a fiber?

a. Temperature changes

b. Pressure changes

c. Sound waves

d. All of the above

e. None of the above

5. Which sort of change in a fiber sensor can be measured by interferometry?

a. Changes in the wavelength of input light

b. Changes in intensity of light

c. Changes in refractive index caused by pressure

d. Changes in optical absorption

6. An example of an interferometric sensor is a

a. punched card reader.

b. microbending sensor of acoustic waves.

c. fiber-optic gyroscope.

d. sensor that measures the height of parts on a production line.

7. How do fiber grating sensors work?

a. Microbending causes increased attenuation

b. They alter wavelengths transmitted and reflected

c. They change polarization

d. They modulate light with a digital code

8. A 1° increase in temperature reduces the refractive index of the glass in a sensing fiber by 0.000005 at a wavelength of 1 μm. Assuming the length of the fiber does not change significantly, how much does a 10° temperature shift the phase of 1-μm light passing through a 10-mm sensor?

a. 1.8°

b. 90°

c. 180°

d. 360°

e. 1800°

9. How do fiber-optic gyroscopes detect rotation?

a. By measuring changes in the wavelength of light in the fiber

b. By interferometrically measuring differences in the paths of light going in opposite directions around a fiber loop

c. By detecting changes in polarization of light caused by inertial changes in the moving fiber loop

d. By measuring intensity changes caused by microbending

e. By detecting changes in the refractive index induced by acceleration

10. What can fiber sensors measure when embedded in a smart structure?

a. Curing conditions of a composite material

b. Internal strain in a composite material

c. Structural integrity of a completed component

d. Stresses on a component during use

e. All of the above.

CHAPTER

30

Imaging and Illuminating Fiber Optics

About This Chapter

Communications and sensing were latecomers in the world of fiber optics. The early developers of fiber optics were interested in transmitting images through bundles of fibers. Fibers developed for imaging differ greatly from those designed for communications, because different considerations affect performance. This chapter is about fiber-optic imaging and the related field of illumination. Some fiber-optic illumination uses bundles of fibers to deliver light, but much illumination is with single fibers. This chapter covers these diverse noncommunication applications of fiber-optic devices in medicine, industry, and military systems.

Basics of Fiber Bundles

As you read in Chapter 1, optical fibers were invented for imaging and were soon applied to illumination as well. Imaging requires a bundle of fibers, one to carry each point on the image. For imaging, the bundle must be *coherent,* which in this case means that the ends of the fibers must be arranged in the same way on both ends of the bundle. Project an image onto one end of a coherent bundle, and the same image appears on the other end.

To visualize how a coherent bundle works, start with a handful of drinking straws all the same length. With a little care, you can hold the straws so they are aligned parallel to each other. Look through the straws at a printed page, and you'll see the words through the array of little pipes. The smaller the straws, the smaller the bit of the page you see through each one. Individual fibers are like individual straws, but they are much thinner and far more flexible. Fibers guide light by total internal reflection, but straws only transmit light straight along their axes.

Coherent bundles have the fibers in the same places on both ends.

There are two basic families of fiber bundles, each developed for distinct applications. Long, thin flexible bundles of loose fibers are used to examine or deliver light to otherwise inaccessible places. Important examples are the flexible endoscope threaded down a patient's throat to examine the stomach, and the flexible colonoscope used to examine the colon. Industrial counterparts are used to inspect the interiors of engines. Imaging requires coherent bundles, but illumination normally is done with bundles in which the fibers are randomly aligned. For most imaging and illumination applications, flexibility is important.

A second family is rigid fiber bundles in which the fibers have been fused together to make a solid block. Processing retains the light-guiding structure of the individual fibers that are aligned end to end in the bundle. Usually they are shorter and fatter than flexible bundles. These fused bundles can be used as optical devices for transmitting or magnifying images piece by piece, as well as for some types of inspection and for some other optical applications.

Both types of fiber bundles are based on step-index multimode fibers with thick cores and thin claddings. This structure means that light reaching the input face of the bundle is most likely to fall on a core, so it is transmitted to the other end of the bundle. Individual fibers may be drawn quite thin, but the ratio of core to cladding thickness remains unchanged. The difference between core and cladding index is larger than in communication fibers, so the cores can be drawn finer and still transmit multiple modes of light. Single-mode transmission is not desirable in imaging or illumination fibers because it limits how much light they can collect at the face of the bundle.

Making Fiber Bundles

Long, thin flexible bundles are made by winding a fiber around a spool and cutting through a glued region.

Figure 30.1 shows one way to make coherent fiber bundles. Start by looping a single long, thin fiber many times around a spool, glue the fibers together in one spot and remove them from the spool. Then cut through the glued region. This gives a flexible bundle, with fibers loose in the middle and fixed on both ends. Because the two ends were originally adjacent to each other, the fibers are all in the same positions.

This approach is simple in concept and dates back to the mid-1950s, when it was used to make the first flexible fiber bundle. However, it is a demanding process and is difficult when using very thin fibers, which are likely to break.

Fused imaging bundles are drawn jointly into solid rods.

An alternative approach is to draw many fibers simultaneously to finer and finer diameters in a series of stages. The first step is to draw a step-index fiber with a diameter about 2.5 mm. These fibers are easy to handle, and a group of them—typically 37 to 169—are grouped together, heated until they soften, and stretched out into a rigid *multifiber* about 2 mm in diameter, as shown in Figure 30.2. Then a number of multifibers (typically 61 to

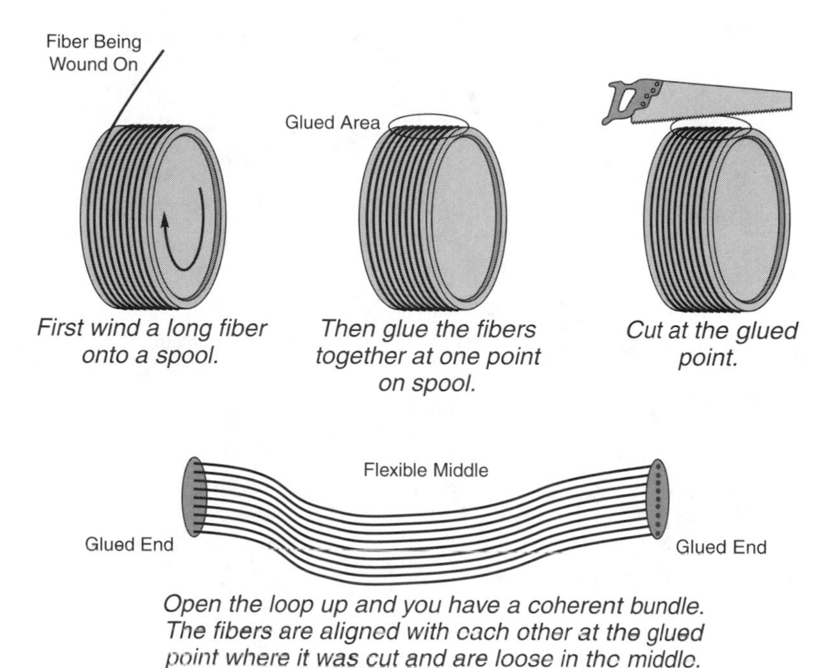

FIGURE 30.1
Flexible bundle made by winding fiber around a spool.

271) are packed together, heated, and drawn again to produce a rigid fiber bundle, containing many thousands of fibers. Each fiber in the final bundle is about 3 to 20 μm in diameter. The number of fibers drawn together in each step is chosen to make patterns that pack together well.

The fused fiber bundle process can be used to make flexible bundles, with a few important changes. Look carefully at Figure 30.2, and you will note that the large core is surrounded by two rings of cladding. One is the conventional low-index cladding that confines light to the core in all fibers. The composition of the other depends on the type of bundle being made.

For rigid bundles, that outer layer is a dark absorptive glass that keeps light from leaking between the cores in the bundle. A certain amount of light always leaks into the fiber cladding. Usually this stays in the inner part of the cladding, but for imaging bundles the cladding is quite thin. If the claddings were all fused together—as they would be without the dark glass—the light could freely disperse through the whole bundle within the fused cladding glass. Then it could leak back into the cores and degrade the image.

For flexible bundles, that outer layer is a glass that is soluble in acid. Manufacturers cover the ends of the rigid rod and then dip it in an acid that dissolves away that leachable layer in the middle of the rod, leaving a flexible bundle of many thin fibers, which are arranged so their ends are aligned for imaging.

Individual fibers in a flexible coherent bundle can be small, but not quite as small as in a fused bundle. Some performance limits of flexible bundles are comparable to those of rigid

FIGURE 30.2
Stages in making a fiber bundle. (Courtesy of Schott Fiberoptics)

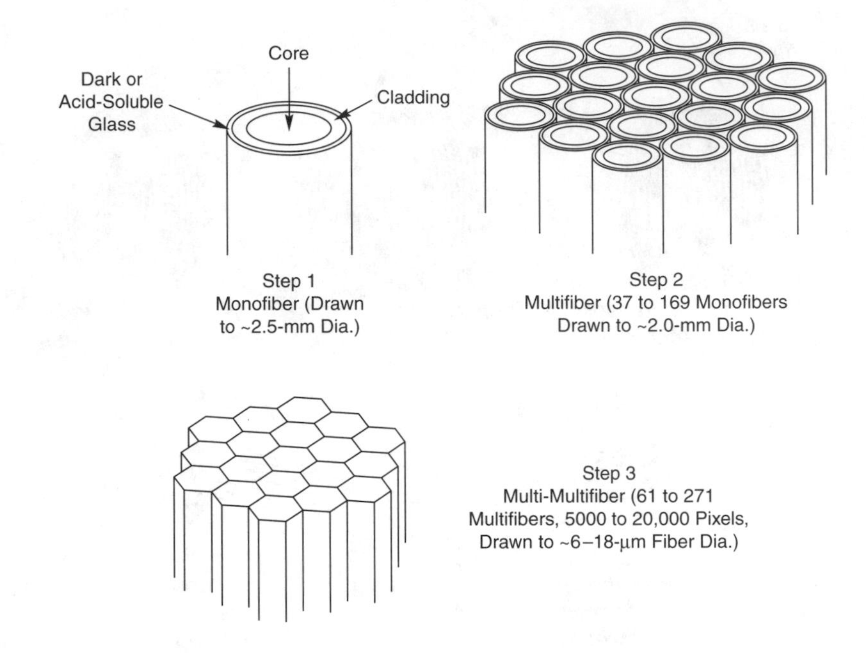

bundles. When flexible bundles are used, an added concern is breakage of individual fibers, which does not occur in fused bundles. Each fiber break prevents light transmission from one spot on the input face. The loss of a single fiber is not critical, but as more fibers break, the transmitted light level drops and resolution can decline as well. Eventually breakage reaches a point where the image-transmitting bundle is no longer usable. Plastic fibers can reduce the breakage problem, but have other limitations.

Randomly aligned bundles serve as "light pipes."

Randomly aligned bundles are made by collecting many fibers into a bundle, much like collecting strands of spaghetti. This would be very difficult if the fibers were as thin as those in imaging bundles, but such fine fibers are not needed because the resolution does not matter; random bundles serve purely as "light pipes." Typically, random bundles are made of fibers with diameters in the 100-μm range, which are flexible enough to bend freely with minimum fiber breakage.

Imaging and Resolution

Resolution is a crucial issue in an imaging fiber bundle. Figure 30.3 shows how an image is carried from one end of the bundle to the other. Each fiber core carries its own segment of the image to the other end, maintaining its alignment.

Bundle resolution depends on the core sizes of the fibers it contains.

To visualize what happens, imagine that each fiber core captures a chunk of the image and delivers it to the other end of the bundle. This process averages out any details that fall within a single core. For example, if the input to a single core is half black and half white, the output will be gray. Thus, the fiber cores must be small to see much detail. For a sta-

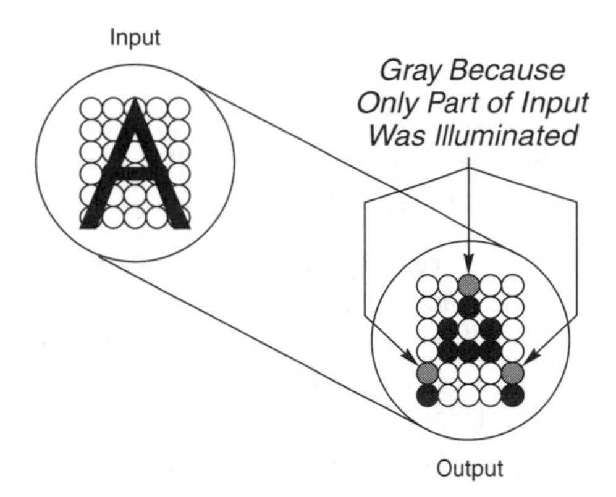

FIGURE 30.3
Image transmission through a fiber bundle.

tionary fiber bundle, the resolution is about half a line pair per fiber core, meaning two fiber core widths are needed to measure a line pair. Numerically, that means 10 μm fiber cores could resolve 50 line pairs per millimeter (1 line pair per 20 μm). Imaging bundles have fiber cores as small as 3 μm. Resolution is significantly higher—about 0.8 line pair per fiber core diameter—if the fiber bundle is moving with respect to the object.

Before you wonder too much about the quality of fiber-bundle images, you should realize that fiber cores typically are about 10 μm. If that was the case, the letter A in Figure 30.3 would be only 60 μm high, less than $\frac{1}{16}$ mm tall. That's many times smaller than the finest of fine prints used in legal documents. You have to look very hard, and may need a strong magnifying lens to see the individual fiber spots on a good imaging bundle.

Cladding Effects

The cores conduct light in fiber bundles, but they are surrounded by cladding layers. Bundles are made with thin cladding layers, but some light must fall onto the cladding rather than the core. The fate of that light depends on the bundle design. Fibers in rigid bundles have a dark outer cladding layer that absorbs light so that little can pass between fiber cores. Light that leaks out of the cores of individual fibers in flexible bundles cannot easily enter other fibers. However, neither type can completely prevent light from leaking between fibers.

Light that falls into fiber claddings in bundles is lost, but typically 90% falls onto fiber cores.

Most light entering the cladding is lost, which can limit transmission efficiency. This makes the fraction of the surface made up by fiber cores an important factor in a bundle's light-collection efficiency. That is, the collection efficiency depends (in part) on the packing fraction, defined as

$$\text{Packing fraction} = \frac{\text{total core area}}{\text{total surface area}}$$

A typical value is around 90%.

Transmission Characteristics

Typical attenuation of bundled fiber is around 1 dB/m.

Fiber bundles need to carry light only a few meters, so they do not have as low attenuation as communication fibers. Typical attenuation of bundled fiber is around 1 dB/m, over a thousand times higher than that of communication fibers at 1300 nm.

Likewise, operating wavelengths differ. Visible light is needed for imaging and illumination, and even for other applications the short distances make it unnecessary to operate at wavelengths where fibers are most transparent. Glass fiber bundles are typically usable at wavelengths of 400 to 2200 nm, and special types made from glass with good ultraviolet transmission are usable at somewhat shorter wavelengths. Plastic fibers are usable at visible wavelengths, 400 to 700 nm. Some special-purpose bundles are made of other materials, but they are not widely used.

Bundled fibers are step-index multimode types with large NA.

Bundled fibers generally have higher numerical apertures than communication fibers, because light-collection efficiency is critical and pulse dispersion is irrelevant. The relatively large difference between core and cladding index gives bundled fibers typical NAs of 0.35 to 1.1. The same holds true for large-core single fibers used in illumination; larger NAs are better because they boost light-collection efficiency.

Optics of Bundled Fibers

Some simplifications valid for communications are not valid for bundled fibers.

The underlying principles of fiber optics are the same if fibers are separate or bundled. However, earlier descriptions of communication fibers relied on some simplifications of optical principles. These simplifications don't always work for bundles or other noncommunication fibers. It's time to go back and face some complications that don't affect communications through single fibers.

Light Rays in Optical Fibers

Light rays are an important concept in understanding how optical devices affect light. Earlier you learned how lenses worked, and saw a simple explanation of how total internal reflection of light rays guide light down a fiber. Those explanations are true as far as they go, but the behavior of light rays in a fiber is a bit more complicated.

As shown in Figure 30.4, a light ray that enters the fiber at an angle θ to the fiber axis will later emerge at roughly the same angle to the fiber axis, as long as the ray is within the fiber's acceptance angle. However, the ray may not emerge in the same direction; it will be part of a ring of light at roughly the original angle to the fiber axis. *Roughly* is the operative word, because imperfections in the fiber and other factors cause the light to emerge in a ring of angles centered on θ.

This does not conflict with what you learned about communication fibers. There the light ray was only an example of the path light could follow. In looking at multimode fibers, we considered the light rays and the modes collectively, never worrying about individual mode patterns. Generally there's no reason to worry about individual modes in multimode fibers.

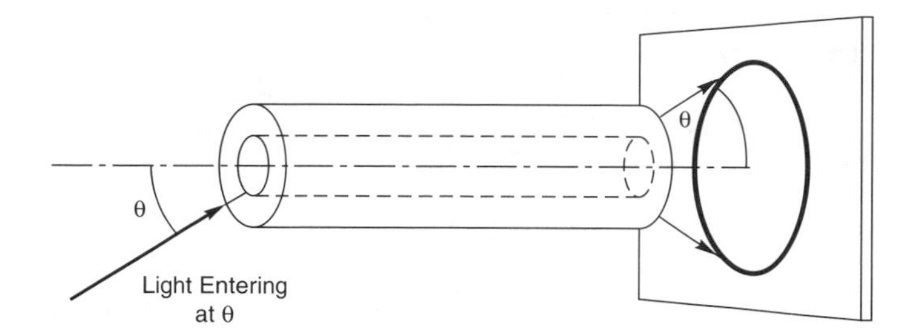

FIGURE 30.4
Light rays emerge from a fiber in a diverging ring.

One other thing should be pointed out: step-index fibers with constant diameter cores do not focus light. (As you will see later, both tapered and graded-index fibers can focus light passing along their lengths.) All light emerges from a step-index multimode fiber at roughly the same angle that it entered, not at a changed angle, as would happen if it did focus light. As long as the fiber's sides and ends are straight and perpendicular to each other, a single step-index multimode fiber or a bundle of them—like a flat window pane—cannot focus light.

Step-index fibers with constant size cores do not focus light.

This has one important practical consequence that you'll discover the first time you look through an imaging bundle. You have to put the distant end up very close to what you want to see, or the image will become blurred. You see the image on the near end of the bundle because light travels straight through each fiber. For the light from the object to enter the right collecting fibers, it must either be focused onto the collecting end or the collecting end must be very close to the object, so light can't slip into other fibers.

If the fiber's output end is cut at an angle not perpendicular to its axis, light entering at an angle θ still emerges in a cone, but the center of the cone is at an angle to the fiber axis. If the slant angle (from the perpendicular) is a small value ϕ, the angle β by which the rays are offset is approximately

$$\beta = \phi(n - 1)$$

where n is the refractive index of the fiber core.

Tapered Fibers

I assumed earlier that fiber cores are straight and uniform, but they could also be tapered (although not over long distances). Figure 30.5 shows what happens to a light ray entering a tapered fiber at an angle θ_1. If the ray meets criteria for total internal reflection, it is confined in the core. However, it meets the core-cladding boundary at different angles on each bounce so each total internal reflection is at different angles from the axis. The result is that it emerges from the fiber at a different angle, θ_2. If input core diameter is d_1 and output core diameter is d_2, the relationship between input and output angles is

Tapered fibers magnify or demagnify objects seen through them. Tapered fibers are used in bundles.

$$d_1 \sin \theta_1 = d_2 \sin \theta_2$$

The same relationship holds for the fiber's outer diameter as long as core and outer diameter change by the same factor, d_2/d_1.

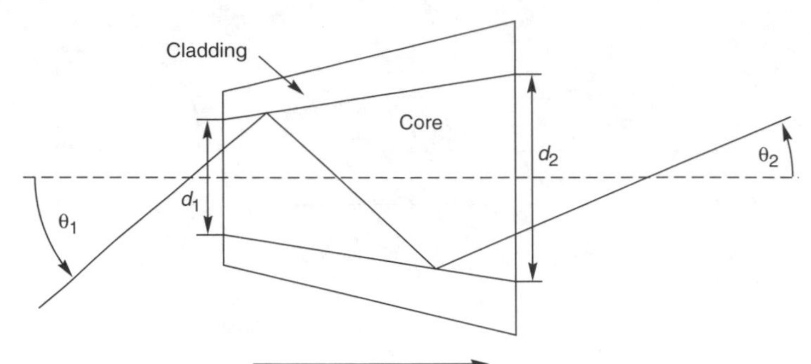

FIGURE 30.5
Light passing from the narrow to the broad end of a tapered fiber.

As a numerical example, suppose the input angle is 30° and the taper expands diameter by a factor of 2. The sine of the output angle θ_2 would be

$$\sin\theta_2 = \frac{d_1}{d_2}\sin\theta_1 = \frac{1}{2}(\sin 30°) = 0.25$$

Thus θ_2 would be about 14.5° and light exiting the broad end of a taper would emerge at a smaller angle to the fiber axis than it entered. Conversely, light going from the broad end to the narrow end would emerge at a broader angle.

Tapered bundles of fused fibers can be used as magnifiers if the narrow end is placed on a page and you look at the top side. Each fiber expands or shrinks the spot of the image it transmits by the same amount. The eye sees this as each spot being spread over a larger area at the large end of the taper. This increases the size of the image, but not the clarity, because the transmitted image has only as many picture elements as the narrow end of the bundle.

Focusing with Graded-Index Fibers

Graded-index fibers can focus light in certain cases.

Unlike step-index fibers, graded-index fibers can focus light in certain cases. This does not make graded-index fibers useful for image transmission or other fiber-bundle applications, but short segments of graded-index fibers can function as focusing components in some optical systems.

In Chapter 4, you saw that light follows a sinusoidal path through graded-index fiber. When you looked at how a cone of light was transmitted through a long fiber, you saw output as a cone of the same angle. Now look instead at the path of an individual ray through a short segment of graded-index fiber, shown in Figure 30.6, and compare that with the path of a light ray in step-index fiber.

There is an important but subtle difference. Total internal reflection from a step-index boundary keeps light rays at the same angle to the fiber axis all along the fiber. However, graded-index fibers refract light rays, so the angle of the ray to the axis is constantly changing as the ray follows a sinusoidal path. If you cut the fiber after the light ray has gone through 180° or 360° of the sinusoid, the light emerges at the same angle to the axis that it entered. However, if the distance the light ray travels is not an integral multiple of 180° of

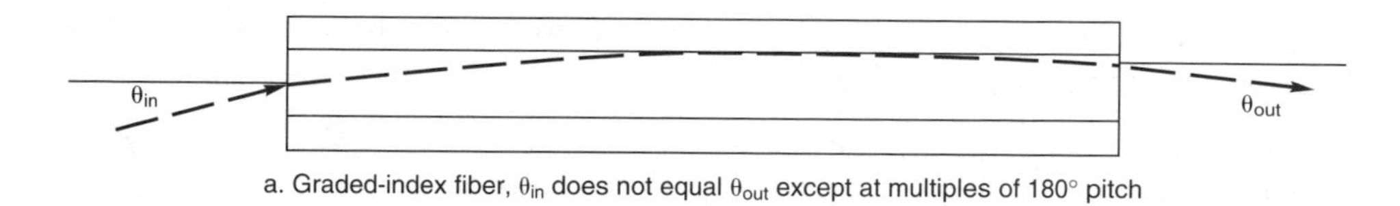

a. Graded-index fiber, θ_{in} does not equal θ_{out} except at multiples of 180° pitch

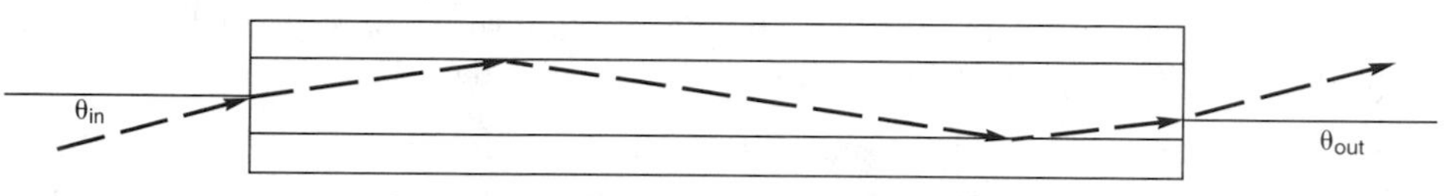

b. Step-index fiber, input and output angles always equal

FIGURE 30.6 *Rays in graded-index and step-index fibers.*

the sinusoid, it emerges at a different angle. This property allows segments of graded-index fiber to focus light.

In the design of graded-index fiber lenses (usually sold under the trade name Selfoc), the key parameter is the fraction of a full sinusoidal cycle that light goes through before emerging. That fraction is called the *pitch.* A 0.23-pitch lens, for instance, has gone through 0.23 of a cycle, or $0.23 \times 360° = 82.8°$. The value of the pitch depends on factors including refractive-index gradient, index of the fiber, core diameter, and wavelength of light.

Pitch is a critical parameter of graded-index lenses.

Although the lenses are segments of fiber, they are short by fiber-optic standards, just a few millimeters long. Thus, they can be considered as rod lenses as well as fiber lenses.

These tiny fiber lenses are used in a variety of applications. Some are used in fiber-optic transmitters, to focus light from an LED or diode laser so that it can be coupled efficiently into a fiber. Others are used in optical systems such as fax machines and scanners. A linear array of fiber-optic microlenses can focus light reflected from a small area of a page onto a linear array of sensors that detect the light. Ideally each sensor collects light focused by one microlens.

Imaging Applications

Imaging covers a broad range of fiber-bundle applications. Most imaging systems use lenses and conventional optics, but fiber bundles do a better job in certain cases. Imaging bundles often are better for reaching into inaccessible places, from the inside of the human body to the interior of machines. Let's look briefly at these applications.

Medical Endoscopes

The most important use of imaging fiber bundles is to allow physicians to look inside the body without surgery. This is done with special-purpose coherent fiber bundles called *endoscopes,* which are up to a couple of meters long. Versions called *gastroscopes* are threaded down the throat to examine the stomach. *Colonoscopes* are versions designed to examine the colon. Short rigid bundles are used for some medical examinations because of their

Endoscopes allow physicians to look inside body cavities.

high resolution, but flexible types are preferred for most purposes because they are easier to insert and manipulate through body orifices.

Traditional fiber-optic endoscopes use one set of fibers to transmit light inside the body and a separate set to collect and view the reflected light. Lenses on the end of the instrument focus light onto the fiber bundle, so it does not have to be pressed against tissue. Endoscopes may include surgical tools to treat lesions in the stomach or colon. Some newer endoscopes use fibers to transmit light into the body, but collect light with a miniature CCD (charge-coupled device) imaging camera that is inserted into the body.

Some endoscopes include fibers capable of transmitting high laser powers as well as illuminating light, so physicians can perform laser surgery. For example, a surgeon performing microsurgery on the knee could make an incision to insert an endoscope. After viewing the area to be treated, the surgeon could look through the viewing fibers to align the instrument, look away, then fire laser pulses to treat the lesion. (Surgeons avoid watching during laser pulses to protect their eyes.) After each set of laser pulses, the surgeon looks back to check progress.

Industrial Inspection Instruments

Fiber-optic imaging instruments also are used in industry to inspect dangerous or otherwise inaccessible areas. Lenses on the end of the instrument focus light onto the end of the bundle. Flexible fiber bundles could be used in this way to examine the inside of a storage tank that has only one small opening. Fiberscopes also could examine the interiors of machinery.

Faceplates

Image transmission does not have to be over a long distance. Another common application of fiber-optic image transmission is the fiber-optic *faceplate,* a thin slice of a coherent bundle in which individual fibers are only a fraction of an inch long. Faceplates are cut from longer, fused coherent bundles like slides of salami, although generally one or both surfaces are not flat.

Fiber-optic faceplates transfer light between surfaces of different shapes.

The job of a faceplate is to transmit an image between two stages of an imaging system that must amplify weak input light to generate a clearly visible image. It is typically used in military systems where faint light is amplified so soldiers can see an image of the scene. Image amplifiers may go through multiple stages, each amplifying the input light by a certain factor. Infrared light is used to generate a visible image. The output stages often are strongly curved screens that can't be focused onto flat input devices without distortion. A faceplate can convert the curved output screen to a flat surface, as shown in Figure 30.7. If the input of the next stage works best with a curved screen, the other side of the faceplate can be curved to match.

The big advantage of the faceplate is that it transfers light very efficiently between two surfaces that otherwise can't be butted face to face. Suppose, for example, you're trying to detect some very weak light from a scene illuminated only by starlight. A single-stage image-intensifier camera makes the image brighter, but not bright enough to see clearly. You

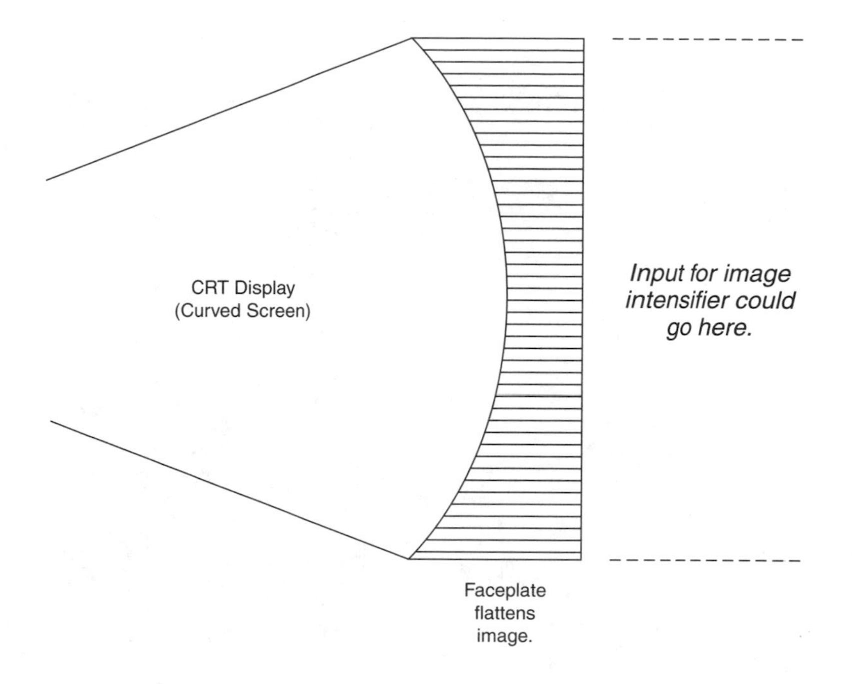

FIGURE 30.7
A fiber-optic faceplate transfers light from a curved display tube.

want to add a second stage, but the output of the first stage is on a curved screen. Put a fiber faceplate between that output and the input of the second-stage tube and you lose very little light. An imaging lens would lose much more light. The first fiber-optic faceplates were developed for such military imaging tubes, and they remain in use for newer equipment.

Faceplates also can help flatten the curved image generated by some display screens, correct for distortion, and make the display appear brighter by concentrating light toward the viewer.

Image Manipulation, Splitting, and Combining

Coherent fiber bundles can manipulate images.

Coherent fiber bundles can do more than just transmit images; they can also manipulate them. Twisting a coherent bundle by 180° inverts the image. You can do the same with lenses, but a fiber-optic image inverter does not require as long a working distance, which is of critical importance in some military systems. (Some image inverters are less than 1 in. long.)

Another type of image manipulation possible with fused fiber optics is the image combiner and splitter shown in Figure 30.8. This is made by laying down a series of fiber-optic ribbons, alternating them as if shuffling a deck of cards. One ribbon goes from the single input to output 1, the next from the input to output 2, the next to output 1, and so on. Put a single image into the input, and you get two identical (but fainter) output images. Put separate images into the two outputs, and you get one combined image.

Similar ideas could be used in other image manipulators or in devices to perform operations on optical signals. However, before you rush out for a patent application on your own

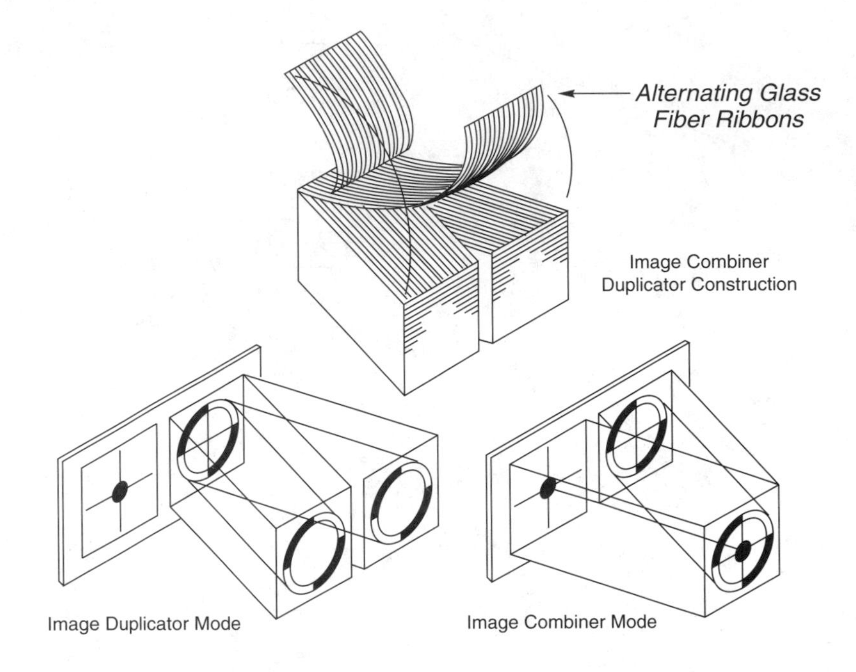

FIGURE 30.8
A fiber-optic image combiner and duplicator. (Courtesy of Galileo Electro-Optics Corp.)

bright idea, you must face the ugly reality of cost. Manufacture of the fiber-optic image combiner in Figure 30.8 requires time and exacting precision, making it too expensive for most uses. Image inverters are used in some systems, but only where less costly lens systems won't do the job.

Light Piping and Illumination

Light piping delivers light through optical fibers.

Illumination and light piping are the simplest applications of optical fibers. *Light piping* is simply the transfer of light from one place to another by guiding it through one or more optical fibers. It doesn't matter how the fibers are arranged, as long as they deliver the light to the desired place. Thus fibers need not be arranged in the same ways at both ends of an illuminating bundle. A single fiber may suffice for many applications.

Light piping for illumination is merely the delivery of light to a desired location. Why bother with optical fibers to do a lightbulb's job? A flexible bundle of optical fibers can efficiently concentrate light in a small area, or deliver light around corners to places it otherwise could not reach, such as inside machinery. A fiber bundle also can deliver light without the heat generated by incandescent bulbs, and without bringing electric current near the illuminated spot. This can be important in locations where bulbs and current can't be used because of explosive vapors or heat-sensitive materials. Fiber bundles also can be divided to deliver light from one bulb to many separate places. Light-piping fibers also can serve as indicators, to verify that an important bulb is operating.

Another important application of light piping is *optical power delivery,* transmitting laser beams for medical treatment or industrial material working. Conventional laser system use lenses or mirrors to focus beams onto the desired spots. These systems use large optics, making them bulky, which is cumbersome for fine tasks such as delicate surgery. Optical-fiber beam delivery systems are much easier to manipulate. Some are designed for surgeons to use with their hands; others are built for robotic control in factories.

Single large-core step-index fibers are best for many power delivery applications as long as the input light can be concentrated into a single core. Low-loss, large-core silica fibers have surprisingly high power transmission capabilities, and can easily carry tens of watts over a few meters. Illuminating bundles transmitting lower powers can use smaller-core, step-index fibers of glass or plastic, as long as light intensities and heat levels are low.

Single large-core fibers can deliver powerful laser beams.

If all fibers in an illuminating bundle go to the same place, they illuminate a single area. If they are directed to different places, they can form a patterned image, such as the fiber-optic sign shown in Figure 30.9. All the fibers collect input light from one bulb, then are splayed out to show the desired pattern. Diffusing lenses at the fiber ends can spread light to make large, easily visible spots. (The WALK sign makes a good example, but it's not widely used.)

FIGURE 30.9
A fiber-optic sign.

A bundle of fibers spreading out from a single illuminating bulb can make a sparkling display, like the one that introduced me to fiber optics nearly 30 years ago. Then they were a rarity, but today they're commonplace. At Christmas, you can buy little fiber splays to attach to Christmas tree lights, and at a recent trade show one company was handing out little plastic flashlights with a splay of plastic fibers that sparkled with colors at their tips. I couldn't resist and picked one up.

What Have You Learned?

1. Rigid or flexible bundles of optical fibers can transmit images if the fibers that make them up are properly aligned at the ends (coherent). Rigid bundles are made of fibers fused together; flexible bundles contain separate fibers bonded at the ends. Resolution is limited by the size of the fiber cores, typically around 10 μm.
2. Bundles of fibers in which the ends are not aligned with one another serve as "light pipes" to illuminate hard-to-reach places. The bundle can be broken up on one end to form an image or display (e.g., a WALK sign).
3. Light that falls into fiber claddings in bundles is lost, but typically 90% falls onto fiber cores.
4. Imaging and other short-distance fibers generally have much higher attenuation than communication fibers. Bundled fibers are step-index multimode types with large NA.
5. Step-index multimode fibers do not focus light, but segments of graded-index fiber do focus light and can serve as lenses. Tapered fibers magnify or demagnify objects seen through them; they are used in bundles.
6. Thin fiber-optic faceplates are used to concentrate light from certain displays in a particular direction. They are used with certain high-performance imaging tubes, but not for ordinary cathode-ray tubes.
7. Coherent fiber bundles can invert, split, and combine images.
8. Endoscopy is the use of coherent fiber bundles to view inside the body without surgery.
9. Large-core fibers can deliver laser power for medicine or materials-working.

Further Reading

Schott Fiberoptics, Introduction to Fiber Optic Imaging, *http://www.schottfiberoptics.com/introfiber.html*

Walter Siegmund, "Fiber Optics," Chapter 13 in Walter G. Driscoll, ed., *Handbook of Optics* (McGraw-Hill, 1978)

Questions to Think About for Chapter 30

1. Standard laser printers used to have resolution of 300 dots per inch. Could you spot those dots with a bundle of 10-μm fibers? (Assume that resolution in line pairs is equivalent to dots per inch.)
2. Using the criteria that the finest possible resolution corresponds to half a line pair per fiber core diameter, what is the largest core fiber that could resolve 300 line pairs per inch?
3. A typical packing fraction for bundled fiber is 90%. Assuming that the fiber cores are 10 μm and you can neglect space between fibers (not a good assumption, but it makes the math manageable) what is the outer diameter of each fiber and how thick is the cladding?
4. Given the same assumptions about neglecting the spacing between fibers, what would be the packing fraction for a bundle assembled from 100/140 step-index multimode fibers?
5. You are trying to deliver a 50-W laser beam through a 3-m length of fiber with attenuation of 10 dB/km. How much power is lost in the fiber?
6. If you put a fiber-optic image inverter flat on a printed sheet of paper, you can read the inverted letters through the taper. You're seeing reflected light. How did it reach the paper?

Quiz for Chapter 30

1. Which of the following statements is false?
   a. Coherent fiber bundles can transmit images.
   b. Coherent fiber bundles can focus light.
   c. Graded-index fiber segments can focus light.
   d. Imaging fiber bundles contain step-index multimode fibers.
2. A graded-index fiber lens has a pitch of 0.45. How much of a sinusoidal oscillation cycle do light rays experience in passing through it?
   a. 27°
   b. 45°
   c. 81°
   d. 162°
   e. 180°
3. What does it mean to say that a fiber bundle has a packing fraction of 90%?
   a. 90% of the fibers are intact.
   b. 90% of the input surface is made up of optical fibers.

c. 90% of the input surface is made up of fiber core.

d. 90% of the input surface is made up of fiber cladding.

e. It transmits 90% of the incident light through its entire length.

4. You want to resolve an image with 8 line pairs per millimeter. In theory, what is the largest fiber core size that you could use in a stationary coherent bundle?

a. 8 μm

b. 50 μm

c. 62.5 μm

d. 100 μm

e. 125 μm

5. Endoscopes used in medicine to view inside the body

a. usually are flexible fiber bundles.

b. sometimes transmit laser beams to treat disease.

c. can examine the stomach or colon.

d. all the above

e. none of the above

6. Fiber-optic faceplates are

a. specialized sensors that detect temperature variations across a surface.

b. thin, rigid fiber bundles used to transfer light efficiently in image intensifiers.

c. assemblies of graded-index fiber lenses that focus light in photocopiers.

d. used on most television sets.

7. Average attenuation of bundled fibers is

a. 0.5 dB/km.

b. 1 to 5 dB/km.

c. 10 to 100 dB/km.

d. around 1 dB/m.

8. What types of fibers are used in imaging bundles?

a. Step-index multimode

b. Graded-index multimode

c. Step-index single-mode

d. All of the above

9. The practical use of fiber-optic bundles to manipulate images is limited by what?

a. Poor resolution

b. High attenuation

c. Fragility

d. High cost

10. Which of the following is the most important advantage of random fiber bundles over coherent bundles for illumination?

a. Flexibility

b. Low cost

c. Size

d. Durability

Appendix A Important Constants, Units, Conversion Factors, and Equations

Table A.1 Metric unit prefixes and their meanings

Prefix	*Symbol*	*Multiple*
peta-	P	10^{15} (quadrillion)
tera-	T	10^{12} (trillion)
giga-	G	10^{9} (billion)
mega-	M	10^{6} (million)
kilo-	k	10^{3} (thousand)
hecto-	h	10^{2} (hundred)
deca-	da	10^{1} (ten)
deci-	d	10^{-1} (tenth)
centi-	c	10^{-2} (hundredth)
milli-	m	10^{-3} (thousandth)
micro-	μ	10^{-6} (millionth)
nano-	n	10^{-9} (billionth)
pico-	p	10^{-12} (trillionth)
femto-	f	10^{-15} (quadrillionth)
atto-	a	10^{-18} (quintillionth)

Table A.2 Important physical constants

Boltzmann constant *(k)*	2.380658×10^{-23} J/K
Charge of electron *(e)*	1.602177×10^{-19} Coulomb
Planck constant *(h)*	4.14×10^{-15} eV/sec
Planck constant *(h)*	$6.6260755 \times 10^{-34}$ J/sec
Speed of light *(c)*	299,792.458 km/sec

Table A.3 Important conversion factors

From	*To*	*Multiply by*
Angstroms (Å)	nm	0.1
electron volts (eV)	J	1.602177×10^{-19}
erg	J	10^{-7}
inch	mm	25.4
J	erg	10^{7}
km	miles (statute)	0.6214
lb	kg	0.453592
miles (statute)	km	1.6093
mils (0.001 inch)	μm	25.4
mm	inch	0.03937

Table A.4 Other useful data and formulas

Wavelength, frequency, and speed of light: $c = \lambda\nu$, $\nu = c/\lambda$, $\lambda = c/\nu$

Wavelength of 1-eV photon = 1.2399 μm

Wavelength of photon with energy E: $\lambda = \frac{hc}{E}$

Photon Energy $E = h\nu$ (Planck constant × frequency)

Photon Energy $E = \nu \times (4.14 \times 10^{-15}\ \text{eV/sec}) = \nu \times 6.6260755 \times 10^{-34}$ J/sec

Photon Energy $E = \frac{hc}{\lambda}$

Refractive index of material $n = \frac{c_{vacuum}}{c_{material}}$

continues

Table A.4 (continued)

Refractive index of air at standard temperature and pressure: 1.000293

Snell's law of refraction $n_i \sin I = n_r \sin R$

Critical Angle: $\theta_c = \arcsin\left(\frac{n_{\text{clad}}}{n_{\text{core}}}\right)$

Numerical Aperture: $\text{NA} = \sqrt{n_{\text{core}}^2 - n_{\text{clad}}^2}$

Decibel attenuation: $\text{dB} = -10 \log_{10}\left(\frac{P_{\text{out}}}{P_{\text{in}}}\right)$

Power output when input and dB loss known: $P_{\text{out}} = P_{\text{in}} \times 10^{(-\text{dB}/10)}$

Bandwidth and data rate relationships:

Maximum NRZ data rate $= \frac{0.7}{\Delta t(\text{total})}$

Maximum RZ data rate $= \frac{0.35}{\Delta t(\text{total})}$

Bandwidth (MHz) $= \frac{350}{\Delta t(\text{ns})}$

Resonant wavelengths in laser cavity of length L and index n: $N\lambda = 2nL$

Table A.5 Standard symbols

Quantity	*Symbol*
Frequency	ν (Greek nu)
Angle	θ (Greek theta)
Angle	ϕ (Greek phi)
Change	Δ (Greek delta)
Planck constant	h
Pulse spreading	Δt
Refractive index	n
Speed of light	c
Wavelength	λ (Greek lambda)
Decibels relative to 1 mW	dBm
Decibels relative to 1 μW	dbμ

Table A.6 Proposed ITU wavelength bands

Name	*Meaning*	*Wavelengths (nm)*
O-band	Original	1260–1360
E-band	Extended	1360–1460
S-band	Short	1460–1530
C-band	Conventional	1530–1565
L-band	Long	1565–1625
U-band	Ultra-long	1625–1675

Appendix B Decibels and Equivalents

Decibels and equivalent power ratios

Loss in Decibels	*Power Ratio*
0.1	0.9772
0.2	0.9550
0.3	0.9333
0.4	0.9120
0.5	0.8913
0.6	0.8710
0.7	0.8511
0.8	0.8318
0.9	0.8128
1	0.7943
2	0.6310
3	0.5012
4	0.3981
5	0.3162
6	0.2512
7	0.1995
8	0.1585
9	0.1259
10	0.1
20	0.01

continues

Decibels and equivalent power ratios (continued)

Loss in Decibels	*Power Ratio*
30	0.001
40	0.0001
50	0.00001
60	0.000001
70	0.0000001
80	0.00000001
90	10^{-09}
100	10^{-10}
200	10^{-20}
300	10^{-30}
400	10^{-40}
500	10^{-50}
600	10^{-60}
700	10^{-70}
800	10^{-80}
900	10^{-90}
1000	10^{-100}

Appendix C Standard Time-Division Multiplexing Rates

North American Digital Telephone Hierarchy (Plesiochronous Digital Hierarchy)

Rate Name	*Data Rate*	*Nominal Voice Circuits*
Single circuit	64,000 bit/s	1
T1 or DS1	1.544 Mbit/s	24
T2 or DS2	6.312 Mbit/s	96
T3 or DS3	44.736 Mbit/s	672
*T3C or DS3C	90 Mbit/s	1,344
*T4 or DS4	274 Mbit/s	4,032

*Obsolete (not used in new equipment)

ITU/European Digital Telephone Hierarchy

Rate Name	*Data Rate*	*Nominal Voice Circuits*
Single Circuit	64,000 bit/s	1
Level 1	2.048 Mbit/s	30
Level 2	8.448 Mbit/s	120
Level 3	34.3 Mbit/s	480
Level 4	139 Mbit/s	1920
*Level 5	565 Mbit/s	7680

*Obsolete (not used in new systems)

SONET/Synchronous Digital Hierarchy (SDH)

SONET Rate Name	*SDH Rate Name*	*Data Rate*	*Nominal Equivalent*
OC-1/STS-1	—	51.84 Mbit/s	672 voice 1 T3
OC-3/STS-3	STM-1	155.52 Mbit/s	2016 voice 1 Level 4
OC-12/STS-12	STM-4	622.08 Mbit/s	8064 voice
OC-48/STS-48	STM-16	2.488 Gbit/s	32,256 voice
OC-192/STS-192	STM-64	9.953 Gbit/s (10 Gbit/s)	129,024 voice
OS-768/STS-768	STM-256	40 Gbit/s	516,096 voice

Appendix D ITU Frequencies and Wavelengths for L- and C-bands, 100-GHz spacing, 100 channels

Frequency (THz)	Wavelength (nm)
186.00	1611.79
186.10	1610.92
186.20	1610.06
186.30	1609.19
186.40	1608.33
186.50	1607.47
186.60	1606.60
186.70	1605.74
186.80	1604.88
186.90	1604.03
187.00	1603.17
187.10	1602.31
187.20	1601.46
187.30	1600.60
187.40	1599.75
187.50	1598.89
187.60	1598.04
187.70	1597.19
187.80	1596.34
187.90	1595.49
188.00	1594.64
188.10	1593.79
188.20	1592.95
188.30	1592.10
188.40	1591.26
188.50	1590.41
188.60	1589.57
188.70	1588.73
188.80	1587.88
188.90	1587.04

continues

ITU Frequencies and Wavelengths (continued)

Frequency (THz)	*Wavelength (nm)*	*Frequency (THz)*	*Wavelength (nm)*
189.00	1586.20	192.50	1557.36
189.10	1585.36	192.60	1556.55
189.20	1584.53	192.70	1555.75
189.30	1583.69	192.80	1554.94
189.40	1582.85	192.90	1554.13
189.50	1582.02	193.00	1553.33
189.60	1581.18	193.10	1552.52
189.70	1580.35	193.20	1551.72
189.80	1579.52	193.30	1550.92
189.90	1578.69	193.40	1550.12
190.00	1577.86	193.50	1549.32
190.10	1577.03	193.60	1548.51
190.20	1576.20	193.70	1547.72
190.30	1575.37	193.80	1546.92
190.40	1574.54	193.90	1546.12
190.50	1573.71	194.00	1545.32
190.60	1572.89	194.10	1544.53
190.70	1572.06	194.20	1543.73
190.80	1571.24	194.30	1542.94
190.90	1570.42	194.40	1542.14
191.00	1569.59	194.50	1541.35
191.10	1568.77	194.60	1540.56
191.20	1567.95	194.70	1539.77
191.30	1567.13	194.80	1538.98
191.40	1566.31	194.90	1538.19
191.50	1565.50	195.00	1537.40
191.60	1564.68	195.10	1536.61
191.70	1563.86	195.20	1535.82
191.80	1563.05	195.30	1535.04
191.90	1562.23	195.40	1534.25
192.00	1561.42	195.50	1533.47
192.10	1560.61	195.60	1532.68
192.20	1559.79	195.70	1531.90
192.30	1558.98	195.80	1531.12
192.40	1558.17	195.90	1530.33

Appendix E
Laser and Fiber Safety

With the exercise of reasonable common sense, fiber-optic systems are reasonably safe. As in any workplace, you should always be careful of chemicals, electrical voltages, and flying fragments that might endanger the eyes. Two potential hazards are specific to fiber-optic systems: sharp fiber fragments and laser light.

Fiber fragments are sharp, very small, and extremely hard to see. They can do serious damage in the wrong place, especially in your eye. They are very light, so they can easily fly into the air when cut, making good safety goggles particularly important when fibers are to be cut. Fragments are hard to see on many surfaces, so you should keep careful track of them. The best procedure is to grasp them with tweezers and dispose of them immediately in a sealed container. (Fiber labs and workshops should have special containers for that purpose.) If you have to lay a fiber fragment down, put it on a flat (nonreflecting) black surface, where it is easiest to see. If a fiber splinter jabs you, pull it out carefully with tweezers if it's accessible. If it gets in your eye, call for medical help.

You may find laser light warnings plastered in many labs and attached to a variety of equipment. The levels of laser light you will encounter in fiber optics are not deadly. The most powerful lasers used in fiber optics are not going to burn holes through you. However, they could cause serious eye damage. Because the beams in fiber-optic systems are invisible to the human eye, they give you no inherent warning that they are turned on, or that they are entering the air in front of you. You have to rely on safety equipment and labels. You should always know what you're working with.

The hazard of laser light comes from the fact that the eye focuses the parallel light rays in a laser beam onto a tiny spot on the retina, the light-sensitive layer at the back of the eyeball. This concentrates the beam from even a milliwatt laser so much that the light intensities on that tiny spot are comparable to that produced from looking directly at the sun. Just as with the sun, a momentary glance into a milliwatt laser beam will not blind you instantly, but you should not look intentionally into it.

Nature is fairly kind in some ways. The eye is full of water, which absorbs light at wavelengths longer than about 1400 nm, blocking it from the retina—the most sensitive part of the eye. Thus, low powers in the erbium-fiber band are less hazardous than those at shorter wavelengths. Light emerging from optical fibers also spreads out much faster than the familiar beams from red laser pointers, so you don't have to worry much about being zapped by beams from loose fiber ends on the other side of the room. However, erbium-fiber amplifiers can generate hundreds of milliwatts, which are not powers you should trifle with.

The most dangerous lasers you are likely to find in modern fiber-optic systems are 980-nm pump lasers for erbium-doped fiber amplifiers, and high-power transmitters in the 1300-nm band for cable television systems. Both can generate hundreds of milliwatts, are invisible to the eye, but can penetrate to the retina.

Remember that the eye hazards presented by different lasers vary widely, so take care to pay close attention to what types are being used where you are working.

Appendix F
Fiber-Optic Resources

No single book can tell you everything about fiber optics. The field is growing and changing too rapidly. A wide variety of resources is available. These are sources that I have used and found helpful. It is far from comprehensive, but should give you a starting point. My comments are in parentheses.

Books

GENERAL FIBER OPTICS

M Bass ed., *OSA Handbook of Optics V.4 Fiber Optics and Nonlinear Optics* (McGraw Hill, New York, 2001) (massive handbook)

John A. Buck, *Fundamentals of Optical Fibers* (Wiley Interscience, New York, 1995) (textbook on fibers and their properties)

Bob Chomycz, *Fiber Optic Installer's Field Manual* (McGraw Hill, New York, 2000) (for the practicing technician)

Dennis Derickson, *Fiber Optic Test and Measurement* (Prentice Hall, Upper Saddle River, NJ, 1998) (excellent coverage)

Jim Hayes, *Fiber Optics Technician's Manual* (Delmar Publishers, Albany, 1996) (for the practicing technician)

Jeff Hecht, *City of Light: The Story of Fiber Optics* (Oxford University Press, New York, 1999) (history, easy reading)

Gerd Keiser, 3rd ed., *Optical Fiber Communications* (McGraw Hill, New York, 2000) (recommended as the next step up from this book)

TELECOMMUNICATIONS IN GENERAL

Walter Ciciora, James Farmer, and David Large, *Modern Cable Television Technology: Video, Voice, and Data Communications* (Morgan Kaufmann, San Francisco, 1999) (very comprehensive)

Alan Freedman, *The Computer Glossary* (American Management Association, New York, 2001), and *Computer Desktop Encyclopedia Software* (Computer Language Company, 2001) (includes many telecommunications terms)

Roger L. Freeman, *Fundamentals of Telecommunications* (Wiley-Interscience, New York, 1999) (oriented toward systems at the engineering level)

Gary M. Miller, *Modern Electronic Communications,* 6th ed. (Prentice Hall, Upper Saddle River, NJ, 1999) (good nuts and bolts introduction)

Laszlo Solymar, *Getting the Message: A History of Communications* (Oxford University Press, 1999) (an entertaining and enjoyable history)

OPTICS, PHOTONICS, AND LASERS

Govind P. Agrawal, ed., *Semiconductor Lasers: Past, Present, and Future* (AIP Press, Woodbury, NY, 1995) (good overview)

J. Warren Blaker and Peter Schaeffer, *Optics, An Introduction for Technicians and Technologists* (Prentice Hall, 2000) (Introductory)

Eugene Hecht, *Optics,* 3rd ed. (Addison Wesley, Reading, MA, 1997) (the standard upper-level undergraduate text; no relation)

Jeff Hecht, *Understanding Lasers* (IEEE Press, Piscataway, NJ, 1994) (introductory level, but a bit dated on semiconductor lasers)

Francis A. Jenkins and Harvey E. White, *Fundamentals of Optics,* 4th ed. (McGraw Hill, New York, 1976) (very old, but a good introduction to classical optics)

B. A. E. Saleh and M. C. Teich, *Fundamentals of Photonics* (Wiley-Interscience, New York, 1991) (wide-ranging upper-level textbook)

Amnon Yariv, *Optical Electronics in Modern Communications,* 5th ed. (Oxford University Press, New York, 1997) (widely used but very advanced textbook)

Periodicals

INDUSTRY MAGAZINES

Fiberoptic Product News (technology)

FibreSystems International (technology and business, from Britain)

Integrated Communications Design (communications technology)

Laser Focus World (for laser and optoelectronics engineers)

Lightwave (fiber-optic communications and business)

Photonics Spectra (optics and photonics)

Telephony (telephone industry and technology)

WDM Solutions (technology)

SCHOLARLY JOURNALS

Bell Labs Technical Journal, free online at: *www.lucent.com/minds/techjournal/*

Electronics Letters (Published by Institution of Electrical Engineers; many cutting-edge fiber-optic papers)

IEEE Communications Magazine

Journal of Lightwave Technology

IEEE Journal on Selected Areas in Communications

IEEE Photonics Technology Letters

ONLINE ONLY PUBLICATIONS

Fiberoptics Online *(www.fiberopticsonline.com)*

Fibers.org *(www.fibers.org)*

Light Reading *(www.lightreading.com)*

Other Resources

PROFESSIONAL SOCIETIES

Institute of Electrical and Electronics Engineers *(www.ieee.org)* (also see IEEE Lasers and Electro-Optics Society)

Optical Society of America: *(www.osa.org)*

SPIE—The International Society for Optical Engineering *(www.spie.org)*

MAJOR NATIONAL AND INTERNATIONAL CONFERENCES

European Conference on Optical Communications (annual, August-October, see *www.ecoc-exhibition.com/*)

National Fiber-Optic Engineers Conference (annual, see *www.nfoec.com*)

Optical Fiber Communications Conference (annual, March, run by Optical Society of America, see *www.osa.org/ofc*)

OTHER WEB SITES

FiberU, on-line education and resources: *www.fotec.com/fiberu-online*

My own now-small fiber site: *www.fiberhome.com*

www.photonicresources.com/

Tutorial links: *//nctt1.stcc.mass.edu/photonic_web/code/tutorialspage.html*

Glossary

Absorption Loss of light energy that is absorbed and converted to heat. Not equal to loss or attenuation, which include other effects.

Acceptance Angle The angle over which the core of an optical fiber accepts incoming light; usually measured from the fiber axis. Related to numerical aperture (NA).

Access Network Part of the telecommunication network that connects to individual and corporate users.

Add-drop multiplexer A device that drops and/or adds one or more optical channels to a signal.

All-Dielectric Cable Cable made entirely of dielectric (insulating) materials without any metal conductors, armor, or strength members.

Analog A signal that varies continuously (e.g., sound waves). Analog signals have frequency and bandwidth measured in hertz.

ATM (Asynchronous Transfer Mode) A digital transmission switching format, with cells containing 5 bytes of header information followed by 48 data bytes.

Attenuation Reduction of signal magnitude, or loss, normally measured in decibels. Fiber attenuation is normally measured per unit length in decibels per kilometer.

Avalanche Photodiode (APD) A semiconductor photodetector with integral detection and amplification stages. Electrons generated at a *p/n* junction are accelerated in a region where they free an avalanche of other electrons. APDs can detect faint signals but require higher voltages than other semiconductor electronics.

Average Power The average level of power in a signal that varies with time.

AWG See *Waveguide Array.*

Axis The center of an optical fiber.

Backbone System A transmission network that carries high-speed telecommunications between regions (e.g., a nationwide long-distance telephone system). Sometimes used to describe the part of a local area network that carries signals between branching points.

Backscattering Scattering of light in the direction opposite to that in which it was originally traveling.

Bandwidth The highest frequency that can be transmitted by an analog system. Also, the information-carrying capacity of a system (especially for digital systems).

Baud Strictly speaking, the number of signal-level transitions per second in digital data. For some common coding schemes, this equals bits per second, but this is not true for more complex coding, and it is often misused. Bits per second is less ambiguous.

Beamsplitter A device that divides incident light into two separate beams.

Bidirectional Operating in both directions. Bidirectional couplers split or combine light the same way when it passes through them in either direction. Bidirectional transmission sends signals in both directions, sometimes through the same fiber.

Birefringent Having a refractive index that differs for light of different polarizations.

Bit Error Rate (BER) The fraction of bits transmitted incorrectly.

Bragg Scattering Scattering of light caused by a change in refractive index, as used in *Fiber Bragg Gratings* and *Distributed Bragg Reflectors.*

Broadband Covering a wide range of frequencies or having a high data rate. The broadband label is sometimes used for a network that carries many different services or for video transmission.

Broadcast Transmission Sending the same signal to many different places, like a television broadcasting station. Broadcast transmission can be over optical fibers if the same signal is delivered to many subscribers.

Bundle (of fibers) A rigid or flexible group of fibers assembled in a unit. Coherent fiber bundles have fibers arranged in the same way on each end and can transmit images.

Byte Eight bits of digital data. (Sometimes parity and check bits are included, so one "byte" may include 10 bits, but only 8 of them are data.)

Carrier In technology, the wave that is modulated with a signal carrying information. In business, a company that provides telecommunication services.

Category 5 A type of twisted-pair copper cable designed to transmit high-speed signals.

CATV An acronym for cable television, derived from Community Antenna TeleVision.

C-Band Wavelengths of about 1530 to 1565 nm, where erbium-doped fiber amplifiers have their strongest gain. Normally erbium-fiber amplifiers operate in either C- or L-band.

Cell A fixed-length data packet transmitted in certain digital systems such as ATM.

Central Office A telephone company facility for switching signals among local telephone circuits; connects to subscriber telephones. Also called a switching office.

Chromatic Dispersion Wavelength-dependent pulse spreading in optical fibers, measured in picoseconds (of pulse spreading) per nanometer (of source bandwidth) per kilometer (of fiber length). It is the sum of waveguide and material dispersion.

Circuit Originally a physical connection that transmits electricity or signals. Now also a communication channel that guarantees a fixed transmission capacity.

Circuit Switching Making temporary physical or virtual connections between two points, which guarantees a fixed transmission capacity.

Cladding The layer of glass or other transparent material surrounding the light-carrying core of an optical fiber. It has a lower refractive index than the core and thus confines light in the core. Coatings may be applied over the cladding.

CLEC (Competitive Local Exchange Carrier) A company that offers local telephone service in competition against dominant phone companies.

Coarse Wavelength-Division Multiplexing Transmitting signals at multiple wavelengths through the same fiber with wide spacing between optical channels. Typical spacing is several nanometers or more. Also called *wide wavelength-division multiplexing.*

Coating An outer plastic layer applied over the cladding of a fiber for mechanical protection.

Coax Coaxial cable—cable with a central metallic conductor surrounded by an insulator that is covered by a metallic sheath that runs the length of the cable.

Coherent Bundle (of fibers) Fibers packaged together in a bundle so they retain a fixed arrangement at the two ends and can transmit an image.

Coherent Communications In fiber optics, a communication system where the output of a local laser oscillator is mixed with the received signal, and the difference frequency is detected and amplified.

Compression Reducing the number of bits needed to encode a digital signal, typically by eliminating long strings of identical bits or bits that do not change in successive sampling intervals (e.g., video frames).

Connector A device mounted on the end of a fiber-optic cable, light source, receiver, or housing that mates to a similar device to couple light into and out of optical fibers. A connector joins two fiber ends, or one fiber end and a light source or detector.

Copper Industry slang for metal wire, either twisted-pair or coaxial cable.

Core The central part of an optical fiber that carries light.

Coupler A device that connects three or more fiber ends, dividing one input between two or more outputs or combining two or more inputs into one output.

Coupling Transfer of light into or out of an optical fiber. (Note that coupling does not require a coupler.)

Critical Angle The angle at which light in a high-refractive-index material undergoes total internal reflection.

Cut-Back Measurements Measurement of optical loss made by cutting a fiber to compare loss of a short segment with loss of a longer one.

Cutoff Wavelength The longest wavelength at which a single-mode fiber can transmit two modes, or (equivalently) the shortest wavelength at which a single-mode fiber carries only one mode.

Cycles per Second The frequency of a wave, or number of oscillations it makes per second. One cycle per second equals one hertz.

Dark Current The noise current generated by a photodiode in the dark.

Dark Fiber Optical fiber installed without transmitter and receiver, usually to provide expansion capacity. Some carriers lease dark fibers to other companies that add equipment to transmit signals through them.

dBm Decibels relative to 1 mW.

dBμ Decibels relative to 1 μW.

DBR See *Distributed Bragg Reflection.*

Decibel (dB) A logarithmic comparison of power levels, defined as ten times the base-10 logarithm of the ratio of the two power levels. One-tenth of a bel.

Demultiplexer A device that separates a multiplexed signal into its original components; the inverse of a multiplexer.

Dense Wavelength-Division Multiplexing (DWDM) Transmitting signals at multiple wavelengths through the same fiber with close spacing. Channel spacing usually is 200 GHz or less in frequency units, corresponding to 1.6 nm in wavelength units at 1550 nm.

Detector A device that generates an electrical signal when illuminated by light. The most common fiber-optic detectors are photodiodes.

DFB See *Distributed Feedback.*

Dielectric Nonconductive.

Dielectric Filter An optical filter that selectively transmits one wavelength and reflects others based on interference effects inside the structure. Also called *interference filter.*

Digital Encoded as a signal in discrete levels, typically binary 1s and 0s.

Digital Subscriber Line (DSL) A service that transmits digital signals to homes at speeds of hundreds of kilobits to tens of megabits per second over twisted-pair wires at higher frequencies than voice telephone signals. There are several variations.

Diode An electronic device that lets current flow in only one direction. Semiconductor diodes used in fiber optics contain a junction between regions of different doping. They include light emitters (LEDs and laser diodes) and detectors (photodiodes).

Diode Laser A semiconductor diode that generates laser light. A current flowing through the diode causes electrons and holes to recombine at the junction layer between *p*- and *n*-doped regions, producing excited states that can release energy in the form of light.

Directional Coupler A coupler in which light is transmitted differently when it goes in different directions.

Dispersion The stretching of light pulses as they travel in an optical fiber, which increases their duration.

Dispersion Compensation Offsetting the dispersion of one fiber by using different fibers or other components that have dispersion of the opposite sign. Usually done for chromatic dispersion; compensation for polarization-mode dispersion is in development.

Dispersion-Shifted Fiber Optical fiber with nominal wavelength of zero chromatic dispersion shifted away from 1310 nm. Often used for zero dispersion-shifted fiber, which has zero chromatic dispersion at 1550 nm and is not used in DWDM systems.

Dispersion Slope The change in dispersion with wavelength.

Distributed Bragg Reflection Reflection of light caused by periodic changes in refractive index in a stack of layers of different composition or—equivalently—by a corrugation at the boundary between two semiconductor layers. The period and the refractive index select one wavelength.

Distributed Feedback Feedback arising from reflection distributed through a structure.

Distributed Feedback Laser A diode laser with a corrugation in the electrically pumped part of the laser, which selects the laser wavelength by reflecting that wavelength back into the active layer.

Drop A cable that delivers service to an individual customer.

DSL See *Digital Subscriber Line.*

DTV Digital television.

Duplex In cables, one that contains two fibers. For connectors, one that connects two pairs of fibers. For data transmission, full-duplex transmitters and receivers simultaneously send and receive signals in both directions, but half-duplex cannot do both at the same time.

DWDM See *Dense Wavelength-Division Multiplexing.*

Edge-Emitting Diode An LED that emits light from its edge, producing more directional output than LEDs that emit from their top surface.

Edge-Emitting Laser A semiconductor laser that emits light in the plane of its junction from the edge of the chip.

Electro-Absorption Modulator A semiconductor diode reverse-modulated so it modulates light passing through it.

Electromagnetic Interference (EMI) Noise generated when stray electromagnetic fields induce currents in electrical conductors.

Electromagnetic Radiation Waves made up of oscillating electrical and magnetic fields perpendicular to one another and traveling at the speed of light. Can also be viewed as photons or quanta of energy. Electromagnetic radiation includes radio waves, microwaves, infrared, visible light, ultraviolet radiation, X rays, and gamma rays.

EMI See *Electromagnetic Interference.*

Endoscope A fiber-optic bundle used for imaging and viewing inside the human body.

Erbium-Doped Fiber Amplifier Optical fiber doped with the rare earth element erbium, which can amplify light at 1530 to 1610 nm when pumped by an external light source.

Ethernet A local-area network standard. The original Ethernet transmits 10 Mbit/s. Other versions are Fast Ethernet at 100 Mbit/s, Gigabit Ethernet at 1 Gbit/s, and 10-Gigabit Ethernet.

Evanescent Wave Guided light waves that extend beyond the boundary of a fiber core into the cladding. Evanescent waves can transfer energy between waveguides.

Excess Loss Loss of a passive coupler above that inherent in dividing light among the output ports.

External Modulation Modulation of output of a light source by an external device.

Extrinsic Loss Splice losses arising from the splicing process itself.

Eye Pattern A pattern formed by overlaying traces of a series of transmitted pulses in a visual display. The more open the eye, the sharper the distinction between on and off pulses.

Ferrule A tube within a connector with a central hole that contains and aligns a fiber.

Fiber Amplifier An optical fiber doped to amplify light from an external source. The most important type is the erbium-doped fiber amplifier.

Fiber Bragg Grating An optical fiber in which the core refractive index varies periodically, causing Bragg scattering at wavelengths selected by the period and refractive index. A fiber Bragg grating reflects the selected wavelength and transmits others.

Fiber Distributed Data Interface (FDDI) A standard for a 100-Mbit/s fiber-optic local-area network.

Fiber-Optic Gyroscope A coil of optical fiber that can detect rotation about its axis.

Fiber to the Curb (FTTC) Fiber-optic service to a node that is connected by wires to several nearby homes, typically on a block.

Fiber to the Home (FTTH) A network in which optical fibers bring signals all the way to homes.

Fibre Channel A standard for transmitting signals at 100 Mbit/s to 4.25 Gbit/s over fiber or (at slower speeds) copper.

FITL Fiber in the loop.

Fluoride Glasses Materials that have the amorphous structure of glass but are made of fluoride compounds (e.g., zirconium fluoride) rather than oxide compounds (e.g., silica).

Frame A fixed-length block of data transmitted as a unit; SONET transmits frames. In video, one of a series of images shown in sequence.

Frequency The number of times an electromagnetic wave oscillates in a second, or the number of wave peaks that pass a point in a second; measured in hertz.

Frequency-Division Multiplexing Combining analog signals by assigning each a different carrier frequency and merging them in a single signal with a broad range of frequencies.

FTTC Fiber to the curb.

FTTH Fiber to the home.

Full Duplex In data transmission, transmitters and receivers that simultaneously send and receive signals in both directions.

Fused Fibers A bundle of fibers melted together so they maintain a fixed alignment with respect to each other in a rigid rod.

Fusion Splice A splice made by melting the tips of two fibers together so they form a solid junction.

Gallium Aluminum Arsenide (GaAlAs) A semiconductor compound used in LEDs, diode lasers, and certain detectors.

Gallium Arsenide (GaAs) A semiconductor compound used in LEDs, diode lasers, detectors, and electronic components.

Gbit/s Gigabits (billion bits) per second.

Glass A solid in which the atoms are arranged randomly instead of ordered in a crystal. In fiber optics, "glass" usually means a silica compound unless otherwise noted.

Graded-Index Fiber A fiber in which the refractive index changes gradually with distance from the fiber axis, rather than abruptly at the core-cladding interface.

Graded-Index Fiber Lens A short segment of graded-index fiber that focuses light passing through it.

Group Delay Time The difference in travel time through a fiber for light of different wavelengths.

Half-Duplex In data transmission, a system in which transmitters and receivers cannot simultaneously send and receive signals.

Hard-Clad Silica Fiber A fiber with a hard plastic cladding surrounding a step-index silica core. (Other plastic-clad silica fibers have a soft plastic cladding.)

HDTV High-definition (or high-resolution) television; digital television with higher resolution than present systems.

Head-End The central facility where signals are combined for distribution in a cable television system.

Hertz Frequency in cycles per second.

Hierarchy A set of transmission speeds arranged to multiplex signals at successively higher data rates.

Hybrid Fiber/Coax The use of fiber to distribute cable-television signals to nodes, which in turn distribute them to homes over coaxial cable.

Index-Matching Gel A gel or fluid with refractive index close to glass that reduces refractive-index discontinuities that can cause reflective losses.

Index of Refraction The speed of light in a vacuum divided by the speed of light in a material, abbreviated *n,* which measures how materials refract light.

Indium Gallium Arsenide (InGaAs) A semiconductor material used in lasers, LEDs, and detectors.

Indium Gallium Arsenide Phosphide (InGaAsP) A semiconductor material used in lasers, LEDs, and detectors.

Infrared Light with wavelengths longer than 700 nm and shorter than about 1 mm, invisible to the human eye, which we can feel as heat. Glass optical fibers transmit infrared signals at 700 to about 1650 nm in the infrared.

Infrared Fiber Colloquially, optical fibers with best transmission at wavelengths of 2 μm or longer, made of materials other than silica glass.

Injection Laser Another name for a semiconductor or diode laser.

Integrated Services Digital Network (ISDN) Originally a standard to transmit two digital voice lines at 64 kbit/s and one 16-kbit/s data channel. Now repackaged as IDSL, a form of DSL, transmitting 128 kbit/s over distances beyond the reach of DSL.

Integrated Optics Optical devices that perform two or more functions and are integrated on a single substrate; analogous to integrated electronic circuits.

Intensity Power per unit solid angle.

Interference For light, the way that waves add together, depending on their phase. *Constructive interference* occurs when the waves are in phase and their amplitudes add. *Destructive interference* occurs when the waves are 180° out of phase and their amplitudes cancel.

Interference Filter An optical filter that selectively transmits one wavelength and reflects others based on interference effects inside the structure. Also called *dielectric filter.*

Interleaver An optical device that separates a series of optical channels so alternating wavelengths emerge out its two ports. The best-known type is a *Mach-Zehnder interferometer.*

Internet Protocol (IP) Standard packet-switched transmission format for the Internet; uses variable-length packets.

Intrinsic Losses Splice losses arising from differences in the fibers being spliced.

Irradiance Power per unit area.

Junction Laser A semiconductor diode laser.

LAN See *Local-Area Network.*

Large-Core Fiber Usually, a fiber with a core of 200 μm or more.

Laser From l*ight* a*mplification by* s*timulted* e*mission of* r*adiation,* one of the wide range of devices that generates light by that principle. Laser light is directional, covers a narrow range of wavelengths, and is more coherent than ordinary light. Semiconductor diode lasers are the usual light sources in fiber-optic systems.

Layer A standard or protocol for signal transmission or processing to perform certain functions. It includes standard interfaces with other layers, which perform other functions.

L-Band Wavelengths of about 1570 to 1625 nm where some erbium-doped fiber amplifiers operate. Separate from the C-band.

LED See *Light-Emitting Diode.*

Legacy Older equipment, generally no longer made.

Light Strictly speaking, electromagnetic radiation visible to the human eye at 400 to 700 nm. Commonly, the term is applied to electromagnetic radiation with properties similar to visible light, including the invisible near-infrared radiation in most fiber-optic communication systems.

Light-Emitting Diode (LED) A semiconductor diode that emits incoherent light at the junction between *p*- and *n*-doped materials.

Lightguide An optical fiber or fiber bundle.

Light Piping Use of optical fibers to illuminate.

Lightwave An an adjective, a synonym for optical, often (but not always) meaning fiber-optic.

Linewidth The range of wavelengths in an optical signal, sometimes called spectral width.

Local-Area Network (LAN) A network that transmits data among many nodes in a small area (e.g., a building or campus).

Local Loop The part of the telephone network extending from the central (switching) office to the subscriber.

Longitudinal Modes Oscillation modes of a laser along the length of its cavity. Each longitudinal mode contains only a narrow range of wavelengths, so a laser emitting a single longitudinal mode has a narrow bandwidth. Distinct from transverse modes.

Loose Tube A protective tube loosely surrounding a cabled fiber, often filled with gel.

Loss Attenuation of optical signal, normally measured in decibels.

Loss Budget An accounting of overall attenuation in a system.

Mach-Zehnder Interferometer An optical device that separates a series of optical channels so alternating wavelengths emerge out its two ports, sometimes called an *interleaver.*

MAN (Metropolitan Area Network) Similar to *Metro Network.*

Margin Allowance for attenuation in addition to that explicitly accounted for in system design.

Material Dispersion Pulse dispersion caused by variation of a material's refractive index with wavelength.

Mbit/s Megabits (million bits) per second.

Mechanical Splice A splice in which fibers are joined mechanically (e.g., glued or crimped in place) but not fused together.

MEMS (Micro-electro-mechanical systems) Tiny moving mirrors fabricated from semiconductor materials.

Mesh A network that makes multiple interconnections between different points.

Metro Network A telecommunication system serving a metropolitan area, typically with cable lengths to 200 km.

Microbending Tiny bends in a fiber that allow light to leak out and increase loss.

Micrometer One-millionth of a meter, abbreviated μm.

Micron Short for the preferred form, micrometer.

Modal Dispersion Dispersion arising from differences in the times that different modes take to travel through multimode fiber.

Mode An electromagnetic field distribution that satisfies theoretical requirements for propagation in a waveguide or oscillation in a cavity (e.g., a laser). Light has modes in a fiber or laser.

Mode-Field Diameter The diameter of the one mode of light propagating in a single-mode fiber, slightly larger than core diameter.

Mode Stripper A device that removes high-order modes in a multimode fiber to give standard measurement conditions.

Multimode Transmits or emits multiple modes of light.

Multiplexer A device that combines two or more signals into a single output.

***n* Region** A semiconductor doped to have an excess of electrons as current carriers.

NA See *Numerical Aperture.*

Nanometer A unit of length, 10^{-9} m. It is part of the SI system and has largely replaced the non-SI Ångstrom (0.1 nm) in technical literature.

Near-Infrared The part of the infrared near the visible spectrum, typically 700 to 1500 or 2000 nm; it is not rigidly defined.

Network A system of cables or other connections that links many terminals or devices, all of which can communicate with each other through the system.

Noise Equivalent Power (NEP) The optical input power to a detector needed to generate an electrical signal equal to the inherent electrical noise.

No Return to Zero (NRZ) A digital code in which the signal level is low for a 0 bit and high for a 1 bit and does not return to 0 between successive 1 bits.

Nonzero Dispersion-Shifted Fiber Single-mode optical fiber with the wavelength of zero chromatic dispersion shifted to just outside the erbium-fiber amplifier region. Some types have zero dispersion near 1500 nm, others near 1625 nm. Types with zero dispersion at 1580 nm are not usable in the L-band of erbium-doped fiber amplifiers.

Normal (angle) Perpendicular to a surface.

NTSC The analog video broadcast standard used in North America, set by the National Television System Committee.

Numerical Aperture (NA) The sine of half the angle over which a fiber can accept light. Strictly speaking, this is multiplied by the refractive index of the medium containing the light, but for air the index is almost equal to 1.

OC-*x* Optical Carrier, a carrier rate specified in the SONET standard.

Optical Amplifier A device that amplifies an input optical signal without converting it into electrical form. The best developed are optical fibers doped with the rare-earth element erbium.

Optical Channel An optical signal transmitted at one wavelength. WDM systems transmit multiple channels at separate wavelengths.

Optical Circulator A device that transmits light only in one direction through a series of ports, so light can go from port 1 to port 2 and port 2 to port 3, but not from port 2 to port 1.

Optical Networking Processing and switching signals in optical form as well as transmitting them optically.

Optical Node The point where signals are transferred from optical fibers to other transmission media, typically twisted-pair wires or coaxial cable.

Optical Performance Monitor A device installed in a WDM system to monitor signals at the transmitted wavelengths.

Optical Spectrum Analyzer An instrument that scans the spectrum to record power as a function of wavelength.

Optical Time-Domain Reflectometer (OTDR) An instrument that measures transmission characteristics by sending a short pulse of light down a fiber and observing backscattered light.

Optical Waveguide Technically, any structure that can guide light. Sometimes used as a synonym for optical fiber, it can also apply to planar light waveguides.

***p* Region** Part of a semiconductor doped with electron acceptors in which holes (vacancies in the valence electron level) are the dominant current carriers.

Packet Switching Organizing signals by dividing them into data packets, each containing a header that specifies its destination and a packet of data intended for that destination. Separate data packets then are directed to their destinations.

Passive Component A component that doesn't require outside power.

Passive Optical Network A fiber-optic distribution network with no active components between the switching point and the customer.

Peak Power Highest instantaneous power level in a pulse.

Phase The position of a wave in its oscillation cycle.

Photodiode A diode that can produce an electrical signal proportional to light falling upon it.

Photonic A term coined for devices that work using photons or light, analogous to "electronic" for devices working with electrons.

Photons Quanta of electromagnetic radiation. Light can be viewed as either a wave or a series of photons.

Picosecond One-trillionth (10^{-12}) second.

***pin* Photodiode** A semiconductor detector with an intrinsic *(i)* region separating the *p*- and *n*-doped regions. It has fast linear response and is used in fiber-optic receivers.

Planar Waveguide A flat waveguide formed on the surface of a flat material. The zone of high refractive index is rectangular in cross-section and guides light in the same way as the core of an optical fiber.

Plastic-Clad Silica (PCS) Fiber A step-index multimode fiber in which a silica core is surrounded by a lower-index plastic cladding.

Plastic Fiber An optical fiber made entirely of plastic compounds.

Plenum Cable Cable made of fire-retardant material that meets electrical code requirements (UL 910) for low smoke generation and installation in air spaces.

Plesiochronous Digital Hierarchy The North American Digital Hierarchy of time-division multiplexing rates.

Point-to-Point Transmission Carrying a signal between two points, without branching to other points.

Polarization Alignment of the electric and magnetic fields that make up an electromagnetic wave; normally refers to the electric field. If all light waves have the same alignment, the light is polarized.

Polarization-Maintaining Fiber Fiber that maintains the polarization of light that enters it.

Polarization-Mode Dispersion Dispersion arising from random fluctuations in how fibers transmit light in vertical and horizontal polarizations.

POP An Internet node called a Point of Presence.

POTS (Plain Old Telephone Service) Analog voice telephone lines.

Preform A cylindrical rod of specially prepared and purified glass from which an optical fiber is drawn.

Provisioning Arranging a network to provide services to customers.

Pulse Dispersion The spreading out of pulses as they travel along an optical fiber.

Pump Laser The semiconductor laser that provides the light that excites atoms in a fiber amplifier, putting them in the right state to amplify light.

Quantum Efficiency The fraction of photons that strike a detector that produces electron-hole pairs in the output current.

Quaternary A semiconductor compound made of four elements (e.g., InGaAsP).

Radiometer An instrument, distinct from a photometer, to measure power (watts) of electromagnetic radiation.

Raman Amplifier A fiber that transfers energy from a strong pump beam to amplify a weaker signal at a longer wavelength, using stimulated Raman scattering.

Rays Straight lines that represent the path taken by light.

Receiver A device that detects an optical signal and converts it into an electrical form usable by other devices.

Recombination Combination of an electron and a hole in a semiconductor that releases energy, sometimes leading to light emission.

Refraction The bending of light as it passes between materials of different refractive index.

Refractive Index The speed of light in a vacuum divided by the speed of light in a material, abbreviated *n,* which measures how materials refract light.

Refractive-Index Gradient The change in refractive index with distance from the axis of an optical fiber.

Regenerator A receiver-transmitter pair that detects a weak signal, cleans it up, then sends the regenerated signal through another length of fiber.

Repeater A receiver-transmitter pair that detects and amplifies a weak signal for retransmission through another length of optical fiber.

Responsivity The ratio of detector output to input, usually measured in units of amperes per watt (or microamperes per microwatt).

Return to Zero (RZ) A digital coding scheme where signal level is low for a 0 bit and high for a 1 bit during the first half of a bit interval and then in either case returns to zero for the second half of the bit interval.

Ribbon Cables Cables in which many parallel fibers are embedded in a plastic material, forming a flat ribbon-like structure.

Ring A cable that forms a closed loop connecting two or more points, so all points remain connected if the cable breaks at one point.

Rise Time The time it takes output to rise from low levels to peak value. Typically measured as the time to rise from 10% to 90% of maximum output.

Router A device that directs data packets to their destinations using information in their headers to pick the best path. Distinct from *wavelength router.*

S-Band A proposed designation for wavelengths of 1460 to 1530 nm, where optical amplifiers based on thulium-doped fibers are in development.

Scattering Loss of light that is scattered off atoms in different directions, so it escapes from the fiber core. A major component of fiber attenuation.

SDH See *Synchronous Digital Hierarchy.*

Selfoc Lens A trade name used by the Nippon Sheet Glass Company for a graded-index fiber lens; a segment of graded-index fiber made to serve as a lens.

Semiconductor Laser A laser in which injection of current into a semiconductor diode produces light by recombination of holes and electrons at the junction between *p*- and *n*-doped regions.

Sheath An outer protective layer of a fiber-optic cable.

SI Units The standard international system of metric units.

Signal-to-Noise Ratio The ratio of signal to noise, measured in decibels; an indication of analog signal quality.

Silica Glass Glass made mostly of silicon dioxide, SiO_2, used in conventional optical fibers.

Simplex Single element (e.g., a simplex connector is a single-fiber connector).

Single-Frequency Laser A laser that emits a range of wavelengths small enough to be considered a single frequency.

Single Mode Containing only one mode. When dealing with lasers, beware of ambiguities because of the difference between transverse and longitudinal modes. A laser operating in a single transverse mode typically does not operate in a single longitudinal mode.

Single-Polarization Fibers Optical fibers capable of carrying light in only one polarization.

Smart Structures (or Smart Skins) Materials containing sensors (fiber-optic or other types) to measure their properties during fabrication and use.

Soliton An optical pulse that naturally retains its original shape as it travels along an optical fiber.

SONET (Synchronous Optical Network) A standard for fiber-optic transmission.

Splice A permanent junction between two fiber ends.

Splitting Ratio The ratio of power emerging from output ports of a coupler.

Standard Single-Mode Fiber Step-index single-mode fiber with zero dispersion at 1310 nm; the first type used in fiber-optic communications, still widely used.

Star Coupler A coupler with more than three or four ports.

Step-Index Multimode Fiber A step-index fiber with a core large enough to carry light in multiple modes.

Step-Index Single-Mode Fiber A step-index fiber with a small core capable of carrying light in only one mode; this type has zero dispersion at 1310 nm.

Stimulated Raman Scattering Interactions between light and atoms in a transparent material that convert energy from one wavelength to another.

Subscriber Loop The part of the telephone network from a central office to individual subscribers.

Surface-Emitting Diode An LED that emits light from its flat surface rather than its side. Simple and inexpensive, with emission spread over a wide angle.

Surface-Emitting Laser A semiconductor laser that emits light from the wafer surface.

Switch A device that directs light or electricity along different paths, such as fibers or wires.

Switched Network A network that routes signals to their destinations by switching circuits, such as the telephone system.

Synchronous Digital Hierarchy (SDH) The international version of SONET, the Synchronous Optical Network standard. The biggest difference is in the names of the transmission rates.

Synchronous Optical Network See *SONET.*

Tbit/s Terabits (trillion, or 10^{12} bits) per second.

T Carrier A system transmitting signals at one of the standard levels in the North American digital hierarchy.

T Coupler A coupler with three ports.

TDM See *Time-Division Multiplexing.*

Thermo-Optic Switches Optical switches controlled by temperature-induced changes in refractive index.

III-V (3-5) Semiconductor A semiconductor compound made of one or more elements from the IIIA column of the periodic table (A1, Ga, and In) and one or more elements from the VA column (N, P, As, or Sb). Used in LEDs, diode lasers, and detectors.

Threshold Current The minimum current needed to sustain laser action in a diode laser.

Tight Buffer A material tightly surrounding a fiber in a cable, holding it rigidly in place.

Time-Division Multiplexing (TDM) Digital multiplexing by taking one bit or byte of data at a time from separate signals and combining them in a single bit stream.

Total Internal Reflection Total reflection of light back into a material when it strikes the interface with a material having a lower refractive index at an angle below a critical value.

Transceiver A combination of transmitter and receiver providing both output and input interfaces with a device.

Transverse Modes Modes across the width of a waveguide, fiber, or laser. Distinct from longitudinal modes, which are along the length of a laser.

Tree A network architecture in which transmission routes branch out from a central point.

Trunk Line A transmission line running between telephone switching offices or from a cable-TV head end to a distribution node.

Twisted Pair Pair of copper wires twisted around each other. The standard way to connect individual voice telephones, widely used for other low-speed communications.

Ultraviolet Electromagnetic waves invisible to the human eye, with wavelengths about 10 to 400 nm, shorter than visible light.

VCSEL (Vertical Cavity Surface Emitting Laser) A semiconductor laser in which light oscillates vertically (perpendicular to the junction plane) and light emerges from the surface of the wafer rather than from the edge of the chip.

Visible Light Electromagnetic radiation visible to the human eye at wavelengths of 400 to 700 nm.

VOA (Variable Optical Attenuator) An attenuator in which the attenuation can be varied.

Voice Circuit A circuit capable of carrying one telephone conversation or its equivalent; the standard subunit in which telecommunication capacity is counted. The U.S. analog equivalent is 4 kHz. The digital equivalent is 64 kbit/s.

WAN Wide-area network.

Waveguide A structure that guides electromagnetic waves along its length. An optical fiber is an optical waveguide.

Waveguide Array An array of curved planar waveguides that separates many optical channels at once. Also called an *Array Waveguide (AWG).*

Waveguide Dispersion The part of chromatic dispersion arising from the different speeds light travels in the core and cladding of a single-mode fiber (i.e., from the fiber's waveguide structure).

Wavelength The distance an electromagnetic wave travels in the time it takes to oscillate through a complete cycle. Wavelengths of light are measured in nanometers (10^{-9} m) or micrometers (10^{-6} m).

Wavelength-Division Multiplexing (WDM) Multiplexing of signals by transmitting them at different wavelengths through the same fiber.

Wavelength Router An optical device that directs input signals according to their wavelength.

Window A wavelength region where fibers have low attenuation, used for transmitting signals.

Zero Dispersion-Shifted Fiber Fiber with zero chromatic dispersion shifted to 1550 nm, used before the advent of DWDM.

Zero-Dispersion Wavelength Wavelength at which net chromatic dispersion of an optical fiber is nominally zero. Arises where waveguide dispersion cancels out material dispersion.

Index